Table of Relative Atomic
(Scaled to the relative atomic mas

The values of $A_r(E)$ given here apply to elements as they exist na elements. Values in parentheses are used for radioactive elements who without knowledge of the origin of the elements; the value given that element of longest known half life.

I0583946

Name	Symbol	Atomic number	Atomic weight	Name	Symbol	Atomic number	Atomic weight
Actinium	Ac	89	227.0278	Mendelevium	Md	101	(258)
Aluminum	Al	13	26.98154	Mercury	Hg	80	200.59
Americium	Am	95	(243)	Molybdenum	Mo	42	95.94
Antimony (Stibium)	Sb	51	121.75	Neodymium	Nd	60	144.24
Argon	Ar	18	39.948	Neon	Ne	10	20.179
Arsenic	As	33	74.9216	Neptunium	Np	93	237.0482
Astatine	At	85	(210)	Nickel	Ni	28	58.70
Barium	Ba	56	137.33	Niobium	Nb	41	92.9064
Berkelium	Bk	97	(247)	Nitrogen	N	7	14.0067
Beryllium	Be	4	9.01218	Nobelium	No	102	(259)
Bismuth	Bi	83	208.9804	Osmium	Os	76	190.2
Boron	B	5	10.81	Oxygen	O	8	15.9994
Bromine	Br	35	79.904	Palladium	Pd	46	106.4
Cadmium	Cd	48	112.41	Phosphorus	P	15	30.97376
Calcium	Ca	20	40.08	Platinum	Pt	78	195.09
Californium	Cf	98	(251)	Plutonium	Pu	94	(244)
Carbon	C	6	12.011	Polonium	Po	84	(209)
Cerium	Ce	58	140.12	Potassium	K	19	39.0983
Cesium	Cs	55	132.9054	Praseodymium	Pr	59	140.9077
Chlorine	Cl	17	35.453	Promethium	Pm	61	(145)
Chromium	Cr	24	51.996	Protactinium	Pa	91	231.0359
Cobalt	Co	27	58.9332	Radium	Ra	88	226.0254
Copper	Cu	29	63.546	Radon	Rn	86	(222)
Curium	Cm	96	(247)	Rhenium	Re	75	186.207
Dysprosium	Dy	66	162.50	Rhodium	Rh	45	102.9055
Einsteinium	Es	99	(252)	Rubidium	Rb	37	85.4678
Erbium	Er	68	167.26	Ruthenium	Ru	44	101.07
Europium	Eu	63	151.96	Samarium	Sm	62	150.4
Fermium	Fm	100	(257)	Scandium	Sc	21	44.9559
Fluorine	F	9	18.998403	Selenium	Se	34	78.96
Francium	Fr	87	(223)	Silicon	Si	14	28.0855
Gadolinium	Gd	64	157.25	Silver	Ag	47	107.868
Gallium	Ga	31	69.72	Sodium	Na	11	22.98977
Germanium	Ge	32	72.59	Strontium	Sr	38	87.62
Gold	Au	79	196.9665	Sulfur	S	16	32.06
Hafnium	Hf	72	178.49	Tantalum	Ta	73	180.9479
Hahnium	Ha	105	(262)	Technetium	Tc	43	(98)
Helium	He	2	4.00260	Tellurium	Te	52	127.60
Holmium	Ho	67	164.9304	Terbium	Tb	65	158.9254
Hydrogen	H	1	1.0079	Thallium	Tl	81	204.37
Indium	In	49	114.82	Thorium	Th	90	232.0381
Iodine	I	53	126.9045	Thulium	Tm	69	168.9342
Iridium	Ir	77	192.22	Tin	Sn	50	118.69
Iron	Fe	26	55.847	Titanium	Ti	22	47.90
Krypton	Kr	36	83.80	Tungsten (Wolfram)	W	74	183.85
Kurchatovium	Ku	104	(261)	(Unnilhexium)	(Unh)	106	(263)
Lanthanum	La	57	138.9055	Uranium	U	92	238.029
Lawrencium	Lr	103	(260)	Vanadium	V	23	50.9415
Lead	Pb	82	207.2	Xenon	Xe	54	131.30
Lithium	Li	3	6.941	Ytterbium	Yb	70	173.04
Lutetium	Lu	71	174.967	Yttrium	Y	39	88.9059
Magnesium	Mg	12	24.305	Zinc	Zn	30	65.38
Manganese	Mn	25	54.9380	Zirconium	Zr	40	91.22

Basic
Physical
Chemistry

Between these two, naked mind and the perceived world, is there then nothing in common? Together they make up the sum total for us; they are all we have. We called them disparate and incommensurable. Are they then absolutely apart? Can they in no wise be linked together? They have this in common—we have already recognized it—they are both concepts; they both of them are parts of knowledge of one mind. They are thus therefore distinguished, but are not sundered. Nature in evolving us makes them two parts of the knowledge of one mind and that one mind our own. We are the tie between them. Perhaps we exist for that.

Charles Sherrington
Man on His Nature *(1940)*

Basic
Physical
Chemistry

Walter J. Moore

University of Sydney

PRENTICE-HALL, INC., *Englewood Cliffs, New Jersey 07632*

Library of Congress Cataloging in Publication Data

MOORE, WALTER JOHN, 1918
    Basic physical chemistry.

    Includes index.
    1. Chemistry, Physical and theoretical. I. Title.
QD453.2.M648 1982   541.3   82-9809
ISBN 0-13-066019-1   AACR2

Editorial/production supervision: Ellen W. Caughey
Interior design: Ellen W. Caughey
Cover design: Marvin Warshaw
Manufacturing buyer: John Hall

Printed in the United States of America

10 9 8 7 6 5 4 3 2 1

ISBN 0-13-066019-1

PRENTICE-HALL INTERNATIONAL, INC., *London*
PRENTICE-HALL OF AUSTRALIA PTY. LIMITED, *Sydney*
EDITORA PRENTICE-HALL DO BRASIL LTDA., *Rio de Janeiro*
PRENTICE-HALL CANADA INC., *Toronto*
PRENTICE-HALL OF INDIA PRIVATE LIMITED, *New Delhi*
PRENTICE-HALL OF JAPAN, INC., *Tokyo*
PRENTICE-HALL OF SOUTHEAST ASIA PTE. LTD., *Singapore*
WHITEHALL BOOKS LIMITED, *Wellington, New Zealand*

Contents

PREFACE   xv

SYMBOLS   xvii

1 DIMENSIONS AND DEFINITIONS   1

1.1. Physical Quantities and Dimensions 1
1.2. Further Definitions of Standards 3   **1.3.** Amount of Substance 3
1.4. Subsidiary Units 4   **1.5.** Pressure 5   **1.6.** Dimensional Analysis 6
1.7. Equations 6   **1.8.** Chemical Reactions 6   **1.9.** Systems 7
1.10. Equilibrium States 8   **1.11.** State Functions 9   *Problems* 9

2 STATES OF MATTER   11

2.1. Equations of State 12   **2.2.** The Ideal-Gas Equation of State 12
2.3. Gases at Low Pressure 13   **2.4.** Gas Mixtures 15
2.5. The Ideal Gas from the Molecular Point of View 16
2.6. Molecular Speeds 18   **2.7.** Condensation of Gases—Critical Points 18
2.8. Real Gases—Virial Equations 20   **2.9.** Corresponding States 22
2.10. Van der Waals Equation 23   **2.11.** Liquids 25
2.12. Compressibility and Expansivity 25   **2.13.** The Solid State 26
2.14. Phases 27   **2.15.** Equilibrium Between Phases 27
2.16. Components 28   **2.17.** Degrees of Freedom 29
2.18. The Phase Rule 29   *Problems* 30

3 MOLECULAR ENERGIES 34

3.1. The Molecular Interpretation of Thermodynamics 35
3.2. Conservation of Energy 35
3.3. The Energy of Molecules: Translation, Rotation, and Vibration 38
3.4. The Energy of a Molecule: Translation 40
3.5. The Energy of a Molecule: Rotation 40
3.6. The Energy of a Molecule: Vibration 43   **3.7.** Normal Modes of Vibration 46
3.8. Classical Equipartition of Energy 47
3.9. Heat Capacity at Constant Volume: An Experimental Measure of Average
Molecular Energy 48   **3.10.** Experimental C_V for Gases 50
Problems 51

4 MOLECULAR QUANTUM LEVELS 53

4.1. Electromagnetic Radiation: Particles and Waves 53
4.2. Spectroscopy: The Experimental View of Molecular Energy Levels 54
4.3. An Example: The Spectrum of CO 55   **4.4.** Wave Properties of Matter 57
4.5. Translational Energy 58
4.6. Wave Number as a Measure of Molecular Energy Level 61
4.7. Rotational Energy 62   **4.8.** Vibrational Energy 64
4.9. A View of Vibrational and Rotational Energy Levels in the CO Molecule
by Infrared Spectroscopy 65   **4.10.** Electronic Energy 68
Problems 68

5 BOLTZMANN DISTRIBUTION AND TEMPERATURE 71

5.1. The Boltzmann Distribution 71
5.2. The Barometric Formula. A Simple Derivation of a Boltzmann Distribution 73
5.3. Sedimentation Equilibrium 74
5.4. A More General Derivation of the Boltzmann Distribution Law 76
5.5. Relative Populations of Molecular Energy Levels 79
5.6. A Molecular Interpretation of Temperature 82
5.7. A Molecular View of Heat Capacities 89
5.8. Distribution of Molecular Speeds 85
5.9. Relation of Maxwell Equation to Gaussian Density Function 86
5.10. Calculation of Average Values 88   **5.11.** Velocity in Three Dimensions 88
Problems 90

6 THE FIRST LAW OF THERMODYNAMICS—ENERGY 92

6.1. The Concept of Work 92   **6.2.** Work in Changes of Volume 93
6.3. Equilibrium Paths and Reversible Processes 95
6.4. Isothermal Reversible Compression of Ideal Gas 96
6.5. General Concept of Work 97   **6.6.** The Concept of Heat 98
6.7. First Law of Thermodynamics for a Closed System 98
6.8. Exact Differentials and State Functions 99   **6.9.** Enthalpy 100

6.10. The Difference Between C_P and C_V 101
6.11. Enthalpy of Phase Changes   102   **6.12.** ΔU in Chemical Reactions   103
6.13. Measurement of ΔU of Reaction   104   **6.14.** Calculation of ΔH from ΔU 105
6.15. The Hess Law   106   **6.16.** Standard States   107
6.17. Enthalpy of Formation of a Compound   107
6.18. ΔH of Reactions in Aqueous Solution   108
6.19. Enthalpy of Formation of Ions   109
6.20. Temperature Dependence of Enthalpy of Reaction   109
6.21. Bond Enthalpies   111   **6.22.** Thermochemistry and Equilibrium   114
  Problems   115

7   THE SECOND AND THIRD LAWS
    OF THERMODYNAMICS—ENTROPY   118

7.1. Entropy and Reversible Heat   119
7.2. Molecular Picture of Heat and Work   120
7.3. Entropy Changes of an Ideal Gas   121   **7.4.** Entropy in Changes of State   122
7.5. ΔS for an Irreversible Process: Heat Conduction   124
7.6. A Supercooled Liquid Freezes—What Is ΔS of the Liquid and of the Universe?   126
7.7. Heat Engines   127   **7.8.** The Carnot Cycle   128
7.9. Can We Use the Energy of the Oceans?   130
7.10. Entropy and the Arrow of Time   131   **7.11.** Entropy of Mixing   132
7.12. Probability of Mixtures   133   **7.13.** Disorder, Probability, and Entropy   135
7.14. Entropies of Chemical Compounds—Calculations from Heat Capacities   136
7.15. Third Law of Thermodynamics   138   **7.16.** Third-Law Entropies   139
7.17. Entropy Changes in Chemical Reactions   140
  Problems   141

8   PHYSICAL AND CHEMICAL EQUILIBRIUM   144

8.1. Entropy and Equilibrium   144   **8.2.** Dynamic Equilibrium   145
8.3. Free-Energy Functions   146
8.4. Interpretation of the Helmholtz Function A 147
8.5. Equation of State Derived from A 148
8.6. Interpretation of the Gibbs Function G 149
8.7. Phase Equilibrium—the Clapeyron–Clausius Equation   151
8.8. How Vapor Pressure Depends on Temperature   153
8.9. Standard States and Gibbs Free Energy Changes in Chemical Reactions   155
8.10. Gibbs Free Energy of an Ideal Gas   157
8.11. How the Gibbs Function Depends on Extent of Reaction   157
8.12. Equilibrium Constant and Gibbs Free Energy   159
8.13. Measurement of K_P 160   **8.14.** How G Varies with T 161
8.15. How K_P Varies with T 163   **8.16.** Gas–Solid Reactions   165
8.17. Effect of Pressure on Equilibrium Constants   165
8.18. The Chemical Potential   166
8.19. Chemical Potential and Chemical Equilibrium   168   *Problems*   168

9 SOLUTIONS—IDEAL AND DILUTE 173

9.1. Measures of Composition 173
9.2. Partial Molar Quantities—Partial Molar Volume 175
9.3. Other Partial Molar Quantities 178
9.4. How Partial Molar Quantities Are Measured 178
9.5. The Ideal Solution—Raoult's Law 179
9.6. Thermodynamics of Ideal Solutions 181
9.7. Solubility of Gases in Liquids—Henry's Law 182
9.8. Mechanism of Anesthesia 182   **9.9.** Pressure–Composition Diagrams 185
9.10. Temperature–Composition Diagrams 186   **9.11.** Fractional Distillation 187
9.12. Solutions of Solids in Liquids 188   **9.13.** Osmotic Pressure 190
9.14. Osmotic Pressure and Vapor Pressure 191
9.15. Osmotic Pressure of Polymer Solutions 193
Problems 194

10 REAL GASES AND SOLUTIONS 198

10.1. Fugacity and Activity 198
10.2. Real Gases—Chemical Potential and Fugacity 199
10.3. How to Calculate the Fugacity of a Gas 200
10.4. Fugacity and Corresponding States 201
10.5. Use of Fugacity in Equilibrium Calculations 202   **10.6.** Activity 204
10.7. Standard States for Components in Solution 205
10.8. Activities of Solvent and Nonvolatile Solute from Vapor Pressure of Solution 206
10.9. Equilibrium Constants in Solution 209
10.10. ΔG_f° of Biochemicals in Aqueous Solution 211
10.11. Deviations of Solutions from Ideality 213
10.12. Boiling-Point Diagrams 214   **10.13.** Solubility of Liquids in Liquids 215
10.14. Distillation of Immiscible Liquids 217
10.15. Mixtures of Oil and Water 219
Problems 220

11 PHASE TRANSITIONS AND PHASE EQUILIBRIA 223

11.1. Conditions for Equilibrium Between Phases 223
11.2. Pure Substances—One Component Systems 224
11.3. How Thermodynamic Functions Behave at Phase Changes 225
11.4. Melting and Vaporization 227   **11.5.** Liquid Crystals 227
11.6. Measurements at High Pressures 229   **11.7.** High-Pressure Systems 230
11.8. An Approach to Absolute Zero: Cooling by Demagnetization 232
11.9. Superconductivity and Superfluidity 233
11.10. Two Component Systems 236
11.11. Solid–Liquid Equilibrium—Simple Eutectic Diagrams 236
11.12. Formation of Compounds 239   **11.13.** Solid Solutions 240
11.14. Partial Miscibility in the Solid State 241
11.15. The Iron–Carbon Diagram 244
Problems 245

12 STATISTICAL THERMODYNAMICS 247

12.1. Ensembles 248   **12.2.** Ensemble Averages 249
12.3. Statistical Calculation of Thermodynamic Energy 250
12.4. Statistical Formula for Entropy 251
12.5. Helmholtz Free Energy and Equation of State 251
12.6. How to Evaluate Z for Noninteracting Particles 252
12.7. Translational Partition Function 254
12.8. Thermodynamic Functions for a Monatomic Gas 254
12.9. Internal Motions—Molecular Partition Function 256
12.10. Rotational Partition Function—Rigid Linear Molecules 256
12.11. Rotational Energy and Entropy—Linear Molecules 257
12.12. How to Calculate Moments of Inertia 258
12.13. Rotational Partition Functions for Nonlinear Molecules 259
12.14. Vibrational Partition Functions 261
12.15. Vibrational Energy and Entropy 262   **12.16.** Heat Capacities 264
12.17. Vibrational Energy and Molecular Dissociation 265
12.18. Statistical Thermodynamics of Crystals 266
12.19. Electronic Partition Function 267
12.20. The Third Law in Statistical Thermodynamics 269
12.21. Equilibrium Constants 270   **12.22.** Statistical Interpretation of K_P 271
12.23. Examples of Calculation of K_P 272
Problems 275

13 CHEMICAL KINETICS 277

13.1. The Rate of Chemical Change 277
13.2. Experimental Methods in Kinetics 279
13.3. The Concept of Order of Reaction 280   **13.4.** Reduced Rate Constants 281
13.5. Reaction Molecularity and Reaction Order 281
13.6. Reaction Mechanisms 282   **13.7.** First-Order Rate Equations 283
13.8. Second-Order Rate Equations 285
13.9. Determination of Reaction Order 287   **13.10.** Opposing Reactions 288
13.11. Consecutive Reactions 289   **13.12.** Parallel Reactions 290
13.13. Chemical Relaxation 291   **13.14.** Reactions in Flow Systems 293
13.15. Stationary States and Dissipative Processes 296
13.16. Chain Reactions: Formation of Hydrogen Bromide 297
13.17. Free-radical Chains 299
13.18. Branching Chains—Explosive Reactions 300
13.19. How Reaction Rates Depend on Temperature 301
13.20. Examples of Temperature Dependence of Reaction Rates 303
Problems 304

14   CATALYSIS   309

14.1. Catalysts Influence Rate but not Equilibrium   309
14.2. Homogeneous Catalysis in Gas Reactions   310
14.3. Acid–Base Catalysis   311   **14.4.** General Acid–Base Catalysis   313
14.5. Catalysis by Enzymes   314
14.6. The Structure of an Enzyme—Carboxypeptidase A   317
14.7. Surface Catalysis   318   **14.8.** Langmuir Adsorption Isotherm   319
14.9. Adsorption on Nonuniform Surfaces   321
14.10. Mechanism of Surface Reactions   322
        Problems   325

15   THEORY OF REACTION RATES   328

15.1. Collision Theory of Gas Reactions—Collision Frequency   328
15.2. Collision Theory of Gas Reactions—The Rate Constant   331
15.3. Molecular Diameters   332   **15.4.** Collision Theory vs. Experiment   334
15.5. Potential Energy Surfaces—Example of $D + H_2$ 335
15.6. Activated-complex Theory of Reaction Rates   338
15.7. Activated-complex Theory in Thermodynamic Terms   339
15.8. The Entropy of Activation   340   **15.9.** Chemical Dynamics   341
15.10. Reactions in Molecular Beams   342
15.11. Theory of Unimolecular Reactions   343   **15.12.** Reactions in Solution   346
15.13. Diffusion-Controlled Reactions   348
        Problems   349

16   ELECTROCHEMISTRY—IONS IN SOLUTION   352

16.1. Electrochemical Equivalence—the Faraday   352
16.2. Conductivity of Solutions   353   **16.3.** Molar Conductivity   355
16.4. Arrhenius Ionization Theory   356
16.5. A High Dielectric Constant of Solvent Facilitates Separation of Ions   357
16.6. Transport Numbers and Mobilities   358
16.7. Transport Numbers—Hittorf Method   359
16.8. Transport Numbers—Moving Boundary Method   360
16.9. Results of Transport Experiments   361
16.10. Electrolytic Dissociation of Water   363
16.11. Mobilities of Hydrogen and Hydroxyl Ions   364
16.12. Diffusion and Ionic Mobility   364
16.13. Activities and Standard States   366   **16.14.** Ion Activities   367
16.15. Experimental Activity Coefficients of Ionic Solutions   368
16.16. The Ionic Strength   370   **16.17.** Debye–Hückel Theory   370
16.18. The Ionic Atmosphere   371   **16.19.** Debye–Hückel Limiting Law   373
        Problems   375

17 ELECTROCHEMICAL CELLS   378

17.1. Metal Electrodes 378   **17.2.** The Electrochemical Potential 380
17.3. Contact Between Two Metals 381   **17.4.** Types of Electrodes 382
17.5. Classification of Cells 382   **17.6.** An Electrochemical Cell 383
17.7. Cell Diagram and Cell Reaction 384
17.8. Equilibrium Condition in an Electrochemical Cell 384
17.9. Electromotive Force (emf) of a Cell 385   **17.10.** A Standard Cell 386
17.11. Reversible Cells 386   **17.12.** Thermodynamics of Cell Reactions 387
17.13. The Standard emf of Cells 389   **17.14.** Standard Electrode Potentials 390
17.15. Calculation of the emf of a Cell 393
17.16. Calculation of Solubility Products 394
17.17. Electrolyte-Concentration Cells 395   **17.18.** Measurement of pH 396
17.19. Biological Membrane Potentials 397   **17.20.** Nerve Conduction 399
    Problems 401

18 SURFACES AND COLLOIDS   404

18.1. Surface Tension 405   **18.2.** Equation of Young and Laplace 406
18.3. Capillarity 407
18.4. Enhanced Vapor Pressure of Small Droplets—Kelvin Equation 409
18.5. Surface Tension of Solutions 411   **18.6.** Insoluble Surface Films 412
18.7. Structure of Surface Films 414   **18.8.** Surfactants and Micelles 415
18.9. Cell Membranes 417   **18.10.** Colloidal Sols—Particle-Size Distribution 418
18.11. Stability of Colloids—The Electric Double Layer 420   *Problems* 422

19 ELECTROCHEMICAL RATE PROCESSES   424

19.1. Electrode Kinetics 424   **19.2.** Polarization 426
19.3. How the Electric Field Controls the Rate of an Electrode Reaction 426
19.4. The Tafel Equations 429   **19.5.** Kinetics of Discharge of Hydrogen Ions 431
19.6. Diffusion Overpotential 431   **19.7.** Fuel Cells 433   *Problems* 435

20 PARTICLES AND WAVES   436

20.1. Wave Motion 437   **20.2.** The Classical Wave Equation 438
20.3. The Time-Independent Classical Wave Equation 439
20.4. The Schrödinger Wave Equation 440   **20.5.** Translational Energy 441
20.6. Statistical Interpretation of Wavefunctions 443
20.7. Further Characteristics of Wavefunctions 444
20.8. Orthogonality of Wavefunctions 445
20.9. Translational Wavefunctions 446   **20.10.** Quantization of Energy 448
20.11. Zero-point Energy and the Uncertainty Principle 449
20.12. The Free Particle 449
20.13. How to Extract Further Information from a Wavefunction 450
20.14. Operators 451   **20.15.** The Hamiltonian Operator 452
20.16. Free Electron Model for Conjugated Dyes 453
20.17. The Box in Three Dimensions 454   **20.18.** The Tunnel Effect 455
    Problems 457

21 ATOMIC STRUCTURE AND SPECTRA 459

21.1. Atomic Spectra 459   **21.2.** Bohr Orbits and Ionization Energies 460
21.3. Schrödinger Equation for the Hydrogen Atom 463
21.4. The Radial Equation Gives the Energy Levels 465
21.5. The Angular Equation Gives the Angular Momenta 465
21.6. The Quantum Numbers 467   **21.7.** The Radial Wavefunctions 469
21.8. Angular Dependence of Hydrogen Orbitals 471
21.9. The Spinning Electron 473   **21.10.** The Pauli Exclusion Principle 473
21.11. Spectrum of Helium 475   **21.12.** Vector Model of the Atom 477
21.13. Atomic Orbitals and Energies—The Variation Method 480
21.14. The Helium Atom 481
21.15. Heavier Atoms—The Self-Consistent Field 482
21.16. Atomic Energy Levels—The Periodic Table 483
Problems 484

22 THE CHEMICAL BOND 486

22.1. The Theory of Valence 487   **22.2.** The Hydrogen-molecule Ion H_2^+ 488
22.3. Born-Oppenheimer Approximation 488   **22.4.** The Chemical Bond in H_2^+ 489
22.5. Angular Momentum of H_2^+ 491   **22.6.** Simple Variation Theory of H_2^+ 491
22.7. The Covalent Bond in H_2 493   **22.8.** The Valence-bond Method 499
22.9. Molecular Orbitals for Homonuclear Diatomic Molecules 499
22.10. Correlation Diagram 502   **22.11.** Heteronuclear Diatomic Molecules 505
22.12. Electronegativity 506   **22.13.** The Ionic Bond 508
22.14. Polyatomic Molecules—H_2O for Example 509
22.15. Calculation of Molecular Geometries 511
22.16. Nonlocalized Molecular Orbitals—Benzene 512
22.17. Photoelectron Spectroscopy 516
Problems 517

23 ELECTRIC AND MAGNETIC PROPERTIES OF MOLECULES 520

23.1. Relative Permittivity 520
23.2. Polarization of Dielectrics—Dipole Moments 521   **23.3.** Polarizability 523
23.4. The Local Field 524   **23.5.** Orientation of Dipoles in an Electric Field 526
23.6. How Relative Permittivity Depends on Frequency 529
23.7. Dipole Moment and Molecular Structure 530
23.8. Magnetic Properties of Molecules 533
23.9. Diamagnetism and Temperature-Independent Paramagnetism 534
23.10. Temperature-Dependent Paramagnetism 535
Problems 535

24 MAGNETIC RESONANCE 537

24.1. Electric and Magnetic Properties of Nuclei 537
24.2. Nuclear Magnetic Resonance 539
24.3. Experimental Apparatus for NMR 541   **24.4.** Spin-Lattice Relaxation 542
24.5. Spin-Spin Relaxation 543   **24.6.** Chemical Shifts 544
24.7. Spin-Spin Splitting 547
24.8. Dynamic NMR—Measurement of Reaction Rates 549
24.9. Fourier Transform NMR 551   **24.10.** Electron Spin Resonance 552
24.11. Nuclear Hyperfine Interaction 555   **24.12.** Spectra of Free Radicals 556
     Problems 557

25 SYMMETRY 559

25.1. Symmetry Operations 559   **25.2.** Definition of a Group 561
25.3. Further Symmetry Operations 562   **25.4.** Notation for Point Groups 563
25.5. Point-groups and Molecular Properties 564
25.6. Transformations of Vectors by Symmetry Operations 565
25.7. Matrix Representation of Group C_{3v} 569
25.8. Irreducible Representations 570
25.9. Characters of Irreducible Representations 571
25.10. Chemical Applications of Group Theory 573
     Chapter Appendix: Character Tables for Some Point Groups 574
     Problems 576

26 ROTATIONAL AND VIBRATIONAL SPECTRA— MICROWAVE, INFRARED, AND RAMAN 577

26.1. Survey of Molecular Spectra 577
26.2. Emission and Absorption of Light 580
26.3. Pure Rotation Spectra—Rigid Rotors 581
26.4. Microwave Spectroscopy 583
26.5. Rotational Spectra of Polyatomic Molecules 584
26.6. Inversions and Internal Rotations 584   **26.7.** The Harmonic Oscillator 586
26.8. The Anharmonic Oscillator 588
26.9. Vibration–Rotation Spectra of Diatomic Molecules 589
26.10. Infrared Spectrum of Carbon Dioxide 591   **26.11.** Lasers 592
26.12. Raman Spectra 594   **26.13.** Molecular Data from Spectroscopy 596
26.14. Normal Modes of Vibration 596
26.15. Symmetry and Normal Vibrations 597   *Problems* 600

27 ELECTRONIC SPECTRA AND PHOTOCHEMISTRY 603

27.1. Absorption of Light 603   **27.2.** Electronic Transitions and Band Spectra 606
27.3. The Franck–Condon Principle 607   **27.4.** Excited States of Oxygen 609
27.5. Excited States of Polyatomic Molecules 611
27.6. Photochemical Principles 612   **27.7.** Pathways of Molecular Excitation 614
27.8. Fluorescence 615   **27.9.** Dissociation and Predissociation 617
27.10. Secondary Photochemical Processes 618   **27.11.** Flash Photolysis 619
27.12. Energy Transfer in Condensed Systems 620   *Problems* 621

28 CRYSTALLOGRAPHY   623

28.1. Crystal Planes and Faces 624   **28.2.** Crystal Systems 625
28.3. Lattices and Unit Cells 626
28.4. Symmetry Properties—Crystal Classes 628   **28.5.** Crystal Structures 629
28.6. Space Groups 631   **28.7.** X-ray Crystallography 631
28.8. The Bragg Analysis of X-ray Diffraction 632
28.9. Structures of Sodium and Potassium Chlorides 633
28.10. The Powder Method 638   **28.11.** Rotating-Crystal Methods 640
28.12. The Structure Factor 640
28.13. Fourier Synthesis of a Crystal Structure 644
28.14. Neutron Diffraction 645
     Problems 646

29 THE SOLID STATE   648

29.1. The Bond Model of Solids 649   **29.2.** Closest Packing of Spheres 651
29.3. Electron-gas Theory of Metals 654   **29.4.** Quantum Statistics 655
29.5. Cohesive Energy of Metals 657   **29.6.** Intrinsic Semiconductors 658
29.7. Impurity Semiconductors 660   **29.8.** Ionic Crystals 660
29.9. Cohesive Energy of Ionic Crystals 662   **29.10.** Crystal Energies 663
29.11. Point Defects 665   **29.12.** Linear Defects—Dislocations 666
29.13. Effects Due to Dislocations 667
     Problems 669

30 THE LIQUID STATE AND INTERMOLECULAR FORCES   671

30.1. Disorder in the Liquid State 672
30.2. X-ray Diffraction of Liquid Structures 674   **30.3.** Liquid Water 675
30.4. Cohesion of Liquids—Internal Pressure 677
30.5. Intermolecular Forces 678   **30.6.** Origins of Intermolecular Forces 681
30.7. Equation of State and Intermolecular Forces 683
30.8. Theory of Liquids 685   **30.9.** Viscosity of Liquids 687
30.10. Poiseuille Equation 688   **30.11.** Viscosity of Polymer Solutions 690
     Problems 692

TABLES   695

INDEX   703

Preface

Basic Physical Chemistry is directed toward science or engineering students who need to understand the basic foundations of physical chemistry. The length of the book is suitable for a typical one-year course. The prerequisites are chemistry, physics, and mathematics through elementary calculus.

A typical first-year chemistry textbook today includes almost all the topics that were covered in the physical chemistry textbook of my own student days. Yet, in discussions with second-year students, I find that few of them have achieved an adequate understanding of concepts such as entropy, free energy, wave functions, Boltzmann factors, and the like. Therefore, I discuss these basic quantitative concepts of physical chemistry without assuming that a student has prior knowledge from a first-year course. On the other hand, I have ommitted pictorial concepts such as hybrid orbitals and valence-shell electron repulsions, which are adequately described in first-year textbooks.

An introduction to the internal motions and energy states of molecules appears in the early chapters, preceding a discussion of the laws of thermodynamics. According to my experience, students follow thermodynamics more easily if they understand what is happening at the molecular level. Learning to think in terms of molecules and their attributes is a vital objective in the study of physical chemistry.

This book contains a considerable number of worked-out examples. Most of these are quite easy, involving little more than substitution of numerical data into equations. This kind of practical application of equations, however, should give students a good idea of the magnitudes of many different physicochemical quantities. It is important for students to obtain an order-of-magnitude feeling for various properties and processes. As such a feeling becomes established, they can begin to think in semiquantitative terms about increasingly complex systems. Another function of these relatively simple examples is to give students practice in using proper units for a great variety of chemical data. From time to time, the text is interrupted by short questions to make students stop and think about the material.

Several colleagues contributed to the writing of this book. Professor Charles Parmenter of Indiana University was involved with the initial planning and writing of the first half of the book. His many important contributions cannot be adequately acknowledged in a few words. Professor Jerry Bell read the entire manuscript and made numerous suggestions designed to improve the exposition. Professor Robert Hunter, Dr. Peter Wright, and Dr. G. L. D. Ritchie of the University of Sydney provided critical readings of individual chapters. To Professor I. M. Ritchie, Dr. R. A. Craig, and Dr. P. J. Thistlethwaite, and their publisher I am indebted for several interesting problems from their book *Problems in Physical Chemistry* (John Wiley & Sons Australasia Pty. Ltd., 1975). I acknowledge the kind permission of the Cambridge University Press for the quotation from the inspiring work of Charles Sherrington.

Elizabeth Perry, chemistry editor for Prentice-Hall, Inc., was helpful in many ways, particularly in finding three experienced teachers of physical chemistry who provided excellent and varied advice concerning the final manuscript. My special thanks are due to the perceptive review of Professor Thomas Dunn of the University of Michigan, Professor Jeff Steinfeld of M.I.T., and Professor Theodore Sakano of Rose-Hulman Institute.

We gratefully acknowledge the following reviewers: Dewey K. Carpenter, Louisiana State University; Thomas R. Dyke, University of Oregon; Donald D. Fits, University of Pennsylvania; L. Peter Gold, Pennsylvania State University; David Harrick, University of Oregon; and George C. Schatz, Northwestern University.

Finally, I should like to thank my wife, Patricia, for her help in preparing the manuscript.

Comments from teachers and students will be received with thanks and applied with care to later editions.

W. J. M.

Symbols

Symbol and meaning		SI unit
a	van der Waals constant	N m^4 mol^{-2}
a	activity	—
a	acceleration	m s^{-2}
a_0	Bohr radius	m
A	amplitude	m
α	area	m^2
A	Helmholtz free energy	J
A	preexponential factor	(varies)
A	absorbance, optical density	—
A	electron affinity	J
A_0	hyperfine coupling constant	Hz
b	van der Waals constant	m^3 mol^{-1}
b	adsorption coefficient	m^2 N^{-1} [Pa^{-1}]
b	thickness of ionic atmosphere	m
b	light-absorption coefficient	m^{-1}
B	rotational constant $(h^2/8\pi^2 I)$	J
$\tilde{B}$	rotational constant $(h/8\pi^2 cI)$	cm^{-1} [non SI]
$\mathbf{B}$	magnetic flux density [induction]	V s m^{-2} [T]

Symbol and meaning		SI unit
$B(T)$	second virial coefficient	m^3 mol^{-1}
c	speed of light in vacuum	m s^{-1}
c	concentration (n/V)	mol m^{-3}
c	number of components	—
$\bar{c}$	average speed	m s^{-1}
C	capacitance	C V^{-1}
C	number concentration (N/V)	m^{-3}
C	heat capacity	J K^{-1}
C_V	heat capacity at constant volume	J K^{-1}
C_P	heat capacity at constant pressure	J K^{-1}
$C(T)$	third virial coefficient	m^6 mol^{-2}
d	molecular diameter	m
d_{hkl}	interplanar spacing	m
D	diffusion coefficient	m^2 s^{-1}
D_e	spectroscopic dissociation energy	J mol^{-1}
D_0	chemical dissociation energy $(D_e - \frac{1}{2}h\nu_0)$	J mol^{-1}
e	base of natural logarithms	—
e	unit of charge	C
$\mathbf{E}$	electric field	V m^{-1}

Symbol and meaning		SI unit
E	emf (electromotive force)	V
E	energy	J
E_k	kinetic energy	J
E_p	potential energy	J
E_a	activation energy (Arrhenius)	$J\,mol^{-1}$
f	number of degrees of freedom (variance)	—
f	surface pressure	$N\,m^{-1}$
f	fugacity	$N\,m^{-2}$
f	oscillator strength	—
$\mathbf{F}$	force	N
F	faraday constant	$C\,mol^{-1}$
$F(hkl)$	structure factor	—
g	gravitational acceleration	$m\,s^{-2}$
g	electronic g factor	—
g_N	nuclear g factor	—
g_j	statistical weight of level j	—
G	Gibbs free energy	J
h	Planck constant	J s
h	height, altitude	m
H	enthalpy	J
$\hat{H}$	Hamiltonian operator	—
i	current density	$A\,m^{-2}$
I	ionic strength [$\Sigma m_i z_i{}^2$]	$mol\,kg^{-1}$
I	moment of inertia	$kg\,m^2$
I	nuclear spin quantum number	—
I	ionization energy	J
I	electric current	A
I_v	luminous intensity	Ca
J	rotational quantum number	—
J	spin–spin coupling constant	Hz
k	Boltzmann constant	$J\,K^{-1}$
k_1	first-order constant	s^{-1}
k_2	second-order rate constant	$m^3\,mol^{-1}\,s^{-1}$
k_r'	reduced rate constant	s^{-1}
K	equilibrium constant	—
K_P	equilibrium constant in partial pressure	—
K_m	Michaelis constant	—
K_a	acid dissociation constant	—
K_b	base dissociation constant	—
K_w	water dissociation constant	—

Symbol and meaning		SI unit
ℓ	length	m
ℓ	azimuthal quantum number	—
L	Avogadro constant	mol^{-1}
$\mathbf{L}$	angular momentum	$kg\,m^2\,s^{-1}$
$m, \mathfrak{m}$	mass	kg
m	molality	$mol\,kg^{-1}$
m'	volume molality	$mol\,m^{-3}$
m_e	mass of electron	kg
m_p	mass of proton	kg
m_ℓ	magnetic quantum number	—
m_s	electron-spin quantum number	—
M	molar mass	$kg\,mol^{-1}$
$\mathbf{M}$	magnetization	$A\,m^{-1}$
M_I	nuclear-spin quantum number	—
n	amount of substance	mol
n	principal quantum number	—
N	number of particles	—
$\mathfrak{N}$	number of systems in ensemble	—
p	number of phases	—
$\mathbf{p}$	momentum	$kg\,m\,s^{-1}$
$p, \mathscr{P}$	probability	—
P	pressure	$N\,m^{-2}\,[Pa]$
P_c	critical pressure	$N\,m^{-2}$
P_r	reduced pressure	—
$\mathbf{P}$	polarization	$C\,m^{-2}$
P_m	molar polarization	m^3
q	heat	J
Q	electric charge	C
$\mathbf{Q}$	quadrupole moment	$C\,m^2$
r	distance, radius	m
r_k	reaction rate per unit volume	$mol\,m^{-3}\,s^{-1}$
R	molar gas constant	$J\,K^{-1}\,mol^{-1}$
R_e	equilibrium internuclear distance	m
$\mathfrak{R}$	Rydberg constant	m^{-1}
s	steric factor	—
S	entropy	$J\,K^{-1}$
S	overlap integral	—
t	time	s
t	transport (transference) number	—
t	centigrade (Celsius) temperature	°C
T	thermodynamic temperature	K

Symbol and meaning		SI unit
T_c	critical temperature	K
T_r	reduced temperature	—
T_1	spin-lattice (longitudinal) relaxation time	s
T_2	spin-spin (transverse) relaxation time	s
$\mathfrak{I}$	transmittance	—
u	velocity component	m s^{-1}
u	electric mobility	m^2 s^{-1} V^{-1}
U	internal energy	J
U	potential energy	J
v	rate of reaction	mol s^{-1}
$\mathbf{v}$	velocity	m s^{-1}
v	velocity component	m s^{-1}
v	speed	m s^{-1}
v	vibrational quantum number	—
V	volume	m^3
V_c	critical volume	m^3

Symbol and meaning		SI unit
V_r	reduced volume (V/V_c)	—
w	work	J
w	velocity component	m s^{-1}
W	number of microstates	—
X	mole fraction	—
x_e	anharmonicity constant	—
z	compression factor (PV/nRT)	
z	molecular partition function	
z	number of charges	—
z_{AB}	collision frequency of one molecule	s^{-1}
Z	nuclear charge (atomic) number	—
Z	canonical ensemble partition function	—
Z_{AB}	collision frequency per unit volume	s^{-1} m^{-3}

Symbol and meaning		SI unit
α	expansivity	K^{-1}
α	degree of dissociation (fractional)	—
α	transfer factor	—
α	Madelung constant	—
α	polarizability	C m^2 V^{-1}
β	compressibility	m^2 N^{-1} [Pa^{-1}]
γ	fugacity coefficient	—
γ	activity coefficient	—
γ	surface tension	N m^{-1}
γ	magnetogyric ratio	s^{-1} T^{-1}
Γ	surface excess (adsorption)	mol m^{-2}
δ	chemical shift	—
ϵ	energy (of a particle)	J
ϵ	molar absorptivity) (extinction coefficient)	m^2 mol^{-1}
ϵ_0	vacuum permeability	C^2 J^{-1} m^{-1}
ϵ_r	relative permeability (dielectric constant)	—
ϵ_F	Fermi energy	J
η	efficiency of heat engine	—
η	viscosity	kg s^{-1} m^{-1}
η	electrochemical polarization (overvoltage)	V

Symbol and meaning		SI unit
θ	diffraction angle	—
θ	fraction of sites covered	—
Θ	characteristic temperature	K
κ	force constant	N m^{-1}
κ	electric conductivity	Ω^{-1} m^{-1}
κ	thermal conductivity	J K^{-1} m^{-1} s^{-1}
λ	wavelength	m
λ	mean free path	m
Λ	molar conductivity	Ω^{-1} m^2 mol^{-1}
Λ	angular momentum about internuclear axis	$h/2\pi$
μ	reduced mass	kg
μ	chemical potential	J mol^{-1}
$\bar{\mu}$	electrochemical potential	J mol^{-1}
$\boldsymbol{\mu}$	dipole moment	C m
μ_m	magnetic dipole moment	A m^2 [J T^{-1}]
μ	permeability	kg m s^{-2} A^{-2}
μ_0	vacuum permeability	kg m s^{-2} A^{-2}
μ_B	Bohr magneton	A m^2 [JT^{-1}]
μ_N	nuclear magneton	A m^2 [JT^{-1}]
ν	frequency	s^{-1} [Hz]

Symbol and meaning		SI unit	Symbol and meaning		SI unit
$\tilde{\nu}$	wave number	cm^{-1} [non-SI]	ϕ	angular variable	—
			Φ	electric potential	V
ν_J	stoichiometric coefficient	—	Φ	quantum yield	—
			χ	magnetic susceptibility	A^2 N^{-1}
ξ	extent of reaction	mol	χ_e	electric susceptibility	—
Π	osmotic pressure	N m^{-2} [Pa]	ψ	wavefunction in stationary state	—
ρ	density	kg m^{-3}			
σ	shielding constant	—	Ψ	wavefunction including time dependence	—
σ	surface charge density	C m^{-2}			
σ	symmetry number	—	ω	angular frequency $(2\pi\nu)$	rad s^{-1}
σ_{AB}	(collision) cross section	m^2			
τ	half-life	s			

Note: Vectors are denoted by bold face roman symbols. The magnitudes of these vectors are denoted by the corresponding italic symbols. For example, **E** is electric field and E is magnitude of electric field.

Basic
Physical
Chemistry

1

Dimensions and Definitions

In the study of physical chemistry we shall discuss many different quantities and their measurements. It is important to understand the distinctions between a physical quantity, its dimensions, and the units in which it is measured. Communication is much easier when everyone agrees to use the same definitions, symbols, and units in describing the results of experiments and calculations. An international system has been devised (called the *Système International* or SI) which, though not perfect, has been accepted by a great majority of scientists. We shall describe this system and use it in this book.

1.1 Physical Quantities and Dimensions

When a new physical quantity is introduced in science, the first step is to give it a name and define it. The definition must be *operational*. The quantity is defined by stating precisely the operations that must be carried out to measure the quantity. Sometimes the physical quantity has a long background in the history of human thought, so that psychological and physiological factors may have originally suggested the concept. For example, "force" may be distantly related to the experience of muscular effort, and "temperature" to feelings of warmth and cold. Such origins are of historical interest, but we must have exact operational definitions for scientific purposes.

The measurement of a physical quantity consists in a determination of the numerical ratio between two examples of the quantity, one of which is the example to be measured and the other a standard example of some kind. For instance, the *length* of a metal rod can be measured by comparison with a standard meter stick. The operations would be: (1) lay the rod parallel to the meter stick, and (2) count the number of markings on the stick between the origin and the end of the rod. Then,

$$\text{length of rod} = \frac{\text{number of markings from origin to end of rod}}{\text{number of markings on meter stick}}$$

Note that the ratio is always a pure cardinal number. Thus physical measurement is always a counting or enumeration process.

The measurement described would be limited in accuracy by our ability to read markings on the meter stick, to interpolate between markings, to align the rod with the stick, and by the accuracy of the stick itself. The wooden meter stick is a rather crude standard of length. More precise results could be obtained with a cathetometer, which is a carefully engraved metal scale with a sliding telescope attached. This instrument in turn could be calibrated against standard platinum-iridium metric bars kept in the national standards laboratories of different countries. These national standards were originally referred to the international standard meter kept at the International Bureau of Weights and Measures in Sèvres, near Paris. All such metal standards, however, are of rather low accuracy. The wavelengths of spectral lines, which can be exactly reproduced in laboratories anywhere in the world, provide a better primary standard. Thus the standard meter was redefined in 1960 as 1 650 763.73 wavelengths *in vacuo* of the radiation corresponding to a transition between two specified energy levels in an atom of krypton-86.

Another example of a physical quantity is *mass*. We would ordinarily measure the mass of an object by comparing it with the mass of standards such as analytical weights. The comparison is made with a balance which compares gravitational forces or weights; since $F = mg$, the forces or weights are directly proportional to the masses. The proportionality factor, the gravitational acceleration, is g. The masses of the analytical weights are calibrated against secondary standards, which in turn have been measured against the standard kilogram, a mass of platinum kept at Sèvres.

Scientists may be interested in the measurements of many different physical quantities. All these quantities can be defined operationally in terms of a small number of fundamental quantities. The exact choice of these fundamental quantities is to some extent arbitrary. We can, however, select a minimal basic set so that all the quantities *in this set* are *dimensionally independent*. This condition means that they cannot be derived from one another by algebraic combinations. Each dimensionally independent quantity must therefore be referred to its own primary standard. The entire set of primary standards thus defines a set of basic physical dimensions. The dimensions of any other physical quantities can be expressed in terms of the basic set.

The basic physical quantities that have been chosen, their symbols, and SI units are summarized in Table 1.1. Note that the symbols for physical quantities are always printed in italic type. In written material an italic symbol is denoted by underlining. Abbreviations for units are always printed in roman type, e.g., meter, m; kilogram, kg. No periods are used after abbreviations.

TABLE 1.1

BASIC PHYSICAL QUANTITIES AND THEIR SI UNITS

Physical Quantity	Symbol	SI Unit	Abbreviation
Length	l	meter	m
Mass	m	kilogram	kg
Time	t	second	s
Electric current	I	ampere	A
Thermodynamic temperature	T	kelvin	K
Amount of substance	n	mole	mol
Luminous intensity	I_v	candela	cd

1.2 Further Definitions of Standards

The SI unit of time t is the second (s). Originally, it was defined as 1/86 400 part of the mean solar day. Since the accuracy of spectroscopic methods is greater than that of astronomical observations, the second has now been redefined as the duration of 9 192 631 770 periods of the radiation corresponding to the transition between the two hyperfine levels in the ground state of the cesium-133 atom.

The SI unit of thermodynamic temperature T is the kelvin (K). The kelvin is 1/273.16 of the thermodynamic temperature of the triple point of water, the point at which liquid water, water vapor, and ice are in equilibrium (see Fig. 2.8).

The SI unit of electric current I is the ampere (A). It is defined as the constant current which, if maintained in two parallel conductors of infinite length, of negligible circular cross section, and placed *in vacuo* one meter apart, would produce between the conductors a force of exactly 2×10^{-7} newton per meter of length.

The SI unit of luminous intensity I_v is the candela (cd). It is defined as the luminous intensity, in the perpendicular direction, of a surface of 1/600 000 square meters of a black body at the temperature of freezing of molten platinum under a pressure of 101 325 newtons per square meter.

1.3 Amount of Substance

The basic quantity known as *amount of substance* would not be required if our interest were only the study of physics. It is when we begin to study chemical changes that the necessity (or at least the strong desirability) for such a quantity becomes evident. Much of the long history of chemistry has been concerned with the quantitative measurement of the products of reactions between chemical compounds. Chemical changes are governed by principles of stoichiometry, which are based upon the atomic and molecular structures of chemically reacting substances.

The General Conference on Weights and Measures decided in 1971 to incorporate the fundamental combining unit in chemical reactions into the set of basic physical quantities. They thus defined a new physical quantity, the *amount of substance n*.

The SI unit of amount of substance is the mole (abbreviation, mol). The mole is the amount of substance in a system that contains as many elementary entities as there are carbon atoms in exactly 0.012 kg of carbon-12. The elementary entity must be specified, and may be an atom, a molecule, an ion, an electron, a photon, etc., *or a specified group of such entities*. Examples are as follows:

1. One mole of HgCl has a mass of 0.23604 kg.
2. One mole of Hg_2Cl_2 has a mass of 0.47208 kg.
3. One mole of Hg has a mass of 0.20059 kg.
4. One mole of $Cu_{0.5}Zn_{0.5}$ has a mass of 0.06446 kg.
5. One mole of $Fe_{0.91}S$ has a mass of 0.08288 kg.
6. One mole of e^- has a mass of 5.4860×10^{-7} kg.
7. One mole of a mixture containing 78.09 mol % N_2, 20.95 mol % O_2, 0.93 mol % Ar, and 0.03 mol % CO_2 has a mass of 0.028964 kg.

Note particularly the last example. It is perfectly correct to specify a mole of air or a mole of a reaction mixture such as $2Cu + \frac{1}{2}O_2$.

The *amount* of substance is thus a measure of the number of elementary units of the substance. If we wanted to buy some apples at the store, we could buy them by the kilogram (a certain mass), by the bushel (a certain volume), or by the dozen (a certain number). In the same way, the chemist can specify a measured mass of a chemical, a measured volume, or a measured amount (number) of chemical units. The number of elementary units in one mole is an experimentally measured quantity called the Avogadro constant $L = 6.022 \times 10^{23}$ mol^{-1}.

The definition of the mole made a concession to history at the expense of consistency. The mole is based upon 12 g rather than 12 kg of carbon-12. Thus "atomic and molecular weights" are numerically equivalent to masses per amount of substance in units of g mol^{-1}. The SI unit of mass per amount of substance, however, is kg mol^{-1}.

The terms "atomic weight" and "molecular weight" have always been anomalous since they are not weights but rather dimensionless ratios of masses to a standard mass. We shall not use these terms but instead always use *mass per amount of substance, M*. When working with consistent sets of SI units, we must express M in units of kg mol^{-1}. The numbers in tables and on reagent bottles, however, correspond to g mol^{-1}.

1.4 Subsidiary Units

The SI system recommends that subsidiary units be confined to positive or negative powers of 10^3 times the basic unit. These powers have special prefixes:

10^{-15}	femto	f	1	—	—
10^{-12}	pico	p	10^3	kilo	k
10^{-9}	nano	n	10^6	mega	M
10^{-6}	micro	μ or u	10^9	giga	G
10^{-3}	milli	m	10^{12}	tera	T

Other prefixes in common use are: 10^{-2}, centi, c; 10^{-1}, deci, d. The prefixes are written in abbreviations without spacing. Examples are: nanometer, nm; microsecond, μs; gigasecond, Gs.

There is a certain awkwardness with units of volume. The cubic decimeter, dm³, is called the *liter*, L. Hence the cubic centimeter, cm³, is a milliliter, mL. Although these volume units are not in strict accord with SI preferences, they will certainly continue to be widely used, since most people buy milk, wine, and gasoline by the liter.

When units are divided, the quotients can be abbreviated as powers or as fractions. Thus meter per second yields m/s or m s⁻¹. One should never write abbreviations with a sequence of two slant lines. Thus mol cm⁻² s⁻¹ might be written mol/cm² s but never mol/cm²/s.

If these rules are accepted early in the study of chemistry, they will soon become second nature as you correctly use the international language of science.

1.5   Pressure

The mechanical quantity *force* **F** can be defined by Newton's equation

$$\mathbf{F} = m\mathbf{a} \tag{1.1}$$

where **a** is the acceleration. The dimensions of force are $m\, \ell\, t^{-2}$ and the SI unit is kg m s⁻², called the *newton* N.

Pressure is a force per unit area. It has the dimensions $m\, \ell\, t^{-2}/\ell^2 = m\, \ell^{-1}\, t^{-2}$. The SI unit of pressure is accordingly N m⁻² = kg m⁻¹ s⁻². This unit, one newton per square meter, is called the *pascal*, Pa. The standard atmosphere equals 101 325 Pa. The kilopascal, kPa, is often used; 1 kPa = 9.87×10^{-3} atm. Thus, roughly, 1 atm = 100 kPa.

The difficulty in a switch to pascals is that many thermodynamic data are referred to a *standard state* of one standard atmosphere. Consider, for example, an equilibrium between gases, $CO + \frac{1}{2}O_2 = CO_2$. The older generation was brought up to write the equilibrium constant for this reaction as

$$K_P = P_{CO_2}/P_{CO}P_{O_2}^{1/2} \tag{1.2}$$

As we shall later prove (page 159), the pressures that appear in this expression are in reality the ratios of the partial pressures of gases in the mixture at equilibrium to a standard pressure $P^\circ = 1$ atm. Thus the K_P in Eq. (1.2) does not have the dimensions of $P^{-1/2}$ but is a dimensionless ratio.

In our work, we use the same standard state but now we write $P^\circ = 101.325$ kPa. The equilibrium constant in Eq. (1.2) thus becomes

$$K_P = \frac{P_{CO_2}/P^\circ}{(P_{CO}/P^\circ)(P_{O_2}/P^\circ)^{1/2}} = \frac{P_{CO_2}(P^\circ)^{1/2}}{P_{CO}P_{O_2}^{1/2}} \tag{1.3}$$

We have thereby lost the simplification obtained by setting $P^\circ = 1$ atm and then dividing out the unities. This is not too high a price to pay to maintain (1) a uniform system of SI units and (2) the correct dimensions in equilibrium constants.

1.6 Dimensional Analysis

Any equation between physical quantities must be dimensionally correct. Quantities that have different dimensions cannot be equated. It is essential to check the dimensional consistency of any equation that you derive or use. For example, suppose that you saw an equation, $\frac{1}{2}mv^2 = \frac{2}{3}PV$, where v is a speed and V a volume. The dimensions on each side can be written out: $m(\ell\, t^{-1})^2 = (m\, \ell^{-1}\, t^{-2})(\ell^3)$, or $m\, \ell^2 t^{-2} = m\, \ell^2\, t^{-2}$. Thus both sides reduce to the same dimensions and the equation is dimensionally correct. We cannot say, however, from the dimensional analysis whether it is a physically meaningful or useful relation, or whether the constants are correct.

Logarithms and exponentials occur frequently in equations of physical chemistry. No physical quantity can have a dimension that is a logarithm or an exponential of basic dimensions. The arguments of all log or exp functions must be dimensionless numbers, since only in this case are the functions mathematically defined.

1.7 Equations

As long as we can use the basic set of physical quantities of the SI system, symbols for quantities in equations do not imply any particular choice of units. Thus the equations are true no matter what the choice of units. You should never read an equation to specify a particular unit. For example, in $PV = nRT$, P denotes the pressure, V denotes the volume, n denotes the amount of substance, and R is a proportionality factor. Do *not* say "where P is pascals, V is cubic meters, and n is number of moles." You can insert values of P, V, and n into the equation in any units. (With electromagnetic quantities the situation is more complicated, but still no problems arise as long as one stays within the SI system.) Of course you must keep track of the units, so that you will know the units of any derived numerical results. Many students find it convenient to do this alongside the numerical calculation.

A list of symbols, alphabetically arranged, for the various quantities used in this book appears following the preface. Almost all these symbols are in accord with international agreements. The dimensions of the quantity are given. The SI unit can readily be obtained from the dimensions, by substituting the SI units of the fundamental quantities.

1.8 Chemical Reactions

A schematic equation for a chemical reaction is $a\mathrm{A} + b\mathrm{B} = c\mathrm{C} + d\mathrm{D}$, or more generally,

$$v_1 A_1 + v_2 A_2 = v_3 A_3 + v_4 A_4 \tag{1.4}$$

The capital letters A, B, C, D or the A_j specify the reactant and product substances. The a, b, c, d or the v_j are the stoichiometric coefficients or *stoichiometric numbers* of the reaction. They are *dimensionless numbers*. They should not be called "numbers of moles."

Equation (1.4) can be rearranged to

$$0 = v_3 A_3 + v_4 A_4 - v_1 A_1 - v_2 A_2 = \sum v_j A_j \qquad (1.5)$$

When a reaction is written in this form a special rule must be remembered: The stoichiometric numbers are positive for products and negative for reactants.

We often need to specify the extent to which a reaction has proceeded. We can focus attention on one particular reactant or product j in the reaction (1.5). Let n_{j0} be the amount of substance j at the beginning of the reaction. The amount n_j of j when the *extent of reaction* is ξ is given by $n_j = n_{j0} + v_j \xi$, and

$$\xi = \frac{n_j - n_{j0}}{v_j} \qquad (1.6)$$

Since v_j is a dimensionless number, the extent of reaction ξ has the same dimensions as n_j, amount of substance. The SI unit of both n_j and ξ is the mole.

Example 1.1 What is the extent of reaction ξ of $CH_4 + 4Cl_2 = CCl_4 + 4HCl$, when 0.25 mol Cl_2 has reacted?

$n_j - n_{j0} = -0.25$ mol, $v_j = -4$, and from Eq. (1.6), $\xi = -0.25$ mol$/-4 = 0.0625$ mol. The value for extent of reaction is independent of the particular reactant or product that is chosen to follow the progress of the reaction. Thus in the reaction above, if we calculate ξ from the CCl_4 formation, we have $n_j - n_{j0} = 0.0625$ mol, $v_j = 1$, and $\xi = 0.0625$ mol$/1 = 0.0625$ mol.

1.9 Systems

A *system* is defined as a part of the universe that is separated from the rest by definite boundaries. The boundaries need not have any physical reality as long as they can be specified by definite geometrical surfaces. The world outside the boundaries of a system is called the *surroundings*.

Consider a volume of gas enclosed in a cylinder with a piston. If the system is defined to consist solely of the gas, then the cylinder and piston are part of the surroundings. The boundary surface of this system is the interface between the walls of the container and the volume of gas. We can, if we wish, define another system to include the gas and the cylinder and piston that contain it. In this case everything outside the gas, cylinder, and piston is the surroundings. It is necessary to be exact in the definition of the system to be considered.

An *isolated system* is one that does not interact *in any way* with its surroundings. Changes in the surroundings cannot result in any changes in the isolated system. Neither matter nor energy can pass across the boundaries of an isolated system. Systems that are completely isolated from the rest of the universe do not really exist, but as limiting concepts they are often used in thermodynamic discussions.

A *closed system* is one in which there is *no transfer of matter* across the boundaries between system and surroundings. The mass of a closed system remains constant. Note that a closed system can still interact with its surroundings in many other ways, for example, by compression or expansion, by transfers of energy. A closed system is by no means *isolated*.

Matter as well as energy can pass across the boundaries of an *open system*. A chemical reactor with reactants entering at one end and products leaving at the other would be an open system.

You could not rely on your ordinary knowledge of the English language to deduce the meanings of the words "isolated," "closed," and "open" as they are used in physical chemistry. Scientific discourse differs from literary discourse in its much greater emphasis on precision of meaning and description. For scientific terms, synonyms usually do not exist. An important aspect of learning to think and reason as a scientist is to develop precision and clarity in the use of language.

Experiments can be performed upon a system to measure its *properties*. The measured properties of the system are the numerical values of physical quantities, such as pressure, density, temperature, and refractive index.

1.10 Equilibrium States

The measurement and description of the properties of a system are simplified when *time* can be excluded as a variable. In physical chemistry we draw a distinction between problems that are concerned with time and those that are not. Rates of chemical reactions, diffusion rates, conductance of electrolytes, flow of fluids—such phenomena are called *rate processes*. The time dependence of the properties of the system are then of primary concern. In a great range of other phenomena, however, the time variable can be excluded.

If a system is isolated and if none of its properties change with time, the system is said to be in a *state of equilibrium*. Note the requirement that the system be isolated. Closed and open systems can exist in states that do not change with time, but which are not equilibrium states. Nonequilibrium states that do not change with time are called *stationary states*. For example, if one end of a thermally insulated metal bar is kept at fixed temperature T_1 and the other end at fixed temperature T_2 less than T_1, the bar will reach a stationary state in which heat flows steadily from one end to the other. At any point along the bar the temperature will be maintained at some stationary value T between T_1 and T_2. Another example is a chemical reaction in a flow system, in which reactants enter and products leave at steady rates. The composition of the reaction mixture anywhere in the system will have a stationary value as long as the flow rate and temperature are held constant.

Changes in a nonequilibrium system, however, may be so slow as to be undetectable, so that we cannot depend solely upon experimental observations over a period of time to define the equilibrium condition. For example, a mixture of methane and oxygen can be stored in a glass bulb at 25°C for many years without detectable reaction, but if a trace of catalyst is present, a rapid reaction occurs: $CH_4 + 2O_2 = CO_2 + 2H_2O$. Similarly, the properties of white phosphorus can be measured without detectable conversion to red phosphorus. Such systems are said to be in states of *metastable equilibrium*. To specify the stable equilibrium state of a system (the *global equilibrium*), we shall need to apply the theory of chemical thermodynamics, as in Chapter 8. [Can a closed system exist in an equilibrium state?]

One of the important achievements of physical chemistry has been to show how equilibrium states can be specified and to derive relations between different properties of a system in a state of equilibrium. The properties of a system in equilibrium are *functions of state*; i.e., they do not depend on the history of the system before it reached its state of equilibrium.

If we have two 1-kg samples of pure methanol at the same pressure and temperature, they have the same volume V, density ρ, the same thermal and electrical conductivities, and so on. The fact that one sample was made in Germany in 1980 and the other made in Canada in 1976 makes no difference to their properties, which are functions only of the state of the methanol here and now and depend in no way on its history. Not all properties of substances, however, are state functions. For instance, the magnetization of a piece of iron that has recently been near a magnet will be quite different from that of a piece of iron that has recently been annealed at 1000 K. Even with the same P, T, and mass, these two systems have properties that depend on their past histories.

Properties that do not depend on the amount of substance or mass of the system are called *intensive properties*. Examples are P and T. If we divide a system at equilibrium into two parts, the P and T of each part is not changed. Properties that depend on the amount of substance or mass of the system are called *extensive properties*. Examples are mass m, amount of substance n, and volume V. The description of any system requires specification of the value of at least one extensive property and one intensive property.

The equilibrium state of the system of pure liquid methanol can be defined by stating the mass m, temperature T, and pressure P. Instead of mass m, amount of substance n can be specified. We need only these few variables because we have decided to eliminate other factors in the experiments at hand. Thus gravitational, electric, and magnetic fields are excluded (or assumed to be constant).

Problems

1. The density of mercury at 0°C is 13.595×10^3 kg m^{-3}. What pressure in pascals is exerted by a column of mercury 1.000 mm high at 0°C if gravitational acceleration is 9.80665 m s^{-2}? This obsolete pressure unit is called the *torr*.

2. The energy unit electron volt (eV) is the energy acquired by the elementary electronic charge e when its electric potential increases by one volt. Calculate J/eV.

3. The pressure unit *bar* is 10^5 N/m^2. Calculate pascals per bar and the pressure of 1 standard atmosphere in bars.

4. Which of the following equations are dimensionally correct?
   (a) $PV = RT$
   (b) $\frac{1}{2}mv^2 = kT \ln (P/P°)$
   (c) $nRT \ln C = E$ (where E is energy)

5. From the Einstein relation $E = mc^2$, calculate the energy corresponding to the rest mass of an electron and of a proton.

6. What is the weight of a 1-kg mass in Quito ($g = 9.780$ m s^{-2}) and in Minneapolis ($g = 9.806$ m s^{-2})?

7. For the reaction $H_2S + \frac{3}{2}O_2 = H_2O + SO_2$, what is the extent of reaction ξ when exactly 1 mol O_2 has reacted? For $2H_2S + 3O_2 = 2H_2O + 2SO_2$, what is ξ when 1 mol H_2S has reacted?

2

States of Matter

This chapter introduces the states of matter: gases, liquids, and solids. Pure substances can usually be assigned unequivocally to one of these states, but they do not cover everything. Is a blade of grass a solid? Is a windowpane a liquid? Is a star a dense gas? What is a rubber band? At least at the beginning, the physical chemist puts aside such disturbing questions and gives his or her attention to more amenable substances, such as nitrogen, water, and diamonds.

Before we can discuss the properties of even such clearly defined states of matter, we need certain rules and definitions. One purpose of this chapter is to outline some of the basic framework that will allow us to construct a scientific account of chemical systems, as we find them in the laboratory, in our environment, or in ourselves.

A good deal of the discussion will be devoted to gases, both their measured properties and theories based on simple molecular models that can explain many of the experimental facts. To a rough first approximation, molecules in a gas keep so far apart, on the average, that the properties of the gas are simply the sum of the properties of individual molecules. Then, in more realistic theories, forces between molecules enter the model, and we can use measured data on gases to obtain information about intermolecular forces.

Why do we begin with a study of gases? One of the principal subjects of this book is the theory of chemical equilibrium and of the equilibrium properties of systems. This subject is the domain of chemical thermodynamics. Thermodynamics is a science of the broadest generality. Nevertheless, to illustrate thermodynamic principles, it is helpful to students to give applications that can be treated by simple mathematical equations. The behavior of gases when intermolecular forces are neglected (ideal gases) provides many examples of this kind. At the ordinary pressures and temperatures on the surface of our planet, the ideal-gas model gives answers within a few percent of experimental values.

2.1 Equations of State

The four basic physical quantities that we shall use to specify the equilibrium state of a system consisting of a single pure substance are n, V, P, and T. The state can be specified by giving the values of any three of these variables, since a relation exists between them of the form

$$F(n, V, P, T) = 0 \qquad (2.1)$$

Here F denotes some relationship between the variables n, V, P, and T. In some cases F can be written as an explicit mathematical function. In other cases, values of the variables in different states may be tabulated without writing the function F explicitly. The fact expressed in Eq. (2.1) is simply that, for a given amount of substance n, if we know two of the variables V, P, T, then the value of the third is unequivocally fixed.

Particular forms of equations like Eq. (2.1), which give a relation between the variables of state, are called *equations of state*. They can be obtained by fitting experimental n, P, V, T data to empirical equations that contain constants that can be adjusted to fit the data. Alternatively, theoretical equations can be derived from models of gases, liquids, and solids, the structures of molecules and the forces between them. Theoretical derivations of equations of state are much further advanced for gases than for liquids or solids.

2.2 The Ideal-Gas Equation of State

An *ideal gas* is defined as a gas that has the following equation of state:

$$PV = nRT \qquad (2.2)$$

Here R is a universal *gas constant*. In SI units, $R = 8.314 \text{ J K}^{-1} \text{ mol}^{-1}$. This value of R is used in Eq. (2.2) with pressure in pascals and volume in cubic meters. The product PV has the dimensions of energy, since pressure is a force per unit area, which, multiplied by volume, gives force times distance. Thus, in SI units, $\text{J} = \text{Pa m}^3$.

Example 2.1 What pressure is exerted by exactly 1 mol of ideal gas at 273.15 K in a volume of exactly 1 m³?

From Eq. (2.2),

$$P = \frac{(1 \text{ mol})(8.314 \text{ J K}^{-1} \text{ mol}^{-1})(273.15 \text{ K})}{1 \text{ m}^3}$$

$$= 2271 \text{ J m}^{-3} = 2271 \text{ Pa} \left(\frac{\text{J}}{\text{m}^3} = \frac{\text{N m}}{\text{m}^3} = \frac{\text{N}}{\text{m}^2} = \text{Pa} \right)$$

The ideal-gas equation includes two fundamental *gas laws*, which are followed quite closely by real gases at low pressures. These laws were discovered by experimental measurements. Robert Boyle found in 1660 that the volume of a given amount of gas at constant temperature varies inversely with pressure. Joseph Gay-Lussac from 1802 to 1808 showed that the volume of a given amount of gas at constant pressure varies directly with temperature. An ideal gas is defined as one that obeys both these laws. Thus, for an ideal gas,

$$\text{Law of Boyle:} \quad PV = \text{const. at const. } T$$
$$\text{Law of Gay-Lussac:} \quad V/T = \text{const. at const. } P$$

These properties of an ideal gas are represented graphically in Fig. 2.1. The curves that show the variation of P with V at constant T are called *isotherms* of the system. The curves for variation of V with T at constant P are called *isobars*.

2.3 Gases at Low Pressure

As real gases become less dense, their actual PVT properties approach more closely to those predicted by the ideal-gas equation. All gases at low enough densities follow the ideal-gas equation.

Density ρ is mass divided by volume:

$$\rho = m/V = nM/V \tag{2.3}$$

where M is the mass per unit amount of substance. Hence, for an ideal gas, from Eq. (2.2),

$$\rho = MP/RT \tag{2.4}$$

Example 2.2 What is the density of N_2 as an ideal gas at 200 K and 100 kPa?

From Eq. (2.4),

$$\rho = m/V = PM/RT$$
$$= \frac{(100 \times 10^3 \text{ Pa})(28.0 \times 10^{-3} \text{ kg mol}^{-1})}{(8.314 \text{ J K}^{-1} \text{ mol}^{-1})(200 \text{ K})} = 1.68 \text{ kg m}^{-3}$$

If we plot ρ vs. P for a real gas, Eq. (2.4) should be followed at sufficiently low pressures. From Eq. (2.4) the slope of the straight line where the linear relation is followed is M/RT, which gives us an experimental determination of the gas constant R, if we know M and T. The value of R is the same for all ideal gases. Hence, once we know R, Eq. (2.4) can be used to measure the molar mass M of any gas.

Example 2.3 The limiting slope of the plot of density vs. pressure for air at 0°C is $1.274 \times 10^{-5} \text{ kg m}^{-3} \text{ Pa}^{-1}$. What is the molar mass of air?

From Eq. (2.4),

$$M = RT\rho/P$$
$$= (8.314 \text{ m}^3 \text{ Pa K}^{-1} \text{ mol}^{-1})(273.15 \text{ K})(1.274 \times 10^{-5} \text{ kg m}^{-3} \text{ Pa}^{-1})$$
$$= 28.93 \times 10^{-3} \text{ kg mol}^{-1}$$

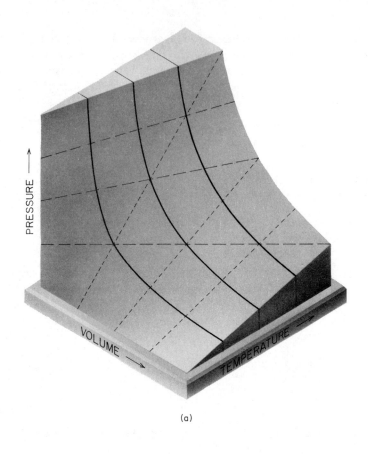

(a)

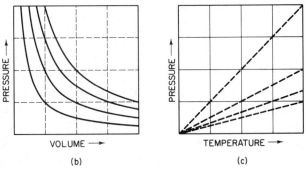

(b)   (c)

FIGURE 2.1 (a) *PVT* surface for an ideal gas. The solid lines are *isotherms,* the broken lines are *isobars,* and the dashed lines are *isometrics.* (b) Projection of *PVT* surface on *PV* plane, showing isotherms. (c) Projection of *PVT* surface on *PT* plane, showing isometrics.

A standard condition of temperature and pressure (STP) for expressing many properties of gases is $T° = 273.15$ K (0°C) and $P° = 101.32$ kPa (1 atm). The volume of one mole of ideal gas under these conditions, from Eq. (2.2), would be

$$V_m^\circ = \frac{nRT^\circ}{P^\circ} = \frac{1 \times 8.3143 \times 273.15}{101.32 \times 10^3} = 0.022415 \text{ m}^3 \text{ mol}^{-1}$$

Comparing this molar volume of an ideal gas with molar volumes V_m° of real gases in Table 2.1, we note that deviations from the ideal volume are easily detectable even at STP.

TABLE 2.1
MOLAR VOLUMES OF GASES IN M³ AT STP
(273.15 K and 101.32 kPa)

Acetylene	0.022085	Ethylene	0.022246
Ammonia	0.022094	Helium	0.022396
Argon	0.022390	Hydrogen	0.022432
Carbon dioxide	0.022263	Methane	0.022377
Chlorine	0.022063	Nitrogen	0.022403
Ethane	0.022172	Oxygen	0.022392

2.4 Gas Mixtures

The composition of a mixture can be specified by stating the amount n_j of each substance that it contains. The total amount of all constituents in the mixture is

$$n = n_1 + n_2 + \cdots = \sum_j n_j$$

The composition of the mixture is then conveniently described by stating the *mole fraction X* of each substance, defined by

$$X_j = \frac{n_j}{n} = \frac{n_j}{\sum n_j} \tag{2.5}$$

Another method of specifying composition is the *concentration*,

$$c_j = \frac{n_j}{V} \tag{2.6}$$

In physical chemistry the term "concentration" always means amount (of a substance) per unit volume of mixture. The SI unit of concentration is mole per cubic meter, but mole per cubic decimeter is more often used. The liter (L) is defined as 10^{-3} m³ or 1 dm³. A solution with a concentration, for example, of 1.63 mol dm⁻³ (mol L⁻¹) is often called a 1.63 molar (1.63 M) solution.

In a mixture of gases, we can define the *partial pressure P_j* of any particular gas as the pressure which that gas could exert if it occupied the total volume all by itself. If we know the concentration of a particular gas in a mixture, we can find its partial pressure from its *PVT* data or equation of state. We *define* an *ideal gas mixture* as one in which the total pressure equals the sum of the partial pressures,

$$P = P_1 + P_2 + \cdots = \sum P_j \tag{2.7}$$

Many gas mixtures follow this equation about as well as individual gases follow the ideal-gas equation. If a mixture of gases obeys Eq. (2.7) and the partial pressure of each gas in the mixture obeys Eq. (2.2), we have an *ideal gas mixture*. In this case, since $P_j = (RT/V)n_j$, we have from Eqs. (2.5) and (2.2),

$$P_j = X_j P \tag{2.8}$$

The partial pressure of each gas in an ideal gas mixture is equal to its mole fraction times the total pressure. Equations (2.7) and (2.8) are versions of *Dalton's Law of Partial Pressures*.

2.5 The Ideal Gas from the Molecular Point of View

We can imagine a gas at moderate pressures to be mostly empty space, in which molecules move at high speeds, making frequent collisions with one another and the walls of the container. For example, the volume of a molecule of H_2 considered as a tiny hard sphere would be about 0.11 nm³, so that the volume of one mole of hard-sphere H_2 molecules would be $(6.02 \times 10^{23}$ mol$^{-1})(1.1 \times 10^{-28}$ m³$) = 6.6 \times 10^{-5}$ m³. Thus only $6.6 \times 10^{-5}/22.4 \times 10^{-3}$ or 0.3 % of the total volume of the gas would be occupied by the hard-sphere volumes at STP.

The pressure of a gas is caused by collisions of molecules with the rigid walls of the container. Pressure is a force per unit area, and from the principles of Newtonian mechanics, the force on a body is equal to the rate of change of its momentum. Consider in Fig. 2.2 a molecule approaching a wall in the positive x direction with a velocity u. If the mass of the molecule is m, its momentum is mu, and its kinetic energy is $\frac{1}{2}mu^2$. The collisions of the molecules with the wall are perfectly elastic; i.e., no energy is transferred to the wall by the collisions. Thus molecules rebound with velocities that are changed in direction but not in magnitude. The velocity of a molecule after collision is $-u$, its momentum is $-mu$, and its kinetic energy is still $\frac{1}{2}mu^2$. The change in momentum of a molecule in the collision is $-2mu$. The momentum transferred to the wall is $2mu$.

To calculate the pressure we find the momentum transferred to unit area of wall by all the collisions in unit time. Since the gas as a whole is at rest in its container, the gas molecules are moving at random in all directions. Let us suppose that half

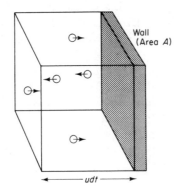

FIGURE 2.2 Collision with an area A normal to the X axis by gas molecules with velocity component in the x direction. The pressure of the gas on the wall is calculated from the rate of transfer of momentum to the wall (force) per unit area.

the molecules have the same velocity u in the direction perpendicular to the wall. Then in unit time half the molecules in a volume Au strike an area A of the wall. (The other half are moving away from the wall.) If there are N molecules in a volume V, the number of collisions with the area A in unit time is $(N/V)Au/2$. Each collision transfers a momentum $2mu$ so that the momentum transfer in unit time to area A is $(N/V)(Au)(mu)$. This rate of transfer of momentum gives the magnitude of the force F on area A, so that the pressure is $P = F/A = Nmu^2/V$.

Actually, the molecules do not all have the *same* velocity component u. Hence we must take the average value of u^2 to obtain

$$P = Nm\overline{u^2}/V \tag{2.9}$$

Here $\overline{u^2}$ is the *mean square* velocity component, i.e., the average value of the squares of the individual values of u.

In reality the molecules in a volume of gas are moving in all directions, not just perpendicular to a given wall as in the calculation above. In Fig. 2.3 we have designated u, v, and w as the components of a molecular velocity in the x, y, and z directions, respectively. The magnitude and direction of any molecular velocity can then be represented by a vector from the origin to a point (u, v, w). The square of the magnitude of the velocity is $c^2 = u^2 + v^2 + w^2$. [Prove this.]

In a volume of gas at equilibrium no particular direction of velocity is preferred to any other. Thus, on average, $\overline{u^2} = \overline{v^2} = \overline{w^2}$, and $\overline{u^2} = \frac{1}{3}\overline{c^2}$. Hence Eq. (2.9) for the pressure becomes

$$P = Nm\overline{c^2}/3V \tag{2.10}$$

The kinetic energy E_k of the gas molecules is $N(\frac{1}{2}m\overline{c^2})$, so that Eq. (2.10) yields

$$PV = \tfrac{2}{3}E_k \tag{2.11}$$

Thus the product PV for an ideal gas is a measure of the kinetic energy of the gas molecules.

Example 2.4 What is the kinetic energy of the molecules in 1 mol of ideal gas at STP?

For $n = 1$, the volume $V = 0.22415$ m³ (STP). $P° = 101.32$ kPa. From Eq. (2.11) the kinetic energy is

$$E_k = \tfrac{3}{2}PV = (\tfrac{3}{2})(101.32 \times 10^3 \text{ Pa})(0.022415 \text{ m}^3) = 3406 \text{ Pa m} = 3406 \text{ J}$$

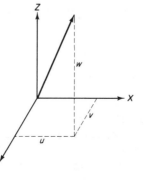

FIGURE 2.3 Components of velocity u, v, w in directions of Cartesian axes X, Y, Z.

For the molecular model of an ideal gas we have shown that $E_k = \frac{3}{2}PV$. For the experimental ideal gas $PV = nRT$. If we now make the assumption that the kinetic-theory model of the ideal gas predicts exactly the experimental behavior, we can write

$$E_k = \tfrac{1}{2}Nm\overline{c^2} = \tfrac{3}{2}nRT \tag{2.12}$$

We see that kinetic theory thus identifies the temperature T as a measure of the average kinetic energy of the gas molecules. With this identification, the molecular model gives the equation of state of the ideal gas.

2.6   Molecular Speeds

It is interesting to use Eq. (2.12) to obtain an idea of the average speeds of gas molecules. Since $M = Nm/n$,

$$\overline{c^2} = 3nRT/Nm = 3RT/M \tag{2.13}$$

The square root of $\overline{c^2}$ is called the *root mean square (rms) speed*. From Eq. (2.13),

$$(\overline{c^2})^{1/2} = c_{rms} = (3RT/M)^{1/2} \tag{2.14}$$

The average speed $\bar{c}$ of the molecules can be calculated from kinetic theory (Section 5.11). It differs slightly from c_{rms}:

$$\bar{c} = (8RT/\pi M)^{1/2} \tag{2.15}$$

Values of $\bar{c}$ for various gas molecules at 273.15 K are given in Table 2.2. The average speed of a methane molecule at 273.15 K is close to the top speed of the Concorde.

TABLE 2.2
AVERAGE SPEEDS OF GAS MOLECULES AT 273.15 K

Gas	$\bar{c}$ (m s^{-1})	Gas	$\bar{c}$ (m s^{-1})
Ammonia	582.7	Helium	1204.0
Argon	380.8	Hydrogen	1692.0
Benzene	272.2	Mercury	170.0
Carbon dioxide	362.5	Methane	600.6
Carbon monoxide	454.5	Nitrogen	454.2
Chlorine	285.6	Oxygen	425.1
Deuterium	1196.0	Water	566.5

2.7   Condensation of Gases—Critical Points

In 1877, Louis Cailletet succeeded in liquefying oxygen and nitrogen by rapid expansion of the cold compressed gases. For every gas there exists a *critical temperature T_c*, above which it cannot be liquefied no matter how great the applied pressure. The pressure that just suffices to liquefy the gas at T_c is called its *critical pressure P_c*.

The volume of the gas at P_c and T_c is its *critical volume* V_c. The state of the gas at P_c, V_c, T_c is called its *critical point*. The values T_c, P_c and V_{mc} are the *critical constants* of the gas. Critical constants for various gases are cited in Table 2.3.

TABLE 2.3

CRITICAL POINT DATA AND VAN DER WAALS[a] CONSTANTS

Formula	T_c (K)	P_c (MPa)	$10^6 V_c$ (m^3 mol^{-1})	$10^3 a$ (m^6 Pa mol^{-2})	$10^6 b$ (m^3 mol^{-1})
He	5.3	0.229	61.6	3.45	23.7
H_2	3.33	1.30	69.7	24.7	26.6
N_2	126.1	3.39	90.0	141	39.1
CO	134.0	3.51	90.0	151	39.9
O_2	154.3	5.04	74.4	138	31.8
C_2H_4	282.9	5.16	127.5	453	57.1
CO_2	304.2	7.38	94.2	364	42.7
NH_3	405.6	11.37	72.0	422	37.1
H_2O	647.2	22.06	55.44	553	30.5
Hg	1735.0	105.0	40.1	820	17.0

[a]See Section 2.10.

The behavior of a gas in the neighborhood of its critical point was first studied by Thomas Andrews in 1869 in a classic series of measurements on carbon dioxide. Results of more recent determinations of the P–V isotherms of CO_2 around the critical temperature of 304.16 K are shown in Fig. 2.4. Consider the isotherm at 303.55 K, which is slightly below T_c. As the vapor is compressed, the P–V curve first follows AB, which is approximately a Boyle's Law isotherm. When point B is reached, a meniscus appears and liquid begins to form. The liquid state and the gaseous state now coexist in equilibrium. The pressure $P = 7250$ kPa is the pressure of gaseous CO_2 existing in equilibrium with liquid CO_2 at 303.55 K. This equilibrium pressure is called the *vapor pressure*. On further compression, the volume decreases at this constant pressure until point C is reached, at which all the vapor has been converted into liquid. The curve CD is then the 303.55 K isotherm of liquid carbon dioxide, its steepness indicating the low compressibility of the liquid.

As isotherms are taken at successively higher temperatures, the points of discontinuity B and C are observed to approach each other gradually, until at 304.16 K they coalesce, and no appearance of a second phase is observable. This isotherm corresponds to the critical temperature of CO_2. Isotherms above this temperature exhibit no formation of a second phase no matter how great the applied pressure. Above its critical temperature, a substance is said to be in a *fluid state*.

There is a *continuity of states* from liquid to fluid to gas. This fact may be demonstrated by following the path $EFGH$. The gas at point E, at temperature below T_c, is warmed at constant volume to point F, above T_c. It is then compressed along the isotherm FG, and finally cooled at constant volume along GH. At point H, below T_c, carbon dioxide exists as a liquid, but at no point along the path $EFGH$ from gas to liquid are two phases simultaneously present. The transformation from gas to liquid has occurred smoothly and continuously.

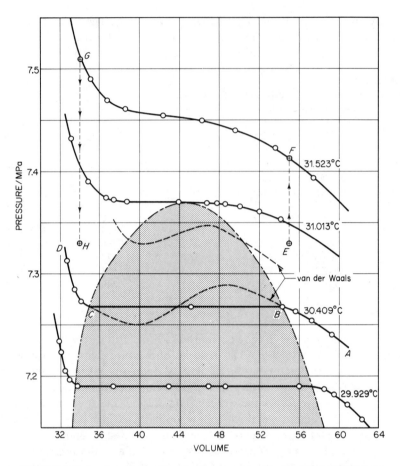

FIGURE 2.4 Isotherms of carbon dioxide near the critical point of 31.013°C. The two-phase liquid–vapor region is shaded. [A. Michels, B. Blaisse and C. Michels, *Proc. R. Soc. Lond. A. 160*, 367 (1937)]

2.8 Real Gases—Virial Equations

Many equations of state have been proposed for real (nonideal) gases, derived from different theoretical models or based on different ideas about how to fit experimental *PVT* data to an empirical equation. The most general way to fit data to an equation is to use a power series. Since deviations from ideality depend on the density of the gas, it would be reasonable to represent the equation of state as a power series in n/V and to include as many terms as may be necessary to represent the experimental *PVT* data with the desired accuracy. The equation so obtained is called a *virial equation* from the Latin *vir*, power. The virial equation can be written as

$$\frac{PV}{nRT} = 1 + \frac{nB(T)}{V} + \frac{n^2C(T)}{V^2} + \frac{n^3D(T)}{V^3} + \cdots \qquad (2.16)$$

The coefficients $B(T)$, $C(T)$, etc., are called the second, third, etc., *virial coefficients*. They are functions of temperature T.

Virial coefficients are used also in other branches of physical chemistry, especially the theory of solutions. The virial equation is also the basis of theoretical calculations from statistical models of nonideal gases, in which it turns out that $B(T)$ represents the deviation from ideality due to interactions between pairs of molecules.

An illustration of how the virial equation fits experimental data is shown in the following results for argon at 298 K.

P (kPa)	$\frac{PV}{nRT}$	$=$	1	$+$	$B\left(\frac{n}{V}\right)$	$+$	$C\left(\frac{n}{V}\right)^2$	$+$	remainder
10^2			1	$-$	0.00064	$+$	0.00000	$+$	0.00000
10^3			1	$-$	0.00648	$+$	0.00020	$-$	0.00007
10^4			1	$-$	0.06754	$+$	0.02127	$-$	0.00036
10^5			1	$-$	0.38404	$+$	0.68788	$+$	0.37232

Note that at 10^5 kPa, the first three terms are no longer providing an adequate approximation.

Some values of second virial coefficients $B(T)$ are plotted in Fig. 2.5 as functions

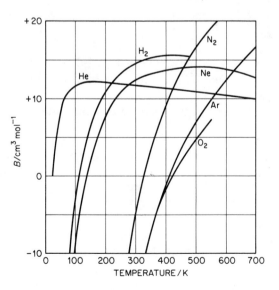

FIGURE 2.5 The second virial coefficients B of a number of gases as functions of temperature.

or T. We note that different gases may vary considerably in their deviations from ideal behavior as measured by the values of $B(T)$.

2.9 Corresponding States

The ratios of P, V, and T to the critical values P_c, V_c, and T_c, respectively, are called the *reduced* pressure, volume, and temperature. These reduced variables are

$$P_R = P/P_c, \qquad V_R = V/V_c, \qquad T_R = T/T_c \qquad (2.17)$$

In 1881, J. H. van der Waals pointed out that, to a fairly good approximation at moderate pressures, all gases follow the same equation of state in terms of reduced variables, P_R, T_R, and V_R. Thus $V_R = F(P_R, T_R)$, where F would be the same function for all the different gases studied. He called this rule the *Principle of Corresponding States*. If this rule were exact, the critical ratio $P_c V_c / nRT_c$ would be the same for all gases. Actually, as can be found from the data in Table 2.3, the ratio varies from 0.20 to 0.33 for the common gases. The Principle of Corresponding States is nevertheless important, since it shows that deviation of a gas from ideality is not determined by the absolute values of P, V, T, but by the reduced values P_R, V_R, T_R, which take into account the nature of the actual gas.

A convenient way to summarize PVT data for real gases is to introduce a *compression factor z*, such that

$$PV = znRT \qquad (2.18)$$

Chemical engineers concerned with the properties of gases at elevated pressures have prepared graphs to show the variation of the z factor with P and T, and they have found to a good approximation, even at fairly high pressures, that z appears to be a universal function of the reduced variables P_R and T_R,

$$z = F(P_R, T_R) \qquad (2.19)$$

This behavior is illustrated in Fig. 2.6 for a number of different gases, where $z = PV/nRT$ is plotted, at various reduced temperatures, against the reduced pressure. At these moderate pressures, the fit is good to within about 1%. These graphs are illustrations of the law of corresponding states.

Example 2.5 Xenon has $T_c = 289.7$ K, $P_c = 5.88$ MPa. From the compression factor in Fig. 2.7, estimate the volume of 1 mol Xe at pressure 12.47 MPa and 320 K. Compare this volume with the ideal-gas value.

$P_R = 2.12$, $T_R = 1.10$. From Fig. 2.7, $z = 0.42$ and hence

$$V = znRT/P = \frac{(0.42)(1 \text{ mol})(8.314 \text{ J K}^{-1} \text{ mol}^{-1})(320 \text{ K})}{12.47 \times 10^6 \text{ Pa}} = 8.96 \times 10^{-5} \text{ m}^3$$

For the ideal gas, $V = 2.13 \times 10^{-4}$ m^3.

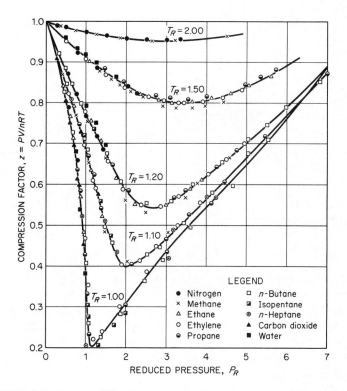

FIGURE 2.6 Compressibility factor as a function of reduced variables of state [Gouq-Jen Su, *Ind. Eng. Chem. 38,* 803 (1946)]

2.10 Van der Waals Equation

Many equations of state have been devised to describe the behavior of real gases. The virial equation (2.16) is an example. The most famous and perhaps most widely used equation is that of J. H. van der Waals who, in 1873, first took up the problem of deriving an equation of state for nonideal gases. He devised a kinetic-theory model for an *imperfect gas*, in which both basic postulates of the perfect-gas model were modified. Instead of treating gas molecules as mass points, he treated them as rigid spheres of diameter *d*. Instead of saying that no forces exist between molecules, he assumed that they exert attractive forces on one another, which lead to condensation under appropriate conditions of *T* and *P*. Figure 2.7 shows the features of the van der Waals model.

The nearest approach of the center of one molecule to the center of another molecule of the same species is *d*. Thus for each pair of molecules a volume of $\frac{4}{3}\pi d^3 = 8v_m$ is excluded from the total gas volume *V*, where v_m is the molecular volume. The total excluded volume for *nL* molecules is thus $nL(4v_m) = nb$, where *b* is a constant for a given species of gas. This correction modifies the ideal-gas equation to

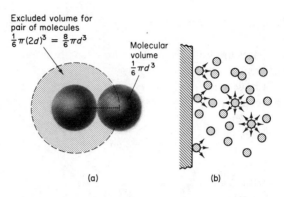

Excluded volume for
pair of molecules
$\frac{1}{6}\pi(2d)^3 = \frac{8}{6}\pi d^3$

Molecular
volume
$\frac{1}{6}\pi d^3$

(a)   (b)

FIGURE 2.7 Van der Waals correction to the model of a perfect gas: (a) excluded volume; (b) intermolecular forces.

$$P(V - nb) = nRT \tag{2.20}$$

Equation (2.20) is useful at high temperatures and high pressures, where the excluded volume is more important than the effect of intermolecular forces.

Van der Waals made the correction for attractive forces as follows. Since attraction occurs between *pairs* of molecules, it must increase as the square of the concentration, $(n/V)^2$. The actual pressure P will be less than the ideal-gas pressure by an amount $a(n/V)^2$, where a is a different constant for each gas. The result of both the van der Waals corrections is, therefore,

$$\left[P + a\left(\frac{n}{V}\right)^2\right](V - nb) = nRT \tag{2.21}$$

Some values of the van der Waals constants were given in Table 2.3. The van der Waals equation represents the PVT behavior of gases quite well when they deviate moderately from ideality. For example, consider the following ratios of PV (calculated) to PV (observed) for carbon dioxide at 313.15 K and various pressures $P/P°$ ($P° = $ 101.32 kPa).

$P/P°$	1	10	50	100	200	500	1000
PV(calc) from Eq. (2.21)/PV(obs)	1.001	1.01	1.04	1.28	1.34	1.35	1.35

In Fig. 2.4, the theoretical isotherms of the van der Waals equation for CO_2 are plotted as dashed lines. Although the van der Waals equation does not represent the P–V behavior in the region where gas and liquid phases coexist, it does display a maximum and a minimum in $P(V)$ in a continuous curve through this region. At the critical point itself, the maximum and minimum converge to a point of inflection. (See Problem 21.)

2.11 Liquids

Liquids and gases are both *fluids*. They flow readily when subject to small shear forces. Above the critical temperature T_c no distinction can be drawn between liquid and gas. Below T_c, however, liquid and gas can coexist as separate phases in equilibrium. When a pure liquid is introduced into an evacuated vessel of fixed volume maintained at a temperature T above its freezing point, some liquid evaporates. Evaporation continues until either (1) all the liquid is converted to vapor, or (2) a limiting pressure is reached, $P^•$, the *vapor pressure* of the liquid at T. At $P^•$ and T, liquid and vapor coexist in equilibrium. The rate of evaporation of liquid to vapor then exactly equals the rate of condensation of vapor to liquid, so that no further change occurs in the relative amounts of the two phases, as long as the temperature and volume of the vessel are kept constant.

The attractive and repulsive intermolecular potential energies are both large compared to the translational kinetic energy of molecules in a liquid. Cohesive forces in liquids cannot even be treated as the sum of interactions between pairs of molecules. For these reasons it has not yet been possible to obtain adequate theoretical equations of state for liquids.

Even though the van der Waals equation does not represent accurately the PVT behavior of liquids, it nevertheless provides a valuable insight into the nature of the liquid state. It displays the continuity of states between liquid and gas and the existence of a critical point. It emphasizes the importance of short-range intermolecular forces. The short-range attractive forces in liquids, represented by the a term in the van der Waals equation, lead to condensation of vapor to liquid when the temperature is below the critical temperature T_c and the density of the system becomes high enough.

2.12 Compressibility and Expansivity

The variation of V with P at constant T defines the *compressibility* β of a substance,

$$\beta = -\frac{1}{V}\left(\frac{\partial V}{\partial P}\right)_T \tag{2.22}$$

The variation of V with P is written as a partial derivative since $V(P, T)$ is a function of both P and T, and T is held constant while the variation of V with P is determined. Usually, the volume decreases as the pressure increases so that $(\partial V/\partial P)_T$ is negative. Therefore, a minus sign is included in the definition since a positive coefficient β is more convenient.

The variation of V with T when P is held constant defines the *expansivity* α of a substance, sometimes called the *coefficient of thermal expansion*:

$$\alpha = \frac{1}{V}\left(\frac{\partial V}{\partial T}\right)_P \tag{2.23}$$

Although α and β are themselves functions of P and T, they can often be taken as constant over moderate ranges of these variables.

There is a useful mathematical relation between the complete differential of a state function (such as V) and the partial coefficients (such as α, β). From $V(T, P)$,

$$dV = (\partial V/\partial T)_P \, dT + (\partial V/\partial P)_T \, dP \tag{2.24}$$

The only way to calculate a change in V when T changes to $T + dT$ and P changes to $P + dP$ is to follow a path in which P is held constant while T is changed, and then T is held constant while P is changed. The expression (2.24) gives dV for a change along such a path. [Please draw a diagram of V as function of T and P, and indicate such a path.]

Suppose that we know α and β for a substance and wish to calculate $(\partial P/\partial T)_V$, i.e., the change in pressure with temperature while volume is held constant. This can be an important calculation from the point of view of safety in storage of chemicals and gases or in the use of closed vessels as reactors. When $V = $ const., $dV = 0$, so that Eq. (2.24) becomes $0 = (\partial V/\partial T)_P \, dT + (\partial V/\partial P)_T \, dP$, and

$$\left(\frac{\partial P}{\partial T}\right)_V = -\frac{(\partial V/\partial T)_P}{(\partial V/\partial P)_T} = \frac{\alpha}{\beta} \tag{2.25}$$

Example 2.6 Suppose that a mercury thermometer in which the column is filled at 50°C is overheated to 52°C. What pressure will develop inside the thermometer?

For mercury at 50°C, $\alpha = 1.77 \times 10^{-4}$ K^{-1} and $\beta = 3.80 \times 10^{-8}$ kPa^{-1}. Thus from Eq. (2.25), $(\partial P/\partial T)_V = \alpha/\beta = 4660$ kPa/K. For $\Delta T = 2$ K, $\Delta P = 9320$ kPa. This pressure is likely to burst the thermometer. This example is a warning never to heat anything in a closed container before a $(\partial P/\partial T)_V$ has been calculated to show that the final pressure is within safe limits.

2.13   The Solid State

The three states of matter are often distinguished by their mechanical properties. Solids and liquids have definite volumes, whereas gases expand to fill any container. Solids have definite shapes; under typical laboratory conditions gases and liquids have shapes determined by their containers. In response to a small applied force, either compression or shear, a solid undergoes an elastic deformation; when the force is removed the solid springs back to its original shape. A liquid responds to a small shear force by a displacement of one layer over another in the process of *viscous flow*.

In a gas, the molecules are moving randomly through space and, at any instant, their arrangement is disordered except for transitory pairing associations. In a liquid, the molecules fall into ordered arrangements, but any particular arrangement persists at most for a fraction of a second, and each ordered region is limited in extent to at most a few hundred molecules. In the solid state, however, an ordered region can extend through a large volume and can persist virtually unaltered for long periods of time.

We have noted (e.g., in Fig. 2.4 for carbon dioxide) the continuity of states between gas and liquid. No such continuity can exist between gas and solid or between liquid and solid. There is no continuous path between disorder and long-range order, and the passage from liquid or gas to solid always occurs discontinuously, with a definite interface between the solid and the fluid states.

Equations of state for solids have been fitted to experimental PVT data, but for most purposes the data are used in tabulated form, usually as compressibilities β and expansivities α.

2.14 Phases

Changes such as the melting of ice, the solution of sugar in water, the vaporization of benzene, or the transformation of graphite to diamond, are called changes in state of aggregation or *phase changes*. They are characterized by discontinuous changes in certain properties of the system at some definite temperature and pressure. The word *phase* is derived from the Greek word for "appearance." We distinguish phase changes from chemical changes, which involve chemical reactions, and physical changes such as expansion or compression that occur continuously with variations in temperature or pressure. In the solid state especially, the distinction between a chemical change and a phase change is not always possible to maintain, since certain solid phases exist over a range of compositions within which the structures may exhibit various degrees of disorder.

As defined by J. Willard Gibbs, a phase is a mechanically separable part of a system that is uniform throughout, not only in chemical composition, but also in physical state. Examples are a volume of air, a noggin of rum, or a cake of ice. Mere difference in subdivision is not enough to determine a new phase; a mass of cracked ice is still only one phase. We are assuming here that a variable surface area has no appreciable effect on the properties of the substance.

A system consisting of a single phase is called *homogeneous*. A system of more than one phase is *heterogeneous*. Thus water containing cracked ice is a heterogeneous system of two phases. [Is a liter of blood a single phase? a glass of champagne?] In systems consisting entirely of gases, only one phase can exist at equilibrium. In systems consisting entirely of liquids, depending on their mutual solubility, one or more phases can coexist at equilibrium. Such systems can even remain heterogeneous above the critical temperatures of the liquids. Solids usually enjoy limited intersolubility, and many different solid phases can coexist in a system at equilibrium.

2.15 Equilibrium between Phases

We shall often consider systems in which solid, liquid, and gaseous phases coexist at equilibrium. A familiar example is illustrated in Fig. 2.8. This diagram shows the behavior of water at relatively low pressures, at which the only form of solid water is ordinary ice I.

The ranges of pressure and temperature over which ice, liquid water, and water

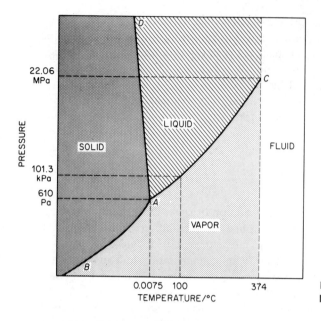

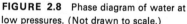

FIGURE 2.8 Phase diagram of water at low pressures. (Not drawn to scale.)

vapor can exist as single phases are represented by areas on the diagram. The lines dividing these areas represent the states of P, T for which two different phases of water can coexist in equilibrium. Thus AB, the coexistence line of ice and water vapor, is the sublimation curve of ice. The line AC is the vapor-pressure curve of liquid water; it terminates at C, which is the critical point of water, $T_c = 647.2$ K, $P_c = 22.06$ MPa. The line AD is the melting-point curve of ice I (or the freezing-point curve of water, which is the same thing for an equilibrium process).

When the value of P is $P°$, 1 standard atmosphere, 101.32 kPa, the corresponding values of $T_f°$ and $T_b°$ are the normal freezing point and boiling point of water, respectively.

The point A is the triple point of ice I–water–water vapor, $P_t = 610$ Pa, $T_t = 273.16$ K. The temperature T_t is the single exact fixed point established by definition as the reference temperature for the Kelvin scale of temperature.

A diagram analogous to Fig. 2.8 can be drawn as the *phase diagram* of any pure substance for which adequate experimental data are available. Phase diagrams can also be constructed for systems consisting of mixtures and solutions of two or more different substances.

2.16 Components

The composition of a system can be completely defined in terms of its *components*. The ordinary meaning of the word *component* is restricted in this technical usage. The components are the minimum number c of chemically distinct constituents needed to define the composition of each phase in the system.

A practical way to define the number of components is to set it equal to the total number of *independent* chemical constituents in the system minus the number of distinct chemical reactions that can occur in the system between these constituents. By number of independent constituents we mean the total number minus the number of any restrictive conditions, such as material balance or charge neutrality. By a distinct chemical reaction we mean one that cannot be written simply as a sequence of other reactions in the system.

Consider, for example, the system consisting of calcium carbonate, calcium oxide, and carbon dioxide. There are three distinct chemical constituents, $CaCO_3$, CaO, and CO_2. One reaction occurs between them: $CaCO_3 = CaO + CO_2$. Hence, the number of components $c = 3 - 1 = 2$. The chemical composition of the solid $CaCO_3$ phase could be described as $CaO \cdot CO_2$.

Another example is the system formed by mixing NaCl, KBr, KCl, and H_2O. The composition of any phase can be specified in terms of four ions (Na^+, K^+, Cl^-, and Br^-) plus H_2O, but the condition of electroneutrality requires that $Na^+ + K^+ = Cl^- + Br^-$. Therefore, $c = 5 - 1 = 4$.

2.17 Degrees of Freedom

Equilibria between phases are independent of the actual amounts of the phases that may be present. Thus the vapor pressure above liquid water in no way depends on the volume of the vessel or on whether a few mm³ or maybe m³ of water are in equilibrium with the vapor phase. Similarly, the concentration of a saturated solution of salt in water is a fixed and definite quantity, regardless of whether a large or a small excess of undissolved salt is present.

In discussing phase equilibria, therefore, we do not consider the extensive variables, which depend on the masses of the phases. We consider only the intensive variables, such as temperature, pressure, and concentrations. Of these variables, a certain number may be independently varied, but the rest are fixed by the values chosen for the independent variables and by the thermodynamic requirements for equilibrium. The number of the intensive state variables that can be independently varied without changing the number of phases is called the number of *degrees of freedom f* of the system, or sometimes, the *variance*.

For example, the state of a certain amount of a pure gas may be specified completely by any two of the variables, pressure, temperature, and density. If any two of these are known, the third can be calculated. Therefore, this system has two degrees of freedom; it is a *bivariant* system. In the system water–water vapor, only one variable need be specified to determine the state. At any given temperature, the pressure of vapor in equilibrium with liquid water is fixed. This system has one degree of freedom, or is said to be *univariant*.

2.18 The Phase Rule

When a system containing one or more components in one or more phases comes to equilibrium, there is an important general relation that must be satisfied among f, c, and p. This is the famous *phase rule* of Willard Gibbs (1878):

$$f = c - p + 2 \qquad (2.26)$$

We can readily derive this rule as follows. Temperature T and pressure P are two intensive state variables that provide the "2" in the phase rule. The remaining $c - p$ degrees of freedom are the concentration variables needed to specify the state of the system at equilibrium.

The concentrations of c components in a phase can be specified by $c - 1$ values. The remaining one is obtained by difference. For example, if a phase contains A, B, C with 50% A and 30% B, we know that it must have 20% C. For p phases, therefore, a total of $p(c - 1)$ concentration variables exist.

At equilibrium the ratios of concentrations of a component in different phases are given by distribution coefficients K_d. For a component j in phases a and b, for example,

$$c_j^a / c_j^b = K_d^{ab} \qquad (2.27)$$

These distribution coefficients are important in many practical extraction and purification procedures in chemistry. At present we wish only to emphasize that at equilibrium the concentrations of components in different phases are governed by conditions of the form of Eq. (2.27). For each component there will be $(p - 1)$ conditions like Eq. (2.27) and thus for c components, $c(p - 1)$ conditions in all.

The total number of concentration variables minus the number of conditions gives $p(c - 1) - c(p - 1) = c - p$. Adding 2 for T and P, we have the phase rule, $f = c - p + 2$.

Example 2.7 The system shown in Fig. 2.9 consists of two solutions, diethyl ether in water and water in ether, in equilibrium with a vapor phase at 293 K and 59.3 kPa. How many degrees of freedom exist for this system?

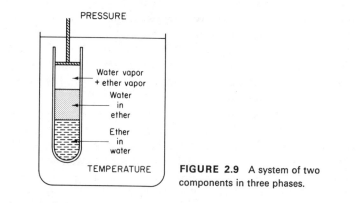

FIGURE 2.9 A system of two components in three phases.

Since $c = 2$ and $p = 3$, $f = c - p + 2 = 1$. Thus once the temperature is specified, the pressure and the compositions of all the phases are fixed.

Problems

1. The volume of one mole of ideal gas at 273.15 K is 0.022414 m³ at 101.32 kPa. What would be the volume at 298.15 K and the same pressure?

2. Calculate the pressure exerted by 1.00 kg of air in a volume of 1.00 m^3 at 300 K, assuming ideal-gas behavior.

3. In an experiment to measure the vapor pressure of naphthalene, a current of air at a flow rate of 1.000 dm^3 min^{-1} at STP (273.15 K and 101.32 kPa) was passed over naphthalene crystals in a tube at 298.15 K. In exactly 1 h, 0.0475 g of naphthalene sublimed and was carried away in the air stream. Calculate the vapor pressure of naphthalene on the assumption that the current of air became saturated with naphthalene vapor.

4. Calculate the volume occupied by 100 g of ethylene C$_2$H$_4$ at 25°C and 300 kPa (a) as an ideal gas; (b) as a van der Waals gas. (See Table 2.3.) (*Hint:* The cubic equation for V can be solved by the Newton–Raphson iterative process.)

5. An evacuated glass bulb weighs 24.8015 g. Filled with dry air at 101.32 kPa and 298.15 K, it weighs 24.9295 g. Filled with a mixture of methane and ethane at the same P and T, it weighs 24.8768 g. Calculate the mole percent of methane in the gas mixture.

6. The density of CH$_3$NH$_2$ at 273.15 K is:

P (kPa)	20.264	50.660	81.056
ρ (g dm^{-3})	0.2796	0.7080	1.1476

Determine the molar mass M of CH$_3$NH$_2$. (Plot ρ/P vs. P.)

7. Henning and Heuse measured the expansivity of helium as a function of pressure.

$\alpha \times 10^6$	3658.9	3660.3	3659.1	3658.2	3658.1
P (mmHg)	504.8	520.5	760.1	1102.9	1116.5

According to the Law of Charles and Gay-Lussac, the volume of an ideal gas is $V = V_0 \alpha T$. Calculate α and $1/\alpha = T_0$.

8. Dinitrogen tetroxide dissociates as N$_2$O$_4$ = 2NO$_2$. 1.0386 g of the compound is introduced into an evacuated bulb of $V = 347.5$ cm^3 at a pressure of 101.45 kPa at 45.0°C. What is the fractional degree of dissociation (ideal-gas approximation)?

9. What are the second and third virial coefficients $B(T)$ and $C(T)$ of a gas that follows the van der Waals equation of state? Expand the van der Waals equation in the form of Eq. (2.16) and compare terms.

10. How many molecules are in 1 cm^3 of an ideal gas at STP?

11. Interstellar gas contains about one atom per cubic centimeter and the temperature is about 3 K. Calculate the pressure.

12. Calculate the volume of a balloon filled with helium if its lifting force at 25°C and atmospheric pressure is to be 50 kg.

13. A human breath has a volume of 6×10^{-4} m^3 and the earth's atmosphere a volume of 4×10^{18} m^3 at STP. If the last breath of Julius Caesar has been mixed completely with the atmosphere, what is the chance that you inhale an atom of argon that was exhaled by Caesar's last breath? (Argon is 1% of the atmosphere.)

14. A sample of natural gas is found on analysis to have the following composition by volume: CH$_4$ = 25%, C$_2$H$_6$ = 35%, C$_3$H$_8$ = 40%. What are the partial pressures

of the gases at 100 kPa total pressure? What is the mass of propane per kilogram of gas?

15. The deepest spot in the ocean is in the Marianas Trench at 11 022 m. Neglecting the change in density with pressure and temperature, assume that $\rho = 1.05$ g cm^{-3} and estimate the pressure at this depth.

16. Radon has $T_c = 377$ K and $P_c = 6.28$ MPa. From the Principle of Corresponding States, estimate V_c per mole and z, the compression factor, at the critical point.

17. A hot-air balloon consists of a light, rigid vessel, open at the bottom where a small gas flame keeps the air within the vessel at constant temperature, say 20 K above that of the surrounding atmosphere. Estimate the size of the balloon needed to lift two men (70 kg each) in a basket, which together with the rest of the balloon weighs 200 kg. Assume that ambient $t = 20°C$ (RTC).

18. What would be the pressure of a 760-mm-high mercury column at 30°C in New York City where $g = 9.803$ m s^{-2}? For Hg, $\alpha = 1.80 \times 10^{-4}$ K^{-1}, $\rho(0°C) = 13.5950$ g cm^{-3}.

19. Draw the van der Waals P–V isotherm at 303.08 K for CO_2 (Table 2.3). Mark the limits of the two-phase region on the diagram given that the vapor pressure of liquid CO_2 at 303 K is 7192 kPa. Calculate the areas between each loop of the van der Waals curve and the flat horizontal portion of the P–V isotherm.

20. For mercury, $\alpha' = (1/V_0)(\partial V/\partial T)_P = 1.817 \times 10^{-4} + 5.90 \times 10^{-9}t + 3.45 \times 10^{-10}t^2$, where t is the Celsius temperature and $V_0 = V$ at $t = 0°C$. If an ideal-gas thermometer and a mercury thermometer agree exactly at 0°C and 100°C, what is the temperature t on the mercury scale when $t = 50°C$ on the ideal-gas scale?

21. Show that for a van der Waals gas, $T_c = 8a/27bR$, $V_c = 3b$, $P_c = a/27b^2$. [*Hint*: The van der Waals equation is a cubic equation in V. At the critical point all roots become the same, so that $(V - V_c)^3 = 0$.]

22. The Berthelot equation of state is

$$P = \frac{nRT}{V - nb} - \frac{n^2a}{TV^2}$$

   (a) Calculate a and b for the Berthelot equation in terms of the critical constants.
   (b) For 1 mol N_2 at 200 K in a volume of 1 dm^3, how do the pressures calculated from the Berthelot and van der Waals equations compare?

23. An unusual gas has been discovered that follows the equation of state $(P + n^2a/V^2)V = nRT$, but a is a function of T such that for $T \leq T_x$, $a = 0$ and for $T > T_x$, $a = a_0/T$, where a_0 is a constant. Plot α and β for this gas as functions of T.

24. The equation of state of Dieterici (1899) is

$$P = \frac{nRT}{V - nb} e^{-na/VRT}$$

   Calculate a and b in terms of the critical constants of a gas.

25. With the results from (24), calculate the volume of 100 g C_2H_4 at 25°C and 300 kPa on the basis of the Dieterici equation and compare result with those in Problem 4.

26. The virial equation of state can be written

$$PV/nRT = 1 + nB'(T)P + n^2C'(T)P^2 + \cdots$$

   Calculate $B'(T)$ and $C'(T)$ in terms of $B(T)$ and $C(T)$ in Eq. (2.16).

27. The radius of Earth is $R = 6.37 \times 10^6$ m and of Mars, $R = 3.38 \times 10^6$ m. For Earth, $g = 9.80$ m s^{-2} and Mars has a mass 0.108 that of Earth. What is the minimum speed

for molecules of (a) hydrogen, (b) helium, and (c) oxygen to escape from the surface of each planet? At what temperature would the rms speed of the molecules equal this escape speed?

28. What is the kinetic energy of 1 mol N_2 at 300 K? Of 1 mol C_2H_4 at 300 K? at 600 K?

29. In the method of Knudsen [*Ann. Phys.*, *29*, 179 (1909)] the vapor pressure is determined by the rate at which the substance, under its equilibrium pressure, diffuses through an orifice. In one experiment, Be powder was placed inside a Mo box having an effusion hole 3.18 mm in diameter. At 1537 K, it was found that 8.88 mg Be effused in 15.2 min. Calculate the vapor pressure of Be at 1537 K. Every molecule striking the orifice effuses and the number striking the unit area is $\frac{1}{4}(N/V)\bar{c}$, where N/V is the number of molecules per unit volume.

30. A mixture of 50% H_2 and 50% D_2 at 25°C and $P = 100$ Pa is allowed to diffuse through a porous filter. Calculate the composition of the initial gas that diffuses through. (At these low pressures, diffusion rate is proportional to molecular speed.)

31. The uranium isotopes ^{235}U and ^{238}U can be separated by diffusion of UF_6 through porous barriers. Calculate the separation factor for one passage through the barrier at 60°C. The ratio of $^{235}U/^{238}U$ in natural U is 0.7/99.3. If the diffusion barriers operate at 50% of maximum theoretical efficiency, calculate the number of passages through the barriers necessary to separate a fraction enriched to 10% ^{235}U.

32. What pressure would have to be applied to a sample of liquid cyclohexane at 25°C to reduce its volume by 2%? The compressibility $\beta = 1.13$ GPa^{-1} (assume constant) and $\rho = 0.7739$ g cm^{-3} at 100 kPa.

33. Sketch the phase diagram (P–T) of CO_2: triple point: 217.7 K and 5.2×10^5 Pa; critical point: 304.3 K and 7.4×10^6 Pa; melting point 293.15 K at 5.0×10^8 Pa.

3

Molecular Energies

The first use of the word "energy" seems to have been by d'Alembert in the French Encyclopedia of 1785. The concept, however, has a much longer history. It first appeared in mechanics, and was then extended to the thermal properties of substances and to electrical and chemical changes. The First Law of Thermodynamics is entirely concerned with energy. It states that energy can be neither created nor destroyed: the Law of Conservation of Energy. Much of thermodynamics is concerned with exchanges of energy that accompany physical and chemical transformations of matter.

In this chapter we review the various forms in which energy appears. Molecular energies are first introduced by using the familiar classical physics of Newton. We then show how chemists can study these energies by measurements of heat capacities. These measurements reveal shortcomings in the classical treatment of energies and indicate the need for a quantum mechanical treatment.

In Chapter 4, we give the quantum description of molecular energy levels and show how the theory explains the experimental data of spectroscopy. Chapter 5 introduces one of the most important principles of physical chemistry, expressed in the Boltzmann distribution equation. With this principle we can understand the concept of thermal equilibrium that is so basic to thermodynamics. We shall learn how molecular populations are distributed over molecular energy levels, and how to understand from a molecular point of view the concept of heat capacity. Finally, we shall come to one of the most important points: The Boltzmann distribution provides a fundamental molecular definition of temperature.

3.1 The Molecular Interpretation of Thermodynamics

The calculations and predictions of thermodynamics are all based on data concerning *bulk* properties of chemical substances. That is, thermodynamics describes the behavior of a mole of CH_4, a micromole of DNA, and so forth. It says nothing whatsoever about individual CH_4 or DNA molecules. Thermodynamics is concerned only with large aggregates (say, masses greater than about 10^{-18} kg). The properties of such large assemblies of atoms or molecules are called *macroscopic* properties. In contrast, a spectroscopist who determines the energy levels of a single atom or molecule is dealing with *microscopic* properties.

This division of systems or of properties into macroscopic and microscopic is important to recognize. Thermodynamics is a macroscopic science, and this fact is one of its basic strengths. Thermodynamics makes detailed and accurate predictions about the equilibrium characteristics of macroscopic chemical systems without any reliance on the properties of individual atoms or molecules. It uses only data concerning bulk properties. Because it is relatively easy to obtain such data even for complex chemical systems, thermodynamics is a practical tool of immense value to chemists.

On the other hand, equilibrium properties can also be approached from the point of view of the microscopic or molecular world. The result is a special branch of theoretical chemistry known as *statistical thermodynamics*, which will be discussed in Chapter 12. We shall borrow some qualitative concepts from statistical thermodynamics ahead of time in order to provide useful insights into our development of macroscopic thermodynamics, or as it is sometimes called, *classical thermodynamics*. We must always remember, however, that classical thermodynamics stands on its own, quite independent of any knowledge of microscopic properties.

Many students find that their first study of thermodynamics is more comprehensible if the abstract thermodynamic variables are related to physical concepts of molecular properties. Molecular energies prove to be the properties most useful for this purpose. Time spent with facts concerning molecular energies and related concepts in this and the following two chapters is repaid by a superior ability to navigate the development of classical thermodynamics in the subsequent chapters.

3.2 Conservation of Energy

The principle of conservation of energy can be expressed with varying degrees of generality. Consider first *mechanical energy*, energy due to the motions and positions of large bodies like crystals, tennis balls, and planets. In the science of mechanics two kinds of energy are distinguished, kinetic energy E_k and potential energy E_p. When mechanical processes operate in the absence of friction, the total mechanical energy $E = E_k + E_p$ always remains constant. This is the mechanical principle of *conservation of energy*. For example, if a body falls through a vacuum in a gravitational field, its gain in kinetic energy is exactly balanced by its loss in gravitational potential energy.

$$E = \tfrac{1}{2}mv_1^2 + mgh_1 = \tfrac{1}{2}mv_2^2 + mgh_2 \qquad (3.1)$$

Here v_1 and v_2 are initial and final magnitudes of velocity, h_1 and h_2 are intitial and final heights, and g is the gravitational acceleration, approximately 9.81 m s^{-2} at the earth's surface.

Example 3.1 At what speed would a diver enter the water from the top of a 10-m tower? Neglect resistance due to air.

In Eq. (3.1), $v_1 = 0$, $h_1 = 10$ m, $h_2 = 0$, $v_2 = (2gh_1)^{1/2}$, and

$$v_2 = [(2)(9.81 \text{ m s}^{-2})(10 \text{ m})]^{1/2} = 14 \text{ m s}^{-1}$$

Now let us move to a more general statement of the principle. When frictional forces occur in a mechanical system (e.g., a bicycle wheel), we notice a rise in temperature. We speak of friction producing "heat," when what is meant is that some of the mechanical energy of motion is being transferred to the molecules of the system (in the case of the wheel to the metal axle and surrounding air). Thus the mechanical energy is not destroyed by frictional processes. It is simply converted into other forms of energy associated with the smallest particles of the substance. This microscopic energy is called the *internal energy U* of the body. The principle of conservation of energy requires that $E_k + E_p + U$ always remain constant for any isolated system.

We can subdivide the internal energy U into several categories. *Thermal energy* is associated with motions of the individual atoms, molecules, or ions that constitute a system. We shall describe these motions in later sections of this chapter. (Note that they do not include the movement of the system as a whole.) *Intermolecular energy* is associated with interactive forces between the molecules of a system. For example, energy must be supplied to vaporize a liquid. If the liquid is insulated from its surroundings, the necessary energy must be supplied by the liquid from its own thermal energy, so it cools. The energy is not lost, however; it has gone into increasing the internal energy of the molecules as they pass from liquid to gas. *Chemical energy* is associated with the making and breaking of chemical bonds due to interactions between the electrons and nuclei of individual atoms, ions, and molecules. Energy is released as atoms bond together to make molecules; this is called chemical energy. A change occurs in the chemical energy as the electrons and nuclei rearrange to form the molecule. Other changes in chemical energy occur as molecules react with one another to make other kinds of molecules.

This dissection of energies can be taken further. For the purposes of chemical studies, we can consider atoms to be the elementary particles that constitute large-scale systems. We know, however, that atoms themselves have a structure of nuclei and electrons, and that nuclei are composed of protons and neutrons. For the purpose of physical studies, therefore, we can carry the analysis of energy back to electrons, protons, and neutrons. For nuclear reactions the equivalence of mass and energy becomes important; this is embodied in the Einstein relation,

$$E = mc^2 \qquad (3.2)$$

where c is the magnitude of the velocity of electromagnetic radiation in a vacuum ($= 2.9979 \times 10^8$ m s^{-1}). Thus we can finally equate the masses of the electrons, protons, and neutrons in the system to an energy E. The elementary particles are ultimately composed of energy. Energy becomes matter when it takes the form of elementary particles.

Example 3.2 Calculate the thermal energy released per mole in the nuclear reaction $2^2H \longrightarrow {}^3H + {}^1H$, where the nuclear masses are ${}^1H = 1.007820$, ${}^2H = 2.014092$, and ${}^3H = 3.016044$, all in g mol^{-1}.

$\Delta m = 3.016044 + 1.007820 - 2(2.014092) = -0.00432$ g mol^{-1}. From Eq. (3.2),

$$\Delta E = c^2 \Delta m = (2.998 \times 10^8 \text{ m s}^{-1})^2(-4.32 \times 10^{-6} \text{ kg mol}^{-1})$$

$$-\Delta E = 38.8 \times 10^{10} \text{ J mol}^{-1} = 3.88 \times 10^8 \text{ kJ mol}^{-1}$$

This compares to $-\Delta E = 242$ kJ mol^{-1} for $H_2 + \frac{1}{2}O_2 = H_2O(g)$.

As a result of this analysis we have a balance sheet for the total energy in the system. This total energy has a definite value. In any isolated system, the energy can be transformed in many ways, but the total energy E never changes. This is the general law of conservation of energy.

A few seconds after the universe was created, about 20×10^9 years ago, it was in the form of an immensely dense mass of energy. As it expanded and cooled, particles such as electrons, protons, and neutrons were formed from radiant energy. On further cooling, nuclei appeared and then atoms. In the universe, all possible transformations of energy have occurred and are occurring. The mass–energy of the universe, however, is a constant.

All conservation laws are derived from underlying laws of symmetry that reflect the scientific mind of the universe. The law of conservation of energy can be derived from the symmetry of scientific observations with respect to displacements in time; i.e., any experiment we perform on an isolated thermodynamic system can depend only on the elapsed time interval of the experiment and not on the absolute time at which the experiment is started. Displacement of the time axis cannot change the result of an experiment in an isolated system.

For the approximately isolated systems that we set up in chemical laboratories, we need not consider the ultimate analysis of energy down through nuclear structure and mass–energy conversion. For chemical purposes, we therefore separate the law of conservation of energy from the law of conservation of mass. Most of our studies will deal only with thermal, intermolecular, and chemical energies. In our discussions of chemical systems, we shall almost always consider them to be at rest and the mechanical potential energy of the systems will be kept constant. The internal energy U is usually, therefore, the energy function employed in physical chemistry. In chemical engineering, however, one often needs to consider also mechanical energies, namely the kinetic energy of moving fluids or the potential energy of materials in gravitational fields.

3.3 The Energy of Molecules:
Translation, Rotation, and Vibration

What happens to energy that is put into a system to raise its temperature? How do the molecules in the system store this added energy? According to the kinetic theory of matter, much of the energy is stored in the motion of molecules and in the internal movements of atoms within molecules. To a good approximation the total thermal energy of most molecules can be separated into three categories: translational, rotational, and vibrational. For example, the rotational energy of a molecule is not affected by its energy of translation, and to a first approximation rotation and vibration are independent.

This separation of molecular motions also occurs conveniently in the mathematical description. We first begin with a molecular model in which the masses of the constituent atoms are concentrated at points. Almost all the atomic mass is concentrated in a tiny nucleus, the radius of which is about 10^{-15} m. Since the overall dimensions of molecules are of the order of 10^{-10} m, it is reasonable to consider a molecule as a group of mass points. Take a molecule composed of N atoms. To represent the instantaneous positions in space of N mass points requires $3N$ coordinates. These can be the values of $x_j y_j z_j$ which locate each of the N masses. The number of coordinates required to specify the positions of all the mass points (atoms) in a molecule is called the number of its *degrees of freedom*. Thus a molecule of N atoms has $3N$ degrees of freedom.

The atoms comprising each molecule move through space as a connected unit and we can represent the translational motion of the molecule as a whole by the motion of its *center of mass*. Consider a collection of mass points m_j located at positions $(x_j y_j z_j)$. If $\sum m_j(x_j + y_j + z_j) = 0$, the origin of coordinates is the *center of mass* of the collection of mass points. Three coordinates (degrees of freedom) are required to represent the instantaneous position of the center of mass, to which the translational motion is fully assigned. The remaining $3N - 3$ coordinates represent the *internal degrees of freedom*.

Example 3.3 Locate the center of mass of the linear molecule OCS in which $R_{OC} = 116$ pm and $R_{CS} = 156$ pm.

We take the relative atomic masses as $m_O = 16$, $m_C = 12$, and $m_S = 32$. Let the X axis coincide with the axis of the molecule. The condition that the origin of coordinates be at the center of mass is $m_O x_1 + m_C x_2 + m_S x_3 = 0$.

$$-16(x_2 + 116) - 12x_2 + 32(156 - x_2) = 0$$
$$x_2 = 52 \text{ pm}$$

The center of mass is 52 pm from the C atom along the CS bond.

The internal degrees of freedom may be subdivided into rotations and vibrations. For a rotational motion there are two degrees of freedom for a linear molecule and three for a nonlinear molecule. We can see these requirements in Fig. 3.1. [Why do we exclude rotation of a linear molecule about its internuclear axis?]

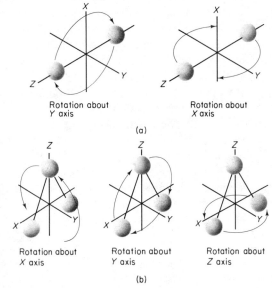

Rotation about
Y axis

Rotation about
X axis

(a)

Rotation about
X axis

Rotation about
Y axis

Rotation about
Z axis

(b)

FIGURE 3.1 (a) Rotations of a linear molecule about two mutually perpendicular axes. (b) Rotations of a nonlinear molecule about three mutually perpendicular axes.

There now remain $3N - 5$ degrees of freedom in linear molecules or $3N - 6$ in nonlinear molecules. These describe the motions of the nuclei with respect to one another. They represent the number of *vibrational degrees of freedom*. Diatomic molecules have $3N - 5 = 1$ vibrational degree of freedom. The linear polyatomic molecule acetylene (HCCH) has $3N - 5 = 7$ vibrational degrees of freedom. The nonlinear polyatomic molecule formaldehyde (HCHO) has $3N - 6 = 6$ vibrational degrees of freedom. In terms of coordinates, these degrees of freedom represent the interatomic distances and angles needed to specify the geometry of the atomic framework. In terms of molecular motions, these degrees of freedom represent the number of (nearly) independent vibrational modes that can occur in the molecule. Thus acetylene has seven normal modes of vibration, and a vibrational energy can be associated with each mode.

In summary, a molecule made up of N atoms has $3N$ degrees of freedom, which can be divided into independent molecular motions:

Linear molecules: 3 translations   Nonlinear molecules: 3 translations
                     2 rotations   3 rotations
              $3N - 5$ vibrations $3N - 6$ vibrations

[How many (1) rotations and (2) vibrations do the following molecules have: (a) NO; (b) NO_2 (bent); (c) $SOCl_2$; (d) CS_2 (linear); (e) C_6H_6?]

3.4 The Energy of a Molecule: Translation

By *translation* we mean the motion of the molecule as a unit through space. We can imagine the total mass m of the molecule to be located at its center of mass, and then follow the motion of this mass point through space. The kinetic energy will be discussed from the standpoint of classical mechanics, which assumes the energy to have a continuous range of allowed values. Actually, we know that energy is divided into discrete packets, called *quanta*, so that it cannot vary continuously but only by definite jumps. In the case of translation, however, the quanta are so small that the variation of energy is effectively continuous.

If we set up a system of three orthogonal axes, XYZ, the velocity of the molecule can be resolved into components along each of the axes as was shown in Fig. 2.3. From Eq. (2.12), the kinetic energy of translation of the molecule can be written as $\epsilon_t = \frac{1}{2}mc^2 = \frac{1}{2}mu^2 + \frac{1}{2}mv^2 + \frac{1}{2}mw^2$. The three energy terms correspond to the three translational degrees of freedom.

We can use the results of the model of a perfect gas from Chapter 2 to calculate the average translational energy of a molecule. Equation (2.12) showed that the translational energy of a mole of perfect-gas molecules is $\frac{3}{2}RT$. This is related to the average of the square of the molecular velocity $\overline{c^2}$ by

$$L(\tfrac{1}{2}m\overline{c^2}) = \tfrac{3}{2}RT \tag{3.3}$$

where L is the Avogadro constant, 6.022×10^{23} mol^{-1}. Thus the average translational energy per molecule $\bar{\epsilon}_t$ becomes

$$\bar{\epsilon}_t = \tfrac{1}{2}m\overline{c^2} = \tfrac{3}{2}\frac{R}{L}T \tag{3.4}$$

The quantity R/L is called the *Boltzmann constant k*. It has the dimensions of energy per molecule divided by temperature, and in SI units $k = 1.381 \times 10^{-23}$ J K^{-1}. Thus Eq. (3.4) yields the average translational energy $\bar{\epsilon}_t$ of a molecule as

$$\bar{\epsilon}_t = \tfrac{1}{2}m\overline{c^2} = \tfrac{3}{2}kT \tag{3.5}$$

If the gas is not flowing, all directions of motion of the molecules are equally probable (in other words, the motion of the molecules is *isotropic*). Hence

$$\overline{u^2} = \overline{v^2} = \overline{w^2} = \overline{c^2}/3 \tag{3.6}$$

Thus the average translational energy of a molecule from Eq. (3.5) is $\frac{1}{2}kT$ for each degree of freedom. Any particular molecule in the gas may have an energy considerably below or above the average value of $\frac{1}{2}kT$. Note that only the *average* energy is related to the temperature T. It does not make sense to speak of the temperature of an individual molecule.

3.5 The Energy of a Molecule: Rotation

To acquire rotational motion about an axis a molecule must possess a *moment of inertia* about that axis. Suppose that a body is composed of a collection of masses m_j, each located at some distance r_j from a given axis. The moment of inertia is defined

in mechanics as
$$I = \sum m_j r_j^2 \tag{3.7}$$
The kinetic energy of rotation about the axis is
$$\epsilon_r = \tfrac{1}{2} I \omega^2 \tag{3.8}$$
where ω is the angular velocity of the rotation. The SI unit of ω is radians per second. The expression for rotational kinetic energy is easy to remember since it is analogous to the expression $\tfrac{1}{2} mc^2$ for translational kinetic energy.

Figure 3.2(a) shows some of the rotational characteristics of CO. The molecule is depicted as two spheres representing the atoms of C and O, with a bond of the length R_e between them. Since R_e has a fixed value, this model is a *rigid rotor*. In the figure, the line of centers (bond) between the atoms coincides with the Z axis, and the X and Y axes are perpendicular to Z and intersect at the center of mass of the molecule. The distances of masses m_O and m_C from the origin of this system of axes are
$$r_O = \frac{m_C}{m_C + m_O} R_e; \qquad r_C = \frac{m_O}{m_C + m_O} R_e \tag{3.9}$$
where R_e is the equilibrium distance between the centers of atoms C and O. [Use the fact that $r_C m_C = r_O m_O$ to derive Eq. (3.9).]

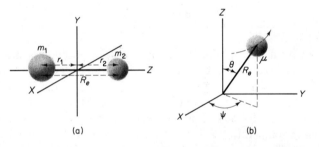

(a)   (b)

FIGURE 3.2 (a) The two rotational axes of CO (X and Y) with the center of mass at the origin of the coordinate system. (b) The rotational motion of CO represented as that of a reduced mass at a distance R_e from the origin.

From Eq. (3.9) the moment of inertia about the X axis is
$$I_x = m_O r_O^2 + m_C r_C^2 \tag{3.10}$$
From the symmetry in Fig. 3.2,
$$I_x = I_y = I \tag{3.11}$$
By combining Eqs. (3.10) and (3.11) with Eq. (3.9), we get
$$I = \frac{m_C m_O}{m_C + m_O} R_e^2 \tag{3.12}$$
The ratio of masses in Eq. (3.12) appears repeatedly in problems involving a pair of atoms. It is called the *reduced mass*, μ, and is defined by
$$\frac{1}{\mu} = \frac{1}{m_C} + \frac{1}{m_O} \qquad \text{or} \qquad \mu = \frac{m_C m_O}{m_C + m_O} \tag{3.13}$$
The beauty of reduced mass is that it changes a two-body problem, such as that of the CO molecule, into a simple one-body problem. The two masses are replaced by the

single reduced mass μ and the two distances from the center of mass are replaced by the single distance of separation between the two bodies. Thus, in the present case, one needs to work only with the simple expression

$$I = \mu R_e^2 \tag{3.14}$$

We can interpret this result physically by saying that the rotational motions of a diatomic molecule can be represented by the rotation of a single mass μ at a fixed distance R_e from the origin of coordinates. This result is shown in Fig. 3.2(b). The two degrees of freedom required to represent the motion of the mass μ are conveniently chosen to be the two angles θ and ϕ of a system of spherical polar coordinates.

Example 3.4   Calculate μ and I for $^{12}C^{16}O$ for which $R_e = 112.8$ pm.

$$m_C = 12.000 \times 10^{-3}/6.022 \times 10^{23} = 1.993 \times 10^{-26} \text{ kg}$$

$$m_O = 15.995 \times 10^{-3}/6.022 \times 10^{23} = 2.656 \times 10^{-26} \text{ kg}$$

$$\mu = \frac{(1.993 \times 10^{-26})(2.656 \times 10^{-26})}{(1.993 + 2.656) \times 10^{-26}} = 1.139 \times 10^{-26} \text{ kg}$$

$$I = \mu R_e^2 = (1.139 \times 10^{-26} \text{ kg})(112.8 \times 10^{-12} \text{ m})^2$$

$$= 1.449 \times 10^{-46} \text{ kg m}^2$$

We can see from Fig. 3.2 that a diatomic molecule (and any linear polyatomic molecule) has only two rotational degrees of freedom. Since the atoms are represented by point masses, I_z is zero because in Eq. (3.7) the r_i's from the Z axis are all zero.

We shall not give the proof here (see Section 12.11), but simply state the result that if a linear molecule follows classical mechanics with rotational energies $\bar{\epsilon}_{x,r} = \bar{\epsilon}_{y,r} = \frac{1}{2}I\omega^2$, then at temperature T its total average rotational energy is $\bar{\epsilon}_r = (\frac{1}{2}kT)_x + (\frac{1}{2}kT)_y = kT$. Note that the average kinetic energy of rotation is $\frac{1}{2}kT$ for each degree of freedom, just as it is for translation.

The rotational kinetic energy of a nonlinear molecule is

$$\epsilon_r = \frac{1}{2}I_A\omega_A^2 + \frac{1}{2}I_B\omega_B^2 + \frac{1}{2}I_C\omega_C^2 \tag{3.15}$$

where the I's and ω's refer to a set of three mutually perpendicular axes of rotation, called the *principal axes*. The way in which these axes are chosen is discussed in Chapter 12.

A typical example, the triangular molecule SO_2, is shown in Fig. 3.3. The shape, and dimensions of this molecule have been determined by spectroscopic methods (Chapter 26). In the case of SO_2, the problems of choosing the principal axes and finding the center of mass are solved by symmetry. The center of mass must lie on the axis that bisects the OSO angle (i.e., the B axis). Since the mass of the sulfur atom is almost equal to the sum of the masses of the oxygen atoms, the center of mass must be midway between the sulfur atom and the projection of the oxygen atoms on the B axis. The solution to the problem is shown in Fig. 3.3 and the axes to which the moments of inertia refer are indicated. The moments I_A, I_B, and I_C are calculated from Eq. (3.7) as before.

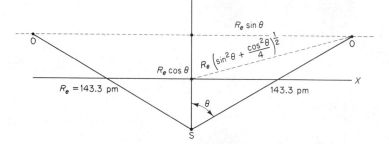

$$\theta = 59°46.5', \sin \theta = 0.864, \cos \theta = 0.503$$
$$R_e \cos \theta = 72.1 \text{ pm}, R_e \sin \theta = 123.8 \text{ pm}$$

$$I_A = I_{xx} = 2\left(\frac{16 \times 10^{-3}}{6.02 \times 10^{23}}\right)(36.1 \times 10^{-12})^2 + \frac{32}{6.02 \times 10^{23}}(36.1 \times 10^{-12})^2$$

$$= 1.39 \times 10^{-46} \text{ kg m}^2$$

$$I_B = I_{yy} = 2\left(\frac{16 \times 10^{-3}}{6.02 \times 10^{23}}\right)(123.8 \times 10^{-12})^2 = 8.15 \times 10^{-46} \text{ kg m}^2$$

$$I_C = I_{zz} = \frac{32 \times 10^{-3}}{6.02 \times 10^{23}}(36.1 \times 10^{-12})^2 + 2\left(\frac{16 \times 10^{-3}}{6.02 \times 10^{23}}\right)(128.9 \times 10^{-12})^2$$

$$= 9.52 \times 10^{-46} \text{ kg m}^2$$

FIGURE 3.3 Calculation of the principal moments of inertia of SO_2. The SO_2 molecule is in the plane of the paper and the Z axis is perpendicular to this plane and passes through the center of mass.

The kinetic energy of rotation is given by Eq. (3.15). According to classical mechanics each of the three rotational degrees of freedom has an average energy $\frac{1}{2}kT$. Thus the average rotational energy of a nonlinear molecule is

$$\bar{\epsilon}_r = 3(\tfrac{1}{2}kT) = \tfrac{3}{2}kT \tag{3.16}$$

3.6 The Energy of a Molecule: Vibration

Consider as an example the diatomic molecule CO as shown in Fig. 3.4. The CO bond between the atoms is represented by an idealized spring. If the spring is stretched from its equilibrium length R_e and then released, the atoms undergo a regular vibratory motion. In order to understand the nature of this motion we must examine the relation between the stretching force and the distance between the atoms. The more the spring is stretched, the more it resists further extension. The spring exerts a restoring force **F**, which acts to return it to its equilibrium length R_e. If the spring is not stretched too far, the magnitude of this restoring force is proportional to the displacement (Hooke's Law):

$$F = -\kappa x \tag{3.17}$$

where $x = R - R_e$ and the proportionality constant κ is called the *force constant*. The stiffer the spring, i.e., the chemical bond, the larger is the force constant.

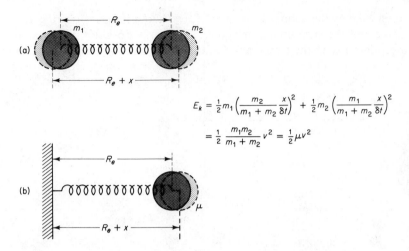

$$E_k = \tfrac{1}{2}m_1\left(\frac{m_2}{m_1 + m_2}\frac{x}{\delta t}\right)^2 + \tfrac{1}{2}m_2\left(\frac{m_1}{m_1 + m_2}\frac{x}{\delta t}\right)^2$$

$$= \tfrac{1}{2}\frac{m_1 m_2}{m_1 + m_2}v^2 = \tfrac{1}{2}\mu v^2$$

FIGURE 3.4 A diatomic molecule such as CO is shown in two models suitable for treating the vibrational motion: (a) the two-body model with masses m_1 and m_2 and equilibrium internuclear distance R_e; (b) a one-body model with mass μ at a distance R_e from a fixed point.

From Newton's Second Law, force = mass × acceleration,

$$\mathbf{F} = m\mathbf{a} = m\frac{d^2x}{dt^2} = -\kappa x \tag{3.18}$$

The last two terms comprise the differential equation for the motion of a vibrating spring, or more generally, any *harmonic oscillator*. The displacement x varies as a sine or cosine function of t—hence the term *harmonic*.

Example 3.5 Integrate Eq. (3.18) with the initial conditions $x = 0$ at $t = 0$ and $dx/dt = 0$ at $x = A$.

Let $v = dx/dt$; then $d^2x/dt^2 = dv/dt = v(dv/dx)$ and Eq. (3.18) becomes $v(dv/dx) + (\kappa/m)x = 0$. Integration gives $v^2 + (\kappa/m)x^2 = \text{const.}$ When $x = A, v = 0$, const. $= (\kappa/m)A^2$, so that

$$v^2 = \left(\frac{dx}{dt}\right)^2 = \frac{\kappa}{m}(A^2 - x^2)$$

$$\frac{dx}{dt} = \left[\frac{\kappa}{m}(A^2 - x^2)\right]^{1/2}$$

$$\frac{dx}{(A^2 - x^2)^{1/2}} = \left(\frac{\kappa}{m}\right)^{1/2}dt$$

Integration gives $\sin^{-1}(x/A) = (\kappa/m)^{1/2}t + \text{const.}$ When $t = 0$, $x = 0$, const. $= 0$. Hence $x = A\sin(\kappa/m)^{1/2}t = A\sin 2\pi\nu t$, where the vibration frequency $\nu = (1/2\pi)(\kappa/m)^{1/2}$. [To verify the solution obtained in Example 3.5, substitute it back into Eq. (3.18).]

If a macroscopic spring is extended and released, its oscillations will decrease in amplitude and fairly soon come to rest. The energy of the large-scale motion has been dissipated by the action of frictional forces, both inside the spring and between the spring and the surrounding medium. The energy does not disappear but rather causes a slight rise in the temperature of the spring and its surroundings. In this way the macroscopic motion is transformed into the internal microscopic motions of the particles that compose the spring and its surroundings. In the case of an isolated vibrating molecule, however, the oscillations, once started, continue as long as the molecule exists. There are no frictional forces within the molecule. If the molecule is not isolated, it can interact with other molecules or with electromagnetic radiation, to emit or absorb energy.

Suppose that we stretch the bond shown in Fig. 3.4 from its equilibrium position ($R = R_e$) until $R = R_e + A$ or $x = A$. The work done on the molecule at each stage dx of the stretching process is equal to force × distance = $F(x)\,dx$, where the force F is itself a function of x. Note that the force applied to the molecular spring is opposite to the restoring force of Eq. (3.17). The total work done is obtained by integration; it is equal to the potential energy stored in the stretched molecule:

$$\epsilon_p = \int_0^A F(x)\,dx = \int_0^A \kappa x\,dx = \tfrac{1}{2}\kappa x^2 \Big|_0^A = \tfrac{1}{2}\kappa A^2 \tag{3.19}$$

Suppose that we now release the spring from $x = A$. The stored potential energy can be converted into kinetic energy of motion. As shown in Fig. 3.4,

$$\epsilon_k(\text{vib}) = \tfrac{1}{2}\mu v^2 \tag{3.20}$$

The velocity of the mass μ increases until it reaches a maximum value as μ passes through the equilibrium position at $x = 0$, where ϵ_p is zero. At each point in the motion the total energy is

$$\epsilon = \epsilon_p + \epsilon_k = \tfrac{1}{2}\kappa x^2 + \tfrac{1}{2}\mu v^2 = \tfrac{1}{2}\kappa A^2 \tag{3.21}$$

A is called the *amplitude* of the vibration.

The total energy ϵ is always constant. When the spring reaches $x = 0$, its momentum carries on the compression until the potential energy of the compressed spring has soaked up all the kinetic energy of motion. This point is reached at $R = R_e - A$ or $x = -A$. The motion then reverses to complete a cycle of the harmonic oscillation.

Figure 3.5 is a plot of the potential energy $\epsilon_p = \tfrac{1}{2}\kappa x^2$ of the vibrating molecule as x is varied. This parabola shows the potential energy of the system at each point in its motion. For each ϵ_p, the kinetic energy ϵ_k can be calculated from Eq. (3.21).

We have seen that the average kinetic energy for each translational or rotational degree of freedom of a molecule is $\tfrac{1}{2}kT$. For each vibrational degree of freedom both kinetic and potential energies contribute to the energy. According to classical mechanics, each vibration has an average kinetic energy of $\tfrac{1}{2}kT$ and an average potential energy of $\tfrac{1}{2}kT$ and thus an average total energy of kT. This classical behavior occurs for molecular vibrations only at high temperatures. Because of the quantized nature of energy (which is particularly important in the case of vibrations), the average energy for each vibrational degree of freedom is usually much less than kT. The proof of these facts about vibrational energies will be given in Chapter 12.

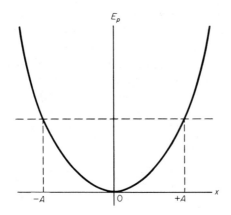

FIGURE 3.5 Potential-energy curve of a harmonic oscillator in one dimension, $E_p = \frac{1}{2}\kappa x^2$. A particular vibration amplitude A is indicated.

3.7 Normal Modes of Vibration

The vibrational motions of a polyatomic molecule can be complicated. However, if the forces between the atoms obey Hooke's Law, the motions can always be represented as the superposition of a number of simple harmonic vibrations. These simple basic motions are called the *normal modes* of vibration. In a given normal mode (j) each vibrating atom of the molecule is oscillating at the same frequency ν_j.

Examples of the normal modes for CO_2 (linear) and SO_2 (nonlinear) are shown in Fig. 3.6. The bent SO_2 molecule has $3N - 6 = 3$ distinct normal modes, each

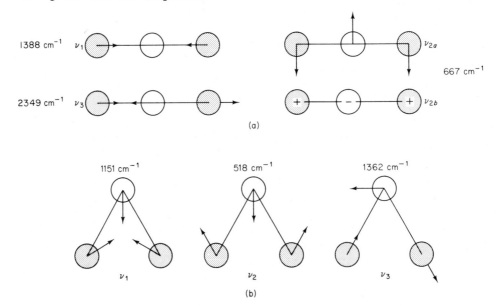

FIGURE 3.6 Normal modes of vibration of (a) CO_2 (linear) and (b) SO_2 (nonlinear). Arrows indicate displacements in the plane of paper; + and − indicate displacements normal to that plane.

with a characteristic frequency. (The frequencies have different values in molecules of different compounds.) The linear CO_2 molecule has $3N - 5 = 4$ normal modes. Two (v_1 and v_3) correspond to stretching the molecule, and two (v_{2a} and v_{2b}) to bending. The two bending vibrations differ only in that one is in the plane of the paper and one (denoted by $+$ and $-$) is normal to the plane. These two vibrations, from symmetry considerations, must have the same frequency.

The separation of polyatomic vibrations into normal modes makes life much easier for the chemist. Fortunately, molecular vibrations are nearly harmonic, so that to a good approximation the energies in these modes can be treated separately. Thus the total vibrational energy in a polyatomic molecule is simply the sum of the separate energies in each of its $3N - 5$ vibrations (linear molecules) or $3N - 6$ vibrations (nonlinear molecules). Each vibrational mode is expected to contribute kT to the classical molecular energy ($\frac{1}{2}kT$ for kinetic energy and $\frac{1}{2}kT$ for potential energy).

3.8 Classical Equipartition of Energy

Our discussion of molecular energies can be summarized as follows: In classical mechanics the average molecular energy is the sum of energies $\frac{1}{2}kT$, with one such term contributed for each squared term in the classical energy expression. This is the theorem of equipartition of energy. Thus for a diatomic molecule, the translations contribute $\frac{3}{2}kT$, since there are three squared terms: $\frac{1}{2}mu^2$, $\frac{1}{2}mv^2$, and $\frac{1}{2}mw^2$. The two rotations contribute kT because of one squared term $\frac{1}{2}I\omega^2$ for each rotation. Finally, the vibration contributes kT since the vibrational energy itself contains two squared terms [Eq. (3.21)]. The theorem thus says that the classical internal energy of a diatomic molecule should be

$$\bar{\epsilon} = \bar{\epsilon}_t + \bar{\epsilon}_r + \bar{\epsilon}_v = \tfrac{3}{2}kT + kT + kT = \tfrac{7}{2}kT$$

For the energy per mole, one uses $R = Lk$ instead of the Boltzmann constant, giving $U_m = \frac{7}{2}RT$.

Table 3.1 summarizes average classical energies for linear and nonlinear molecules. These are the energies we should observe if molecules behaved in accordance with Newtonian mechanics.

TABLE 3.1
CLASSICAL EQUIPARTITION OF ENERGY

	Linear Molecules		Nonlinear Molecules	
	Degrees of freedom	Average classical energy	Degrees of freedom	Average classical energy
Translation	3	$\frac{3}{2}kT$	3	$\frac{3}{2}kT$
Rotation	2	$\frac{2}{2}kT$	3	$\frac{3}{2}kT$
Vibration	$3N - 5$	$(3N - 5)kT$	$3N - 6$	$(3N - 6)kT$

Example 3.6 What is the average energy of the molecule C_2H_6 as predicted by the classical theorem of equipartition of energy?

C_2H_6 has $N = 8$ atoms and $3N = 24$ degrees of freedom. There are 3 translations, 3 rotations (nonlinear molecule), and $3N - 6 = 18$ vibrations. Thus

$$\bar{\epsilon} = 3(\tfrac{1}{2}kT) + 3(\tfrac{1}{2}kT) + 18(kT) = 21kT$$

3.9 Heat Capacity at Constant Volume:
An Experimental Measure of Average Molecular Energy

How can we test the concepts and models for molecular energies outlined in the previous sections? Is the average energy of an ethane molecule really $21kT$? There are several solutions to this problem. For example, rotational and vibrational energies of molecules can be measured by spectroscopic methods as described in Chapter 26. Translational energies of molecules can be measured by experiments with beams of molecules or by subtle spectroscopic effects.

Measurements of heat capacities of gases provide another approach. In this chapter we have seen how the internal energy U of a gas can be contained in the various motions of the gas molecules. While it is difficult to measure the average energy content of molecules, it is relatively easy to measure any change in temperature of a substance as a consequence of energy added to it. All the energy added to a substance kept at constant volume must increase the internal energy U of the substance. The ratio of energy added ΔU to temperature increase ΔT measures what is called the *heat capacity* of the substance. The heat capacity of a substance is its capacity to store energy.

We measure the uptake of energy and the resulting temperature rise in a *heat-capacity calorimeter*. Figure 3.7 shows a schematic version of this instrument. A closed insulated container is equipped with (1) an internal coil that can be electrically heated and (2) some device (such as a thermocouple) for measuring the temperature of the fluid surrounding the coil. The container is filled with the substance to be studied. Then

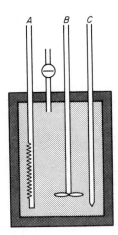

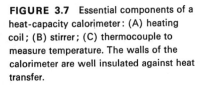

FIGURE 3.7 Essential components of a heat-capacity calorimeter: (A) heating coil; (B) stirrer; (C) thermocouple to measure temperature. The walls of the calorimeter are well insulated against heat transfer.

a source of electrical energy is connected to the heating coil. The electrical energy introduced into the system is I^2Rt, where I is the constant current, R the resistance of the coil, and t the time during which current flows. The rise in temperature is measured as ΔT. Some of this energy is used to increase the internal energy of the sample by ΔU and some is used to raise the temperature of the calorimeter itself. The latter value can be found by making an experiment with an empty calorimeter. It is thus possible to measure with considerable precision the increase in internal energy ΔU of the sample. [Calculate the electrical energy input to a calorimeter if a voltage source of 6.00 V delivers a current of 0.200 A for 1.00 min.]

The two quantities ΔU and ΔT are proportional to each other provided that the temperature change is small: $\Delta U = C \Delta T$. The proportionality factor C is the *heat capacity*. Actually, since C itself is a function of T, the value of the ratio $\Delta U/\Delta T$ in the limit of small ΔT is taken to define the heat capacity:

$$\lim_{\Delta T \to 0} \frac{\Delta U}{\Delta T} = C \qquad (3.22)$$

This limit, which can be measured experimentally, is written as $C = dU/dT$.

We need to stress the condition that the measurement is made with the sample at constant volume so that no energy is expended in pushing back the surroundings by expansion of the sample as its temperature rises. This condition is especially important for gases, owing to their large change of volume with temperature. Thus we specify a particular type of heat capacity, namely C_V, at constant volume, and we use the notation for a partial differential in its definition:

$$C_V = (\partial U/\partial T)_V \qquad (3.23)$$

This notation states that C_V measures how internal energy U changes with temperature T, when one of the system variables, namely volume V, is kept constant. Thus $(\partial U/\partial T)_V$ represents a physical concept based on a definite experimental operation. The SI unit of heat capacity is joule per kelvin, $J\ K^{-1}$.

Note that C_V is an extensive quantity; that is, it depends on the amount of substance. Doubling the amount of substance doubles C_V. To obtain a quantity that is characteristic of each substance we tabulate the value of C_V for one mole, C_V/n, measured in $J\ K^{-1}\ mol^{-1}$.

Historically, the concept of *heat capacity* was related to the concept of *heat*. The experiment above would have been described as follows: In the resistance wire, the electric energy I^2Rt is converted to heat, and the heat passes from the wire into the surrounding medium, thereby raising the temperature of the contents of the calorimeter.

How does heat capacity C_V tell us about the internal energy U? It does so by its intimate connection with U, via its definition, Eq. (3.23). Heat capacity measures the capacity of a substance to store energy as its temperature is raised. Consider, for example, the internal energy of ethane as calculated by classical equipartition of energy. From Example 3.6, for n moles of ethane, the predicted energy is $U = 21nRT$. The predicted heat capacity is, therefore,

$$C_V = (\partial U/\partial T)_V = (\partial[21nRT]/\partial T)_V = 21nR \qquad \text{or} \qquad C_V/nR = 21$$

Thus a measure of C_V will tell us directly if the theoretical prediction about the energy is correct. The predicted $C_V = 21nR$ is independent of temperature.

3.10 Experimental C_V for Gases

When the temperature of a gas at constant volume is increased, the gas gains energy U in the translational, rotational, and vibrational motions of its molecules. The energy gained per unit increase in temperature is C_V. Experimental C_V data for various gases are collected in Table 3.2. Note that C_V/nR is a dimensionless number.

How well do the classical mechanics of molecular motions and the principle of equipartition of energy explain these values? The inert gases, He, Ne, Ar, etc., all have $C_V/nR = 1.500$, independent of T. See, for example, the values for He in Table 3.2. These gases are monatomic, so that they have three degrees of freedom of translation, but no rotation or vibration. Consequently, the predicted average energy is $\bar{\epsilon} = \frac{3}{2}kT$ per molecule or $U = \frac{3}{2}RT$ per mole. The predicted heat capacity is $C_V = (\partial U/\partial T)_V = \frac{3}{2}nR$ or $C_V/nR = \frac{3}{2}$. This exactly matches the observed value. Experimental results and classical theory are in perfect agreement for translational motion.

The situation changes dramatically for diatomic and polyatomic molecules. The experimental heat capacities in Table 3.2 are below the theoretical values except for Cl_2 at the highest temperatures. In addition, the heat capacities are temperature dependent.

TABLE 3.2
HEAT CAPACITIES C_V/nR OF GASES

Gas	Classical value	Observed values at temperature $T(K)$					
		298.15	400	600	800	1000	1500
He	1.500	1.500	1.500	1.500	1.500	1.500	1.500
H_2	3.500	2.468	2.510	2.527	2.562	2.633	2.882
O_2	3.500	2.532	2.621	2.860	3.059	3.195	3.398
Cl_2	3.500	3.071	3.246	3.403	3.475	3.511	3.571
N_2	3.500	2.503	2.519	2.622	2.781	2.934	3.192
H_2O	6.000	3.037	3.119	3.365	3.652	3.956	4.651
CO_2	6.500	3.465	3.970	4.691	5.185	5.530	6.020

For all diatomic molecules the classical theory predicts $C_V/nR = 3.50$ independent of temperature. The experimental data for N_2, O_2, and H_2 indicate that C_V/nR is about 2.5 at room temperature. Only at higher temperatures does C_V approach the predicted value. A possible explanation would be that only translational and rotational degrees of freedom can acquire their equipartition quota of energy, giving $C_V/nR = \frac{3}{2} + \frac{2}{2} = \frac{5}{2}$. This explanation is reinforced by the values for H_2O and CO_2 in Table 3.2. They are approximately $C_V/nR = 3$ at low temperatures, the value expected for translations and rotations alone. For ethane at 100 K, $C_V/nR = 3.35$, nowhere near the classical prediction of 21.

Thus evidence is found in these C_V measurements that molecules do not behave completely in accord with the predictions of classical mechanics. Vibrational energy in particular appears to participate in equipartition of energy only at higher temperatures. Put another way, the figures suggest that molecules are unable to store added energy in vibrational modes at ordinary temperatures but only at high temperatures. The development of quantum theory showed this explanation to be correct and brought calculated C_V values for gases into precise agreement with the experimental data [Can you suggest any reason why Cl_2 at 1000 K has C_V/nR actually higher than the classical value?]

Problems

1. The energy released in the combustion of 1.00 mol of n-octane is 5450 kJ. Calculate the decrease in mass of Δm in the combustion of 10^3 kg of n-octane.

2. Calculate the classical C_V (on equipartition principle) of the linear molecule acetylene C_2H_2 and the nonlinear benzene C_6H_6.

3. From the values of C_V in Table 3.2, calculate the internal energy that must be added to a mole of helium to raise its temperature from 300 K to 1000 K.

4. Repeat the calculation in Problem 3 for 1 mol of H_2 and of CO_2. A rough graphical integration can be made.

5. Locate the center of mass of the molecule H_2O given that the internuclear distance O—H is 96.0 pm and the HOH bond angle is 105°.

6. P_4 is a tetrahedral molecule with a P—P distance 221 pm. Locate the center of mass of the P_4 molecule and calculate its moment of inertia.

7. A thermally insulated constant-volume heat-capacity calorimeter is filled with 20.950 g of liquid benzene at 298.15 K. An electric current of 0.125 A is passed through the internal resistance wire of 5.27 Ω for 10.0 min. The final temperature is 299.04 K. If the effective heat capacity of the calorimeter is 35.0 J K^{-1}, what is C_V per mole of benzene?

8. In a calorimeter similar to that in Problem 7 but of different dimensions, an energy input of 72.0 J into 32.07 g of liquid benzene at 298.15 K produced a $\Delta T = 1.09$ K. When calorimeter is filled with 30.0 g of n-octane at 298.15 K, an energy input of 61.5 J gave the same ΔT. What is the molar heat capacity C_V of n-octane?

9. Calculate the reduced mass μ and moment of inertia I of $^{13}C^{18}O$ for which $R_e = 112.8$ pm. Compare the results with those in Example 3.4.

10. The equation of state of a monatomic solid is

$$PV + nf(V/n) = BU$$

where B is a constant and f is some function of molar volume. Prove that

$$B = \alpha V/\beta C_V$$

(α is thermal expansivity and β is isothermal compressibility). This is the Gruneisen equation used in solid-state studies.

11. Show that $(\partial U/\partial P)_V = \beta C_V/\alpha$. For water at 20°C, $\alpha = 206.6 \times 10^{-6}$ K^{-1}, $\beta = 45.9 \times 10^{-11}$ Pa^{-1}. and $C_V = 4.184$ J K^{-1}g^{-1}. Calculate the ΔU on isothermal compression of 1 kg of water at 20°C from 10^2 to 10^4 kPa.

12. The CH bond in the CH radical has a force constant of 4.490×10^2 N m^{-1}. Calculate the fundamental vibration frequency. What would be the frequency in the CD radical?

13. The bond distance in the NaCl molecule is 251 pm. What is the moment of inertia?

14. The bond length in CH is 112 pm. Calculate the energy necessary to stretch or compress the bond by 10%, assuming that it behaves as a harmonic oscillator. (See Problem 12.)

15. The fundamental vibration frequency of F_2 is 2.67×10^{13} Hz, $R_e = 141.8$ pm; of Cl_2, 1.69×10^{13} Hz, $R_e = 198$ pm. Sketch the potential-energy curves $U(R)$ for these molecules (on same plot) on a harmonic-oscillator model.

4

Molecular Quantum Levels

The classical mechanics of Newton is entirely satisfactory for large-scale objects moving at speeds well below the speed of light. Consistent with our observations of falling stones and planetary motions, it treats energy as a continuous variable. The theory fails, however, when we try to use classical mechanics to describe the world of small masses such as atoms and molecules. Our discussion of the heat capacity of gases in Chapter 3 showed that the failure of classical mechanics is sometimes quite obvious. In other instances one must look more carefully. For example, the treatment of the translation of molecules by classical mechanics predicts translational heat capacities that are precisely in accord with experiment. In some cases, however, even translational energies fail to follow the predictions of classical theory—electrons in metals are a famous example.

The theoretical mechanics that applies to the molecular world is called *quantum mechanics* or *wave mechanics*. One of its chief distinctions from classical mechanics is that energy is no longer a continuous variable but usually appears in discrete packets or *quanta*. We shall discuss the basis of quantization in later chapters. Here we present some facts about discrete molecular energy levels. We introduce the discussion with a review of some properties of light.

4.1 Electromagnetic Radiation: Particles and Waves

In the history of light, particle theory and wave theory contended for supremacy, until finally a synthesis was achieved that allowed both representations to play complementary roles. A beam of light is a stream of particles called *photons*. A beam of light is also an electric and magnetic field moving through space as a wave motion (Fig .4.1). The energy ϵ of the photon is related to the frequency ν of the electromagnetic wave by

$$\epsilon = h\nu \tag{4.1}$$

This relation was discovered by Max Planck in 1900. The Planck constant h has the dimensions of energy $\times$ time (a quantity called *action*). In SI units, $h = 6.626 \times 10^{-34}$ J s.

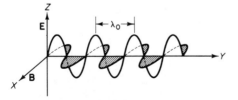

FIGURE 4.1 A plane wave of electromagnetic radiation at one instant in time. The magnetic induction **B** is in the xy plane and the electric field **E** is in the yz plane. In a vacuum, the frequency $\nu = c/\lambda_0$ where λ_0 is the wavelength and c the speed.

Light travels through a vacuum at a speed $c = 2.998 \times 10^8$ m s⁻¹. Measurements of this speed have a history extending to the seventeenth century, and its value is presently known to an accuracy of about 1 part in 10^9. [How fast does light ($\lambda = 600$ nm) go through (a) air; (b) water of 298 K?]

The frequency ν, wavelength λ, and speed c are related by

$$\nu = c(1/\lambda) = c\tilde{\nu} \tag{4.2}$$

The quantity $1/\lambda = \tilde{\nu}$ is called the *wave number*. Almost everyone gives wave numbers in the non-SI units of cm⁻¹.

A photon has no rest mass and its energy ϵ is related to its mass by the Einstein equation,

$$\epsilon = mc^2 \tag{4.3}$$

This equation couples the energy of a photon to its mechanical properties mass and velocity. The Planck equation $\epsilon = h\nu$ relates the energy of a photon to its wave property, frequency. The two equations are combined to give $h\nu = mc^2$ or, with $\lambda = c/\nu$, $\lambda = h/mc$. This equation relates a wave property of the photon (λ) to a particle property (mc).

The product mc is the magnitude of the momentum p of the photon. Hence

$$\lambda = h/p \tag{4.4}$$

Just as gas molecules colliding with a wall exert a pressure due to their transfer of momentum to the wall, so photons colliding with matter exert a radiation pressure. The pressure of radiation drives clouds of interstellar dust through the vast spaces of the universe.

4.2 Spectroscopy: The Experimental View
of Molecular Energy Levels

The particle and wave natures of light are both revealed by a study of the interaction of light with atoms and molecules. The wave nature is most often used to explain light scattering, and the particle nature to explain absorption and emission. Measurements of absorption or emission of radiation are the most widely used methods to determine atomic and molecular energy levels.

The study of the interaction between radiation and matter is called *spectroscopy*. Its principal product is a *spectrum*, which is a representation of the emission, absorption, or scattering of light as a function of its frequency v. Molecular spectroscopy covers a broad range of frequencies, ranging from the high frequencies of soft X rays ($v \approx 10^{16}$ s^{-1} or Hz) to the low frequencies of radiowaves ($v \approx 10^8$ Hz). Visible light spans only the small region from 4.3×10^{14} Hz to 7.5×10^{14} Hz, but we refer to ultraviolet or infrared radiation as "light" even though we cannot see it. The range in spectroscopic wavelengths is from about 10^{-8} m (soft X rays) to about 3 m (radio waves), with visible light being in the region from about 4×10^{-7} to 7×10^{-7} m. The wavelengths used to study various aspects of atomic and molecular properties are summarized in Table 26.1. For comparison, recall that an atomic radius or a molecular bond length is of the order of 10^{-10} m.

Molecules absorb or emit energy only in discrete amounts or packets called *quanta*. The essential fact about atomic or molecular spectroscopy is that absorption or emission of radiation by an atom or a molecule can occur only if the energy of the radiation quantum equals the energy difference $\Delta\epsilon$ between two energy levels in the atom or molecule. (This condition is *necessary* but other conditions must also be satisfied for absorption or emission actually to happen.) Thus

$$\Delta\epsilon = \epsilon_2 - \epsilon_1 = hv \qquad (4.5)$$

Whenever it gives or takes a quantum of energy, an atom or molecule must make a transition from one definite energy level ϵ_1 to another ϵ_2. Quantization is not restricted to transfer of energy by electromagnetic radiation. It applies equally well to the transfer of energy between a pair of molecules when they collide and to the energy transferred to a substance when it is placed in a thermal bath or heated by a resistance wire in a calorimeter. Molecules can exist only in definite, discrete energy states. Energy levels between these states are not permitted at all.

4.3 An Example: The Spectrum of CO

A demonstration of the quantized interaction of light with molecules is shown by the spectrum of carbon monoxide gas in Fig. 4.2. The figure shows light absorption by CO over a small range of frequencies in the infrared region. Consider first the absorption in part (a) of the figure. There is a region or *band* of pronounced absorption near $v = 6.4 \times 10^{13}$ Hz, or wavelength $\lambda = 4.7$ μm. This absorption results in an increase in the vibrational energy of CO. Given this fact, one knows that a separation between two vibrational levels in CO is

$$\epsilon_2 - \epsilon_1 = \Delta\epsilon_v = hv = (6.63 \times 10^{-34} \text{ J s})(6.4 \times 10^{13} \text{ s}^{-1}) = 4.2 \times 10^{-20} \text{ J}$$

Note that regions to higher and lower energy are essentially free of absorption. These photon energies do not correspond to differences between any energy levels in the CO molecule and hence there is no absorption.

A weak absorption occurs near $v = 12.8 \times 10^{13}$ Hz or about $\lambda = 2.3$ μm.

WAVELENGTH, λ/μm

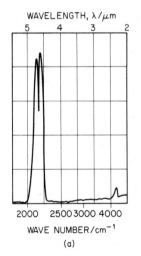

WAVE NUMBER/cm⁻¹

(a)

FIGURE 4.2 Vibration–rotation absorption bands of CO gas in the infrared: (a) the spectrum at low resolution, showing two regions of absorption; (b) the stronger absorption band at a resolution high enough to see the rotational structure. [*Pure Appl. Chem. 1*, 537 (1961)]

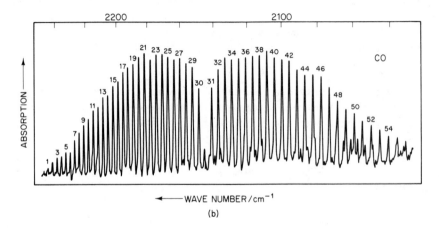

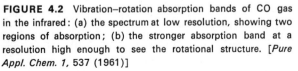

(b)

This region corresponds to twice the photon energy of the strong absorption and shows that a difference between another pair of vibrational energy levels in CO is about 8.4×10^{-20} J.

The band at 4.7 μm upon examination at higher resolution in (b) of Fig. 4.2 reveals a beautiful regularity of finer details. These features occur because the light absorption causes small changes in the rotational energy of the molecule along with the larger changes in its vibrational energy. We can tell from the presence of many lines in this spectrum that rotations as well as vibrations are quantized. The nature of the quantization and an understanding of the finer structure in this spectrum will come later in this chapter when we consider the quantum mechanical expression for the rotational energy of a molecule.

The absorptions in Fig. 4.2 change vibrational and rotational energies of CO, but they do not affect translational energy. Neither do they affect electronic energy, i.e., the energy due to the position and motion of electrons in the CO molecule. If we search in the microwave region ($\nu \approx 10^{11}$ Hz), we find absorptions that increase *only* the rotational energy of CO. An absorption that principally changes the electronic energy of CO can be found at the comparatively high frequencies of the ultraviolet ($v \approx 10^{15}$ Hz).

We now consider the molecular energy levels that are predicted by quantum mechanics. The predictions have been verified repeatedly by analysis of spectra such as those in Fig. 4.2.

4.4 Wave Properties of Matter

We described in Section 4.1 the dual particle–wave nature of light. In 1923 this concept was extended by the French physicist Louis de Broglie to particles of matter such as electrons, protons, neutrons, atoms, and molecules. He proposed that the relationship of Eq. (4.4), $\lambda = h/p$, applies not only to photons but to all particles. For particles traveling at less than the speed of light, the magnitude of the momentum is $p = mv$, where v is the speed of the particle. From Eq. (4.4), by analogy, for any material particle,

$$\lambda = h/mv \qquad (4.6)$$

The Broglie relation (4.6) embodies the wave–particle duality at the heart of the physical world. It relates a wave property of an entity, its wavelength λ, to a mechanical property, the magnitude mv of its momentum.

Example 4.1 An electron is accelerated in an electric field to a speed of 1.00×10^6 m s^{-1}. What is the wavelength of the electron?

The mass of the electron is $m_e = 9.11 \times 10^{-31}$ kg. From Eq. (4.6),

$$\lambda = \frac{6.63 \times 10^{-34} \text{ J s}}{(9.11 \times 10^{-31} \text{ kg})(1.00 \times 10^6 \text{ m s}^{-1})} = 728 \times 10^{-12} \text{ m} = 728 \text{ pm}$$

The wavelength of the electron in Example 4.1 is of the same order of magnitude as distances between atoms in molecules or crystals. It was therefore predicted that beams of electrons would experience diffraction effects on passing through material substances: gases, liquids, or solids. Figure 4.3 is a diffraction pattern obtained by passing a beam of electrons through a thin gold foil and recording the positions where the electrons strike a photographic plate perpendicular to the beam direction. Such diffraction patterns provide experimental proof of the wave properties of electrons. Analysis of the spacings of diffraction maxima permits us to calculate the wavelength λ of the electrons if we know the distances between atoms in solid gold. The experimental λ agrees perfectly with the λ from Eq. (4.6).

Beams of charged particles such as electrons and protons can be focused by lenses based on suitable dispositions of electric and magnetic fields. Thus electron microscopes are extensively used to examine the fine structure of materials with electron beams. Figure 4.4, a picture of molecules in a crystal obtained with an electron microscope, is a beautiful demonstration of the wave properties of electrons.

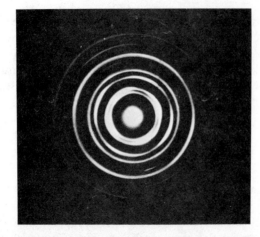

FIGURE 4.3 One of the electron diffraction pictures that first showed the wave nature of electrons, obtained by G. P. Thomson (1928) from gold foil.

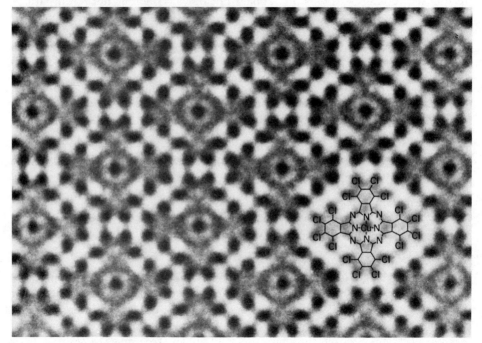

FIGURE 4.4 An electron microscope picture that shows molecules of copper phthalocyanine chloride in a crystal. [N. Uyeda, Institute for Chemical Research, University of Kyoto]

4.5 Translational Energy

The kinetic energy of translation also has discrete energy levels, but separations between these levels are so small that for practical purposes translational energy can be regarded as continuous. Nevertheless, the theory of translational energy levels is interesting and important for the insight it gives into the wave mechanical basis for quantization of energy.

The restriction of allowed energy levels of particles only to certain discrete values is closely related to the wave aspect of particles. Consider a one-dimensional example of translational motion as shown in Fig. 4.5. A particle is confined to move between

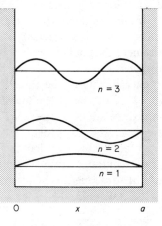

FIGURE 4.5 Waves representing a particle confined to move in one dimension between fixed barriers.

barriers fixed at $x = 0$ and $x = a$. When we consider the wave aspect of the particle, the amplitude of the associated wave that represents the particle must vanish at $x = 0$ and at $x = a$ since the particle cannot penetrate into regions beyond these points. A wave restricted in this way to a certain region of space is called a *standing wave*. Only certain discrete wavelengths are allowed for a standing wave. In this case the necessary condition is that $a = n(\lambda/2)$, where n is any *integer*. An integral number of half wavelengths must fit between $x = 0$ and $x = a$. The number n is called a *quantum number*. It can have values $n = 1, 2, 3, \ldots$. This requirement allows the *particle* to be associated with a *standing wave* within the confines of $0 < x < a$. If the requirement is not satisfied, no standing wave can exist.

A particle moving in one dimension and confined within definite limits must have a wavelength specified by one of the values of n that satisfy the condition $n(\lambda/2) = a$. From Eq. (4.6), $\lambda = h/mv$ and thus $nh/mv = 2a$, and the magnitude of the velocity of the particle is restricted to values $v = nh/2ma$. The allowed values of the kinetic energy ϵ_k thus depend on the square of the translational quantum number n,

$$\epsilon_k = \tfrac{1}{2} mv^2 = n^2 h^2 / 8ma^2 \tag{4.7}$$

The corresponding formula for a particle confined within a rectangular box of sides a, b, c is readily derived by quantum mechanics (see Chapter 20). The result is

$$\epsilon_k = \frac{h^2}{8m} \left(\frac{n_1^2}{a^2} + \frac{n_2^2}{b^2} + \frac{n_3^2}{c^2} \right) \tag{4.8}$$

Since translational motion in three dimensions has three degrees of freedom, three quantum numbers n_1, n_2, n_3 are required to specify the allowed energy levels. The quantum numbers can take any integral values. If $a = b = c$, the container is a cube and the allowed energy levels are $\epsilon_k = (h^2/8ma^2)(n_1^2 + n_2^2 + n_3^2)$. The volume of the cubical box is $V = a^3$, so that

$$\epsilon_k = \frac{h^2}{8mV^{2/3}}(n_1^2 + n_2^2 + n_3^2) \qquad (4.9)$$

Note that as the volume increases, the spacing of the energy levels decreases. In the limit as the volume becomes infinite, the spacing of allowed energy levels becomes zero and the energy then varies continuously and not in quantum jumps. In other words, classical behavior is followed when the particle is not restricted to a confined space. It is the restriction of the particle to a finite volume that leads to quantized energy levels. The quantization is a necessary consequence of the wave aspect of the particle and the requirement that only standing waves of appropriate wavelengths can exist in any confined space.

The lowest allowed energy state of the three-dimensional box is designated as (111) with $n_1 = n_2 = n_3 = 1$. [Can one of the quantum numbers be zero?] The next state is (211), but there are three states of the same energy for a cubical box, (211), (121), and (112). This energy level is said to have a *statistical weight* or *degeneracy* $g = 3$. Note that levels but not states may have degeneracy. The degeneracy of an energy level is the number of quantum states with exactly that energy. The degeneracy of energy levels has important consequences in later calculations of average molecular properties. [What is g for the state (123)?]

Example 4.2 We saw in Chapter 3 that the average translational energy of a molecule is $\frac{3}{2}kT$ as predicted by classical mechanics. What is the average quantum number of an energy level in O_2 corresponding to this energy at 300 K in a volume of 1 L?

Assume that $n_1^2 + n_2^2 + n_3^2 = 3n^2$, so that we can estimate an average quantum number. From Eq. (4.9),

$$\epsilon_t = \tfrac{3}{2}kT = 3h^2n^2/8mV^{2/3}$$

$$n^2 = 4kTmV^{2/3}/h^2 = \frac{(4)(1.4 \times 10^{-23}\ \text{J K}^{-1})(300\ \text{K})(5.3 \times 10^{-26}\ \text{kg})(10^{-3}\ \text{m}^3)^{2/3}}{(6.6 \times 10^{-34}\ \text{J s})^2}$$

$$n \approx 4.5 \times 10^9$$

Example 4.3 What is the lowest allowed translational energy of an O_2 molecule in a volume of 1.0 L?

From Eq. (4.9) with $n_1 = n_2 = n_3 = 1$,

$$\epsilon_t(111) = \frac{(6.6 \times 10^{-34}\ \text{J s})^2(1 + 1 + 1)}{(8)(5.3 \times 10^{-26}\ \text{kg})(1.0 \times 10^{-3}\ \text{m}^3)^{2/3}}$$

$$= 3.1 \times 10^{-40}\ \text{J}$$

(Recall that J is $\text{kg m}^2\ \text{s}^{-2}$.)

Pay special attention to the magnitudes of the numbers that have been calculated in the examples above. The translational quantum numbers of molecules of average thermal energy are of the order of 10^9. The average energy $\frac{3}{2}kT$ is 6×10^{-21} J at

300 K. Thus the average energy per level is about $10^{-20}/10^9 = 10^{-29}$ J. Translational quanta are extremely small compared to kT. As a consequence, molecules use many translational energy levels that lie closely together, and the discrete or quantized character of the energy is usually undetectable. Translational energies therefore appear to have a continuous range of values, just as would be given by classical mechanics. We saw in Chapter 3 that the measured translational heat capacity (e.g., the heat capacity of a monatomic gas) is in precise accord with the classical prediction of $\frac{3}{2}R$ per mole.

This classical behavior of translational energy is an example of a general rule: *Quantum mechanics gives results corresponding to classical mechanics in the limit of large quantum numbers.* Classical mechanics is, in a sense, a special case of quantum mechanics applicable to large quantum numbers. The emergence of classical behavior from quantum mechanics in this limit is called the *correspondence principle.*

Quantized translational energy can be observed by infrared spectroscopy under special circumstances. Consider the factors necessary to place molecules in levels with low quantum numbers so that quantum behavior might be detected. Equation (4.9) shows that large separations between levels occur when both particle mass and available volume are small. In Chapter 3 we found that average molecular energies are low at low temperatures. Thus we need to examine small molecules at low temperatures in a small volume. Hydrogen is a logical choice. A small volume can be created by trapping an H_2 molecule in a solvent cavity in a liquid at low temperature. Under these conditions, one can observe excitation of H_2 from one translational level to another by absorption of light in the far infrared ($v \approx 6 \times 10^{12}$ Hz or $\lambda \approx 50$ μm). Figure 4.6 shows the absorption spectrum of isolated H_2 molecules trapped in liquid-argon solvent cages at 90 K.

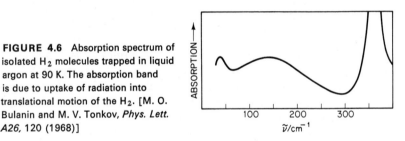

FIGURE 4.6 Absorption spectrum of isolated H_2 molecules trapped in liquid argon at 90 K. The absorption band is due to uptake of radiation into translational motion of the H_2. [M. O. Bulanin and M. V. Tonkov, *Phys. Lett. A26,* 120 (1968)]

4.6 Wave Number as a Measure of Molecular Energy Level

The joule of the SI system could be used as the energy unit for molecular quantum levels. However, another quantity, the wave number $\tilde{v} = \lambda^{-1}$ in units of cm^{-1}, is more widely used in spectroscopy, The photon energy in absorption or emission of radiation is equal to the separation of a pair of molecular energy levels, $hv = \epsilon_2 - \epsilon_1 = \Delta\epsilon$. The wavelength λ is more easily measured than the photon frequency v. With $v = c/\lambda$, where c is the speed of light in a vacuum. $\Delta\epsilon = hv = hc(1/\lambda) = hc\tilde{v}$. Thus the wave number $\tilde{v} = 1/\lambda$ is proportional to the energy quantum $\Delta\epsilon$ through the factor *hc*. The conversion between joules and cm^{-1} is

$$(\text{energy in J}) = (1.986 \times 10^{-23} \text{ J cm})(\text{wave number in cm}^{-1})$$

The spectroscopic unit of frequency, the hertz (Hz), equal to one cycle per second (s^{-1}), was named after Heinrich Hertz, who first demonstrated experimentally the existence of electromagnetic waves, at the Karlsruhe Technische Hochschule in 1888.

4.7 Rotational Energy

The classical rotational motion of a diatomic molecule was considered in Section 3.5. Let us now consider how the energy levels are restricted by the requirements of wave mechanics. Figure 4.7 represents schematically a wave associated with an allowed rotation of a particle with a reduced mass μ at a fixed distance r from the origin. The moment of inertia of the molecule is $I = \mu r^2$. If v is the magnitude of the linear velocity of μ, the angular velocity $\omega = v/r$, and the kinetic energy is $\epsilon_r = \frac{1}{2}\mu v^2 = \frac{1}{2}I\omega^2$.

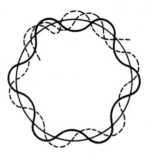

FIGURE 4.7 Model of standing wave representing a rotating particle. Unless $n\lambda = 2\pi r$, the wave is destroyed by interference (dashed line).

A standing wave pattern for the rotating particle requires that an integral number J of whole wavelengths be fitted into the circumference $2\pi r$ of its orbit. Thus $J\lambda = 2\pi r$ or $\lambda = 2\pi r/J$. J is called the *rotational quantum number*; it can have values $J = 0, 1, 2, \ldots$. From the Broglie relation, $\lambda = h/\mu v$. Hence $v = hJ/2\pi\mu r$, and $\frac{1}{2}\mu v^2 = \epsilon_r = J^2 h^2/8\pi^2 I$. [Why can J be 0, whereas n cannot?]

The rotation represented by Fig. 4.7 is confined to two dimensions. An actual molecule can rotate in three-dimensional space. Instead of J^2 the correct factor in the energy equation turns out to be $J(J + 1)$. Thus the allowed rotational energy levels are

$$\epsilon_r = J(J + 1)h^2/8\pi^2 I \qquad (4.10)$$

Note in Eq. (4.10) that a small moment of inertia for a molecule corresponds to a large quantum of rotational energy. From Eq. (4.10), since $\epsilon = \frac{1}{2}I\omega^2$, the rotational frequency ω of the molecule increases as the quantum number J of the rotational energy level increases. Remember that the angular frequency $\omega = 2\pi v$. Experimental proof of the formula (4.10) is obtained from measurements of the absorption of microwaves or of infrared radiation by molecules.

Example 4.4 The internuclear distance in the molecule $^{16}O_2$ is 120.80 pm. What is the spacing $\Delta\epsilon$ between the rotational energy levels $J = 0$ and $J = 1$?

The reduced mass of $^{16}O_2$ is

$$\mu = \frac{(16.0^2/32.0) \times 10^{-3} \text{ kg mol}^{-1}}{6.02 \times 10^{23} \text{ mol}^{-1}} = 1.33 \times 10^{-26} \text{ kg}$$

$$I = \mu r_e^2 = (1.33 \times 10^{-26} \text{ kg})(121 \times 10^{-12} \text{ m})^2$$

$$= 19.4 \times 10^{-47} \text{ kg m}^2$$

For $J = 0$, $\epsilon_r = 0$. For $J = 1$, Eq. (4.10) yields

$$\epsilon_r = \frac{(6.63 \times 10^{-34} \text{ J s})^2(2)}{8\pi^2(19.4 \times 10^{-47} \text{ kg m}^2)} = 5.75 \times 10^{-23} \text{ J} = \Delta\epsilon$$

Example 4.5 What is the approximate quantum number J of a $^{16}O_2$ molecule whose rotational energy is approximately equal to kT, the classical average value, at 350 K?

$$kT = (h^2/8\pi^2 I)(J(J+1)) \approx (h^2/8\pi^2 I)(J^2)$$

$$J \approx (8\pi^2 IkT)^{1/2}/h$$

$$= \frac{(8\pi^2)^{1/2}(19.4 \times 10^{-47} \text{ kg m}^2)^{1/2}(1.38 \times 10^{-23} \text{ J K}^{-1} \times 350 \text{ K})^{1/2}}{6.63 \times 10^{-34} \text{ J s}}$$

$$\approx 13$$

Both examples show that rotational quanta are larger than translational quanta by many orders of magnitude, but they are still small relative to average classical energies ($kT = 4 \times 10^{-21}$ J at 300 K). The quanta are large enough, however, for quantum effects to be easily observed. At the same time quantum numbers of occupied levels at room temperature are often sufficiently high for average values of rotational energy to approach the classical prediction. At low temperature, however, the classical treatment can be in serious error.

The pattern of rotational energy levels is shown in Fig. 4.8. Because the energy

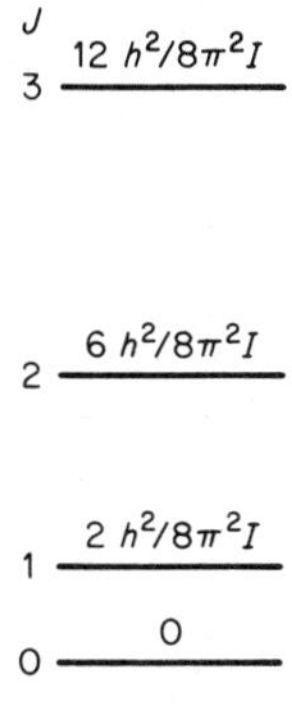

J

3    $12\ h^2/8\pi^2 I$ ——

2    $6\ h^2/8\pi^2 I$ ——

1    $2\ h^2/8\pi^2 I$ ——

0   0 ——

FIGURE 4.8 Energy levels of a rigid linear rotor.

depends on $J(J+1)$, the spacing between levels increases as J increases. In other words, the size of the quantum of rotational energy grows as J increases.

The statistical weight (or degeneracy) g_J of a rotational level of quantum number J is $2J + 1$. This result is derived from the quantum mechanics of rotation.

4.8   Vibrational Energy

The quantum treatment of molecular vibrations uses the model of the harmonic oscillator as a first approximation. The quantum mechanical expression for the energy levels of a harmonic oscillator is

$$\epsilon_v = h\nu(v + \tfrac{1}{2}) \tag{4.11}$$

Here ν is the frequency of the vibration and v is the *vibrational quantum number*, with allowed values $v = 0, 1, 2, \ldots$. Equation (4.11) gives the vibrational energy of a diatomic molecule or of a particular normal mode of vibration in a polyatomic molecule.

From Eq. (4.11) you can see that the vibrational energy levels of a harmonic oscillator are evenly spaced, as shown in Fig. 4.9. The vibrational quantum has energy $h\nu$. As the quantum number increases, the vibrational energy increases but the fundamental vibration frequency does not change. (Note the contrast with rotational motion where the rotational frequency increases with rotational quantum number and energy).

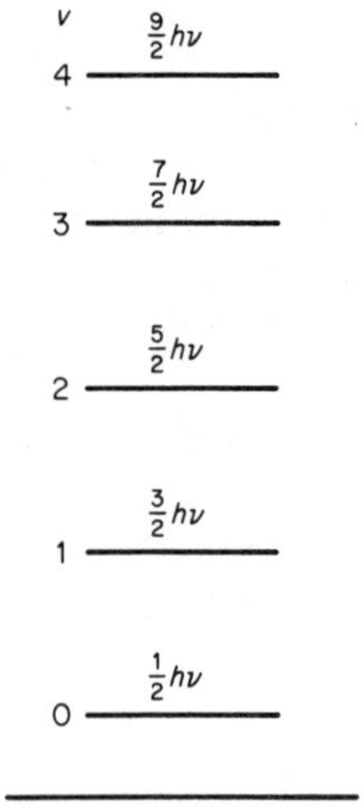

$$
\begin{array}{ll}
v & \\
4 \quad & \tfrac{9}{2}h\nu \\[1.5em]
3 \quad & \tfrac{7}{2}h\nu \\[1.5em]
2 \quad & \tfrac{5}{2}h\nu \\[1.5em]
1 \quad & \tfrac{3}{2}h\nu \\[1.5em]
0 \quad & \tfrac{1}{2}h\nu
\end{array}
$$

FIGURE 4.9   Equally spaced energy levels of a harmonic oscillator are specified by the quantum number v.

Equation (4.11) differs from the original Planck hypothesis in that $(v + \tfrac{1}{2})$ takes the place of v. Therefore, the lowest possible vibrational level $v = 0$ still has considerable vibrational energy. This is called the *zero-point energy* and amounts to

$$\epsilon_0(v = 0) = \tfrac{1}{2}h\nu \tag{4.12}$$

Even in the limit of absolute zero, when all translational and rotational motion has ceased in a frigid crystal, the molecules would still be oscillating internally about their equilibrium internuclear distance R_e. The zero-point energy is by no means small or negligible. [Calculate the zero-point energy of a vibration with $v = 10^{13}$ Hz and compare it to kT at 300 K.]

For a diatomic molecule, the vibration frequency v in Eq. (4.11) is given by the classical expression (see Example 3.5)

$$v = \frac{1}{2\pi}\sqrt{\kappa/\mu} \qquad (4.13)$$

where κ is the force constant and μ the reduced mass. This equation is used to calculate force constants from spectroscopic data. [Calculate κ for the vibration of O_2 given $v = 4.74 \times 10^{13}$ Hz.]

Vibrational frequencies (energies) are reliably obtained from experiments in spectroscopy, as described in Chapters 26 and 27. The example of CO in Fig. 4.2 showed a difference between two vibrational energy levels of about 4.2×10^{-20} J. In Chapter 26 we show that the absorption of radiation occurs between two *adjacent* levels, so that the quantum number v changed by one unit: $v' - v'' = \Delta v = 1$. Thus the vibrational quantum in CO is $hv = 4.2 \times 10^{-20}$ J, and the wave number of the absorption is $\tilde{v} = 2100$ cm^{-1}.

The average (classical) vibrational energy of a diatomic molecule is kT, which corresponds to 200 cm^{-1} at 300 K. The vibrational quantum of CO at 2100 cm^{-1} is thus 10 times larger than kT. The probability is close to zero that any particular CO molecule at 300 K can have even one quantum of vibrational energy. The classical prediction is therefore grossly in error. It is also evident that the vibrational quantum levels occupied by CO molecules at 300 K almost always have a low quantum number, $v = 0$ or 1. The correspondence principle indicates classical behavior for translational and (to a lesser extent) rotational motions, but only quantum theory is able to deal with vibrations.

The quantum treatment of vibrations of polyatomic molecules retains the normal-mode description of Chapter 3. A polyatomic molecule thus has $3N - 5$ (linear molecule) or $3N - 6$ (nonlinear molecule) modes, each with possible energy $\epsilon_i = hv_i(v_i + \frac{1}{2})$, where i designates the particular mode. The modes are distinguished by the frequency v_i or the energy of the vibrational quantum hv_i. The wave numbers of the various modes i in a polyatomic molecule range from about 100 to 3000 cm^{-1}. The vibrational energies are so large relative to kT that quantum effects are dominant. As an example, the failure of the heat capacity of ethane to reach the classical prediction is almost entirely due to the absence of contributions from vibrational energies.

4.9 A View of Vibrational and Rotational Energy Levels in the CO Molecule by Infrared Spectroscopy

Quantum mechanics tells us that restrictions exist on the magnitude of changes in quantum number that occur as molecules emit or absorb light. These restrictions are called *selection rules* (see Chapter 26). The selection rule for the rotational quantum number J of CO for the infrared absorption in Fig. 4.2 is $\Delta J = \pm 1$. In any spectroscopic absorption or emission of radiation by CO, J must either increase or decrease by one unit. ΔJ itself is defined by the convention

$$\Delta J = J(\text{upper}) - J(\text{lower}) = J' - J''$$

where upper and lower (or prime and double prime) refer to J states of higher and lower energy, respectively.

In the vibration–rotation (infrared) spectrum of Fig. 4.2, the main energy change of the molecule is established by the transition between two vibrational levels with a selection rule $\Delta v = 1$. In this case, absorption occurs between vibrational levels $v'' = 0$ and $v' = 1$. The changes in rotational energy, with $\Delta J = \pm 1$, that accompany the vibrational change now refer to rotational levels $J(\text{upper}) = J'$ in the $v' = 1$ state, and $J(\text{lower}) = J''$ in the $v'' = 0$ state.

Figure 4.10 is a diagram of some absorptions allowed by the selection rules. Every absorption causes the change $v = 0 \rightarrow v = 1$. The selection rules preclude any absorptions with $\Delta J = 0$.* The rules thus require that a change in J accompanies the change in v; i.e., the rotational level must change when the vibrational level changes in CO. A transition that is not allowed by the selection rules is said to be "forbidden." Two types of J changes occur: Those with $\Delta J = +1$ form the *R branch* of the absorption band in the spectrum; those with $\Delta J = -1$ form the *P branch* of the absorption band. Four possible transitions are shown for each branch.

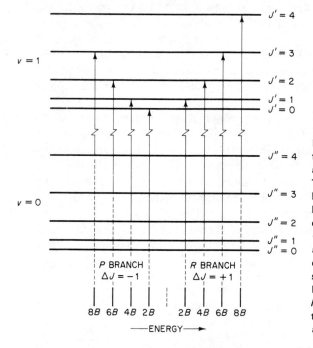

FIGURE 4.10 A schematic diagram of the lower rotational levels in the $v = 0$ and $v = 1$ levels of a diatomic molecule. The allowed absorptions form a series of lines at increasing energies (the *R* branch, $\Delta J = +1$) and a series at decreasing energies (the *P* branch, $\Delta J = -1$). The energy scale shows the positions at which these transitions would be observed in an absorption spectrum on a scale that is linear in energy. The scale has been marked in units of energy $B = h^2/8\pi^2 I$ from an origin at the position of the forbidden $J'' = 0 \rightarrow J' = 0$ absorption.

Calculation of the photon energies is straightforward. The energies have a common contribution of $h\nu$ from the $\Delta v = 1$ vibrational change (where ν in this case

*For diatomic molecules having an odd number of electrons (NO, for example) a transition with $\Delta J = 0$ is allowed.

is the CO fundamental vibration frequency). In addition, a rotational contribution is added to or subtracted from this energy $h\nu$ to establish the total energy of the absorption. It is convenient to calculate the rotational part by first rewriting Eq. (4.10) as

$$\epsilon_r = BJ(J + 1) \tag{4.14}$$

where $B = h^2/8\pi^2 I$ is called the *rotational constant*. The difference $\Delta\epsilon_r$ between the levels now becomes

$$\Delta\epsilon_r = \epsilon'_r - \epsilon''_r = B[J'(J' + 1) - J''(J'' + 1)]$$

For the R branch ($\Delta J = +1$), we have $J' = J'' + 1$, so that

$$\Delta\epsilon_r = 2B(J'' + 1)$$

For the P branch ($\Delta J = -1$), we have $J' = J'' - 1$, so that

$$\Delta\epsilon_r = -2BJ''$$

The energies $\Delta\epsilon_{abs}$ of the infrared vibrational–rotational absorptions of CO thus become

$$\text{R branch:} \quad \Delta\epsilon_{abs} = h\nu_v + 2B(J'' + 1) \tag{4.15}$$

$$\text{P branch:} \quad \Delta\epsilon_{abs} = h\nu_v - 2BJ'' \tag{4.16}$$

where ν_v is the *vibrational frequency* of CO, and B is the *rotational constant* of CO determined by its moment of inertia $I = \mu R_e^2$.

Observe how these simple formulas from quantum mechanics describe the CO spectrum:

1. The R branch should be a series of lines beginning at $\Delta\epsilon_{abs} = h\nu_v + 2B$ (for $J'' = 0$) and going to higher energies with equal separations of $2B$ in energy units.

2. The P branch should begin at $\Delta\epsilon_{abs} = h\nu_v - 2B$, with a series of lines to lower energies, again with equal separations of $2B$. [Note that the lowest J'' must be $J'' = 1$ rather than $J'' = 0$. Why?]

3. There should be a gap between the P and R branches of energy $4B$. The center of this gap, $2B$ from the first line of each branch, marks the position of the forbidden $J'' = 0 \rightarrow J' = 0$ absorption and lies at an energy of $\Delta\epsilon_{abs} = h\nu_v$.

Study carefully these predictions of quantum mechanics to see how well they fit the CO spectrum of Fig. 4.2. It is hard to imagine any better confirmation of the validity of the quantum mechanical formulas for molecular energy levels.

Two features of the CO spectrum remain unexplained. The first concerns the intensities of the spectral lines. These are determined by the fraction of CO molecules in the various rotational levels $J'' = 0, 1, 2, \ldots$ of the lowest vibrational level $v = 0$ from which absorption occurs. The relative intensities will soon be explained quantitatively (Section 5.5). The second problem appears to be an imperfection in our theory. As opposed to the predictions of Eqs. (4.15) and (4.16), the spacing between lines is not quite constant, but grows slightly in the P branch and shrinks in the R branch. It turns out that the rotational constants $B = h^2/8\pi^2 I$ are not the same in both v levels, as we presumed, but actually differ slightly. They differ because of an increase in bond length R_e, as one goes from the vibrational state $v = 0$ to $v = 1$. The R_e appears in the moment of inertia $I = \mu R_e^2$ and hence affects the rotational constant B. The change in R_e is less than 1%, but even that small detail concerning the CO molecule is readily apparent in the spectrum. Additional discussion of the dependence of R_e on v is given in Chapter 25, where we shall see that the effect arises mainly from the fact that vibrations in molecules are not strictly harmonic.

4.10   Electronic Energy

Atoms and molecules contain a fourth type of energy, which was not discussed in Chapter 3. All readers are familiar with the energy levels of the hydrogen atom representing occupation of different atomic orbitals by the single electron. We speak of the hydrogen atom with its electron in a 1s, a 2p, or a 3s orbital. In each case, the atom has a different *electronic energy* and is in a different electronic energy level. Such electronic levels occur for molecules as well as atoms. Simple formulas generally are not available for these electronic levels, the hydrogenlike atoms being exceptions. The separation between electronic levels is generally large, being of the order of $100\,kT$ or more. This is 10 to 100 times larger than typical separations between vibrational levels. Clearly, quantum theory must *always* be used to describe electronic energies. The levels can be studied by absorption or emission spectroscopy, using methods similar to those discussed in the preceding section. Visible or ultraviolet light can be used to excite an electron to a higher electronic level.

With few exceptions, molecules are found only in their lowest electronic energy levels at temperatures below 1000 K. For this reason, electronic energies were ignored in our discussion of internal energies and heat capacities in Chapter 3. It is quite unusual to find any electronic contribution to the heat capacity of a molecule. We shall postpone any further discussion of electronic energies until Chapter 12.

Problems

1. Calculate the energy of 1 mole of photons for (a) soft X-rays at $\lambda = 10^{-8}$ m; (b) visible light at 5×10^{-7} m; (c) radio waves at 3 m.

2. What is the Broglie wavelength of a helium atom moving at its average speed at 300 K? at 1 K? at 10^{-6} K?

3. A beam of neutrons is obtained from a nuclear reactor at an equilibrium temperature of 100 K. Would this beam be suitable for diffraction studies on metal foils?

4. Advanced electron microscopes now operate at 1000 kV. What is the corresponding wavelength of the electron beam?

5. An electron is confined in an extended linear molecule 2 nm long. Estimate the minimum energy the electron could have, and the energy to excite the electron to the next highest state. (The model of a particle in one-dimensional box is used, a fair approximation.)

6. Calculate the wave number $\tilde{v}$ (cm^{-1}) of a spectral line that arises from a transition between two energy levels 1 eV apart.

7. A CO_2 laser gives an infrared beam at $\lambda = 10.6$ μm and a power of 1 kW. How many quanta per second are emitted? If the laser output is all absorbed in 1 L of water, how long would it take to raise the water temperature from 20°C to the boiling point?

8. Consider a molecule $^{12}C^{16}O$ (Example 3.4) with energy $\frac{1}{2}kT$ in each rotational degree of freedom. At 25°C, what is (a) its angular momentum; (b) its angular velocity; (c) its rotational frequency (revolutions per second); (d) what is the velocity of the rotating ^{16}O atom and how does it compare with its average translational velocity at 25°C?

9. The wave number of the fundamental vibration in the molecule Na^{35}Cl is 380 cm^{-1}. Calculate the vibration frequency; what is the wavelength measured in a vacuum? What would be the wavelength measured in air if the refractive index n(air) $= 1.000286$. What is the energy of the photon emitted or absorbed (a) in J; (b) in eV; (c) in kJ mol^{-1}?

10. Repeat the calculations in Problem 9 for HBr for which $\tilde{v} = 2649.67$ cm^{-1}. Note that $\tilde{v}$ almost always refers to a vacuum value, whereas measured λ are usually in air. ($c' = c/n$.)

11. Suppose that a H atom is confined in a cubic box 1 nm long. What is the lowest allowed energy level (a) in J; (b) in eV? What would be the difference between the lowest and next lowest levels (a) in eV; (b) in cm^{-1}?

12. The O—H stretching vibration in CH_3OH is at $\tilde{v} = 3300$ cm^{-1}. Calculate $\tilde{v}$ for the O—D stretch in CH_3OD. (The force constants are the same. Why?)

13. In the microwave spectrum of $^{12}C^{16}O$, the $J = 0 \longrightarrow 1$ transition occurs at 115 271.204 MHz. Calculate the moment of inertia and the internuclear distance R_e in $^{12}C^{16}O$.

14. For H^{35}Cl, $R_e = 127.5$ pm; calculate the wave number of rotational transition $J = 0 \longrightarrow J = 1$. Calculate it for D^{35}Cl, which has the same R_e.

15. Calculate the width (cm^{-1}) of the gap between the P and R branches of the fundamental infrared absorption band of H^{35}Cl. (See Problem 14.)

16. Vibrational force constants do not change with isotopic substitution. For H^{35}Cl, $\tilde{v} = 2989$ cm^{-1}. Calculate $\tilde{v}$ for H^{37}Cl.

17. Calculate the rotational constant B for Na^{35}Cl, the moment of inertia of which is 1.453×10^{-45} kg m^2. Sketch the positions (cm^{-1}) of the first four lines of the P and R branches of the fundamental infrared absorption band. (See Problem 9.)

18. For large J values the frequency of absorption of a quantum of rotational energy approaches the classical rotation frequency of the molecule. Derive this instance of the correspondence principle.

19. (a) OCS is a linear triatomic molecule. Show that its moment of inertia is

$$I = \frac{m_c m_o R_{co}^2 + m_c m_s R_{cs}^2 + m_o m_s (R_{co} + R_{cs})^2}{m_o + m_c + m_s}$$

(b) For $^{16}O^{12}C^{32}S$, $B/hc = 0.202964$ cm^{-1}, $^{16}O^{12}C^{34}S$, $B/hc = 0.197910$ cm^{-1}. Calculate R_{co} and R_{cs}.

20. Calculate the number of energy states N_ϵ less than a specified value ϵ for a particle in a three-dimensional box. Consider Eq. (4.9) and draw a three-dimensional space with axes n_1, n_2, n_3. Each set of integral values (n_1, n_2, n_3) corresponds to an allowed quantum state. For fairly large ϵ, the N_ϵ is the volume of positive octant of sphere of radius R with $R^2 = n_1^2 + n_2^2 + n_3^2$.

21. From the result derived in Problem 20, calculate the number of electron states with energy less than 1 eV in a cube of metal 10 cm³ in volume (the box).

22. The density of gold at 20°C is $\rho = 19.3$ g cm⁻³. If one valence electron in each atom of Au is essentially free to move within the metal, estimate the width in eV of the band of energy levels occupied by these free electrons. (Two electrons with opposite spins can enter each level.)

5

Boltzmann Distribution
and Temperature

We have described the general features of the energy levels of individual molecules. How are molecules in a gas distributed among these energy levels? This is one of the most important questions in physical chemistry. If we can find the answer, we shall be able to calculate all the thermodynamic equilibrium properties of ideal gases. It is worth underlining this remarkable fact: For systems composed of ideal gases, all large-scale properties, including energies, heat capacities, and equilibrium constants, can be calculated from a knowledge of how the molecules are distributed among their energy levels.

We shall first state the answer to the distribution problem. Two ways of obtaining the answer are then shown along with an important historical application. Finally, we use the distribution law to show how our knowledge of energy levels can provide a molecular interpretation of temperature and heat capacity.

5.1 The Boltzmann Distribution

Let us consider a collection of a large number of molecules, such as a volume of gas at a specified temperature and pressure.

Example 5.1 Calculate the number N of molecules in 1 m³ of O_2 at 101 kPa and 298 K.

$$N = (PV/RT)L = \frac{(101 \times 10^3 \text{ Pa})(1 \text{ m}^3)}{(8.31 \text{ J K}^{-1} \text{ mol}^{-1})(298 \text{ K})}(6.02 \times 10^{23} \text{ mol}^{-1})$$

$$= 2.46 \times 10^{25}$$

We assume that the molecules are in *thermal equilibrium* at constant temperature T in a container of fixed volume V. This condition can be achieved by immersing a container of gas with rigid but heat-conducting walls in a large bath of fluid at temperature T and waiting until the temperature is uniformly equal to T in all parts of the gas. Let N_0 be the number of molecules in the lowest energy level ϵ_0. The number N_i in any other energy level ϵ_i is given by

$$\frac{N_i}{N_0} = \frac{g_i}{g_0} e^{-(\epsilon_i - \epsilon_0)/kT} \tag{5.1}$$

The factors g_0, g_i are the *statistical weights* (or *degeneracies*) of the respective levels. Recall that the statistical weight is the number of distinct quantum states that have the same molecular energy. Although the ratio in Eq. (5.1) has been written with reference to the lowest state, a similar formula can easily be obtained from Eq. (5.1) for any two states with energies ϵ_i and ϵ_j:

$$\frac{N_i}{N_j} = \frac{g_i}{g_j} e^{-(\epsilon_i - \epsilon_j)/kT} \tag{5.2}$$

Equation (5.1) was first derived by Ludwig Boltzmann in 1868. It is called the *Boltzmann distribution law* and is one of the most important equations in physical chemistry.

[A gas molecule has two energy levels of equal statistical weight separated by $\epsilon_i - \epsilon_j = kT$. What is the ratio of the populations in these levels at temperature T?]

The fraction of molecules out of the total number N that is in a specified energy level can be obtained as follows. From Eq. (5.1),

$$N_i = N_0 (g_i/g_0) e^{-(\epsilon_i - \epsilon_0)/kT}$$

The total number of molecules N is the sum of N_i over all energy levels:

$$N = \sum_i N_i = \frac{N_0}{g_0} \sum g_i e^{-(\epsilon_i - \epsilon_0)/kT}$$

Hence

$$\frac{N_i}{N} = \frac{g_i e^{-\epsilon_i/kT}}{\sum_{i=0}^{\infty} g_i e^{-\epsilon_i/kT}} \tag{5.3}$$

The sum over energy levels in the denominator is called the *molecular partition function*. The molecules are partitioned among the various energy levels and each term in the sum is proportional to the number of molecules N_i in a particular level ϵ_i.

If we write for the molecular partition function,

$$z \equiv \sum_{i=0}^{\infty} g_i e^{-\epsilon_i/kT} \tag{5.4}$$

Eq. (5.3) becomes

$$\frac{N_i}{N} = \frac{g_i e^{-\epsilon_i/kT}}{z} \tag{5.5}$$

The partition function is the key function of statistical thermodynamics. It is the link between microscopic molecular energies and macroscopic variables of a chemical system. Chapter 12 will show how z can be used to calculate molecular energies, entropies, heat capacities, free energies, equilibrium constants, and other thermodynamic functions.

5.2 The Barometric Formula.
A Simple Derivation of a Boltzmann Distribution

There are many approaches to the Boltzmann distribution, some general and others based on specific experimental situations. This section gives a derivation by way of an old problem, first approached experimentally by Blaise Pascal. In 1648 he arranged to have a barometer carried up the mountain at Le Puy de Dome so as to record the change in atmospheric pressure with altitude.

Imagine (Fig. 5.1) a column of the earth's atmosphere of cross section A extending upward in height h from sea level $h = 0$. Suppose that temperature T does not change with height. (This assumption would not be true of any real column of the atmosphere.) At height h let the barometric pressure be P and the density be ρ. Recall

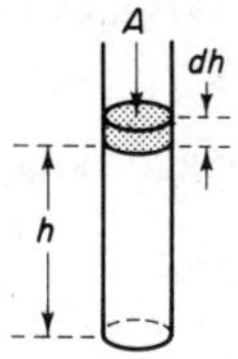

FIGURE 5.1 Sketch of a column of the earth's atmosphere used in derivation of the barometric formula.

that weight mg is the gravitational force on a mass m and that pressure is the force exerted on a unit area. The difference in pressure dP between levels h and $h + dh$ is then the weight of the slice of air of thickness dh divided by the area A of the slice.

$$dP = -g \, dm/A = -g\rho A \, dh/A = -g\rho \, dh \tag{5.6}$$

For an ideal gas, $\rho = PM/RT$ [Eq. (2.4)]. Hence

$$dP = -\frac{gPM}{RT} \, dh, \qquad dP/P = -\frac{gM}{RT} \, dh \tag{5.7}$$

On integration, $\ln P = -(gM/RT)h + \text{const.}$ When $h = 0, P = P_0$, the pressure at sea level, so that const. $= \ln P_0$. Therefore, $\ln (P/P_0) = -Mgh/RT$ or

$$P/P_0 = e^{-Mgh/RT} \tag{5.8}$$

If we substitute $R = Lk$ and $M = Lm$, the equation becomes

$$P/P_0 = e^{-mgh/kT} \tag{5.9}$$

The potential energy of a molecule at h in a gravitational field is $\epsilon_p = mgh$. Equation (5.9) is an example of the Boltzmann formula (5.1), written as

$$N/N_0 = e^{-\epsilon_p/kT} \tag{5.10}$$

Here it is understood that $\epsilon_p = 0$ for N_0 molecules at sea level, and the degeneracies g are all equal. The ratio of pressures P/P_0 in Eq. (5.9) is equal to the ratio of the numbers of molecules N/N_0 in the very thin slices of equal volume, since $P = nRT/V = NkT/V$.

The barometric formula is thus a special case of the Boltzmann distribution, in which the energy levels are the potential energies in a gravitational field. Since various forms of potential energy are interconvertible, Eq. (5.10), though derived for gravitational potential energy, applies to any kind of potential energy. Furthermore, since potential energy and kinetic energy are interconvertible, Eq. (5.10) should be applicable to any kind of energy.

Example 5.2 What would be the barometric pressure at 1700 m, the approximate altitude of Boulder, Colorado? Assume a standard sea level pressure of 100 kPa at 300 K.

Use Eq. (5.9). The energies mgh and kT must be in identical units. It is convenient to use the mass of 1 mole of air and this requires use of RT rather than kT in the denominator. The gravitational acceleration must be in m s^{-2} rather than in cm s^{-2}, as found in many tables. Since air is approximately 20% O_2 and 80% N_2, the molar mass of air is $M \approx (0.20)(0.032) + (0.80)(0.028) = 0.0288$ kg mol^{-1}.

$$\frac{P_0}{P} = \exp\left[-\frac{(0.0288 \text{ kg mol}^{-1})(9.80 \text{ m s}^{-2})(1700 \text{ m})}{(8.31 \text{ J K}^{-1} \text{ mol}^{-1})(300 \text{ K})}\right] = 0.825$$

The pressure at 1700 m is $P = (0.825)(100 \text{ kPa}) = 83$ kPa.

5.3 Sedimentation Equilibrium

The barometric formula may be of limited practical use as applied to Earth's atmosphere, but it can be applied exactly to the sedimentation of microscopic particles suspended in a fluid medium at constant temperature. In 1909, Jean Perrin described a beautiful set of experiments on such a sedimentation equilibrium. They illustrated the Boltzmann distribution and, perhaps surprisingly, provided a means to estimate the Avogadro constant.

Gamboge, which is used as a water color, is the dried latex of a Vietnamese tree, *Garcinia morella*. When a piece of this substance is rubbed under water it forms a brilliant yellow emulsion, which appears under the microscope as a swarm of perfectly spherical yellow globules. The spheres have different sizes, but they all have the same density, 1.207 g cm^{-3} at 20°C. They can be fractionated by careful centrifugation to yield a suspension of spheres with uniform diameters.

To measure the distribution of globules in a gravitational field, an observation chamber was formed by boring a hole in a glass plate about 100 μm thick and cementing this plate to a flat glass base. A microscope with a small depth of focus was used so that only globules in a thin layer about 1 μm thick were seen at any one focus. When a water-immersion objective was used, no correction for refractive index was necessary and the height of the level being observed could be read directly to better than 0.25 μm from a scale attached to the micrometer screw on the microscope.

The globules were counted by taking an observation every 15 s of the field at each level and averaging the numbers per unit area. The radius a of the globules was measured by several different methods. The weight of a particle in water is $\frac{4}{3}\pi a^3(\rho_g - \rho_w)g$. The Boltzmann equation (5.9) in this instance can be written

$$N/N_0 = \exp\left[-\tfrac{4}{3}\pi a^3 gh(\rho_g - \rho_w)/kT\right]$$

or

$$\ln\frac{N}{N_0} = \frac{-\frac{4}{3}\pi a^3(\rho_g - \rho_w)gh}{kT} \tag{5.11}$$

Figure 5.2 shows the results of some of the experiments of Perrin. Drawings at five different levels 10 μm apart are mounted vertically so as to give a schematic picture of the distribution of particle numbers with height.

FIGURE 5.2 Drawing of results of one of the experiments of Perrin: distribution of gamboge globules in a gravitational field.

In one of the most careful experiments the particle radius $a = 0.212$ μm and 13 000 globules were counted, giving the following relative numbers at different levels:

Height h (μm)	5	35	65	95
N	100	47	22.6	12

When $\ln N$ is plotted vs. h in accord with Eq. (5.11), the slope of the best straight line gives

$$\frac{\frac{4}{3}\pi a^3(\rho_g - \rho_w)gL}{RT} = 2.42 \times 10^4 \text{ m}^{-1}$$

On inserting the data,

$$a = 0.212 \ \mu\text{m} \qquad\qquad g = 9.8067 \text{ m s}^{-2}$$
$$\rho_g = 1.207 \times 10^3 \text{ kg m}^{-3} \qquad R = 8.3143 \text{ J K}^{-1} \text{ mol}^{-1}$$
$$\rho_w = 1.00 \times 10^3 \text{ kg m}^{-3} \qquad T = 293.15 \text{ K}$$

the equation yields a value of the Avogadro constant, $L = 7.3 \times 10^{23}$ mol^{-1}.

The agreement with values from other data is good, considering the accuracy of the experiment. For example, the ratio of the faraday F to the unit electric charge e (Section 16.1) gives

$$L = F/e = \frac{9.64867 \times 10^4 \text{ C mol}^{-1}}{1.602192 \times 10^{-19} \text{ C}} = 6.0222 \times 10^{23} \text{ mol}^{-1}$$

Before Perrin's experiments were performed, some scientists, in particular Wilhelm Ostwald, the doyen of physical chemists, still had doubts about the "real existence" of molecules and atoms. The value of L obtained by Perrin from his gamboge particles was in good agreement with that calculated for the number of molecules in a mole of gas. Philosophers could thus admit that the ontological status of visible microscopic particles and of molecules was the same. (Ontology is the branch of metaphysics that studies the nature of existence or being.) It was argued that if the gamboge particles are real material entities, then molecules in a gas must also be real material entities. If molecules are mental concepts or ideals, then so are gamboge particles. [Do you find this argument convincing?]

5.4 A More General Derivation of the Boltzmann Distribution Law

The Boltzmann distribution equation (5.3) also emerges from a derivation based on an analysis of probabilities. Consider a system consisting of a very large number of particles in a container of constant volume V maintained at some constant temperature T. Suppose that each particle i has a certain energy ϵ_i. Assume that the particles can interact by colliding with one another to exchange energy, but that the potential energy due to forces between particles is, on average, negligible. The particles in the system must be able to exchange energy with one another so that an input of energy into one part of the system can spread uniformly throughout the entire system. Then the system can reach equilibrium with a uniform temperature T.

We withdraw a particle from the container. Let p_a be the probability that it has energy ϵ_a. Nothing distinguishes the particles from one another except the energy levels they happen to be in, so that p_a must be simply some function of ϵ_a, that is, $p_a(\epsilon_a)$. We now withdraw a second particle. The probability that we now have a particle with energy ϵ_b is $p_b(\epsilon_b)$. The combined probability that the two particles withdrawn have energy ϵ_a for one and ϵ_b for the other is the product of the two independent separate probabilities, $p_a(\epsilon_a)p_b(\epsilon_b)$. The energy of the pair of particles that we have withdrawn is $\epsilon_{ab} = \epsilon_a + \epsilon_b$. Since nothing distinguishes this pair from any other pair except its energy ϵ_{ab}, the probability of withdrawing such a pair is $p(\epsilon_{ab})$. Thus

$$p(\epsilon_{ab}) = p(\epsilon_a + \epsilon_b) = p_a(\epsilon_a)p_b(\epsilon_b) \tag{5.12}$$

Now comes the key question. What is the mathematical consequence of the relationship in Eq. (5.12)? The answer is that the individual probabilities $p_j(\epsilon_j)$ must have the form

$$p_j(\epsilon_j) = Ae^{-\beta\epsilon_j} \tag{5.13}$$

where j stands for a or b, and A and β are constants.

The proof is as follows. Equation (5.12) is equivalent to

$$f(w) = g(x)h(y) \tag{5.14}$$

where

$$w = x + y \tag{5.15}$$

Equation (5.14) holds for all possible pairs of x and y and for all w. Differentiate both sides of Eq. (5.14) with respect to x:

$$\frac{df}{dx} = \frac{df}{dw}\frac{dw}{dx} = \frac{dg(x)}{dx}h(y)$$

But from Eq. (5.15), $dw/dx = 1$. Divide the left side by $f(w)$ and the right by $g(x)h(y) = f(w)$, to obtain

$$\frac{1}{f}\frac{df}{dw} = \frac{1}{g}\frac{dg}{dx} \quad \text{or} \quad \frac{d\ln f}{dw} = \frac{d\ln g}{dx} \tag{5.16}$$

If Eq. (5.16) is to be valid for all values of x consistent with Eq. (5.15) and for all w, each side must equal the same constant. We can call the constant $-\beta$. Thus

$$\frac{d\ln f}{dw} = -\beta = \frac{d\ln g}{dx}$$

Hence

$$\ln f(w) = -\beta w + \text{constant}, \quad f(w) = Ae^{-\beta w}$$

and

$$\ln g(x) = -\beta x + \text{constant}, \quad g(x) = A'e^{-\beta x}$$

In terms of the probability functions of Eq. (5.12), these are equivalent to the general form given in Eq. (5.13), $p(\epsilon) = Ae^{-\beta\epsilon}$.

The constant A is readily evaluated from the condition that the sum of the individual probabilities p_i must equal unity:

$$\sum p_i = 1 = \sum_i Ae^{-\beta\epsilon_i}$$

$$A = 1/\sum_i e^{-\beta\epsilon_i} \tag{5.17}$$

Equation (5.13) thus becomes $p_i(\epsilon_i) = e^{-\beta\epsilon_i}/\sum e^{-\beta\epsilon_i}$.

If we now *postulate* that there is equal probability of withdrawing any individual particle regardless of its energy, we must then have

$$p_i(\epsilon_i) = \frac{N_i}{N} = \frac{e^{-\beta\epsilon_i}}{\sum_i e^{-\beta\epsilon_i}} \tag{5.18}$$

where N_i is the number of particles of energy ϵ_i and $N = \sum N_i$, the total number of particles. Equation (5.18) may be compared with Eq. (5.3). Aside from the degeneracy factors, the Boltzmann distribution law has been derived if we can show that $\beta = 1/kT$. This equality can be established by showing that Eq. (5.18) is consistent with experiment when the equality is assumed. But what sort of experiment can be used? An experiment already discussed can do the job. It is the determination of the heat capacity of a monatomic gas, say, neon or helium. In Chapter 3 the heat capacity was found to be an indication of average molecular energy. In the case of a monatomic gas the energy is due to translation only, and heat-capacity measurements show that the average molecular translational energy is $\bar{\epsilon}_t = \frac{3}{2}kT$, or for one degree of freedom

$\bar{\epsilon}_t = \frac{1}{2}kT$. Let us calculate $\bar{\epsilon}_t$ from Eq. (5.18) and compare it to this experimental result. The basic definition of the average energy is

$$\bar{\epsilon} = \sum_i p_i \epsilon_i \qquad (5.19)*$$

From Eqs. (5.18) and (5.19),

$$\bar{\epsilon} = \frac{\sum_i \epsilon_i e^{-\beta \epsilon_i}}{\sum_i e^{-\beta \epsilon_i}} \qquad (5.20)$$

The quantum mechanical expression for the translational energy in one dimension in a level with quantum number n is, from Eq. (4.7),

$$\epsilon_t = \frac{h^2}{8ma^2} n^2 = cn^2 \qquad (5.21)$$

where $c = h^2/8ma^2$. We use this expression to calculate $\bar{\epsilon}_t$.

The key to evaluating the summations in Eq. (5.20), which run over the levels in Eq. (5.21), is to realize that many levels contribute to the sums, with each level differing by an extremely small energy from its nearest neighbors (see Section 4.5). Thus the quantum number n can be treated as a continuous variable, and the sums in Eq. (5.20) can be replaced by integrals. By combining Eqs. (5.20) and (5.21), we then obtain

$$\bar{\epsilon}_t = \frac{c \int_0^\infty n^2 e^{-\beta c n^2}\, dn}{\int_0^\infty e^{-\beta c n^2}\, dn} \qquad (5.22)$$

The $n = \infty$ in the upper limit of the integrals presents no problem because contributions to the integrals vanish at high n, owing to the $e^{-\beta c n^2}$ factors. Both the numerator and denominator can be found in standard integral tables:

$$\int_0^\infty x^2 e^{-bx^2}\, dx = \frac{1}{4}\sqrt{\pi/b^3} \quad \text{and} \quad \int_0^\infty e^{-bx^2}\, dx = \frac{1}{2}\sqrt{\pi/b}$$

Equation (5.22) thus yields

$$\bar{\epsilon}_t = \frac{(c/4)[\pi/(\beta c)^3]^{1/2}}{(\frac{1}{2})(\pi/\beta c)^{1/2}} = \frac{1}{2\beta}$$

We thus see that our calculation of $\bar{\epsilon}_t$ with Eq. (5.18) gives agreement with the experimental value $\bar{\epsilon}_t = \frac{1}{2}kT$, provided that

$$\beta = 1/kT \qquad (5.23)$$

With the identification of $\beta = 1/kT$, Eq. (5.18) is seen to be the Boltzmann distribution of Eq. (5.3), with the exception of the degeneracy factors g_i. These factors did not appear since there is only one state per energy level, as specified by the quantum number n.

$$* \qquad \bar{\epsilon} = \frac{E}{N} = \frac{\sum_i N_i \epsilon_i}{N} = \sum \frac{N_i}{N} \epsilon_i = \sum_i p_i \epsilon_i$$

where E is the total energy of the system with N particles.

5.5 Relative Populations of Molecular Energy Levels

The Boltzmann equation (5.2) describes the relative populations of two molecular energy levels. It can be written with $\Delta\epsilon = \epsilon_i - \epsilon_j$ as

$$\frac{N_i}{N_j} = \frac{g_i}{g_j}e^{-\Delta\epsilon/kT} \tag{5.24}$$

Because questions concerning the relative populations of molecular energy levels arise so frequently, this is one of the basic equations in physical chemistry. The exponential itself often has dominant control over the populations: It is called the *Boltzmann factor*. Table 5.1 cites values of e^{-x} a few values of $x = \Delta\epsilon/kT$.

TABLE 5.1
NUMERICAL VALUES OF
THE BOLTZMANN FACTOR e^{-x}

x	e^{-x}
0	1
1	0.37
2	0.14
3	0.05
4	0.02
5	0.007

The factor kT in the exponential is the universal yardstick of molecular energy since we must always pay attention to the size of $\Delta\epsilon$ *relative to* kT when gauging relative populations. When $\Delta\epsilon \ll kT$, $x \ll 1$ and the Boltzmann factor is nearly unity. Usually, degeneracies are small so that the ratios of populations will not be far from unity. When $\Delta\epsilon$ approaches kT, however, the exponential factor falls rapidly from unity and damps out the upper-level populations. When $\Delta\epsilon \gg kT$, the fraction of molecules in the higher level becomes virtually nil.

It is useful to recall the value of kT at 300 K, 4.1×10^{-21} J. The spectroscopic wave number corresponding to this energy is $1/\lambda = 200$ cm^{-1} (from $\epsilon = h\nu = hc/\lambda = kT$). If energies E are expressed per mole, Eq. (5.24) becomes

$$\frac{N_i}{N_j} = \frac{g_i}{g_j}e^{-\Delta E/RT}$$

At 300 K, $RT = 2.5$ kJ mol^{-1}.

We can now use Eq. (5.24) to learn about the relative populations of energy levels in an ensemble of molecules at thermal equilibrium. Relative translational populations are independent of relative rotational or vibrational populations. This separation of population distributions follows directly from the separation of molecular energies and the exponential form of Eq. (5.24) since $e^{x+y} = e^x e^y$. Let us first get a qualitative picture of the distributions at 300 K by comparing the sizes of the energy quanta to kT.

Translational distributions. In Section 4.5 we worked out the magnitude of the lowest translational energy level (111) of O_2. Similar calculations determine the energy of the next higher level (211), which is triply degenerate with states (211), (121) and (112) of equal energy. The result shows that the two levels are separated by $\Delta \epsilon = \epsilon_{211} - \epsilon_{111} = 9 \times 10^{-40}$ J. Comparing this $\Delta \epsilon$ to $kT = 4 \times 10^{-21}$ J, we see that the translational quantum is about $2 \times 10^{-19} kT$, so that the Boltzmann factor $\exp(-10^{-19})$ will be unity, and the populations will be in the ratio of the level degeneracies, $3:1$. With the translational quantum so much smaller than kT, it is evident that a population of molecules must be distributed over an enormous number of levels.

Rotational distributions. The size of the rotational quantum of O_2 worked out in Section 4.7 is typical for many small molecules. The quantum size, $\Delta \epsilon_r = 6 \times 10^{-23}$ J is quite small compared to $kT = 4 \times 10^{-21}$ J. Thus rotational populations are spread over a considerable number of levels, of the order of 10 to 100 at 300 K. Recall from Section 4.7 that an O_2 molecule with average rotational energy at 350 K is in a level with $J = 13$.

Vibrational distributions. Vibrational quanta are usually 2 to 10 times larger than kT at 300 K. In the case of O_2, for example, the vibrational wave number of 1580 cm^{-1} exceeds the value for kT of 200 cm^{-1} by a factor of nearly 8. The Boltzmann factor for the ratio of populations is small ($N_1/N_0 = e^{-\Delta \epsilon / kT} = e^{-1580/200} = 4 \times 10^{-4}$). Most of the molecules are in the lowest vibrational level, $v = 0$. This situation is typical of most diatomic molecules. Polyatomics often have a few low-frequency modes with vibrational quanta at less than 500 cm^{-1}. As the factors in Table 5.1 show, these modes have modest populations (a few percent) in $v = 1$ at 300 K.

Electronic distributions. With few exceptions, the ground or lowest electronic level is separated from the first excited electronic level by energies of at least $100 kT$. The Boltzmann factor is thus nearly zero, and essentially the entire population of molecules is in the ground electronic state at 300 K.

We can summarize a typical situation. The molecules in a liter of O_2 at 300 K have a broad distribution over an enormous number of translational levels, of the order of 10^{30}. The rotational energy is distributed over a smaller array of levels, the first 50 or so. In contrast, over 99% of the molecules are in their lowest vibrational level and virtually all of the molecules are in their lowest electronic state.

A typical pattern of the relative populations of vibrational energy levels is shown in Fig. 5.3. The relative rotational populations at 298 K are shown in Fig. 5.4(a). The lowest level ($J = 0$) has only a modest population, with a maximum occurring at $J = 7$. The maximum is a consequence of competing factors in the Boltzmann

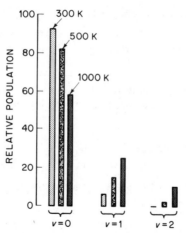

FIGURE 5.3 Relative populations of vibrational energy levels in Cl_2 at three different temperatures.

distribution. While the exponential factor damps out higher-J populations, the degeneracy factor $g_J = (2J + 1)$ favors them.

We now return to an unfinished problem concerning the infrared absorption spectrum of CO, Fig. 4.4: Why do the absorption lines in the P and R branches vary in intensity from the center to the extremes of the branches? The answer is found in the Boltzmann distribution. Intensities of absorption lines to a first approximation are determined by the relative populations of the energy levels from which the absorption originates. The infrared spectral intensities thus give a direct view of the relative rotational-level populations in CO.

Figure 5.4(b) reproduces the R branch of the absorption spectrum with the

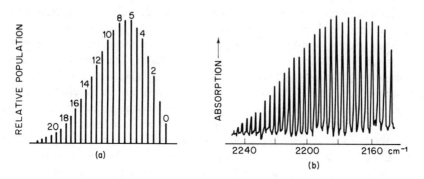

FIGURE 5.4 A comparison of (b) the R branch ($J'' = 0 \rightarrow J' = 1$, $J'' = 1 \rightarrow J' = 2$, etc.) of the $v'' = 0 \rightarrow v' = 1$ absorption spectrum of CO with (a) the calculated relative populations of the rotational levels from which the absorption lines originate. The populations are for a CO gas temperature of 298 K. The J'' quantum number of the level from which each absorption line originates is shown. Owing to the manner in which absorption is usually measured, the apparent absorption intensities give only a rough representation of the relative numbers of absorbing molecules responsible for each line. When the intensities are corrected to indicate accurately the relative numbers of absorbing molecules, the match with calculated level populations is extremely good. The absorption spectrum was taken directly from Fig. 4.2.

relative populations of rotational levels in CO at 298 K calculated from Eq. (5.24), with $g_J = 2J + 1$. The correspondence between experimental and calculated values is another piece of evidence attesting to the validity of the Boltzmann distribution.

Example 5.3 The fundamental vibration frequency of O_2 is 4.738×10^{13} Hz. What fraction of the total number of O_2 molecules have two quanta of vibrational energy at 1000 K?

The vibrational quantum

$$h\nu = (6.626 \times 10^{-34} \text{ J s})(4.738 \times 10^{13} \text{ s}^{-1}) = 3.139 \times 10^{-20} \text{ J}$$

$$kT = (1.381 \times 10^{-23} \text{ J K}^{-1})(1000 \text{ K}) = 1.381 \times 10^{-20} \text{ J}$$

$$h\nu/kT = 2.273$$

From Eq. (5.3),

$$\frac{N_2}{N} = \frac{e^{-2(2.273)}}{1 + e^{-2.273} + e^{-2(2.273)} + e^{-3(2.273)} + \cdots}$$

$$= \frac{0.0106}{1.1148} = 0.0095$$

Thus, even at 1000 K, less than 1% of O_2 molecules have two quanta of vibrational energy.

5.6 A Molecular Interpretation of Temperature

Sensations of hot and cold inspired an intuitive concept of temperature. Measurement was first based on the expansion of liquids, as in mercury thermometers. The ideal-gas law $PV = nRT$ was a step closer to a more fundamental definition of temperature, but the Boltzmann distribution law provides the deepest understanding.

We have so far treated temperature as a rather undistinguished parameter in the Boltzmann law. The Boltzmann distribution, however, actually allows us to see that *temperature* is the crucial factor that establishes the energy content of a system of N molecules at equilibrium. Temperature defines not only the energy per unit amount of the system, but also how the energy is distributed among the molecular energy levels. A simple example can illustrate this point. Consider the vibrational energy of N Cl_2 molecules, with vibrational quanta at 465 cm^{-1}. We can calculate the total vibrational energy by direct summation, $E = N_0\epsilon_0 + N_1\epsilon_1 + N_2\epsilon_2 + \cdots$. The average energy of a molecule is

$$\frac{E}{N} = \bar{\epsilon} = \frac{N_0}{N}\epsilon_0 + \frac{N_1}{N}\epsilon_1 + \frac{N_2}{N}\epsilon_2 + \cdots = \sum_{i=0}^{\infty} \frac{N_i}{N}\epsilon_i \tag{5.25}$$

N_i is the population of level i and $N = \sum N_i$. With Eq. (5.3) for N_i/N, Eq. (5.25) becomes

$$\bar{\epsilon} = \frac{N \sum \epsilon_i e^{-\epsilon_i/kT}}{\sum e^{-\epsilon_i/kT}} \tag{5.26}$$

All the g_i factors are unity for diatomic vibrational levels. The evaluation of Eq. (5.26) with the vibrational energies of Eq. (4.7) is straightforward since only the first few terms contribute significantly to the summations.

Table 5.2 shows the calculated vibrational energies over a range of temperatures, 300 to 1000 K, for one mole of Cl_2 ($N = L$). The total vibrational energy rises steadily

TABLE 5.2

VIBRATIONAL ENERGIES AND LEVEL POPULATIONS OF Cl_2
AT TEMPERATURES BETWEEN 300 AND 1000 K

T (K)	Total vibrational energy (J mol^{-1})	Level populations as percent of total[a]				
		$v = 0$	$v = 1$	$v = 2$	$v = 3$	$v = 4$
300	3859	93	6	—	—	—
400	4390	87	11	1	—	—
500	5039	81	16	3	1	—
600	5693	74	19	5	1	—
1000	7970	58	25	10	5	2

[a]Round-off errors cause sums not exactly 100%. Absence of an entry indicates a population less than 0.5%.

with temperature. As the temperature rises, lower levels are emptied and higher levels become increasingly occupied. Such behavior is characteristic of all systems and of all types of molecular energies.

The Boltzmann distribution thus leads to the concept of a *molecular thermometer*. Since populations of molecular energy levels are uniquely dependent on temperature, why not turn the tables and use a measurement of these populations to determine the temperature of a system? Such a procedure is in fact widely used to measure both high and low temperatures in systems where other methods are impractical. We have already seen an example of how these measurements work in Section 5.5, where we were able to reproduce the appearance of the CO infrared absorption spectrum with a calculation of the distribution at 298 K of rotational-level populations in CO. By using temperature as an adjustable parameter to match the experimental absorption-line strengths with calculations of level populations, the temperature can be established to within a few kelvins.

Such temperature measurements are possible whenever distributions of translational, rotational, vibrational, or in some cases even electronic-level populations can be measured. Vibrational distributions such as those shown in Fig. 5.3 are commonly used to measure temperatures in hot gases (flames, exhausts, plasmas, and explosions). Rotational distributions have been used to learn about the low temperatures of interstellar space. High stellar temperatures can sometimes be gauged by the distribution of electronic-level populations in atoms or ions.

Example 5.4 Molecular N_2 is heated in an electric arc. Spectroscopic measurements show that the relative numbers of molecules in excited vibrational states are:

v	0	1	2	3
N_v/N_0	1.00	0.26	0.068	0.018

Show that the gas is in thermal equilibrium with respect to distribution of vibrational energy and calculate the temperature of the gas.

If gas follows the Boltzmann distribution, it is in equilibrium. Since $\epsilon = (v + \frac{1}{2})h\nu$,

$$N_v/N_0 = \exp(-vh\nu/kT) = [\exp(-h\nu/kT)]^v$$

$$v = 1, \; N_1/N_0 = 0.26$$

$$v = 2, \; N_2/N_0 = (0.26)^2 = 0.068$$

$$v = 3, \; N_3/N_0 = (0.26)^3 = 0.018$$

Thus the gas does follow the Boltzmann distribution. The vibration of N_2 is at 2331 cm^{-1}. $h\nu/kT = hc\tilde{\nu}/kT$. From $\exp(-h\nu/kT) = 0.26$,

$$-\ln(0.26) = \frac{(6.63 \times 10^{-34} \text{ J s})(3.00 \times 10^8 \text{ m s}^{-1})(2331 \times 10^2 \text{ m}^{-1})}{(1.381 \times 10^{-23} \text{ J K}^{-1})T}$$

$$T = 2490 \text{ K}$$

One final point must be made concerning this fundamental interpretation of temperature. Temperature is defined only for a *large number of particles in thermal equilibrium*. The conditions emphasized in italics are essential. The concept of temperature has been based upon the Boltzmann distribution. As the derivation in Section 5.4 shows, this distribution is statistical, since it is derived in terms of probabilities. Hence it applies only to large ensembles of particles for which a statistical treatment is valid. Accordingly, the concept of temperature has significance only for a large number of particles.

5.7 A Molecular View of Heat Capacities

The Boltzmann distribution explains the temperature dependence of heat capacities. As shown in Section 3.8, heat-capacity measurements provide one of the clearest views of molecular energies. The definition $C_V = (\partial U/\partial T)_V$ indicates that the heat capacity C_V measures the change in energy that occurs when a system changes temperature at constant V. At constant volume the energy levels of the system remain fixed. As temperature rises, the Boltzmann distribution requires that the populations of lower energy levels be depleted in order to increase the populations of higher levels to give the distribution characteristic of the higher T. The heat capacity simply gives the energy required to attain equilibrium by promoting appropriate numbers of molecules in the system to higher levels as the temperature rises by 1 K.

In the case of electronic energies and vibrations, where the levels are far apart relative to kT, so few molecules change energy levels that little energy is absorbed; C_V for these degrees of freedom is near zero at 300 K. Translations and rotations, however, have levels more closely spaced relative to kT, so that substantial readjustment of populations occurs for a 1-K rise in temperature. These arguments are of course always dependent on the size of $\Delta\epsilon$ relative to kT. Thus at high T, the energy kT can equal or exceed the vibrational spacings $\Delta\epsilon$ and the average vibrational energy per degree of freedom can approach kT. Conversely, at low T, kT may become much smaller than rotational spacings so that the rotational contribution to C_V drops to nearly zero. For these reasons, molecular heat capacities depend strongly on temperature.

5.8 Distribution of Molecular Speeds*

The Boltzmann distribution law can be used to obtain the equation for the distribution of molecular speeds in a gas. This problem was originally solved by James Clerk Maxwell in 1860.

A molecule in a gas is moving in three dimensions and we can consider that its velocity† vector c has components u, v, and w along a set of orthogonal axes X, Y, and Z, as shown in Fig. 2.4. The square of the magnitude of velocity is

$$c^2 = u^2 + v^2 + w^2 \qquad (5.27)$$

The problem of the distribution of velocities in three dimensions is simplified by considering first one particular component, say u. The kinetic energy associated with the velocity component u is $\frac{1}{2}mu^2$, where m is the mass of the molecule.

The probability that any molecule picked at random has a velocity component between u and $u + du$ is denoted $p(u)$. Since u is a continuous variable, we cannot define a probability that a molecule has a velocity of exactly u; we must allow a certain small range of velocities du about the value of interest. The kinetic energy of a molecule is $\frac{1}{2}mu^2$, and from the general Boltzmann formula (5.2) it is evident that $p(u)$ must be proportional to $e^{-\epsilon/kT} = e^{-(1/2)mu^2/kT}$. Thus

$$p(u)\,du = A e^{-mu^2/2kT}\,du \qquad (5.28)$$

where A is a factor that converts the proportionality into an equality.

The constant A is evaluated from the condition that the probabilities must add up to unity. Thus

$$\int_{-\infty}^{+\infty} p(u)\,du = A \int_{-\infty}^{+\infty} e^{-mu^2/2kT}\,du = 1$$

This integral is a standard form (with $a = m/2kT$),

$$\int_{-\infty}^{+\infty} e^{-ax^2}\,dx = (\pi/a)^{1/2} \qquad (5.29)$$

---

*Without serious loss of continuity, the material in Sections 5.8 to 5.11 can be postponed until the beginning of Chapter 15 if so desired.

†Recall that *velocity* is a vector quantity and the magnitude of the velocity is the *speed*.

Thus $A(2\pi kT/m)^{-1/2} = 1$ or $A = (m/2\pi kT)^{1/2}$, and Eq. (5.28) becomes

$$p(u)\, du = (m/2\pi kT)^{1/2}e^{-mu^2/2kT}\, du \tag{5.30}$$

This is the Maxwell equation for velocities in one dimension.

Example 5.5 Apply Eq. (5.30) to calculate the probability that the velocity component of a N_2 molecule lies between 999.5 and 1000.5 m s^{-1} at 300 K.

We can set $du = \Delta u = 1$ m s^{-1}, the small velocity range required. In a large number N_0 of molecules, the fraction $\Delta N/N_0$ with a velocity component between u and $u + \Delta u$ is simply $p(u)\,\Delta u$ from Eq. (5.30). In this case,

$$p(u)\,\Delta u = \left(\frac{28 \times 10^{-3} \text{ kg mol}^{-1}}{2\pi \times 8.314 \text{ J K}^{-1} \text{ mol}^{-1} \times 300 \text{ K}}\right)^{1/2}$$

$$\times \exp\left(\frac{-28 \times 10^{-3} \text{ kg mol}^{-1} \times 10^6 \text{ m}^2 \text{ s}^{-2}}{2 \times 8.314 \text{ J K}^{-1} \text{ mol}^{-1} \times 300 \text{ K}}\right)$$

$$= 1.337 \times 10^{-3}\, e^{-5.613} = 4.88 \times 10^{-6}$$

Note that instead of m/kT, we have used M/RT, where M is the molar mass of N_2 and R is the gas constant per mole. We compute, therefore, that about five molecules out of each million will have a velocity component in the specified range.

5.9   Relation of Maxwell Equation to Gaussian Density Function

In statistics, a function like $p(u)$, which gives the probability of finding the variable u in a specified range, is called a *density function*. The Maxwell density function is mathematically identical with one of the most important functions in statistics, the *Gaussian density function*, sometimes called *normal density function*.

If we define a new variable s by $s^2 = mu^2/kT$, the function $p(u)$ becomes $(m/kT)^{1/2}f(s)$, where

$$f(s) = (2\pi)^{-1/2}e^{-s^2/2} \tag{5.31}$$

This is the usual form of the normal density function.

In the language of statistics, the density function $f(s)$ gives the fraction of a population that lies between s and $s + ds$. The distribution function $F(s)$ gives the cumulative fraction of the population that lies between the lower limit $-\infty$ and an upper value of s. Thus

$$F(s) = \int_{-\infty}^{s} f(s')\, ds' = (2\pi)^{-1/2}\int_{-\infty}^{s} e^{-s'^2/2}\, ds' \tag{5.32}$$

This function is called the *normal distribution function* or the *Gaussian distribution function*.

The normal density function and the normal distribution function are plotted in Fig. 5.5. Extensive tables of these functions (or of the closely related *error functions*) are available in all scientific libraries. A few values of $F(s)$ are given in Table 5.3. These functions are used in many problems involving the distribution of molecular

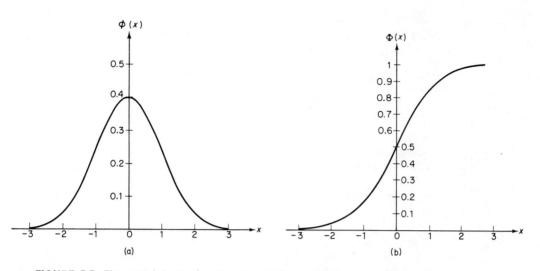

FIGURE 5.5 The normal density function (a) and the normal distribution function (b) as defined in statistics.

TABLE 5.3
THE NORMAL DISTRIBUTION FUNCTION

s	$F(s)$	s	$F(s)$
0.0	0.500 000	2.3	0.989 276
0.1	0.539 828	2.4	0.991 802
0.2	0.579 260	2.5	0.993 790
0.3	0.617 911	2.6	0.995 339
0.4	0.655 422	2.7	0.996 533
0.5	0.691 462	2.8	0.997 445
0.6	0.725 747	2.9	0.998 134
0.7	0.758 036	3.0	0.998 650
0.8	0.788 145	3.1	0.999 032
0.9	0.815 940	3.2	0.999 313
1.0	0.841 345	3.3	0.999 517
1.1	0.864 334	3.4	0.999 663
1.2	0.884 930	3.5	0.999 767
1.3	0.903 200	3.6	0.999 841
1.4	0.919 243	3.7	0.999 892
1.5	0.933 193	3.8	0.999 928
1.6	0.945 201	3.9	0.999 952
1.7	0.955 435	4.0	0.999 968
1.8	0.964 070	4.1	0.999 979
1.9	0.971 283	4.2	0.999 987
2.0	0.977 250	4.3	0.999 991
2.1	0.982 136	4.4	0.999 995
2.2	0.986 097	4.5	0.999 997

speeds. For example, when we consider the kinetics of gas reactions, we shall want to know the fraction of collisions that have energies greater than a critical value, the activation energy E_a, needed for reactions to occur.

Example 5.6 In a sample of CO_2 at 1000 K, what fraction of the molecules has a kinetic energy greater than $5(\frac{1}{2}kT)$ in one degree of freedom?

In terms of the variable s in Eq. (5.31), the required fraction is that with $s^2 > 5$. This is

$$2 \int_{\sqrt{5}}^{\infty} f(s)\, ds = 2[1 - F(\sqrt{5})]$$

The factor 2 is due to the fact that s can be positive or negative (two directions of a velocity component). From Table 5.3, $F(\sqrt{5}) = 0.987$, so that the answer is 0.026. Only a small fraction of any lot of molecules can have kinetic energy much greater than kT per degree of freedom.

5.10 Calculation of Average Values

The density function $p(s)$ for a variable s allows us to evaluate the average value of any function of s, say $g(s)$. We use the *mean value theorem*, which states: If $p(s)$ is the density function for any variable s, so that $p(s)\, ds$ is the probability that the variable s lies between s and $s + ds$, the mean value of any function $g(s)$ of s is given by

$$\overline{g(s)} = \int_{-\infty}^{\infty} p(s)g(s)\, ds \qquad (5.33)$$

We can derive this theorem from the ordinary definition of the mean value in a discrete distribution. Suppose that in a set of trials or experiments to find the value of $g(x)$, the value for x_1 occurs n_1 times; for x_2, n_2 times; and so on. Then the mean value of $g(x)$ is

$$\overline{g(x)} = \frac{\sum n_j g(x)}{\sum n_j}$$

But $n_j / \sum n_j$ is simply the probability p_j for the value x_j, so that

$$\overline{g(x)} = \sum_{j=1}^{\infty} p_j(x_j)g(x_j)$$

where the sum is over all the discrete values of x_j. If we now let the separation between the discrete values pass in the limit to zero, the corresponding limit of the sum becomes an integral over dx, or $\overline{g(x)} = \int_{-\infty}^{+\infty} p(x)g(x)\, dx$. The formula (5.33) can be used to calculate the mean values of many interesting properties of gas molecules.

5.11 Velocity in Three Dimensions

We can obtain the three-dimensional density function for molecular speeds by multiplying together three one-dimensional functions. The fraction of molecules having simultaneously a velocity component between u and $u + du$, v and $v + dv$, and w and $w + dw$, is

$$p(u, v, w)\, du\, dv\, dw = \left(\frac{m}{2\pi kT}\right)^{3/2} \exp\left[\frac{-m(u^2 + v^2 + w^2)}{2kT}\right] du\, dv\, dw \qquad (5.34)$$

We wish an expression for the fraction of molecules with a speed between c and $c + dc$, regardless of direction, where $c^2 = u^2 + v^2 + w^2$. These are the molecules whose velocity points lie within a spherical shell of thickness dc at a distance c from the origin (Fig. 5.6). The volume of this shell is $4\pi c^2\, dc$. This is the integral of $du\, dv\, dw$ in Eq. (5.34) over the spherical shell. Therefore, the desired density function is

$$p(c)\, dc = 4\pi \left(\frac{m}{2\pi kT}\right)^{3/2} \exp\left(\frac{-mc^2}{2kT}\right) c^2\, dc \qquad (5.35)$$

This is the expression derived by Maxwell in 1860.

Equation (5.35) is plotted in Fig. 5.7 at several different temperatures. The curve becomes broader and less peaked at the higher temperatures, as the average speed

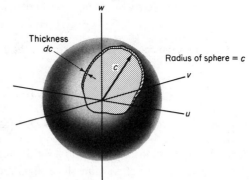

FIGURE 5.6 Components of molecular velocity, u, v, w, along axes X, Y, Z, and a spherical shell in velocity space between c and $c + dc$. The volume of the shell is $4\pi c^2\, dc$.

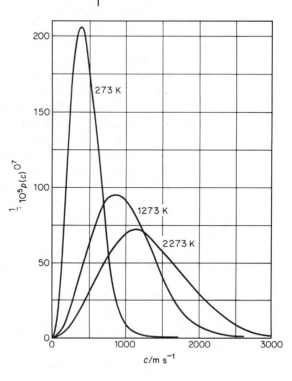

FIGURE 5.7 Relative probabilities of molecular speeds in nitrogen at three different temperatures. The density function $p(c)$ of Eq. (5.35) is plotted vs c. The most probable speed at any temperature is at the maximum of the curve.

increases and the distribution about the average becomes wider. Experimental measurements of molecular velocity distributions are in accord with the Maxwell equations. [Can you devise an experimental apparatus to measure a velocity distribution?]

We can now calculate the average molecular speed $\bar{c}$. Using Eqs. (5.33) and (5.35), we have

$$\bar{c} = \int_0^\infty p(c)c \, dc = 4\pi\left(\frac{m}{2\pi kT}\right)^{3/2} \int_0^\infty \exp\left(\frac{-mc^2}{2kT}\right)c^3 \, dc$$

The evaluation of this integral can be obtained from

$$\int_0^\infty e^{-ax^2}x^3 \, dx = 1/2a^2$$

Making the appropriate substitutions, we find that

$$\bar{c} = (8kT/\pi m)^{1/2} \qquad (5.36)$$

Note that the average speed is somewhat less than the rms speed, given by Eq. (2.14).

Problems

1. Consider a set of three energy levels with equal spacings ϵ. Suppose that identical particles occupy these levels in a geometric progression; e.g., $N_J = 2000, 200, 20$. Does this occupation correspond to a Boltzmann distribution?

2. The number of ways of realizing the assignment in (1) is $W = N!/N_1! \, N_2! \, N_3!$. Calculate W. Show that for any other assignment W will be lower (e.g., shift 20 particles from the lowest level to the highest level).

3. The vibration frequency of O_2 is 4.74×10^{13} Hz. Assuming a harmonic oscillator, what would be the average number of O_2 molecules in 1 mol O_2 in the state $v = 4$ at 300 K? at 1000 K?

4. The fundamental vibration frequency of N_2 corresponds to $\tilde{v} = 2360$ cm^{-1}. Assuming a harmonic oscillator, what fraction of N_2 molecules possess no vibrational energy except their zero-point energy at 500 K?

5. The ground electronic state of Cl is a doublet with separation 881 cm^{-1}. Plot the relative population N_1/N_0 of the two states of the doublet from $T = 100$ to 1000 K.

6. An electron spin can have two different directions in a magnetic field of flux density B, having energies $\pm\mu_B B$, where μ_B is the Bohr magneton, 9.27410×10^{-24} J T^{-1}. Calculate the relative electron population in the two energy levels at equilibrium in a field of 5 tesla (T), at 5 K, at 300 K.

7. Plot the partition function of Eq. (5.4) for an electron in a magnetic field of 5 T as a function of T.

8. The quantity $\tau = -dT/dx$, where x is height in the atmosphere, is called the *temperature lapse rate*. Show that if τ is a constant, $P/P_0 = (T/T_0)^{gM/R\tau}$, where P_0 and T_0 are sea-level values.

9. Derive an expression for the average speed, the rms speed, and the most probable speed for a molecule moving in two dimensions (e.g., adsorbed on a plane surface).

10. Show that the fraction of molecules in a gas having kinetic energies between ϵ and $\epsilon + d\epsilon$ is

$$f(\epsilon) \, d\epsilon = [2\pi/(\pi \, kT)^{3/2}]e^{-\epsilon/kT}\epsilon^{1/2} \, d\epsilon$$

11. What fraction of the molecules in (a) helium and (b) radon have a kinetic energy between 0.9 and 1.1 kT at 300 K and at 1000 K? (See Problem 10.)

12. Calculate the average speed, the rms speed, and the most probable speed c_{mp} of CH_4 molecules at 300 K and at 600 K. First derive the expression $c_{mp} = (2kT/m)^{1/2}$.

13. In the N_2 gas at 2490 K in Example 5.4, what fraction of the total energy (translation, rotation, vibration) of the N_2 is of each category?

14. The moment of inertia of CO is 14.48×10^{-47} kg m^2. Recalling that the statistical weight of a rotational level is $2J + 1$, draw the relative intensities of the first 10 rotational absorption lines ($\Delta J = 1$) at 200 K and at 400 K.

15. The maximum intensity in the microwave spectrum of a sample of CO occurs for absorption from the rotational level with $J = 6$. What is the temperature of the gas? [Problem 14]

16. A sample of sodium vapor is heated in a platinum cell with a small orifice from which a beam of sodium atoms emerges. Velocity analysis shows that the maximum number of sodium atoms occurs in the velocity range at 1000 m s^{-1}. What is the temperature of the sodium vapor?

6

The First Law
of Thermodynamics—Energy

The preceding three chapters have been concerned with energy at the molecular level. Now we return to macroscopic systems to begin our development of classical thermodynamics by considering applications of the concept of energy to large-scale processes. We shall first discuss the concept of *work* and discover how to measure or calculate *work* in various processes. We shall look at the historically important concept of *heat* and its relation to work. We shall then see how work and heat are related to the change in the internal energy of a system. This discussion will provide a new formulation of the principle of conservation of energy, the First Law of Thermodynamics. Finally, we shall see how the First Law can be applied to the energetics of chemical reactions.

6.1 The Concept of Work

In mechanics, if the point of application of a force **F** moves, the force is said to "do work." For instance, for a body of mass m subject to a gravitational force, the point of application can be taken to be the center of mass of the body. The work done by a force of magnitude F whose point of application moves a distance dr along the direction of the force is

$$dw = F\,dr \tag{6.1}$$

For the case of a force that is constant in direction and magnitude, Eq. (6.1) can be integrated between an initial position r_1 and a final position r_2 to give

$$w = \int_{r_1}^{r_2} F\,dr = F(r_2 - r_1) \tag{6.2}$$

An example of such a constant force is the weight of a body in the gravitational field of the earth. Over distances small compared to the diameter of the earth, the gravitational force acting on a body of mass m is $F = mg$, where g is the acceleration due to gravity. The value of g varies from point to point over the surface of the earth, but a standard value is $g = 9.80665$ m s^{-2}. To lift a body of mass m against the force of gravitation, one must apply an external force equal to mg.

Example 6.1   What is the work done when a mass of exactly 1 kg is lifted a distance of exactly 1 m?

From Eq. (6.2), $w = mgr_2 = (1 \text{ kg})(9.80665 \text{ m s}^{-2})(1 \text{ m})$

$\qquad = 9.80665 \text{ (kg m s}^{-2}) \text{ m} = 9.80665 \text{ newton meter (N m)}$

$\qquad = 9.80665 \text{ joule (J)}$

The example illustrates the essential features of mechanical work, namely, the force **F** and the movement of its point of application. It then became customary to say that "the force does the work." When the work involved motion of a body, a further extension of the concept was to consider that "the force does work on the body." Sometimes one sees a more dramatic statement, "the force performs work on the body." In thermodynamics the massive "bodies" of mechanics are generalized to become "systems," which include solids, liquids, and gases with defined boundaries. The concept of work is then extended to the system. The work is still always calculated as the product of a force and the displacement of its point of application, but now it is said that "the force does work on the system."

By international convention, it has been decided to define work done *upon* a system as *positive*. We can imagine an observer located within the system. Anything that enters the system from the surroundings or is done on the system by the surroundings is given a positive sign.

Although a force can "do work on" a system, the force does *not* "give work to" or "add work to" a system, nor does a system "hold," "contain," or "receive work." Any expression, therefore, is incorrect if it implies that work behaves like a substance or like a function of the state of the system.

6.2   Work in Changes of Volume

Figure 6.1 depicts a gas of volume V confined in a cylinder by a piston of area A. The external pressure on the piston is

$$P'_{ex} = F/A \tag{6.3}$$

where F is the magnitude of the force **F** due to the weight of the piston itself plus any extra weight placed on it. At equilibrium the pressure of the fluid, $P' = P'_{ex}$, so that

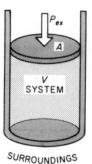

FIGURE 6.1 A system that consists of a volume of gas and its surroundings.

SURROUNDINGS

there is no net force upon the piston and it does not move. If we add a weight to the piston, the new $P_{ex} > P'$, and the piston moves in the direction of the gravitational force, compressing the gas to a smaller volume until the pressure of the gas increases to a new pressure $P = P_{ex}$ and equilibrium is established again.

As the piston moves downward, work is done *on the gas*. The element of work is

$$dw = F \, dr = \frac{F}{A} A \, dr = -P_{ex} \, dV \tag{6.4}$$

where A is the area of the piston and $dV = -A \, dr$ is the element of volume change. Note that as the piston moves a distance dr, the volume of the gas is decreased. Thus dV is negative and the minus sign must be included in Eq. (6.4) so the dw, the work done on the system, is positive.

If the external pressure is kept constant during a volume change from V_1 to V_2, the work done can be calculated by integration of Eq. (6.4),

$$w = -\int_{V_1}^{V_2} P_{ex} \, dV = -P_{ex}(V_2 - V_1) = -P_{ex} \, \Delta V \tag{6.5}$$

Example 6.2 What is the work done on the gas when 10 m³ of gas is compressed to 5 m³ under a constant pressure of 10^3 kPa?

$$w = (-10^3 \times 10^3 \text{ Pa})(-5 \text{ m}^3) = 5 \times 10^6 \text{ Pa m}^3 = 5 \times 10^6 \text{ J}.$$

If a change in volume is carried out in such a way that the external pressure is known at each successive state, we can plot the process on a graph of P_{ex} vs V. Such a plot is called an *indicator diagram*; an example is shown in Fig. 6.2(a). The work done on the system in a compression is equal to the area under the curve.

It is evident that the work done in going from point A to point B in the P_{ex}–V diagram depends on the particular path that is traversed. Consider, for example, two alternative paths from A to B in Fig. 6.2(b). More work will be done in going by path ADB than by path ACB, as is evident from the greater area under curve ADB. If we proceed from state A to state B by path ADB and return to A along BCA, we shall have completed a *cyclic process*. The net work done by the system during this cycle is seen to be equal to the difference between the areas under the two paths, which is the shaded area in Fig. 6.2(b).

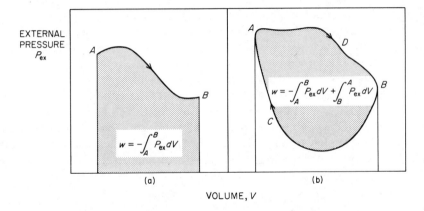

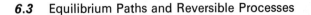

FIGURE 6.2 Indicator diagrams for *PV* work: (a) a general process from *A* to *B*; (b) a cyclic process *ADBCA*.

6.3 Equilibrium Paths and Reversible Processes

In the language of physical chemistry, we carefully distinguish between a *change* and a *process*. When a system passes from a state *A* to a state *B*, it undergoes a certain *change* $A \rightarrow B$. To define a change we need to specify only the final state and the initial state. The change in any state function $\mathfrak{F}$ when the system changes from *A* to *B* is $\Delta\mathfrak{F} = \mathfrak{F}_B - \mathfrak{F}_A$ (final value of $\mathfrak{F}$ — initial value of $\mathfrak{F}$). The system can change from *A* to *B* in many different ways. Any particular path that leads from *A* to *B* is called a *process*. The *process* from *A* to *B* is specified by the exact succession of states of the system along the path between *A* and *B*, whereas the *change* from *A* to *B* is specified by the initial and final states only. Processes that result in *PVT* changes in fluids are often represented by curves on indicator diagrams like those in Fig. 6.2.

If each successive point along the P_{ex}–*V* curve is an equilibrium state of the system, we have the very special case that P_{ex} always equals *P*, the pressure of the fluid itself. The indicator curve then becomes an *equilibrium curve* for the system. Such a case is shown in Fig. 6.3. Only when equilibrium is maintained can the work be

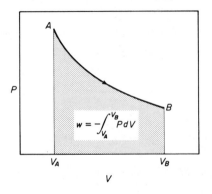

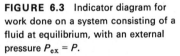

FIGURE 6.3 Indicator diagram for work done on a system consisting of a fluid at equilibrium, with an external pressure $P_{ex} = P$.

calculated from functions of the state of the substance itself, P and V. When the change from A to B proceeds (or can be imagined to proceed) through a succession of equilibrium states of the system, the process from A to B is called a *reversible process*.

To expand a gas by a reversible process as indicated in Fig. 6.3, the pressure must be released so slowly that at every stage the pressure within the gas volume is just equal to the pressure on the piston. If the pressure is infinitesimally decreased below the equilibrium P, the gas will expand by dV; if the pressure is infinitesimally increased above the equilibrium P, the gas will contract by dV. Thus the equilibrium path is in principle a reversible path, but a reversible process is really only a limiting ideal case, since it would take an infinite time to carry out an actual change reversibly.

In the equilibrium case, $P_{ex} = P$ and Eq. (6.4) becomes

$$dw = -P\,dV \quad \text{(reversible)} \tag{6.6}$$

6.4 Isothermal Reversible Compression of Ideal Gas

For an ideal gas at constant temperature Eq. (6.6) takes an especially simple form. Since $P = nRT/V$,

$$dw = -nRT\frac{dV}{V} = -nRT\,d\ln V \tag{6.7}$$

This expression can be integrated to yield

$$w = \int_{V_1}^{V_2} - nRT\,d\ln V$$

where V_1 and V_2 are the initial and final volumes, respectively. Thus,

$$w = -nRT(\ln V_2 - \ln V_1) = -nRT\ln\frac{V_2}{V_1} \tag{6.8}$$

Or, since $P_1 V_1 = P_2 V_2$ for an ideal gas,

$$w = nRT\ln\frac{P_2}{P_1} \tag{6.9}$$

Example 6.3 What is the work done on 1 mol of an ideal gas at 300 K when it is isothermally compressed from 100 to 400 kPa by a reversible process?

$$w = (1\text{ mol})(8.314\text{ J K}^{-1}\text{ mol}^{-1})(300\text{ K})\ln\left(\frac{400\text{ kPa}}{100\text{ kPa}}\right)$$
$$= 3460\text{ J}$$

Isothermal compression of a gas can be carried out if a cylinder of gas and piston are surrounded by a large volume of water at constant temperature T. When the gas is compressed reversibly, work w is done on the gas, given by Eq. (6.9). The internal energy of an *ideal* gas depends only on its temperature, as we have seen in the description of the translational, rotational, and vibrational energies of gas molecules. Thus there is no change in the internal energy of an ideal gas when it is compressed at constant T from V_1 to V_2. When the gas is compressed, the large-scale motion of the piston is converted into the microscopic motions of the gas molecules. We would expect the temperature of the gas to rise as a consequence, but the collisions of the gas molecules with the walls of the piston transfer energy to the walls and in turn this energy is transferred to the molecules of water in the huge water bath that surrounds the system. It is assumed that this bath is so huge that no appreciable rise in temperature T is produced in it. Thus the condition of constant T is imposed.

6.5   General Concept of Work

In mechanical systems, work is always formulated as the product of two terms, an intensive factor, which is a force, and an extensive factor, which is a displacement. A similar formulation applies also to nonmechanical work.

In physical chemistry, we are often interested in changes carried out in electrical cells. In the case of electrical work, the force becomes the electromotive force (emf) E of the cell, and the differential displacement becomes the charge dQ transferred through the external circuit as the cell discharges ($dQ < 0$). The element of work done on the cell is $dw = E \, dQ$.

We summarize various examples of work in Table 6.1. [Why does a negative sign occur for PV work only?]

TABLE 6.1
EXAMPLES OF WORK

Intensive factor	Extensive factor	Element of work dw
Tension F	Distance ℓ	$F \, d\ell$
Surface tension γ	Area A	$\gamma \, dA$
Pressure P	Volume V	$-P \, dV$
Electromotive force E	Charge Q	$E \, dQ$
Magnetic flux density B	Magnetization M	$B \, dM$

Example 6.4   The surface tension γ of liquid mercury at 273 K is 4.70 N m^{-1}. Calculate the minimum work required to divide 1.00 kg of mercury into spherical droplets 1.00 μm in diameter. The density ρ of mercury at 273 K is 13.6 $\times$ 10^3 kg m^{-3}.

$dw = \gamma \, dA$, $w = \gamma(A_2 - A_1) \approx \gamma A_2$. The number of droplets is the volume of mercury (m/ρ) divided by the volume of one droplet $(\frac{1}{6}\pi d^3)$.

$$A_2 = (\pi d^2)(m/\rho)(\tfrac{1}{6}\pi d^3)^{-1} = 6m/\rho d$$
$$= 6(1.00 \text{ kg})/(13.6 \times 10^3 \text{ kg m}^{-3})(1.00 \times 10^{-6} \text{ m}) = 441 \text{ m}^2$$
$$w = (4.70 \text{ N m}^{-1})(441 \text{ m}^2) = 2070 \text{ N m} = 2070 \text{ J}$$

6.6   The Concept of Heat

The experimental observations that led to the concept of temperature also led to the concept of heat. For many centuries, however, natural philosophers did not distinguish clearly between these two concepts, and often used the same name, *calor* or *caloric*, for both. In English the word *warmth* conveys a similar ambiguity. Joseph Black of Edinburgh, late in the eighteenth century, was one of the first to explain the distinction between heat and the intensive factor, temperature. Black showed that thermal equilibrium between two bodies *a* and *b* requires that their temperatures be equal, $T^a = T^b$.

Suppose that two bodies *a* and *b* with $T^a > T^b$ are placed in contact. Then heat will flow from *a* to *b* until $T^a = T^b$ and thermal equilibrium is reached. Heat can be transferred from one system to another by *conduction* as in this example, and also by *radiation* (as electromagnetic energy) and by *convection* (associated with a transfer or circulation of material substances).

We must not say that a system "contains heat." Heat is a mode of transfer of energy that can occur when temperature gradients exist between systems or when there is a transfer of mass between systems. After the transfer of energy has occurred, we must refer only to the *energy* of the system, and never to the *heat* in the system.

Even primitive peoples knew the connection via friction between heat and large-scale motion. Philosophers were quite clear about the relation of heat to small-scale motions. Thus Francis Bacon wrote in 1620:

> Heat itself, its essence and quiddity, is motion and nothing else. . . . Heat is a motion of expansion, not uniformly of the whole body together, but in the smaller parts of it. . . . The body acquires an alternating motion, perpetually quivering, striving, and struggling, and initiated by repercussion, whence springs the fury of fire and heat.

Although there was no difficulty in the concept of heat as motion, the correct understanding of the relation of heat to work had a long and arduous history. Originally heat was wrongly conceived to be a weightless, indestructible fluid, whereas work was something done by men and horses with sweat and toil. In 1798, Benjamin Thompson studied the boring of cannon by horse power in the Munich arsenal, and calculated a *mechanical equivalent of heat* $J = w/q = 5.46$ J cal^{-1} (in our units, not his). Julius Robert Mayer in 1852 made a calculation from the fall of weights and heating of water, and obtained $J = 3.56$ J cal^{-1}. Today, the calorie is no longer an independent unit and by international definition, $J = 4.184$ J cal $^{-1}$ exactly.

6.7   First Law of Thermodynamics for a Closed System

A closed system is defined by boundaries that permit the passage of energy but not of massive particles. The surroundings can do work *w* on the system and heat *q* can flow through the walls of the system. Both the work *w* and the heat *q* contribute to a change ΔU in the internal energy of the system, so that if U_B is the final energy and U_A the initial energy:

$$U_B - U_A = \Delta U = q + w \qquad (6.10)$$

Equation (6.10) is a mathematical statement of the First Law of Thermodynamics. It may at first appear to be an obvious and inconsequential statement. Why did it take so much effort through the first half of the nineteenth century by scientific giants like Joule and Helmholtz to establish it?

The importance of Eq. (6.10) lies in the inner meaning of the symbols we have written. ΔU is the change in a function of the state of the system—a quantity that depends not on past history or details of processes but simply and only on the final and initial states. The quantities q and w, however, are not state functions. We saw in Fig. 6.2, for example, that the work done on a system depends on the particular path (on the P–V diagram) from initial to final state. The inner significance of Eq. (6.10), therefore, is that although q and w are not state functions, their sum is always a state function.

6.8 Exact Differentials and State Functions

In the limit of smaller and smaller changes, Eq. (6.10) becomes

$$dU = dq + dw \qquad (6.11)$$

Although the quantities dU, dq, and dw are written in the same notation, they are not mathematical functions of the same kind. We must distinguish two classes of differentials. Those that are obtained by differentiation of state functions are called *exact differentials*. Examples are dU, dV, and dT. To integrate an exact differential we need to know only the initial and final state of the system. Consider, for example, the differential dU of internal energy. The change in internal energy between two states A and B is

$$\Delta U = \int_A^B dU = U_B - U_A$$

We do not need to know the particular process that leads from A to B.

In contrast, consider the quantity dw. This is an inexact differential. It is not the differential of a state function, since w is not a state function. We cannot integrate dw between an initial state A and a final state B unless we know the exact path of the process that leads from A to B. We saw an illustration of this requirement in the indicator diagram of Fig. 6.2. The work done on the system in going from A to B depends on the particular curve that is followed on the P_{ex}–V diagram.

Now we can appreciate the significance of Eq. (6.11). Although dq and dw are not exact differentials, their sum $dq + dw$ is always an exact differential dU and thus defines a state function U. This fact is an important mathematical consequence of the principle of conservation of energy applied to changes in a system due to heat transfers and work. Equation (6.11) is an important mathematical statement of the First Law of Thermodynamics.

When a state function is a function of two or more independent variables, its complete differential can always be written as a sum of terms each due to changes in one of the variables. For example, if U is a function of V and T, $U(V, T)$, its complete differential is

$$dU = (\partial U/\partial V)_T \, dV + (\partial U/\partial T)_V \, dT \tag{6.12}$$

The function $(\partial U/\partial V)_T$ is the partial derivative of U with respect V at constant temperature. It gives the variation of U with V when T is held constant. The function $(\partial U/\partial T)_V$ is the partial derivative of U with respect to T when V is held constant. (We have already met this function as C_V, the heat capacity at constant volume.) Figure 6.4 shows the geometrical significance of Eq. (6.12). The $U(V, T)$ is a surface in (U, V, T) space. The partial derivatives are slopes along curves of constant T or V.

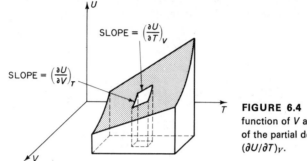

FIGURE 6.4 Internal energy U as function of V and T. Geometrical meaning of the partial derivatives $(\partial U/\partial V)_T$ and $(\partial U/\partial T)_V$.

6.9 Enthalpy

The internal energy U is an especially suitable state function for conditions of constant V. Most experiments in the chemical laboratory, however, are not carried out in vessels at constant volume. It is more usual to operate under a condition of constant (or almost constant) pressure. The natural independent variables are then P and T. For use with these variables, a new state function called (from the Greek, "in warmth") the *enthalpy*, H, has been defined:

$$H = U + PV \tag{6.13}$$

[Why is H a state function?]

For a change from an initial state (1) of a system to a final state (2),

$$(H_2 - H_1) = (U_2 - U_1) + (P_2 V_2 - P_1 V_1)$$

or

$$\Delta H = \Delta U + \Delta(PV) \tag{6.14}$$

Consider in Fig. 6.5 an experiment similar to that shown in Fig. 3.11, but this time at constant pressure P instead of constant volume V. When the electrical energy is put into the fluid through the heating coil, part of it goes into the internal energy of the fluid and part into the work of expanding the fluid at constant pressure, $-dw = P \, dV$. From Eq. (6.13) *at constant pressure, $dH = dU + P \, dV$.*

The temperature coefficient of H at constant P is

$$C_P = (\partial H/\partial T)_P \tag{6.15}$$

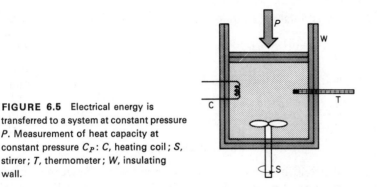

FIGURE 6.5 Electrical energy is transferred to a system at constant pressure P. Measurement of heat capacity at constant pressure C_P: C, heating coil; S, stirrer; T, thermometer; W, insulating wall.

C_P is called the *heat capacity at constant pressure*. We can measure C_P simply by measuring the electrical energy input into the system and the consequent rise in temperature. If we know C_P, we can calculate the change in enthalpy H with change in T at constant P. At constant P,

$$\Delta H = \int_{T_1}^{T_2} (\partial H/\partial T)_P \, dT = \int_{T_1}^{T_2} C_P \, dT \tag{6.16}$$

Example 6.5 From 180 to 310 K, the C_P in J K^{-1} mol^{-1} of CS$_2(\ell)$ at 100 kPa fits the empirical equation, $C_P = A + BT + CT^2$, as

$$C_P = 77.28 - 2.07 \times 10^{-2}T + 5.15 \times 10^{-5}T^2$$

The C_P were measured experimentally and the data were fitted to the equation cited with an accuracy of better than 2%. What is the ΔH of 1 mol CS$_2$ when it is heated from 180 K to 310 K at 100 kPa?

From Eq. (6.16), at constant pressure,

$$\Delta H = \int_{180}^{310} (A + BT + CT^2) \, dT = \left[AT + \tfrac{1}{2}BT^2 + \tfrac{1}{3}CT^3 \right]_{180}^{310}$$

$$= 77.28(310 - 180) - \frac{2.07 \times 10^{-2}}{2}(310^2 - 180^2)$$

$$+ \frac{515 \times 10^{-5}}{3}(310^3 - 180^3)$$

$$= 10\,050 - 660 + 410 = 9800 \text{ J}$$

6.10 The Difference between C_P and C_V

The heat capacity at constant pressure $C_P = (\partial H/\partial T)_P$ is usually larger than that at constant volume $C_V = (\partial U/\partial T)_V$, because at constant P some of the heat transferred to a substance may be used in the work of expanding it rather than in heating it, whereas at constant V all the heat produces only a rise in temperature. An equation

for the difference $C_P - C_V$ is obtained as follows: $C_P - C_V = (\partial H/\partial T)_P - (\partial U/\partial T)_V$.
Since $H = U + PV$,

$$C_P - C_V = (\partial U/\partial T)_P + P(\partial V/\partial T)_P - (\partial U/\partial T)_V$$

From Eq. (6.12), by dividing by dT at constant P,

$$(\partial U/\partial T)_P = (\partial U/\partial V)_T(\partial V/\partial T)_P + (\partial U/\partial T)_V$$

Substituting this into the equation above gives

$$C_P - C_V = [P + (\partial U/\partial V)_T](\partial V/\partial T)_P \qquad (6.17)$$

The term $P(\partial V/\partial T)_P$ is the contribution to C_P that is caused by the change in volume of the system against the external pressure P. The other term, $(\partial U/\partial V)_T(\partial V/\partial T)_P$, is the contribution to C_P due to the energy required to increase the volume against the internal cohesive forces of the substance. The term $(\partial U/\partial V)_T$ is called the *internal pressure*. In the case of liquids and solids, which have strong internal cohesive forces, this term is usually large. In the case of gases, on the other hand, $(\partial U/\partial V)_T$ is generally small compared to P.

For ideal gases $(\partial U/\partial V)_T = 0$. There are no internal cohesive forces between molecules in an ideal gas. Also, for an ideal gas, with $V = nRT/P$, $(\partial V/\partial T)_P = nR/P$ and Eq. (6.17) becomes

$$C_P - C_V = nR \qquad \text{(ideal gas)} \qquad (6.18)$$

[C_V for argon is $\frac{3}{2}R$ per mole. What is C_P?]

6.11 Enthalpy of Phase Changes

Changes such as fusion of a solid, vaporization of a liquid, sublimation of a solid, or change of a solid from one crystal structure to another are called *phase changes* or *changes in state of aggregation*.

Changes in the internal energy U and the enthalpy H of a substance accompany a phase change. Consider, for instance,

$$\text{liquid water} \rightleftharpoons \text{water vapor}$$

$$\Delta H(\text{vaporization}) = H(H_2O, g) - H(H_2O, \ell)$$

The internal kinetic and potential energies of the water molecules are quite different in the two states of aggregation. The intermolecular potential energies are much more negative in a condensed phase than in a gas. This cohesive energy due to net attractive forces between the molecules is responsible for the condensation of gas to liquid when the temperature falls below the boiling point of the liquid.

If a substance is heated from T_1 to T_2 and a change in phase occurs in the interval, the ΔH of the change in phase must be included in the computation of the total ΔH for the change in temperature. The enthalpy of H_2O per mole is plotted vs. T in Fig. 6.6. Note the jumps in H at the melting and boiling points. The ΔH of phase changes are sometimes called *latent heats*, in accord with the historic terminology of Joseph Black (1760).

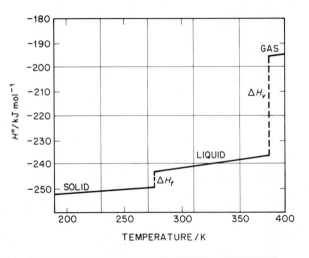

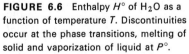

FIGURE 6.6 Enthalpy $H°$ of H_2O as a function of temperature T. Discontinuities occur at the phase transitions, melting of solid and vaporization of liquid at $P°$.

TEMPERATURE / K

Example 6.6 What is the ΔH when 1 mol H_2O at 101 kPa is heated from 353 to 393 K? The following data are available.

$$C_P(\ell, H_2O) = 75.0 \text{ J K}^{-1} \text{ mol}^{-1}$$

$$\Delta H(\text{vaporization}) = 47.3 \text{ kJ mol}^{-1} \text{ at } 373 \text{ K}$$

$$C_P(g, H_2O) = 35.4 \text{ J K}^{-1} \text{ mol}^{-1}$$

$$\Delta H = \int_{353}^{373} 75.0 \, dT + 47\,300 + \int_{373}^{393} 35.4 \, dT = 49\,500 \text{ J}.$$

6.12 ΔU in Chemical Reactions

We shall now consider the internal energies U and enthalpies H of systems in which chemical reactions occur. This is the subject that has traditionally been called *thermochemistry*. If a chemical reaction occurs in an isolated system, the internal energy U of the system is constant, but some chemical energy may be converted into thermal energy or intermolecular energy, or vice versa.

As an example of a chemical reaction consider $2H_2 + O_2 = 2H_2O$. The internal energy of the product at a specified temperature and pressure is not the same as the sum of internal energies of the reactants at that same temperature and pressure. We define

$$\Delta U(\text{reaction}) = U(\text{products}) - U(\text{reactants})$$
$$T_1, P_1 \qquad\qquad T_1, P_1 \qquad\qquad T_1, P_1 \qquad\qquad\qquad (6.19)$$

The balanced reaction equation indicates that

$$\Delta U = 2U_{H_2O} - 2U_{H_2} - U_{O_2} \qquad\qquad (6.20)$$

Here U_j denotes the internal energy per mole for each substance ($j = H_2O, H_2, O_2$). The standard notation uses a subscript or parentheses to denote an extensive thermodynamic quantity per unit amount of substance. Thus U_{H_2} or $U(H_2)$ is the internal energy of H_2 per unit amount of H_2, with SI unit joules per mole.

We write the ΔH of the reaction in the same way,

$$\Delta H = 2H_{H_2O} - 2H_{H_2} - H_{O_2}$$

Note that for a chemical reaction the symbol ΔX always means "value of X for products — value of X for reactants."

In generalized form, we can write

$$\Delta U = \sum v_j U_j \quad \text{and} \quad \Delta H = \sum v_j H_j \qquad (6.21)$$

where v_j are the stoichiometric coefficients in the reaction equation, positive for products, negative for reactants. Note that v_j is a dimensionless number and in SI units U_j or H_j is in joules per mole, so that $v_j H_j$ and $v_j U_j$ are also in joules per mole. The ΔU or ΔH of a reaction is always expressed in units of joules per mole.

Per mole of what? One answer would be per mole of reactants or products in the stoichiometric mixture as given by the balanced equation for the reaction. Recall (page 4) that the mole is an Avogadro number of any specified group of entities. Thus for $2H_2 + O_2 = 2H_2O$, the ΔH of reaction would refer to reaction of 1 mole of a mixture of $2H_2 + O_2$, i.e., to $2L$ molecules of H_2 plus L molecules of O_2.

Another way of defining the ΔU or ΔH of a reaction is in terms of the extent of reaction, ξ (page 7). Then ΔU (reaction) $= \Delta U/\xi$. Since the SI unit of ξ is the mole, the ΔU of reaction again is seen to have SI units of joules per mole.

6.13   Measurement of ΔU of Reaction

The ΔU of a chemical reaction can be measured in a *bomb calorimeter*, as shown in Fig. 6.7. The reaction vessel (bomb) is constructed of a strong alloy. It is equipped with electrical leads and an ignition wire to initiate the reaction. The bomb is surrounded by a volume of water, in which heating coils and a device for precise temperature measurement are installed. This water jacket is thermally insulated, so that during an experiment practically no energy is lost from the calorimeter to its surroundings.

In many cases the reaction studied is a combustion. The sample of reactant is placed in a holder into which the ignition wire is fitted. The bomb is filled with oxygen at elevated pressure. When equilibrium temperature is established throughout the apparatus, combustion is started by heating the ignition wire, and it proceeds rapidly to completion. The temperature rise ΔT of the water jacket is recorded.

After the system has cooled back to its original temperature, carefully measured electrical energy I^2Rt is introduced into the system via the heating coil until a ΔT equal to that in the reaction experiment is produced. During the reaction experiment, the water jacket + bomb + reaction mixture constitute an isolated system, and thus for the entire system, $\Delta U = 0$. The ΔU of the chemicals that react must therefore be exactly equal and opposite to ΔU of the calorimeter system of water and bomb. If the extent of reaction (determined by analysis of the product mixture) is ξ,

$$\Delta U = 0 = \xi\,\Delta U(\text{reaction}) + \Delta U(\text{calorimeter})$$
$$\Delta U(\text{reaction}) = -\Delta U(\text{calorimeter})/\xi \qquad (6.22)$$

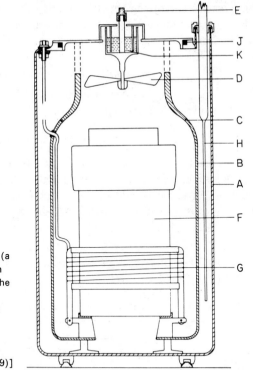

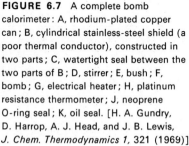

FIGURE 6.7 A complete bomb calorimeter: A, rhodium-plated copper can; B, cylindrical stainless-steel shield (a poor thermal conductor), constructed in two parts; C, watertight seal between the two parts of B; D, stirrer; E, bush; F, bomb; G, electrical heater; H, platinum resistance thermometer; J, neoprene O-ring seal; K, oil seal. [H. A. Gundry, D. Harrop, A. J. Head, and J. B. Lewis, *J. Chem. Thermodynamics 1*, 321 (1969)]

The electrical energy measured in the second part of the experiment equals ΔU(calorimeter), giving ΔU(reaction) $= \Delta U_r = -I^2 R t/\xi$.

The value of ΔU_r for a chemical reaction depends on the temperature at which it is measured. Consider, for example,

$$2H_2 + O_2 = 2H_2O, \qquad \Delta U_r(T)$$

Since ΔU_r depends only on the specified initial and final states, we can consider that the reaction occurs at the constant temperature T, and then let the products be heated from T to $T + \Delta T$ by the heat evolved in the reaction. The ΔU in this latter step is C_V (products) ΔT. The ΔU_r that we measure for the reaction is therefore ΔU_r at the specified initial temperature T of the reactants. Usually, a temperature of 25°C (298.15 K) is chosen as the reaction temperature in tables of thermodynamic data.

6.14 Calculation of ΔH from ΔU

Many reactions are studied in bomb calorimeters. These measurements give ΔU for the reaction, but we often want to know ΔH. From Eq. (6.13), $H = U + PV$. Thus

$$\Delta H = \Delta U + \Delta(PV) \tag{6.23}$$

By $\Delta(PV)$ is meant PV(products) $- PV$(reactants).

For a reaction involving only liquids and solids, $\Delta(PV)$ is so small as to be usually negligible. For reactions with gases as products or reactants, however, $\Delta(PV)$ can be appreciable. Consider a generalized reaction in which all reactants and products are gases,

$$aA + bB = cC + dD \tag{6.24}$$

We write $\Delta v = c + d - a - b$. For ideal gases $PV = nRT$, so that, per mole of reaction, we have at constant T, $\Delta(PV) = (\Delta v)RT$. Thus, for ideal-gas reactions, Eq. (6.23) becomes

$$\Delta H_r = \Delta U_r + (\Delta v)RT \tag{6.25}$$

If any of the reactants or products in Eq. (6.24) are liquids or solids, their contribution to $\Delta(PV)$ can be ignored compared to that of the gases. Many calorimetric studies are carried out in reaction calorimeters at constant pressure. With this equipment, one can measure the reaction enthalpy ΔH directly. Then we can calculate ΔU, if it is needed, from Eq. (6.25).

Example 6.7 For the reaction S(rhombic crystal) $+ O_2 = SO_2$, $\Delta U_r(298\ \text{K}) = -298\ \text{kJ mol}^{-1}$. What is ΔH_r?

If we ignore the solid reactant, $\Delta v = 0$ and hence from Eq. (6.25), $\Delta H_r = \Delta U_r$.

Example 6.8 For the reaction $\frac{3}{2}H_2 + \frac{1}{2}N_2 = NH_3$, $\Delta U_r(298\ \text{K}) = -43.5\ \text{kJ mol}^{-1}$. What is ΔH_r?

$\Delta v = -1$. From Eq. (6.25),

$$\Delta H_r = \Delta U_r - RT$$
$$= -43\ 500 - (8.314\ \text{J K}^{-1}\ \text{mol}^{-1})(298\ \text{K}) = -46.0\ \text{kJ mol}^{-1}$$

6.15 The Hess Law

As an immediate consequence of the First Law of Thermodynamics, ΔU or ΔH for any chemical reaction is independent of the path, i.e., independent of any particular intermediate reactions that may occur. This principle, first established experimentally by G. H. Hess (1840), was called the Law of Constant Heat Summation. If this law were not true, one could carry out a reaction in an isolated system by one path with ΔU_1 and then carry out the reverse reaction by a different path with ΔU_2 and thereby create or destroy energy $\Delta U = \Delta U_1 - \Delta U_2 \neq 0$.

It is often possible, by means of Hess's Law, to calculate the ΔH or ΔU of a reaction from measurements on other reactions. For example,

(1) $COCl_2(g) + H_2S(g) = 2HCl(g) + COS(g)$	$\Delta H(298\ \text{K}) =$	$-78\ 705\ \text{J mol}^{-1}$
(2) $COS + H_2S = H_2O(g) + CS_2(\ell)$	$\Delta H(298\ \text{K}) =$	$3\ 420\ \text{J mol}^{-1}$
(3) $COCl_2 + 2H_2S = 2HCl + H_2O(g) + CS_2(\ell)$	$\Delta H(298\ \text{K}) =$	$-75\ 285\ \text{J mol}^{-1}$

(We often write 298 K to stand for 298.15 K or 25°C.) The Hess Law greatly extends the range of thermochemical data. For example, we can calculate the ΔH of any reaction provided that we know the ΔH of combustion of its reactants and products.

6.16 Standard States

To specify the ΔH (or ΔU) of a reaction it is necessary to write the exact chemical equation and to specify the states of all reactants and products. The thermodynamic state of a pure substance is specified by naming its temperature and pressure. Two examples follow, in which $P° = 101.32$ kPa (1 standard atmosphere).

$$CO_2(P°) + H_2(P°) = CO(P°) + H_2O(g, P°), \quad \Delta H°(298.15 \text{ K}) = 41\ 160 \text{ J mol}^{-1}$$
$$AgBr(s, P°) + \tfrac{1}{2}Cl_2(P°) = AgCl(s, P°) + \tfrac{1}{2}Br_2(\ell, P°),$$
$$\Delta H°(298.15 \text{ K}) = 28\ 670 \text{ J mol}^{-1}$$

By international agreement, the standard state of all substances for the purpose of recording thermodynamic data is their state at $P° = 101.32$ kPa. Furthermore, if the substance is a gas, its standard state is that of the ideal gas. (The values of thermodynamic data in the ideal-gas state may be calculated from the equation of state of the gas.) The superscript $°$ on a symbol denotes a value taken at the standard pressure $P°$ and (for a gas) at the ideal-gas condition. It is important to remember that the superscript $°$ does *not* specify the temperature.

6.17 Enthalpy of Formation of a Compound

In chemical energetics the identities of chemical elements never change—we do not deal with nuclear reactions. It is therefore convenient—and ordained by another international convention—that the enthalpies of the chemical elements in their most stable states at $P° = 101.32$ kPa and 298.15 K are set equal to zero. This state is the reference standard state for enthalpy data. Thus the reference state for oxygen is O_2 gas as an ideal gas at $P°$ and 298.15 K. In this state $H°_{298}(O_2) = 0$ J mol^{-1}. For O_3 or for O atoms, $H°$ at $P°$ and 298.15 K is not set equal to zero. For carbon, two crystalline states exist at $P°$ and 298.15 K. The more stable solid is graphite, which therefore is assigned $H°_{298} = 0$ in the reference state. For diamond $H°_{298} = 1900$ J mol^{-1}. [If 1.00 mg of diamond and 1.00 mg of graphite are each burned in a bomb calorimeter, which combustion yields the more heat?]

Thus,

$$H°_j(298 \text{ K}) = 0 \quad \textit{for elements in reference states}$$

At any other temperature $H°_j$ will not be zero but some other value, which can be computed if $C°_P$ is known for the element. For an element in the same form as in its reference state,

$$H°_j(T) = \int_{298}^{T} C°_P(T')\, dT'$$

The integration variable is primed (T') to distinguish it from the upper limit (T) [Does this equation hold for $Br_2(g)$?]

The *standard enthalpy of formation* ΔH_f° of any compound is the ΔH_r° of the reaction by which it is formed from its elements. All the reactants and products are at $P^\circ = 101.32$ kPa. We can, however, have standard enthalpies of formation at different temperatures, so that the temperature of the reaction of formation must be separately specified. The superscript $^\circ$ does not specify the temperature but only P° and the fact that gases are ideal. The elements are in their reference forms, e.g., graphite not diamond. We shall later describe the standard state for components in solution, but so far we have dealt only with pure substances. For example, at 298.15 K,

$$S(\text{rhombic}) + O_2 = SO_2 \qquad \Delta H_f^\circ(298 \text{ K}) = -296.9 \text{ kJ mol}^{-1}$$

$$2Al + \tfrac{3}{2}O_2 = Al_2O_3 \qquad \Delta H_f^\circ(298 \text{ K}) = -1669.8 \text{ kJ mol}^{-1}$$

Thermochemical data are conveniently tabulated as standard enthalpies of formation ΔH_f°. Some examples, selected from a compilation of the National Bureau of Standards, are given in Table A.1 of the Appendix. The standard enthalpy of any reaction at 298.15 K is readily found as the difference between the tabulated ΔH_f° of products and reactants.

Many thermochemical data have been obtained from measurements of heats of combustion. If the heats of formation of all its combustion products are known, the heat of formation of a compound can be calculated from its heat of combustion. For example, all at 298.15 K:

$$
\begin{array}{ll}
(1) \ -1[C_2H_6 + \tfrac{7}{2}O_2 = 2CO_2 + 3H_2O(\ell)] & \Delta H_f^\circ = -1560.1 \text{ kJ mol}^{-1}(-1) \\
(2) \ 2[C(\text{graphite}) + O_2 = CO_2] & \Delta H_f^\circ = -\ 393.5 \text{ kJ mol}^{-1}(2) \\
(3) \ 3[H_2 + \tfrac{1}{2}O_2 = H_2O(\ell)] & \Delta H_f^\circ = -\ 285.8 \text{ kJ mol}^{-1}(3) \\
\hline
(4) \ 2C + 3H_2 = C_2H_6 & \Delta H_f^\circ = -\ \ 84.3 \text{ kJ mol}^{-1}
\end{array}
$$

6.18 ΔH of Reactions in Aqueous Solution

For chemical reactions in aqueous solution (which include many inorganic and biochemical reactions) we cannot use the ΔH_f° values of compounds as pure crystals, liquids, or gases. The ΔH for solution of a compond in water (or in any other solvent) depends on the final concentration in the solution. The data most often tabulated, however, are for an "infinitely dilute solution," which in practice means a solution at about 0.01 mol L^{-1} or less. These data are reported as $\Delta H_f^\circ(aq)$.

The $\Delta H(aq)$ are obtained by combining data on ΔH_f° for the pure compounds with data obtained by experimental measurement of the ΔH of solution of the compound in a large excess of water. For example,

$$
\begin{array}{ll}
\tfrac{1}{2}H_2(g) + \tfrac{1}{2}Cl_2(g) = HCl(g) & \Delta H_{298}^\circ = -\ 92.31 \text{ kJ mol}^{-1} \\
HCl(g) = HCl(aq) & \Delta H_{298}^\circ = -\ 75.14 \text{ kJ mol}^{-1} \\
\hline
\tfrac{1}{2}H_2(g) + \tfrac{1}{2}Cl_2(g) = HCl(aq) & \Delta H_f^\circ(298 \text{ K, aq}) = -167.45 \text{ kJ mol}^{-1}
\end{array}
$$

The value -167.45 kJ mol^{-1} is tabulated as the standard enthalpy of formation of HCl in aqueous solution (at infinite dilution). The values of $\Delta H_f^\circ(aq)$ can be combined in the usual way to give the $\Delta H°(aq)$ of a great number of reactions in aqueous solution.

6.19 Enthalpy of Formation of Ions

Dilute solutions of electrolytes in water are dissociated into ions. For example,

$$HCl(aq) = H^+(aq) + Cl^-(aq)$$
$$Na_2SO_4(aq) = 2Na^+(aq) + SO_4^{2-}(aq)$$

Thus we would save a good deal of space if we could tabulate $\Delta H_f°(aq)$ for the individual ions. The data for the ions could then be combined to yield $\Delta H_f°(aq)$ for all possible compounds derived from these ions.

There is no way to measure the ΔH of solution of 1 mole of gaseous ions all bearing like charges. We therefore adopt a convention that arbitrarily sets the $\Delta H_f°(aq)$ for the H^+ ion at 298.15 K equal to zero. Values for all the other ions can then be tabulated relative to this standard. For example, consider HCl:

$$\Delta H_f(HCl, aq) = \Delta H_f(H^+) + \Delta H_f(Cl^-)$$

From the experimental $\Delta H_f(HCl, aq) = -167.4 \text{ kJ mol}^{-1}$ and the convention $\Delta H_f(H^+ aq) = 0$, we obtain $\Delta H_f°(Cl^-, aq) = -167.4 \text{ kJ mol}^{-1}$. Then we can take an experimental value for NaCl of $\Delta H_f°(aq) = -407.1 \text{ kJ mol}^{-1}$ and the value for Cl^-, to obtain $\Delta H_f°(Na^+) = -239.7 \text{ kJ mol}^{-1}$. In this way we can build up a table for all ions of interest. The values for some common ions are given in Table A.3.

Example 6.9 From standard ionic enthalpies and the $\Delta H_f°$, calculate $\Delta H°$ at 298 K for the reaction,

$$CaSO_4(s) = Ca^{2+}(aq) + SO_4^{2-}(aq)$$

$$\Delta H° = \Delta H_f°(Ca^{2+}) + \Delta H_f°(SO_4^{2-}) - \Delta H_f°(CaSO_4)$$
$$= -543.0 - 907.5 - (-1432.7) = -17.8 \text{ kJ mol}^{-1}$$

6.20 Temperature Dependence of Enthalpy of Reaction

Often the ΔH of a reaction is measured at one temperature and we need to know its value at a different temperature. We write $\Delta H = H(\text{products}) - H(\text{reactants})$, and differentiate both sides with respect to T at constant P:

$$\left(\frac{\partial \Delta H}{\partial T}\right)_P = \left(\frac{\partial H}{\partial T}\right)_P (\text{products}) - \left(\frac{\partial H}{\partial T}\right)_P (\text{reactants})$$

$$\left(\frac{\partial \Delta H}{\partial T}\right)_P = C_P(\text{products}) - C_P(\text{reactants}) = \Delta C_P$$

(6.26)

This equation was first obtained by G. R. Kirchhoff in 1858. It shows that the rate of change of enthalpy of reaction with temperature is equal to the difference in heat capacities C_P of products and reactants.

The heat capacities themselves vary with temperature. Often, however, it is sufficiently accurate to use an average value of the heat capacity over the range of temperature considered.

Example 6.10 $H_2O(g) = H_2 + \frac{1}{2}O_2$, and $\Delta H^\circ = 241\ 750\ \text{J mol}^{-1}$ at 291 K. What is ΔH° at 308 K? Over the small temperature range, the effectively constant C_P values per mole are $C_P(H_2O, g) = 33.56$, $C_P(H_2) = 28.83$, and $C_P(O_2) = 29.12\ \text{J K}^{-1}\ \text{mol}^{-1}$.

We use Eq. (6.26) with
$$\Delta C_P = C_P(H_2) + \tfrac{1}{2}C_P(O_2) - C_P(H_2O, g)$$
$$= 28.83 + 14.56 - 33.56 = 9.83\ \text{J K}^{-1}\ \text{mol}^{-1}$$

From Eq. (6.26),
$$\int_{291}^{308} d(\Delta H^\circ) = \int_{291}^{308} \Delta C_P\,dT = 9.83 \int_{291}^{308} dT$$

Thus, $\Delta H^\circ(308\ \text{K}) - \Delta H^\circ(291\ \text{K}) = 9.83(17) = 170\ \text{J mol}^{-1}$

$$\Delta H^\circ\ (308\ \text{K}) = 241\ 920\ \text{J mol}^{-1}$$

Experimental heat-capacity data can be represented by a power series:

$$C_P = a + bT + cT^2 \tag{6.27}$$

Examples of such heat-capacity equations are given in Table 6.2. These three-term equations fit the experimental data to within about 0.5% over a temperature range

TABLE 6.2
HEAT CAPACITY OF GASES AT PRESSURE $P^\circ = 101.32\ \text{kPa}$
(273 to 1500 K)

Gas	$C_P = a + bT + cT^2$ (C_P in J K^{-1} mol^{-1})		
	a	$b \times 10^3$	$c \times 10^7$
H_2	29.07	−0.836	20.1
O_2	25.72	12.98	−38.6
Cl_2	31.70	10.14	−2.72
Br_2	35.24	4.075	−14.9
N_2	27.30	5.23	−0.04
CO	26.86	6.97	−8.20
HCl	28.17	1.82	15.5
HBr	27.52	4.00	6.61
H_2O	30.36	9.61	11.8
CO_2	26.00	43.5	−148.3
Benzene	−1.71	326.0	−1100
n-Hexane	30.60	438.9	−1355
CH_4	14.15	75.5	−180

from 273 to 1500 K. The series expression for ΔC_P is substituted into Eq. (6.26), and the integration is carried out. Thus at a constant pressure P°, for the standard enthalpy change,

$$d(\Delta H^\circ) = \Delta C_P \, dT = (A + BT + CT^2) \, dT$$
$$\Delta H_T^\circ = \Delta H_0^\circ + AT + \tfrac{1}{2}BT^2 + \tfrac{1}{3}CT^3 \tag{6.28}$$

where A, B, C, etc., are the sums of the individual a, b, c in Eq. (6.27). ΔH_0° is the constant of integration. Any one measurement of ΔH° at a known temperature T makes it possible to evaluate the constant ΔH_0° the standard enthalpy of reaction at 0 K. Then the ΔH° at any other temperature (within the range of validity of the heat-capacity equations) can be calculated from Eq. (6.28).

6.21 Bond Enthalpies

In many cases, to a good approximation, it is possible to express the enthalpy of formation of a molecule as an additive property of the bonds forming the molecule. This formulation has led to the concepts of *bond energy* and *bond enthalpy*.

Consider a reaction in which a bond A—B is broken between atom A and atom B: A—B $\rightarrow$ A + B. In terms of this reaction, the bond energy or enthalpy of A—B has been variously defined as:

1. The energy of the reaction at absolute zero, $\Delta U^\circ(0 \text{ K})$
2. The enthalpy of the reaction at absolute zero, $\Delta H^\circ(0 \text{ K})$
3. The enthalpy of the reaction at 298.15 K, $\Delta H^\circ(298.15 \text{ K})$

The first two definitions are useful in discussions of molecular structure, which often refer to spectroscopic data on dissociation of molecules at 0 K. We shall use the last definition because it is most convenient for thermochemical calculations. The entities A and B need not be atoms; they may be fragments of molecules. For example, the ΔH° of the C—C bond in ethane would be the ΔH° (298 K) of the reaction $C_2H_6 = 2CH_3$.

For a given type of bond, the bond enthalpy $\Delta H^\circ(A\text{—}B)$ depends on the particular molecule in which the bond occurs and on its particular situation in that molecule. Consider, for example, CH_4 and suppose that the H atoms are removed from it one at a time:

(1) $CH_4 = CH_3 + H$, $\quad \Delta H^\circ \sim 422$ kJ mol^{-1}

(2) $CH_3 = CH_2 + H$, $\quad \Delta H^\circ \sim 364$ kJ mol^{-1}

(3) $CH_2 = CH + H$ $\quad \Delta H^\circ \sim 385$ kJ mol^{-1}

(4) $CH = C + H$, $\quad \Delta H^\circ \sim 335$ kJ mol^{-1}

For many purposes, less detailed information is adequate. Thus the four C—H bonds in methane are all equivalent, and if we imagine the dissociation of CH_4 to a C atom and four H atoms, we can set one-fourth of the enthalpy of this reaction equal to the average $\Delta H°$ of a C—H bond in CH_4. The reaction would be $CH_4 = C(g) + 4H$.

To calculate such average bond-enthalpy values, therefore, we need the enthalpies of formation of molecules from their atoms. The $\Delta H°$ of atomization of an element is the $\Delta H°$ of conversion of the element in its standard state to its atoms in their standard states. We can use the $\Delta H°$ of atomization of the elements to calculate bond enthalpies from standard enthalpies of formation. In most cases, it is not difficult to obtain the $\Delta H°$ for converting the elements to monatomic gases. In the case of metals, this is simply the $\Delta H_s°$ for sublimation to the monatomic form. For example,

$$Mg(c) \longrightarrow Mg(g) \qquad \Delta H_s°(298 \text{ K}) = 150.2 \text{ kJ mol}^{-1}$$

$$Ag(c) \longrightarrow Ag(g) \qquad \Delta H_s°(298 \text{ K}) = 298.2 \text{ kJ mol}^{-1}$$

In other cases, the $\Delta H°$ of atomization can be obtained from the dissociation energies of diatomic gases. For example,

$$\tfrac{1}{2}Br_2(g) \longrightarrow Br(g) \qquad \Delta H°(298 \text{ K}) = 111.9 \text{ kJ mol}^{-1}$$

$$\tfrac{1}{2}O_2(g) \longrightarrow O(g) \qquad \Delta H°(298 \text{ K}) = 249.2 \text{ kJ mol}^{-1}$$

A most important $\Delta H°$ of atomization is the heat of sublimation of graphite since all bond enthalpies of organic molecules depend on it.

$$C(\text{graphite}) \longrightarrow C(g) \qquad \Delta H_s°(298 \text{ K}) = 716.68 \text{ kJ mol}^{-1}$$

Some enthalpies of atomization of elements are given in Table 6.3.

TABLE 6.3
STANDARD ENTHALPIES OF ATOMIZATION OF ELEMENTS
$\Delta H°(298 \text{ K})$

Element	$\Delta H_{298}°$ (kJ mol^{-1})	Element	$\Delta H_{298}°$ (kJ mol^{-1})
H	217.97	N	472.70
O	249.17	P	333.8
F	78.99	C	716.68
Cl	121.68	Si	455.6
Br	111.88	Hg	60.84
I	106.84	Ni	425.14
S	278.81	Fe	404.5

As an example, let us use thermochemical data to determine the average $H°$ of the two O—H bonds in water.

$$H_2 = 2H, \qquad \Delta H°(298 \text{ K}) = \quad 436.0 \text{ kJ mol}^{-1}$$

$$O_2 = 2O, \qquad \Delta H°(298 \text{ K}) = \quad 498.3 \text{ kJ mol}^{-1}$$

$$H_2 + \tfrac{1}{2}O_2 = H_2O, \qquad \Delta H°(298 \text{ K}) = -241.8 \text{ kJ mol}^{-1}$$

Therefore,

$$2H + O = H_2O, \qquad \Delta H^\circ(298\ K) = -927.2\ \text{kJ mol}^{-1}$$

This is the $\Delta H^\circ(298\ K)$ for the formation of two O—H bonds, so that the average ΔH° for the O—H bond in water is $927.2/2 = 436.6\ \text{kJ mol}^{-1}$. This value is quite different from ΔH° for $HOH \rightarrow H + OH$, which is $498\ \text{kJ mol}^{-1}$.

The ΔH of a bond A—B is approximately constant in a series of similar compounds. Thus tables of average bond enthalpies can be used to estimate ΔH° values for chemical reactions. Different bond types must be distinguished, e.g., single, double, and triple bonds in the case of C—C bonds. Table 6.4 is a summary of average single-bond enthalpies. Table 6.5 gives a few individual values for specified molecules.

TABLE 6.4
AVERAGE SINGLE-BOND ENTHALPIES[a] (kJ mol^{-1})

	A	Si	I	Br	Cl	F	O	N	C	H
H	339	339	299	366	432	563	463	391	413	436
C	259	290	240	276	328	441	351	292	348	
N					200	270		161		
O		369			203	185	139			
F	250	541	258	237	254	153				
Cl	250	359	210	219	243					
Br		289	178	193						
I		213	151							
Si	227	177								
S	213									

[a]After L. Pauling, *Nature of the Chemical Bond*, 3rd ed. (Ithaca, N.Y.: Cornell University Press, 1960).

TABLE 6.5
SINGLE- AND MULTIPLE-BOND ENTHALPIES (kJ mol^{-1})

Triple bonds	ΔH°	Double bonds	ΔH°	Single bonds	ΔH°	Single bonds	ΔH°
N≡N	946	$CH_2{=}CH_2$	682	CH_3—CH_3	368	CH_3—H	435
HC≡CH	962	$CH_2{=}O$	732	H_2N—NH_2	243	NH_2—H	431
HC≡N	937	O=O	498	HO—OH	213	OH—H	498
C≡O	1075	HN=O	481	F—F	159	F—H	569
		HN=NH	456	CH_3—Cl	349	CH_3—NH_2	331
		$CH_2{=}NH$	644	NH_2—Cl	251	CH_3—OH	381
				HO—Cl	251	CH_3—F	452
				F—Cl	255	CH_3—I	234
						F—I	243

Example 6.11 Estimate the standard enthalpy of formation of C_2H_5OH by means of the table of bond energies.

	Bonds	ΔH° (kJ mol^{-1})
	1 C—C	348
	5 C—H	5×413
	1 C—O	351
	1 O—H	463
		3227 kJ mol^{-1}

Structure:
$$H-\underset{\underset{H}{|}}{\overset{\overset{H}{|}}{C}}-\underset{\underset{H}{|}}{\overset{\overset{H}{|}}{C}}-O-H$$

Hence,

$$2C(g) + O(g) + 6H(g) \longrightarrow C_2H_5OH(g), \quad \Delta H^\circ(298 \text{ K}) = -3227 \text{ kJ mol}^{-1}$$

From the enthalpies of atomization in Table 6.3,

$$
\begin{aligned}
2C(\text{graphite}) &\longrightarrow 2C(g) & 2 \times 717 &= 1434 \\
\tfrac{1}{2}O_2 &\longrightarrow O & &249 \\
3H_2 &\longrightarrow 6H & 6 \times 218 &= 1308 \\
& & &\overline{2991 \text{ kJ mol}^{-1}}
\end{aligned}
$$

Therefore,

$$2C(\text{graphite}) + \tfrac{1}{2}O_2 + 3H_2 \longrightarrow C_2H_5OH(g),$$

$$\Delta H^\circ(298 \text{ K}) = -236 \text{ kJ mol}^{-1}$$

The experimental value is $\Delta H^\circ(298 \text{ K}) = -237 \text{ kJ mol}^{-1}$, so that the estimate is good to within about 0.5%, which is better than usual.

To obtain the standard ΔH°_{298} of ethanol (C_2H_5OH) we must remember that this compound is a liquid in its standard state. For $C_2H_5OH(g) = C_2H_5OH(\ell)$, $\Delta H^\circ_v(298 \text{ K}) = 43.5 \text{ kJ mol}^{-1}$. Hence the estimated $\Delta H^\circ_f(298 \text{ K})$ for ethanol(ℓ) is $-236 - 43.5 = -279.5 \text{ kJ mol}^{-1}$.

6.22 Thermochemistry and Equilibrium

One of the major problems of physical chemistry is to understand chemical equilibrium. We should like to be able to calculate the equilibrium constant of any chemical reaction from a knowledge of the properties of the reactants and products. At one time it was thought that measurement of the ΔH of a reaction would allow one to calculate its equilibrium constant. Marcelin Berthelot and Julius Thomsen, the two great thermochemists of the late nineteenth century, were inspired to carry out a vast program of thermochemical measurements by their belief that the ΔH of a reaction fixed its position of equilibrium.

We know that their idea was wrong; the existence of spontaneous endothermic reactions suffices to disprove it. Nevertheless, the equilibrium constant of a chemical reaction can indeed be calculated from calorimetric measurements on reactants and products. The value of ΔH brings us only halfway to understanding chemical equilibrium. In addition to ΔH we must know the heat capacities C_P of reactants and products over the whole range of temperature from 0 K to the temperature of the reaction.

Thus the calorimeter is still the experimental key to the thermodynamic equilibrium problem. We shall see in Chapter 7 that the energy (or enthalpy) must be combined with another important thermodynamic function, the entropy, before we can finally solve the problem of chemical equilibrium.

Problems

1. An average man produces about 10^4 kJ of heat a day through metabolic processes. If the man were a closed system of 70 kg mass with $C_P = 4.2$ kJ K^{-1} kg^{-1}, estimate his temperature rise in 1 day. The man is actually an open system. The main mechanism of heat loss is evaporation of water. How much water must be evaporated per day to maintain his body temperature constant? The ΔH(vaporization) of water is 2405 kJ kg^{-1} at 27°C.

2. A lead bullet is fired into a soft wood plank. At what speed must it be traveling to melt on impact if initial temperature is 25°C and heating of wood is neglected. The melting point of lead is 327°C, its $\Delta H_f = 5.19$ kJ mol^{-1}, and its heat capacity follows the rule of Dulong and Petit, $C_V = 3R$ per mole.

3. The density of aluminum at 20°C is 2.70 g cm^{-3} and that of the liquid at 660°C is 2.38 g cm^{-3}. Calculate the work done on surroundings when 1 kg of Al is heated under $P = 100$ kPa from 20°C to 660°C.

4. Calculate the work $-w$ done on the surroundings when 1 mole of H_2O (a) freezes at 0°C; (b) boils at 100°C.

$$\Delta H_f = 6.03 \text{ kJ mol}^{-1}, \qquad \Delta H_v = 40.67 \text{ kJ mol}^{-1}$$

$$\rho_s(0°C) = 0.9170, \qquad \rho_\ell = (0°C) = 0.9999, \qquad \rho_\ell(100°C) = 0.9583 \text{ g cm}^{-3}$$

Compare the work values to the ΔH_f and ΔH_v.

5. One kilogram of solid CO_2 is placed in a 0.1-m^3 steel cylinder with a pressure release valve. The ambient temperature is 25°C. What pressure will build up in the cylinder when all the CO_2 has vaporized? If the valve is opened to the atmosphere, what is the work done on the surroundings as the compressed gas is released?

6. One mole of ideal gas is expanded from 10 L to 100 L (a) at a constant $P = 100$ kPa; (b) in three steps: 1000 → 500 → 200 → 100 kPa; (c) isothermally and reversibly at 300 K; (d) isothermally and reversibly at 500 K. For each of these cases, calculate the work $-w$ done on the surroundings.

7. One mole of ideal gas with $C_V = \frac{3}{2}R$ is confined under a pressure of 1000 kPa at 300 K. The pressure is suddenly released to 100 kPa and the gas expands adiabatically. What is the final temperature of the gas? What is the work done on the surroundings, $-w$?

8. For argon, $C_P = 20.79$ J K^{-1} mol^{-1}.
   (a) One mole of argon is expanded isothermally and reversibly from 22.414 L to 50.000 L at 273.15 K. Calculate q, w, ΔU, and ΔH.
   (b) Calculate q, w, ΔU, ΔH, and ΔT for an adiabatic reversible expansion between the same volumes starting at 273.15 K.
   Sketch the expansions on a P–V plot.

9. 0.100 kg of CO_2 is held in an insulated container at constant volume, which has an internal electric heater. What is the ΔU of the gas when its temperature is increased from 300 K to 400 K? (See Table 6.2.)

10. Vinyl chloride can be made by the reaction

$$C_2H_2(g) + HCl = CH_2{=}CHCl(g), \qquad \Delta H^\circ(298\ K) = -100\ kJ\ mol^{-1}$$

Estimate the mass of cooling water at 15°C needed to keep the reaction vessel at 25°C per kilogram of HCl used in a stoichiometric reaction.

11. Ammonia at 25.0°C and 100 kPa flows through an insulated tube at 80 cm³ s⁻¹. It passes over a heated wire that dissipates 0.50 W and the NH_3 leaves the tube at 29.1°C. Calculate C_P and C_V per mole of NH_3.

12. WC is burned with excess O_2 in a bomb calorimeter: $WC(c) + \frac{5}{2}O_2 = WO_3(c) + CO_2$. $\Delta U^\circ(298\ K) = -1192\ kJ\ mol^{-1}$. What is $\Delta H^\circ(298\ K)$? What is $\Delta H_f^\circ(298\ K)$ of WC if the $\Delta H^\circ(298\ K)$ of combustion of pure W is $-837.5\ kJ\ mol^{-1}$? (See Table A.1.)

13. Calculate $\Delta H^\circ(298\ K)$ for the reaction $2CO_2 + 2NH_3 = 2H_2O + 2NO + C_2H_2$.

14. The Victoria Falls are 120 m high. If all the kinetic energy of the water going over the falls is converted into heat, estimate the temperature difference between the water at the top and that at the bottom of the falls. If the water power were harnessed at 100% efficiency, what flow rate (kg s⁻¹) would give 1 kW of power?

15. If $f(x, y, z, \ldots)$ is a function of $x, y, z, \ldots$, then

$$(\partial f/\partial x)_y = -(\partial f/\partial y)_x(\partial y/\partial x)_f$$

Show that for a van der Waals gas, $R\beta = \alpha(V_m - b)$, where V_m is the molar volume, α the expansivity, and β the isothermal compressibility.

16. In Problem 9 the resistance of the electric heater is 10.0 Ω. What current must be passed through the heater to raise the temperature of the 0.100 kg CO_2 from 300 K to 400 K in 1.00 min?

17. If the gas in Problem 9 is kept at constant pressure of 1.00 atm, what is ΔH when T is raised from 300 K to 400 K? What is ΔU if gas is ideal?

18. Show that for an adiabatic reversible expansion of an ideal gas from V_1 to V_2 for which C_V is independent of T, and $\gamma = C_P/C_V$,

$$(V_2/V_1)^{\gamma-1} = T_1/T_2$$

19. An ideal gas at 300 K with C_V 10.0 J K⁻¹ mol⁻¹ is expanded reversibly and adiabatically to twice its initial volume. What is its final temperature? (See Problem 18.)

20. Compute the specific energy change in J kg⁻¹ for the following processes:
(a) Dry air is heated at constant volume from 0°C to 10°C. (See Table 6.2.)
(b) The horizontal speed of a current of air is increased from 10 m s⁻¹ to 25 m s⁻¹.

21. Hydrogen is burned in a torch with a stoichiometric amount of oxygen. Calculate the maximum possible flame temperature. (See Table 6.2.)

22. The adult human brain can oxidize 10 g of glucose (or its equivalent) per hour. The ΔH_f of glucose at 37°C is $-1274\ kJ\ mol^{-1}$. Estimate the power output of the brain.

23. Show that $(\partial U/\partial P)_V = \beta C_V/\alpha$. Calculate ΔU when pressure of 1 mol N_2 is increased from 100 kPa to 1000 kPa at 298 K assuming that gas follows van der Waals equation. (See Tables 2.3 and 6.2.)

24. For N_2 at 273 K, $(C_P - C_V)/nR = 2.00$ at 2×10^4 kPa, and $\alpha = 4.73 \times 10^{-3}$ K⁻¹, $V_m = 1.16 \times 10^{-4}$ m³ mol⁻¹. Calculate the internal pressure $(\partial U/\partial V)_T$.

25. Estimate the $\Delta H_f(298\ K)$ of $C_2H_5NH_2(g)$ from the bond enthalpies. The experimental value is $-48.5\ kJ\ mol^{-1}$.

26. Plot ΔH° for the reaction $H_2 + Br_2(g) = 2HBr$ vs. T from 300 K to 1000 K.

27. For silver, $C_P(Ag) = 21.3 + 8.54 \times 10^{-3}T + 1.51 \times 10^5 T^{-2}$ (J K^{-1} mol^{-1}). If 100 g of silver at 800 K is immersed in 1 L of water at 300 K, calculate the equilibrium temperature (neglect heat loss).

28. When a nonideal gas expands at constant enthalpy, its temperature changes. The Joule–Thomson coefficient $\mu_{JT} = (\partial T / \partial P)_H$. Show that $\mu_{JT} = -(1/C_P)(\partial H / \partial P)_T$. For CO_2 at 300 K, $\mu_{JT} = 1.105 \times 10^{-5}$ K Pa^{-1} at 100 kPa and 1.084×10^{-5} kPa^{-1} at 1000 kPa. Estimate the change in molar enthalpy CO_2 on compression from 100 kPa to 1000 kPa at 300 K. (See Table 6.2.)

29. The $\Delta H°(298$ K) for combustion of cyanamide is -741 kJ mol^{-1}:

$$CH_2N_2(s) + \tfrac{3}{2}O_2(g) = CO_2(g) + H_2O(\ell) + N_2(g)$$

Calculate $\Delta H_f°(298$ K) for cyanamide. (See Table A.1.)

30. A jet engine on an aircraft takes in cold air at low velocity, transfers heat from the combustion of fuel, and expels hot gases at higher velocity. Let the inlet temperature be 200 K and the outlet 900 K. Heat is produced in the combustion chamber at 40 MW. Neglect the increase in mass flow due to the fuel supply and calculate the mass flow of gas through the engine. Assume that $C_P = 1.0$ kJ K^{-1} kg^{-1} for the gases.

7

The Second and Third Laws of Thermodynamics—Entropy

The First Law of Thermodynamics tells us that energy can be changed from one form to another but can be neither created nor destroyed in any process. This law, however, tells us nothing about the direction a process will take, yet we know that there are particular directions in which natural processes occur. Heat flows from hotter to colder bodies, gases mix by interdiffusion, and chemical reactions proceed from reactant mixtures to well-defined products. For example, if we have two blocks of metal at different temperatures and we place them in contact, we know that heat will flow from the block at higher temperature to the one at lower temperature, until ultimately the temperature of both blocks is the same. All we can say from the First Law is that the energy lost as heat by the hot block is the same as the energy gained by the cold block, assuming that the system is insulated from its surroundings. It would not in any way contradict the First Law if heat flowed from the colder to the hotter body, provided that energy was conserved in the process. Yet we know that such a process never occurs without outside intervention.

If we wish to understand the direction of physical and chemical processes, we must have a thermodynamic function that allows us to predict the direction of the change that can be expected if we specify the initial conditions of a system and the constraints imposed on it. The energy function cannot help us, since it always remains constant in any isolated system or in the universe as a whole. Isolated systems change spontaneously from nonequilibrium states toward equilibrium states. We want a function that changes when the system changes and stays constant when the system rests at equilibrium. The Greek for "in change" is *en tropos*. The function we want was discovered by Rudolf Clausius in 1850, and he called it the *entropy*. The energy function U is defined by the First Law of Thermodynamics. The entropy function S is defined by the Second Law of Thermodynamics.

This chapter introduces the concept of entropy and shows how to measure it quantitatively by calorimetric determination of heat capacity over a range of temperature extending down close to absolute zero. The molecular interpretation of entropy is described, based on the original work of Boltzmann. Chapter 8 will show how entropy data can be combined with energy data to calculate the equilibrium constants of chemical reactions.

7.1 Entropy and Reversible Heat

The processes of change that appear to occur in the world can be classified as reversible or irreversible. Suppose that a motion picture is taken of a process and the film is run in reverse. If the process we see when the film is run in reverse also actually occurs in the world, then the original process was a reversible process. If, however, the backward movie does not depict an actual process that occurs in the world, then the original process was irreversible. For example, the oscillation of a pendulum (free of friction) or the vibration of a molecule is a reversible process. The fall of a teacup and its smashing to bits on the kitchen floor is an irreversible process. A vibrating molecule is a microscopic system—there are no real reversible processes in the macroscopic world. Nevertheless, we can make calculations for reversible processes if we set them up as limiting cases of real processes in which the system never departs appreciably from a succession of equilibrium states.

One mathematical expression for the First Law of Thermodynamics was given in Eq. (6.11), $dU = dq + dw$. This equation holds for any transfer, reversible or irreversible, of heat or work to a closed system. If, however, the transfers are reversible, we can write

$$dU = dq_{rev} + dw_{rev} \tag{7.1}$$

In Section 6.3 we found that the element of reversible PV work is

$$dw_{rev} = -P\,dV \tag{7.2}$$

The differential of reversible work is the product of an intensity factor P and the differential dV of a capacity factor V that is a state function. We can write Eq. (7.2) as

$$\frac{dw_{rev}}{-P} = dV \tag{7.3}$$

Division of dw_{rev}, which is not an exact differential, by $-P$, an intensive state function, gives dV, which is the exact differential of an extensive state function, the volume V. Thus $(-1/P)$ is called an *integrating factor* for the element of reversible work. From Eqs. (7.1) and (7.2),

$$dU = -P\,dV + dq_{rev} \tag{7.4}$$

What about the element of reversible heat dq_{rev}? Can we also find an integrating factor for it? The intensity factor for the transfer of heat is obviously the temperature T. Thus by analogy with Eq. (7.3), let us write

$$\frac{dq_{rev}}{T} = dS \qquad (7.5)$$

We now state that dS is the exact differential we are seeking and that S is a state function, the *entropy*. The integrating factor for the reversible heat is $1/T$. Equation (7.5) also *defines* T as the *thermodynamic temperature*.

Note carefully that we have not *derived* Eq. (7.5). The existence of a state function S defined by Eq. (7.5) is a *postulate*. The definition of a new state function, the entropy, in Eq. (7.5) is the most basic statement of the Second Law of Thermodynamics. Just as the First Law defined a state function U through its exact differential $dU = dq + dw$, so the Second Law defines a state function S through its exact differential $dS = dq_{rev}/T$. Note especially the important fact that an entropy change must be calculated from a *reversible heat transfer*. We cannot write simply $dS = dq/T$ but must specify dq_{rev}. Many alternative statements of the Second Law can be derived from this simple but powerful equation. In particular, we shall see how it governs the directions of chemical reactions, and indeed of all changes in the universe.

In terms of dS, Eq. (7.1) can be now written

$$dU = -P\,dV + T\,dS \qquad (7.6)$$

This is the most important single equation of thermodynamics. It combines in one expression the First and Second Laws. We shall be able to derive many useful equations from Eq. (7.6). All these derived equations give results that are confirmed by experimental measurements, and none of the equations has ever made a false prediction.

7.2 Molecular Picture of Heat and Work

In Section 5.6 we saw that when heat is transferred to a system and the temperature rises, the molecules redistribute themselves among their allowed energy levels, in accord with the Boltzmann distribution. The energy levels themselves do not change, but the chances of finding molecules in various higher levels are increased.

When work is done on a system, on the other hand, the actual energy levels themselves change. For a gas this effect can be seen in the translational energy levels as given by Eq. (4.9), $\epsilon_k = (h^2/8mV^{2/3})(n_1^2 + n_2^2 + n_3^2)$. As work is done on a gas, it is compressed, its volume is decreased, and in accord with Eq. (4.9) all its energy levels are raised. With liquids and solids, the effect of compression is also to raise the energy levels, but the detailed theory is more difficult.

Figure 7.1 summarizes this molecular interpretation of heat and work.

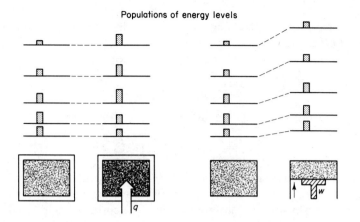

Populations of energy levels

FIGURE 7.1 Molecular pictures of heat q and of work w.

7.3 Entropy Changes of an Ideal Gas

Before proceeding to more general applications of the entropy function, we shall calculate the change in entropy ΔS for changes in volume and in temperature of an ideal gas. These calculations provide simple practical examples of the measurement of entropy changes.

From Eqs. (7.6) and (6.12),

$$dU = \left(\frac{\partial U}{\partial T}\right)_V dT + \left(\frac{\partial U}{\partial V}\right)_T dV = -P\,dV + T\,dS \qquad (7.7)$$

At constant temperature, $dT = 0$ and $(\partial U/\partial V)_T\,dV = -P\,dV + T\,dS$. For an ideal gas, $(\partial U/\partial V)_T = 0$, and thus

$$dS = \frac{P\,dV}{T} = nR\frac{dV}{V} \qquad \text{(ideal gas)} \qquad (7.8)$$

For a change in volume from V_1 to V_2 integration of Eq. (7.8) gives

$$\Delta S = \int_{V_1}^{V_2} nR\frac{dV}{V} = nR\ln\frac{V_2}{V_1} \qquad \text{(ideal gas)} \qquad (7.9)$$

Example 7.1 Calculate the increase in entropy of 1 mol of an ideal gas when it expands from 100 L to 200 L.

From Eq. (7.9), $\Delta S = (1\text{ mol})(8.314\text{ J K}^{-1}\text{ mol}^{-1})\ln\dfrac{200}{100}$

$$= 5.76\text{ J K}^{-1}$$

Note that the dimensions of entropy are those of energy/temperature and the SI unit is J K^{-1}. The dimensions and SI units of entropy are the same as those of heat capacity, but this fact does *not* imply that they are the same physical quantities.

To calculate ΔS for a change in temperature at constant volume, we set $dV = 0$ in Eq. (7.6):

$$dU = \left(\frac{\partial U}{\partial T}\right)_V dT = C_V\, dT = T\, dS$$

$$dS = C_V\, dT/T = C_V\, d\ln T \tag{7.10}$$

For a change from T_1 to T_2, integration of Eq. (7.10) gives

$$\Delta S = \int_{T_1}^{T_2} C_V \frac{dT}{T} = \int_{T_1}^{T_2} C_V\, d\ln T \tag{7.11}$$

This equation is valid for a change in temperature of any substance at constant volume. It is not restricted to ideal gases since in its derivation no use is made of ideal-gas properties. For a change at constant pressure, the corresponding equation is

$$\Delta S = \int_{T_1}^{T_2} C_P \frac{dT}{T} = \int_{T_1}^{T_2} C_P\, d\ln T \tag{7.12}$$

To perform the integration in Eq. (7.11) we must know C_V as a function of T. A simple case occurs if C_V is effectively constant over the temperature range from T_1 to T_2.

Example 7.2 What is the increase in entropy of 1 mole of argon heated from 300 K to 600 K at constant volume?

Argon is a monatomic gas with $C_V = n(\frac{3}{2}R)$, independent of T. From Eq. (7.11),

$$\Delta S = (1 \text{ mol} \times \tfrac{3}{2} \times 8.314 \text{ J K}^{-1} \text{ mol}^{-1}) \ln \frac{600}{300} = 8.64 \text{ J K}^{-1}$$

By comparison with Example 7.1 we see that, for the same amount of ideal gas, doubling T increases S by somewhat more than doubling V.

Examples 7.1 and 7.2 illustrate the fact that S is a state function. Expressions for ΔS were derived in terms of the values of other state functions such as T and V in the initial and final states of the system. The equation (7.7) that was used to evaluate ΔS was derived for a reversible process between initial and final states, but once the result for ΔS is obtained, it depends only on the initial and final states and in no way on the particular path between these states.

When the volume of a gas increases, the molecules of the gas become less localized or restricted in space, and the entropy of the gas increases. When the temperature of a gas increases, the molecules of the gas become less restricted in their velocities (or momenta) and the entropy of the gas increases.

7.4   Entropy in Changes of State

If we wish to consider changes in entropy at constant pressure, Eq. (7.6) should be transformed so as to substitute the enthalpy H for the internal energy U. Since $H = U + PV$, taking the complete differential gives $dH = dU + P\, dV + V\, dP$.

Adding this equation to Eq. (7.6) gives

$$dH = V\,dP + T\,dS \qquad (7.13)$$

This equation can be applied to calculate the change in entropy of a substance when it undergoes a change in phase. For example, consider in Fig. 7.2 a pure liquid in equilibrium with its vapor at some temperature T_b, the boiling point, and pressure P. If P is the standard atmospheric pressure $P^\circ = 101.32$ kPa, T is the *normal boiling point* of the liquid T_b°. When energy is added to the system through the heating coil while the pressure is maintained constant, some of the liquid is converted into vapor. At constant pressure, Eq. (7.13) becomes $dH = T\,dS$, or $dS = dH/T$. As a finite amount of energy is added to the system at constant pressure, liquid is converted to gas as the temperature remains constant at T_b. The equation for dS can be integrated, to give

$$S^g - S^\ell = \Delta S_v = \frac{\Delta H_v}{T_b} = \frac{H^g - H^\ell}{T_b} \qquad (7.14)$$

where H^g and H^ℓ are the enthalpies of gas and liquid, respectively. The difference, $H^g - H^\ell$, is the *enthalpy of vaporization* ΔH_v of the liquid. Since liquid and vapor are always at equilibrium at temperature T_b, ΔH_v is the reversible heat transferred to the system.

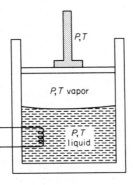

FIGURE 7.2 A liquid and its vapor in equilibrium at P and T. Reversible transfer of heat to the system at equilibrium causes transformation of liquid to vapor.

By exactly similar reasoning, for the fusion of a solid at its melting point T_f, the entropy of fusion is

$$\Delta S_f = \Delta H_f / T_f \qquad (7.15)$$

Solid and liquid are in equilibrium at the melting point, so that ΔH_f is the reversible heat added to the system at T_f.

Example 7.3 The enthalpy of vaporization of ethanol is 43.5 kJ mol^{-1} at its normal boiling point of 351.5 K. What is the entropy of vaporization per mole of ethanol at 351.5 K?

From Eq. (7.14),

$$\Delta S_v = \frac{43\,500 \text{ J mol}^{-1}}{351.5 \text{ K}} = 124 \text{ J K}^{-1} \text{ mol}^{-1}$$

Note that ΔS_v is considerably greater than ΔS_f. This result is typical. The
increase in molar entropy of a substance is much greater when the liquid vaporizes
than when the solid melts. Our evidence that entropy increases as a system becomes
less restricted is borne out by these results for ΔS_v and ΔS_f. When a solid melts, its
atoms, ions, or molecules become less restricted and more disordered in their locations and motions, and the entropy increases. When a liquid vaporizes, the molecules find much greater freedom of motion in the gaseous state, and correspondingly
a large increase in entropy occurs on vaporization.

An interesting general rule about ΔS_v was given by Trouton in 1884. For many
liquids ΔS_v, when calculated at the normal boiling point as $\Delta H_v/T_b^\circ$, is in the range
90 to 100 J K⁻¹ mol⁻¹. There are exceptions (e.g., water, molten salts), but the rule is
still quite useful. If you know the normal boiling point of a liquid, you can estimate
its ΔH_v to be about 100 T_b° joules. The ΔH_v of a liquid is quite a strong function of the
temperature, decreasing with rise in T and going to zero at the critical temperature T_c.

7.5 ΔS for an Irreversible Process: Heat Conduction

We have a prescription for calculating ΔS for a change via a reversible process,
Eq. (7.5), but how do we calculate ΔS for a change brought about by an irreversible
process? We remember that S is a state function, and thus for a change $A \rightarrow B$,
$\Delta S = S_B - S_A$ depends only on the final and initial states and not on the path between
them. Therefore, we devise a reversible path between A and B and calculate ΔS along
this path.

An example of an irreversible process is the transfer of heat from a warmer to
a colder body. We can use an ideal gas to carry out the transfer reversibly, and thereby
calculate the entropy change. The reversible process is summarized in Fig. 7.3a.

1. The gas is placed in thermal contact with the warm reservoir at T_2 and expanded
   reversibly and isothermally until it takes up heat equal to q. To simplify the
   argument, we assume that the reservoirs have heat capacities so large that
   changes in their temperatures on adding or withdrawing heat are negligible.
2. The gas is then removed from contact with the hot reservoir and allowed to
   expand reversibly and adiabatically until its temperature falls to T_1.
3. Next, it is placed in contact with the colder reservoir at T_1 and compressed
   isothermally until it gives up heat equal to q.

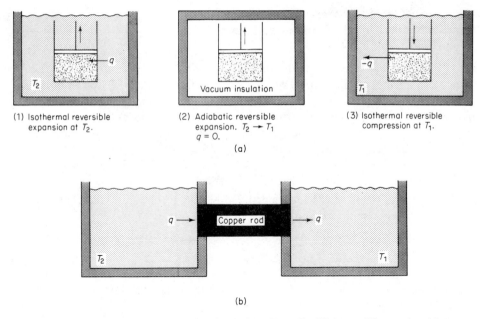

(1) Isothermal reversible expansion at T_2.

(2) Adiabatic reversible expansion. $T_2 \rightarrow T_1$ $q = 0$.

(3) Isothermal reversible compression at T_1.

(a)

(b)

FIGURE 7.3 (a) Reversible transfer of heat from T_2 to T_1. (b) Irreversible transfer of heat from T_2 to T_1.

The hot reservoir has now lost entropy equal to q/T_2, and the cold reservoir has gained entropy equal to q/T_1. The net entropy change of the reservoirs has therefore been $\Delta S = q/T_1 - q/T_2$. Since $T_2 > T_1$, $\Delta S > 0$, and their entropy has *increased*. The entropy of the ideal gas, however, has *decreased* by an exactly equal amount $(q/T_2 - q/T_1)$, so that for the entire isolated system of ideal gas plus heat reservoirs, $\Delta S = 0$ for the reversible process. If the heat transfer is carried out irreversibly, as in Fig. 7.3(b), by placing the two reservoirs in thermal contact and allowing heat q to flow along the finite temperature gradient thus established, there is no compensating entropy decrease in the surroundings. The entropy of the entire isolated system (or indeed the entropy of the universe) would increase during the irreversible process, by the amount $\Delta S = q/T_1 - q/T_2$.

Example 7.5 A huge copper block at 1000 K is joined to another huge copper block at 500 K by a copper rod. The rate of heat conduction is 10^4 J s^{-1}. What is the rate of entropy increase of the universe due to this process?

By carrying out the heat transfer reversibly as shown in Fig. 7.3(a), we can calculate that

$$\Delta S = q/T_1 - q/T_2 = 10^4 \text{ J s}^{-1}(1/500 \text{ K} - 1/1000 \text{ K}) = 10 \text{ J K}^{-1} \text{ s}^{-1}$$

Thus the entropy of the universe increases by 10 J K^{-1} per second.

7.6 A Supercooled Liquid Freezes—
What Is ΔS of the Liquid and of the Universe?

It is possible to cool liquids well below their normal freezing points. Let us imagine that 1 mol of supercooled water is kept in a refrigerator at $-10°C$ and 101 kPa but one day it suddenly freezes. What is ΔS for the change, $H_2O(\ell, P°, 263\text{ K}) \rightarrow H_2O(\text{ice}, P°, 263\text{ K})$? This is an irreversible process: It is impossible to remelt the ice at 263 K and 101 kPa. To calculate ΔS for the change of state, we devise the following reversible path between the initial and final states:

1. Warm the water from 263 K to 273 K, at constant pressure $P°$.
2. Freeze the water at 273 K to ice at 273 K at $P°$. This is a phase change under equilibrium conditions and hence a reversible process.
3. Cool the ice from 273 K to 263 K at constant pressure $P°$.

Example 7.6 Carry out the calculation of ΔS for the water and for the universe, given $C_P(\text{water}) = 75.3$ J K^{-1} mol^{-1}, $C_P(\text{ice}) = 36.9$ J K^{-1} mol^{-1} and $\Delta H_f(273\text{ K}, P°) = 5950$ J mol^{-1}.

The processes are as follows:

$$\text{water(263 K)} \xrightarrow{\text{(irreversible)}} \text{ice(263 K)}$$

$$\downarrow \text{(reversible)} \qquad\qquad \uparrow \text{(reversible)}$$

$$\text{water(273 K)} \xrightarrow{\text{(reversible)}} \text{ice(273 K)}$$

From Eqs. (7.12) and (7.15),

$$\Delta S = \int_{263}^{273} C_P(\ell)\frac{dT}{T} + \frac{\Delta H_f}{T_f} + \int_{273}^{263} C_P(\text{ice})\frac{dT}{T}$$

$$= 75.3 \ln\frac{273}{263} + \frac{5950}{273} + 36.9 \ln\frac{263}{273}$$

$$= 2.81 - 21.80 - 1.38 = -20.4 \text{ J K}^{-1} \text{ mol}^{-1}$$

This is the decrease in entropy of the supercooled water on freezing to ice at 263 K. To find the ΔS of the surroundings, we can assume that the large heat reservoir of the surroundings remains at constant T during the essentially reversible transfer to it of ΔH_f at 263 K. We calculate $\Delta H_f(263\text{ K})$ from the Kirchhoff equation (since the pressure is constant at $P°$):

$$d(\Delta H_f)/dT = \Delta C_P = C_P(\ell) - C_P(\text{ice}) = 38.4 \text{ J K}^{-1} \text{ mol}^{-1}$$

$$\Delta H_f(273\text{ K}) - \Delta H_f(263\text{ K}) = 38.4(273 - 263) = 384 \text{ J mol}^{-1}$$

Hence, $\Delta H_f(263) = 5950 - 384 = 5566$ J mol and $\Delta S_f = 5566/263 = 21.2$ J K^{-1} mol^{-1}. The ΔS of the universe is $21.2 - 20.4 = +0.8$ J K^{-1} mol^{-1}. The entropy of the universe has increased during the irreversible freezing of the supercooled water.

We have given two quantitative examples of a general law that can be derived from the Second Law of Thermodynamics as stated in Eq. (7.5): *In any irreversible process, the entropy of the universe always increases.* The entropy function S thus governs the direction of changes in the universe.

7.7 Heat Engines

The development of thermodynamics during the nineteenth century was closely linked to the industrial and engineering problems of steam engines. As Lloyd Henderson once remarked: "Science owes more to the steam engine than the steam engine owes to science." In particular, it owes to the steam engine the Second Law of Thermodynamics and the discovery of the entropy function.

A steam engine operates essentially as follows. A fire is used to heat the *working substance* steam, causing it to expand through a valve into a cylinder fitted with a piston. The expansion drives the piston forward, and, by suitable coupling, work can be obtained from the engine. The steam, which has been cooled by the expansion, is withdrawn from the cylinder through a valve. A flywheel returns the piston to its original position, in readiness for another expansion stroke. The great advance made by James Watt in 1769 was to provide a second chamber to receive the expanded working substance so that heat could be recovered from it.

The steam engine is an example of the general class of heat engines. In simplest terms, any heat engine withdraws heat q_2 from a source at higher temperature T_2, converts some of this heat into work w, and discards the remaining heat q_1 to a sink at lower temperature T_1. In practice, frictional losses of work occur in moving parts of the engine. A schematic outline of a heat engine is shown in Fig. 7.4.

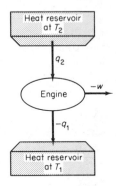

FIGURE 7.4 The basic components of a heat engine. The system to which the heat and work terms refer is the engine.

The basic theoretical analysis of the heat engine was published in 1824 in a work of remarkable originality by a young French engineer Sadi Carnot, "Reflections on the Motive Power of Fire." At that time it was believed that heat was the result of the presence in a substance of a weightless element called *caloric.* Carnot understood perfectly that work can be obtained from a heat engine only when heat is transferred from a hot source to a cold sink. He saw in his mind the caloric flowing like a fluid from source to sink, and he compared it to water flowing over a waterfall. Just as water in its fall can turn a millwheel and yield useful work, so caloric in its fall can yield work in the steam engine. Carnot reasoned that neither the water in the mill-stream nor the caloric in the engine would be lost. Therefore, his only mistake was to set the caloric taken from the hot source equal to the caloric added to the cold sink. So persuasive was this model that even William Thomson (Kelvin), the greatest thermodynamicist of the nineteenth century, in 1849 needed several more years of intensive work on the theory of heat engines before he could shake off the concept of an indestructible caloric. If you have difficulties today with thermodynamics, you may be consoled by the intellectual struggles of its pioneers.

7.8 The Carnot Cycle

Carnot introduced the important idea of analyzing the operation of an engine by means of a cyclic process, in which the working substance was returned to exactly its original state. In this way the relation of heat to work in operating the engine can be obtained without complications due to other changes in the system. In what follows, the signs of heat and work terms are given with reference to the working substance. Since at the end of a cycle it is in the same state as at the beginning, for the working substance $\Delta U = 0$ for each cycle. Hence $\Delta U = 0 = q_2 + q_1 + w$,

$$-w = q_2 + q_1 \tag{7.16}$$

The efficiency η of the heat engine is defined as the ratio of work output to heat taken by the working substance from the fire:

$$\eta = \frac{-w}{q_2} = \frac{q_2 + q_1}{q_2} \tag{7.17}$$

The Carnot cycle consists of the following four steps:

1. Place the working substance (e.g., steam) in contact with the hot reservoir at T_2 and allow it to expand isothermally and reversibly until it has absorbed heat q_2. Work during expansion is w_1.
2. Remove the working substance from contact with the hot reservoir, insulate it adiabatically, and allow it to expand reversibly until its temperature falls to T_1. In this step $q = 0$, work is w_2.
3. Place the working substance in contact with the cold reservoir at T_1 and compress it isothermally and reversibly. Heat $-q_1$ is given up to the cold reservoir; work is w_3.
4. Remove the working substance from the cold reservoir, insulate it adiabatically, and compress it reversibly until it reaches its original state to complete the cycle. Heat $q = 0$; work is w_4.

The steps in the Carnot cycle described above are shown on a P–V indicator diagram in Fig. 7.5(a). In carrying out the cycle it is necessary to stop the isothermal compression in step 3 at exactly the right state to allow a final adiabatic compression to restore the original state. The area of the P–V cycle measures the work $-w$ done in one cycle of operation by the heat engine.

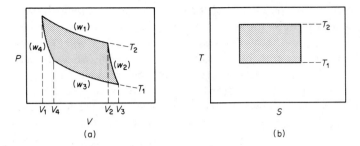

FIGURE 7.5 (a) Carnot cycle on *P–V* diagram. (b) Carnot cycle on *T–S* diagram.

In Fig. 7.5(b) the cycle is shown on a *T–S* diagram. The isotherms are straight lines of constant *T*. From Eq. (7.5) when $dq_{rev} = 0$, $dS = 0$ or $S =$ const.; therefore, the reversible adiabatics are straight lines of constant *S*. The area of the *T–S* cycle measures the heat added to the working substance.

For the cyclic process, since *S* is a state function, $\Delta S = 0$ or $(S_2 - S_1) + (S_1 - S_2) = 0$. From Eq. (7.5), however,

$$\text{At } T_2, \quad S_2 - S_1 = q_2/T_2; \quad \text{at } T_1, \quad S_1 - S_2 = q_1/T_1$$

so that

$$\frac{q_2}{T_2} + \frac{q_1}{T_1} = 0 \qquad (7.18)$$

The efficiency of the cycle, from Eq. (7.17) and (7.18), becomes

$$\eta = \frac{T_2 - T_1}{T_2} \qquad (7.19)$$

This is the famous *Carnot Theorem* for the efficiency of heat engines. The efficiency depends only on the two temperatures and not on the choice of working substance. No other engine operating between the same two temperatures can possibly have any higher efficiency. If in Eq. (7.19) $\eta > (T_2 - T_1)/T_2$, we could show by working backward that $\Delta S < 0$ for the cyclic process, which would contradict the basic postulate that *S* is a state function.

In this discussion we have reversed the historical development. The entropy function *S* was not used by Carnot, but was introduced later by Clausius. Its use, however, simplifies the discussion of the Carnot cycle and leads directly to the basic theorem of Eq. (7.19).

Example 7.7 If the heat source is at 500 K and the heat sink at 300 K, what is the maximum thermal efficiency of the heat engine?

From Eq. (7.19), $\eta = (500 - 300)/500 = 0.40$.

The thermodynamic efficiency in Eq. (7.19) is the maximum possible, since all the steps in operation of the engine are reversible. In practice, the efficiency of conversion of heat into electric energy in a steam turbine can achieve about 85% of the theoretical thermodynamic maximum. One of the problems with current designs of nuclear-fission power plants is that the reactors are cooled with liquid water and the operating temperature is too low to achieve a thermodynamic efficiency better than about 32%. Reactors that use helium gas as coolant can operate at higher temperatures with efficiencies about 39%. Because of their low efficiencies, liquid-water reactors produce about 40% more waste heat (q_1) than modern fossil-fuel power stations. Since the waste heat is usually discharged into rivers or oceans, ecological problems may result. In some places, however, like the icy waters off the Maine coast where even lobsters feel too cold, some local warming of the sea may not be a disadvantage.

7.9   Can We Use the Energy of the Oceans?

A man takes his motor boat out to sea on a fishing trip. He runs out of fuel and has forgotten to bring a spare can. If he is lucky, he may be rescued by another boat; otherwise, he may never get back. His boat is surrounded by the ocean, containing a vast amount of thermal energy. Yet this energy is of no use to the boatsman. He cannot utilize any of the vast energy of the ocean because it is at a uniform temperature, and hence from Carnot's Theorem (7.19) he cannot convert it to work.

Kelvin's favorite statement of the Second Law of Thermodynamics was: "It is impossible by any cyclic process, no matter how idealized, to transform heat into work, without transferring a certain amount of heat from a hotter to a colder reservoir." The First Law of Thermodynamics is no threat to the drifting fisherman, but the Second Law of Thermodynamics could prove fatal to him.

Actually, however, there is marked temperature difference in the tropics between the surface of the ocean (about 25°C) and a depth of 300 m (about 5°C). Experimental power plants have been designed to tap this enormous source of useful energy. A pilot plant operating off the coast of Hawaii is outlined in Fig. 7.6. The maximum thermal efficiency between 25°C and 5°C from Eq. (7.19) is only

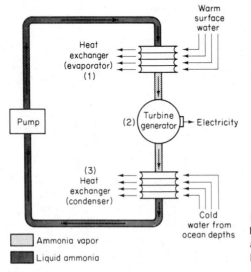

FIGURE 7.6   A heat engine operating between T_2 at surface of ocean and T_1 at a depth of 300 m. Pilot plant designed by Lockheed Company.

$$\eta = \frac{T_2 - T_1}{T_2} = \frac{20}{298} = 0.067$$

but there is no cost for fuel. Clearly, thermodynamics alone cannot decide between various methods of power generation. Each method has its geographical, technological, economic, and political factors.

7.10 Entropy and the Arrow of Time

In an isolated system the energy U is constant, and any process that occurs spontaneously must increase the entropy S of the system. Thus gases mix by interdiffusion, heat flows from hotter to colder bodies, and so on. Of course, any *part* of the system can decrease in entropy, provided that the rest of the system increases by a greater amount. Overall, however, the S of the system can never decrease. It can only stay constant or increase. Thus for the isolated system,

$$\Delta S \geq 0 \tag{7.20}$$

The equality sign applies to the equilibrium condition of the system. For any spontaneous process that actually occurs in the system, the entropy must increase, $\Delta S > 0$. We saw a numerical example of this law in the freezing of supercooled water in Section 7.6. In that case the isolated system was the water plus the surrounding refrigerator.

In isolated systems, entropy always tends to increase. The only truly isolated system is the entire universe. Clausius was thus led to his famous statement of the First and Second Laws of Thermodynamics: *The energy of the universe is constant; the entropy of the universe strives always toward a maximum.* The First Law of Thermodynamics is a conservation law and therefore a symmetry law. The Second Law is not a symmetry law; in fact, it is a principle that breaks the symmetry of the universe by indicating a definite direction of change.

Arthur Eddington proposed that "entropy is the arrow of time." If we examine the entropy of the universe at two different times, the point of higher entropy corresponds to the point of later time. The direction of time is governed by the increase in the entropy of the universe. Yet this argument is somewhat circular. We must have some physical measurement on the universe that changes with time, other than its entropy, before we can say that increasing entropy measures increasing time. Modern cosmology supports the theory that the universe has been expanding uniformly since the moment of its creation. Thus distant galaxies are receding rapidly from any point of observation. The entropy of the universe therefore is increasing as the universe expands.

The entropies of so-called isolated systems, such as we study in the laboratory, also increase as they tend toward equilibrium. Consequently, they are not really "isolated" from the universe. The fact that any change in such an isolated system increases its entropy can not be separated from the fact that the universe is expanding. We might go so far as to say that the bits of our smashed teacup never spontaneously come together and leap back onto the table because the universe is expanding and time is unidirectional. [Can we imagine a universe in which entropy is decreasing?]

Figure 7.7 represents two different gases in two separate containers with a removable barrier between them. The gases are assumed to be ideal. They are kept at some constant T and P. For example, one gas might be N_2 and the other O_2. The gas molecules are denoted by small black or white circles, but of course in an actual gas the number of molecules would be enormously higher.

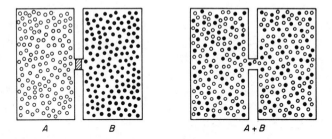

A   B   A + B

FIGURE 7.7 A picture of the increase in disorder on mixing two gases by diffusion, which results in an increase in entropy.

If the barrier is removed, the gases will mix by diffusion. Since the gases are ideal, there are no intermolecular forces between them. The driving force that causes the gases to mix can certainly not be any redistribution of the energy of the universe. The driving force is entirely due to the increase in entropy as the gases mix. It is the same driving force that causes an ideal gas to expand into a larger volume, as in Eq. (7.9).

Suppose that we mix an amount n_A moles of gas A with an amount n_B of gas B at constant pressure and temperature. The ΔS of mixing is equal to the ΔS of expanding each gas from its initial pressure P to its partial pressure P_A or P_B in the mixture. From Eq. (7.9), for an expansion,

$$\Delta S = nR \ln \frac{V_2}{V_1} = nR \ln \frac{P_1}{P_2}$$

In the case of mixing,

$$\Delta S = n_A R \ln \frac{P}{P_A} + n_B R \ln \frac{P}{P_B}$$

From Eq. (2.8), we set $P_A = X_A P$ and $P_B = X_B P$, where X_A and X_B are the mole fractions, to give

$$\Delta S = n_A R \ln \frac{1}{X_A} + n_B R \ln \frac{1}{X_B}$$

or

$$\Delta S = -n_A R \ln X_A - n_B R \ln X_B$$

Dividing both sides by $n_A + n_B$, we obtain the ΔS per unit amount of substance,

$$\frac{\Delta S}{n_A + n_B} = -RX_A \ln X_A - RX_B \ln X_B \qquad (7.21)$$

For any number c of components, this result would become

$$\Delta S / \sum n_j = -R \sum_{j=1}^{c} X_j \ln X_j \qquad (7.22)$$

This expression gives the ΔS of mixing per unit amount of substance (SI units: $J\ K^{-1}\ mol^{-1}$).

Although the expression (7.22) was derived for ideal gases, it also is valid for liquid and solid solutions in which the intermolecular forces between the different components are uniform. In particular, the ideal entropy of mixing will always be valid for mixtures of isotopic species.

Example 7.8 The isotopic composition of natural lead in atom percent is ^{204}Pb, 1.5%; ^{206}Pb, 23.6%; ^{207}Pb, 22.6%; ^{208}Pb, 52.3%. Calculate the entropy of mixing of the isotopes per mole of natural Pb.

Atom percent is simply $100X$, so that Eq. (7.22) gives

$$\Delta S/1 = -R(0.015 \ln 0.015 + 0.236 \ln 0.236 + 0.226 \ln 0.226 + 0.523 \ln 0.523)$$

$$= -R(-0.063 - 0.341 - 0.336 - 0.339)$$

$$= -8.314(-1.079) = 8.97 \ J\ K^{-1}\ mol^{-1}$$

7.12 Probability of Mixtures

To obtain a deeper insight into the increase in entropy when two substances mix by interdiffusion, let us consider a mixing process that involves a much smaller number of particles. Consider two adjoining vessels of equal volume separated by a barrier, with 10 black balls in one vessel and 10 white balls in the other one. Suppose that we remove the barrier and shake the vessels thoroughly. The result will be a mixture of white and black balls randomly distributed between the two vessels. Experience has taught us that we might shake the containers many times but the likelihood of unmixing the balls and regaining the original ordered arrangement is very small. Like the interdiffusion of two gases, the mixing of the balls appears to be an essentially irreversible process.

The reason why the mixed arrangement persists is that there are very many ways to achieve a mixture whereas there is only one way to achieve the completely separated state. An analogy that will appeal to card players is the occurrence of 13 hearts in a bridge hand. Most bridge players have never been dealt a hand of 13 hearts, but this hand is not any more or less probable than any other *specified* hand of 13 cards. There are, however, so many possible hands not containing 13 hearts that it is extremely unlikely in random dealings that a hand containing 13 hearts will occur. (The actual probability is once in $52!/39!\ 13! \approx 6.35 \times 10^{11}$ deals.)

We start with N black balls and N white balls. The probability of any arrangement that places B black balls and $W = N - B$ white balls in one vessel is the probability of one particular example of such an arrangement $p_i(B, W)$ times the number of ways g_i of assigning the balls that lead to this arrangement. Remember that the arrangement in the right container is a sort of "mirror image" of that in the left: If the left contains 6 black and 4 white, the right must contain 4 black and 6 white. Thus it will suffice to consider one container only, since the other one must adjust automatically. Suppose that we consider the balls in pairs, therefore, one in left and one in right vessel. The chance that the first ball we look at on the left is black is $\frac{1}{2}$. If it is black, we assign a white ball on the right. The chance that the second ball on the left is white is also $\frac{1}{2}$; we assign it to white and add a black to the right vessel. We can see that any particular sequence of black and white balls in the left container will have a probability

$$p_i(B, W) = \left(\frac{1}{2}\right)^N$$

The statistical weight of this arrangement will be the number of ways of achieving it (i.e., the total number of permutations of the N balls divided by the permutation of the blacks and whites); this factor is

$$g_i = \frac{N!}{B!W!} = \frac{N!}{B!(N-B)!}$$

The probability of any arrangement is then

$$p(B, W) = g_i p_i = \frac{N!}{B!(N-B)!}\left(\frac{1}{2}\right)^N \tag{7.23}$$

From Eq. (7.23), the probability of placing 10 blacks on the left and 10 whites on the right is

$$p(10, 0) = \frac{10!}{10!\,0!}\left(\frac{1}{2}\right)^{10} = \left(\frac{1}{2}\right)^{10}$$

Since any other particular sequence of 10 choices has the same probability, the total number of possible arrangements is $1/p(10, 0) = 2^{10} = 1024$.

The probability of the mixed configurations with $W = 5$, $B = 5$ is

$$p(5, 5) = \frac{10!}{5!\,5!}\left(\frac{1}{2}\right)^{10} = 252\left(\frac{1}{2}\right)^{10} = \frac{252}{1024} = 0.246$$

There are 252 ways of distributing 5 blacks and 5 whites on each side, compared to only one way of distributing 10 whites in the left container.

Twenty balls are not very many. With 200 balls, the total number of possible arrangements would be $1/(\frac{1}{2})^{100} = 1.268 \times 10^{30}$. Only one of these would place 100 white balls in the left container. The number of arrangements with 50 whites and 50 blacks in each container would be

$$g(50, 50) = \frac{100!}{50!\,50!} = 3.70 \times 10^{27}$$

There are a lot of partially mixed states that are also important and will occur frequently, for example $p(51, 49)$, $p(49, 51)$, etc.

If we considered a mole of N_2 mixing with a mole of O_2, there would be $6.022 \times 10^{23} = L$ molecules of each gas to distribute between the two containers. Probabilities like $p(L/2, L/2)$ become huge compared to $p(L, 0)$.

A convenient formula was derived by Stirling for the factorials of large numbers. This was used to evaluate 100! in the example above.

$$\ln N! \approx N \ln N - N$$

or $$N! \approx (N/e)^N \qquad (7.24)$$

7.13 Disorder, Probability, and Entropy

We have seen that when the barrier is removed between the two containers in Fig. 7.7, the system will spontaneously change into a disordered state since disordered states have a much higher probability than unmixed ordered states. The increase in entropy that occurs when two gases mix by interdiffusion is simply the movement by the system from a state of lower probability to a state of higher probability. The result of the mixing is as follows:

1. Decrease in order
2. Increase in randomness or disorder
3. Increase in probability
4. Increase in entropy

A quantitative relation between entropy S and probability p can be obtained from the fact that entropy is an additive function, whereas probability is a multiplicative function. If two systems are considered in states with probabilities p_1 and p_2, and entropies S_1 and S_2, the probability of the two systems together will be $p_1 p_2$ and their entropy $S_1 + S_2$. Thus the relation between S and p must be logarithmic, $S \propto \ln p$.

When a system like the mixture of black and white balls (or N_2 and O_2 molecules) is considered, there will be many different arrangements of the molecules between the two containers once the barrier is removed. Some of these arrangements are more probable than others, since from formula (7.23) there are more ways of achieving these arrangements. We call any particular arrangement of the particles in a system a *microstate* of the system. For example, in the case of 10 black and 10 white balls, there were 252 microstates that placed 5 white and 5 black balls in each container. When the particles can be distributed among different energy levels, each way of assigning the particles to the allowed energy levels is also a particular microstate of the system. Boltzmann showed that the entropy should be related to the average probability of the various microstates of the system. Thus S is proportional to $\overline{\ln p}$ the average of the natural logarithm of p.

$$S = -k \overline{\ln p} \qquad (7.25)$$

The negative sign is necessary because S is positive, and $\overline{\ln p}$ is negative since p is a fraction. The proportionality constant is the Boltzmann constant k (as will be demonstrated a little later).

If we consider all the different microstates of a system and denote each one by an index $j = 1, 2, 3, \ldots$ the average value of $\ln p$ can be calculated as

$$\overline{\ln p} = \sum p_j \ln p_j \qquad (7.26)$$

This is another example of our general averaging formula of Section 5.10, where now we are calculating the average of $\ln p$. From Eq. (7.25), the general Boltzmann expression for the entropy is, therefore,

$$S = -k \sum p_j \ln p_j \qquad (7.27)$$

The use of this formula to calculate the entropy will be discussed in Chapter 12.

A special case of Eq. (7.27) occurs when all the microstates have the same probability. This would be the situation for the black and white balls in Fig. 7.7. Then every $p_j = p = 1/W$, where W is the total number of possible microstates, so that the sum over j is just W times $p \ln p$.

$$S = -k \sum p_j \ln p_j = -kW \frac{1}{W} \ln \frac{1}{W}$$

$$S = k \ln W \qquad (7.28)$$

Suppose that we use this formula to calculate the entropy of mixing of 1 mol N_2 and 1 mol O_2. From our previous consideration of the mixing problem, $p_j = (\frac{1}{2})^{2L}$ or $W = 2^{2L}$. Hence, from Eq. (7.28), we have $\Delta S = k \ln 2^{2L} = 2L\, k \ln 2$. The entropy of mixing per mole is $\Delta S/2 = Lk \ln 2 = R \ln 2$.

If, on the other hand, we calculate ΔS of mixing per mole from the thermodynamic formula (7.22),

$$\Delta S = -R \sum X_j \ln X_j = -R(\tfrac{1}{2} \ln \tfrac{1}{2} + \tfrac{1}{2} \ln \tfrac{1}{2})$$

$$= R \ln 2 = Lk \ln 2$$

This calculation shows that the proportionality constant in the Boltzmann formula (7.27) is indeed the Boltzmann constant k.

7.14   Entropies of Chemical Compounds— Calculations from Heat Capacities

We have now seen how to calculate ΔS for changes in ideal gases, for phase changes, and for heat flows. We have explored the interpretation of entropy in terms of probability. How are we going to obtain numerical values for the entropies of the vast number of chemical compounds of interest in the laboratory and in technical operations? But first—why should we want such data? Entropy is the thermodynamic function that governs change and chemists want to understand how chemicals change and react under all possible conditions. This is the problem of the position of equilibrium of chemical systems under different conditions of temperature and pressure. The entropy S will provide the answer to this problem, but only after some further thermodynamic theory to be given in Chapter 8. Now we shall show how the entropy of a substance can be calculated by measuring its heat capacity over a range of temperatures from close to absolute zero to the temperature of interest. The basic instrument for measuring entropy is the calorimeter, just as it is for measuring energies and enthalpies.

Heat capacities are usually measured at constant pressure as C_P. From Eq. (7.13) at constant pressure, $dH = C_P \, dT = T \, dS$. Thus $dS = C_P \, dT/T$. If we know C_P as a function of T, we can integrate this equation to find the change in S with T constant P. Thus

$$\Delta S = \int_{T_1}^{T_2} C_P \, dT/T = \int_{T_1}^{T_2} C_P \, d \ln T \qquad (7.29)$$

If a phase transition occurs at T_t, the corresponding $\Delta S_t = \Delta H_t/T_t$ must be included.

Example 7.9 Calculate ΔS when 1 mol H_2O is heated from 263 to 283 K at $P^\circ = 101.32$ kPa. $C_P(\text{ice}) = 2.09 + 0.126 \, T$ (J K^{-1} mol^{-1}), $\Delta H_f(273 \text{ K}) = 6000$ J mol^{-1}, $C_P(\text{water}) = 75.3$ J K^{-1} mol^{-1}.

From Eq. (7.29),

$$\Delta S = \int_{263}^{273} (2.09 + 0.126T) \frac{dT}{T} + \frac{6000}{273} + \int_{273}^{283} 75.3 \frac{dT}{T}$$

$$= 2.09 \ln \frac{273}{263} + 0.126(273 - 263) + \frac{6000}{273} + 75.3 \ln \frac{283}{273}$$

$$= 26.0 \text{ J K}^{-1} \text{ mol}^{-1}$$

If the heat-capacity measurements are carried to sufficiently low temperatures, the C_P data can be extrapolated to 0 K. We can then write a general equation for the difference between the entropy of a substance at 0 K and the entropy at any other temperature T', say 298.15 K. From Eq. (7.29),

$$S - S_0 = \int_0^{T_f} \frac{C_P(c)}{T} \, dT + \frac{\Delta H_f}{T_f} + \int_{T_f}^{T_b} \frac{C_P(\ell)}{T} \, dT + \frac{\Delta H_v}{T_b} + \int_{T_b}^{T'} \frac{C_P(g)}{T} \, dT \qquad (7.30)$$

This equation shows how to calculate the entropy of a substance from calorimetric measurements of its heat capacity C_P over a range of T starting near 0 K. The ΔH of all phase changes that occur between 0 K and the upper temperature limit T' must also be measured. These can be measured in the same calorimeter, from the input of electrical energy at constant T during the phase changes. Thus all the terms in Eq. (7.30) are measured in a calorimeter, except S_0, the limiting value of the entropy as T approaches 0 K. A graphical representation of this integration is shown in Fig. 7.8, where the C_P of oxygen is plotted vs. $\ln T$. The area under the curve plus the ΔS of any phase transitions give ΔS for the change in temperature between the limits $T_1 = 0$ and $T_2 = 298$ K.

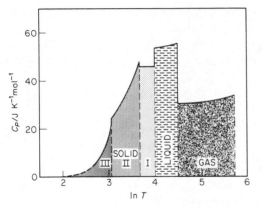

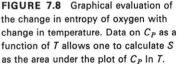

FIGURE 7.8 Graphical evaluation of the change in entropy of oxygen with change in temperature. Data on C_P as a function of T allows one to calculate S as the area under the plot of C_P ln T.

7.15 Third Law of Thermodynamics

The limiting value of the entropy at absolute zero, S_0 in Eq. (7.30) is given by the Third Law of Thermodynamics. As stated in 1923 by G. N. Lewis and M. Randall: "If the entropy of each *element* in some crystalline state be taken as zero at the absolute zero of temperature, each substance has a finite positive entropy; but at the absolute zero of temperature the entropy may become zero, and does so become in the case of perfect crystalline substances." There are two parts to this statement: (1) the arbitrary setting of $S_0 = 0$ for all elements in their standard state at $P°$, and (2) the statement that in the limit of 0 K the entropy of any element or compound in any perfect crystalline state is then also zero. This statement is equivalent to saying that $\Delta S_0 = 0$ for any physical or chemical change involving perfect crystals at absolute zero.

The convention $S_0 = 0$ for elements is similar to that which sets $H°_{298} = 0$ for the elements. Chemical reactions never transmute the elements so that arbitrary zero levels for the $H°$ and $S°$ of the elements can have no effect on the calculated $\Delta H°$ and $\Delta S°$ values for chemical reactions. The fact that 0 K is chosen as the reference point for $S°$ and 298 K for $H°$ is simply a matter of experimental convenience, in that the $S°$ data are obtained by extrapolating C_P to 0 K. The convention $S_0 = 0$ for the elements is therefore not itself part of the Third Law of Thermodynamics.

The Third Law states that for any physical or chemical change of perfect crystalline substances at 0 K, $\Delta S_0 = 0$. Given the *convention* $S°_0 = 0$ for the elements, the Third Law thus states that $S_0 = 0$ for the elements in any state, not merely their standard state of $P°$. If the S_0 of gold at $P° = 101.32$ kPa is zero, the Third Law states that $S_0 = 0$ for gold at $P = 10^5$ kPa or any other pressure. If $S°_0 = 0$ for graphite, then $S_0 = 0$ for diamond also. Furthermore, for any chemical reaction in the limit of 0 K, $\Delta S_0 = 0$. For instance, since $S_0 = 0$ for Ag and I_2, $S_0 = 0$ for AgI.

The Third Law was discovered by Walther Nernst in his investigations of how the ΔS values of chemical reactions behave as the limit of 0 K is approached.

Many checks of the Third Law have been made for both elements and crystalline compounds. We must not forget, however, that $S_0 = 0$ is restricted to *perfect crystal-*

line substances. Thus glasses, solid solutions, and crystals retaining structural disorder near absolute zero, are excluded from the rule $S_0 = 0$.

7.16   Third-Law Entropies

When $S_0 = 0$ in Eq. (7.30) we can evaluate standard entropies per mole of substances at 298.15 K from measured C_P data. An example of such a calculation is summarized in Table 7.1. The values of entropy obtained in this way from C_P data are called *Third-Law entropies*. Examples of Third-Law entropies are included in Table A.2.

TABLE 7.1

THE THIRD-LAW ENTROPY OF HCl FROM MEASUREMENTS
OF ITS HEAT CAPACITY

Contribution	J K mol⁻¹
1. Extrapolation from 0–16 K (Debye Theory, Section 29.10)	1.3
2. $\int C_P \, d \ln T$ for solid I from 16–98.36 K	29.5
3. Transition, solid I $\longrightarrow$ solid II, 1190/98.36	12.1
4. $\int C_P \, d \ln T$ for solid II from 98.36–158.91 K	21.1
5. Fusion, 1992/158.91	12.6
6. $\int C_P \, d \ln T$ for liquid from 158.91–188.07 K	9.9
7. Vaporization, 16 150/188.07	85.9
8. $\int C_P \, d \ln T$ for gas from 188.07–298.15 K	13.5
	$S^{\circ}_{298.15} = 185.9$

The statistical theory of entropy leads directly to the conclusion that $S_0 = 0$ for perfect crystals. In a perfect crystal at 0 K every particle (atom, ion, or molecule) would be in its lowest vibrational energy state, and would have no rotational or translational energy. Each particle would occupy a specific site on a crystal lattice which is perfectly ordered. Thus, in Eq. (7.28) the number of microstates for the crystal at 0 K would be $W = 1$ and $S = k \ln W = 0$. Alternatively, in Eq. (7.27) only one distribution exists, the perfectly ordered one with $p_1 = 1$ and all other $p_j = 0$, so that $S = -\sum p_j \ln p_j = 0$.

In some cases, the molecules in a crystal may persist in more than one arrangement even in the limit of absolute zero. The Lewis and Randall condition for $S_0 = 0$ is therefore not satisfied. An example is nitrous oxide. Two adjacent molecules of N_2O can be oriented in the crystal either as (ONN NNO) or as (NNO NNO) as shown in Fig. 7.9. The difference in energy $\Delta \epsilon$ between these two alternatives is so slight that their relative probability $e^{\Delta \epsilon / kT}$ is virtually unity even at very low temperatures. When C_P of crystalline N_2O is being measured in a calorimeter, the crystals at extremely low temperatures always contain residual disorder due to the two possible orientations of the molecules in the crystal structure. At very low temperatures the disorder is "frozen in" as the rate of reorientation in the frigid crystal becomes extremely low.

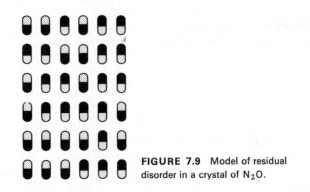

FIGURE 7.9 Model of residual disorder in a crystal of N_2O.

As a consequence of this residual disorder, the limiting value of S_0 is not zero, since there always remains an entropy of mixing of the two arrangements. Since the probability of either arrangement is $\frac{1}{2}$, the residual entropy of mixing from Eq. (7.21) is, per mole,

$$S_0^m = -R(\tfrac{1}{2} \ln \tfrac{1}{2} + \tfrac{1}{2} \ln \tfrac{1}{2}) = R \ln 2 = 5.77 \text{ J K}^{-1} \text{ mol}^{-1}$$

The entropy of N_2O calculated from the statistical formula in Chapter 12 is found to be 4.8 J K^{-1} mol^{-1} higher than the Third-Law value. Since the experimental uncertainties in the entropy measurements are about 1.0 J K^{-1} mol^{-1}, the residual randomness of the N_2O crystal provides a quantitative explanation of the discrepancy in this case between Third-Law and statistical entropies. In most cases, however, the agreement between statistical and Third-Law values is within experimental errors, and then we can be confident that the Lewis and Randall requirement of a perfect crystal has been satisfied at very low temperatures.

7.17 Entropy Changes in Chemical Reactions

Third-Law entropies calculated from heat-capacity data and statistical entropies calculated from spectroscopic data (Chapter 12) have now been obtained for many chemical compounds. They are usually tabulated as $S°(298 \text{ K})$, the values in the standard state at 25°C. From such data the $\Delta S°(298 \text{ K})$ of a vast number of chemical reactions can be calculated.

Example 7.10 From the Third-Law entropies in Table A.2, calculate $\Delta S_{298}^°$ for the reaction $CH_4 + 2O_2 = CO_2 + 2H_2O(g)$.

$$\Delta S_{298}^° = 2S°(H_2O) + S°(CO_2) - S°(CH_4) - 2S°(O_2)$$
$$= 2(188.7 \text{ J K}^{-1} \text{ mol}^{-1}) + 213.7 - 186.2 - 2(205.1)$$
$$= -5.3 \text{ J K}^{-1} \text{ mol}^{-1}$$

Note that the units of $\Delta S°$ of reaction are J K^{-1} mol^{-1} (see page 104).

When any chemical reaction occurs, the ΔS of the universe must increase, in strict accord with the Second Law of Thermodynamics. This law does not mean, of course, that the ΔS for every reaction must be positive, $\Delta S_r^\circ > 0$. It simply means that ΔS for the reacting system plus its surroundings must be positive. In the next chapter we shall see how to combine the ΔS_r° values with ΔH_r° values to obtain a criterion for the direction of any chemical reaction. [For the reaction in Example 7.10, $\Delta H^\circ(298 \text{ K}) = -803 \text{ kJ mol}^{-1}$. Estimate the ΔS of the universe.]

Problems

1. Calculate ΔS when 1 mol of water at 288 K is mixed with 1 mol of water at 370 K and the system comes to equilibrium in an adiabatic enclosure. Assume that $C_P = 77$ J K^{-1} mol^{-1} independent of T.

2. For 1.000 kg CH$_4$, calculate S(101 kPa, 1000 K) $-$ S(101 kPa, 300 K). (See Table 6.2.)

3. An electric current of 5 A flows through a resistor of 30 Ω, which is kept at a temperature of 300 K by running water. In 30 min, what is the ΔS of (a) the resistor; (b) the water?

4. An electric current of 5 A flows through a thermally insulated resistor of 30 Ω for 30 sec. The initial T of the resistor is 300 K, its mass is 10 g, and its specific heat $c_P = 1.00$ J K^{-1} g^{-1}. What is the ΔS of (a) the resistor; (b) its surroundings?

5. Mercury boils at P° and 630 K, with $\Delta H_v = 64.9 \text{ kJ mol}^{-1}$, $C_P(\text{Hg}, \ell) = 28.0$ J K^{-1} mol^{-1}, and $C_P(\text{Hg, g}) = 20.8$ J K^{-1} mol^{-1}. Calculate the difference in molar entropies $S_m(\text{Hg}, \ell, 300 \text{ K}) - S_m(\text{Hg, g}, 700 \text{ K})$.

6. One mole of superheated water is evaporated at 383 K and 101 kPa. Calculate ΔS of water and of the surroundings. ΔH_v (373 K) $= 47.3 \text{ kJ mol}^{-1}$, $C_P(\text{H}_2\text{O, g})$ (Table 6.2), and $C_P(\text{H}_2\text{O}, \ell) = 75.4$ J K^{-1} mol^{-1}.

7. Calculate ΔS_m for mixing 79% N$_2$, 1% Ar, and 20% O$_2$.

8. One mole of superheated water is evaporated at 383 K and 101 kPa. Calculate ΔS of the water and the surroundings from the following data. Water boils at 383 K under a pressure of 142.8 kPa. Its ΔH_v at this T and P is 40.15 kJ mol^{-1}. Assume constant: $\alpha(\ell) = 7.8 \times 10^{-4}$ K^{-1}, $\rho(\ell) = 0.955$ g cm^{-3}. [Use $(\partial S/\partial P)_T$ from Table 8.1.]

9. A sample of water is suddenly (and adiabatically) compressed from 100 kPa to 1000 kPa. Initial $T = 300$ K. What is final T? [First show that $(\partial T/\partial P)_S = \alpha TV/C_P$. See Problem 8.]

10. The thermal expansivity α' of mercury from 0 to 100°C,

$$\alpha' = (1/V_0)(\partial V/\partial T)_P = 1.817 \times 10^{-4} + 5.90 \times 10^{-9}t + 3.45 \times 10^{-10}t^2$$

when t is °C and V_0 is the volume at 0°C. The $S^\circ(298 \text{ K})$ of Hg is 75.9 J K^{-1} mol^{-1}. Calculate S(298 K) at 10^3 kPa, assuming that α' is independent of P. [Problem 8.]

11. For AgCl(c), $S^\circ(298 \text{ K}) = 96.2$ J K^{-1} mol^{-1} and $C_P = (62.3 + 4.18 \times 10^{-3}T - 11.2 \times 10^5 T^{-2})$ J K^{-1} mol^{-1}. Draw a graph of S° from 298 K to 728 K (melting point).

12. A steam engine operates between 140 and 30°C. What is minimum heat withdrawal from heat source to yield 1 kJ of work?

13. In regions with mild winters, heat pumps can be used for space heating in winter and cooling in summer. Assuming ideal thermodynamic efficiency for the pump, compare the cost of keeping a room at 25°C in winter and in summer when outside temperatures are 12 and 38°C, respectively. What assumptions did you make in this calculation? Suppose that the heat pump were running 50% of the time when the outside tempera-

ture was 12°C. If the outside temperature fell to 0°C, could this heat pump maintain the inside at 25°C?

14. A refrigerator maintains a temperature of 2°C in a room at 30°C. Heat transfer to the refrigerator is 10^4 J min^{-1}. If the unit can operate at 50% of its maximum thermal efficiency, what is its power requirement?

15. Prove that it is impossible for two reversible adiabatics on a P–V diagram to intersect.

16. Derive a formula for $(\partial S/\partial V)_T$ for a van der Waals gas. Calculate the ΔS of compressing 1 mol CO_2 at 298 K from 0.1 m^3 to 0.001 m^3 assuming (a) an ideal gas; (b) a van der Waals gas.

17. Benzene and toluene are miscible in all proportions with no evolution of heat; calculate the *minimum* work in any isothermal process to separate an equimolar mixture of benzene and toluene at 300 K.

18. Calculate $S°(298.15\text{ K})$ of silver from the heat-capacity data.

T (K)	15	30	50	70	90	110	130	150
C_P (J K^{-1} mol^{-1})	0.67	4.77	11.65	16.33	19.13	20.96	22.13	22.97
T	170	190	210	230	250	270	290	300
C_P	23.61	24.09	24.42	24.73	25.03	25.31	25.44	25.50

Below 15 K, $C_P = AT^3$, where T is a constant.

19. The crystal structure of ice is shown in Fig. 29.3. Show that the residual entropy of ice at 0 K due to the different possible arrangements W of the water molecules is

$$S_0 = R \ln W = R \ln \tfrac{3}{2} = 3.37 \text{ J K}^{-1} \text{ mol}^{-1}$$

20. For an equation of state, $PV = nRT - n^2a/V$, show that $(\partial S/\partial P)_T = nRV^2/(n^2a - PV^2)$. Calculate $S_m(300\text{ K}, 10^3\text{ kPa}) - S_m°(300\text{ K})$ for NH_3, with $a = 0.417$ m^6 Pa mol^{-2}. [See Problem 8]

21. In 1871, J. C. Maxwell created the sorting demon, "a being whose faculties are so sharpened that he can follow every molecule in its course, and would be able to do what is at present impossible to us. . . . Let us suppose that a vessel is divided into two portions, A and B by a division in which there is a small hole, and that a being who can see the individual molecules opens and closes this hole, so as to allow only the swifter molecules to pass from A to B, and only the slower ones to pass from B to A. He will, thus, without expenditure of work, raise the temperature of B and lower that of A, in contradiction to the second law of thermodynamics." Can you save the Second Law from the demon?

22. The enthalpy of ionization of water was measured [*J. Chem. Thermodyn.*, **9**, 65 (1977)]

T (K)	273.70	284.85	298.15	305.65	313.15	323.15
$\Delta H_i°$ (kJ mol^{-1})	62.13	58.94	55.84	54.28	52.81	51.01

Fit these data to an equation of the form

$$\Delta H_i° = A(T - 298.15) + B(298.15)^2$$

and evaluate $\Delta C_P°$ for the reaction at 298.15 K.

23. Exactly 1 mol H_2 at 400 K is compressed adiabatically and reversibly from 100 to 1000 kPa. Assume ideal-gas behavior and $C_P = 28.9$ J K^{-1} mol^{-1}. What is ΔU, ΔH, ΔS for the change?

24. The isentropic (adiabatic reversible) compressibility is $\beta_S = -V^{-1}(\partial V/\partial P)_S$. Show that $\beta - \beta_S = \alpha^2 TV/C_P$.

25. The speed of sound u is related to β_S by $u^2 = (\beta_S \rho)^{-1}$. For cyclohexane at 100 kPa and 298 K, calculate β and β_S.

$$u = 1250 \text{ m s}^{-1}, \qquad \rho = 0.7739 \text{ g cm}^{-3}$$
$$\alpha = 1.215 \times 10^{-3} \text{ K}^{-1}, \qquad c_P = 1.86 \text{ J K}^{-1} \text{ g}^{-1}$$

8

Physical and Chemical Equilibrium

The world is a theatre in which the scene is always changing. Our existence depends on these physical and chemical changes, those in our environment and those in the living cells of our own bodies. We kindle a fire to keep warm, burn fuel in an engine to carry us to and fro, while internally the oxidation of food maintains body temperature and yields biochemical fuel (ATP, adenosine triphosphate) for muscular work.

The aim of chemical thermodynamics is to provide a general theory of the changes in the world, the directions in which they proceed, and the equilibrium points they can reach under given conditions. Many of these changes and equilibria occur in systems such as liquid solutions, nonideal gases, and crystalline structures, in which the forces between molecules are large and have important effects on the equilibrium. In such systems the basic bare bones of the equilibrium theory are complicated by problems of the thermodynamic theory of nonideal systems. Therefore, in this chapter, we shall discuss equilibrium theory as applied to pure substances and ideal gases, and in Chapter 10 we shall consider extensions of the theory to nonideal gases and solutions.

8.1 Entropy and Equilibrium

The entropy function S provides the unifying principle that governs physico-chemical change. For any change that occurs in an isolated system,

$$\Delta S \geq 0 \tag{8.1}$$

An isolated system is a system in which energy U and volume V are constant. Thus Eq. (8.1) states that for a system at constant U and V, the entropy S can only increase or remain constant. When S remains constant, the system is at equilibrium. If the state of the system changes, its entropy must increase. The natural law embodied in Eq. (8.1) is simply a consequence of the fact that the large-scale world is constructed of a vast multitude of small-scale particles: molecules, atoms, and ions. Entropy spontaneously increases in an isolated system undergoing change because the particles of the microscopic world in their natural motions and wanderings must move toward arrangements of increasing disorder and probability.

The fact that we are here at all, with our complex, highly ordered, and most improbable bodies and brains, shows that the rule of entropy applies only to entire isolated systems. It is possible to pay for a decrease in S in one part of a system by having a greater increase in S in another part. In other words, if the particular system in the world that interests us is not isolated, the inexorable law of increasing entropy applies to the system plus its surroundings, but not to the system alone. Living organisms can maintain their ordered structures or grow in complexity because they are open systems, taking in organized matter from their surroundings and giving forth less organized matter.

8.2 Dynamic Equilibrium

Equilibrium requires that a system no longer changes with time in a *macroscopic sense*. That is, no measurement made on the system will detect any difference in a macroscopic property measured at times t and $t + \Delta t$. At the microscopic level, however, the system is certainly not static. The rapid motions of the molecules in gases and liquids continue, the particles in crystals vibrate about equilibrium positions. Consider, for example, a typical phase equilibrium, a liquid in equilibrium with its vapor at the boiling point. In a closed container, molecules leave the liquid surface at the same rate as molecules from the vapor enter the liquid surface.

In a chemical equilibrium also, the interchange of molecular partners continues at a rapid rate. For instance, consider the reaction $H_2 + I_2 = 2HI$. At equilibrium H_2 and I_2 molecules are still rapidly being converted to HI molecules, while HI molecules are rapidly decomposing back to $H_2 + I_2$. The rates of forward and reverse reactions are the same, so that no change in composition of the system occurs and $\Delta S = 0$, in accord with Eq. (8.1).

Physicochemical equilibrium is always dynamic and never static. The macroscopic properties of the system exhibit no change, but at the microscopic level the molecules are a beehive of activity. The Norwegian chemists Guldberg and Waage gave the first mathematical exposition of dynamic chemical equilibrium in 1863. In the HI reaction, they wrote the forward rate as

$$v_f = k_f\{H_2\}\{I_2\}$$

where $\{H_2\}$ and $\{I_2\}$ denote what they called the "active masses" of H_2 and I_2, and

k_f is a rate constant. The active masses are proportional to concentrations. The reverse rate was written as

$$v_r = k_r\{HI\}\{HI\} = k_r\{HI\}^2$$

At equilibrium, the forward and reverse rates are equal, $v_f = v_r$, or $k_f\{H_2\}_{eq}\{I_2\}_{eq} = k_r\{HI\}_{eq}^2$. Thus

$$\frac{\{HI\}_{eq}^2}{\{H_2\}_{eq}\{I_2\}_{eq}} = \frac{k_f}{k_r} = K \tag{8.2}$$

The ratio of forward and reverse rate constants is an equilibrium constant K.

8.3 Free-Energy Functions

The only thermodynamic driving force for changes in the universe is the increase in entropy. Often, however, it is not convenient to consider the entire universe (or even the system of interest plus its immediate surroundings). We want criteria for equilibrium that can be applied to the system alone. In other words, it is often convenient to relate the $\Delta S > 0$ of the universe to a change in some function that can be calculated for the system only.

The two most important conditions for studies of chemical systems are (1) constant T and V, and (2) constant T and P. The condition of constant T and V is often useful for theoretical interpretation of the properties of substances, since it allows us to study effects of temperatures without complications due to PV work on the system. The condition of constant T and P is especially useful for discussions of the equilibrium conditions met in chemical reactions and phase changes.

Hermann von Helmholtz in 1882 introduced the thermodynamic function for which the natural variables are T and V, and he called it the *free energy*. We shall call it the *Helmholtz free energy* and assign the symbol A. It is defined by

$$A = U - TS \tag{8.3}$$

In this A function, the S is multiplied by T, as is necessary to give proper dimensionality, since TS has the dimensions of energy.

Willard Gibbs introduced in 1876 the function whose natural variables are P and T. We call it the *Gibbs free energy G*. It is defined by

$$G = A + PV \tag{8.4}$$

Since $H = U + PV$,

$$G = H - TS \tag{8.5}$$

The basic thermodynamic functions and some important relations (to be derived later in this chapter) are summarized in Table 8.1.

TABLE 8.1
THERMODYNAMIC FUNCTIONS

Name of function	Symbol and natural variables	Definition	Differential expression	Corresponding Maxwell relation
Internal energy	$U(S, V)$		$dU = T\,dS - P\,dV$	$\left(\dfrac{\partial T}{\partial V}\right)_S = -\left(\dfrac{\partial P}{\partial S}\right)_V$
Enthalpy	$H(S, P)$	$H = U + PV$	$dH = T\,dS + V\,dP$	$\left(\dfrac{\partial T}{\partial P}\right)_S = \left(\dfrac{\partial V}{\partial S}\right)_P$
Helmholtz free energy	$A(T, V)$	$A = U - TS$	$dA = -S\,dT - P\,dV$	$\left(\dfrac{\partial S}{\partial V}\right)_T = \left(\dfrac{\partial P}{\partial T}\right)_V$
Gibbs free energy	$G(T, P)$	$G = H - TS$	$dG = -S\,dT + V\,dP$	$\left(\dfrac{\partial S}{\partial P}\right)_T = -\left(\dfrac{\partial V}{\partial T}\right)_P$

8.4   Interpretation of the Helmholtz Function A

We shall now show how the change ΔA in the Helmholtz free energy denotes the direction of spontaneous change in a system at constant T and V. Consider in Fig. 8.1 an isolated part of the universe that consists of system a maintained at constant

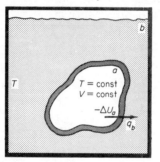

FIGURE 8.1   An isolated part of the universe consisting of a system a at constant T and V immersed in a large heat bath, b, which forms the surroundings of the system.

T and V and its surroundings b, which consist of a huge heat bath at constant T. Any spontaneous change that occurs within system a must be accompanied by an increase in entropy of the universe. Therefore,

$$\Delta S = (\Delta S_a + \Delta S_b) \geq 0 \qquad (8.6)$$

The equality sign corresponds to the condition of equilibrium, i.e., reversible processes.

Since the volume of a is constant, no PV work can be done on b due to any change that occurs in a. If no other work is done on b, the only interaction between a and b is a transfer of heat. Since $\Delta U_a = q + w = q$, this transfer of heat results in a change ΔU_a in the internal energy of a. The heat received by b is then $-\Delta U_a$. Since b is a huge reservoir at constant temperature, the heat transfer to it is reversible, and

$$\Delta S_b = q_{\text{rev}}/T = -\Delta U_a/T$$

Hence Eq. (8.6) becomes

$$\Delta S = \Delta S_a - \frac{\Delta U_a}{T} \geq 0$$

or

$$\Delta U_a - T \, \Delta S_a \leq 0 \tag{8.7}$$

In terms of the Helmholtz function A, Eq. (8.7) for the process at constant T becomes

$$\Delta A_a = \Delta U_a - T \, \Delta S_a \leq 0$$

Thus the function A must decrease in any spontaneous change in a system at constant T and V. In the discussion above, note that the spontaneous change increases the entropy of the universe, but the use of the Helmholtz function A allows us to relate this $\Delta S > 0$ of the universe to a $\Delta A < 0$ for the system at constant T and V.

The Helmholtz function was given the symbol A from the German *Arbeit*, work. For an *isothermal process*, $-\Delta A$ for a system measures the maximum work that can be obtained from the system in any change. From Eq. (8.3),

$$dA = dU - TdS \qquad \text{(const. } T) \tag{8.8}$$

But $dU = dq_{\text{rev}} + dw_{\text{rev}}$, and $dq_{\text{rev}} = TdS$. Thus from Eq. (8.8),

$$dA = dw_{\text{rev}} \qquad \text{or} \qquad w_{\text{rev}} = \Delta A \tag{8.9}$$

The reversible work is the maximum work that can be obtained, since it is done against the maximum opposing force, which is just at the equilibrium value.

8.5 Equation of State Derived from A

From $A = U - TS$, the complete differential of A is

$$dA = dU - T \, dS - S \, dT \tag{8.10}$$

On substituting $dU = T \, dS - P \, dV$ from Eq. (7.6), we have

$$dA = -P \, dV - S \, dT \tag{8.11}$$

At constant T,

$$\left(\frac{\partial A}{\partial V} \right)_T = -P \tag{8.12}$$

This equation allows us to calculate the pressure of any substance as a function of V and T, provided that we know the Helmholtz free energy $A(V, T)$. The equation of state of a substance is exactly the function $P(V, T)$. Equation (8.12) is completely general. It applies to liquids and solids as well as to gases.

If we substitute Eq. (8.3), $A = U - TS$, into Eq. (8.12), we get

$$\left(\frac{\partial U}{\partial V} \right)_T - T \left(\frac{\partial S}{\partial V} \right)_T = -P \tag{8.13}$$

Since we can show* that $(\partial S/\partial V)_T = (\partial P/\partial T)_V$, Eq. (8.13) becomes

$$P = T\left(\frac{\partial P}{\partial T}\right)_V - \left(\frac{\partial U}{\partial V}\right)_T \tag{8.14}$$

This equation is sometimes called a *thermodynamic equation of state*. We recall that $(\partial U/\partial V)_T$ is the *internal pressure*, since it measures the intermolecular forces in a substance. $(\partial P/\partial T)_V$ measures the rate of change of pressure with temperature, and $T(\partial P/\partial T)_V$ is called the *thermal pressure*. If we have an ordinary equation of state for a substance, e.g., of form $P = F(V, T)$, we can use Eq. (8.14) to calculate the internal pressure.

Example 8.1   What is the internal pressure in a gas that follows the van der Waals equation?

For a van der Waals gas,

$$P = \frac{nRT}{V - nb} - \frac{n^2a}{V^2}$$

Hence $(\partial P/\partial T)_V = nR/(V - nb)$. From Eq. (8.14),

$$(\partial U/\partial V)_T = -P + \frac{nRT}{V - nb} = \frac{n^2a}{V^2}$$

The answer is consistent with the origin of the van der Waals a term as a measure of cohesive forces in the gas. For 1 mol CO_2 at 101.32 kPa and 273.15 K, $V = 22.263 \times 10^{-3}$ m^3. From $a = 0.3637$ m^6 Pa mol^{-2}, the internal pressure is 734 Pa, $\sim 0.7\%$ of the external pressure of 1 atm.

8.6   Interpretation of the Gibbs Function G

By an argument similar to that in Section 8.4 we can show that ΔG indicates the direction of any spontaneous change in a system at constant T and P. In this case the heat transferred to the surroundings at constant pressure is $-\Delta H_a$ (instead of $-\Delta U_a$ at constant volume). Thus $\Delta S_a = -\Delta H_a/T$ and the analog of Eq. (8.7) is

$$\Delta H_a - T\,\Delta S_a \leq 0$$

or, from Eq. (8.5),

$$\Delta G_a \leq 0 \qquad \text{(const. } T, P) \tag{8.15}$$

From $G = A + PV$, the complete differential is $dG = dA + P\,dV + V\,dP$. Substitution of dA from Eq. (8.11) yields

*From $dA = (\partial A/\partial V)_T\,dV + (\partial A/\partial T)_V\,dT$ and $dA = -P\,dV - S\,dT$, $(\partial A/\partial V)_T = -P$ and $(\partial A/\partial T)_V = -S$. Since the order of differentiation makes no difference, $\partial/\partial T\,(\partial A/\partial V)_T = \partial/\partial V\,(\partial A/\partial T)_V$. Therefore, $(\partial P/\partial T)_V = (\partial S/\partial V)_T$ (one of the *Maxwell relations*). This derivation is a special application of Euler's rule: If $dF = M\,dx + N\,dy$ is the perfect differential of a function $F(x, y)$, then $(\partial M/\partial y)_x = (\partial N/\partial x)_y$.

$$dG = -S\,dT + V\,dP \qquad (8.16)$$

This expression is the basis for the use of G as the criterion for equilibrium in systems at constant T and P. We should remember that equations such as Eq. (8.16) are relations between differentials of state functions for systems *at equilibrium*. They are derived ultimately from Eq. (7.6), $dU = dw_{rev} + dq_{rev} = -P\,dV + T\,dS$, the combined first and second laws applied to equilibrium (reversible) processes. In a system at constant T and P, $T = $ const., $dT = 0$; $P = $ const., $dP = 0$. Hence Eq (8.16) becomes

$$dG = 0 \qquad (\text{const. } T \text{ and } P) \qquad (8.17)$$

In any system at equilibrium at constant T and P, any change in the system must be such that $dG = 0$. In other words, as the system at constant T and P approaches equilibrium, G must approach a maximum or a minimum. It is evident from Eq. (8.15) that the *minimum* in G is the equilibrium condition, since G *decreases* in any spontaneous change in a system at constant T and P.

If one plots the G of a system at constant T and P against any parameter ξ that measures a change in the system, for example the extent of a chemical reaction, the minimum in G occurs at the equilibrium value of the parameter ξ. The equilibrium condition in mathematical terms is the condition for a minimum in G:

$$\left(\frac{\partial G}{\partial \xi}\right)_{T,P} = 0; \qquad \left(\frac{\partial^2 G}{\partial \xi^2}\right)_{T,P} > 0 \qquad (8.18)$$

An example of this criterion for equilibrium applied to a chemical reaction will be considered somewhat later.

Consider the application of the Gibbs free energy G to a change in phase of a substance, for instance the melting of a pure solid. Figure 8.2 shows the Gibbs free energies per unit amount of substance (G_i) for solid and liquid lead at $P° = 101.32$ kPa plotted as functions of temperature. (Note that G denotes an extensive state

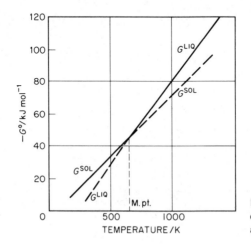

FIGURE 8.2 Molar Gibbs free energy of solid and liquid lead as a function of T at 101.3 kPa.

function, whereas G_i denotes an intensive state function. In SI units, G_i is the molar Gibbs free energy measured in J mol^{-1}.) The curve for solid intersects the curve for liquid at the normal melting point $T_f = 600$ K. At T_f, therefore,

$$G_i(\text{solid}) = G_i(\text{liquid})$$

The curves for solid lead above T_f and for liquid lead below T_f are extrapolated as dashed lines.

When a liquid is supercooled below its freezing point or superheated above its boiling point, it is said to be in a *metastable state*. Superheated solids can exist for only very short times—less than a second—but superheated liquids can often be maintained by carefully heating a liquid kept free of suspended particles. Even when experimental measurements on substances in metastable states are not feasible, their thermodynamic properties, such as G, H, S, and C_P, can be calculated by extrapolation from empirical equations or from the theoretical formulas of statistical thermodynamics. Thus the values of G_i shown in Fig. 8.2 for lead under metastable conditions are quite reliable provided that they are not extrapolated too far beyond the normal melting point.

Below the melting point, we can see that $G_i(\text{liq}) > G_i(\text{solid})$. Hence if a metastable liquid freezes, the G of the system is lowered. Similarly, above the melting point, $G_i(\text{solid}) > G_i(\text{liq})$, and melting a superheated solid will lower the G of the system. If the system is maintained at constant T and P, the phase change will always occur in such a way as to achieve the minimum possible final G for the system.

8.7 Phase Equilibrium—the Clapeyron–Clausius Equation

The thermodynamic condition for equilibrium between two different phases of a pure substance is that the molar Gibbs free energies be equal for the two phases. For two phases a and b of a substance denoted by the subscript i,

$$G_i^a(T, P) = G_i^b(T, P) \tag{8.19}$$

We have explicitly written the G_i as functions of their natural variables T and P. The two phases denoted by superscripts a and b in Eq. (8.19) may be solid and liquid, solid and gas, liquid and gas, two solid phases of the same substance having different crystal structures, or in the unique case of liquid helium, two different liquid phases. If we visualize $G_i^a(T, P)$ and $G_i^b(T, P)$ as two surfaces in a three-dimensional G–T–P space, the intersection of these surfaces is given by the condition (8.19). The points specified by Eq. (8.19) define a curve $P(T)$ or $T(P)$. This curve is the equilibrium P–T curve for the phase change $a \longrightarrow b$. The equation for such a phase equilibrium was first proposed in 1834 by the French engineer Clapeyron, and a rigorous derivation based on the entropy function was provided by Clausius about 30 years later.

If T and P change along the equilibrium curve to $T + dT$ and $P + dP$, the equality of G_i^a and G_i^b must be maintained, so that Eq. (8.19) becomes

$$G_i^a + dG_i^a = G_i^b + dG_i^b \qquad (8.20)$$

where dG_i^a and dG_i^b are the changes in G_i^a and G_i^b that result from dT and dP. From Eqs. (8.19) and (8.20), $dG_i^a = dG_i^b$. The general equation for dG in Eq. (8.16) then gives

$$-S_i^a\, dT + V_i^a\, dP = -S_i^b\, dT + V_i^b\, dP$$

On rearrangement,

$$\frac{dP}{dT} = \frac{S_i^b - S_i^a}{V_i^b - V_i^a} = \frac{\Delta S}{\Delta V} \qquad (8.21)$$

This is the most general form of the Clapeyron–Clausius equation.

As shown in Eq. (7.14), the ΔS for a phase transition under equilibrium conditions is $\Delta H/T$, so that Eq. (8.21) may be written

$$\frac{dP}{dT} = \frac{\Delta H}{T\, \Delta V} \qquad (8.22)$$

The integration of this equation gives an explicit expression for the P–T curve of the transition. To perform the integration, we need to know the dependence of ΔH and ΔV on T and P. For example, Fig. 8.3 shows the variation in the enthalpy of

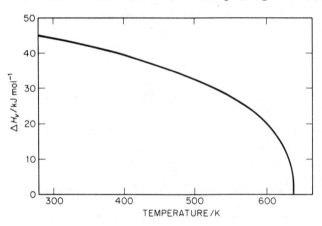

FIGURE 8.3 The enthalpy of vaporization of water as a function of temperature. ΔH_v goes to zero at the critical temperature T_c.

vaporization of water, ΔH_v, from 273 K to the critical point at 647.2 K. Data on the densities ρ of the two phases over the range of T and P of interest are equivalent to data on $\Delta V(T, P)$.

$$\Delta V_m = V_m^b - V_m^a = M\left(\frac{1}{\rho_b} - \frac{1}{\rho_a}\right) \qquad (8.23)$$

where M is the molar mass and V_m the molar volume.

Usually, we do not have complete data on ΔH and ΔV for the transition, but suitable approximations can be made. Over short ranges of T and P, Eq. (8.22) can be integrated with ΔH and ΔV taken to be constant.

Example 8.2 It is often stated that in ice skating the pressure on the blades of the skates melts the ice underneath, forming a thin slippery layer of water. Consider an 80-kg man wearing skates with runners 200×1.00 mm. If the pressure were uniformly distributed, what would be the melting point of ice under these skate blades?

The densities of ice and water at 101.32 kPa and 273.15 K are $\rho(\text{ice}) = 0.917 \times 10^3$ kg m^{-3}, $\rho(\text{water}) = 0.998 \times 10^3$ kg m^{-3}. The ΔH_f is 333.5 kJ kg^{-1}. The increase in pressure under the skate blades is

$$\Delta P = \frac{80 \text{ kg} \times 9.81 \text{ m s}^{-2}}{0.200 \text{ m} \times 0.001 \text{ m}} = 3.92 \times 10^6 \text{ Pa}$$

From Eq. (8.23),

$$\Delta V_f = 18.0 \times 10^{-3} \text{ (kg mol}^{-1})\left(\frac{1}{0.998 \times 10^3} - \frac{1}{0.917 \times 10^3}\right)\frac{1}{\text{kg m}^{-3}}$$

$$= -1.59 \times 10^{-6} \text{ m}^3 \text{ mol}^{-1}$$

Thus, from Eq. (8.22), at 273.15 K,

$$\frac{dP}{dT} = \frac{333.5 \times 10^3 \times 18.0 \times 10^{-3} \text{ (J mol}^{-1})}{(273.15 \text{ K})(-1.59 \times 10^{-6})(\text{m}^3 \text{ mol}^{-1})} = -1.38 \times 10^7 \text{ Pa K}^{-1}$$

We assume this dP/dT to be constant over a short range of temperature. Then, for $\Delta P = 3.92 \times 10^6$ Pa, $\Delta T = -0.28$ K. It would thus appear that a uniform pressure on the skates could hardly lower the melting point of ice enough to create a water layer. Despite this calculation, people do ice-skate: Something must be wrong with the theoretical model: The blades of skates are not flat but concave and the actual area of contact is 10 to 100 times lower than that used in our calculation.

For most solid $\rightleftharpoons$ liquid transitions, the density of the solid is greater than the density of the liquid at the melting point. In such cases, ΔV_f in Eq. (8.22) is positive. Since ΔH_f is always positive, the derivative $dP/dT > 0$, and the melting point of the solid increases with pressure. Ice and solid bismuth are exceptional, in that $\rho(\text{liq}) > \rho(\text{solid})$ at T_f. Ice has an unusually open structure (see Fig. 29.3) and when ice melts there is actually a contraction in volume as H_2O molecules can get closer together in the liquid state.

8.8   How Vapor Pressure Depends on Temperature

For a liquid $\rightleftharpoons$ vapor transition, the equilibrium P–T curve gives the dependence of vapor pressure on temperature. For equilibrium between liquid and vapor, Eq. (8.22) becomes

$$\frac{dP}{dT} = \frac{\Delta H_v}{T_b \Delta V_v} = \frac{\Delta H_v}{T_b(V^g - V^\ell)} \tag{8.24}$$

Here T_b is the boiling point and ΔH_v is the enthalpy of vaporization. In applying Eq. (8.24) it is usually a good approximation to neglect V^ℓ compared to V^g, and if the vapor is assumed to behave as an ideal gas, $V^g = nRT/P$. Thus, Eq. (8.24) becomes

$$\frac{dP}{dT} = \frac{\Delta H_v}{T(RT/P)} \qquad \text{or} \qquad \frac{1}{P}\frac{dP}{dT} = \frac{\Delta H_v}{RT^2} \qquad \text{or} \qquad \frac{d\ln P}{dT} = \frac{\Delta H_v}{RT^2} \qquad (8.25)$$

Over a moderate range of T, say 30 or 40 K, we may consider ΔH_v to be constant. With this approximation, integration of Eq. (8.25) gives

$$\ln P = \frac{-\Delta H_v}{RT} + \text{const.} \qquad (8.26)$$

Or, integration between definite limits gives

$$\ln\frac{P_2}{P_1} = \frac{-\Delta H_v}{R}\left(\frac{1}{T_2} - \frac{1}{T_1}\right) \qquad (8.27)$$

This equation allows us to calculate the vapor pressure P_2 at T_2 if we know the value P_1 at T_1 and the enthalpy of vaporization.

Example 8.3 The normal boiling point of benzene (at $P° = 101.3$ kPa) is 353.2 K. Estimate the reduced pressure at which benzene would boil at 330 K.

In approximate calculations of vapor pressures or boiling points, as in this example, Trouton's rule can be used to estimate ΔH_v as $90T_b$. In this case $\Delta H_v = (90)(353) = 31\,800$ J mol^{-1}. From Eq. (8.27),

$$\ln\frac{P_2}{101.3\text{ kPa}} = -\frac{31\,800\text{ J}}{8.134\text{ J/K}}\left(\frac{1}{330\text{ K}} - \frac{1}{353\text{ K}}\right) = -0.755$$

Hence $P_2 = 47.6$ kPa.

Example 8.4 The normal boiling point of diethyl ether is 307.6 K. Ether is to be stored in aluminum drums that can withstand a pressure of 10^3 kPa. What is the maximum temperature to which the drums of ether could be safely exposed? The ΔH_v of ether is 27.0 kJ mol^{-1} at its boiling point.

From Eq. (8.25) one can see that the greater ΔH_v, the greater the rise of vapor pressure P with T. As the temperature of the ether is raised above T_b, its ΔH_v will decrease. Thus if we use the value of ΔH_v at T_b in this problem, we will overestimate the increase of P with T. Since such an overestimate will improve the safety factor, we have no qualms about using $\Delta H_v = 27\,000$ J mol^{-1} in this situation. Thus, from Eq. (8.27),

$$\ln\frac{10^3}{101.32} = \frac{-27\,000}{8.314}\left(\frac{1}{T_2} - \frac{1}{307.6}\right)$$

The result is $T_2 = 393$ K. Of course, we would not wish to come close to the bursting pressure, and it would be reasonable to allow a safety factor of at least 2. Then,

$$\ln\frac{5 \times 10^2}{101.32} = \frac{-27\,000}{8.314}\left(\frac{1}{T_2} - \frac{1}{307.6}\right)$$

and $T_2 = 362$ K. The ether drums should not be allowed to exceed this temperature.

Equation (8.26) suggests a form often used to tabulate vapor-pressure data, $\ln P = A - B/T$. This form implies that ΔH_v is constant. Since we know that ΔH_v actually varies appreciably with T, this equation cannot represent vapor-pressure data accurately. An additional term is therefore often added to obtain an improved empirical equation,

$$\ln P = A - \frac{B}{T} - C \ln T \tag{8.28}$$

Some examples of vapor-pressure data fitted to this equation are given in Table 8.2.

TABLE 8.2
VAPOR PRESSURES OF LIQUIDS FROM MELTING POINT TO BOILING POINT
$\ln P(Pa) = A - B/T - C \ln T$

Liquid	A	B	C	Liquid	A	B	C
Ar	37.27	1 147	2.814	CH_4	39.36	1 378	3.283
Na	31.23	13 290	1.178	CH_3Cl	55.59	3 917	5.133
K	31.56	10.980	1.37	$CHCl_3$	49.32	5 019	3.916
Zn	32.53	15 360	1.127	CF_4	28.73	1 687	1.132
Hg	29.19	7 710	0.840	CCl_4	60.88	5 856	5.669
Pb	32.76	25 100	1.05	CS_2	41.56	4 247	2.90
N_2	34.99	861	2.833	C_2H_4	45.12	2 328	3.865
O_2	39.25	1 147	3.334	C_2H_6	42.09	2 430	3.332
Cl_2	41.71	3 258	3.017	$(CH_3)_2CO$	49.93	5 074	3.966
Br_2	40.88	4 603	2.661	C_4H_{10}	50.67	3 982	4.376
HCN	30.41	3 687	1.148	CH_3OH	56.55	6 128	4.634
H_2O	60.32	6 851	5.138	C_6H_6	52.84	5 501	4.779
NH_3	45.70	3 714	3.406	$(CH_3)_3N$	74.20	4 933	7.978

8.9 Standard States and Gibbs Free Energy Changes in Chemical Reactions

The standard international conventions for enthalpy data are outlined in Chapter 6. The conventions for recording values for the Gibbs free energy are similar. The standard states are defined as follows for pure substances:

1. For a gas: an ideal gas at $P° = 101.32$ kPa
2. For a liquid: the pure liquid at $P°$
3. For a solid: the solid in its most stable crystal structure at $P°$

The Third Law of Thermodynamics specifies for all substances a standard level for the entropy S, so that we can set $S_0° = 0$ at 0 K. No similar specification exists for G and H, and their zero levels cannot be fixed in any physically determined way. We are, therefore, free to adopt a standard convention by international agreement. It was decided to set the standard enthalpy of the elements at 298.15 K equal to zero,

$H°(298.15 \text{ K}) = 0$. Since we have fixed both $H°$ and $S°$, we are no longer free to choose $G°$ arbitrarily for the elements, but must calculate it from $G° = H° - TS°$ at each temperature of interest, including 298.15 K.

The standard Gibbs free energy of formation $\Delta G_f°$ of a compound is the $\Delta G°$ of the reaction by which it is formed from its elements, when all reactants and products are in their standard states. For example, S(rhombic crystal) $+ 3F_2 = SF_6$; F_2 and SF_6 are gases at $P° = 101.32$ kPa. $\Delta G_f° = G°(SF_6) - G°(S, \text{rh}) - 3G°(F_2)$; $\Delta G_f°$ (298.15 K) $= -893.2$ kJ mol^{-1}. The $\Delta G°$ values are usually determined from $\Delta G° = \Delta H° - T \Delta S°$, where $\Delta H°$ is found by calorimetric measurements and $\Delta S°$ is found either by heat-capacity measurements (Third-Law entropies) or by statistical thermodynamic calculations from spectroscopic data.

Standard Gibbs free energies (of formation) for a number of compounds are given in Table A.2, which is based on a publication of the National Bureau of Standards. From these $\Delta G_f°$ values one can determine the $\Delta G_r°$ for a large number of chemical reactions.

$$\Delta G_r° = \Delta G_f°(\text{products}) - \Delta G_f°(\text{reactants}) = \sum v_i \Delta G_{fi}° \qquad (8.29)$$

where v_i is the stoichiometric coefficient in the reaction equation. The elements are not formed from any reactants so that $\Delta G_f°$ is, of course, always zero for the elements when they are in their equilibrium states at any specified temperature.

Example 8.5 Calculate $\Delta G_{298}°$ for the reaction in dilute aqueous solution:

$$NH_3 + H_2O = NH_4^+ + OH^-$$

From Eq. (8.29) and Tables A.1, A.3,

$$\Delta G_{298}° = -79.5 - 157.3 - (-26.6) - (-237.0) \text{ kJ mol}^{-1}$$
$$= 26.8 \text{ kJ mol}^{-1}$$

The dependence of the Gibbs function on temperature is usually tabulated in somewhat different form, as shown in Table A.2. The function $(G° - H_0°)/T$ is listed for different values of T. This function gives the standard Gibbs function at T relative to the standard enthalpy at 0 K. Division by T results in a slower variation with T for $(G° - H_0°)/T$ than for $G°$ itself. Consequently, interpolation is easier when one needs to obtain a value for $G° - H_0°$ at a T that is not tabulated. The function $(G° - H_0°)/T$ can also be calculated by statistical methods to be described in Chapter 12. When we use this function to calculate $\Delta G°$ for a chemical reaction, we need also to know $\Delta H_0°$, the enthalpy of reaction at 0 K. As explained in Section 6.20, $\Delta H_0°$ can be obtained from thermochemical and heat-capacity data.

Example 8.6 What is the standard Gibbs free energy of 1 mol N_2 gas at 1000 K?

From Table A.3, $(G° - H_0°)/T = -197.9$ J mol^{-1} K^{-1}, so that at 1000 K, $G° - H_0°$ $= -197\,900$ J mol^{-1}. From the table, $H_0° = -8670$ J mol^{-1}, so that $G_{1000}°(N_2) =$ $-206\,600$ J mol^{-1}.

8.10 Gibbs Free Energy of an Ideal Gas

From Eq. (8.16), $dG = V\,dP - S\,dT$. At constant temperature, $dT = 0$ and $dG = V\,dP$. For an ideal gas, $V = nRT/P$, and thus $dG = nRT\,dP/P = nRT\,d\ln P$. On integration,

$$G = nRT \ln P + \text{const.} \qquad (8.30)$$

When $P = P^\circ$, the standard pressure of 101.32 kPa, $G = G^\circ$, and from Eq. (8.30), $G^\circ = nRT \ln P^\circ + \text{const.}$, or $\text{const.} = G^\circ - nRT \ln P^\circ$. On substitution for the constant in Eq. (8.30),

$$G = nRT \ln P + G^\circ - nRT \ln P^\circ = G^\circ - nRT \ln \frac{P}{P^\circ} \qquad (8.31)$$

Example 8.7 What is the Gibbs free energy of N_2 as an ideal gas at 1000 K and $P = 800$ kPa?

From Example 8.6, $G^\circ_{1000}(N_2) = -206\,600$ J mol^{-1}. Then, from Eq. (8.31),

$$G = -206\,600 \text{ J mol}^{-1} + (8.314 \text{ J K}^{-1}\text{ mol}^{-1})(1000 \text{ K}) \ln \frac{800}{101.32}$$

$$= -206\,600 + 17\,200 = -189\,400 \text{ J mol}^{-1}$$

For a mixture of ideal gases, when Dalton's Law of Partial Pressures holds, Eq. (8.31) is valid provided that the partial pressure P_i of the gas i in the mixture is substituted for P. Thus for 1 mol of gas i in an ideal-gas mixture,

$$G_i = G_i^\circ + RT \ln \frac{P_i}{P^\circ} \qquad (8.32)$$

Recall the notation that the subscript i on any thermodynamic function such as G indicates the quantity per unit amount of substance (SI, per mole).

8.11 How the Gibbs Function Depends on Extent of Reaction

Let us review the extent of reaction concept of Section 1.8 by an example.

Example 8.8 100 g Cl_2 is converted to SCl_6 by the reaction $S_8 + 24Cl_2 \longrightarrow 8SCl_6$. What is the extent of reaction ξ?

100 g Cl_2 is 100 g/70.90 g mol^{-1} = 1.410 mol. The extent of reaction is

$$\xi = \frac{\Delta n(Cl_2)}{\nu(Cl_2)} = \frac{-1.410 \text{ mol}}{-24} = 0.0588 \text{ mol}$$

To calculate the extent of reaction, we must specify a particular reaction equation. What would be the extent of reaction for $\frac{1}{8}S_8 + 3Cl_2 = SCl_6$? In this case, for conversion of 100 g Cl_2, $\xi = -1.410 \text{ mol}/-3 = 0.470 \text{ mol}$.

The Gibbs function G governs the direction of change and position of equilibrium in any closed physicochemical system maintained at constant T and P. Any change in such a system that would decrease G is thermodynamically possible. Any change that would increase G of such a system is impossible. If the system, subject to the particular constraints imposed upon it, is in a state in which no change can occur that will lower its Gibbs free energy G, the state so defined is an equilibrium state. The minimum in the G function at constant T and P defines the state of equilibrium.

It is interesting to see how G behaves as a chemical reaction occurs in a system. We shall consider a reaction vessel at $T = 1000$ K and $P^\circ = 101.32$ kPa. In the initial state of the system the vessel contains 1 mol H_2 and 1 mol I_2 gas. These gases can react, $H_2 + I_2 = 2HI$. Note that in this case the reaction occurs without any change in pressure.

Let us first calculate the G of the system in its initial condition, when no reaction has yet occurred and hence $\xi = 0$. From Table A.2, at 1000 K.

$$G^\circ_{1000}(H_2) = -137.0 \text{ kJ mol}^{-1}$$
$$G^\circ_{1000}(I_2) = -203.9 \text{ kJ mol}^{-1}$$
$$G^\circ_{1000}(HI) = -184.9 \text{ kJ mol}^{-1}$$

We assume that the gases behave ideally. Thus the initial partial pressures of H_2 and I_2 are both $0.5(101.32) = 50.66$ kPa. From Eq. (8.32), therefore,

$$G_{1000}(H_2) = G^\circ + RT \ln \frac{P}{P^\circ} = -137.0 + (8.314 \times 10^{-3})(1000) \ln \frac{1}{2}$$

$$= -137.0 - 5.8 = -142.8 \text{ kJ mol}^{-1}$$

$$G_{1000}(I_2) = -203.9 - 5.8 = -209.7 \text{ kJ mol}^{-1}$$

Since there is one mole of each, for the reaction system,

$$G(\xi = 0) = -142.8 + (-209.7) = -352.5 \text{ kJ}$$

Now let us calculate G for the system when reaction has occurred to the extent $\xi = 0.2$. The system then contains

$$n(H_2) = 0.8 \text{ mol}; \qquad P(H_2) = \left(\frac{0.8}{2.0}\right)(101.32) \text{ kPa}$$

$$n(I_2) = 0.8 \text{ mol}; \qquad P(I_2) = \left(\frac{0.8}{2.0}\right)(101.32) \text{ kPa}$$

$$n(HI) = 0.4 \text{ mol}; \qquad P(HI) = \left(\frac{0.4}{2.0}\right)(101.32) \text{ kPa}$$

The corresponding Gibbs free energies are

$$G(H_2) = -137.0 + (8.134 \times 10^{-3})(10^3) \ln \frac{0.8}{2.0} = -144.6 \text{ kJ}$$

$$G(I_2) = -203.9 - 7.6 = -211.5 \text{ kJ}$$

$$G(HI) = -184.9 + (8.314 \times 10^{-3})(10^3) \ln \frac{0.4}{2.0} = -198.3 \text{ kJ}$$

Thus, for the system,

$$G(\xi = 0.2) = 0.8(-144.6) + 0.8(-211.5) + 0.4(-198.3)$$
$$= -364.2 \text{ kJ}$$

In the same way we can calculate G at other values of ξ; the results are plotted in Fig. 8.4. The minimal $G(\xi)$ is the value of G for the system at equilibrium. The minimum of G occurs at $G = -374.9$ kJ for $\xi = 0.735$. This is the equilibrium value

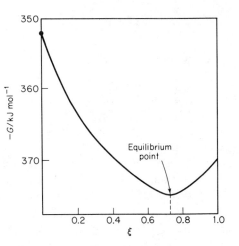

FIGURE 8.4 Gibbs free energy G as a function of extent of reaction ξ for $H_2 + I_2 = 2HI$ at P° and 2000 K.

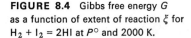

for the extent of reaction; it satisfies the equilibrium condition in Eq. (8.18). The composition of the reaction mixture at equilibrium is

$$n^{eq}(H_2) = 0.265 \text{ mol}; \quad P^{eq}(H_2) = (0.133)(101.32) \text{ kPa}$$
$$n^{eq}(I_2) \ = 0.265 \text{ mol}; \quad P^{eq}(I_2) \ = (0.133)(101.32) \text{ kPa}$$
$$n^{eq}(HI) = 1.47 \text{ mol}; \quad P^{eq}(HI) = (0.735)(101.32) \text{ kPa}$$

At this equilibrium composition $\Delta G = 0$ for the reaction, i.e.,

$$\Delta G = 0 = 2G(HI) - G(H_2) - G(I_2).$$

8.12 Equilibrium Constant and Gibbs Free Energy

We shall now derive an important relation between ΔG°, the standard change in Gibbs free energy for a reaction, and K_P, the equilibrium constant in terms of partial pressures. For the general reaction $aA + bB = cC + dD$,

$$\Delta G = dG_D + cG_C - aG_A - bG_B$$

The expressions for G_i are taken from Eq. (8.32) to give

$$\Delta G = cG_C^\circ + cRT \ln \frac{P_C}{P^\circ} + dG_D^\circ + dRT \ln \frac{P_D}{P^\circ} - aG_A^\circ - aRT \ln \frac{P_A}{P^\circ} - bG_B^\circ - bRT \ln \frac{P_B}{P^\circ}$$

On rearrangement, we have

$$\Delta G = \Delta G^\circ + RT \ln \frac{(P_C/P^\circ)^c (P_D/P^\circ)^d}{(P_A/P^\circ)^a (P_B/P^\circ)^b} \qquad (8.33)$$

where

$$\Delta G^\circ = cG_C^\circ + dG_D^\circ - aG_A^\circ - bG_B^\circ$$

As equilibrium, $\Delta G = 0$, and then

$$0 = \Delta G^\circ + RT \ln \frac{(P_C^{eq}/P^\circ)^c (P_D^{eq}/P^\circ)^d}{(P_A^{eq}/P^\circ)^a (P_B^{eq}/P^\circ)^b} \qquad (8.34)$$

Now, the *standard* free-energy change ΔG° is a function only of T, so that the expression in the argument of the logarithm in Eq. (8.34) is also a function of T, which can be written

$$K_P(T) = \frac{(P_C^{eq}/P^\circ)^c(P_D^{eq}/P^\circ)^d}{(P_A^{eq}/P^\circ)^a(P_B^{eq}/P^\circ)^b} = \frac{(P_C^{eq})^c(P_D^{eq})^d}{(P_A^{eq})^a(P_B^{eq})^b}(P^\circ)^{a+b-c-d} \tag{8.35}$$

Thus Eq. (8.34) becomes

$$-\Delta G^\circ = RT \ln K_P \tag{8.36}$$

The derivation of Eq. (8.36) is a thermodynamic proof that an equilibrium constant exists. Since ΔG° is a function of temperature alone, it cannot depend on the individual values of the partial pressures in Eq. (8.35). No matter how the individual partial pressures may vary, depending on the composition of the system at equilibrium, the combination of P_i in Eq. (8.35) must be $K_P(T)$, a constant at constant T. Equation (8.36) not only proves the *existence* of an equilibrium constant K_P, it also gives us an explicit formula for calculating K_P from the ΔG° data.

Example 8.9 Calculate ΔG° and K_P at 2000 K for the reaction

$$H_2O(g) = H_2(g) + \tfrac{1}{2}O_2(g)$$

From Table A.2, at 2000 K,

$$-\Delta(G^\circ - H_0^\circ)/T = 157.6 + \tfrac{1}{2}(234.7) - 223.1 = 51.85 \text{ J K}^{-1} \text{ mol}^{-1}$$

$$\Delta G^\circ - \Delta H_0^\circ = -(2000 \text{ K})(51.85 \text{ J K}^{-1} \text{ mol}^{-1}) = -103\ 700 \text{ J mol}^{-1}$$

$\Delta H_0^\circ = 238\ 900 \text{ J mol}^{-1}$. Hence, $\Delta G^\circ = 135\ 200 \text{ J mol}^{-1}$.

From Eq. (8.36),

$$\ln K_P = -135\ 200 \text{ J mol}^{-1}/8.314 \text{ J K}^{-1} \text{ mol}^{-1} (2000 \text{ K}) = -8.131$$

$$K_P = 2.94 \times 10^{-4}$$

Example 8.10 If steam at pressure of 200 kPa is passed through a furnace tube at 2000 K, what will be the percent oxygen in the exit stream?

$H_2O = H_2 + \tfrac{1}{2}O_2$. From Eq. (8.35), with $P = P(H_2) = 2P(O_2)$,

$$K_P = \frac{\sqrt{1/2}\, P^{3/2}}{P_{H_2O}}(P^\circ)^{-1/2} = 2.94 \times 10^{-4}$$

Dissociation of H_2O is so small that we can set $P_{H_2O} \approx 200$ kPa. Hence

$$P^{3/2} = \sqrt{2}\,(200)(2.94 \times 10^{-4})(101.3)^{1/2} = 0.837$$

$$P = 0.888 \text{ kPa}$$

Therefore,

$$\% O_2 = \frac{0.888/2}{200} \times 100 = 0.22$$

8.13   Measurement of K_P

The measurement of K_P is simple enough in principle, but may involve experimental difficulties in some cases. A mixture of gases of known composition is prepared and allowed to come to equilibrium at some measured constant temperature T_1.

The exact composition of the reaction mixture in the equilibrium state must then be determined. It would be desirable to analyze the mixture in the reaction vessel at T_1, but this procedure is not always possible. One can try to quench the reaction mixture rapidly to a much lower temperature (example: pull a quartz reaction vessel from a red hot furnace and plunge it into ice water). There is a possibility that the gases will react appreciably during the quench, so that the measured equilibrium composition corresponds not to T_1 but to some lower temperature. If a suitable catalyst is available, one can allow the reaction mixture to reach equilibrium in the presence of the catalyst and then quickly separate the mixture from the catalyst before making the analysis. Flow methods are often useful. The reaction mixture can be passed over the catalyst in a tube heated in a furnace. The exit gases can be quickly cooled in the absence of catalyst without disturbing the equilibrium.

In some cases the reaction at constant volume causes a change in pressure, for example, $SO_2 + \frac{1}{2}O_2 = SO_3$. The pressure in the reaction vessel decreases as this gas reaction progresses. If the composition of the initial reaction mixture is known, a measurement of the pressure at equilibrium allows us to calculate the equilibrium composition. Suppose that the initial gas mixture contained a mol SO_2 and b mol O_2. The initial pressure (for an ideal gas mixture) would be $P_1 = (a + b)RT/V$. If x mol SO_2 is formed at equilibrium, the total number of moles is $(a - x) + (b - \frac{1}{2}x) + x = a + b - \frac{1}{2}x$. The pressure *at equilibrium* is $P_2 = (a + b - \frac{1}{2}x)RT/V$. Hence the ratio of initial pressure to equilibrium pressure is $P_1/P_2 = (a + b)/(a + b - \frac{1}{2}x)$. When x has been determined in this way, the equilibrium constant can be calculated. At equilibrium,

$$n(SO_2) = a - x; \qquad n(O_2) = b - \frac{x}{2}; \qquad n(SO_3) = x$$

Hence

$$X(SO_2) = \frac{a - x}{a + b - \dfrac{x}{2}}; \qquad X(O_2) = \frac{b - \dfrac{x}{2}}{a + b - \dfrac{x}{2}}; \qquad X(SO_3) = \frac{x}{a + b - \dfrac{x}{2}}$$

The partial pressures of each gas in the equilibrium mixture are given by $P_j = X_j P$. Thus

$$K_P = \frac{P(SO_3)}{P(SO_2)P^{1/2}(O_2)}(P^\circ)^{1/2} = \frac{X(SO_3)}{X(SO_2)X^{1/2}(O_2)}\left(\frac{P^\circ}{P}\right)^{1/2}$$

8.14 How G Varies with T

The dependence of G on T is obtained directly from Eq. (8.16) as

$$\left(\frac{\partial G}{\partial T}\right)_P = -S \qquad (8.37)$$

Figure 8.5 shows graphically how G and H vary with T.

An important derivation from Eq. (8.37) is based on the substitution from Eq. (8.5) of $-S = (G - H)/T$, whence

$$\left(\frac{\partial G}{\partial T}\right)_P = \frac{G - H}{T} \qquad (8.38)$$

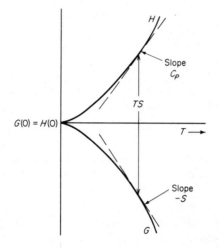

FIGURE 8.5 Variation with T at constant P of Gibbs free energy G and enthalpy H of a pure substance. The limiting slopes as $T \to 0$ are both zero, so that both C_P and $S \to 0$ as $T \to 0$.

This is one form of the *Gibbs–Helmholtz equation,* Now,

$$\left(\frac{\partial (G/T)}{\partial T}\right)_P = \frac{1}{T}\left(\frac{\partial G}{\partial T}\right)_P - \frac{G}{T^2}$$

Substitution of $(\partial G/\partial T)_P$ from Eq. (8.38) gives

$$\left(\frac{\partial (G/T)}{\partial T}\right)_P = \frac{-H}{T^2} \qquad (8.39)$$

These equations are especially useful when applied to $\Delta G°$ for chemical reactions, since they allow us to calculate the $\Delta G°$ at various temperatures if we know the $\Delta G°$ at one temperature (say 298 K) and the $\Delta H°$ over the range of temperatures.

Example 8.11 Estimate $\Delta G°$ at 500 K for the reaction,

$$CaCl_2(c) + 2H_2O(g) = Ca(OH)_2(c) + 2HCl(g)$$

From Table A.1,

$$\Delta G°(298) = -898 - 2(95.3) - (-748) - 2(-228.6) = 116.6 \text{ kJ mol}^{-1}$$

We take $\Delta H°$ as constant over the temperature range. From Table A.1,

$$\Delta H°(298) = -986 - 2(92.3) - (-796) - 2(-241.8) = 109 \text{ kJ mol}^{-1}$$

From Eq. (8.39) at constant P,

$$\int_{T_1}^{T_2} d(\Delta G°/T) = \int_{T_1}^{T_2} -(\Delta H°/T^2)\, dT$$

On integration between 298 and 500 K,

$$\frac{\Delta G°(500)}{500} = \frac{\Delta G°(298)}{298} + \Delta H°(298)\left[\frac{1}{500} - \frac{1}{298}\right]$$

$$\Delta G°(500) = 500\left[\frac{116.6}{298} + 109\left(\frac{1}{500} - \frac{1}{298}\right)\right]$$

$$= 121 \text{ kJ mol}^{-1}$$

It would be better not to assume that $\Delta H°$ is constant, but to use the Kirchhoff equation (6.27) and data on C_P to obtain $\Delta H°(T)$ over the temperature range.

We already have the equation to relate K_P to $\Delta G°$ and the equation to calculate the variation of $\Delta G°$ with T. Thus it is a simple matter to put these two together to obtain an equation for variation of equilibrium constant K_P with T. From Eq. (8.36), $\Delta G°/T = -R \ln K_P$. From Eq. (8.39), $[\partial(\Delta G°/T)/\partial T]_P = -\Delta H°/T^2$. Therefore,

$$\left(\frac{\partial \ln K_P}{\partial T}\right)_P = \frac{\Delta H°}{RT^2} \tag{8.40}$$

This is the *van't Hoff equation*. It shows that if $\Delta H° > 0$ (endothermic reaction), K_P increases with T, whereas if $\Delta H° < 0$ (exothermic reaction), K_P decreases with increasing T.

These results are examples of the *Principle of Le Chatelier* (1888) as applied to the effect of temperature on a chemical equilibrium. When the reaction is endothermic, an increase in temperature moves the equilibrium point toward the product side, so that more heat is absorbed by the reacting system, thus tending to counteract the original rise in T.

Integration of Eq. (8.40) with $\Delta H°$ assumed to be constant gives

$$\ln K_P = -\frac{\Delta H°}{RT} + \text{const.} \tag{8.41}$$

This equation indicates that a plot of $\ln K_P$ vs. $1/T$ is a straight line if $\Delta H°$ is constant over the range of T considered. An example is shown in Fig. 8.6 for the reaction $PCl_5 = PCl_3 + Cl_2$,

$$K_P = \frac{P(PCl_3)P(Cl_2)}{P(PCl_5)}\frac{1}{P°}$$

The slope of the line is $-\Delta H°/R = -11\,000$, and $\Delta H° = 91\,450$ J mol^{-1}

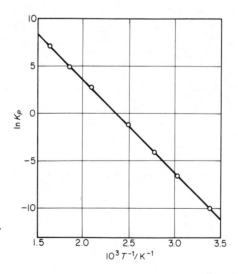

FIGURE 8.6 Equilibrium constant for $PCl_5 = PCl_3 + Cl_2$ plotted as $\ln K_P$ vs. T^{-1}.

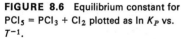

Example 8.12 For the reaction $2NO_2 = N_2O_4$,

$$\Delta G^\circ(298) = -4000 \text{ J mol}^{-1} \quad \text{and} \quad \Delta H^\circ(298) = -57\,000 \text{ J mol}^{-1}$$

Estimate K_P at 500 K.

At 298 K from Eq. (8.36), $\ln K_P = -\Delta G^\circ/RT = 1.61$. Inserting values into Eq. (8.41), we obtain $\ln K_P = 57\,000/(8.314)(298) + \text{const.} = 1.61$, or const. $= -21.4$. Thus for this reaction, $\ln K_P = (57\,000/RT) - 21.4$. At 500 K,

$$\ln K_P = \frac{57\,000}{(8.314)(500)} - 21.4 = -7.7$$

$$K_P = 4.5 \times 10^{-4}$$

Over a more extended range of T, we cannot take ΔH° to be constant in Eq. (8.40). We should instead use Eq. (6.29) based on the Kirchhoff equation. When this expression for ΔH° is substituted into Eq. (8.40), we have (at constant P)

$$\frac{d \ln K_P}{dT} = \frac{1}{RT^2}(\Delta H_0^\circ + AT + \tfrac{1}{2}BT^2 + \tfrac{1}{3}CT^3 + \cdots)$$

On integration,

$$\ln K_P = -\frac{\Delta H_0^\circ}{RT} + A' \ln T + B'T + C'T^2 + \cdots + I \tag{8.42}$$

The value of the integration constant I can be obtained if the value of K_P is known at any one T, either experimentally or by calculation from ΔG°. One value of ΔH° is needed to fix the value of ΔH_0°, the integration constant of the Kirchhoff equation. Therefore, from a knowledge of the heat capacities C_P of products and reactants, and one value each of ΔH° and ΔG° (or K_P), one can calculate the equilibrium constant of an ideal-gas reaction at any temperature.

Example 8.13 For the important gas reaction $CO + H_2O(g) = H_2 + CO_2$, set up an equation in the form of Eq. (8.42) and from this calculate K_P at 800 K. For this reaction, $K_P = P_{H_2}P_{CO_2}/P_{CO}P_{H_2O}$.

From Table A.1, the standard Gibbs free-energy change at 298 K is

$$\Delta G_{298}^\circ = -394.36 - (-228.59 - 137.15) = -28.62 \text{ kJ mol}^{-1}$$

Thus

$$\ln K_P(298) = \frac{28\,620}{298R} = 11.54 \quad \text{or} \quad K_P(298) = 10.3 \times 10^4$$

From the enthalpies of formation in Table A.1,

$$\Delta H_{298}^\circ = -393.50 - (-241.83 - 110.41) = -41.26 \text{ kJ}$$

The heat capacity Table 6.2 yields, for this reaction,

$$\Delta C_P = C_P(CO_2) + C_P(H_2) - C_P(CO) - C_P(H_2O)$$

$$= -2.15 + 26.1 \times 10^{-3}T - 13.2 \times 10^{-6}T^2 \text{ J K}^{-1}$$

From Eq. (6.28), $\Delta H^\circ = \Delta H_0^\circ - 2.15T + 13.1 \times 10^{-3}T^2 - 4.40 \times 10^{-6}T^3$. Substituting $\Delta H^\circ = -41.26$ kJ, $T = 298$ K, and solving for ΔH_0°, we get $\Delta H_0^\circ = -41.67$ kJ. Then the temperature dependence of the equilibrium constant, Eq. (8.42), becomes

$$\ln K_P = \frac{41\,670}{RT} - \frac{2.15}{R}\ln T + \frac{13.1 \times 10^{-3}}{R}T - \frac{4.40 \times 10^{-6}}{2R}T^2 + I$$

By inserting the value of $\ln K_P$ at 298 K, and $R = 8.314$ J K^{-1} mol^{-1}, we can evaluate the integration constant as $I = -4.26$. Now K_P can be readily calculated at any temperature. For example, at 800 K, $\ln K_P = 1.73$, $K_P = 5.64$.

8.16 Gas–Solid Reactions

In reactions between gases and pure solids, only the partial pressures of the gases occur in the equilibrium constants K_P, since the pure solid phases have a constant composition. Consider, for example, the reduction of nickel oxide by carbon monoxide,

$$NiO(c) + CO(g) = Ni(c) + CO_2(g)$$

The equilibrium constant is $K_P = (P_{CO_2})^{eq}/(P_{CO})^{eq}$.

Example 8.14 For the reaction, $NiO(c) + CO(g) = Ni(c) + CO_2(g)$ ΔG° (J mol^{-1}) $= -20\,700 - 11.97T$. Nickel is exposed to a current of hot carbon dioxide. At what temperature will the product gases at equilibrium at 1 atm pressure contain 400 ppm (parts per million) of carbon monoxide?

Since $P_{CO} \ll P_{CO_2}$,

$$K_P = P_{CO_2}/P_{CO} = (400 \times 10^{-6})^{-1} = 2500$$

$$\ln K_P = -\Delta G^\circ/RT = R^{-1}\left(\frac{20\,700}{T} + 11.97\right) = 7.82$$

$$T = 390 \text{ K}$$

8.17 Effect of Pressure on Equilibrium Constants

To discuss the effect of pressure on equilibrium in gas reactions, it is convenient to represent the partial pressures as $P_i = X_iP$. Here, X_i is the mole fraction of component i and P is the total pressure. The expression in Eq. (8.35) for K_P of a general reaction becomes

$$K_P = \frac{(P_C/P^\circ)^c(P_D/P^\circ)^d}{(P_A/P^\circ)^a(P_B/P^\circ)^b} = \frac{X_C^c X_D^d}{X_A^a X_B^b}\left(\frac{P}{P^\circ}\right)^{c+d-a-d} = K_X(P/P^\circ)^{\Delta\nu} \tag{8.43}$$

K_X is the equilibrium constant in terms of mole fractions,

$$K_X = X_C^c X_D^d / X_A^a X_B^b \tag{8.44}$$

and Δv is the sum of the stoichiometric coefficients of products minus the sum for reactants.

For reactions of ideal gases, K_P is independent of pressure. Unless $\Delta v = 0$, K_X will depend on the pressure, as seen in Eq. (8.44). In such cases the composition of the equilibrium mixture will depend on the pressure. In case an inert (nonreacting) gas is present, we must include it in calculating the mole fractions of the reacting species.

Example 8.15 For the reaction $N_2O_4 = 2NO_2$, $K_P = 0.167$ at 300 K. Calculate the fractional dissociation a of N_2O_4 at 300 K and pressures (a) 100 kPa; (b) 400 kPa; (c) with 1 mol of argon added to equilibrium mixture in (a) with P kept at 100 kPa; (d) with 1 mol of argon added to the equilibrium mixture and the volume kept constant.

Let us take as a basis for calculation 1 mol N_2O_4. At equilibrium there will be $1 - a$ mol N_2O_4 and $2a$ mol NO_2. Hence $X(N_2O_4) = (1 - a)/(1 + a)$ and $X(NO_2) = 2a/(1 + a)$. Equation (8.43) becomes:

(a) $\quad 0.167 = \dfrac{[(2a/(1 + a)]^2}{(1 - a)/(1 + a)} \left(\dfrac{100}{101.3} \right)$

$$\frac{4a^2}{1 - a^2} = 0.169, \qquad a = 0.201 \text{ at } 100 \text{ kPa}$$

(b) $\quad \dfrac{4a^2}{1 - a^2} = 0.0423, \qquad a = 0.102 \text{ at } 400 \text{ kPa.}$

(c) $\quad 0.167 = \dfrac{[(2a/(2 + a)]^2}{(1 - a)/(2 + a)} \left(\dfrac{100}{101.3} \right)$

$$\frac{4a^2}{(1 - a)(2 + a)} = 0.169, \qquad a = 0.265 \text{ at } 100 \text{ kPa}$$

(d) This case gives exactly the same result as part (a) since the partial pressures are not changed.

8.18 The Chemical Potential

In 1875, J. Willard Gibbs introduced a new thermodynamic function, the *chemical potential μ*, which has become important both in the theory of equilibrium and in the theory of some rate processes in chemical systems.

We shall make frequent use of the chemical potential in subsequent chapters. Just as masses fall from higher to lower gravitational potential, and electric charges move from higher to lower electrical potential, so chemical substances move from higher to lower chemical potential, by diffusional transport and by chemical reactions. The basic drive in physicochemical systems at constant T and P is down the gradient of chemical potential. The general equilibrium theory as outlined in this section and the next may seem rather abstract and austere. In these qualities the theory reflects the mind of J. Willard Gibbs. Over the years, many attempts have been made to express chemical thermodynamics in different ways, but in the end we usually come back to Gibbs and his chemical potentials.

We must often deal with systems in which the amounts of various components $(n_A, n_B, n_C \ldots)$ change. In other words, the chemical composition of the system changes. For instance, in a closed system, chemical reactions can occur, which alter the various n_j. In an open system, fluxes of various components can cross the boundaries of the system, thereby altering the n_j. In all such cases, the extensive thermodynamic state functions of the system are functions of the amounts n_j of the various components. Thus the Gibbs function is $G(T, P, n_j)$, where n_j stands for the whole set of $n_A, n_B, n_C, \ldots$. The Helmholtz function is $A(T, V, n_j)$, the enthalpy $H(S, P, n_j)$, and the internal energy $U(S, V, n_j)$.

The expressions for the complete differentials of these functions must include terms in dn_j. For example,

$$dG = \left(\frac{\partial G}{\partial T}\right)_{P, n_j} dT + \left(\frac{\partial G}{\partial P}\right)_{T, n_j} dP + \sum_j \left(\frac{\partial G}{\partial n_j}\right)_{T, P, n_i} dn_j \qquad (8.45)$$

The expression $(\partial G/\partial n_j)_{T, P, n_i}$ is the partial derivative of G with respect to the amount of a particular substance n_j, when T, P, and all the other n_i $(i \neq j)$ are held constant. If there are only two components, A and B, for example, Eq. (8.45) would be

$$dG = \left(\frac{\partial G}{\partial T}\right)_{P, n_A, n_B} dT + \left(\frac{\partial G}{\partial P}\right)_{T, n_A, n_B} dP + \left(\frac{\partial G}{\partial n_A}\right)_{T, P, n_B} dn_A + \left(\frac{\partial G}{\partial n_B}\right)_{T, P, n_A} dn_B \qquad (8.46)$$

From Eq. (8.16), in a system of constant composition (i.e., when all the $dn_j = 0$), $dG = -S\, dT + V\, dP$. Hence Eq. (8.45) becomes

$$dG = -S\, dT + V\, dP + \sum_j \left(\frac{\partial G}{\partial n_j}\right)_{T, P, n_i} dn_j \qquad (8.47)$$

By the same argument that led to Eq. (8.45), we obtain from $A(T, V, n_j)$,

$$dA = \left(\frac{\partial A}{\partial T}\right)_{V, n_j} dT + \left(\frac{\partial A}{\partial V}\right)_{T, n_j} dV + \sum_j \left(\frac{\partial A}{\partial n_j}\right)_{T, V, n_i} dn_j$$

From Eq. (8.11), this equation then becomes

$$dA = -S\, dT - P\, dV + \sum_j \left(\frac{dA}{\partial n_j}\right)_{T, V, n_i} dn_j \qquad (8.48)$$

Similar expressions can be obtained for the complete differentials of $H(P, S, n_j)$ and $U(V, S, n_j)$, each state function being represented in terms of its natural variables.

The chemical potential of Gibbs is defined by

$$\mu_j = \left(\frac{\partial G}{\partial n_j}\right)_{T, P, n_i} = \left(\frac{\partial A}{\partial n_j}\right)_{T, V, n_i} = \left(\frac{\partial H}{\partial n_j}\right)_{P, S, n_i} = \left(\frac{\partial U}{\partial n_j}\right)_{V, S, n_i} \qquad (8.49)$$

We shall show the equality of the first two partial derivatives and leave the other two as an exercise. From $G = A + PV$, $dG = dA + P\, dV + V\, dP$. When this dG is substituted into Eq. (8.47), we have

$$dA + P\, dV + V\, dP = -S\, dT + V\, dP + \sum_j \left(\frac{\partial G}{\partial n_j}\right)_{T, P, n_i} dn_j$$

$$dA = -P\, dV - S\, dT + \sum_j \left(\frac{\partial G}{\partial n_j}\right)_{T, P, n_i} dn_j$$

Comparison with Eq. (8.48) shows that

$$\left(\frac{\partial G}{\partial n_j}\right)_{T,P,n_i} = \left(\frac{\partial A}{\partial n_j}\right)_{T,V,n_i} = \mu_j \tag{8.50}$$

8.19 Chemical Potential and Chemical Equilibrium

Let us consider a system at constant T and P. Then Eq. (8.47) becomes

$$dG = \sum_j \mu_j \, dn_j \tag{8.51}$$

Suppose that the system consists of substances A, B, C, D, and a chemical reaction can occur, $aA + bB = cC + dD$. In this case, on the basis of the usual convention that product terms are positive and reactant terms negative, Eq. (8.51) is

$$dG = \mu_C dn_C + \mu_D dn_D - \mu_A dn_A - \mu_B dn_B$$

The extent of reaction ξ is introduced from Eq. (1.6), $dn_j = v_j \, d\xi$, where v_j is the stoichiometric coefficient. Hence Eq. (8.51) can be written

$$dG = \sum_j v_j \mu_j \, d\xi \tag{8.52}$$

For the typical reaction, we have

$$dG = v_C \mu_C \, d\xi + v_D \mu_D \, d\xi - v_A \mu_A \, d\xi - v_B \mu_B \, d\xi$$

Now let us apply the condition that equilibrium has been reached in the reaction mixture at constant T and P. From Eq. (8.17), Eq. (8.52) gives $dG = 0 = \sum_j v_j \mu_j \, d\xi$, and hence, dividing through by $d\xi$, we obtain

$$\sum_j v_j \mu_j = 0 \tag{8.53}$$

This general equation for equilibrium becomes, in the typical reaction,

$$c\mu_C + d\mu_D - a\mu_A - b\mu_B = 0 \tag{8.54}$$

This equation will provide the basis for our derivation in Chapter 12 of a powerful statistical formula for the equilibrium constant.

Problems

1. Calculate ΔU, ΔH, ΔS, ΔA, and ΔG in compressing 1 mol of an ideal gas at 300 K from 100 to 1000 kPa.

2. Prove that a gas that has equation of state $PV = nRT$ must have $(\partial U/\partial V)_T = 0$.

3. At $P = 100$ kPa, the molar volume of mercury at 273 K is 14.73 cm³ mol⁻¹, and its compressibility is $\beta = 3.88 \times 10^{-11}$ Pa⁻¹. Assuming that β is constant over the temperature range, calculate ΔG_m for compression of mercury from 100 to 1000 kPa.

4. At 350 K the molar volume V_m of CO_2 and the expansivity $\alpha = (1/V)(\partial V/\partial T)_P$ are:

P (MPa)	1	2	3	5	7	9	10
V_m (cm^3 mol^{-1})	2823	1367	881	490	321	226	192
$10^3\alpha$ (K^{-1})	3.17	3.53	3.95	5.04	6.64	9.03	10.7

Evaluate:

$$G_m(350 \text{ K}, 10 \text{ MPa}) - G_m(350 \text{ K}, 1 \text{ MPa})$$

$$S_m(350 \text{ K}, 10 \text{ MPa}) - S_m(350 \text{ K}, 1 \text{ MPa})$$

$$H_m(350 \text{ K}, 10 \text{ MPa}) - H_m(350 \text{ K}, 1 \text{ MPa})$$

5. Show that $(\partial H/\partial P)_T = V(1 - \alpha T)$. Calculate $(\partial H/\partial P)_T$ for CO_2 at 350 K and 5 MPa. (See Problem 4.)

6. Derive an equation for the Joule–Thomson coefficient $\mu_{JT} = (\partial T/\partial P)_H$ for a gas that follows the van der Waals equation of state.

7. Mercury has been spilled in an unventilated room at 25°C. The vapor pressure of Hg is 0.23 Pa. Does the maximum level in parts per million of mercury vapor in the atmosphere of the room exceed the safe limit of 0.02 ppm?

8. The vapor pressure of diethyl ether at 0°C is 133.3 Pa. You are considering a setup to trap ether from a flowing gas with a cold trap at dry-ice temperature (-78°C). Suppose that a flow of N_2 saturated with ether enters the trap at 10 L min^{-1}, 100 kPa, and 0°C. Calculate the mass of ether lost through the trap in 10 h and the mass collected in the trap. The ΔH_v of ether can be taken as 420 J g^{-1} between 0 and -78°C.

9. Assume that N_2 follows van der Waals equation of state and calculate the change in temperature when N_2 at 200 K expanded through a throttling valve from 10^3 to 10^2 kPa. [Joule–Thomson expansion: assume that $\mu_{JT} = (\Delta T/\Delta P)_H$.]

10. Calculate the internal pressure $(\partial U/\partial V)_T$ of a gas that follows a virial equation, $PV/nRT = 1 + B(T)n/V + C(T)n^2/V^2$.

11. Suppose that the vapor pressure of NH_3 is given accurately by the equation in Table 8.2.
    (a) What pressure would be required to liquefy NH_3 at 260 K?
    (b) What would be the ΔH_v of liquid NH_3 at 230 K and at 260 K?
    (c) If ΔH_v is constant in the Clapeyron equation (8.24), calculate its value from the vapor pressures between 230 and 260 K.

12. For UF_6 the vapor pressures (Pa) are given by:

    Solid: $\log P^s = -3312T^{-1} - 5.33 \log T + 28.97$

    Liquid: $\log P^\ell = -1502T^{-1} + 9.624$

    Calculate the T and P at triple point (s, ℓ, g).

13. The vapor pressures of solid and liquid UCl_4 fit the equations

    $$\log_{10}(P/P°) = A + B/T$$

    Solid: $A = 10.43 \pm 0.10$, $B = -10\,412 \pm 82$ K

    Liquid: $A = 7.24 \pm 0.13$, $B = -7649 \pm 121$ K

Calculate (a) the melting point of UCl_4(s); (b) the ΔH_f of UCl_4(s) at the melting point; (c) from the estimated errors in the vapor-pressure parameters, estimate the errors in your results.

14. The density ρ of diamond at 298 K and 101.3 kPa is 3.513 g cm^{-3}, and of graphite, 2.260 g cm^{-3}. Assuming that ρ and $\Delta H_t = 1900$ J mol^{-1} are independent of pressure, estimate the pressure at which diamond and graphite would be in equilibrium at 298 K and at 1300 K.

15. For each of the following processes, state which of the quantities, ΔU, ΔH, ΔS, ΔG, and ΔA, are equal to zero for system specified.
    (a) An ideal gas is adiabatically expanded through a throttling valve.
    (b) A nonideal gas is adiabatically expanded through a throttling valve.
    (c) A nonideal gas is taken around a Carnot cycle.
    (d) H_2 and O_2 react to form H_2O in a thermally isolated bomb.
    (e) Liquid water is vaporized at 100°C and 1 atm.
    (f) HCl and NaOH react to form H_2O and NaCl in an aqueous solution at constant T and P.

16. The vapor pressure of solid CO_2 is 58.5 kPa at 188 K and 135 kPa at 198 K. Calculate ΔH(sublimation) and the sublimation temperature at $P° = 1$ atm.

17. Max Planck derived an equation for the temperature variation of the enthalpy ΔH_t of a phase transition along the equilibrium P–T curve. [Note that the Kirchhoff equation (6.26) applies only to a change at constant pressure.]

$$\frac{d(\Delta H_t)}{dT} = \Delta C_P + \frac{\Delta H_t}{T} - \Delta H_t(\partial \ln \Delta V/\partial T)_P$$

Derive the Planck equation starting from

$$dH = (\partial H/\partial T)_P \, dT + (\partial H/\partial P)_T \, dP$$

18. Show that for a solid $\longrightarrow$ gas or liquid $\longrightarrow$ gas transition, the Planck equation to a good approximation is $d(\Delta H_t)/dT \approx \Delta C_P$.

19. Pure gold melts at 1336 K and boils at 3133 K. $\Delta H_v(3133$ K$) = 343$ kJ mol^{-1}, $C_P(\ell) = 29.3$ J K^{-1} mol^{-1}. Calculate the vapor pressure of liquid gold from melting point to boiling point and plot the results on a $\ln P$ vs. T^{-1} diagram.

20. A student measures the vapor pressures of an organic compound over a range of temperatures and calculates $\Delta H_v = 30.0$ kJ mol^{-1} from the Clapeyron equation. There is a systematic error of 0.50 K in the temperature readings. What error would this cause in the ΔH_v?

21. At the normal melting point of ice, $\alpha^\ell = -6.0 \times 10^{-5}$ K^{-1} and $\alpha^s = 11.0 \times 10^{-5}$ K^{-1}. $\Delta H_f = 5.983$ kJ mol^{-1}, $c_P^\ell = 4.184$ J K^{-1} g^{-1}, $c_P^s = 2.02$ J K^{-1} g^{-1}, v(ice) $= 1.093$ cm^3 g^{-1}. Calculate $d(\Delta H_f)/dT$ from the Planck equation and compare this correct value with that from the Kirchhoff equation (6.26).

22. From the $\Delta G_f°$ (Table A.1), calculate $\Delta G°$ and K_P for following reactions at 298 K:
    (a) $N_2O(g) + 3H_2O(g) = 2NH_3(g) + 2O_2$
    (b) $HCl(g) + NaBr(c) = NaCl(c) + HBr(g)$
    (c) $N_2O(g) + 2Cu(c) = N_2(g) + Cu_2O(c)$
    (d) $H_2O_2(\ell) + SO_2(g) = SO_3(g) + H_2O(\ell)$
    (e) $NH_3(g) + CO(g) = HCN(g) + H_2O(g)$

23. For reactions (c) and (d) in Problem 22, what is ΔG_{298} when extent of reaction $\xi = 0.50$?

24. From the $\Delta G_f°(298$ K$)$ in Table A.3, calculate $\Delta G_{298}°$ for the following reactions in dilute aqueous solution.

(a) $H_2SO_4 = H^+ + HSO_4^-$

(b) $Ag^+ + Cl^- = AgCl(c)$

(c) $3Ca^{2+} + 2PO_4^{3-} = Ca_3(PO_4)_2(c)$

25. Over a short range of temperature, ΔS° can be taken to be constant. For the reactions in Problem 22, estimate $\Delta G^\circ(350 \text{ K})$. (See Table A.1.)

26. For $\frac{1}{2}SnO_2(c) + H_2(g) = \frac{1}{2}Sn(\ell) + H_2O(g)$, the equilibrium ratios P_{H_2O}/P_{H_2} are 923 K, 1.66; 973 K, 2.22; 1023 K, 2.84; 1073 K, 3.53. Calculate ΔG°, ΔH°, and ΔS° for the reaction at 1000 K.

27. Iodine (I_2) vapor is suddenly heated to 2000 K in a shock tube loaded with argon. If the partial pressure of I_2 was 100 Pa, calculate the extent of dissociation of I_2 after the shock wave has passed and the overall pressure is restored.

28. Attendants at swimming pools who smoke inhale both CO and Cl_2. If the reaction $CO + Cl_2 = COCl_2$ occurs in their lungs, they will be exposed to deadly phosgene (permitted limit 0.1 ppm). Thus besides the 10 years of life that they can expect to lose to cigarettes, they may be subject to acute $COCl_2$ poisoning. From a *thermodynamic viewpoint*, is such an increased danger possible? Assume partial pressures $P_{CO} = 10^{-6}$ atm, $P_{Cl_2} = 10^{-6}$ atm. (See Table A.2.) (RCT)

29. For the reaction $6CH_4 = C_6H_6(g) + 9H_2$, $\Delta H^\circ(298 \text{ K}) = 531$ kJ mol^{-1}, $\Delta S^\circ(298 \text{ K}) = 322$ J K^{-1} mol^{-1}, $\Delta C_P = 176 - 134 \times 10^{-3}T + 16.0 \times 10^{-6}T^2$ J K^{-1} mol. If other reactions could be ignored, would this reaction be suitable for synthesis of benzene from methane?

30. For the reaction, $2SO_3(g) = 2SO_2(g) + O_2(g)$:

(a) Calculate ΔG° and K_P at 298 K.

(b) Estimate ΔG° and K_P at 600 K assuming that ΔH° is constant from 298 to 600 K.

(c) Show that $K_P = a^3P/(2 + a)(1 - a)^2$, where a is the fractional extent of dissociation of SO_3. Calculate a at 600 K and $P = 100$ kPa, 500 kPa.

31. The equilibrium constant of the reaction (cyclohexane $\longrightarrow$ methylcyclopentane) $C_6H_{12}(g) \longrightarrow C_5H_9CH_3(g)$ was measured over a range of T as ln $K_P = 4.814 - 2059/T$. Calculate ΔG°, ΔS°, and ΔH° for the reaction at 1000 K.

32. Eastman gave K_P for the following sequence of reactions:

$$Fe_3O_4(s) + CO(g) = 3FeO(s) + CO_2(g) \quad K_{P_2}$$

$$FeO(s) + CO(g) = Fe(s) + CO_2(g) \quad K_{P_1}$$

T (K)	873	973	1073	1173
K_{P_1}	0.871	0.678	0.552	0.466
K_{P_2}	1.15	1.77	2.54	3.43

At what temperature could FeO, Fe, Fe$_3$O$_4$, CO, and CO$_2$ all coexist at equilibrium? What would be the ratio P_{CO_2}/P_{CO} at this point? If T were raised, what substance(s) would disappear?

33. $HD(g) + H_2O(g) = H_2(g) + HDO(g)$, $K_P = 2.6$ at 373 K. The percent HD in natural hydrogen is 0.0298%. You wish to remove the D from a supply of hydrogen gas by successive equilibrations of 50–50 mixtures of steam and hydrogen at 373 K. How many stages would be needed to reduce the HD to 0.003%? (Assume that H_2, D_2, and HD are always at equilibrium.)

34. For $Cl_2 = 2Cl$,

$T(K)$	600	800	1000	1200
K_P	4.80×10^{-16}	1.04×10^{-10}	2.45×10^{-7}	2.38×10^{-5}

Calculate $\Delta H°(298 \text{ K})$ for the reaction and the bond enthalpy of Cl—Cl. (See Table 6.4.)

35. Deoxygenated nitrogen is often prepared in the laboratory by passing tank N_2 over hot copper turnings: $2Cu(c) + \frac{1}{2}O_2(g) = Cu_2O(c)$, $\Delta G°(\text{kJ mol}^{-1}) = -195.4 - 0.0164T \log T + 0.1427T$. What would be the residual concentration of O_2 in N_2 with the copper at 600 C? Why is the copper heated?

9

Solutions—
Ideal and Dilute

Life on earth probably originated in aqueous solutions, and most biochemical reactions occur in water as a solvent or at interfaces between membranes and the watery medium in which cells live. On the other hand, most synthetic organic reactions are carried out in organic solvents. The theory of solutions is well developed on the thermodynamic side, that is, as far as large-scale properties and their relationships are concerned. The theory that would allow us to calculate the properties of solutions from molecular structures and interactions is still at a rather primitive stage.

In this chapter we shall first discuss the general description of solutions, and then develop equations for ideal solutions and for dilute solutions. In Chapter 10 we shall consider some of the theory of nonideal solutions.*

9.1 Measures of Composition

Table 9.1 is a summary of various ways to measure the composition of a solution.

Suppose that a solution contains n_A moles of component A, n_B moles of B, n_C moles of C, and so on. Then the mole fraction of A is

$$X_A = \frac{n_A}{n_A + n_B + n_C + \cdots}$$

Or, in general,

$$X_J = n_J / \sum_J n_J \tag{9.1}$$

*J. H. Hildebrand, "A History of Solution Theory," *Ann. Rev. Phys. Chem. 32*, 1 (1981).

TABLE 9.1
COMPOSITION OF SOLUTIONS

Name	Symbol	Definition	Usual SI unit
Molality	m	Amount of solute in unit mass of solvent	mol kg^{-1}
Concentration	c	Amount of solute in unit volume of solution	mol dm^{-3}
Volume molality	m'	Amount of solute in unit volume of solvent	mol dm^{-3}
[Mass] percent	%	Mass of solute in 100 unit masses of solution	Dimensionless
Mole fraction	X_A	Amount of component A divided by total amount of all components	Dimensionless

The *concentration* c_B of a component B in solution is the amount of B in unit volume of solution,

$$c_B = n_B/V \qquad (9.2)$$

Strictly speaking, concentration has only this meaning, but we often extend the concept to the number N of particles in a volume of solution, particularly when dealing with solutions of gases. We should then speak of a "number concentration, $C_B = N_B/V$, or perhaps a "number density." In nonscientific language one might say, "the concentration of the solution is 10% glucose in water," but we shall always use *concentration* as defined in Eq. (9.2). The SI unit of concentration is mole per cubic meter. In practice, mole per cubic decimeter (mol/dm^3) is often used. In these units, and only in these units, c is called the *molarity*.

For a solution of two components A and B, whose molar masses are M_A and M_B, if ρ is the density of the solution,

$$X_B = \frac{c_B M_A}{\rho + c_B(M_A - M_B)} \qquad \text{[Please prove this.]} \qquad (9.3)$$

Example 9.1 What are the concentration and mole fraction of sucrose in a solution in water at 25°C that is 20.0% by weight? The density of the solution is $\rho = 1.0794$ g cm^{-3}.

$M_A = 18.0$ g mol^{-1}, $M_B = 342$ g mol^{-1}. Take as basis 100 g of solution:

Concentration: $c_B = \dfrac{n_B}{100/\rho} = \dfrac{(20/342)10^3}{100/1.0794} = 0.631$ mol dm^{-3}

Mole fraction: $X_B = \dfrac{n_B}{n_A + n_B} = \dfrac{20/342}{20/342 + 80/18} = 0.0130$

The concentration c of a solution depends on temperature, since, from Eq. (9.2) $c = n\rho/\mathfrak{m}$, where $\mathfrak{m}$ is the mass and ρ the density of solution, and the latter depends on T. For example, at 5°C, where the density of the solution in Example 9.1 is

$\rho = 1.0843$ g cm^{-3}, the concentration of the same solution of sucrose would be 0.634 compared to 0.631 mol dm^{-3} at 25°C. This built-in dependence of c on T makes c useful in some theoretical analyses, but it can be inconvenient in practical applications.

The *molality* m_B of a component B in a solution is defined as the amount of B divided by the mass m_A of some other component A, which is denoted as the solvent,

$$m_B = n_B/m_A = n_B/n_A M_A \tag{9.4}$$

where M_A is the molar mass of A. Note that molality does not depend on the density of the solution and hence does not vary with T. In SI, molality m has units of mol kg^{-1}.

For a solution of two components, A and B, in which the solvent A has molar mass M_A, the relation between molality m_B and mole fraction X_B is

$$X_B = \frac{m_B}{m_B + 1/m_A} \tag{9.5}$$

Example 9.2 What is the molality of sucrose in the solution described in Example 9.1?

From Eq. (9.5),

$$m_B = \frac{X_B}{(1 - X_B)M_A} = \frac{0.0130}{(0.987)(18.0 \times 10^{-3} \text{ kg mol}^{-1})}$$

$$= 0.732 \text{ mol kg}^{-1}$$

9.2 Partial Molar Quantities—Partial Molar Volume

The equilibrium properties of solutions are described in terms of thermodynamic functions of state. How do extensive functions such as V, U, S, G, and H depend on the composition of a solution? As an example of how this problem is handled, consider the volume V. Suppose that we have large amounts of pure water and pure ethanol at $T = 298.15$ K and $P° = 101.32$ kPa. The density of water under these conditions is $\rho = 0.997$ g cm^{-3}, and the density of ethanol is $\rho = 0.785$ g cm^{-3}. The molar volumes of the pure liquids are

$$V_w^\bullet = M_w/\rho_w = 18.0 \text{ g mol}^{-1}/0.997 \text{ g cm}^{-3} = 18.1 \text{ cm}^3 \text{ mol}^{-1}$$

$$V_e^\bullet = M_e/\rho_e = 46.1 \text{ g mol}^{-1}/0.785 \text{ g cm}^{-3} = 58.7 \text{ cm}^3 \text{ mol}^{-1}$$

where M_w and M_e are the molar masses of water and ethanol. A subscript on any extensive state function for a pure substance means the value of the function for unit amount of the substance (SI, per mole). Thus $V_j^\bullet$ is the molar volume of j for a pure substance j. $V_j^\bullet$ is an intensive and not an extensive function.

If we add one mole of water to a large volume of water at 25°C the volume will increase by 18.1 cm^3. If we add one mole of ethanol to a large volume of ethanol, the volume will increase by 58.7 cm^3. If, however, we add one mole of ethanol to a large volume of water, the volume will increase by 54.2 cm^3, and if we add one mole of water to a large volume of ethanol, the volume will increase by 14.1 cm^3. These volume changes are the *partial molar volumes* of ethanol in water and of water in ethanol *in the limit of extreme dilution*. The partial molar quantity is an *intensive state function*, as denoted by a subscript on the symbol for an extensive state function.

$$V_e(X_e \to 0) = 54.2 \text{ cm}^3 \text{ mol}^{-1}$$

(This is the volume of a mole of ethanol when the ethanol molecules are surrounded by water.)

$$V_w(X_w \to 0) = 14.1 \text{ cm}^3 \text{ mol}^{-1}$$

(This is the volume of a mole of water when the water molecules are surrounded by ethanol.)

Partial molar volumes depend on the composition of the solution. They can always be defined in the following way. The partial molar volume of a component is the change in volume of solution when one mole of the component is dissolved at constant T and P in a very large volume of solution of specified composition. Thus if we wanted to measure the partial molar volume of ethanol in a 50 mol % solution of ethanol in water, we would add one mole of ethanol to a very large volume of the 50 mol % solution. (A "very large volume" would be about 100 times the molar volume of any component.) Figure 9.1 shows the partial molar volumes of ethanol and of water as measured over the entire range of concentrations of the solutions, from $X_e = 0$ to $X_e = 1$.

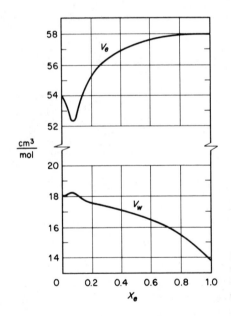

FIGURE 9.1 Partial molar volumes of water and of ethanol in solutions at 20°C: V_w (water), V_e (ethanol), X_e (mole fraction ethanol).

The partial molar volume V_A of a component A in any solution is the increase in volume of the solution per mole of A at the specified P, T, and composition. Since V_A is the change of volume with amount n_A of A at constant P, T, and amount n_B of B, it is defined as

$$V_A = (\partial V/\partial n_A)_{T,P,n_B} \qquad (9.6)$$

Partial molar volumes are useful since, if we know their values, we can calculate the molar volume of a solution of any specified composition. At constant T and P, the volume of a solution of A and B (a *binary solution*) is a function of n_A and n_B, $V(n_A, n_B)$. If amounts dn_A of A and dn_B of B are added to the solution, the increase in volume is given by the complete differential of V,

$$dV = (\partial V/\partial n_A)_{T,P,n_B}\, dn_A + (\partial V/\partial n_B)_{T,P,n_A}\, dn_B \qquad (9.7)$$

From Eq. (9.6),

$$dV = V_A dn_A + V_B dn_B \qquad (9.8)$$

This expression can be integrated, which corresponds physically to increasing the volume of the solution without changing its composition, V_A and V_B hence being kept constant. The result is

$$V = n_A V_A + n_B V_B \qquad (9.9)$$

This equation tells us that the volume of the solution equals the amount of A times the partial molar volume of A, plus the amount of B times its partial molar volume.

Example 9.3 From the data in Fig. 9.1, calculate the volume of a solution at 25°C that contains exactly 10 mol of ethanol and 6 mol of water. What is the change in volume when 10 mol of ethanol is mixed with 6 mol of water?

The mole fraction of ethanol in the solution is $X_e = 10/16 = 0.625$. From Eq. (9.9), the final volume of solution is

$$V = n_e V_e + n_w V_w = (10)(57.6) + 6(16.5) = 675 \text{ cm}^3$$

The volume of the liquids before mixing was $10(58.7) + 6(18.1) = 696 \text{ cm}^3$. The volume change is final volume minus initial volume, $\Delta V = -21 \text{ cm}^3$.

The complete differential of Eq. (9.9), in which both n_J and V_J are allowed to change, is

$$dV = n_A dV_A + V_A dn_A + n_B dV_B + V_B dn_B \qquad (9.10)$$

For both Eqs. (9.8) and (9.10) to hold, one must have, therefore,

$$n_A dV_A + n_B dV_B = 0 \qquad (9.11)$$

This is one example of the *Gibbs–Duhem equation*. It can also be written as

$$dV_A = -\frac{n_B}{n_A} dV_B = \frac{X_B}{X_B - 1} dV_B \qquad (9.12)$$

This useful formula allows us to calculate one partial molar volume V_A of a binary solution if we know the other one as a function of composition, $V_B(X_B)$. From Eq. (9.12)

$$V_A(X_B) = \int_0^{X_B} \frac{X'_B}{X'_B - 1} \frac{dV_B}{dX'_B} dX'_B \qquad (9.13)$$

Example 9.4 The partial molar volume of K_2SO_4 in aqueous solutions at 298 K is given by $V_2(\text{cm}^3) = 32.280 + 18.216m^{1/2}$, where m is the molality of K_2SO_4. Use the Gibbs–Duhem relation to obtain an equation for V_1, the partial molar volume of water in the solutions. The molar volume of pure water at 298 K is $V_1^{\bullet} = 18.070$ cm³ mol⁻¹.

From Eq. (9.12), $dV_1 = -(n_2/n_1)\, dV_2$; $dV_2 = (9.108m^{-1/2})\, dm$. When $n_2 = m$, $n_1 = 10^3/M_1$. Thus,

$$dV_1 = -(mM_1/1000)(9.108m^{-1/2})\, dm$$

$$V_1 = -\frac{M_1}{1000} \int 9.108m^{1/2}\, dm + C = -\frac{M_1}{1000}(6.072m^{3/2}) + C$$

When $m = 0$, $V_1^{\bullet} = C = 18.070$. Thus $V_1 = 18.070 - 0.1094m^{3/2}$. For example, when $m = 2.000$ mol/kg, $V_1 = 17.760$ cm³.

9.3 Other Partial Molar Quantities

We have presented partial molar functions in terms of partial molar volumes, but similar equations can be written for all the other partial molar functions. The partial molar Gibbs function is the chemical potential,

$$\mu_A = (\partial G/\partial n_A)_{T,P,n_B} \tag{9.14}$$

The partial molar enthalpy is

$$H_A = (\partial H/\partial n_A)_{T,P,n_A} \tag{9.15}$$

H_A and H_B are used to calculate the enthalpy changes when components mix to form solutions. Thus, for a solution of A and B,

$$\Delta H(\text{solution}) = H(\text{solution}) - H(\text{components})$$

$$\Delta H(\text{solution}) = n_A H_A + n_B H_B - n_A H_A^{\bullet} - n_B H_B^{\bullet} \tag{9.16}$$

$$= n_A(H_A - H_A^{\bullet}) + n_B(H_B - H_B^{\bullet})$$

All the thermodynamic relations derived in earlier chapters can also be applied to the partial molar functions; for example,

$$\left(\frac{\partial G_A}{\partial T}\right)_P = \left(\frac{\partial \mu_A}{\partial T}\right)_P = -S_A \tag{9.17}$$

$$\left(\frac{\partial G_A}{\partial P}\right)_T = \left(\frac{\partial \mu_A}{\partial P}\right)_T = V_A \tag{9.18}$$

9.4 How Partial Molar Quantities Are Measured

Let us again take the volume as an example. The partial molar volume V_A, defined by Eq. (9.6), is equal to the slope of the curve obtained when the volume of the solution is plotted against the molality m_A of A. This result follows since m_A is the amount n_A of A in a *constant* mass, say 1 kg, of component B, Determinations of partial molar quantities by this *slope method* are rather inaccurate but are convenient for quick surveys.

The *method of intercepts* is generally preferred. A quantity is defined, called the *mean molar volume* of the solution $\bar{V}_m$, which is the volume of the solution divided by the total number of moles of components. For a binary solution, $\bar{V}_m = V/(n_A + n_B)$. Then $V = \bar{V}_m(n_A + n_B)$ and

$$V_A = \left(\frac{\partial V}{\partial n_A}\right)_{n_B} = \bar{V}_m + (n_A + n_B)\left(\frac{\partial \bar{V}_m}{\partial n_A}\right)_{n_B} \tag{9.19}$$

Now, the derivative with respect to n_A is transformed into a derivative with respect to mole fraction X_B,

$$\left(\frac{\partial \bar{V}_m}{\partial n_A}\right)_{n_B} = \frac{d\bar{V}_m}{dX_B}\left(\frac{\partial X_B}{\partial n_A}\right)_{n_B}$$

Since $X_B = n_B/(n_A + n_B)$, $(\partial X_B/\partial n_A)_{n_B} = -n_B/(n_A + n_B)^2$, and thus Eq. (9.19) becomes

$$V_A = \bar{V}_m - \frac{n_B}{n_A + n_B}\frac{d\bar{V}_m}{dX_B}$$

$$\bar{V}_m = X_B\frac{d\bar{V}_m}{dX_B} + V_A \tag{9.20}$$

The application of this equation is illustrated in Fig. 9.2, where the mean molar volume $\bar{V}_m$ of a solution is plotted against the mole fraction X_B. (Recall the standard slope–intercept form of the straight line: $y = mx + b$.) The line S_1S_2 is drawn tangent to the curve at point P, corresponding to a definite mole fraction X_B. The intercept O_1S_1 at $X_B = 0$ is V_A, the partial molar volume of A at the particular composition X_B. It can readily be shown that the intercept on the other axis, O_2S_2, is the partial molar volume of B, V_B. [Please show it.]

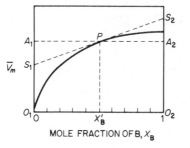

FIGURE 9.2 Determination of partial molar volumes by the intercept method. The dashed line is the tangent to the curve of $\bar{V}_m$ as a function of X_B at a particular mole fraction X'_B. The intercept O_1S_1 gives V_A, the partial molar volume of A at X'_B, and the intercept of O_2S_2 gives V_B, the partial molar volume of B.

9.5 The Ideal Solution—Raoult's Law

The most important observable quantity in the theory of solutions is the vapor pressure of a component above a solution. This partial vapor pressure is a measure of the tendency of the given species to escape from solution into the vapor phase. The tendency of a component to escape from solution is a direct indication of its chemical potential within the solution. The lower the chemical potential, the lower the vapor pressure. By studying the escaping tendencies, or partial vapor pressures, as functions of temperature, pressure, and concentration, we obtain a description of the thermodynamics of the solution. (To be precise, however, we need to consider small corrections to the partial vapor pressure due to the fact that the vapor is not an ideal gas.)

The concept of the ideal gas has played an important role in the thermodynamics of gases. Many cases of practical interest are treated adequately by means of the ideal-gas approximation, and systems deviating from ideality are described by comparing them to the ideal case. It would be helpful to find some similar concept to act as a guide in the theory of solutions, and fortunately it is possible to do so. Ideality in a gas implies a complete *absence* of cohesive forces. Ideality in a solution is defined by complete *uniformity* of cohesive forces. For two components A and B, the intermolecular forces between A and B, B and B, and A and A are all the same in an ideal solution.

For an ideal solution, the tendency of A to pass into the vapor is proportional to X_A, the mole fraction of A in solution:

$$P_A = k'X_A \tag{9.21}$$

where k' is a proportionality constant. When $X_A = 1$, $P_A = P_A^\bullet$, the vapor pressure of pure A. Thus Eq. (9.21) becomes

$$P_A = X_A P_A^\bullet \tag{9.22}$$

In 1886, François Raoult reported data on the partial pressures of components in some solutions that closely followed Eq. (9.21), which is therefore called *Raoult's Law*. An ideal solution is defined as one that follows Raoult's Law.

The vapor pressures of the system ethylene bromide + propylene bromide are plotted in Fig. 9.3. The experimental results almost coincide with the theoretical curves predicted by Eq. (9.22). In this instance, the agreement with Raoult's Law is excellent. It is unusual, however, to find solutions that follow Raoult's Law so closely over the whole range of concentrations. Ideality demands a uniformity of intermolecular

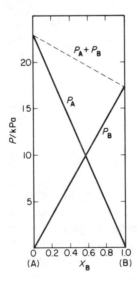

FIGURE 9.3 Vapor pressures in system $C_2H_4Br_2$(A) + $C_3H_6Br_2$(B) at 85°C closely follow Raoult's Law.

forces between two different components, and this can be achieved only when the components are chemically very much alike.

If the component B added to pure A lowers the vapor pressure, Eq. (9.22) can be written an terms of a relative vapor-pressure lowering,

$$\frac{P_A^{\bullet} - P_A}{P_A^{\bullet}} = 1 - X_A = X_B \qquad (9.23)$$

This form of the equation is especially useful for solutions of a nonvolatile solute in a volatile solvent, and it can sometimes be used to determine the molar mass of a solute.

Example 9.5 The vapor pressure of CS_2 at 293 K is 11.386 kPa. When 2.000 g of sulfur is dissolved in 100.0 g CS_2, the vapor pressure falls to 11.319 kPa. Calculate the molar mass of sulfur in the solution.

From Eq. (9.23),

$$X_B = \frac{2.000/M_B}{2.000/M_B + 100.0/76.13} = \frac{11.386 - 11.319}{11.386} = 0.00588$$

$M_B = 257$. This corresponds to a formula S_8 for the sulfur molecule in CS_2 solution.

9.6 Thermodynamics of Ideal Solutions

Equation (8.19) stated the requirement for equilibrium of a pure substance A in two phases a and b at some specified T and P, $G_A^a = G_A^b$. In the case of components in solution, the equilibrium condition is written in terms of chemical potentials, μ_A and μ_B. For equilibrium between a solution and its vapor,

$$\mu_A = \mu_A^g, \qquad \mu_B = \mu_B^g \qquad (9.24)$$

If the vapor is an ideal mixture of ideal gases, we have from Eq. (8.31),

$$\mu_A^g = \mu_A^\circ + RT \ln P_A/P^\circ, \qquad \mu_B^g = \mu_B^\circ + RT \ln P_B/P^\circ \qquad (9.25)$$

where μ_A° and μ_B° are the chemical potentials in the standard state and P_A and P_B are the partial vapor pressures of A and B above the solution. From Raoult's Law, $P_A = X_A P_A^{\bullet}$ and $P_B = X_B P_B^{\bullet}$, so that Eqs. (9.24) and (9.25) yield

$$\mu_A = \mu_A^\circ + RT \ln P_A^{\bullet}/P^\circ + RT \ln X_A$$
$$\mu_B = \mu_B^\circ + RT \ln P_B^{\bullet}/P^\circ + RT \ln X_B \qquad (9.26)$$

At any specified temperature and pressure $P_A^{\bullet}$ and $P_B^{\bullet}$ are definite values of vapor pressures of pure components A and B. We therefore can combine the first two terms in Eq. (9.26) as $\mu_A^\circ + RT \ln P_A^{\bullet}/P^\circ = \mu_A^{\bullet}(T)$ so that

$$\mu_A = \mu_A^{\bullet}(T) + RT \ln X_A \qquad (9.27)$$

Note that $\mu_A^{\bullet}(T)$ is the value of μ_A when $X_A = 1$, or in other words, the chemical potential of pure liquid A at P° and the specified temperature. An equation like Eq. (9.27) holds for each component A, B, C, . . . in an ideal solution. It is the basic thermodynamic definition of an ideal solution.

When components are mixed to form an ideal solution there is no ΔV and there is no ΔH of mixing. In other words, for each component in an ideal solution, the partial molar volume equals the molar volume of pure component, $V_j = V_j^{\bullet}$, and the partial molar enthalpy equals the molar enthalpy of pure component, $H_j = H_j^{\bullet}$. These results are consistent with the fact that the intermolecular forces in an ideal solution are uniform. A molecule of A experiences the same intermolecular attractions and repulsions whether it is surrounded entirely by B, partly by B and partly by A, or entirely by other A molecules. A molecule could not tell who its neighbors were if it kept its eyes closed in an ideal solution.

The entropy change on mixing components to form an ideal solution is the same as the expression obtained in Section 7.11 for the mixing of molecules in an ideal gas. The entropy of mixing per mole is

$$\Delta S_m = -R \sum X_i \ln X_i \qquad (9.28)$$

The entropy of mixing in an ideal solution is purely statistical. There is no ΔS caused by effects of mixing on the structure of the liquid, although such effects would occur in a nonideal solution.

Since $\Delta H_m = 0$ for mixing components to form an ideal solution, the free energy of mixing $\Delta G_m = \Delta H_m - T \Delta S_m = -T \Delta S_m = -RT \sum X_i \ln X_i$ per mole of solution. The thermodynamic quantities for formation of 1 mole of ideal solution are summarized in the diagram of Fig. 9.4.

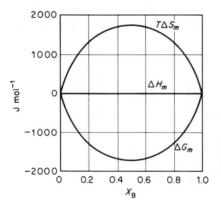

FIGURE 9.4 Thermodynamic quantities, $T\Delta S_m$, ΔH_m, and ΔG_m for formation of one mole of ideal solution of A and B by mixing the pure substance to give a final mole fraction X_B.

9.7 Solubility of Gases in Liquids—Henry's Law

Consider a solution of component B, which may be called the solute, in A, the solvent. If the solution is sufficiently diluted, a condition ultimately is attained in which each molecule of B is surrounded completely by component A. The solute B is then in a uniform environment irrespective of the fact that A and B may form solutions that are far from ideal at higher concentration. In such a very dilute solution, the escaping tendency of B from its uniform environment is proportional to its mole fraction, but the proportionality constant k_H no longer is $P_B^{\bullet}$ as it is for the ideal solution in Eq. (9.22). We may write

$$P_B = k_H X_B \quad \text{or} \quad X_B = P_B/k_H \tag{9.29}$$

William Henry, in 1803, established this equation by extensive measurements of the dependence on pressure (P_B) of the solubility (X_B) of gases in liquids. Henry's Law is not restricted to gas–liquid systems, however, but is followed by a wide variety of fairly dilute solutions, and *by all solutions in the limit of extreme dilution*. The form of the law for dissociated solutes, such as electrolytes, is discussed in Section (10.12).

Some data on the solubility of gases in water over a wide range of pressures are shown in Fig. 9.5. If Henry's Law were obeyed exactly, these solubility curves would all be straight lines. Actually, the curves for H_2, H_2, and N_2 are quite linear up to about 10^7 Pa, but for O_2 we can see deviations even in this range.

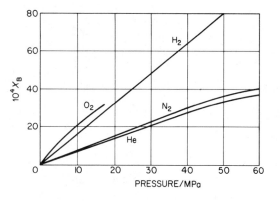

FIGURE 9.5 Solubility of gases in water at 298.15 K as a function of pressure.

Example 9.6 A diver descends to a depth of 200 m in a suit filled with 4% O_2–96% He at an equivalent pressure. Based on Henry's Law, what would be the equilibrium mole fraction of He in his body tissues? What volume of gas would be released per cm^3 of tissue if he underwent a sudden decompression?

The pressure at 200 m is $(200 \text{ m})(10^3 \text{ kg m}^{-3})(9.81 \text{ m s}^{-2}) = 1.96 \times 10^6$ Pa. From Fig. 9.5, the Henry's Law constant for He in water is 1.52×10^{10} Pa. Hence, from Eq. (9.29), $X_B = 0.96 \times 1.96 \times 10^6 \text{ Pa}/1.52 \times 10^{10} \text{ Pa} = 1.24 \times 10^{-4}$. At atmospheric pressure, $X_B = 0.96 \times 101.3 \times 10^3 \text{ Pa}/1.52 \times 10^{10} \text{ Pa} = 0.64 \times 10^{-5}$. Assume tissue to be essentially water; 1 cm^3 of H_2O amounts to 1/18 mol, so that 1.18×10^{-4} (1/18) mol He released would give 0.15 cm^3 at STP. Such a sudden release of He in the tissues could cause a fatal attack of "the bends."

9.8 Mechanism of Anesthesia

One of the most fascinating problems in medical physiology is the mechanism by which gases produce anesthesia and narcosis. Many anesthetics, such as krypton and xenon, are inert chemically; in fact, all gases have an anesthetic effect at high enough pressures. Jacques Cousteau, in *The Silent World*, gives a memorable account of the nitrogen narcosis experienced at great depths, which has claimed the life of more than one diver. Inert-gas narcosis should not be confused with decompression sickness ("the bends"), which is caused by rapid release of gas dissolved in the tissues with consequent bubble formation and damage to cells (Example 9.6). Table 9.2 summarizes some anesthetic data on mice. Even helium produces narcosis at high pressures.

TABLE 9.2
ANESTHETIC PRESSURES FOR MICE (RIGHTING REFLEX)

Key no.	Gas	Pressure (kPa)	Key no.	Gas	Pressure (kPa)
1	He	19 300	10	C_2H_4	100
2	Ne	11 000	11	C_2H_2	90
3	Ar	2 400	12	Cyclo-C_3H_6	10
4	Kr	400	13	CF_4	1 900
5	Xe	100	14	SF_6	700
6	H_2	8 500	15	CF_2Cl_2	40
7	N_2	3 500	16	$CHCl_3$	0.8
8	N_2O	150	17	Halothane	1.7
9	CH_4	600	18	Ether	3.2

Early attempts to understand the causes of anesthesia were made by Meyer (1899) and Overton (1901), who found a good correlation between the solubility of a gas in a lipid (olive oil) and its narcotic efficacy. Examples are given in Fig. 9.6.

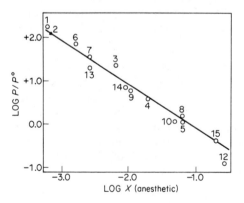

FIGURE 9.6 Graph of log anesthetic pressure of a gas vs. log solubility in olive oil at 37°C with $P^\circ = 101.3$ kPa. For key to numbers, see Table 9.2. The line has unit slope.

Since nerve-cell membranes are composed mainly of lipids and proteins, it was suggested that the anesthetic molecules dissolve in the membranes and block the process of nerve conduction in some way as yet unknown. The mole fraction of dissolved anesthetic necessary to produce anesthesia is in the range 0.02 to 0.05, so that the anesthetic can certainly alter the properties of the membrane to some considerable extent.

9.9 Pressure–Composition Diagrams

The example of a *P–X* diagram in Fig. 9.7 shows the system 2-methylpropanol-1 + propanol-2, which obeys Raoult's Law quite closely over the entire range of compositions. The straight upper line (*liquidus curve*) represents the dependence of the total vapor pressure above the solution on composition of the liquid. The curved lower line represents the dependence of the total vapor pressure on composition of the vapor.

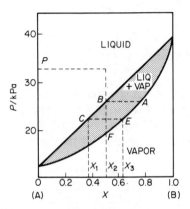

FIGURE 9.7 Pressure–composition (mole fraction) diagram at 333 K for the system 2-methylpropanol-1 (A) + propanol-2 (B), which form practically ideal solutions.

Consider a liquid of composition X_2 at a pressure P_2. This point G lies in a one-phase all-liquid region, in which there are three degrees of freedom (from the phase rule, $f = c - p + 2 = 2 - 1 + 2 = 3$). One of the degrees of freedom is used by the requirement of constant temperature for the diagram. Hence, for any arbitrary composition X_2, the liquid solution at constant T can exist over a range of different pressures.

As pressure is decreased along the dashed line of constant composition, nothing happens until the liquidus curve is reached at B. At this point, liquid begins to vaporize. The vapor that is formed is richer than the liquid in the more volatile component, propanol-2. The composition of the first vapor to appear is given by point A on the vapor curve.

As pressure is further reduced below B, a two-phase region on the diagram is entered. This represents the region of stable coexistence of liquid and vapor. The dashed line passing horizontally through a typical point D in the two-phase region is called a *tie-line;* it connects liquid and vapor compositions that are in equilibrium. In the two-phase region, the system is bivariant. One of the degrees of freedom is used by the requirement of constant temperature, and only one remains. When the pressure is fixed in this region, therefore, the compositions of both the liquid and the vapor phases are also definitely fixed. They are given, as we have seen, by the end points of the tie-line.

The overall composition of the system at point D in the two-phase region is X_2. This is made up of liquid having a composition X_1, and vapor having a composition X_3. We can calculate the relative amounts of liquid and vapor required to yield the overall composition. Let n_ℓ and n_v be the sums of the numbers of moles of both components A and B in liquid and in vapor, respectively. From a material balance applied to component B, $X_2(n_\ell + n_v) = X_1 n_\ell + X_3 n_v$, or

$$\frac{n_\ell}{n_v} = \frac{X_3 - X_2}{X_2 - X_1} = \frac{DE}{DC} \tag{9.30}$$

This expression is called the *lever rule*. It applies to the amount of any two phases in equilibrium connected by a tie-line in a phase diagram of a two-component system.

As pressure is still further decreased along BF, more liquid is vaporized, until finally, at F, no liquid remains. Further decrease in pressure then proceeds in the one-phase, all-vapor region.

9.10 Temperature–Composition Diagrams

The temperature–composition diagram of a liquid–vapor equilibrium is the boiling-point diagram of the solution at the constant pressure chosen. If the pressure is $P° = 101.32$ kPa, the boiling points are the *normal* ones. The T–X diagram for the 2-methylpropanol-1 + propanol-2 system is shown in Fig. 9.8.

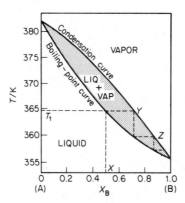

FIGURE 9.8 Boiling point vs. composition diagram for 2-methylpropanol-1 (A) + propanol-2 (B), which form practically ideal solutions.

The boiling-point diagram for an ideal solution can be calculated if the vapor pressures of the pure components are known as functions of temperature. The two end points of the boiling point diagram shown in Fig. 9.8 are the temperatures at which the pure components have vapor pressures of 101.32 kPa (i.e., 82.3 and 108.5°C). The composition of the solution that boils anywhere between these two temperatures, say at 100°C, is found as follows: If X_A is the mole fraction of C_4H_9OH, from Raoult's Law we have $101.3 = P_A^\bullet X_A + P_B^\bullet(1 - X_A)$. At 100°C, the vapor pressure of C_3H_7OH is 192 kPa; of C_4H_9OH, 76.0 kPa. Thus $101.3 = 76.0\,X_A + 192(1 - X_A)$, or $X_A = 0.781$, $X_B = 0.219$. This gives one intermediate point on the liquidus curve (boiling point vs. composition of solution); the others are calculated in the same way.

The composition of the vapor is given by Dalton's Law of Partial Pressures:

$$X_A^{\text{vap}} = P_A/101.3 = X_A^\ell P_A^\bullet/101.3 = 0.781 \times 76.0/101.3 = 0.586$$

$$X_B^{\text{vap}} = P_B/101.3 = X_B^\ell P_B^\bullet/101.3 = 0.219 \times 192/101.3 = 0.415$$

The curve for vapor composition vs. pressure is therefore readily constructed from the liquidus curve.

9.11 Fractional Distillation

The application of the boiling-point diagram to distillation is shown in Fig. 9.8. The solution of composition X begins to boil at temperature T_1. The first vapor formed has a composition Y, richer in the more volatile component. If this vapor Y is condensed and reboiled, vapor of composition Z is obtained. This process is repeated until the distillate is composed of pure component B. In practical cases, the successive fractions each cover a range of compositions, but the vertical lines in Fig. 9.8 may be taken as average compositions within these ranges.

A fractionating column is a device that carries out automatically the successive condensations and vaporizations required for fractional distillation. An especially clear example is the bubble-cap type of column in Fig. 9.9. As vapor ascends from the boiler, it bubbles through a film of liquid on the first plate. This liquid is somewhat cooler than that in the boiler, so that a partial condensation takes place. The vapor that leaves the first plate is therefore enriched in the more volatile component compared to vapor from the boiler. A similar enrichment takes place on each successive plate. Each attainment of equilibrium between liquid and vapor corresponds to one of the steps in Fig. 9.8. Each stage is called a *theoretical plate*. Let us suppose, for

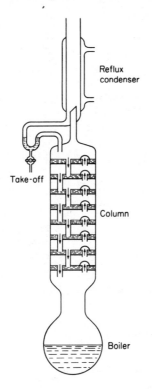

Reflux condenser

Take-off

Column

Boiler

FIGURE 9.9 A bubble-cap fractionating column.

example, that we start with a solution having mole fraction $X_B = 0.500$ of propanol (B) in 2-Me propanol-1 (A) and distill it in a column with three theoretical plates. The first distillate that is taken, point D in the figure, will have a composition of $X_B = 0.952$ propanol.

The efficiency of a distilling column is measured by the number of equilibration stages that it achieves. In a well-designed bubble-cap column, each unit acts as nearly one theoretical plate. The performance of various types of packed columns is also described in terms of theoretical plates. The separation of liquids whose boiling points lie close together requires a column with a considerable number of plates. The number depends on the *reflux ratio*, the ratio of distillate returned to the column to that taken off as product.

9.12 Solutions of Solids in Liquids

Solubility curve and *freezing-point depression curve* are two different names for the same thing, that is, a temperature vs. composition curve for a solid–liquid equilibrium at some constant pressure, usually chosen as $P°$. Such a diagram is shown in Fig. 9.10 for the system naphthalene + menthol. The curve CE may be interpreted

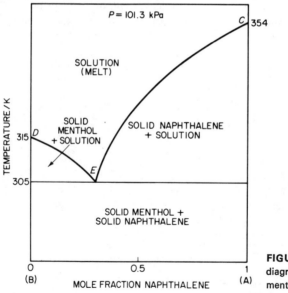

FIGURE 9.10 Temperature–composition diagram for the system naphthalene + menthol.

as either (1) the depression of the freezing point of naphthalene by the addition of menthol, or (2) the solubility of solid naphthalene in the solution. In one case, we regard T as a function of X; in the other, X as a function of T. The lowest point E on the solid–liquid diagram is called the *eutectic point* (from the Greek, *eutektos*, easily melted).

In this diagram, the solid phases that separate are shown as pure naphthalene

(A) on one side and pure menthol (B) on the other. This representation is not exactly correct, since there must be a small extent of solid solution of B in A and of A in B. Nevertheless, the absence of any solid solution is in many cases an excellent approximation.

For a pure solid A to be in equilibrium with a solution containing A, it is necessary that the chemical potential of A be the same in the two phases, $\mu_A^s = \mu_A^l$. From Eq. (9.27), the chemical potential of component A in an ideal solution is $\mu_A^l = \mu_A^{\bullet l} + RT \ln X_A$, where $\mu_A^{\bullet l}$ is the chemical potential of pure liquid A. Thus the equilibrium condition can be written

$$\mu_A^{\bullet s} = \mu_A^{\bullet l} + RT \ln X_A$$

Now $\mu_A^{\bullet s}$ and $\mu_A^{\bullet l}$ are simply the molar free energies of pure solid and pure liquid. Hence

$$\frac{G_A^{\bullet s} - G_A^{\bullet l}}{RT} = \ln X_A \tag{9.31}$$

Since $\partial(G/T)\partial T = -H/T^2$ from Eq. (8.39), differentiation of Eq. (9.31) with respect to T yields (with ΔH_f the enthalpy of fusion)

$$\frac{H_A^{\bullet l} - H_A^{\bullet s}}{RT^2} = \frac{\Delta H_f}{RT^2} = \frac{d \ln X_A}{dT} \tag{9.32}$$

It is a good approximation to take ΔH_f independent of T over moderate ranges of temperature. Integrating Eq. (9.32) from $T_f^\bullet$, the freezing point of pure A, mole fraction unity, to T, the temperature at which pure solid A is in equilibrium with solution of mole fraction X_A, we obtain

$$\ln X_A = \frac{\Delta H_f}{R}\left(\frac{1}{T_f^\bullet} - \frac{1}{T}\right) \tag{9.33}$$

This is the equation for the temperature variation of the solubility X_A of a pure solid in an ideal solution. The plot of $\ln X_A$ vs. T^{-1} for an ideal solution is linear.

Example 9.7 Calculate the solubility of naphthalene in an ideal solution at 298 K.

Naphthalene melts at 353 K, and its ΔH_f at the melting point is 19.29 kJ mol^{-1}. Thus, from Eq. (9.33),

$$\frac{19\,290 \text{ (J mol}^{-1})}{8.314 \text{ (J K}^{-1} \text{ mol}^-)^1}(353^{-1} - 298^{-1})(\text{K}^{-1}) = \ln X_A, \quad X_A = 0.297$$

This is the mole fraction of naphthalene in any ideal solution, whatever the solvent may be. Actually, the solution will approach ideality only if the solvent is rather similar in chemical and physical properties to the solute. Typical experimental values for the solubility X_A of naphthalene in various solvents at 298 K are: chlorobenzene, 0.317; benzene, 0.296; toluene, 0.286; acetone, 0.224; hexane, 0.125.

We can rewrite Eq. (9.33) in terms of $X_B = 1 - X_A$, the mole fraction of solute:

$$\frac{\Delta H_f}{R}\frac{(T - T^\bullet)}{TT^\bullet} = \ln(1 - X_B)$$

For depressions of the freezing point $(T_f^\bullet - T) = \Delta T_f$ that are small compared to T, we can set $TT^\bullet = T^{\bullet 2}$, and then, on expanding the logarithm in a power series.

$$\frac{\Delta H_f(\Delta T_f)}{RT_f^{\bullet 2}} = -\ln(1 - X_B) = X_B + \tfrac{1}{2}X_B^2 + \tfrac{1}{3}X_B^3 + \cdots$$

For dilute solutions, $X_B \ll 1$, and from Eq. (9.5), with $1/M_A \gg m_B$, the molality of B, we obtain

$$\Delta T_f \approx \frac{RT_f^{\bullet 2}}{\Delta H_f} X_B \approx \frac{RT_f^{\bullet 2}}{\Delta H_f} M_A m_B = K_F m_B \qquad (9.34)$$

The *molal freezing-point depression* is

$$K_F = \frac{RT_f^{\bullet 2} M_A}{(\Delta H_f)} \qquad (9.35)$$

where M_A is the molar mass of solvent.

For example: for water, $K_F = 1.855$ K kg mol^{-1}; for camphor, 40.0 K kg mol^{-1}. Because of its exceptionally large K_F, camphor is used in a micromethod for molecular weight determination by freezing-point depression.

Example 9.8 For benzene, $T_f^\bullet = 278.6$ K, $\Delta H_f = 9.83$ kJ mol^{-1}, and $M_A = 78 \times 10^{-3}$ kg mol^{-1}. What is K_F?

$$K_F = \frac{(8.314 \text{ J K}^{-1} \text{ mol}^{-1})(278.6 \text{ K})^2(78.0 \times 10^{-3} \text{ kg mol}^{-1})}{9830 \text{ J mol}^{-1}}$$

$$= 5.12 \text{ K kg mol}^{-1}$$

9.13 Osmotic Pressure

Colligative properties are those properties of dilute solutions that depend on the *number* of solute molecules per unit volume of solution.

Colligative properties can all be related to the lowering of the vapor pressure when a nonvolatile solute is dissolved in a solvent. Since the vapor pressure P_A of the solvent A is a measure of the chemical potential μ_A of the solvent, it is evident that colligative properties are all determined by the chemical potential of the solvent. Such properties include boiling-point elevation, freezing-point depression, and osmotic pressure.

In 1748, J. A. Nollet described an experiment in which brandy was placed in a large vial, the mouth of which was closed with an animal bladder and immersed in water. The bladder gradually became greatly swollen and sometimes even burst. The animal membrane is *semipermeable*; water can pass through it, but alcohol cannot. The increased pressure in the vial, caused by diffusion of water into the solution, is called the *osmotic pressure* (from the Greek, *osmos*, impulse). Osmotic pressure can be measured by the hydrostatic pressure of the column of solution that it can support. Typical precise osmotic-pressure data are summarized in Table 9.3.

In 1885, J. H. van't Hoff pointed out that in dilute solutions the osmotic pressure Π obeyed the relationship $\Pi V = nRT$, or

$$\Pi = cRT \qquad (9.36)$$

where $c = n/V$ is the concentration of solute. The validity of the equation can be judged by comparison of calculated and experimental values of Π in Table 9.3.

TABLE 9.3
OSMOTIC PRESSURES OF SOLUTIONS OF SUCROSE
IN WATER AT 20°C

Molality, m (mol kg^{-1})	Concentration, c (mol dm^{-3})	Observed osmotic pressure (kPa)	Calculated osmotic pressure (kPa)		
			Eq. (9.36)	Eq. (9.40)	Eq. (9.38)
0.1	0.098	262	239	243	247
0.2	0.192	513	469	487	553
0.3	0.282	771	689	731	792
0.4	0.370	1027	902	975	1035
0.5	0.453	1292	1100	1220	1279
0.6	0.533	1559	1300	1460	1520
0.7	0.610	1837	1490	1700	1763
0.8	0.685	2119	1670	1945	2003
0.9	0.757	2403	1840	2190	2244
1.0	0.825	2693	2010	2430	2480

An osmotic pressure arises when two solutions of different concentrations (or a pure solvent and a solution) are separated by a semipermeable membrane. A difference in solubility is responsible for many cases of semipermeability. For example, protein membranes, like those employed by Nollet, can dissolve water but not alcohol. In other cases, the membrane may act as a molecular sieve; the cross sections of the channels through the membrane may be so small that they can be permeated by small molecules, such as water, but not by large molecules, such as carbohydrates or proteins. Irrespective of the mechanism by which the semipermeable membrane operates, the final result is the same. Osmotic flow continues until the chemical potential of the diffusing component has the same value on both sides of the barrier. If the flow takes place into a closed volume, the pressure inside must increase.

9.14 Osmotic Pressure and Vapor Pressure

The equilibrium osmotic pressure can be calculated by thermodynamic methods. Consider in Fig. 9.11 a pure solvent A, which is separated from a solution of B in A by a membrane permeable to A alone. At equilibrium, an osmotic pressure Π has developed. The condition for equilibrium is that the chemical potential of A is the same in the phases a and b on the two sides of the membrane, $\mu_A^a = \mu_A^b$. Thus at equilibrium,

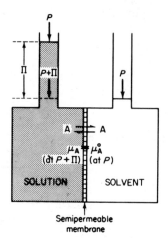

FIGURE 9.11 A solution of A and B separated from pure A by a membrane permeable only to A. At equilibrium an osmotic pressure Π exists.

μ_A in the solution must equal $\mu_A^\bullet$ of pure A. There are two factors that cause the value of μ_A in the solution to depart from that of pure A. These factors must therefore have exactly equal and opposite effects on μ_A. The first is the change in μ_A caused by dilution of A in the solution. From Eq. (9.25) this change is a decrease in μ_A given by $\Delta\mu_A = RT \ln P_A/P_A^\bullet$, where P_A is the partial vapor pressure of A in the solution and $P_A^\bullet$ is the vapor pressure of pure A. Exactly counteracting this effect is the increase in μ_A in the solution due to the imposed osmotic pressure Π. From Eq. (9.18), $d\mu_A = V_A dP$, so that $\Delta\mu_A = V_A\Pi$. (It has been assumed that the solution is incompressible, so that V_A is independent of P.) Equating the equal and opposite changes in μ_A, we find that

$$V_A\Pi = RT \ln \frac{P_A^\bullet}{P_A} \tag{9.37}$$

The significance of this equation can be stated as follows: The osmotic pressure is the external pressure that must be applied to the solution to raise the vapor pressure of the solvent A to that of pure A.

In most cases, the partial molar volume V_A of solvent in solution can be well approximated by the molar volume of pure liquid $V_A^\bullet$. In the special case of an ideal solution, Eq. (9.37) then becomes

$$\Pi V_A^\bullet = -RT \ln X_A \tag{9.38}$$

Replacing X_A by $(1 - X_B)$ and expanding $-\ln(1 - X_B)$ as in Section 9.12 we obtain when $X_B \ll 1$ the formula for a dilute solution,

$$\Pi V_A^\bullet = RT X_B \tag{9.39}$$

Since the solution is dilute, $X_B \approx n_B/n_A$ and

$$\Pi = \frac{RT}{V_A^\bullet} \frac{n_B}{n_A} \approx RTm' \tag{9.40}$$

As the solution becomes very dilute, m', the volume molality, approaches c, the molar concentration, and we find as the end product of the series of approximations the van't Hoff equation, $\Pi = cRT$. The adequacy with which Eqs. (9.38), (9.40), and (9.36) represent the experimental data can be judged from the comparisons in Table 9.3.

Example 9.9 A solution of 54.1 g of mannitol ($M = 182.2$) per 1000 g of water at 20°C has a vapor pressure 2.326 kPa, whereas that of pure water is 2.338 kPa. Calculate the osmotic pressure from Eqs. (9.37), (9.38), and (9.40).

Equation (9.37):

$$\Pi = \frac{(8.314 \text{ J K}^{-1} \text{ mol}^{-1})(293 \text{ K})}{18.0 \times 10^{-6} \text{ m}^3 \text{ mol}^{-1}} \ln \frac{2.338}{2.326} = 696 \text{ kPa}$$

Equation (9.38) with $X_A = 0.9947$:

$$\Pi = \frac{-(8.314 \text{ J K}^{-1} \text{ mol}^{-1})(293 \text{ K})}{18.0 \times 10^{-6} \text{ m}^3 \text{ mol}^{-1}} \ln 0.9947 = 719 \text{ kPa}$$

Equation (9.40):

$$\Pi = (8.314 \text{ J K}^{-1} \text{ mol}^{-1})(293 \text{ K})[(54.1)/(182.2)](10^3) = 723 \text{ kPa}$$

Note that volume molality must be expressed in mol m^{-3} for consistent SI units.

9.15 Osmotic Pressure of Polymer Solutions

The osmotic pressures of solutions of high polymers and proteins provide some of the best data on the thermodynamic properties of these macromolecules. Special membranes are available that allow small molecules and electrolytes to pass freely, but are impermeable to polymers above a defined range of molecular mass.

Solutions of small molecules follow the van't Hoff equation, $\Pi = cRT$, up to a concentration of about 0.1 molar (about 1%), but large deviations occur in polymer solutions at much lower concentrations. In 1914, Caspari first reported this behavior, finding that the osmotic pressure of a 1% solution of rubber in benzene was about twice the value calculated from the van't Hoff formula.

In 1945, McMillan and Mayer showed that the osmotic pressure of nonelectrolytes could be represented by a power series in the concentrations, exactly like the virial equation for a nonideal gas. Their equation is usually written:

$$\Pi = RT\left[\frac{m}{V\bar{M}_N} + B\left(\frac{m}{V}\right)^2 + C\left(\frac{m}{V}\right)^3 + \cdots\right] \tag{9.41}$$

Here m is the mass of solute (polymer) in volume V of solution, $\bar{M}_N$ is the number average molar mass of the polymer molecules in solution, and B and C are functions of temperature, the second and third virial coefficients, respectively. In fairly dilute solutions, the third term in Eq. (9.41) can be neglected, giving

$$\frac{\Pi}{(m/V)} = \frac{RT}{\bar{M}_N} + RTB\left(\frac{m}{V}\right) \tag{9.42}$$

Thus a plot of $\pi/(m/V)$ vs. (m/V) should give a straight line in the region of low concentration. The intercept of the line extrapolated to $(m/V) = 0$ yields $(RT/\bar{M}_N)$

and hence the number average molar mass $\bar{M}_N$ of the polymer molecules. [$\bar{M}_N = \sum N_j M_j / \sum N_j$, where N_j is the number of molecules with molar mass M_j in the sample.] Some data on polypyrolidones in water plotted in accord with Eq. (9.42) are shown in Fig. 9.12.

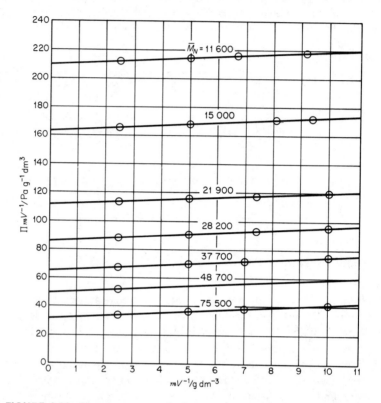

FIGURE 9.12 Osmotic pressures of solutions of polypyrrolidones in water at 298 K plotted in accord with Eq. (9.42), where $\mathfrak{m} V^{-1}$ is the mass of polymer per unit volume of solution. [J. Hengstenberg, *Makromol. Chem.* **7**, 236 (1952)] The number average molar masses of the polymer samples, M_N in g mol^{-1}, are noted on the plots.

Problems

1. A solution of 4.450 g H_2SO_4 in 82.20 g H_2O had a density $\rho = 1.029$ g cm^{-3} at 25°C and 1 atm pressure. Calculate (a) wt %; (b) mole fraction; (c) molality; (d) molarity of the solution.

2. A solution of sucrose in water is prepared that is 10.00% by weight. The density of the solution at 0°C is $\rho = 1.04135$ g cm^{-3} and at 25°C is $\rho = 1.03679$ g cm^{-3}. Calculate (a) the molarity and (b) the molality of the solution at the two temperatures.

3. The densities of solutions of NH_3 in H_2O at 288.1 K:

wt % NH_3	0	4.68	9.80	15.40	21.20	27.80	35.50
ρ (g cm^{-3})	1.000	0.980	0.960	0.940	0.920	0.900	0.880

Calculate the partial molar volumes of H_2O and NH_3 (by the method of intercepts) at the compositions given.

4. Solutions are prepared at 25°C containing 1.000 kg of water and n moles of NaCl. The volume in cm^3 is found to vary with n as

$$V = 1001.38 + 16.6253n + 1.7738n^{3/2} + 0.1194n^2$$

Draw a graph of the partial molar volumes of H_2O and of NaCl in the solution as a function of molality m from 0 to 2 m.

5. In a solution of acetone (1) and chloroform (2) at 298.16 K, $V_1 = 74.108$ cm^3 and $V_2 = 80.309$ cm^3 at $X_2 = 0.38526$. The molar volumes of the pure liquids are $V_1^\bullet = 73.993$ and $V_2^\bullet = 80.665$ cm^3. What is the volume of solution when 500 cm^3 acetone is mixed with 342 cm^3 of chloroform?

6. The partial specific volume $v_2 = V_2/M_2$, where M_2 is the molar mass. Determine v_2 for hemoglobin in aqueous solution at 275 K from the measured densities p.

c_2 (g cm^{-3} solution)	0.07712	0.15479	0.24998	0.33682	0.42108
p (g cm^{-3})	1.0195	1.0394	1.0635	1.0856	1.1068

Note in this case v_1 and v_2 do not depend on concentration. p_1(pure water) = 0.999854 g cm^{-3}.

7. When 3.000 g of a nonvolatile hydrocarbon containing 94.4% C is dissolved in 100 g of benzene, the vapor pressure of benzene is lowered from 9.953 to 9.823 kPa. Calculate the molecular formula of the hydrocarbon.

8. The vapor pressure of a solution containing 11.94 g of a nonvolatile solute X in 100 g H_2O is 98.8 kPa at 373 K. Estimate the molar mass of X.

9. Liquids A and B form an ideal solution. At 50°C the total vapor pressure of a solution containing 1 mol A and 2 mol B is 33.3 kPa. On addition of one more mole of A to solution the vapor pressure increases to 40.0 kPa. Calculate $P_A^\bullet$ and $P_B^\bullet$.

10. At 20°C the vapor pressure of benzene is 13.3 kPa and that of octane is 2.66 kPa. If 1 mol of octane is dissolved in 4 mol of benzene, and the solution is ideal, calculate the total vapor pressure of the solution and the composition of the vapor.

11. Roughly plot $\mu_A - \mu_A^\circ$ vs. X_A for an ideal solution. What can you conclude about the possibility of preparing a perfectly pure substance?

12. Chlorobenzene A and bromobenzene B form an ideal solution. At 400 K, $P_A^\bullet = 115$ kPa and $P_B^\bullet = 60.4$ kPa. Calculate the composition of the solution that boils at 400 K at 1 atm pressure and the composition of the vapor in equilibrium with this solution at 400 K.

13. Pure water is saturated with a 2:1 gas mixture of H_2/O_2 at 500 kPa total P. The water is then boiled to remove all dissolved gases. What is the composition of the gases driven off after they are dried? (See Fig. 9.5.)

14. Suppose that you have a mixture of n-propane ($X_A = 0.4$) and n-butane ($X_B = 0.6$) in an ideal solution at 77 K. Devise an isothermal reversible process to separate the solution into pure components. What practical limitations would there be to the use of your process? Calculate the minimum work needed to separate 1 mol of the solution into its pure components.

15. A certain mass of substance X in 100 g of benzene lowers the freezing point 1.280 K. The same mass of X in 100 g of water lowers the freezing point 1.395 K. If X does not dissociate in benzene, to how many ions does it dissociate in water?

16. Ethylene glycol, $HO \cdot CH_2 \cdot CH_2 \cdot OH$, is used as an antifreeze. Estimate the composition of a glycol–water solution that just begins to form ice at $-30°C$.

17. The solubility of picric acid in benzene is:

$t\ (°C)$	5	10	15	20	25	35
$(g/100\ g\ C_6H_6)$	3.70	5.37	7.29	9.56	12.66	21.38

The melting points are 121.8°C and 5.5°C, respectively. Estimate the enthalpy of fusion of picric acid based on ideal solution.

18. The melting points and ΔH_f of o, p, m-dinitrobenzenes are: 116.9°C, 173.5°C, 89.8°C; and 16 340, 14 000, and 17 900 J mol^{-1}. Based on the ideal solubility law, calculate the ternary eutectic temperature and composition for mixtures of the o,m,p compounds.

19. The following boiling points T_b are found for solutions of O_2 and N_2 1 atm:

$T_b\ (K)$	77.3	78	79	80	82	84	86	88	90.1
mol $\%\ O_2(\ell)$	0	8.1	21.6	33.4	52.2	66.2	77.8	88.5	100
mol $\%\ O_2(vap)$	0	2.2	6.8	12	23.6	36.9	52.2	69.6	100

Draw the T–X diagram. If 90% of a mixture containing 20% O_2 is distilled, what will be the composition of the residual liquid and its T_b?

20. A solution contains 1.000 g of sucrose and an unknown mass x of glucose in 1 kg of water. At 298 K its osmotic pressure is 3.00×10^4 Pa. Calculate x.

21. The following osmotic pressures were observed for bovine serum albumin (BSA) in 0.15 M NaCl at 273 K. [G. Scatchard, *J. Am. Chem. Soc.*, *68*, 2320 (1946)]

BSA (g/cm^3 solution)	0.020	0.030	0.040	0.060
Π (Pa)	820	1360	1900	3230

Calculate the molar mass of BSA.

22. The ΔH_f of ice is 6020 J mol^{-1}. What mole fraction nonionic impurity in water would lower the freezing point by 1.00°C?

23. The solubility of orthorhombic sulfur in CCl_4 at 298 K is 8.52 g/kg CCl_4. The Gibbs free energy of formation $\Delta G_f^\circ(298\ K)$ of monoclinic sulfur is 120 J mol^{-1}. Both forms exist as S_8 rings in the solids and after solution. What is the solubility of monoclinic S in CCl_4 at 298 K?

24. The solubilities of I_2 in CS_2 at various T are:

$T\ (K)$	253	273	293	313
$X(I_2)$	0.0122	0.0232	0.0418	0.0710

Calculate the average ΔH° for $I_2(s) \longrightarrow I_2$(solution). Actually, ΔH° varies with T. If $\Delta H^\circ = \Delta H^\circ_{253} + AT$, estimate the change in ΔH° over the temperature range.

25. The molar enthalpy of solution of thymine is $\Delta H^\circ = 24.32 \text{ kJ mol}^{-1}$ at infinite dilution and 298 K. The solubility of thymine at 298 K is 32.0 mmol/kg water. Estimate ΔS° (298 K) for solution of thymine in water. Mention any assumptions made. [M. V. Kilday, *J. Res. Nat. Bur. Stds.* **83**, 529 (1978)]

10

Real Gases and Solutions

For ideal gases and ideal solutions, the thermodynamic equations are greatly simplified. At the molecular level, the reason for this simplicity is the absence of intermolecular forces in ideal gases and the uniformity of such forces in ideal solutions. Most of the systems that interest the physical chemist depart appreciably from ideal behavior, and we must now face the problem of developing a more general thermodynamic theory to deal with real gases and solutions.

10.1   Fugacity and Activity

A clever way has been devised to treat problems of nonideality. We keep the forms of the ideal equations but introduce new functions to take the places of the partial pressure P_A in a gas and the concentration c_A or mole fraction X_A of a component in solution. For a gas, instead of P_A, we use a function called the *fugacity* f_A, an idealized partial pressure. For a component in solution, instead of c_A or X_A, we use a function called the *activity* a_A.

We then write for a real gas,

$$f_A = \gamma_A P_A \tag{10.1}$$

where γ_A is the *fugacity coefficient,* a factor that converts a pressure into an idealized pressure or fugacity. Similarly, for a nonideal solution, we write

$$^c a_A = {}^c\gamma_A c_A \quad \text{or} \quad {}^x a_A = {}^x\gamma_A x_A \tag{10.2}$$

where $^c\gamma_A$ and $^x\gamma_A$ are *activity coefficients* that convert concentrations or mole fractions into activities. Note that the activity based on c is not the same as the activity based on X. When it is clear from the context which activity is being used, the superscript may be omitted.

This chapter will describe how the thermodynamics of nonideal gases and solutions can be formulated in terms of the fugacity and activity functions.

10.2 Real Gases—Chemical Potential and Fugacity

The basic thermodynamic function for equilibrium studies is the chemical potential μ_A of a component A. In an ideal-gas mixture,
$$\mu_A = \mu_A^\circ + RT \ln (P_A/P_A^\circ) = \mu_A^\circ + RT \ln X_A \qquad (10.3)$$
For a nonideal gas mixture, instead of the ideal equation (10.3) we have
$$\mu_A = \mu_A^\circ + RT \ln (f_A/f_A^\circ) \qquad (10.4)$$
The dimensions of fugacity are the same as those of pressure, and the unit of fugacity in SI is therefore the pascal. The fugacity of a gas in the standard state is $f_A^\circ = 101.32$ kPa.

The standard state of a real gas is that state in which $f_A^\circ = 101.32$ kPa and in which, furthermore, the gas behaves as if it were ideal. From Eq. (10.1) it therefore follows that in the standard state the fugacity coefficient $\gamma = 1$ and $f_A^\circ = P_A^\circ$. Substituting $f_A = \gamma_A P_A$ into Eq. (10.4) gives
$$\mu_A = \mu_A^\circ + RT \ln (P_A/P_A^\circ) + RT \ln \gamma_A \qquad (10.5)$$
Now, by the definition of the standard state, μ_A° is the chemical potential of the gas as an ideal gas at $P_A^\circ = 101.32$ kPa and the specified temperature. All the effects of nonideality of the gas are included in the fugacity coefficient γ_A. When the gas is ideal $\gamma_A = 1$. Departures from ideality are measured by the deviation of γ_A from unity.

The definition of the standard state is subtle and curious, since it is not any real state of the gas, but an imaginary (or hypothetical) state. To visualize this state, imagine the gas at P_A° and then imagine that all attractive and repulsive intermolecular forces between the gas molecules are made to disappear while pressure is still maintained at P_A°. The resulting hypothetical state is the defined standard state of the gas. The advantage of this choice of standard state is that it requires all gases to be in the same ideal condition when they are in their standard states.

But, you might well ask, how can we find the properties of a gas in an imaginary state? This determination is quite easy, and the procedure can be seen from Fig. 10.1,

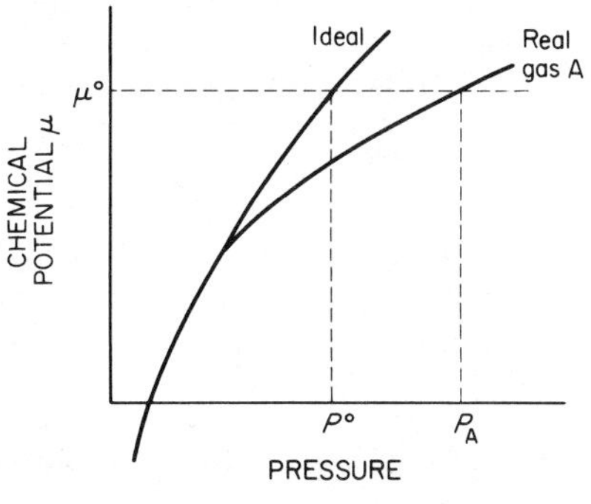

FIGURE 10.1 Definition of the standard state of a gas at which $f^\circ = P^\circ$ and $\gamma = 1$.

where chemical potential is plotted vs. pressure for a real gas and for an ideal gas. This would be a plot of Eq. (10.5). In the limit of sufficiently low pressures, all gases behave ideally, and as $f \rightarrow P, \gamma \rightarrow 1$. To get the property of a gas in its standard state, we must move back along the experimental curve (as a function of pressure) until it joins the theoretical ideal curve, and then move along the ideal curve until we reach the point of standard fugacity ($f^\circ = P^\circ$). There is no difficulty in calculating the change in a property along the ideal curve, since we have simple equations for all the properties of an ideal gas.

10.3 How to Calculate the Fugacity of a Gas

The fugacity of a pure gas or of a gas in a mixture can be evaluated if sufficiently detailed PVT data are available. We shall consider only the case of a pure gas. At constant temperature, from Eqs. (8.49) and (10.4),

$$dG = n\, d\mu = V\, dP = nRT\, d\ln f \tag{10.6}$$

If the gas is ideal, $V = nRT/P$. For a nonideal gas, we define a quantity α by

$$n\alpha = V(\text{ideal}) - V(\text{real}) = \frac{nRT}{P} - V \tag{10.7}$$

whence $V = n(RT/P) - n\alpha$. Substituting this expression into Eq. (10.6), we find that

$$d\mu = RT\, d\ln f = RT\, d\ln P - \alpha\, dP \tag{10.8}$$

The equation is integrated from $P' = 0$ to $P' = P$,

$$RT \int_{f, P=0}^{f} d\ln f' = RT \int_{P=0}^{P} d\ln P' - \int_{0}^{P} \alpha\, dP'$$

As its pressure approaches zero, a gas approaches ideality, and for an ideal gas, the fugacity equals the pressure, $f = P$. The lower limits of the first two integrals must therefore be equal, so that we obtain

$$RT \ln f = RT \ln P - \int_{0}^{P} \alpha\, dP' \tag{10.9}$$

This equation enables us to evaluate the fugacity at any pressure and temperature, provided that PVT data for the gas are available. If the deviation from ideality of the gas volume α is plotted against P, the integral in Eq. (10.9) can be evaluated graphically. Alternatively, an equation of state can be used to calculate an expression for α as a function of P, making it possible to evaluate the integral by analytic methods.

In terms of the fugacity coefficient $\gamma = f/p$, Eq. (10.9) is

$$RT \ln \gamma = - \int_{0}^{P} \alpha\, dP' \tag{10.10}$$

Example 10.1 Calculate the fugacity and the fugacity coefficient of Xe at 1000 K and 10^4 kPa if the gas follows an equation of state $P(V - nb) = nRT$, with $b = 5 \times 10^{-5}$ m^3 mol^{-1}.

In this case $\alpha = (RT/P) - (V/n) = -b$. From Eq. (10.9),

---

$$RT \ln f = RT \ln P - \int_0^P -b \, dP' = RT \ln P + bP$$

$$RT \ln \gamma = bP, \qquad \gamma = e^{bP/RT}$$

Thus when $P = 10^4$ kPa, $\gamma = 1.06$, $f = 1.06 \times 10^4$ kPa.

10.4   Fugacity and Corresponding States

In Chapter 2, we saw that different gases display nearly the same deviations from ideality when they are in corresponding states. This rule is illustrated by the fact that different gases have almost the same fugacity coefficient γ when they are at the same reduced temperature and pressure, $T_R = T/T_c$ and $P_R = P/P_c$.

Figure 10.2 shows a family of curves* relating the fugacity coefficient of a gas

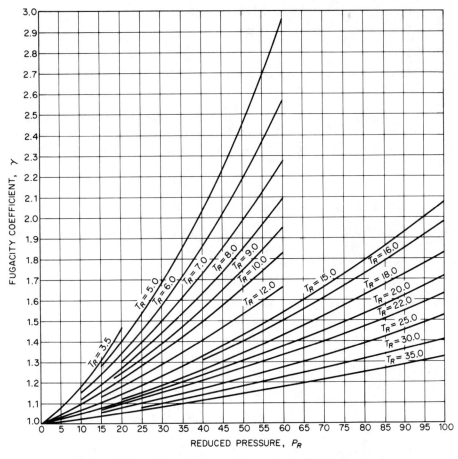

FIGURE 10.2   (a) Fugacity coefficients γ of gases in the high-temperature range. Each curve gives γ vs. P_R for a particular value of reduced temperature T_R.

*R. H. Newton, *Ind. Eng. Chem.*, **27**, 302 (1935).

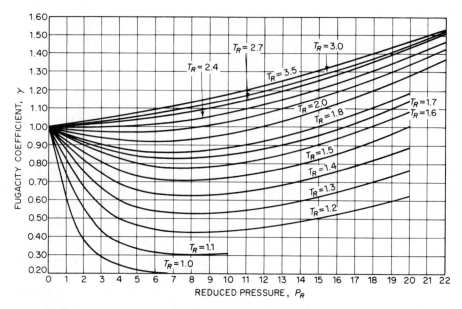

FIGURE 10.2 (b) Fugacity coefficients of gases in the intermediate temperature range.

to P_R at various values of T_R. To the approximation that the Law of Corresponding States is valid, all gases will fit this single set of curves. We thus can estimate the fugacity of a gas solely from its critical constants, T_c and P_c, data that are almost always available.

Example 10.2   Estimate the fugacity of C_2H_4 at 10^4 kPa and 300 K.

From Table 2.3, $P_c = 5160$ kPa and $T_c = 283$ K, hence $P_R = 1.94$, $T_R = 1.06$. From Fig. 10.2(b), $\gamma = 0.49$ and $f = 4900$ kPa.

10.5   Use of Fugacity in Equilibrium Calculations

From the same kind of derivation used in Section 8.12, but with fugacities instead of partial pressures, we can derive an equilibrium constant for reaction between nonideal gases, $aA + bB = cC + dD$:

$$\Delta G° = -RT \ln K_f \qquad (10.11)$$

$$K_f = \frac{(f_C/f°)^c (f_D/f°)^d}{(f_A/f°)^a (f_B/f°)^b} = \frac{\gamma_C^c \gamma_D^d}{\gamma_A^a \gamma_B^b} \frac{P_C^c P_D^d}{P_A^a P_B^b} (f°)^{a+b-c-d}$$

where $f° = 101.32$ kPa, the fugacity in the standard state. If we write $\gamma_C^c \gamma_D^d / \gamma_A^a \gamma_B^b = K_\gamma$, Eq. (10.11) becomes

$$K_f = K_\gamma K_P \qquad (10.12)$$

Note that K_y is not an equilibrium constant but simply the ratio of fugacity coefficients needed to convert partial pressures in K_P into fugacities in K_f.

As an example of the use of fugacities in equilibrium problems, consider the synthesis of ammonia, $\frac{1}{2}N_2 + \frac{3}{2}H_2 = NH_3$. This industrially important reaction is carried out under high pressures, at which the ideal-gas approximation fails badly. The reaction has been investigated up to 3.5×10^8 Pa.* The NH_3 percents in equilibrium with a 3:1 H_2/N_2 mixture at 723 K and various total pressures are shown in Table 10.1. In the third column of the table are the values calculated from these data of $K_P = (P_{NH_3}/P_{N_2}^{1/2}P_{H_2}^{3/2})\,(P^\circ)$. Since for ideal gases K_P should be independent of the total pressure, these results show a large departure from ideal behavior.

Let us calculate the equilibrium constants K_f by means of fugacity coefficients from Newton's graphs. We thereby adopt the approximation that the fugacity coefficient of a gas in a mixture is determined only by the temperature and by the total pressure. This approximation ignores specific interactions between components in the mixture of gases. The values of γ are read from the graphs, at the proper values of reduced pressure P_R and reduced temperature T_R and we calculate $K_y = \gamma_{NH_3}/\gamma_{N_2}^{1/2}$ $\gamma_{H_2}^{3/2}$. The values of K_y and K_f are shown in Table 10.1. There is a marked improvement in the constancy of K_f as compared to K_P. Only at 10^5 kPa and above does the approximate treatment of fugacities fail. If we had the exactly correct K_y, K_f would of course remain constant even at the highest pressures. To carry out such an exact thermodynamic treatment, we must use the fugacity of each gas in the particular mixture under study. Extensive PVT data on the mixture are needed for such a calculation.

TABLE 10.1

EQUILIBRIUM IN THE AMMONIA SYNTHESIS AT 723 K
WITH 3:1 RATIO OF H_2 TO N_2

Total pressure (10^6 Pa)	Percent NH_3 at equilibrium	K_P	K_y (approximate)	K_f (approximate)
1.0	2.04	0.00659	0.995	0.00655
3.0	5.80	0.00676	0.975	0.00659
5.0	9.17	0.00690	0.945	0.00650
10	16.36	0.00725	0.880	0.00636
30	35.5	0.00884	0.688	0.00608
60	53.6	0.01294	0.497	0.00642
100	69.4	0.02496	0.434	0.01010
200	89.8	0.1337	0.342	0.0458
350	97.2	1.0751	—	—

We often wish to calculate the equilibrium concentrations in a reaction for which ΔG° is known. The procedure is to obtain K_f from $-\Delta G^\circ = RT \ln K_f$, to estimate K_y from the graphs, and then to calculate the partial pressure from $K_P = K_f/K_y$.

*L. J. Winchester and B. F. Dodge, *Am. Inst. Chem. Eng. J.*, **2**, 431 (1956).

Example 10.3 The synthesis of methanol is carried out at high pressures by the reaction $CO + 2H_2 = CH_3OH$. Calculate the mol percent CH_3OH in the equilibrium mixture at 1000 K and 10^5 kPa starting with a $2:1$ ratio of H_2/CO (a) assuming ideal gases; (b) using fugacities from Newton's graphs. For reaction at 1000 K, $\Delta G^\circ = 143.9$ kJ mol^{-1}.

For the critical data and fugacity coefficients, we find:

	T_c (K)	P_c (kPa)	T_R	P_R	$\gamma = f/P$
CH_3OH	513.2	7950	1.95	12.6	1.02
CO	134.0	3550	7.48	28.2	1.48
H_2	33.0	1330	30.1	77.0	1.30

Thus $K_\gamma = \gamma_{CH_3OH}/\gamma_{CO}\gamma_{H_2}^2 = 0.408$. From $\Delta G^\circ = -RT \ln K_f$, $K_f = 3.04 \times 10^{-8}$ $= K_P(\text{ideal})$. $K_P(\text{real}) = K_f/K_\gamma = 7.45 \times 10^{-8}$. Suppose that the reactant mixture contains 1 mol CO and 2 mol H_2. At equilibrium x mol CH_3OH is formed. Then the total number of moles is $(1 - x) + 2(1 - x) + x = 3 - 2x$. Then

$$K_P = \frac{\left(\dfrac{x}{3 - 2x}\right)P/P^\circ}{\left(\dfrac{1 - x}{3 - 2x}\right)(P/P^\circ)\left[\dfrac{2(1 - x)}{3 - 2x}\right]^2 (P/P^\circ)^2} = \frac{x(3 - 2x)^2}{4(1 - x)^3}\left(\frac{P^\circ}{P}\right)^2$$

Ideal case: $(3.04 \times 10^{-8})(10^5/101.3)^2 = 0.0296$

$$= x(3 - 2x)^2/4(1 - x)^3$$

Real case: $(7.45 \times 10^{-8})(10^5/101.3)^2 = 0.0726$

$$= x(3 - 2x)^2/4(1 - x)^3$$

By trial and error with a hand calculator:

Ideal case: $x = 0.013$; mol $\% = \dfrac{0.013}{3 - 2(0.013)} \times 100 = 0.44\%$

Real case: $x = 0.031$; mol $\% = \dfrac{0.031}{3 - 2(0.031)} \times 100 = 1.06\%$

10.6 Activity

The fugacity function is related to the chemical potential through Eq. (10.6) and, in principle, we could use fugacities to discuss equilibrium in liquid and solid solutions. In practice, however, a new function has been introduced, the activity a. The activity is the ratio of fugacity to fugacity in a defined standard state,

$$a_A = f_A/f_A^\circ \tag{10.13}$$

The standard state may be chosen in various ways, as we shall describe shortly. By its definition, activity a is seen to be a dimensionless ratio. Whenever we use an activity, we must know the standard state that has been selected. In terms of activity, Eq.

(10.4) becomes

$$\mu_A = \mu_A^\circ + RT \ln a_A \tag{10.14}$$

From Eqs. (10.11) and (10.14), we can readily derive an equilibrium constant in terms of activities,

$$K_a = a_C^c a_D^d / a_A^a a_B^b \tag{10.15}$$

with

$$\Delta G^\circ = -RT \ln K_a \tag{10.16}$$

These expressions are always valid, even for the most nonideal systems. In themselves, however, they do not allow us to calculate the equilibrium composition of a reaction mixture from the thermodynamic data for $\Delta G^\circ(T)$. To translate K_a back into measurable terms, we must have some way of computing the actual concentrations in the equilibrium reaction mixture from the value of K_a calculated from Eq. (10.16).

10.7 Standard States for Components in Solution

Before we can apply Eq. (10.16) to practical problems, we must define standard states for components in a solution. There are two different standard states in common use. With increasing dilution the solvent always approaches the ideal behavior specified by Raoult's Law, and the solute always approaches the behavior specified by Henry's Law. One standard state (I) is therefore based on Raoult's Law as a limiting law, and the other standard state (II) is based on Henry's Law. We may select whichever definition seems more convenient for any component in a particular solution.

Case I. Standard State (I) for a component considered as solvent. In this case, the standard state of component A in a solution is taken to be the pure liquid or pure solid at P° and at the temperature in question. Thus the activity becomes

$$a_A = \frac{f_A}{f_A^\circ} \approx \frac{P_A}{P_A^\bullet} \tag{10.17}$$

where $P_A^\bullet$ is the vapor pressure of pure A at 101.32 kPa total pressure. It is almost always sufficiently accurate to take the activity equal to the ratio of the partial pressure P_A of A above the solution to the vapor pressure of pure A. Should the vapor depart appreciably from ideal-gas behavior, we can use f_A instead of P_A.

With this choice of standard state, Raoult's Law, Eq. (9.22), becomes $a_A = P_A/P_A^\bullet = X_A$. Thus, for an ideal solution, or for any solution in the limit as $X_A \to 1$, $X_A = a_A$.

We define an activity coefficient $^x\gamma_A$ by

$$a_A = {}^x\gamma_A X_A \tag{10.18}$$

As $X_A \to 1$, $^x\gamma_A \to 1$.

Case II. Standard State (II) for a component considered as a solute. In this case, we choose the standard state so that in the limit of extreme dilution, as $X_B \to 0$, $a_B \to X_B$. As long as Henry's Law is obeyed, as shown in Fig. 10.3,

$$f_B = k_H X_B \tag{10.19}$$

The standard state is obtained by extrapolating the Henry's Law line to $X_B = 1$.

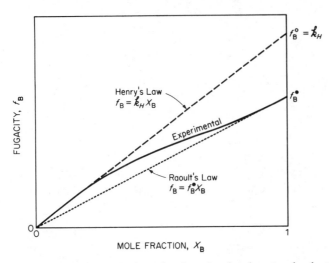

FIGURE 10.3 Definition of the standard state $[X_B = 1, f_B^\circ = k_H]$ for a solute B based upon Henry's Law for dilute solutions.

Thus we see that the fugacity in the standard state, f_B°, is simply equal to k_H, the constant of Henry's Law:

$$f_B^\circ = k_H \tag{10.20}$$

As in the case of the nonideal gas, this standard state is a hypothetical state. We can think of it in physical terms as a state in which the pure solute B ($X_B = 1$) has the properties it would have in an infinitely dilute solution in the solvent A. In this case

$$a_B = f_B/f_B^\circ = f_B/k_H = {}^x\gamma_B X_B \tag{10.21}$$

The composition of a solution is often expressed in terms of molality m or (molar) concentration c. We can define activities and activity coefficients for use with these quantities; for a component B:

$${}^m a_B = {}^m\gamma(m_B/m_B^\circ) \tag{10.22}$$

$${}^c a_B = {}^c\gamma(c_B/c_B^\circ) \tag{10.23}$$

In Eq. (10.22) m_B° is the molality in a defined standard state of unit molality (1 mol B per kg solvent). In Eq. (10.23) c_B° is the concentration in a defined standard state, usually 1 mol B per dm³ solution (1 molar). Note that the activity coefficients are dimensionless numbers. Relations between the three activity coefficients $^x\gamma$, $^m\gamma$, and $^c\gamma$ are readily derived.*

10.8 Activities of Solvent and Nonvolatile Solute from Vapor Pressure of Solution

As an example of this important method, let us consider how the activities of water A and of sucrose B are determined from data on vapor pressures of the solution. The same method has been applied to obtain activities of amino acids, peptides, and other solutes of biochemical interest.

*S. Glasstone, *An Introduction to Electrochemistry* (Princeton, N. J.: D. Van Nostrand Company, 1942), p. 134.

Sucrose is nonvolatile, so that the total vapor pressure above the solution of B in A equals the partial vapor pressure of water, P_A. If we neglect a small correction for nonideality of the water vapor, we can readily tabulate the activities a_A of the water from $a_A = P_A/P_A^{\bullet}$. The results are shown in Table 10.2 for the particular temperature of 323.2 K, where the vapor pressure of pure water is $P_A^{\bullet} = 12.33$ kPa. The vapor pressures of a sucrose solution in which the mole fraction of water is $X_A = 0.9665$ is $P_A = 11.86$ kPa. Hence $a_A = 11.86/12.33 = 0.9619$. The activity coefficient $^x\gamma_A = 0.9619/0.9665 = 0.9952$ at this concentration. Note that the standard state for the solvent, water, is chosen as the pure liquid at external pressure P°. As $X_A \rightarrow 1$, $a_A \rightarrow X_A$ and $^x\gamma_A \rightarrow 1$.

TABLE 10.2

ACTIVITIES OF WATER AND SUCROSE IN THEIR SOLUTIONS
AT 323.2 K OBTAINED FROM VAPOR-PRESSURE LOWERING
AND THE GIBBS–DUHEM EQUATION

Water			Sucrose	
Mole fraction X_A	Vapor pressure (kPa)	Activity a_A	Mole fraction X_B	Activity a_B
1.0000	12.333	1.0000	0.0000	0.0000
0.9940	12.258	0.9939	0.0060	0.0060
0.9864	12.252	0.9934	0.0136	0.0136
0.9826	12.085	0.9799	0.0174	0.0197
0.9762	11.959	0.9697	0.0238	0.0302
0.9665	11.863	0.9619	0.0335	0.0481
0.9559	11.688	0.9477	0.0441	0.0716
0.9439	11.468	0.9299	0.0561	0.1037
0.9323	11.153	0.9043	0.0677	0.1390
0.9098	10.801	0.8758	0.0902	0.2190
0.8911	10.539	0.8545	0.1089	0.3045

The activity of the sucrose obviously cannot be determined from its partial vapor pressure because this is immeasurably low. If we had a volatile solute, e.g., ethanol, we would doubtless take pure ethanol to be its standard state and compute its activity from Eq. (10.17). In the case of sucrose, on the other hand, we choose the second definition of standard state, that which is based on Henry's Law for the solute.

The activity of the sucrose can be calculated provided that we know the vapor pressure of the solvent (water) over the whole range of solute concentration from $X_B = 0$ up to the highest concentration of interest. From a Gibbs–Duhem equation like Eq. (9.11), $n_A \, d\mu_A + n_B d\mu_B = 0$. From Eq. (10.14), this becomes

$$n_A \, d\ln a_A + n_B \, d\ln a_B = 0$$

On division by $(n_A + n_B)$, we have

$$X_A \, d\ln a_A + X_B \, d\ln a_B = 0$$

Dividing by X_B and integrating, we obtain

$$\int_1^2 d \ln a_B = -\int_1^2 (X_A/X_B) \, d \ln a_A = -\int_1^2 X_A/(1 - X_A) \, d \ln a_A \qquad (10.24)$$

We wish to calculate a_B from the measurements giving a_A as a function of X_A. There appears to be some difficulty in using Eq. (10.24), since as $X_B \rightarrow 0$, $X_A \rightarrow 1$, and the integrals approach ∞. This difficulty is easily avoided by starting the integration, not at $X_A = 1$, but at a value of X_A at which the solvent begins to follow Raoult's Law, i.e., at which $X_A = a_A$. At this value, $X_B = a_B$, where a_B is defined on the basis of Henry's Law. Therefore, the integrals in Eq. (10.24) have lower limits corresponding to extremely dilute solutions. The results of such a calculation of the activities of sucrose in aqueous solution are shown in Table 10.2.

Example 10.4 The vapor pressures of water above sucrose solutions at 323.2 K are given in Table 10.2. Calculate the activity and activity coefficient of sucrose B in the solution at $X_B = 0.0561$ from the Gibbs–Duhem equation.

In Fig. 10.4 we plot $X_A/(1 - X_A)$ vs. $\ln a_A$ in accord with Eq. (10.24). As an example of the integration of the Gibbs–Duhem equation, we take an initial point where solution is very dilute and $a_B = X_B = 0.0060$. We take (as an example) a final concentration with $X_B = 0.0561$. The area under the curve between these two points is 2.850. Hence $\ln [a_B(0.0561)/a_B(0.0060)] = 2.850$ or $a_B(0.0561) = 17.283(0.0060) = 0.1037$. Other points are calculated in the same way.

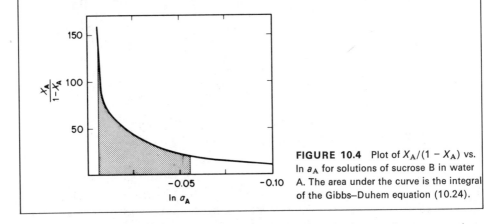

FIGURE 10.4 Plot of $X_A/(1 - X_A)$ vs. $\ln a_A$ for solutions of sucrose B in water A. The area under the curve is the integral of the Gibbs–Duhem equation (10.24).

A convenient way to determine the activity of water for use in computations based on Eq. (10.24) is the *isopiestic method*.* We take a set of reference standards for the activity of water. Sucrose solutions are convenient for this purpose. We place a reference solution and the solution of unknown activity in open vessels in an evacuated chamber such as a vacuum desiccator (Fig. 10.5). Water evaporates from the solution of higher vapor pressure and condenses into the solution of lower vapor pressure, until equilibrium is reached when the water vapor pressure is the same for the two solutions. At this *isopiestic point*, the chemical potential of water and hence the activity

*For experimental details see R. F. Platford, "Experimental Methods—Isopiestic" in *Activity Coefficients of Electrolyte Solutions*, ed. R. M. Pytkowicz (Boca Raton, Florida: CRC Press, 1979).

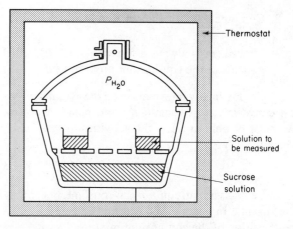

FIGURE 10.5 Isopiestic method for determining the activity of a volatile solvent in solution with a nonvolatile solute.

of water is the same in the two solutions. We measure the composition of the two solutions. We then know that the activity of water at the measured composition of solution X is the same as that of water in the sucrose solution. By repeating the experiment for different compositions, we determine the values of a_A over the range of composition required to calculate the activity a_B of the solute from the Gibbs–Duhem equation (10.24). Data on molal activity coefficients for compounds of biochemical interest measured in this way are summarized in Table 10.3.

TABLE 10.3
MOLAL ACTIVITY COEFFICIENTS OF SOME AMINO ACIDS
AND PEPTIDES IN AQUEOUS SOLUTION AT 298.15 K

Compound	Molality, m					
	0.2	0.3	0.5	1.0	1.5	2.0
Glycine	0.961	0.944	0.913	0.854	0.812	0.782
Alanine	1.005	1.007	1.012	1.024	1.027	—
Threonine	0.989	0.984	0.975	0.959	0.951	0.944
Proline	1.019	1.028	1.048	1.097	1.149	1.205
ϵ-Aminocaproic acid	0.971	—	0.951	0.942	1.002	1.072
Glycylglycine	0.912	0.879	0.828	0.745	0.697	—
Glycylalanine	0.935	0.912	0.883	0.855	—	—

10.9 Equilibrium Constants in Solution

The relation $\Delta G° = -RT \ln K_a$ is universally valid, but it simply summarizes the mathematical analysis of the equilibrium problem. We might paraphrase the content of this equation as follows:

1. Define a standard state for each of the reactants and products in a chemical equilibrium, $aA + bB = cC + dD$.

2. Compute $\Delta G°$ for the reaction in which all the components are in this standard state.

3. Then there is always a function $K_a(T, P)$, which is related to activities of components in the equilibrium mixture by $K_a = a_C^c a_D^d / a_A^a a_B^b$.

To obtain any information on the actual composition of this equilibrium mixture, or, conversely, to calculate K_a, and hence $\Delta G°$, from the equilibrium composition, we must be able to relate the activities to some composition variables. For example, since $^x a = {}^x \gamma X$,

$$K_a = \left(\frac{\gamma_C^c \gamma_D^d}{\gamma_A^a \gamma_B^b}\right)\left(\frac{X_C^c X_D^d}{X_A^a X_B^b}\right) = K_\gamma K_x \qquad (10.25)$$

Note that K_γ is not an equilibrium constant, but simply the indicated product of activity coefficients. In general, K_x will not be constant at constant T and P as we vary the composition of the equilibrium mixture. In some cases, however, it may happen that K_γ does not change much as we vary the composition. In particular, in dilute solutions, in which the solutes approximately follow Henry's Law and the solvent approximately follows Raoult's Law, we can choose standard states (as shown in the preceding section), so that all the γ's approach unity. In this case, $K_a \rightarrow K_x$, and $\Delta G° = -RT \ln K_x$.

To the extent that these approximations are satisfactory, we shall be able to use an equilibrium constant K_x for reactions in solution. Let us not disdain such an approximate equilibrium constant, because often the experimental data do not justify a more elaborate treatment. We must, however, keep clearly in mind the choice of standard states for $\Delta G°$. For all the reactants, the standard states would be those at $X = 1$. In the case of the solvent, this would be the pure liquid. In the case of the solute, this would be a hypothetical state in which $X = 1$, while the intermolecular forces are like those in an extremely dilute solution.

An example of measurement of equilibrium in solution is the early work (1895) of Cundall on the dissociation $N_2O_4 = 2NO_2$ in chloroform. Some of his data are shown in Table 10.4, with the calculated K_x.

TABLE 10.4
DISSOCIATION OF N_2O_4 IN CHLOROFORM SOLUTION AT 281.4 K

$10^2\ X(N_2O_4)$	$10^6\ X(NO_2)$	$10^{11}\ K_x$	$c(N_2O_4)$ mol dm^{-3}	$10^3\ c(NO_2)$ mol dm^{-3}	$10^5\ K_c$
1.03	0.93	8.37	0.129	1.17	1.07
1.81	1.28	9.05	0.227	1.61	1.14
2.48	1.47	8.70	0.324	1.85	1.05
3.20	1.70	9.04	0.405	2.13	1.13
6.10	2.26	8.35	0.778	2.84	1.04
		Mean 8.70			Mean 1.09

We have also included the computed values of an equilibrium constant in terms of concentrations c,

$$K_c = \frac{c_{NO_2}^2}{c_{N_2O_4}} \frac{1}{c°}$$

Note carefully that the use of K_c implies a derivation from $\Delta G^\circ = -RT \ln K_a$ based on a new and distinctive choice of standard states. We then must have a new set of activity coefficients $^c\gamma$, so that $a = {}^c\gamma c$, and as $c \rightarrow 0$, $^c\gamma \rightarrow 1$, and $K_a \rightarrow K_c$. We define the corresponding standard state as the hypothetical state of the solute at a concentration $c^\circ = 1$ mol dm^{-3}, but with an environment the same as that in an extremely dilute solution.

For the choice of standard states consistent with K_X, we find from K_X in Table 10.4 for the reaction $N_2O_4 = 2NO_2$,

$$\Delta G^\circ_{(X)} = -RT \ln K_X = 54.20 \text{ kJ mol}^{-1}$$

For the choice of standard states consistent with K_c, we find from K_c in Table 10.4,

$$\Delta G^\circ_{(c)} = -RT \ln K_c = 26.73 \text{ kJ mol}^{-1}$$

Having seen this example, a sagacious chemist will remember the following warning: It is folly to use ΔG° values for reactions in solution unless you are sure you understand the exact standard state upon which they are based.

10.10 ΔG°_f of Biochemicals in Aqueous Solution

Biochemical reactions proceed in an aqueous medium at rather closely controlled pH and ionic concentration. These conditions are quite different from the usual standard states for reactions in gases or nonpolar liquids, and there is a problem of translating the thermodynamic data from the standard condition that is usual in calorimetric work to conditions of physiological interest. For example, we can obtain ΔG°_f for biochemicals in the crystalline state at 298.15 K from enthalpies of formation and Third-Law entropies. Examples of such thermodynamic data are given in Table 10.5.

TABLE 10.5
THERMODYNAMIC DATA FOR AMINO ACIDS AND PEPTIDES AT 298.15 K[a]

Compound	Crystalline state			Aqueous solution		
	ΔH°_f (kJ mol^{-1})	ΔS°_f (J K^{-1} mol^{-1})	ΔG°_f (kJ mol^{-1})	Solubility (molality sat. soln.)	$^m\gamma$	$\Delta G^{\circ b}_f$ (kJ mol^{-1})
DL-Alanine	−563.6	−644	−372.0	1.9	1.046	−373.6
DL-Alanyglycine	−777.8	−967	−489.5	3.161	0.73	−491.6
L-Aspartic acid	−973.6	−812	−731.8	0.0377	0.78	−723.0
Glycine	−528.4	−431	−370.7	3.33	0.729	−372.8
Glycylglycine	−745.2	−854	−490.4	1.7	0.685	−490.8
DL-Leucine	−640.6	−975	−349.4	0.0756	1.0	−343.1
DL-Leucylglycine	−860.2	−1310	−469.9	0.126	1.0	−468.4

[a]From F. H. Carpenter, *J. Am. Chem. Soc.*, *82*, 1120 (1960), where references to original sources are given.
[b]For the dipolar-ionic form at standard state $^m a = 1$.

The affinity in biochemical reactions will be determined by the ΔG of the reaction in an aqueous physiological medium. Thus, instead of the $\Delta G_f^\circ(C)$ of the crystalline compounds, we wish to know the $\Delta G_f^\circ(W)$ of the compounds as solutes in aqueous solution. The appropriate standard state will usually be at unit activity on the molality scale, i.e., at $^m a = 1$. We can readily calculate $\Delta G_f^\circ(W)$ from $\Delta G_f^\circ(C)$ by a two-step process:

1. Dissolve the crystals in water to form a saturated solution. Since crystals and solute are in equilibrium in a saturated solution, for this step $\Delta G = 0$.

2. Calculate ΔG for the change in solute activity from its value at saturation a_{sat} to its value $a = 1$ in the standard state:

$$\Delta G = RT \ln \frac{1}{a^{sat}} = -RT \ln (^m \gamma m)^{sat} \qquad (10.26)$$

To carry out this computation, we need only know, besides the molality of the saturated solution, the activity coefficient at that molality. Hence

$$\Delta G_f^\circ(W) = \Delta G_f^\circ(C) - RT \ln (^m \gamma m)^{sat} \qquad (10.27)$$

Example 10.5 The solubility of glycine in water at 298.15 K is 3.33 molal, and $\Delta G_f^\circ(c) = -370.7$ kJ mol^{-1}. Calculate $\Delta G_f^\circ(w)$.

From Table 10.5, $^m \gamma = 0.729$. Hence, from Eq. (10.27),

$\Delta G_f^\circ(w) = -370.7$ kJ mol^{-1} $- (8.314 \times 10^{-3}$ kJ K^{-1} mol$^{-1})(298.15$ K)

$$\ln (3.33 \times 0.729)$$

$$= -370.7 - 2.2 = -372.9 \text{ kJ mol}^{-1}$$

We can use $\Delta G_f^\circ(w)$ values to calculate equilibrium constants K_a for reactions of interest. Actually, the physiological solvent medium is not pure water, and for precise calculations we should like to know the various activity coefficients in particular solutions containing inorganic salts as well as organic solutes. Furthermore, a temperature of about 310 K is more relevant than 298 K for biochemical reactions *in vivo* in warm-blooded animals.

Many important biochemicals are acids or bases, and the pH of the medium can have a considerable effect on affinities and equilibrium constants. Fortunately, the physiological medium is well buffered close to pH 7.0, so that it is usually safe to assume a hydrogen ion concentration of 10^{-7} mol dm^{-3}. If, however, in a given reaction, H$^+$ ions are used up or set free, the driving force ΔG will be sensitive to pH. The ΔG for dilution of H$^+$ from unit activity to 10^{-7} mol dm^{-3} at 300 K would be about $\Delta G = RT \ln (10^{-7}/1) = -40.2$ kJ mol^{-1}.

Example 10.6 From data in Table 10.5, calculate ΔG° and K_a for the synthesis of a simple dipeptide at 298 K by the reaction:

$$\text{alanine} + \text{glycine} = \text{alanylglycine} + \text{H}_2\text{O}$$

An amino acid NH_2—CHR—COOH exists in neutral solutions principally as a dipolar ion NH_3^+—CHR—COO^-, where R indicates the side chain of the amino acid. We shall ignore the effect of pH and assume that all reactants and products are completely in the form of dipolar ions. The ΔG_f° for liquid water is -237.0 kJ mol^{-1}. Hence

$$\Delta G^\circ = (-491.6 - 237.0) - (-373.6 - 372.8) = 17.8 \text{ kJ mol}^{-1}$$

$$K_a = \exp(-\Delta G^\circ/RT) = 7.60 \times 10^{-4} \text{ at } 298.15 \text{ K}$$

We can conclude that the synthesis of peptide bonds will not occur spontaneously to an appreciable extent from amino acids in aqueous solution. In living cells, the synthetic reaction is coupled with a reaction that has a large negative ΔG, the hydrolysis of adenosine triphosphate (ATP).

10.11   Deviations of Solutions from Ideality

Only a few solutions follow Raoult's Law over the complete range of concentrations. For this reason, most practical applications of the ideal equations are restricted to dilute solutions. As a solution becomes more dilute, the behavior of the solute B approaches more closely to that given by Henry's Law. Henry's Law is a limiting law that is followed by all solutes in the limit of extreme dilution, as $X_B \rightarrow 0$. The behavior of the solvent, as the solution becomes more dilute, approaches more closely to Raoult's Law. In the limit of extreme dilution, as $X_A \rightarrow 1$, all solvents obey Raoult's Law as a limiting law.

The properties of nonideal solutions are conveniently discussed in terms of their deviations from ideality. The first extensive measurements of vapor pressures of solutions were made by Jan von Zawidski around 1900. The partial vapor pressures of components in a solution provide a direct measure of their chemical potentials or activities. Two types of deviation from ideality can be distinguished: cases in which $a_A > X_A$ or $\gamma_A > 1$ are *positive deviations*, and cases in which $a_A < X_A$ or $\gamma_A < 1$ are *negative deviations*. In some cases a solution may have positive deviation in one range of concentration and negative deviation in another range.

A system exhibiting a positive deviation from Raoult's Law is water + dioxane. The partial vapor pressures of water and dioxane above the solutions are shown in Fig. 10.6(a). In an ideal solution the partial pressures would follow the dashed lines. The positive deviation is shown by partial vapor pressures higher than the values calculated for ideal solutions. The escaping tendencies of the components in solution are accordingly higher than the escaping tendencies in the individual pure liquids. This effect has been ascribed to cohesive forces between unlike components in the solution that are smaller than those within the pure liquids. In metaphoric terms, the components are happier by themselves than when they are mixed together; they are unsociable. A scientific translation is obtained by equating a happy component to one in a state of low chemical potential. When positive deviations from Raoult's Law occur, volume and enthalpy increase when the solution is formed, $\Delta V > 0$, $\Delta H > 0$. Positive deviations are often observed in aqueous solutions. Pure water is itself strongly

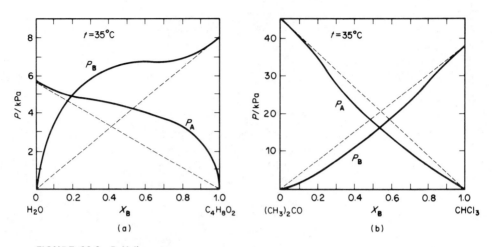

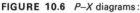

(a)

(b)

FIGURE 10.6 *P–X* diagrams:
(a) positive deviation from Raoult's Law—the system water + dioxane.
(b) negative deviation from Raoult's Law—the system acetone + chloroform.

associated, and addition of a second component may break down the water structure to some extent, causing an increased partial vapor pressure.

Example 10.7 Estimate the activity coefficients of water and of dioxane in their solution at 35°C at a mole fraction of 0.50.

From Fig. 10.6(a), at $X_B = 0.50$, $P_B = 6.53$, $P_A = 4.20$, $P_A^{\bullet} = 5.63$, $P_B^{\bullet} = 8.00$, all in kPa. Thus,

$$\gamma_A = a_A/X_A = (P_A/P_A^{\bullet})/X_A = 1.49$$
$$\gamma_B = a_B/X_B = (P_B/P_B^{\bullet})/X_B = 1.63$$

A system exhibiting a negative deviation from Raoult's Law is chloroform + acetone. The partial vapor pressures are shown in Fig. 10.6(b). In this case, the escaping tendency of a component from solution is less than it would be from the pure liquid. This fact may be the result of attractive forces between the unlike molecules in solution greater than those between the like molecules in the pure liquids. In some cases, actual association or compound formation may occur between the unlike components in the solution. In all cases of negative deviation, we observe a contraction in volume and a decrease in enthalpy on mixing.

10.12 Boiling-Point Diagrams

A sufficiently great positive deviation from ideality may lead to a maximum in the *P–X* diagram, and a sufficiently great negative deviation, to a minimum. An illustration of such behavior is shown in Fig. 10.7(a). At a maximum or minimum in the vapor-pressure curve, vapor and liquid must have the same composition.

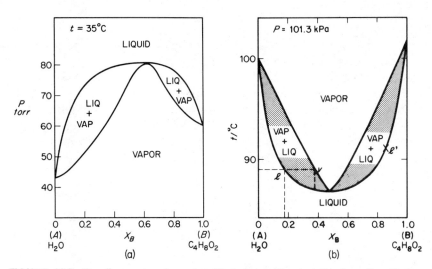

FIGURE 10.7 The dioxane + water system illustrates positive deviation from Raoult's Law. (a) P–X diagram at 35°C; (b) T–X diagram at 1 atm (normal boiling-point diagram).

The *P–X* diagram in Fig. 10.7(a) has its counterpart in the boiling-point (*T–X*) diagram in Fig. 10.7(b). A maximum in the *P–X* curve corresponds to a minimum in the *T–X* curve.

A solution with the composition corresponding to a maximum or minimum point on the boiling-point diagram is called an *azeotropic solution* (from the Greek *zein* to boil, and *a-tropos* unchanging), since there is no change in composition on boiling. Such solutions cannot be separated by distillation at constant pressure. In fact, at one time it was thought they were real chemical compounds. If the overall pressure is changed, however, the composition of an azeotropic solution is changed. Thus an azeotrope does not obey the law of definite composition, which holds for all gaseous compounds.

The distillation of a system with a maximum or minimum boiling point can be discussed by reference to Fig. 10.7(b). If the temperature of a solution having the composition ℓ is raised, it begins to boil at $t = 89°C$. The first vapor that distills has the composition v, richer than the original liquid in component B. The residual solution therefore becomes richer in A; and if the vapor is continuously removed, the boiling point of the residue rises, as its composition moves along the liquidus curve from ℓ toward pure A. If a fractional distillation is carried out, a final separation into pure A and the azeotropic solution is achieved. Similarly, a solution of original composition ℓ' can be separated into pure B and azeotrope.

10.13 Solubility of Liquids in Liquids

If positive deviation from Raoult's Law becomes sufficiently large, the components may no longer form a continuous series of solutions. As successive portions of one component are added to the other, a limiting solubility is finally reached, beyond which two distinct liquid phases separate. Usually, but not always, increasing temperature tends to promote solubility, as the thermal kinetic energy overcomes the reluctance of the components to mix freely. In other words, the $T \, \Delta S$ term in $\Delta G = \Delta H -$

$T \Delta S$ becomes more important. A solution that displays a large positive deviation from ideality will frequently split into two phases when it is cooled.

An example is the *n*-hexane + nitrobenzene system shown in the *T–X* diagram of Fig. 10.8(a). At the temperature and composition indicated by the point *x*, two phases coexist, the conjugate solutions represented by *y* and *z*. The relative amounts of the phases are proportional, as usual, to the segments of the tie-line. As the temperature is increased along the isopleth *XX'*, the amount of hexane-rich phase decreases and the amount of nitrobenzene-rich phase increases. Finally, at *Y*, the hexane-rich phase disappears completely.

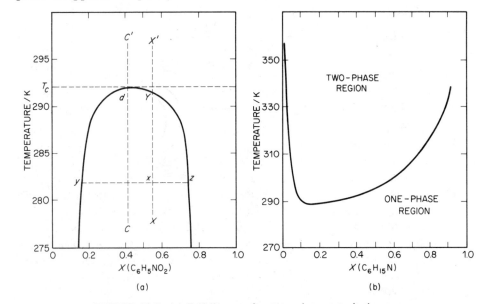

FIGURE 10.8 (a) *T–X* diagram of system *n*-hexane + nitrobenzene.
(b) *T–X* diagram of system triethylamine + water.

The composition corresponding to the maximum in the *T–X* curve is called the *critical composition* and the temperature at the maximum is the *critical solution temperature* or upper *consolute temperature*. As a two-phase system having the critical composition is gradually heated [line *CC'* in Fig. 10.8(a)], there is no gradual disappearance of one phase. Even in the immediate neighborhood of the maximum *d*, the ratio of the segments of the tie-line remains practically constant. The compositions of the two conjugate solutions gradually approach each other until, at the point *d*, the boundary line between the two phases suddenly disappears and a single phase remains.

As the critical temperature is slowly approached from above, a curious phenomenon is observed. Just before the single homogeneous phase passes over into two separate phases, the solution is suffused by a pearly opalescence. This critical opalescence is caused by the scattering of light from small regions of slightly differing density, which appear and disappear in the liquid as microscopic fluctuations of separate phases. X-ray studies have revealed that such regions may persist even several degrees above the critical point.

Some systems exhibit a lower consolute temperature. At higher temperatures, two partially miscible solutions are present, which become completely intersoluble when sufficiently cooled. An example is the triethylamine + water system in Fig. 10.8(b), with a lower consolute temperature of 290 K, at $P°$. Note that great increase in solubility as the temperature decreases to this point.

Finally, systems have been found with both upper and lower consolute temperatures. These are more common at elevated pressures, and we might expect all systems with a lower consolute temperature to display an upper one at sufficiently high temperature and pressure. An example is the butanol-2 + water system shown in Fig. 10.9 at different pressures.

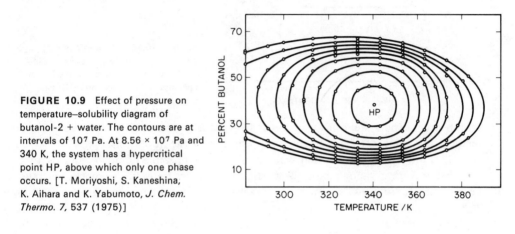

FIGURE 10.9 Effect of pressure on temperature–solubility diagram of butanol-2 + water. The contours are at intervals of 10^7 Pa. At $8.56 × 10^7$ Pa and 340 K, the system has a hypercritical point HP, above which only one phase occurs. [T. Moriyoshi, S. Kaneshina, K. Aihara and K. Yabumoto, *J. Chem. Thermo.* **7**, 537 (1975)]

10.14 Distillation of Immiscible Liquids

The boiling-point diagram at $P°$ of a pair of partially miscible liquids, isobutanol and water, is shown in Fig. 10.10. To appreciate the interesting information contained in such a diagram, let us suppose that we make up a mixture of 20 g of isobutanol and 80 g of water and follow its behavior as we heat the system at the constant pressure $P°$. Consider first the system at 83.5°C, point P on the phase diagram. By the lever rule the ratio of the masses of the two phases of composition X and Y will be

$$\frac{\text{water in isobutanol}}{\text{isobutanol in water}} = \frac{PX}{PY}$$

The system thus consists of 19.5 g of composition Y and 80.5 g of composition X.

When the temperature reaches 89°C, vapor first begins to appear. The vapor composition is given by point Z, 61.5% isobutanol. Three phases now coexist at equilibrium. In accord with the phase rule $f = c − p + 2$; with $c = 2$ and $p = 3$, $f = 1$. There is one degree of freedom and this has been used by the condition of constant pressure $P°$. The pressure $P°$ equals the sum of the vapor pressures of water and isobutanol. The temperature is fixed at 89°C and cannot be altered without causing the disappearance of one of the phases.

As soon as temperature exceeds 89°C, however, the phase rich in isobutanol

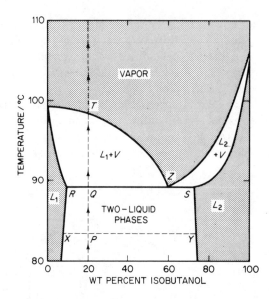

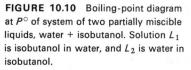

FIGURE 10.10 Boiling-point diagram at P° of system of two partially miscible liquids, water + isobutanol. Solution L_1 is isobutanol in water, and L_2 is water in isobutanol.

disappears, as the system moves into the two-phase region at Q. Further increase in temperature along QT leads to the disappearance of the liquid phase at T.

It is interesting to note that when two partially miscible liquids A and B are in equilibrium with a vapor phase, the partial vapor pressure of each component must be the same for the two solutions. For example, if benzene and water are mixed at 25°C, two immiscible layers are formed, one containing 0.09% C_6H_6 and 99.91% H_2O, the other 99.81% C_6H_6 and 0.19% H_2O. The partial pressure of benzene is the same above either of these solutions, namely 11.3 kPa.

If two partially miscible liquids are distilled at P°, the system boils when the sum of their vapor pressures equals the ambient pressure P°. Consequently, a rather nonvolatile liquid can be distilled at a much lower temperature if it is mixed with a more volatile liquid. Steam distillation of organic liquids is an application of this principle.

Example 10.8 A liquid A that is immiscible with water was steam distilled, giving 200 ml of a distillate that contained 57.2 ml of A. The boiling point of the distillation was 98.2°C and the barometric pressure was 100 kPa. The vapor pressure of water at 98.2°C is 94.9 kPa. The density of liquid A is 1.83 relative to water. What is the molar mass M_A of A?

At 98.2°C, the partial pressure of A is $P_A = 100 - 94.9 = 5.1$ kPa. Now

$$\frac{P_A}{P_w + P_A} = X_A^v = \frac{m_A/M_A}{m_A/M_A + m_w/M_w}$$

$$m_A = 57.2 \times 1.83 = 104.7 \text{ g}$$

$$m_w = 142.8 \times 1.00 = 142.8 \text{ g}$$

Hence

$$\frac{5.1}{100} = \frac{104.7/M_A}{104.7/M_A + 142.8/18.0}; \qquad M_A = 246 \text{ g mol}^{-1}$$

Creatures on other worlds may swim in seas of liquid ammonia with muscles made of silicone polymers, but life on Terra is firmly based on oil and water. Life depends on the existence of cells and cells depend on the existence of cell membranes. The structural basis of the cell membrane is a layer of lipid only two molecules thick (a lipid bilayer). Various protein molecules are incorporated in the lipid layer and are attached to its surface. The evolution of life became possible when these oily membranes enclosed small volumes of aqueous media in which processes of decreasing entropy could take place at the expense of increases in the entropy of the surroundings.

Despite the adage that oil and water do not mix, hydrocarbons can dissolve in water to a limited extent. The thermodynamic study of such solutions has helped to elucidate the behavior of proteins and lipids in the aqueous media of biological systems. As an example, consider the solubility of some hydrocarbons, (1) in water (2) in n-heptane. If we denote the chemical potential of the solute in water as μ_i^w and in heptane as μ_i^ℓ, from Eq. (10.14),

$$\mu_i^w = \mu_i^{\circ w} + RT \ln X_i^w + RT \ln \gamma_i^w$$

$$\mu_i^\ell = \mu_i^{\circ l} + RT \ln X_i^\ell + RT \ln \gamma_i^\ell$$

X_i^w, X_i^ℓ and γ_i^w, γ_i^ℓ are the mole fractions and activity coefficients of the hydrocarbon in the two solvents. The difference in chemical potential of the solute in the two solvents is $\mu_i^w - \mu_i^\ell$. For present purposes, we can assume that the activity coefficients are approximately unity. (The water solution is very dilute and the hydrocarbon solution consists of two chemically similar components). The terms $RT \ln X_i^w$ and $RT \ln X_i^\ell$ are called the *cratic* parts of the chemical potentials, from the Greek word for *mixing bowl*. These terms are due to the entropy of mixing, a statistical effect, which should be the same in both solutions and hence can be omitted from our comparisons.

Thus we can use the difference in *standard* chemical potential to compare the two solutions. Table 10.6 gives $\mu_i^{\circ w} - \mu_i^{\circ l}$ for several hydrocarbons in water and heptane. These values are the $\Delta \mu_i^\circ$ for transfer of the solute from heptane to water.

TABLE 10.6
THERMODYNAMIC DATA FOR TRANSFER OF HYDROCARBONS
FROM n-HEPTANE TO WATER AT 298 K

Solute	$\mu_i^{\circ w} - \mu_i^{\circ l}$ (J mol^{-1})	$H_i^{\circ w} - H_i^{\circ l}$ (J mol^{-1})	$S_i^{\circ w} - S_i^{\circ l}$ (J K^{-1} mol^{-1})
C_2H_6	16 300	$-10\,500$	-88
n-C_3H_8	20 500	$-7\,100$	-92
n-C_4H_{10}	24 700	$-3\,350$	-96
C_6H_6	19 200	2 500	-54

The values are all positive, denoting that the transfer is a thermodynamically unfavorable process. At 300 K a $\Delta \mu_i^\circ$ of 24 700 J mol^{-1} would correspond to an equilibrium constant (in this case the partition coefficient of solute between the two solvents) of exp $(24\,700/8.314 \times 300) = 2 \times 10^4$ for butane between n-heptane and water. It is clear that n-butane would prefer to be in n-heptane rather than in water, and we might go so far as to call the n-butane molecule *hydrophobic* or "water hating."

The surprise comes when we analyze the $\Delta\mu_i^\circ$ into its entropic and enthalpic contributions,

$$\Delta\mu_i^\circ = \Delta H_i^\circ - T\,\Delta S_i^\circ$$

From Eq. (8.38), $(\partial\mu_i^\circ/\partial T)_P = -\Delta S_i^\circ$, so that ΔS_i° can be obtained from data on the temperature dependence of the distribution coefficient of n-butane between n-heptane and water. The results in Table 10.6 show that the transfer of n-butane from n-heptane to water is actually exothermic; i.e., the energy state of the butane is lower in water than in n-heptane. On the other hand, the entropy decreases markedly for the transfer of n-butane from heptane to water, and the $T\,\Delta S_i^\circ < 0$ more than counterbalances the negative ΔH_i° terms, so that the $\Delta\mu_i^\circ$ is positive for the transfer reaction.

What could cause this large decrease in entropy? It cannot be due to restrictions on the motion of butane molecules in water but not in heptane; actually, the motions of the butane molecule (rotations, vibrations, translation) are much the same in both solvents. The explanation most favored is that the structure of water is unusually sensitive to local modification by dissolved solute molecules. Normally, water has an open hydrogen-bonded structure (see Fig. 30.5). The hydrogen bonds between neighboring water molecules can take many different orientations. In the neighborhood of a dissolved butane molecule, the water structure collapses to one of considerably lower entropy as configurations of the hydrogen bonds become more restricted.

The entropy factors that we have found in these solutions are also important in many biochemical reactions. Consider a protein molecule in two different conformations, a random-coil and an ordered globular structure. The random structure would have *in vacuo* the advantage of a higher entropy. Nevertheless, the coil can fold into a globular structure in such a way that nonpolar side chains are brought together, thereby lowering the area of contact between oily side chains and water. When this happens, the effect of the solute on the water structure is reduced and more of the water can resume its normal higher-entropy structure. The net effect of the folding is then to increase the entropy, and the conformation change, random coil → globular structure, will be favored by a net $\Delta S > 0$. This effect is called a *hydrophobic interaction*.

Problems

1. The molar volumes of CO_2 at 333 K have been measured:

P (kPa)	1318	3589	5436	7567	8648
V_m (cm^3 mol^{-1})	2000	666.7	400	250	200

Calculate the fugacity f and fugacity coefficient γ for CO_2 at 333 K and $P = 10^3$, 2×10^3, 4×10^3, and 8×10^3 kPa. Compare with results from Newton's graphs.

2. For a gas that follows Berthelot's equation of state (page 000), show that

$$\ln \gamma = \frac{9}{128} \frac{T_c}{P_c T}\left(1 - \frac{6T_c^2}{T^2}\right)P$$

Calculate γ and the fugacity of C_2H_4 at 300 K and 10^4 kPa. (See Table 2.3.) Calculate $\mu(300 \text{ K}, 10^4 \text{ kPa}) - \mu^\circ (300 \text{ K})$.

3. Show that a gas that follows the virial equation, $PV_m/RT = 1 + B'P + C'P^2$, has $\ln \gamma = B'P + \frac{1}{2}C'P^2$. Hence show that the difference, $\Delta G_m = G_m(\text{real}) - G^m(\text{ideal}) = RT(B'P + \frac{1}{2}C'P^2)$. Example: For N_2 at 273 K, $B' = -4.59 \times 10^{-4}$ atm^{-1}, $C' = 2.31 \times 10^{-6}$ atm^{-2}. Calculate ΔG_m at 273 K and 4.00 MPa.

4. For the reaction $BeO(c) + Cl_2 = BeCl_2(g) + O_2$, $\Delta G^\circ = 4.184(123\,200 - 47.8T)$ J mol^{-1}. What temperature would be necessary to get a partial pressure of $BeCl_2$ of 1 kPa on treating $BeO(c)$ with Cl_2 at 100 kPa?

5. $3C_2H_2(g) = C_6H_6(g)$. Calculate the equilibrium conversion of C_2H_2 to C_6H_6 at 100 kPa, 1000 kPa, 10 000 kPa at 1000 K (a) as ideal gas reaction; (b) with fugacity coefficients from Newton's graphs.

6. $SO_3(g) = SO_2(g) + \frac{1}{2}O_2$. The equilibrium constants are:

T (K)	800	900	1000	1100	1170
$\log K_P$	-1.494	-0.816	-0.268	0.199	0.446

(a) Calculate ΔG°, ΔH°, and ΔS° at 1000 K.

(b) SO_3 gas is passed over a catalyst at 1000 K and $P^\circ = 1$ atm. Calculate the mol % SO_2 in the product gases.

(c) SO_3 gas diluted to 5 mol % with inert N_2 at total pressure of 1 atm and treated as in part (b). Calculate mol % SO_2 in product.

(d) O_2 and SO_2 in molar ratio $O_2/SO_2 = 2$ (i.e., excess O_2) are passed over catalyst at a pressure of 1 atm and 500 atm. Calculate the mol % SO_3 in the product (1) assuming ideal gases, (2) with fugacities based on Newton's graphs. Critical constants T_c, P_c: SO_2(431 K, 77.7 atm), SO_3(490 K, 83.8 atm).

7. At 661 K the vapor pressure of $K(\ell)$ is 0.433 kPa and that of Hg is 170.6 kPa. Over a solution 50 mol % K in Hg, the partial vapor pressures are $P(K) = 0.142$ kPa, $P(Hg) = 1.73$ kPa. Calculate the activities and activity coefficients of K and Hg in the melt. Calculate ΔG_m of mixing 0.5 mol K + 0.5 mol Hg at 661 K. If ΔS_m of mixing is ideal, calculate ΔH_m of mixing.

8. For solutions of acetone and chloroform at 50°C, the mole fractions of acetone in liquid X_A^ℓ, in vapor X_A^v, and the total vapor pressure of the solution P (kPa) are:

X_A^ℓ	0	0.100	0.200	0.300	0.400	0.500	0.600	0.700	0.800	0.900	1.000
X_A^v	0	0.071	0.165	0.279	0.408	0.550	0.684	0.789	0.890	0.955	1.000
P	69.5	66.0	63.2	61.7	61.3	62.5	65.2	68.1	72.0	76.8	81.6

Calculate the activities and activity coefficients of both components and plot them as functions of X_A^ℓ. Show that the Gibbs–Duhem equation (10.24) applies, by calculating a value of $a(CHCl_3)$ from the results for a_A.

9. When FeO dissolves in molten CaF_2 at 1450°C its activity relative to pure solid FeO is 0.60 when $X(FeO) = 0.02$. Calculate $\mu(FeO$ in melt$) - \mu(FeO)$ at 1450°C.

10. At what molality would a solution of glycylglycine in water be in isopiestic equilibrium at 298 K with a 0.200 molal solution of glycine? (See Table 10.3.)

11. For the hydrolysis of ATP at 37°C, at pH 7, ATP + H_2O = ADP + P_i (where P_i is inorganic phosphate) $K_c = 1.30 \times 10^5$ (standard state $c^\circ = 1$ mol L^{-1}). If $\Delta H^{\circ\prime} = -20.1$ kJ mol^{-1}, calculate K_c at 25°C.

12. What is the ΔG° at 50°C per mole of sucrose B when 100 cm^3 of a solution with $X_B = 0.100$ is diluted to 1000 cm^3 with water? (See Table 10.2.)

13. For the reaction

$$NH_4^+ + \begin{matrix} CH-COO^- \\ \| \\ CH-COO^- \end{matrix} = \begin{matrix} NH_3^+ \cdot CH \cdot COO^- \\ | \\ CH_2COO^- \end{matrix}$$

the equilibrium constant at 300.5 K is 1.95×10^2 when $c^\circ = 1$ mol L^{-1}. Calculate the equilibrium concentration of aspartate when initial concentration of NH_4^+ and fumarate are both 0.30 M; 0.03 M.

14. For the hydrolysis of fumarate to malate,

$$\begin{matrix} CH \cdot COO^- \\ \| \\ CH \cdot COO^- \end{matrix} + H_2O = \begin{matrix} CHOH \cdot COO^- \\ | \\ CH \cdot COO^- \end{matrix}$$

the $\Delta G^\circ(298$ K$) = -3.68$ kJ mol^{-1} and $\Delta H^\circ = -14.9$ kJ mol^{-1}. (Standard state $c^\circ = 1$ mol L^{-1}.) Calculate K_c at 37°C.

15. The activity coefficient of glycylglycine G_2 in a 1.00 molal aqueous solution at 25°C is given by log $\gamma = -0.128$. Calculate $\mu - \mu^\circ$ for G_2 under these conditions.

16. Nitrobenzene and water are to be steam distilled at a pressure of 0.963 atm, $T = 371.4$ K. The vapor pressure of H_2O at this T is 0.937 atm. Calculate the mass of nitrobenzene that will distill per 1 kg of water.

11

Phase Transitions and Phase Equilibria

This chapter describes a variety of systems in which two or more phases are in equilibrium. Such systems have great practical as well as theoretical interest. Phase equilibria control the geochemical structures of the planets; they govern the properties of industrial materials such as alloys and ceramics; and they provide techniques that lead to temperatures close to absolute zero and all the amazing phenomena that occur there. The discussion of phase equilibria is based on three basic principles: (1) the Gibbs phase rule $f = c - p + 2$, (2) the use of the Gibbs function G and the chemical potential μ to specify equilibrium conditions, and (3) the graphical representation of the properties of systems by *phase diagrams* showing the equilibrium relations of state variables P, T, and c. These principles will be applied to various systems in the following discussions.

11.1 Conditions for Equilibrium between Phases

Suppose that a system contains several species i ($= 1, 2, 3,$ etc.) distributed in two phases a and b. The conditions for equilibrium are: $T^a = T^b$, $P^a = P^b$, $\mu_i^a = \mu_i^b$ (for each i). If the first condition failed to hold, there would be a flow of heat between a and b and equilibrium would not exist. Hence $T^a = T^b$ is the condition for *thermal equilibrium* between phases. If the second condition failed to hold, one phase would expand into the other. Hence $P^a = P^b$ is the condition for *mechanical equilibrium*. If the third condition failed to hold for any component i, there would be a flow of that component from one phase to the other by diffusion or chemical reaction. Hence $\mu_i^a = \mu_i^b$ is the condition for *chemical equilibrium*.

From the phase rule, when $c = 1, f = 3 - p$, and three different cases are possible: $p = 1, f = 2$, bivariant system; $p = 2, f = 1$, univariant system; $p = 3, f = 0$, invariant system.

Since the maximum number of degrees of freedom is two, the equilibrium conditions for a one-component system can be represented by a phase diagram in two dimensions, the most convenient choice of variables being P and T. If we wish also to display the volume changes in the system, we can construct a model in three dimensions of the complete PVT surface. Every point on this surface denotes a set of equilibrium values for the substance. We usually plot the volume per mole (V_m) or per gram (v).

Figure 11.1 shows such a PVT diagram for carbon dioxide, a substance that contracts on freezing. Since the density of this solid at the freezing point is greater than that of liquid, in accord with the Clapeyron–Clausius equation (8.22) the freezing point increases with pressure. In the case of a substance that expands on freezing, like water, the solid–liquid surface slopes in the opposite direction.

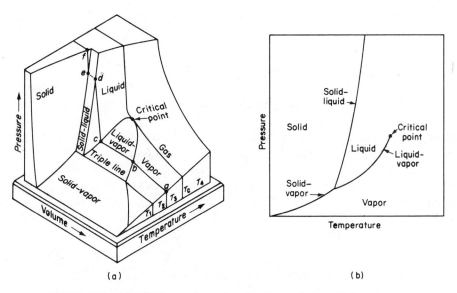

FIGURE 11.1   (a) PVT diagram for carbon dioxide, a substance that contracts on freezing. (b) Projection of the PVT surface on the PT plane. [After F. W. Sears, *An Introduction to Thermodynamics, The Kinetic Theory of Gases and Statistical Mechanics* (Cambridge, MA: Addison Wesley Publishing Co., 1953)]

Let us follow an isotherm on the PVT diagram, by increasing P at constant T. We begin at point a, which corresponds to CO_2 gas at $P = 100$ kPa, $T = 293$ K, and $V_m = 24\,600$ cm^3 mol^{-1}. As pressure increases, volume decreases along line ab until at point b liquid CO_2 begins to form. The pressure at this point is 5.70 MPa and the molar volume of the vapor in equilibrium with liquid is $V_m = 230$ cm^3 mol^{-1}.

The molar volume of the liquid is given by point c as $V_m = 56.5 \text{ cm}^3 \text{ mol}^{-1}$. The line bc is a tie-line; it connects points representing phases in equilibrium with each other. On the 293 K isotherm, the pressure remains constant at 5.70 MPa until vapor is completely converted to liquid at c. Increase in the pressure beyond c is applied to the pure liquid phase, which has a low compressibility, so that the isotherm rises steeply until it intercepts the melting point curve at d, which is very close to 502 MPa at 293 K. The densities of liquid and solid CO_2 at this pressure have not been directly measured, although Bridgman determined the decrease in volume on freezing, ΔV_m from d to e on diagram, to be 3.94 $\text{cm}^3 \text{ mol}^{-1}$. We can estimate V_m of liquid CO_2 at d to be about 35 $\text{cm}^3 \text{ mol}^{-1}$, so that V_m of solid CO_2 at e would be about 31 $\text{cm}^3 \text{ mol}^{-1}$. The isotherm continues with further compression of solid CO_2 along the line ef and beyond.

Figure 11.1 also shows the projections of the PVT surface on the PT plane. The PT projection is the one usually used as a *phase-rule diagram*. On the PT diagram, states with two phases in equilibrium are represented by lines.

At sufficiently high pressures, solid CO_2 can exist well above the critical temperature of the liquid–vapor transition. There has been a lively debate as to whether a critical point ever exists for a solid–liquid transition. At present, the "nays" seem to have the better of it, on the basis of the argument that the solid–liquid transition requires a change in the symmetry of the structure of matter, so that continuity of states between a symmetrical crystal structure and an isotropic liquid would not be possible.

11.3 How Thermodynamic Functions Behave at Phase Changes

For pure substances, the chemical potential μ_i becomes simply the Gibbs free energy per unit amount of substance G_i. Figure 11.2 shows how G_i and other thermodynamic functions behave when a change in phase occurs. To make the example realistic, we plot H, G, V, S and C_P for one mole of pure tin as functions of T at the standard $P° = 101.32$ kPa. At 286.4 K, the transition temperature of gray (cubic) Sn^α to white (tetragonal) Sn^β, there are discontinuities in H, S, V, and C_P. The $\Delta_\alpha^\beta H = 2.5 \text{ kJ mol}^{-1}$ is the enthalpy of transformation and the $\Delta_\alpha^\beta V = -3.34 \text{ cm}^3 \text{ mol}^{-1}$ is the volume change. Usually the high-temperature form has the higher volume, but in the case of tin the structure of white Sn is more dense than that of gray Sn.

The existence of two different forms of an element is called *allotropy*. The transformation of tin is an example of an *enantiotropic* change, which is defined as a change between two different crystalline forms of the same pure substance when both forms have a definite range of thermodynamic stability. When Sn^β is cooled below 286.4 K, it does not rapidly change to Sn^α. Indeed one can make measurements on Sn^β all the way down to the neighborhood of absolute zero. In olden times, the change of Sn^β to Sn^α was sometimes observed in cold winters. The gray tin would appear as a scaly eruption on the surface of tin objects such as church organ pipes, and it was called "tin plague" or "tin pest."

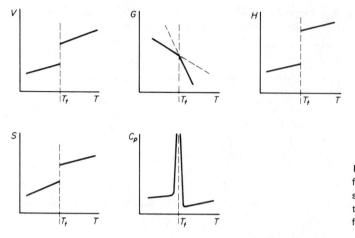

FIGURE 11.2 Thermodynamic functions for one mole of a pure substance that has a phase transition at T_t. Dependence of functions on T at constant P°.

The transition Sn^α (gray) $\rightarrow$ Sn^β (white) is a change in the electronic structure of the solid element. When dissolved in acid, white tin yields $Sn(II)$ salts and gray tin, $Sn(IV)$ salts. Gray Sn has the same crystal structure as diamond—it is not a metal but a semiconductor (like germanium and silicon). White Sn is a good metal, bright, shiny, and an excellent conductor of electricity and heat. At high pressures, silicon and germanium also undergo transitions from a diamond structure to a white-tin structure and then change from semiconductors to metals.

In Fig. 11.2 the entropy S also undergoes an abrupt change at the transition temperature, T_t,

$$\Delta_\alpha^\beta S = \Delta_\alpha^\beta H_t/T_t = \frac{2500 \text{ J mol}^{-1}}{286.4 \text{ K}} = 8.7 \text{ J K}^{-1} \text{ mol}^{-1}$$

We note also that there must be a discontinuity in C_P at the transition temperature.

The molar Gibbs free energies are equal at the transition temperature, $G_i^\alpha = G_i^\beta$. We recall that $(\partial G/\partial T) = -S$. Since S is always a positive quantity, the slopes of G vs. T curves are always negative. The curves for Sn^α and Sn^β are shown in Fig. 11.2. The more stable form is that with the lower G_i, and the intersection of the $G_i(T)$ curves for the two phases determines the transition point.

Phase changes that display the thermodynamic behavior shown in Fig. 11.2 are called *first-order changes.* They include all solid–liquid, liquid–gas, and solid–gas changes, and a large number of solid–solid changes (but not all).

Sometimes two different structures of a solid can coexist, but one of them is always metastable with respect to the other. The G_i vs. T curve of the metastable form thus lies above that of the stable form. This phenomenon is called *monotropy.* A common example of monotropy is in the phosphorus system. White phosphorus is metastable and violet phosphorus is the stable form. One cannot convert P(violet) to P(white) directly under any conditions of P and T. If P(violet) is vaporized, however, the vapor can be condensed to yield P(white), which can then be kept for years at room temperature without reversion to P(violet).

11.4 Melting and Vaporization

Table 11.1 contains data on melting points, boiling points, enthalpies, and entropies of melting and vaporization for a number of substances. The enthalpy of melting a substance is much lower than its enthalpy of vaporization. It requires much less energy to convert a crystal to liquid than to vaporize a liquid. Entropies of melting are also considerably lower than the entropies of vaporization. The latter are roughly constant, from 90 to 125 J K^{-1} mol^{-1} for many liquids (Trouton's Rule). The ΔS (melting) display more variation.

TABLE 11.1
DATA ON MELTING AND VAPORIZATION

Substance	Normal melting point (K)	Enthalpy of melting (kJ mol^{-1})	Entropy of melting (J K^{-1} mol^{-1})	Normal boiling point (K)	Enthalpy of vaporization (kJ mol^{-1})	Entropy of vaporization (J K^{-1} mol^{-1})
Hg	234	2.43	10.4	1165	64.9	103
K	336	2.43	7.20	1047	91.6	87.5
Na	371	2.64	7.11	1165	103	88.3
Al	932	10.7	11.4	2740	283	103
Ag	1234	11.3	9.16	2485	290	117
Fe	1802	14.9	8.24	3273	404	123
Pt	2028	22.3	11.0	4100	523	128
NaCl	1073	30.2	28.1	1686	766	454
KCl	1043	26.8	25.7	1500	690	460
H$_2$	14	0.12	8.4	20.7	0.92	44.4
Ar	83	1.17	14.1	87.5	7.87	89.9
C$_2$H$_5$OH	156	4.60	29.7	351.7	43.5	124
NH$_3$	198	7.70	38.9	240	29.9	125
H$_2$O	273	5.98	22.0	373	47.3	126.8
C$_6$H$_6$	278	9.83	35.4	353.3	34.7	98.2

11.5 Liquid Crystals

In some substances, the crystalline form does not melt directly to a liquid phase but first passes through an intermediate stage (the *paracrystalline* state), which only at a higher temperature undergoes transition to the liquid state. These intermediate states have been called *liquid crystals*, since they display some properties of both liquids and crystals. Thus, some paracrystalline substances flow in a gliding stepwise fashion and form graded droplets having terracelike surfaces; other varieties flow quite freely but are not isotropic, exhibiting interference figures when examined with polarized light. An example is shown in Fig. 11.3.

Liquid crystals tend to occur when molecules are markedly unsymmetrical in shape. For example, in the crystalline state, long-chain molecules may be lined up in

FIGURE 11.3 Textures in a thin layer of nematic liquid crystals. [G. H. Brown and J. J. Wolken, *Liquid Crystals and Biological Structures* (New York: Academic Press, Inc., 1979)]

a regular array. On raising the temperature two types of anisotropic melt might be obtained, shown in Fig. 11.4(b) and (c). In the *smectic* (Greek, *soap*) state, the molecules are oriented in well-defined planes. When a stress is applied, one plane glides over another. In the *nematic* (Greek, *thread*) states, the planar structure is lost, but the orientation is preserved. With some substances, notably the soaps, several different phases, differentiated by optical and flow properties, can be distinguished between typical crystal and typical liquid.

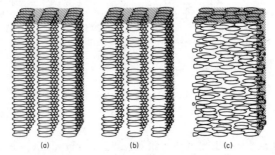

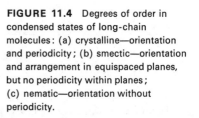

FIGURE 11.4 Degrees of order in condensed states of long-chain molecules: (a) crystalline—orientation and periodicity; (b) smectic—orientation and arrangement in equispaced planes, but no periodicity within planes; (c) nematic—orientation without periodicity.

Liquid crystals undoubtedly exist in living cells. The doubly refracting portion of striated muscle fiber is the classical instance of this arrangement, but there are many others, such as cephalopod spermatozoa, the axons of nerve cells, cilia, and birefringent phases in molluscan eggs. "The paracrystalline state seems the most suited to biological functions, as it combines the fluidity and diffusibility of liquids while preserving the possibilities of internal structure characteristic of crystalline solids."*

*Joseph Needham, *Biochemistry and Morphogenesis* (Cambridge, England: Cambridge University Press, 1942). The ideas of Needham about the importance of paracrystalline states in biological systems are borne out by current work on phase changes in cell membranes [*Liquid Crystals and Ordered Fluids*, J. F. Johnson and R. S. Porter, eds. (New York: Plenum Press, 1970)].

11.6 Measurements at High Pressures

We tend to classify pressures and temperatures as high or low by comparing them to the 10^5 Pa and 293 K of a spring day, despite the fact that almost all matter in the universe exists under vastly different conditions. Even at the center of Earth, by no means a large astronomical body, the pressure is about 4×10^{11} Pa, so that the core material must have properties quite unlike those with which we are familiar. At the center of a comparatively small star, like the Sun, the pressure is about 10^{15} Pa.

The pioneer work of Gustav Tammann on high-pressure measurements was extended by P. W. Bridgman and coworkers. Pressures up to 4×10^{10} Pa were achieved, and methods were developed for measuring the properties of substances at 10^{10} Pa. Most modern apparatus for ultrahigh pressures is based on the idea of using some of the mechanical force that produces the pressure also to support the apparatus. A high-pressure apparatus that provides considerable support to the inner piston is the tetrahedral anvil, shown in Fig. 11.5. This apparatus was used for the first commercial production of synthetic diamonds.

FIGURE 11.5 The tetrahedral anvil, designed by Tracy Hall, used to achieve high pressures and temperatures.

The highest laboratory pressures (up to about 2×10^{11} Pa) have been achieved by dynamic methods in which a shock wave, produced by compressed gas or explosives, travels through the specimen. High-pressure, high-velocity gases accelerate the material of the specimen, and a high pressure is produced owing to the inertia of the material that has just been accelerated. Within a few microseconds, the entire specimen reaches a high pressure, and as the shock front passes through the material, the rarefaction wave spreads back, reducing the pressure again. In 1961, B. J. Alder and R. M.

Christian found evidence for diamond formation in graphite that was shock loaded at high temperatures, and Alder made his famous remark, "We were millionaires for a microsecond."

In the Geophysical Laboratory of the Carnegie Institution (Washington, D.C.) static pressures up to 1.7×10^{11} Pa have been achieved by H. K Mao and coworkers. They use a pressure cell constructed from two diamonds, and the apparatus shown in Fig. 11.6. Optical and X-ray measurements are made on the compressed specimens. The pressure is measured from the wavelength of a line in the fluorescent spectrum of ruby. The ruby wavelength scale is calibrated against X-ray measurements of the volume changes of metals, which were previously determined from accurate shock-wave experiments.

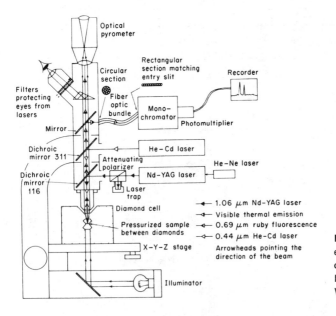

FIGURE 11.6 Diamond pressure cell and equipment for spectroscopic observation of material at high pressure. [Geophysical Laboratory, Carnegie Institution, Washington, D.C.]

11.7 High-Pressure Systems

Measurements on water at high pressures have yielded the results shown in the phase diagram of Fig. 11.7. The melting point of ordinary ice (ice I) falls on compression, until a value of 251 K is reached at 2×10^8 Pa. Further increase in pressure results in the transformation of ice I into a new modification, ice III, whose melting point increases with pressure. Altogether six different polymorphic forms of ice have been found. There are five triple points shown on the water diagram. At a pressure of about 2×10^9 Pa, liquid water freezes to ice VII at about 373 K. Ice IV is not shown. Its existence was indicated by the work of Tammann, but it was not confirmed by Bridgman.

Answers to major geochemical problems will require further data on the properties of minerals at high pressures. Just as icebergs float in the oceans, mountains float

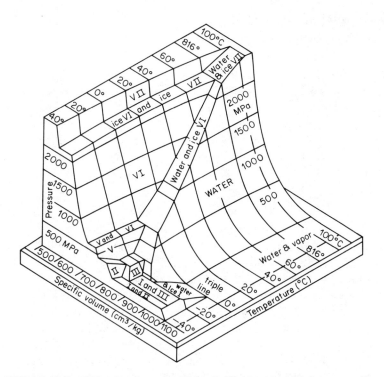

FIGURE 11.7 *PVT* surface for water. [After D. Eisenberg and W. Kauzmann, *The Structure and Properties of Water* (London: Oxford University Press, 1969)]

in a sort of sea of plastic rock that flows readily under pressure. The discontinuity between lighter minerals of the crust and underlying denser minerals is the famous Mohorovičić or M discontinuity. Under the continents, this is about 40 km below the surface, but under the deep ocean floor, it is only 7 to 10 km down. The M discontinuity was detected by a sudden increase at a certain depth below the earth's surface in the velocity of seismic waves. Early theories postulated a difference in chemical composition at the M discontinuity, but the current hypothesis is that the discontinuity marks the locus of phase transformations from low-density crystalline forms of silicate minerals like albite and calcium feldspars to high-density forms like jadeite and garnet. The density changes from about 2.95 to 3.50 g cm^{-3}, and the high-density material is the form stable at the higher pressures. The transformation pressure is around 2×10^8 Pa, depending on temperature. According to this theory, the M discontinuity is simply a natural expression of the *P–T* transformation curve of these two classes of minerals, a sort of large-scale plot of a phase diagram. Mountains arise when a temperature fluctuation beneath the surface causes a fall in the stable level of the phase transformation and formation of a large amount of new material of lower density. The fluctuation is not sudden—it may take several million years. A temperature fluctuation of 10% at the M discontinuity would displace the phase equilibrium sufficiently to elevate peaks to the heights of the Rocky Mountains.

11.8 An Approach to Absolute Zero: Cooling by Demagnetization

Phase changes can be used to cool substances to low temperatures. An example is the demonstration in which water is frozen to ice by evaporation of ether. A test tube of water is immersed in a beaker of ether, which is placed in a bell jar. A vacuum pump exhausts the system and as pressure falls, the boiling point of the ether is decreased. The rapid evaporation of ether produces a drastic cooling and the water freezes. The reason why an evaporating liquid cools can be seen from the way in which thermodynamic functions change on vaporization of a liquid:

$$\textit{Liquid state} \quad \longrightarrow \quad \textit{Vapor state}$$
$$\text{Lower } S_i \qquad\qquad \text{Higher } S_i$$
$$\text{Lower } H_i \qquad\qquad \text{Higher } H_i$$

Since the system is insulated, the only source of the enthalpy of vaporization ΔH_v must be the thermal energy of the liquid itself. Thus as some liquid is evaporated, the temperature of the remaining liquid must fall.

This evaporation method for producing low temperatures was applied to liquid helium by Kammerlingh-Onnes in Leiden and George Dewar in London. A temperature of 0.84 K was reached, but enormous pumps were needed to carry off the gaseous helium, and there was little hope for much further cooling by this method.

In 1933, William Giauque at Berkeley made a great experimental advance in the science of low temperatures. His method was analogous to cooling by evaporation of a liquid, but it was based upon a different kind of phase change, the demagnetization of a magnetic solid. Many transition-metal ions have net electron spins, which act as little magnets that can become aligned in the direction of an imposed magnetic field. Rare-earth salts such as gadolinium sulfate have particularly high magnetic moments.

The transition, magnetized crystal $\longrightarrow$ demagnetized crystal, is analogous to the transition, liquid $\longrightarrow$ vapor. Thus

$$\textit{Magnetized state} \quad \longrightarrow \quad \textit{Demagnetized state}$$
$$\text{Lower } S_i \qquad\qquad\qquad \text{Higher } S_i$$
$$\text{Lower } H_i \qquad\qquad\qquad \text{Higher } H_i$$

The magnetized state is the more ordered state; the magnetic field acting on a magnetized solid is analogous to the pressure acting on a liquid. When the pressure is released, liquid evaporates. When the magnetic field is released, the ordered array of magnetic moments in the solid relaxes back into random orientations. If demagnetization occurs under adiabatic conditions, the increase in enthalpy of the demagnetized state is at the expense of thermal energy of the crystals, and the temperature of the substance falls when the magnetic field is turned off.

The magnetic-cooling experiment is shown in Fig. 11.8. The sample is cooled while the magnetic field is applied and thermal contact between the sample and liquid helium maintained by a jacket filled with helium gas. The helium gas is then pumped away from the jacket leaving the sample thermally insulated by a good vacuum. The field is then turned off and as adiabatic demagnetization occurs, the sample sponta-

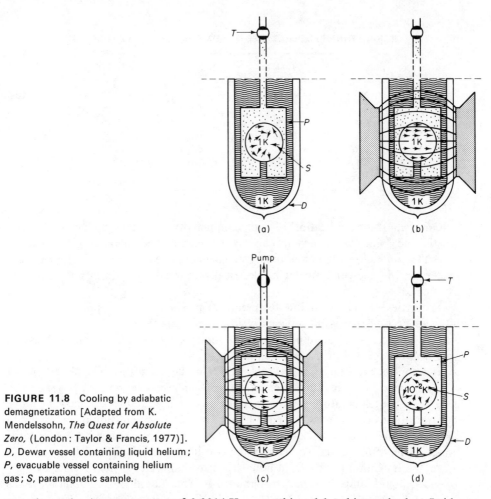

FIGURE 11.8 Cooling by adiabatic demagnetization [Adapted from K. Mendelssohn, *The Quest for Absolute Zero,* (London: Taylor & Francis, 1977)]. *D*, Dewar vessel containing liquid helium; *P*, evacuable vessel containing helium gas; *S*, paramagnetic sample.

neously cools. A temperature of 0.0014 K was achieved by this method at Leiden. Even lower temperatures can be reached by an adiabatic demagnetization based upon nuclear spins instead of electron spins. Nuclear cooling to 0.0003 K in a copper specimen was achieved by Lounasmaa in Finland.

11.9 Superconductivity and Superfluidity

In 1911, Kammerlingh-Onnes discovered that the electrical resistance of pure solid mercury suddenly drops to an immeasurably small value when the metal is cooled below 4.1 K. This phenomenon is called *superconductivity*. Other metals also become superconducting below characteristic transition temperatures. Examples are given in Table 11.2. The conductivity of a superconductor is of the order of 10^{10} times higher than that of a normal metal. The superconducting transition is not a phase transition in the usual sense, since an equilibrium between metal and superconductor never occurs.

TABLE 11.2

TRANSITION TEMPERATURES OF SOME SUPERCONDUCTORS

Substance	T_c (K)	Substance	T_c (K)
W	0.01	Sn	3.72
Ir	0.14	Hg	4.15
Cd	0.52	Pb	7.19
Zn	0.85	Nb	9.10
Mo	0.92	Nb_3Al	18.7
Th	1.37	Nb_3Ge	23.7
InSb	1.90		

One important application of superconductivity is in the production of high magnetic fields, since large currents can pass through a superconducting coil without heating the conductor. Superconducting magnets are used in high-field NMR spectrometers and in the containment of electrically charged particles in nuclear-fusion experiments.

Properties of helium at low temperatures provide many surprises. All other substances become solids at sufficiently low temperatures under their own vapor pressures. In the case of helium, even in the limit as $T \to 0$, a solid phase does not form unless an elevated pressure is applied.

Helium has two stable isotopes, ^{4}He and ^{3}He, the latter having an abundance of only about 1 part in 10^6 in atmospheric helium. The phase diagram of ^{4}He is shown in Fig. 11.9. As the temperature is reduced along an isobar at 100 kPa, somewhat above 2 K a transition occurs from ordinary liquid He-I to a second liquid phase, liquid He-II. This is the only known system in which two liquid phases coexist for the same substance. The transition curve between the two liquids is called the λ line.

As the transition occurs from liquid He-I to liquid He-II, the thermal conductivity of the liquid jumps by a factor of about 10^6. Thus liquid He-II conducts heat as well as solid copper. The transition He-I $\longrightarrow$ He-II was first observed as an instanta-

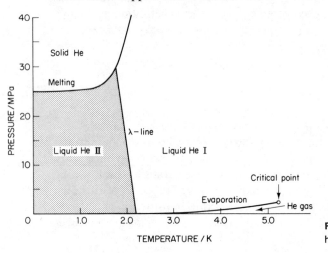

FIGURE 11.9 Phase diagram of helium-4.

neous change from a violently boiling liquid to an utterly calm one. (Bumps and bubbles in a boiling liquid are due to uneven heating and when thermal conductivity is high, uneven heating does not occur.)

Helium-II has a very low viscosity η, and in fact the colder it gets, the lower is its viscosity, so that in the limit as $T \rightarrow 0$, viscosity $\eta \rightarrow 0$ also. Viscosity is a measure of internal friction between molecules in a fluid. Superfluid helium is thus a frictionless state of matter. In this way superfluidity is analogous to superconductivity. The latter is a frictionless state of electrons in a crystal; the former is a frictionless state of atoms in a fluid. In a crystal each atom or molecule is located in space on a regular lattice. A crystal is an ordered structure in position space. By analogy, a superfluid can be considered to be an ordered structure in momentum space.

In Chapter 4 we found that at ordinary temperatures and densities the number of quantum states available to the particles in a system is much greater than the number of particles to be assigned to these states. When matter becomes very dense or the temperature becomes very low, this condition no longer applies, and a quantum-mechanical condensation occurs. This causes drastic changes in the thermodynamic properties of matter. At the densities prevailing on Earth, we can find these effects, such as superconductivity and superfluidity, only at very low temperatures. In the dense states of matter in the interior of stars, however, it is possible that superconductivity and superfluidity may occur even at very high temperatures. The density of a white dwarf star may reach 200 kg cm^{-3}.

The transition liquid He-II $\rightarrow$ liquid He-I does not behave like an ordinary first-order phase transition, with a latent heat ΔH and change in volume ΔV. If we plot the heat capacity C_V vs. T on both sides of the transition, as shown in Fig. 11.10, there is a singularity at the lambda (λ) point, at which $C_V \rightarrow \infty$. The shape of the resultant curve resembles a Greek λ, and this is the origin of the name *lambda transition*. In a λ transition, the heat capacity begins to rise as the temperature approaches

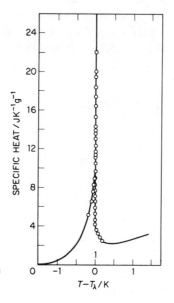

FIGURE 11.10 The specific heat of ^{4}He under its saturated vapor pressure, as a function of T.

the transition temperature. The transition does not occur sharply, and the change occurs over a range of temperature given by the λ region. The increasing value of $C_V = (\partial U/\partial T)_V$ indicates that there is a rearrangement of the internal structure of the material accompanied by an absorption of energy from the surroundings. Lambda transitions are not restricted to liquid helium but are observed in solid-state transitions whenever the change from one form to another is thermodynamically continuous rather than discontinuous as in a first-order phase transition.

11.10 Two-Component Systems

For systems in which the number of components $c = 2$, the Gibbs phase rule, $f = c - p + 2$, becomes $f = 4 - p$. The following cases are then possible: $(p = 1, f = 3)$, $(p = 2, f = 2)$, $(p = 3, f = 1)$, $(p = 4, f = 0)$. The three degrees of freedom are pressure P, temperature T, and the composition of the system, usually measured as mole fraction X or weight percent.

The graphical representation of the phase diagram of a two-component system therefore requires a three-dimensional model or perspective drawing. Often, we are interested mainly in the system at standard atmospheric pressure $P°$, and in this case a temperature–composition $(T-X)$ diagram suffices. We should keep in mind, however, that these $T-X$ diagrams are only slices at constant P through the complete three-dimensional $T-X-P$ diagram.

11.11 Solid–Liquid Equilibrium—Simple Eutectic Diagrams

Two-component systems with solid and liquid phases, in which the liquids are intersoluble at all compositions but in which there is no appreciable solid–solid solubility, are represented by the diagram for Cd–Bi of Fig. 11.11. Examples of systems of this kind are listed in Table 11.3. We previously met a diagram like Fig. 11.11 when we discussed the solubility of ideal solutions in Section 9.12, but the simple eutectic diagram is not restricted to ideal solutions.

TABLE 11.3
SYSTEMS WITH SIMPLE EUTECTIC DIAGRAMS, SUCH AS FIG. 11.11

Component A	Melting point of A (K)	Component B	Melting point of B (K)	Eutectic T(K)	Eutectic Mol % B
$CHBr_3$	280.5	C_6H_6	278.5	247	50
$CHCl_3$	210	$C_6H_5NH_2$	267	202	24
Picric acid	395	TNT	353	333	64
Sb	903	Pb	599	519	81
Cd	594	Bi	444	417	55
KCl	1063	AgCl	724	579	69
Si	1685	Al	930	851	89
Be	1555	Si	1685	1363	32

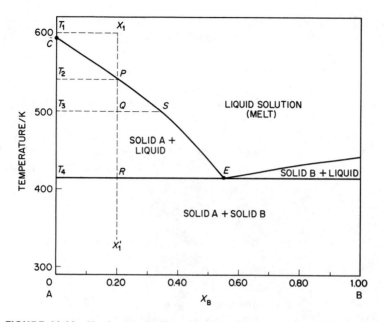

FIGURE 11.11 Simple eutectic diagram for two components A and B which are completely intersoluble as liquids but have negligible solid–solid solubility. The data are for the system cadiumbismuth.

Let us consider what happens, according to the diagram, if we cool a solution of A and B with mole fraction $X_B = X_1$ and temperature T_1. Since the overall composition of the system does not change, the cooling process can be represented by the line of constant composition, or *isopleth*, $X_1 X_1'$. Nothing happens on cooling along $X_1 X_1'$ until the temperature T_2 is reached where the isopleth intersects the solid–liquid equilibrium curve CE. At this point, P, solid first begins to crystallize from the solution (or melt). The composition of the solid is given by the other end of the tie-line PT_2, which is pure A ($X_B = 0$). As pure A crystallizes from the solution, the composition of the liquid that remains must become richer in B. The composition of solution thus moves along the curve PE as the temperature falls and more pure A crystallizes out.

As cooling continues, pure solid A continues to separate, and the composition of the solution continues to move along PE until a temperature T_4 is reached. This is the *eutectic point*, the lowest temperature at which solid and liquid phases can exist in equilibrium in this system. The composition of the solution at the eutectic point is $X_B = 0.55$. As heat is withdrawn from the system, the temperature of the system remains at the eutectic temperature T_4 until all the remaining liquid has solidified. The solid that crystallizes at the eutectic has therefore the same composition as the residual liquid at the eutectic. The solid is a mixture of two phases, pure solid A and pure solid B. The eutectic point is an invariant point on a constant-pressure diagram. Since three phases are in equilibrium, solid A, solid B, and liquid solution, $f = c - p + 2 = 4 - 3 = 1$. The single degree of freedom is used by the condition of constant pressure.

If we retrace the isopleth $X_1' X_1$ starting with a mixture of solids, the first liquid appears at T_4 and it has the eutectic composition. This diagram shows why impure crystals do not melt sharply. No matter how close to $X_B = 0$ the isopleth $X_1' X_1$ is drawn, some liquid appears in the system as soon as the temperature is raised above the eutectic temperature.

Microscopic examination of alloys often reveals a structure indicating that they have been formed from a melt by a cooling process similar to that considered along the isopleth $X_1 X'$ of Fig. 11.11. Crystallites of pure metal are found dispersed in a matrix of finely divided eutectic mixture. An example is shown in Fig. 11.12. We must not forget, however, that the eutectic mixture itself consists of two separate phases.

FIGURE 11.12 Photomicrograph at 50× of 80% Pb + 20% Sb showing crystals of Sb in a eutectic matrix. [Arthur Phillips, Yale University]

Example 11.1 68.27 g Cd and 31.73 g Bi are melted together in a crucible and cooled slowly. Describe the composition of the solid material in the crucible at room temperature.

To use Fig. 11.11 we convert wt % to mol %. $n(\text{Cd}) = 68.27$ g$/112.5$ g mol$^{-1} = 0.6074$ mol; $n(\text{Bi}) = 31.73$ g$/209.0$ g mol$^{-1} = 0.1523$ mol. Thus mol% Cd $= (0.6074/0.7597) \times 100 = 80.0$. As the crucible is cooled, pure Cd separates from melt until T_4 is reached when solid eutectic E begins to separate. The relative amounts of pure Cd and eutectic mixture are given by $RE/RT_4 = 35/20$. Thus the solidified system contains $(35/55)(0.760) = 0.48$ mol of pure Cd, and $(20/55)(0.760) = 0.28$ mol of eutectic mixture containing 45 mol % Cd.

Example 11.2 Suppose that the crucible containing the molten mixture of Cd and Bi of Example 11.1 is heated to 600 K and then placed in a quite well insulated box and allowed to cool slowly. Describe the way in which temperature T decreases with time t.

See Fig. 11.11. The temperature falls at a nearly constant rate until $T_2 = 540$ K, where solid Cd begins to form. The liberation of the enthalpy of fusion ΔH_f causes the rate of fall in T to decrease. The new rate is maintained until the eutectic $T_4 = 417$ K is reached. The temperature remains constant at T_4 (*eutectic halt*) until all sample is solidified, whereupon slow cooling is resumed.

If aniline and phenol are melted together in equimolar proportions, a definite compound crystallizes on cooling, $C_6H_5OH \cdot C_6H_5NH_2$. Pure phenol melts at 313 K, pure aniline at 267 K, and the compound melts at 304 K. The *T–X* diagram in Fig. 11.13 is typical of a system in which a stable compound occurs as a solid phase. A

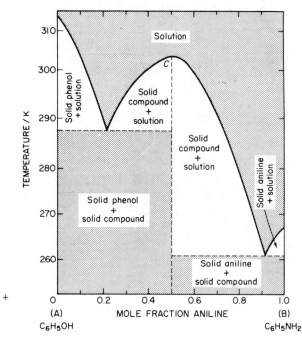

FIGURE 11.13 The system phenol + aniline shows the formation of an intermediate compound.

convenient way to look at such a diagram is to imagine that it is made up of two diagrams of the simple eutectic type placed side by side. In this case, one such diagram would be for phenol + compound, and the other for aniline + compound. The phases corresponding with the various regions of the diagram are labeled. A maximum, such as the point *C*, indicates the formation of a compound with a *congruent* melting point, since, if a solid having the composition $C_6H_5OH \cdot C_6H_5NH_2$ is heated to 304 K, it melts to a liquid of identical composition.

In some systems, solid compounds are formed that do not melt to a liquid having the same composition, but instead decompose before such a melting point is reached. An example is the silica + alumina system (Fig. 11.14), which includes a compound, $3Al_2O_3 \cdot 2SiO_2$, called *mullite*. If a melt containing 40% Al_2O_3 is prepared and cooled slowly, solid mullite begins to separate at about 2053 K. If some of this solid compound is removed and reheated along the line $X'X$, it decomposes at 2080 K into solid corundum and a liquid solution (melt) having the composition *P*. Thus $3Al_2O_3 \cdot 2SiO_2 \rightarrow Al_2O_3$ + solution. Such a change is called *incongruent melting*, since the composition of the liquid differs from that of the solid.

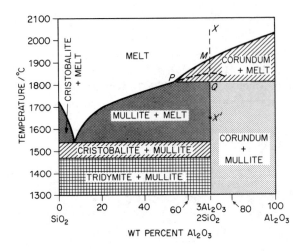

FIGURE 11.14 The system silica + alumina displays a peritectic point at P, above which the compound mullite $3Al_2O_3 \cdot 2SiO_2$ melts incongruently to yield solid corundum (Al_2O_3) + a liquid phase.

The point P is the incongruent melting point or *peritectic point* (Greek, melting around). The suitability of this name becomes evident if one follows the course of events as a solution with composition $3Al_2O_3 \cdot 2SiO_2$ is gradually cooled along XX'. When the point M is reached, solid corundum (Al_2O_3) begins to separate from the melt, the composition of which therefore becomes richer in SiO_2, falling along the line MP. When the temperature falls below that of the peritectic at P, the following change occurs: liquid + corundum $\longrightarrow$ mullite. The solid Al_2O_3 that has separated reacts with the surrounding melt to form the compound mullite. If a specimen taken at a point such as Q is examined, the solid material is found to consist of two phases, a core of corundum surrounded by a coating of mullite. It was from this characteristic appearance that the term *peritectic* originated.

11.13 Solid Solutions

Solid solutions are solid phases containing more than one component. The phase rule makes no distinction between the kind of phase (gas, liquid, or solid) that occurs, being concerned only with how many phases are present. Therefore, the phase diagrams typical of liquid–vapor and liquid–liquid systems have counterparts among solid–liquid and solid–solid systems.

Two general classes of solid solution can be distinguished on structural grounds. A *substitutional solid solution* is one in which solute atoms or molecules are substituted for solvent atoms or molecules in the crystal structure. For example, nickel has a face-centered cubic structure; if some nickel atoms are replaced at random by copper atoms, a solid solution is obtained. This substitution of one group for another is possible only when the substituents do not differ greatly in size. An *interstitial solid solution* is one in which the solute atoms or groups occupy interstices in the crystal structure of the solvent. For example, carbon atoms may occupy some of the interstices in the nickel structure. Interstitial solid solution can occur to an appreciable extent only when the solute atoms are small compared to the solvent atoms.

An example of a system with a continuous series of solid solutions is copper + nickel (Fig. 11.15). Important copper alloys such as Constantan (60Cu, 40Ni) and Monel (60Cu, 35Ni, 5Fe) are solid solutions.

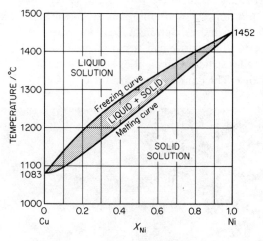

FIGURE 11.15 The copper + nickel system—a continuous series of solid solutions.

11.14 Partial Miscibility in the Solid State

For many systems, especially intermetallic systems, the simple eutectic diagram of Fig. 11.11 does not apply. There is usually some solubility of one component in the crystal structure of the other. Thus the limiting solid phases in Fig. 11.11 would not be pure A and pure B, but solid solutions of B in A and of A in B. Usually, the solid solubility increases with temperature.

We thus typically obtain a diagram similar to that shown in Fig. 11.16, which is part of the phase diagram for the Cu–Al system. Only a portion of the system extending from pure Al to the intermetallic compound $CuAl_2$ is covered. The solid solution

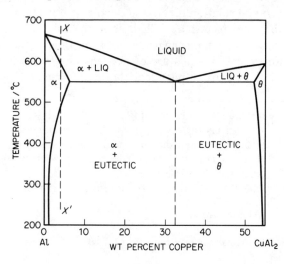

FIGURE 11.16 Part of phase diagram for the system Cu + Al.

of copper in aluminum is called the α phase, and the solid solution of aluminum in the compound $CuAl_2$ is called the θ phase. When we speak of a compound $CuAl_2$ in the solid state we do not necessarily imply that distinct molecules of $CuAl_2$ exist in the crystal structure. We will find, however, that there is a definite crystal structure based on a unit cell containing atoms in the ratio 2Al/1Cu. The range of existence of the θ phase in the diagram of Fig. 11.16 shows the extent to which the ratio of Al/Cu in this crystal structure can depart from exactly 2 : 1. Figure 11.17 shows a remarkable lamellar structure in an alloy of 33% Cu and 67% Al examined under the electron microscope. The dark ribbons are the θ phase, the lighter ones, the α phase.

FIGURE 11.17 Direct transmission electron micrograph of 33% Cu + 67% Al, the eutectic composition in Fig. 11.16. The light ribbons are α phase, the dark ones are θ phase. [N. Takahashi and K. Ashinuma, University of Yamanashi]

The phenomenon of age hardening of alloys is interpreted in terms of the effect of temperature on the range of existence of a solid–solution phase. If a melt containing about 4% Cu and 96% Al is cooled along XX', it first solidifies to a solid solution α. This solid solution is soft and ductile. If it is quenched rapidly to room temperature, it becomes metastable. Changes in the solid state are usually sluggish, so that the metastable solution can persist for some time. It changes slowly, however, to the stable form, which is a mixture of two phases, solid solution α and solid solution θ. This two-phase alloy is much less plastic than the homogeneous solid solution α.

Example 11.3 Plaster of Paris, $CaSO_4 \cdot \frac{1}{2}H_2O$ (density $\rho = 2.80$ g cm^{-1}) is used extensively for making casts. An important property of such a casting material is that on setting it should expand slightly so as to press up to and obtain an accurate impression of the object from which a cast is being taken. When mixed with water plaster of Paris reacts to form gypsum, $CaSO_4 \cdot 2H_2O$ ($\rho = 2.30$ g cm^{-3}). Calculate the volume change ΔV of the setting reaction of plaster of Paris. [RTC]

The reaction is

$$CaSO_4 \cdot \tfrac{1}{2}H_2O(s) + \tfrac{3}{2}H_2O(\ell) = CaSO_4 \cdot 2H_2O(s)$$

Molar volumes: 145.1/2.80   18.0   172.2/2.30   cm^3 mol^{-1}

$$\Delta V = 74.9 - 51.8 - 27.0 = -3.9 \text{ cm}^3 \text{ mol}^{-1}$$

$$\text{Fractional } \Delta V/V = -3.9/(51.8 + 27.0) = -0.05 \, (-5\%)$$

As a matter of fact, plaster of Paris expands slightly on setting. The gypsum formed is somewhat porous. In many practical problems, there may be such a hidden variable and a good chemist should be alert to this possibility.

11.15 The Iron–Carbon Diagram

No discussion of phase diagrams should omit the iron–carbon system, which is the principal theoretical basis for the iron and steel industry. The part of the diagram of greatest interest extends from pure iron to the compound iron carbide, or *cementite*, Fe_3C. This section is reproduced in Fig. 11.18.

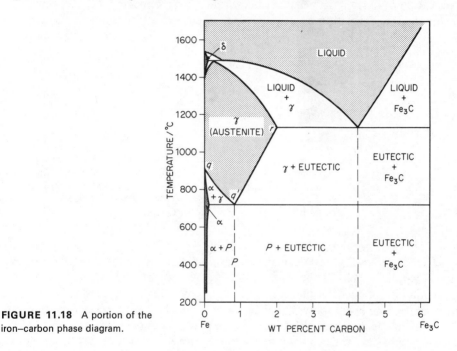

FIGURE 11.18 A portion of the iron–carbon phase diagram.

Pure iron exists in two different structures. The stable crystalline form up to 910°C, called α iron, has a body-centered cubic structure. At 910°C, transition occurs to a face-centered cubic structure, γ iron; but at 1401°C, γ iron transforms back to a body-centered cubic structure, now called δ iron. This is an allotrope that is stable, at constant pressure, both below and above a certain temperature range. The solid solutions of carbon in the iron structures are called *ferrite*.

Apart from the small section concerned with δ ferrite, the upper portion of the diagram is a typical example of limited solid-solid solubility. The curve qq' shows how the transformation temperature of α to γ ferrite is lowered by interstitial solution of carbon in iron. The region labeled α represents the range of solid solutions of C in α iron. The region labeled γ represents the range of solid solutions of C in γ iron, which are given a special name, *austenite*. The decrease in the transition temperature α → γ is terminated at q', where the curve intersects the solid solubility curve rq' of carbon in iron. A point such as q', which has the properties of a eutectic but occurs in a completely solid region, is called a *eutectoid*.

The two phases formed by the eutectoid decomposition of austenite are α ferrite and cementite. These phases form a lamellar structure of alternate bands called *pearlite*. If the composition is close to eutectoid, the steel is composed entirely of pearlite. If the composition is richer in carbon, or *hypereutectoid*, it may contain other grains of cementite in addition to those occurring in the pearlite. If the composition is poorer in carbon, or *hypoeutectoid*, and the steel is cooled slowly, it may contain additional grains of ferrite. Figure 11.19 shows the formation and appearance of pearlite. The first stage in the formation seems to be the nucleation of a crystallite of cementite. As this grows, it removes carbon diffusing from the surrounding austenite. Nucleation of ferrite then occurs at the surface of the cementite, because low carbon favors the transformation of γ to α. Ferrite imparts ductility to the steel.

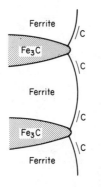

FIGURE 11.19 Formation and appearance of pearlite (alternate bands of Fe_3C and α-ferrite). Photomicrograph 125×. [U.S. Steel Corporation Research Center]

The diagram Fig. 11.18 explains the distinction between the *steels* and *cast irons*. Any composition below 2% carbon can be heated until a homogeneous solid solution (austenite) is obtained. In this condition, the alloy is readily hot rolled or submitted to other forming operations. On cooling, segregation of two phases occurs. Cementite is a hard, brittle material, and its occurrence in pearlitic steels is responsible for their high strength. The way in which the cooling is carried out determines the rate of segregation of the two phases and their grain sizes, and provides many possibilities for obtaining different mechanical properties by annealing and tempering. Compositions above 2% in carbon belong to the general class of cast irons. They cannot be brought into a homogeneous solid solution by heating, and therefore cannot be mechanically worked. They are formed by casting from the molten state, and used where hardness and corrosion resistance are desirable and where brittleness due to high content of cementite is not deleterious.

Problems

1. Sketch the phase diagram of acetic acid from the following data. There are two solid forms, α and β, which are both denser than the liquid, and β is denser than α. The α form melts at 290 K under its own vapor pressure of 1210 Pa. The normal boiling point of liquid is 391 K. Liquid, α, and β are all in equilibrium at 328 K and 200 MPa.

2. Pure lead melts at 600 K. With addition of $X = 0.0228$ of silver the melting point (liquidus point) is lowered by 13.9 K. Estimate the ΔH_f of lead. (Solid Pb does not dissolve Ag.)

3. Sketch the phase diagram of sulfur from following data. (Use a log P scale.) There are two solid forms, orthorhombic S^O and monoclinic S^M, with a transition point at 369 K at $P°$. S^M melts at 393 K under its own vapor pressure of 3.3 Pa. The triple points are (S^O, S^M, S^v) 368 K, 1.0 Pa; (S^O, S^M, S^ℓ) 428 K and 131 MPa. Density $\rho^M < \rho^O$. If S^O is heated quite rapidly, it melts at 387 K. Explain.

4. Sketch the phase diagram of N_2 at low temperatures. There are three crystal forms, which coexist at 44.5 K and 471 MPa. At this triple point, the volume change ΔV in cm³ mol⁻¹ are: $\gamma \longrightarrow \alpha$, 0.165; $\gamma \longrightarrow \beta$, 0.208; $\alpha \longrightarrow \beta$, 0.043. At $P°$ and 36 K, $\alpha \longrightarrow \beta$ with $\Delta V = 0.22$. The ΔS values of transitions cited are 1.25, 5.88, 4.59, and 6.52 J K⁻¹ mol⁻¹, respectively.

5. 1.000 kg of a Ni–Cu alloy with $X_{Ni} = 0.300$ is heated to 1473 K. What is the mass of the unmelted alloy, and what is its composition? (See Fig. 11.15.)

6. The boiling point of liquid nitrogen at $P°$ is 78 K. What is the lowest temperature that could be obtained by boiling liquid N_2 under a reduced pressure of 100 Pa? If the N_2 is kept in a thermally insulated container with a heat leak of 10⁻¹ W, what is the pumping speed required (m³ s⁻¹ at 100 Pa)? (Assume Trouton's rule.)

7. A Coolgardie safe is a device for keeping food cool in hot arid regions of Australia. It consists of a box with sides covered in muslin, down which water creeps by capillary action from a reservoir on top of box. As water evaporates from the muslin, the food safe is cooled. What is the lowest temperature attainable by such a device on a hot day when $t = 45°C$ and relative humidity is 20%? The ΔH_v of water is 43.8 kJ mol⁻¹ and $P_{H_2O} = 3.17$ kPa at 25°C. [RTC]

8. Sketch the phase diagram of the CaF_2–$CaCl_2$ system at $P°$. CaF_2 melts at 1360°C and $CaCl_2$ at 772°C. There is a eutectic at 625°C and 78 mol % $CaCl_2$, where solid $CaCl_2$ and solid compound $CaCl_2 \cdot CaF_2$ separate from melt. At 737°C there is a peritectic where $CaCl_2 \cdot CaF_2 \longrightarrow CaF_2 + $ a solution containing 58 mol % $CaCl_2$. Label regions on diagram.

9. Estimate ΔH_f of CCl_4 at 10⁸ Pa.

P (Pa)	10⁵	10⁸	2×10^8
T_f (K)	250.6	288.3	321.0
ΔV_f (cm³ g⁻¹)	0.0258	0.0199	0.0163

10. For $Cu(c) \longrightarrow Cu(g)$, $\Delta G°(1000 \text{ K}) = 206$ kJ mol⁻¹. You wish to deposit a film of

copper by evaporation of the metal in a high-vacuum chamber. If your system can reach a pressure of 10^{-1} Pa, would the process be feasible at 1000 K?

11. Estimate the temperature at which pure Sb is in equilibrium with a Pb — Sb solution containing 75 mol % Sb. (See Table 11.3.)

12. Monoclinic and orthorhombic sulfur both dissolve in CCl_4 as S_8 molecules. At 298 K the solubility of S(ortho) is 8.52 g per kg CCl_4 and that of S(mono) is 10.9 g per kg CCl_4. Calculate $\Delta G°(298)$ for S(mono) $\longrightarrow$ S(ortho). If an excess of a mixture of monoclinic and orthorhombic sulfur crystals is allowed to stand in CCl_4 at 298 K, describe what happens.

12

Statistical Thermodynamics

A central question in physical chemistry has always been how to calculate the thermodynamic properties of systems from the properties of the molecules of which they are composed. Any large-scale system contains an enormous number of molecules. For example, there are about 1.82×10^{22} molecules in a liter of oxygen at 100 kPa and 300 K. We could not possibly keep track of all the variables needed to describe so many individual molecules. Any theory that seeks to explain the behavior of large-scale systems in terms of molecular properties must therefore rely on statistical methods, which are designed to describe the *average behavior* of large collections of objects. For instance, nobody can determine in advance whether a given radium nucleus will disintegrate within the next 10 minutes, the next 10 days, or the next 10 centuries. If a milligram of radium is considered, however, we know that close to 2.23×10^{10} nuclei will disintegrate in any 10-minute period.

The discipline that allows us to make the theoretical connection between microscopic mechanical properties and macroscopic thermodynamic properties is called *statistical mechanics*. We can symbolize the connection as follows:

Mechanical Properties of Molecules i		*Thermodynamic Properties of Systems*
Positions x_i, y_i, z_i		Temperature T
Momenta p_{xi}, p_{yi}, p_{zi}	$\longrightarrow$	Pressure P
Masses m_i	[*Statistical mechanics*]	Mass m
Kinetic energies E_{ki}		Entropy S
Potential energies E_{pij}		Internal energy U
between molecules i and j		Gibbs free energy G

In addition to the variables listed above, we must specify certain external variables, which are common to both the microscopic and the thermodynamic descriptions of the system. In most cases that we consider, the only external variable is the volume V. In problems dealing with surface films, we have also the surface area α. Other specifications might include an external electric field $\mathbf{E}$, and so on.

Our problem now is to find out how to make the arrow work in the above diagram.

12.1 Ensembles

Suppose that we consider a system, 1 liter of oxygen at 300 K and 100 kPa, for example. Suppose that we have all the information we need about oxygen molecules: mass, rotational and vibrational energy levels, translational energy levels in a volume V, and so on. Actually, we can obtain these data quite readily from spectroscopic measurements and basic quantum mechanical theory. Note that we do not have detailed information about particular individual oxygen molecules, but simply the allowable values of the various energy levels for the general species called the *oxygen molecule*. How can we now go about calculating the thermodynamic properties (U, S, H, P, etc.) of the liter of oxygen?

If we prepared a liter of O_2 in the lab and considered its properties in a certain short interval in time, its molecules would exist in certain energy states allowable in the gas volume V. A short time later, however, the values of the states of the individual molecules would have changed, as a result of intermolecular collisions, energy transfers and motions through space. Even over the short period of time necessary to measure, say, the pressure of the gas, the energy states of individual molecules would change many times.

Well, you will say, we must therefore take an *average over time* of the expressions for properties of the gas in terms of states of the molecules. This average will give us the calculated value of the thermodynamic property, such as pressure P, in which we are interested. We soon discover, however, that such an average over time is impossible to carry out in practice, since it would require us to keep track of the motions and energy states of the 1.82×10^{22} molecules.

Faced with the impossible task of time averaging for a system containing of the order of an Avogadro number L of molecules, J. Willard Gibbs in 1900 introduced the idea of an *ensemble* of systems. An ensemble is a mental construction that consists of a large number of exact replicas of the system under discussion, each member of the ensemble being subject to exactly the same thermodynamic constraints as the original system. If the original system is an isolated mass of matter, the number N of molecules it contains and its volume V are fixed, and its energy E is fixed within an extremely narrow range. We can construct an ensemble of $\mathfrak{N}$ such systems, each member having the same, N, V, and E. This ensemble is shown schematically in Fig. 12.1. It is called a *microcanonical ensemble*. The example in the figure contains only $\mathfrak{N} = 36$ members. The ensembles used in statistical mechanics will have very large $\mathfrak{N}$, and for the purposes of some calculations and theorems, can be taken in the limit to have $\mathfrak{N} \to \infty$.

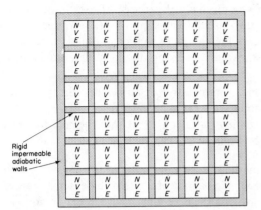

FIGURE 12.1 Representation of a microcanonical ensemble of $\mathfrak{N}$ systems each with same N, V, E.

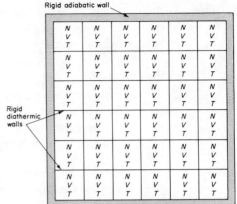

FIGURE 12.2 Representation of a canonical ensemble of $\mathfrak{N}$ systems each with same N, V, T.

A second type of ensemble introduced by Gibbs is the *canonical ensemble* shown schematically in Fig. 12.2. This ensemble consists of a large number of systems, each having the same N, V, T. The individual members of the canonical ensemble, therefore, are not isolated systems, as in the microcanonical ensemble, but closed systems in the thermodynamic sense. The members are separated by *diathermic* walls, which permit the passage of energy but not of material particles. At equilibrium, therefore, each member of the canonical ensemble has the same T, but not necessarily the same E. Its value of E fluctuates about some average value $\bar{E}$. At equilibrium, this average value of E is the same for all members of the ensemble and the $\bar{E}$ determines the constant temperature T of the ensemble. The canonical ensemble is especially useful in statistical thermodynamics.

A third type of ensemble introduced by Gibbs is based on members that are open systems, in which the number of molecules is no longer held fixed. At equilibrium, the V, T, and chemical potential μ of each component is the same for each member of the ensemble. These ensembles, called *grand canonical ensembles*, are useful for more advanced calculations, but we shall not require them in this book.

12.2 Ensemble Averages

The average value of any property M of a system taken over all the systems that are members of the ensemble is called the *ensemble average* $\bar{M}$ of the property, so that

$$\bar{M} = \sum_{\substack{\text{all members} \\ \text{of ensemble}}} M_j / \mathfrak{N}$$

There are two basic postulates required to relate the concept of the ensemble to practical calculations of average properties:

1. *First postulate.* The average of any mechanical variable M over a long time in the actual system is equal to the ensemble average of M, provided that the systems of the ensemble replicate the thermodynamic state and surroundings of the actual system of interest. Strictly speaking, this postulate holds only in the limit as $\mathfrak{N} \longrightarrow \infty$.

2. *Second postulate.* In an ensemble representative of an isolated thermodynamic system (microcanonical ensemble), the members are distributed with equal probability over the possible *quantum states* consistent with the specified values of N, V, E. This is the *principle of equal a priori probabilities.*

The power of the ensemble method is that it replaces the impossible time average by a feasible ensemble average. We cannot prove that the ensemble average is the same as the time average (if we could prove this equivalence, obviously we could not have said that the time average is mathematically impossible). Nevertheless, Postulate 1 has an appealing air of reasonableness. Better still is the fact that when we apply the postulate to calculate thermodynamic properties of systems, we obtain results that agree with experimental measurements.

Postulate 2 also is not subject to proof, but seems logical to us since we can conceive of no reason why one state of given N, V, E should have a probability different from that of another state of the same N, V, E.

We shall now use the ensemble average to derive formulas for the thermodynamic properties of one mole of a substance. First we mentally construct a canonical ensemble similar to that in Fig. 12.2, in which $N = L$, the Avogadro number. A member j of the ensemble then has a molar volume V, is at temperature T, and has an energy E_j. Suppose that we pluck at random one member out of the ensemble. In accord with the general Boltzmann principle, the probability that this member be in the state j with energy E_j is

$$p_j = \alpha e^{-E_j/kT}$$

where α is a proportionality factor. Since the sum of p_j over all the members of the ensemble must equal unity,

$$\sum p_j = 1 = \alpha \sum_j e^{-E_j/kT}$$

and

$$p_j = e^{-E_j/kT} / \sum e^{-E_j/kT} \tag{12.1}$$

The denominator of this expression is the *canonical ensemble partition function.* $Z(V, T)$.

$$Z(V, T) = \sum_j e^{-E_j/kT} \tag{12.2}$$

As indicated, Z is a function of the volume V and temperature T that are common to all members of the ensemble.

12.3 Statistical Calculation of Thermodynamic Energy

The ensemble average value of the energy E can be calculated from the definition of the mean value (page 88) $\bar{E} = \sum p_j E_j$, as

$$\bar{E} = \frac{\sum E_j e^{-E_j/kT}}{\sum e^{-E_j/kT}} = \frac{\sum E_j e^{-E_j/kT}}{Z} \qquad (12.3)$$

From Eq. (12.2) we can show that $(\partial Z/\partial T)_V = \sum (E_j/kT^2)e^{-E_j/kT}$, whence Eq. (12.3) gives

$$\bar{E} = \frac{kT^2(\partial Z/\partial T)_{N,V}}{Z} = kT^2\left(\frac{\partial \ln Z}{\partial T}\right)_{N,V} \qquad (12.4)$$

By virtue of Postulate 1, this $\bar{E}$ is now stated to be identical with the thermodynamic internal energy U of the system upon which the ensemble is based. If $N = L$, it is the molar internal energy U_m.

$$U = \bar{E} \qquad (12.5)$$

12.4 Statistical Formula for Entropy

The formula for the ensemble average entropy can be obtained from the general relation between entropy and probability that was given in Eq. (7.27), $S = -k \sum p_j \ln p_j$. From Eq. (12.1) and (12.2) the probability p_j that a system in the ensemble occurs in a state j with energy E_j is given by $\ln p_j = (-E_j/kT - \ln Z)$. Hence

$$S = k \sum \frac{p_j E_j}{kT} + k \sum p_j \ln Z$$

However, $\sum p_j \ln Z = \ln Z \sum p_j = \ln Z$, since $\sum p_j = 1$, and $\sum p_j E_j = \bar{E}$. Thus $S = \bar{E}/T + k \ln Z$, or

$$S = U/T + k \ln Z \qquad (12.6)$$

12.5 Helmholtz Free Energy and Equation of State

From the definition of the Helmholtz function, $A = U - TS$, and Eq. (12.6) for S, we find that

$$A = -kT \ln Z \qquad (12.7)$$

Thus the canonical ensemble partition function Z is related very simply to the Helmholtz free energy A.

If we know A as a function of T and V, its natural variables, we can calculate the equation of state for a substance from Eq. (8.12), $P = -(\partial A/\partial V)_T$. Thus if we can find $Z(V, T)$ for a substance, we can calculate all its thermodynamic properties, including its equation of state. This result would be equally true for solid, liquid, and gaseous states.

The partition function Z is thus the basis of a powerful statistical method for obtaining thermodynamic properties. Archimedes may have said "Give me a fulcrum and I can lift the world"; we can say "Give me a partition function and I can calculate all the equilibrium properties in the world." The general formulas of statistical thermodynamics are summarized in Table 12.1.

TABLE 12.1

SUMMARY OF GENERAL FORMULAS OF STATISTICAL THERMODYNAMICS

Function	Thermodynamic formula	Statistical formula
A	$U - TS$	$-kT \ln Z$
P	$-(\partial A/\partial V)_T$	$kT (\partial \ln Z/\partial V)_T$
S	$-(\partial A/\partial T)_V$	$k \ln Z + kT (\partial \ln Z/\partial T)_V$
U	$A + TS$	$kT^2(\partial \ln Z/\partial T)_V$
C_V	$(\partial U/\partial T)_V$	$2kT(\partial \ln Z/\partial T)_V + kT^2(\partial^2 \ln Z/\partial T^2)_V$
G	$A + PV$	$kT[V(\partial \ln Z/\partial V)_T - \ln Z]$
H	$U + PV$	$kT[(\partial \ln Z/\partial \ln T)_V + (\partial \ln Z/\partial \ln V)_T]$

12.6 How to Evaluate Z for Noninteracting Particles

To calculate Z for a system of strongly interacting particles (as in a real gas or a liquid) is a formidable problem. You will probably not be surprised to learn that the evaluation of Z for many real systems is not yet possible with the techniques now available.

In the case of noninteracting particles (as in an ideal gas), however, Z can be calculated quite readily. Another case with a straightforward solution is the perfect crystal considered as an array of atoms vibrating in simple harmonic motions about their equilibrium positions. In these cases of noninteracting particles, the energy of a system in the ensemble can be written as the sum of the energies of the independent particles,

$$E = \epsilon_a + \epsilon_b + \epsilon_c + \cdots \tag{12.8}$$

We begin by assuming that the individual particles are distinguishable from one another, so that the subscripts a, b, c, etc., imply that a system with particle a in energy state 1 and particle b in energy state 2 can be physically distinguished from a system with b in state 1 and a in state 2.

Let us recall, from Eq. (5.4), the definition of particle partition functions,

$$z_a = \sum e^{-\epsilon_{aj}/kT}, \qquad z_b = \sum e^{-\epsilon_{bj}/kT}, \qquad \text{etc.}$$

We now see that

$$Z = \sum e^{-E_j/kT} = (\sum e^{-\epsilon_{aj}/kT})(\sum e^{-\epsilon_{bj}/kT}) \cdots = z_a z_b \cdots \tag{12.9}$$

This result will be evident in a simple example. Suppose that there are two particles, a and b, with three energy states 1, 2, and 3. Then the possible states of the system for physically distinguishable particles are

$$\epsilon_{a1} + \epsilon_{b1} \qquad \epsilon_{a1} + \epsilon_{b2} \qquad \epsilon_{a1} + \epsilon_{b3}$$
$$\epsilon_{a2} + \epsilon_{b1} \qquad \epsilon_{a2} + \epsilon_{b2} \qquad \epsilon_{a2} + \epsilon_{b3}$$
$$\epsilon_{a3} + \epsilon_{b1} \qquad \epsilon_{a3} + \epsilon_{b2} \qquad \epsilon_{a3} + \epsilon_{b3}$$

In this case,

$$Z = \sum e^{-E_j/kT} = e^{-(\epsilon_{a1}+\epsilon_{b1})/kT} + e^{-(\epsilon_{a2}+\epsilon_{b1})/kT} + e^{-(\epsilon_{a3}+\epsilon_{b1})/kT} + \cdots$$

where the sum is continued over the nine possible distinguishable states. But this sum is equal to $z_a z_b$, as you can see by multiplying the expressions for z_a and z_b, where

$$z_a = e^{-\epsilon_{a1}/kT} + e^{-\epsilon_{a2}/kT} + e^{-\epsilon_{a3}/kT}$$

$$z_b = e^{-\epsilon_{b1}/kT} + e^{-\epsilon_{b2}/kT} + e^{-\epsilon_{b3}/kT}$$

How can we use Eq. (12.9) to obtain Z for an ideal gas? All the molecules in a volume of gas have exactly the same set of allowable energy states. Thus the molecular partition functions z_a, z_b, z_c, etc. must be exactly the same, $z_a = z_b = z_c = z$. For a system containing N molecules, therefore, Eq. (12.9) would become $Z = z^N$. However, the molecules are not distinguishable particles and permuting the individual molecules among their occupied energy states does not produce a different state of the gas. If each molecule occupied a distinct energy state, there would be $N!$ ways of permuting N molecules among N states, and hence we could correct $Z = z^N$ to obtain for the indistinguishable molecules

$$Z = \frac{1}{N!} z^N \tag{12.10}$$

For gases at ordinary temperatures and pressures, it is an excellent approximation to assume that *each molecule is in a separate energy state*. The reason is that the number of available translational energy states is enormously higher than the number of molecules. Equation (4.8) for the translational energy levels of a particle in a box allows us to calculate (Problem 4.20 of Chapter 4) that the number of available energy states up to an energy ϵ is

$$N(\epsilon) = \frac{\pi}{6} (8m\epsilon)^{3/2} V/h^3 \tag{12.11}$$

For an oxygen molecule at 300 K in a volume of 1 liter, with $\epsilon = \frac{3}{2}kT = 6.21 \times 10^{-21}$ J and molecular mass $m = 5.32 \times 10^{-26}$ kg, $N(\epsilon) = 2.44 \times 10^{29}$. This number is so much larger than the number of O_2 molecules per liter at 10^5 Pa (1.82×10^{22}) that the chance of finding more than one molecule in the same energy state is negligibly small. Therefore, our use of $N!$ to obtain the partition function for indistinguishable molecules is physically correct for all gases at ordinary temperatures and pressures. For matter in the interior of stars, the problem is not so easy, but that is another story.

Armed with Eq. (12.10) we can successfully attack the problem of finding Z for a perfect gas, provided that we can find the molecular partition function z. To calculate z, we need only know the allowed energy levels of the molecules. In Chapter 5 we showed how these energy levels are calculated from quantum theory and experimental spectroscopic data. More detail about the theory and the spectroscopic methods will be given in Chapter 26.

As discussed in Section 3.3, the energy of a molecule can be divided into the translational kinetic energy ϵ_t of the center of mass of the molecule and energy terms ϵ_I associated with internal degrees of freedom, such as rotation and vibration. Hence, we can write $\epsilon = \epsilon_t + \epsilon_I$. From $z = \sum e^{-\epsilon_i/kT}$, it follows that $z = \epsilon_t \epsilon_I$. Thus the translational part of the molecular partition function can be factored from the internal part.

12.7 Translational Partition Function

In Eq. (4.7) the translational energy levels for motion of a particle in one dimension were given as $\epsilon_n = n^2h^2/8ma^2$. The corresponding molecular partition function is

$$z_a = \sum \exp\left(-n^2h^2/8ma^2\right)/kT$$

The energy levels are so closely spaced (owing to the smallness of h^2) that the sum can be replaced by an integral. Thus

$$z_a = \int_0^\infty \exp\left(-n^2h^2/8ma^2kT\right) dn \tag{12.12}$$

To evaluate the integral, let $x^2 = n^2h^2/8ma^2kT$, so that, $dx = (h^2/8ma^2kT)^{1/2}\,dn$. With these substitutions into Eq. (12.12), we have

$$z_a = \left(\frac{8ma^2kT}{h^2}\right)^{1/2} \int_0^\infty e^{-x^2}\,dx$$

The integral is now a standard form, $\int_0^\infty e^{-x^2}\,dx = (\pi/4)^{1/2}$, so that

$$z_a = (2\pi mkT)^{1/2}\frac{a}{h} \tag{12.13}$$

For the three translational degrees of freedom, $z = z_a z_b z_c$, which is the product of three terms like Eq. (12.13). Thus, since $abc = V$, we have from Eq. (12.13),

$$z_t = (2\pi mkT)^{3/2}V/h^3 \tag{12.14}$$

The ensemble partition function for one mole of gas containing L molecules is

$$Z_t = \frac{1}{L!}z^L = \frac{1}{L!}\left[\frac{(2\pi mkT)^{3/2}V}{h^3}\right]^L \tag{12.15}$$

12.8 Thermodynamic Functions for a Monatomic Gas

Let us consider a gas made up of atoms. Examples would be the inert gases helium through radon, monatomic metal vapors such as mercury, and other elements such as the halogens at temperatures so high that the molecules are completely dissociated into atoms. Except for electronic energy in some cases, the gaseous atoms have only translational energies. We can calculate all the translational contributions to the thermodynamic functions of an ideal monatomic gas from the translational partition function Z_t of Eq. (12.15). We first take the logarithm of Eq. (12.15),

$$\ln Z_t = \ln\frac{1}{L!} + L\ln\left[\frac{(2\pi mkT)^{3/2}V}{h^3}\right] \tag{12.16}$$

The internal energy per mole U_m is obtained by substituting Eq. (12.16) for $\ln Z$ into Eq. (12.4) for U_m.

$$U_m = \bar{E} = kT^2\left(\frac{\partial \ln Z_t}{\partial T}\right) = LkT^2\left(\frac{\partial \ln T^{3/2}}{\partial T}\right) = RT^2\left(\frac{3}{2}\right)\frac{1}{T} = \frac{3}{2}RT \tag{12.17}$$

This is the simple result expected from the principle of equipartition of energy. The internal energy U_m of an ideal gas is the sum of the energies of the individual molecules. Since the average energy of a molecule is $\frac{3}{2}kT$, the internal energy per mole U_m is simply $L \cdot \frac{3}{2}kT = \frac{3}{2}RT$.

Unlike energy, entropy is not a property of individual molecules, but depends on the probabilities of states that contain many molecules. The entropy of an ideal monatomic gas can be calculated from Eq. (12.6) now that we have a formula for $\ln Z$ in Eq. (12.16). We use the Stirling approximation formula for $\ln L!$

$$\ln L! \approx L \ln L - L = \ln (L/e)^L = L \ln (L/e) \tag{12.18}$$

or

$$L! \simeq (L/e)^L$$

where e is the base of natural logarithms, 2.718 ... From Eq. (12.18), Eq. (12.15) becomes $Z_t = [(2\pi mkT)^{3/2}eV/Lh^3]^L$, and

$$\ln Z_t = L \ln \left[\frac{(2\pi mkT)^{3/2}eV}{Lh^3} \right] \tag{12.19}$$

From Eq. (12.6) the entropy per mole is

$$S_m = R \ln \frac{eV}{Lh^3}(2\pi mkT)^{3/2} + \frac{3}{2}R$$

$$S_m = R \ln \frac{e^{5/2}V}{Lh^3}(2\pi mkT)^{3/2} \tag{12.20}$$

Sackur and Tetrode first obtained this equation in 1913.

Example 12.1 Calculate the entropies of the inert gases He, Ne, Ar, Kr, Xe, Rn, at 298.15 K and $P = 101.32$ kPa.

The following data are needed to apply the Sackur–Tetrode equation.

$R = 8.314$ J K^{-1} mol^{-1} $\qquad$ $\pi = 3.1416$

$V = 24.465 \times 10^{-3}$ m^3 $\qquad$ $k = 1.381 \times 10^{-23}$ J K^{-1}

$L = 6.022 \times 10^{23}$ mol^{-1} $\qquad$ $T = 298.15$ K

$h = 6.626 \times 10^{-34}$ J s $\qquad$ $m = M/6.022 \times 10^{23}$ (where M is the atomic mass in kg)

On substitution of the numerical values into Eq. (12.20), we obtain

$$S = 195 + \tfrac{3}{2}R \ln M \text{ (J K}^{-1}\text{ mol}^{-1}).$$

Gas	He	Ne	Ar	Kr	Xe	Rn
$10^3 M$ (kg mol^{-1})	4.00	20.18	39.95	83.8	131.3	222
S_m (J K^{-1} mol^{-1})	126	146	155	164	170	176

Notice the increase in entropy with molar mass of the gas. The larger the mass, the more densely packed are the translational energy levels given from Eq. (4.8). The greater the number of available energy levels, the greater the number of possible arrangements of atoms among the levels, and thus the greater the probability of the macroscopic equilibrium state and the greater the entropy of the gas.

12.9 Internal Motions—Molecular Partition Function

If we know the internal energy states of a molecule from spectroscopic data, we can calculate a molecular partition function for internal molecular motions. Then we can use Eqs. (12.10) and (12.14) to calculate Z for an ideal gas. If ϵ_j is the internal energy level of a state j, and g_j is its statistical weight,

$$z_I = \sum_{j=0}^{\infty} g_j e^{-\epsilon_j/kT} \tag{12.21}$$

To a quite good approximation we can take the internal energy of a molecule to be the sum of rotational, vibrational, and electronic contributions,

$$\epsilon_I = \epsilon_r + \epsilon_v + \epsilon_e \qquad [\text{so that } z_I = z_r z_v z_e] \tag{12.22}$$

12.10 Rotational Partition Function—Rigid Linear Molecules

We can calculate the contribution of rotational degrees of freedom to the entropy of an ideal gas if we have a formula for the rotational energy levels, or if we have measured the levels by spectroscopic methods.

A simple result is obtained for the case of rigid diatomic and linear molecules. In this case there is only one value for the moment of inertia I, but there are two degrees of freedom of rotation. The energy levels were given in Eq. (4.10) as

$$\epsilon_j = J(J+1)h^2/8\pi^2 I \tag{12.23}$$

If the moment of inertia is sufficiently high, these energy levels are so closely spaced that the $\Delta\epsilon$ between adjacent levels is much less than kT, even at temperatures of a few kelvin. This condition is, in fact, realized for all diatomic molecules except H_2, HD, and D_2. For example, the moment of inertia I of F_2 is 32.5×10^{-47} kg m^2; and for N_2, $I = 13.9 \times 10^{-47}$ kg m^2. For H_2, however, the moment of inertia is only 0.46×10^{-47} kg m^2. From Eq. (12.23), the height of the level $J = 1$ above the zero level $J = 0$ is 8.00×10^{-23} J for N_2, and 2.42×10^{-21} J for H_2. The value of kT at 10 K is 1.38×10^{-22} J.

To calculate the molecular partition function for rotation we substitute the expression Eq. (12.23) for ϵ_j into $z_r = \sum g_j e^{-\epsilon_j/kT}$. The statistical weight g_j for the rotational energy levels of a linear molecule is $g_j = (2J + 1)$. This result is obtained from the quantum mechanics of rotational motion as in Section 21.6. The rotational partition function then becomes

$$z_r = \sum_J (2J+1) \exp\left[\frac{-J(J+1)h^2}{8\pi^2 IkT}\right] \tag{12.24}$$

When the temperature is high enough that $\epsilon_J \ll kT$, the steps between successive terms in this sum over discrete values of J become small enough to allow us to replace the sum over J by an integration over dJ. In this case,

$$z_r = \int_0^\infty (2J + 1) \exp\left[\frac{-J(J + 1)h^2}{8\pi^2 IkT}\right] dJ$$

and

$$z_r = 8\pi^2 IkT/h^2 \tag{12.25}$$

A symmetry number σ is introduced, which is either $\sigma = 1$ (heteronuclear) or $\sigma = 2$ (homonuclear).* Thus Eq. (12.25) becomes

$$z_r = \frac{8\pi^2 IkT}{\sigma h^2} \tag{12.26}$$

If ϵ_J is not much less than kT, the rotational partition function can be obtained from Eq. (12.24) by the Euler–Maclaurin summation formula. We define a characteristic rotational temperature by

$$\Theta_r = \frac{h^2}{8\pi^2 Ik} \tag{12.27}$$

It can be shown that

$$z_r = \frac{I}{\sigma\Theta_r}\left[1 + \frac{1}{3}\frac{\Theta_r}{T} + \frac{1}{15}\left(\frac{\Theta_r}{T}\right)^2 + \frac{4}{315}\left(\frac{\Theta_r}{T}\right)^3 + \cdots\right] \tag{12.28}$$

This is good to within 1% as long as $T > \Theta_r$.

Example 12.2 Calculate z_r for F_2 at 298 K. $I = 32.5 \times 10^{-47}$ kg m^2.

From Eq. (12.26),

$$z_r = \frac{8\pi^2(32.5 \times 10^{-47} \text{ kg m}^2)(1.38 \times 10^{-23} \text{ J K}^{-1})(298 \text{ K})}{2(6.63 \times 10^{-34} \text{ J s})^2}$$

$$= 120$$

12.11 Rotational Energy and Entropy—Linear Molecules

From Eqs. (12.4) and (12.10), we can calculate the rotational energy per mole.

$$U_m = kT^2\left(\frac{\partial \ln z_r^L}{\partial T}\right) = RT^2\left(\frac{\partial \ln z_r}{\partial T}\right) = \frac{RT^2}{T} = RT \tag{12.29}$$

This is just the classical value, in accord with the equipartition principle, for two rotational degrees of freedom.

*See D. A. McQuarrie, *Statistical Thermodynamics* (New York: Harper & Row Publishers, 1973), p. 101. The symmetry number originates in the nuclear-spin contributions to the partition function. There is no satisfactory explanation except the quite detailed one given in references such as that cited.

We calculate the rotational part of the entropy as

$$S_r = k \ln z_r^L + \frac{U}{T} \tag{12.30}$$

On substitution of z_r from Eq. (12.26), and setting $U = RT$, we obtain

$$S_r = R \ln \frac{8\pi^2 IkT}{\sigma h^2} + R \tag{12.31}$$

Example 12.3   Calculate the rotational contribution to the entropy of F_2 at 298 K.

Substitution of the rotational partition function $z_r = 120$, as found in Example 12.2, into Eq. (12.30) gives

$$S_r = (8.314 \text{ J K}^{-1} \text{ mol}^{-1}) \ln (120) + (8.314 \text{ J K}^{-1} \text{ mol}^{-1})$$
$$= 48.1 \text{ J K}^{-1} \text{ mol}^{-1}$$

From Eq. (12.31), rotational entropy becomes larger as the moment of inertia I becomes larger. The larger I, the more closely packed are rotational energy levels and hence the more microstates are available for occupation by the molecules at any given T. Since more states are available, the randomness and disorder of the partition of molecules among the states is greater, and consequently the entropy of the system is higher.

We can calculate the rotational energy and entropy of any molecule if we know its structure and hence can calculate its moments of inertia, from which its energy levels and partition function can be obtained. Except for the hydrogen molecules, H_2, HD, D_2, rotational motion is effectively classical above 15 K.* For hydrogen and some hydrides at low temperatures we cannot use Eq. (12.26) and must carry out the summation in Eq. (12.24) to obtain the rotational partition function.

12.12   How to Calculate Moments of Inertia

In the rotational motion of molecules a fundamental quantity is the moment of inertia I. For a linear molecule, there are two equal moments of inertia, about two mutually perpendicular axes of rotation that pass through the center of mass of the molecule. There is, however, no moment of inertia about the internuclear axis of the linear molecule because virtually all the masses of the atoms are concentrated in tiny nuclei, which behave effectively as point masses lying on the internuclear axis.

For nonlinear molecules, in the general case, there are three principal moments of inertia, defined by

$$I_a = I_{xx} = \sum_{j=1}^{n} m_j[(y_j - y_0)^2 + (z_j - z_0)^2]$$

$$I_b = I_{yy} = \sum_{j=1}^{n} m_j[(x_j - x_0)^2 + (z_j - z_0)^2] \tag{12.32}$$

$$I_c = I_{zz} = \sum_{j=1}^{n} m_j[(x_j - x_0)^2 + (y_j - y_0)^2]$$

*For effects due to nuclear-spin isomers (ortho and para), see McQuarrie, loc. cit.

Here x_0, y_0, and z_0 are the coordinates of the center of mass of the molecule, and x_j, y_j, and z_j are the distances of the atoms of mass m_j from the three mutually perpendicular principal axes of rotation of the molecule. We can calculate the moments of inertia of any molecule from Eq. (12.32) if we know the structure of the molecule, with all the bond distances and bond angles. It is more convenient, however, to use special formulas derived for molecules of different symmetries. These formulas are given in Table 12.2. The case of the asymmetric top with three different moments of inertia does not yield any simple general formula.

Example 12.4 NF_3 is a symmetric-top molecule. The N—F bond length is 140 pm and the bond angle FNF is $\Theta = 110°$. Calculate the three principal moments of inertia of NF_3.

For a symmetric top, from Table 12.2, there is an $I_\parallel$ parallel to the threefold axis of the top and two equal $I_\perp$ normal to this axis.

$$I_c = I_\parallel = 2\left(\frac{19.0 \times 10^{-3} \text{ kg mol}^{-1}}{6.02 \times 10^{23} \text{ mol}^{-1}}\right)(1.40 \times 10^{-10} \text{ m})^2(1 - \cos 110°)$$

$$= 16.6 \times 10^{-46} \text{ kg m}^2$$

$$I_a = I_b = I_\perp = \frac{19.0 \times 10^{-3} \text{ kg mol}^{-1}}{6.02 \times 10^{23} \text{ mol}^{-1}}(1.40 \times 10^{-10} \text{ m})^2(1 - \cos 110°)$$

$$+ \frac{14.0 \times 19.0 \times 10^{-6}}{6.02 \times 10^{23}(71 \times 10^{-3})}(1.40 \times 10^{-10})^2(1 + 2\cos 110°)$$

$$= (8.30 + 0.39)10^{-46} = 8.69 \times 10^{-46} \text{ kg m}^2$$

12.13 Rotational Partition Functions for Nonlinear Molecules

The energy levels for spherical tops, symmetric tops, and asymmetric tops can all be computed by quantum mechanics. For these polyatomic molecules it is always satisfactory to use a classical approach to the partition functions. The details can be found in standard reference works. The final result for the most general case, the asymmetric top, is

$$z_r = \frac{\pi^{1/2}}{\sigma}\left(\frac{8\pi^2 I_a kT}{h^2}\right)^{1/2}\left(\frac{8\pi^2 I_b kT}{h^2}\right)^{1/2}\left(\frac{8\pi^2 I_c kT}{h^2}\right)^{1/2} \tag{12.33}$$

The special cases of higher summetry are readily obtained,

$$\text{Spherical top: } I_a = I_b = I_c$$

$$\text{Symmetric top: } I_a = I_b \neq I_c$$

Special attention should be paid to the symmetry number σ. It is the number of ways that the molecule can be "rotated into itself" by rotations about symmetry axes in the molecule. For example, $\sigma = 2$ for H_2O, $\sigma = 3$ for NH_3, $\sigma = 4$ for C_2H_4, $\sigma = 12$ for benzene and methane. The inclusion of the symmetry number in the partition function (sum over states) prevents us from including identical rotational states twice.

TABLE 12.2

FORMULAS FOR MOMENTS OF INERTIA
FOR DIFFERENT TYPES OF MOLECULES

1. Diatomic

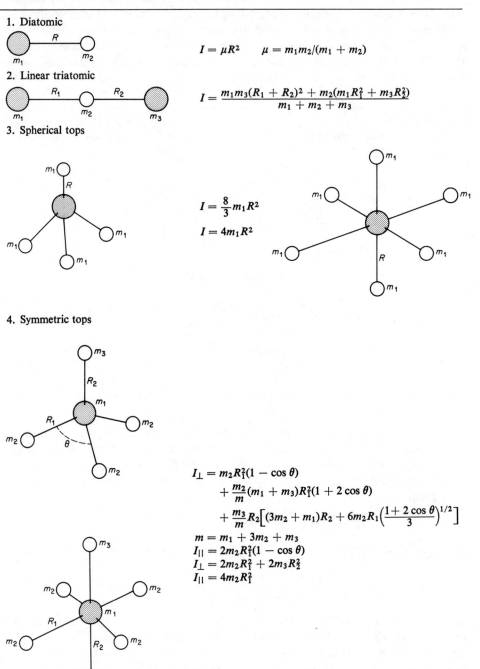

$$I = \mu R^2 \qquad \mu = m_1 m_2 / (m_1 + m_2)$$

2. Linear triatomic

$$I = \frac{m_1 m_3 (R_1 + R_2)^2 + m_2 (m_1 R_1^2 + m_3 R_2^2)}{m_1 + m_2 + m_3}$$

3. Spherical tops

$$I = \frac{8}{3} m_1 R^2$$

$$I = 4 m_1 R^2$$

4. Symmetric tops

$$I_\perp = m_2 R_1^2 (1 - \cos\theta)$$
$$+ \frac{m_2}{m}(m_1 + m_3) R_1^2 (1 + 2\cos\theta)$$
$$+ \frac{m_3}{m} R_2 \left[(3m_2 + m_1) R_2 + 6 m_2 R_1 \left(\frac{1 + 2\cos\theta}{3} \right)^{1/2} \right]$$
$$m = m_1 + 3m_2 + m_3$$
$$I_{\|} = 2 m_2 R_1^2 (1 - \cos\theta)$$
$$I_\perp = 2 m_2 R_1^2 + 2 m_3 R_2^2$$
$$I_{\|} = 4 m_2 R_1^2$$

Example 12.5 Calculate the rotational contribution to the entropy of NF_3 at 400 K from its moment of inertia as found in Example 12.4.

From Eq. (12.33), with $\sigma = 3$,

$$z_r = \frac{\pi^{1/2}}{3}\left(\frac{8\pi^2 kT}{h^2}\right)^{3/2}(I_a I_b I_c)^{1/2}$$

$$= \frac{\pi^{1/2}}{3}\left[\frac{8\pi^2(1.38 \times 10^{-23}\text{ J K}^{-1})(400\text{ K})}{(6.63 \times 10^{-34}\text{ J s})^2}\right]^{3/2}(16.6 \times 10^{-46}\text{ kg m}^2)^{1/2}$$

$$\times\,(8.69 \times 10^{-46}\text{ kg m}^2)$$

$$= 20\ 700$$

$$S = R(\tfrac{3}{2} + \ln z_r) = 95.1\text{ J K}^{-1}\text{ mol}^{-1}$$

12.14 Vibrational Partition Functions

The vibrational partition function of a molecule is readily calculated when the vibrations can be considered as simple harmonic oscillations. The energy levels are then given by Eq. (4.11) as $\epsilon_v = (v + \tfrac{1}{2})h\nu$, where v is a vibrational quantum number with integral values $v = 0, 1, 2, 3$, etc. Each vibrational degree of freedom corresponds to a normal mode of vibration in which all the atoms in the molecule are vibrating in phase with the same frequency ν. The total vibrational energy of the molecule is the sum of the energies in the different degrees of freedom. The vibrational partition function of the molecule is the product of the partition functions for each vibrational degree of freedom, $z_v = z_1 z_2 z_3 \cdots z_n$. This product can be written as

$$z_v = \prod_{i=1}^{n} z_{vi} \tag{12.34}$$

Each partition function has the form

$$z_{vi} = \sum e^{-\epsilon_i/kT} = \sum_v e^{-(v+1/2)h\nu_i/kT} = e^{-h\nu_i/2kT}\sum e^{-vh\nu_i/kT}$$

Since $\sum y^i = 1 + y + y^2 + \cdots = (1-y)^{-1}$, it follows that $\sum e^{-iy} = (1 - e^{-y})^{-1}$. Thus the vibrational partition function for each degree of freedom is simply

$$z_{vi} = e^{-h\nu_i/2kT}(1 - e^{-h\nu_i/kT})^{-1} \tag{12.35}$$

Example 12.6 The fundamental vibrational frequency of F_2 is 2.676×10^{13} Hz. What is the vibrational partition function of F_2 at 298.15 K?

$$\frac{h\nu}{kT} = \frac{(6.626 \times 10^{-34}\text{ J s})(2.676 \times 10^{13}\text{ s}^{-1})}{(1.381 \times 10^{-23}\text{ J K}^{-1})(298.15\text{ K})} = 4.306$$

Then, from Eq. (12.35),

$$z_v = e^{-2.153}(1 - e^{-4.306})^{-1} = 0.1177$$

A typical vibrational partition function is much smaller than a typical rotational one, which in turn is much smaller than a typical translational one. The partition functions give a rough indication of the number of energy levels that are apt to be occupied at a given temperature, since the terms $e^{-\epsilon_i/kT}$ vanish as the probability that a level ϵ_i is occupied becomes vanishingly small. The formulas for molecular partition functions are summarized in Table 12.3.

TABLE 12.3
MOLECULAR PARTITION FUNCTIONS

Motion	Degrees of freedom	Partition function	Order of magnitude (~ 300 K)
Translational	3	$\dfrac{(2\pi mkT)^{3/2}}{h^3} V$	10^{30}–$10^{32} V$ [m^3]
Rotational (linear molecule)	2	$\dfrac{8\pi^2 IkT}{\sigma h^2}$	10–10^2
Rotational (nonlinear molecule)	3	$\dfrac{8\pi^2(8\pi^3 ABC)^{1/2}(kT)^{3/2}}{\sigma h^3}$	10^2–10^3
Vibrational (per normal mode)	1	$\dfrac{1}{1 - e^{-h\nu/kT}}(e^{-h\nu/2kT})$	10^{-2}–10

12.15   Vibrational Energy and Entropy

Always for translation and almost always for rotation, the energy has its classical value of $\frac{1}{2}RT$ per mole in each degree of freedom. For vibrations, however, the classical value is reached only in the limit of high temperatures. From Eqs. (12.4) and (12.35), the vibrational energy per mole for each degree of freedom is

$$U_m = kT^2 \frac{\partial \ln z^L}{\partial T} = RT^2 \frac{\partial \ln z}{\partial T}$$

$$U_m = L\frac{h\nu}{2} + \frac{Lh\nu e^{-h\nu/kT}}{1 - e^{-h\nu/kT}} \tag{12.36}$$

The zero-point energy per mole $L(h\nu/2)$ may be written as U_{m0}, and $h\nu/kT$ as x. Then Eq. (12.36) becomes

$$U_m - U_{m0} = \frac{Rxe^{-x}}{1 - e^{-x}} \tag{12.37}$$

The vibrational heat capacity per mole is

$$C_{Vm} = \left(\frac{\partial U_m}{\partial T}\right)_V = \frac{Rx^2}{2(\cosh x - 1)} \tag{12.38}$$

The vibrational contribution to the molar Gibbs and Helmholtz energies is

$$G_m = A_m = U_{m0} + RT\ln(1 - e^{-x})$$

or $(G_m - U_{m0})/T = R\ln(1 - e^{-x})$. (Note that $G = A + PV$ and $PV = 0$. There is no vibrational contribution to pressure since z_v is not a function of V.)

The harmonic-oscillator contribution to the entropy is found from Eqs. (12.6) and (12.36). For each degree of freedom,

$$S_m = R\left[\frac{x}{e^x - 1} - \ln(1 - e^{-x})\right] \tag{12.39}$$

These functions have been recorded in convenient tables to facilitate calculation of the vibrational contributions. An abridged set of values is given in Table 12.4. The entropy is not tabulated separately since

$$S = \frac{U - U_0}{T} - \frac{G - U_0}{T} \tag{12.40}$$

TABLE 12.4
MOLAR THERMODYNAMIC FUNCTIONS OF A HARMONIC OSCILLATOR
(ENERGY UNITS IN JOULES PER MOLE)

$x = \dfrac{h\nu}{kT}$	C_V	$\dfrac{U - U_0}{T}$	$\dfrac{-(G - U_0)}{T}$	$x = \dfrac{h\nu}{kT}$	C_V	$\dfrac{U - U_0}{T}$	$\dfrac{-(G - U_0)}{T}$
0.10	8.305	7.912	19.56	1.70	6.573	3.159	1.677
0.15	8.297	7.707	16.39	1.80	6.393	2.958	1.502
0.20	8.289	7.510	14.20	1.90	6.209	2.778	1.347
0.25	8.272	7.318	12.55	2.00	6.021	2.603	1.209
0.30	8.251	7.130	11.23	2.20	5.640	2.279	0.976
0.35	8.230	6.945	10.14	2.40	5.255	1.991	0.7907
0.40	8.205	6.761	9.230	2.60	4.870	1.734	0.6418
0.45	8.176	6.586	8.439	2.80	4.494	1.507	0.5213
0.50	8.142	6.410	7.753	3.00	4.125	1.307	0.4246
0.60	8.071	6.067	6.615	3.50	3.270	0.9062	0.2552
0.70	7.983	5.740	5.708	4.00	2.528	0.6204	0.1535
0.80	7.883	5.427	4.962	4.50	1.913	0.4204	0.0933
0.90	7.774	5.125	4.339	5.00	1.420	0.2920	0.0556
1.00	7.657	4.841	3.816	5.50	1.036	0.1878	0.0338
1.10	7.523	4.565	3.366	6.00	0.7455	0.1238	0.0209
1.20	7.385	4.301	2.982	6.50	0.5296	0.0815	0.0125
1.30	7.234	4.049	2.645	7.00	0.3723	0.0531	0.0075
1.40	7.079	3.810	2.355	8.00	0.1786	0.0221	0.0025
1.50	6.916	3.582	2.099	9.00	0.0832	0.0092	0.0016
1.60	6.745	3.365	1.875	10.00	0.0376	0.0037	0.0004

Example 12.7 For one mole of F_2 at 298 K, calculate the vibrational contributions to U, S, G, and C_V, given that the fundamental vibration frequency is $\nu_0 = 2.676 \times 10^{13}$ Hz.

As found in Example 12.6, $x = h\nu/kT = 4.306$. By interpolation, from Table 12.4: $(U - U_0)/T = 0.498$ J K^{-1} mol^{-1}; $-(G - U_0)/T = 0.117$ J K^{-1} mol^{-1}; $U_0 = \frac{1}{2}Lh\nu_0 = 5330$ J; hence $U_m = 5480$ J mol^{-1}. Note that most of the vibrational energy is zero-point energy. $G_m = 5300$ J mol^{-1}. Hence, from Eq. (12.40), $S_m = 0.6$ J K^{-1} mol^{-1}. We can now add the various contributions to find the entropy of F_2 at 298 K. $S_m = S_t + S_r + S_v = 154.7 + 48.1 + 0.6 = 203.4$ J K^{-1} mol^{-1}. This statistical value is in excellent agreement with the Third Law entropy of 203.2 J K^{-1} mol^{-1} obtained from measurements of heat capacity. The vibrational C_V is obtained from Eq. (12.38),

$$C_{Vm} = \frac{(8.314 \text{ J K}^{-1} \text{ mol}^{-1})(4.306)^2}{2(37.24 - 1)} = 2.13 \text{ J K}^{-1} \text{ mol}^{-1}$$

The classical value would be $R = 8.314$ J K^{-1} mol^{-1}.

12.16 Heat Capacities

The value of C_V is readily calculated by taking the derivative with respect to T of the expression for U. $C_V = (\partial U/\partial T)_V$. The value of C_P for an ideal gas is then obtained as $C_P = C_V + nR$.

The translational contribution to C_V is always the classical value $\frac{3}{2}nR$. For diatomic molecules, the rotational and vibrational C_V can be expressed conveniently in terms of *characteristic temperatures* for rotation and for vibration, defined by

$$\Theta_r = h^2/8\pi^2 Ik \tag{12.41}$$

$$\Theta_v = hv/k \tag{12.42}$$

Table 12.5 gives characteristic temperatures for a number of molecules.

TABLE 12.5
CHARACTERISTIC ROTATIONAL AND VIBRATIONAL
TEMPERATURES

Molecule	Θ_r (K)	Θ_v (K)	Molecule	Θ_r (K)	Θ_v (K)
H_2	85.3	6215	N_2	2.88	3374
D_2	42.7	4394	O_2	2.07	2256
Cl_2	0.351	808	HCl	15.02	4227
Br_2	0.116	463	HBr	12.02	3787
I_2	0.0537	308	HI	9.06	3266

Figure 12.3 shows the rotational and vibrational heat capacities for a diatomic molecule as functions of T/Θ_r and T/Θ_v, respectively. The classical values $C_V/R = 1$ are approached when T reaches about 2Θ. [How would you explain the fact that $C_V(\text{rot})$ actually exceeds the classical value over a certain range of temperatures?]

Example 12.8 From the Θ_r and Θ_v values in Table 12.5, estimate the temperatures at which $C_V(\text{vib})$ and $C_V(\text{rot})$ for N_2 have half the classical value, R per mole.

The values are read directly from the graphs of Fig. 12.3. For N_2, $\Theta_r = 2.88$ K and $\Theta_v = 3374$ K. $C_V/nR = \frac{1}{2}$ at $T/\Theta_v = 0.38$, $T/\Theta_r = 0.43$, that is, at $T = 1280$ K, and $T = 1.24$ K, respectively.

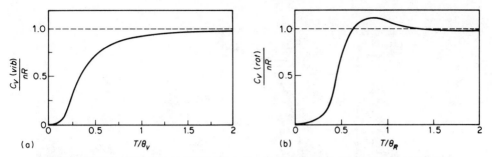

FIGURE 12.3 (a) Vibrational contribution to C_V of a diatomic gas. (b) Rotational contribution to C_V of a diatomic gas. The dashed lines are predictions based on classical equipartition of energy.

12.17 Vibrational Energy and Molecular Dissociation

If a vibrating diatomic molecule really behaved as a harmonic oscillator with a potential energy vs. internuclear distance function like the parabola in Fig. 3.9, the molecule would never dissociate, no matter how much vibrational energy it acquired. We know, however, that molecules like I_2, Cl_2, and N_2 dissociate into atoms at elevated temperatures. This dissociation is the result of an increase in the amplitude of vibration beyond the limit at which the restoring force can bring the atoms together again. A vibrating molecule, therefore, behaves as an *anharmonic oscillator*, in which the restoring force (and force constant κ) decrease with increasing internuclear separation.

The actual potential-energy curve of a typical diatomic molecule, I_2, is shown in Fig. 12.4. Some of the vibrational-energy levels are superimposed on the curve. The restoring force of the vibrational motion is the slope of the curve $\partial U/\partial r$, which decreases at higher vibrational energies and eventually goes to zero as the curve becomes parallel to the internuclear-distance axis. At this point, the vibrational energy is just sufficient to dissociate the molecule into free atoms. The energy level of the separated atoms of iodine at rest is 2.49×10^{-19} J (12 540 cm^{-1}) above the minimum of the potential-energy curve.

At low vibrational-energy levels the curve is close to a parabola, and hence the energy levels are close to those of the harmonic-oscillator model. This agreement of the lower vibrational-energy levels with the harmonic-oscillator model explains why this model is so good for calculation of thermodynamic properties at temperatures where dissociation is not appreciable. We should remember, however, that the statistical thermodynamic method is not tied to any particular model of the internal energies of molecules. If adequate spectroscopic data are available, we can always calculate the molecular partition function directly by summing $e^{-\epsilon_i/kT}$ over the experimental energy levels ϵ_i.

In the diagram of Fig. 12.4, the minimum of the potential-energy curve is taken as the zero of energy, so as to be in accord with the zero used in the vibrational partition function. In calculations of energies of molecules from quantum mechanics, it is more usual to take the zero of energy to correspond to the separated atoms. Then the various vibrational levels shown in Fig. 12.4 would have negative energies.

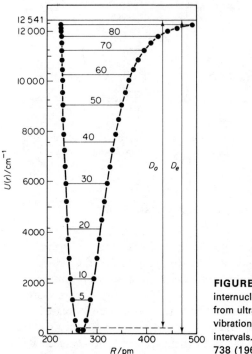

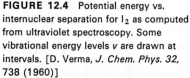

FIGURE 12.4 Potential energy vs. internuclear separation for I_2 as computed from ultraviolet spectroscopy. Some vibrational energy levels v are drawn at intervals. [D. Verma, *J. Chem. Phys. 32*, 738 (1960)]

The energy D_e is called the *spectroscopic dissociation energy* of the molecule. It is the energy difference between the minimum of the potential-energy curve and the level of the separated atoms.

The lowest vibrational energy is not zero, but $\frac{1}{2}hv_0$, the zero-point energy of the vibration of frequency v_0. The difference between this lowest vibrational level and the level of the separated atoms is D_0, the *chemical dissociation energy*. This name is used because in chemical reactions the molecules always have a vibrational energy equal to at least the zero-point energy $\frac{1}{2}hv_0$. Therefore,

$$D_e = D_0 + \tfrac{1}{2}hv_0 \qquad (12.42)$$

12.18 Statistical Thermodynamics of Crystals

In 1819, Dulong and Petit noted that the heat capacities of solid elements at room temperature are usually close to the value given by the formula $C_V/nR = 3$. This result could be explained by the principle of equipartition of energy. A crystal containing n moles of an element is composed of nL atoms, which have $3nL$ degrees of freedom. All except six of these are vibrations, and since $3nL - 6 \approx 3nL$, the classical vibrational energy of the crystal should be $3nL(kT) = 3nRT$, so that $C_V = 3nR$. Actually, in some cases the Dulong–Petit rule fails badly, especially for light, hard crystals such as diamond. Data on C_V as a function of T give results like those in Fig. 12.5, which of course are not consistent with a classical equipartition of energy.

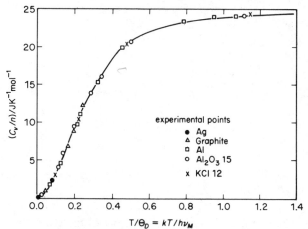

FIGURE 12.5 Heat capacities C_V of solids as functions of T/Θ_D.

$T/\Theta_D = kT/h\nu_M$

In 1906, Einstein devised a quantum theory for the heat capacities of crystals. He used a model in which vibrational energy is partitioned among $3nL$ independent harmonic oscillators, each having the same frequency ν_E. Then Eq. (12.10) without the $N!$ factor gives $Z = z^{3nL}$ for the crystal, with z given by Eq. (12.35). The formulas in Section 12.15 and the functions in Table 12.4 can then be applied directly to the Einstein-model crystal. The calculated values for C_V are good except at temperatures below about 20 K, where the experimental C_V approaches zero as $C_V = aT^3$, whereas the Einstein C_V falls much more sharply.

Debye realized that the Einstein model, by taking a single ν to fit the data at higher temperatures, fails to consider vibrations of low frequency that are important for uptake of energy by the crystal at the lowest temperatures. Debye assumed that the vibrations of the crystal are distributed over a wide range of frequencies from ~ 0 up to a maximum at ν_D. The density function is $g(\nu) = (9nL/\nu_D^3)\nu^2$. Any thermodynamic function, including C_V, for example, can then be calculated by averaging the harmonic-oscillator value in Section 12.15 over the density function $g(\nu)$. For example,

$$C_V = \int_0^{\nu_D} C_V(\nu)g(\nu)\,d\nu$$

Figure 12.5 shows the values of C_V calculated from the Debye theory as a solid curve on which experimental points are superimposed. The agreement between theory and experiment is excellent. In particular, the theory gives correctly the low-temperature dependence of C_V on T^3, and the high-temperature Dulong–Petit limit, $C_V = 3nR$. Some Debye characteristic temperatures are given in Table 12.6. [What would be the classical (Dulong and Petit) value of C_V for CaF_2?]

12.19 Electronic Partition Function

When we discussed internal motions of molecules in Chapter 5, we assumed that the molecules are always in their lowest electronic states, called *ground states*. Since almost all the masses of atoms are concentrated in their nuclei, the translational,

TABLE 12.6

VALUES OF THE DEBYE TEMPERATURE Θ_D (K)

Cs	40	KCl	227
Pb	88	Ca	230
Hg	72	Zn	235
K	91	NaCl	281
Na	156	Cu	343
Au	180	Al	428
AgCl	183	Fe	432
Ag	226	CaF_2	474
Pt	228	Be	1000

rotational, and vibrational energies of molecules can be described as motions of these nuclei. A molecule can also acquire discrete quanta of energy if an electron in the molecule makes a transition from its ground state to an *excited electronic state*. Electronic energies are usually much larger than vibrational energies, so that we can depict the vibrational energies as discrete levels superimposed on the potential energy curve of each electronic state. Sometimes an electronic level consists of more than one state with identical energy. In such instances we say that the electronic state in question has a multiplicity or degeneracy g.

If the electronic states have energies $\epsilon_1, \epsilon_2, \ldots,$ and multiplicities $g_1, g_2, \ldots,$ the electronic partition function is

$$z_e = g_0 + g_1 e^{-\epsilon_1/kT} + g_2 e^{-\epsilon_2/kT} + \cdots = \sum g_i e^{-\epsilon_i/kT} \qquad (12.44)$$

In most diatomic molecules the lowest energy state is a spin *singlet* ($g_0 = 1$) and the next lowest electronic state is so high that $e^{-\epsilon_1/kT} \approx 0$ except at very high temperatures. In such cases $z_e = 1$.

If a molecule has one or more low-lying electronic states, we must calculate the internal partition functions ($z_I = z_v z_r z_e$) separately for each such state. [Why?] For example, with a ground state A and excited state B,

$$z = z_t(g_A z_{IA} + g_B z_{IB} e^{-\epsilon_B/kT}) \qquad (12.45)$$

Example 12.9 The ground state of O_2 is a spin triplet ($^3\Sigma$) and the first excited state is a spin singlet ($^1\Delta$) at 7880 cm^{-1} above the ground state. Apply Eq. (12.45) to O_2 at 1000 K.

Except for Σ states, which have no angular momentum, the orbital multiplicity is $g = 2$. Hence $g_0 = 3$, $g_1 = 2$. $\epsilon_1/kT = hc\tilde{v}/kT = 11.4$, so that $e^{-\epsilon_1/kT} = e^{-11.4}$. Thus $z = z_t(3z_{I0} + 2z_{I1}e^{-11.4}) \simeq 3z_t z_{I0}$. Even at 1000 K, the first excited state will not contribute to z.

Example 12.10 The ground state of F has a degeneracy $g_0 = 4$ and the first excited state lies at $\tilde{\nu} = 404.0 \text{ cm}^{-1}$ above the ground state and has a degeneracy $g_1 = 2$. What fraction of F atoms will be in the first excited state at 1000 K?

The required fraction is $f_1 = g_1 e^{-\epsilon_1/kT}/(g_0 + g_1 e^{-\epsilon_1/kT} + \cdots)$.

$$\frac{\epsilon_1}{kT} = \frac{hc\tilde{\nu}}{kT} = \frac{(6.63 \times 10^{-34} \text{ J s})(3.0 \times 10^8 \text{ m s}^{-1})(404 \times 10^2 \text{ m}^{-1})}{(1.38 \times 10^{-23} \text{ J K}^{-1})(10^3 \text{ K})} = 0.582$$

$$e^{-x} = e^{-0.582} = 0.559$$

$$f_1 = \frac{2(0.559)}{4 + 2(0.559)} = 0.218$$

12.20 The Third Law in Statistical Thermodynamics

The expression for the canonical ensemble partition function in Eq. (12.2) is a sum over all allowable states of a system. It is possible, however, that more than one state has exactly the same energy level E_j. In such a case, the energy level has a degeneracy g_j, where g_j is the number of states with energy E_j. The partition function $Z(V, T)$ can be written also as a sum over energy levels if we include the degeneracy factors. Thus

$$Z(V, T) = \sum g_j e^{-E_j/kT} \tag{12.46}$$

There should be no confusion over the two alternative ways of writing $Z(V, T)$ if you remember that when g_j appears in the sum, we are summing over energy levels and when g_j does not appear, we are summing over states.

The expression in Eq. (12.46) is convenient for obtaining the limiting value of entropy S as temperature T approaches the limit of absolute zero. As $T \to 0$ we can certainly neglect all terms in Z except for the first two, so that Eq. (12.6) becomes

$$S = \frac{1}{T} \frac{g_0 E_0 e^{-E_0/kT} + g_1 E_1 e^{-E_1/kT}}{g_0 e^{-E_0/kT} + g_1 e^{-E_1/kT}} + k \ln (g_0 e^{-E_0/kT} + g_1 e^{-E_1/kT})$$

Near the limit $T = 0$, $e^{-E_1/kT} \ll e^{-E_0/kT}$. The second term is expanded as $\ln (1 + x) \simeq x$, so that

$$S = \frac{E_0}{T} + \frac{E_1}{T} \frac{g_1}{g_0} e^{-(E_1-E_0)/kT} + k \ln g_0 - \frac{E_0}{T} + k \frac{g_1}{g_0} e^{-(E_1-E_0)/kT}$$

In the limit as $T = 0$, the remaining exponential terms decrease much more rapidly than T, so that we are left with only

$$S_0 = k \ln g_0 \tag{12.47}$$

If the degeneracy of the lowest energy state of the ensemble $g_0 = 1$, then $S_0 \to 0$ as $T \to 0$ and all systems of the ensemble go into its lowest energy state. This is the case with perfect crystals in the limit of $T \to 0$, and corresponds to the situation in the Lewis and Randall statement of the Third Law (page 138).

There are several reasons why g_0 may not equal unity; some of them are considered as special cases in chemical thermodynamics, whereas others are excluded by convention. The molecules in a crystal may persist in more than one structural arrangement even at the lowest possible temperatures. Indeed, the low temperature may effectively "freeze in" a disordered arrangement in the crystal. An example is nitrous oxide, as shown in Fig. 7.9.

Example 12.11   What would be the residual entropy S_0 in crystalline CH_3D?

CH_3D is a tetrahedral molecule, and in the crystal the C—D bond could have four different orientations relative to neighboring molecules. Hence

$$S_0 = k \ln g_0 = k \ln (4)^L = R \ln 4 = 11.5 \text{ J K}^{-1} \text{ mol}^{-1}$$

Another source of residual entropy at 0 K arises from the isotopic composition of the elements. Unless we are explicitly concerned with reactions that cause separation or enrichment of isotopes, we neglect the small changes in isotopic composition that occur in chemical reactions. In chemical thermodynamics, we adopt the convention that the elements in their crystalline states at 0 K have a limiting entropy $S_0 = 0$ irrespective of their natural isotopic compositions. Thus a crystal of lead at 0 K is assigned an entropy $S_0 = 0$ despite the fact that it is a mixture of four isotopes.

Another source of residual entropy at 0 K is the nuclear-spin degeneracy of the atoms of the chemical elements. These give a $g_0 \neq 1$, but this effect is ignored by convention in setting $S_0 = 0$.

We can see, therefore, that the Third Law has a simple basis in statistical thermodynamics, but for practical purposes of chemical thermodynamics we do not try to include all the contributions to g_0, but rather adopt a conventional $S_0 = 0$ that is most useful for the thermodynamic treatment of actual chemical reactions.

12.21   Equilibrium Constants

Statistical thermodynamics gives a simple and direct method for calculating equilibrium constants of ideal-gas reactions in terms of the partition functions of reactant and product molecules. Consider a perfectly general reaction in an ideal gas mixture, denoted by $aA + bB = cC + dD$. The condition for equilibrium in terms of chemical potentials was given in Eq. (8.54) as

$$c\mu_C + d\mu_D - a\mu_A - b\mu_B = 0 \tag{12.48}$$

The most direct way to get the statistical expression for the chemical potential is through the Helmholtz free energy, $A = -kT \ln Z$, and Eq. (8.49), $\mu_A = (\partial A/\partial n_A)_{T,V,n_B,n_C,n_D}$. Similar expressions hold for μ_B, μ_C, and μ_D.

The number N_A of molecules of A in the system is related to the number n_A of moles by $N_A = Ln_A$, where L is the Avogadro constant. Hence $\mu_A = L(\partial A/\partial N_A)_{T,V,N_B,N_C,N_D}$. Since $A = -kT \ln Z$, $\mu_A = -kTL(\partial \ln Z/\partial N_A)_{T,V,N_B,N_C,N_D}$. From Eq. (12.10), $Z_A = z_A^{N_A}/N_A!$, $\ln Z_A = N_A \ln z_A - \ln N_A!$. From the Stirling formula for $N!$, $\ln Z_A = N_A \ln z_A - N_A \ln N_A + N_A$. Therefore, $\partial/\partial N_A(\ln Z_A) = \ln z_A - \ln N_A - 1 + 1 = \ln (z_A/N_A)$, and $\mu_A = -LkT \ln (z_A/N_A)$. The equilibrium condition Eq. (12.48) then becomes

$$d \ln \frac{z_D}{N_D} + c \ln \frac{n_C}{N_C} - a \ln \frac{z_A}{N_A} - b \ln \frac{z_B}{N_B} = 0$$

On rearrangement, we obtain

$$\frac{N_D^d N_C^c}{N_A^a N_B^b} = \frac{z_D^d z_C^c}{z_A^a z_B^b} \tag{12.49}$$

The molecular partition function z is a function only of T and V, $z(T, V)$. Therefore, the ratio of numbers of molecules at equilibrium, given in Eq. (12.49), is some function of T and V alone, i.e., an equilibrium constant. This derivation has therefore provided a new statistical thermodynamic proof of the existence of an equilibrium constant and in addition an explicit expression for the constant in terms of molecular partition functions.

It is convenient to divide out the volume V from the z function, by writing $z'V = z$. We then introduce the number concentration, $C = N/V$, into Eq. (12.49):

$$C_D^d C_C^c / C_A^a C_B^b = z_D'^d z_C'^c / z_A'^a z_B'^b$$

Or, in terms of the molar concentration, $c = C/L$,

$$c_C^c c_D^d / c_A^a c_B^b = L^{(a+b-c-d)} z_D'^d z_C'^c / z_A'^a z_B'^b \tag{12.50}$$

Since $P = cRT$, $c = P/RT = P/LkT$, Eq. (12.50) becomes

$$\frac{P_C^c P_D^d}{P_A^a P_B^b} = (kT)^{c+d-a-b} \frac{z_D'^d z_C'^c}{z_A'^a z_B'^b} \tag{12.51}$$

From Eq. (8.30), the equilibrium constant in terms of partial pressures is $K_P = (P_C^c P_D^d / P_A^a P_B^b)(P^\circ)^{a+b-c-d}$. Therefore, from Eq. (12.51), we have

$$K_P = (kT/P^\circ)^{c+d-a-b} z_D'^d z_C'^c / z_A'^a z_B'^b \tag{12.52}$$

This is a marvelously useful formula, since it allows us to calculate the K_P of any ideal-gas reaction from the properties of the reactant and product molecules. These properties are mostly obtained from spectroscopic data, which give the energy levels of the molecules. The data needed to calculate the partition functions are the masses, internuclear distances, vibrational frequencies, and electronic energy levels of the molecules. From this basic information on molecular structure, we can now obtain the equilibrium constants of all possible reactions between gaseous molecules. This simple theory is restricted to ideal gases, because, as has been emphasized, only for ideal gases can we express the properties of a system in terms of the properties of individual isolated molecules. In nonideal gases, liquids, and solids, we must consider the interactions of many molecules. The general theory of equilibrium still holds good, but the calculations become much more difficult.

12.22 Statistical Interpretation of K_P

Let us consider the application of the statistical theory to a simple reaction, $A \rightleftharpoons B$. A and B might represent two isomers, e.g., butane and isobutane. Figure 12.6 shows two sets of energy levels, $\epsilon_j(A)$ belonging to A and $\epsilon_j(B)$ belonging to B. In each case, the reference level of energy in this diagram corresponds to complete dissociation of the molecule into atoms in the ground state. The difference in the

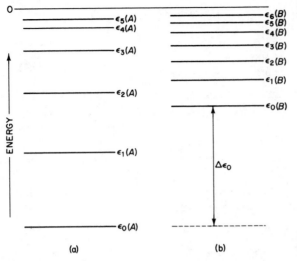

FIGURE 12.6 Sets of energy levels for two isomeric molecules A and B.

lowest energy levels ($j = 0$) of the two compounds is

$$\Delta\epsilon_0 = \epsilon_0(B) - \epsilon_0(A) \tag{12.53}$$

When the reaction $A \rightleftharpoons B$ comes to equilibrium, the molecules of A are distributed in accord with a Boltzmann distribution in the energy levels $\epsilon_j(A)$, and the molecules of B similarly distributed in their levels $\epsilon_j(B)$.

If we take the zero level of energy for the entire system as the lowest energy level of the A molecules, the partition function for A is simply

$$z_A = \sum_{j=0}^{\infty} \exp\left[\frac{-\epsilon_j(A)}{kT}\right]$$

To discuss equilibrium between A and B, we must reckon the energy levels of B from the same zero level as that used for A. Thus from Eq. (12.53)

$$z_B = \sum_{j=0}^{\infty} \exp\left(\frac{-[\epsilon_j(B) + \Delta\epsilon_0]}{kT}\right) = \exp\left(\frac{-\Delta\epsilon_0}{kT}\right) \sum_{j=0}^{\infty} \exp\left(\frac{-\epsilon_j(B)}{kT}\right)$$

Having made such an assignment of a common zero level for the energies, we can obtain the statistical expression for the equilibrium constant from Eq. (12.52). For the reaction $A \rightleftharpoons B$,

$$K_P = \frac{z_B'}{z_A'} \exp\left(\frac{-\Delta\epsilon_0}{kT}\right) \tag{12.54}$$

The equilibrium constant for $A \rightleftharpoons B$ thus has a simple statistical interpretation. It is the sum of the probabilities that the system be found at equilibrium in the energy levels of B divided by the sum of the probabilities that it be found in the energy levels of A.

12.23 Examples of Calculation of K_P

An interesting type of reaction for which K_P is readily calculated from the statistical formula (12.52) is the dissociation of diatomic gases.

Example 12.12 Calculate K_P for the reaction $^{35}Cl_2 = 2^{35}Cl$ at 2000 K. The dissociation energy $D_0 = 238.9$ kJ mol^{-1}. The ground state of Cl is $^2P_{3/2}$ with multiplicity $g_0 = 4$ and there is a low-lying excited state 881 cm^{-1} above the ground state $^2P_{1/2}$ with multiplicity $g_1 = 2$. The vibration frequency of Cl_2 is $\nu = 1.694 \times 10^{13}$ Hz and the moment of inertia $I = 1.16 \times 10^{-45}$ kg m^2. The mass of a Cl atom is $m = 0.035/6.023 \times 10^{23} = 5.81 \times 10^{-26}$ kg.

We use Eqs. (12.45) and (12.52) and calculate each part of the equilibrium constant separately. The atoms, of course, have only translational and electronic partition functions.

$$K_P = \frac{P_{Cl}^2}{P_{Cl_2}P^\circ} = \frac{kT}{P^\circ}\frac{z_t'^2(Cl)}{z_t'(Cl_2)}\frac{1}{z_r(Cl_2)}\frac{1}{z_v(Cl_2)}\frac{z_e^2(Cl)}{z_e(Cl_2)}$$

$$K_P = \frac{kT}{P^\circ}\frac{(\pi mkT)^{3/2}}{h^3}\frac{\sigma h^2}{8\pi^2 IkT}(1 - e^{-h\nu/kT})(4 + 2e^{-\epsilon_1/kT})^2 e^{-\Delta\epsilon_0/kT}$$

$$K_P = (2.72 \times 10^{-25})(1.23 \times 10^{33})(3.48 \times 10^{-4})(0.334)(25.6)(5.76 \times 10^{-7}) = 0.573$$

We now give another example of the calculation of an equilibrium constant. In a couple of hours (including the time needed to find the data) one can calculate this K_P over a range of temperatures. The experimental measurements required over a year and gave less accurate results.

Example 12.13 Calculate the equilibrium constant K_P, at 700 K, for the reaction $H_2 + I_2 = 2HI$.

To calculate the partition functions z and hence K_P, we must know the masses m, moments of inertia I, symmetry numbers σ, vibration frequencies ν, and dissociation energies D_0 of the molecules. These data are as follows:

	H_2	I_2	HI
m (10^{-26} kg)	0.335	42.15	21.24
R_e (pm)	74.0	267	160
$\mu = m_1 m_2/(m_1 + m_2)$ (10^{-27} kg)	0.837	105.4	1.661
$I = \mu R_e^2$ (10^{-46} kg m^2)	0.0458	75.14	0.425
σ	2	2	1
ν (10^{13} Hz)	12.95	0.642	6.805
D_0 (kJ mol^{-1})	431.8	149.0	295.0

From Eq. (12.52) and the equations for the partition functions,

$$K_P = \left(\frac{m_{HI}^2}{m_{H_2}m_{I_2}}\right)^{3/2}\left(\frac{4I_{HI}^2}{I_{H_2}I_{I_2}}\right)\frac{(1 - e^{-h\nu_{H_2}/kT})(1 - e^{-h\nu_{I_2}/kT})}{(1 - e^{-h\nu_{HI}/kT})^2} \times e^{(2D_{0HI}-D_0H_2-D_0I_2)/kT}$$

$$= (32)^{3/2}(0.210)\frac{(1)(0.356)}{(0.991)^2}e^{1.58} = 66.9$$

The experimental value is $K_P = 60.3$.

We shall give one more example of calculation of an equilibrium constant, that of an isotopic exchange reaction. This will indicate the importance of the difference in zero-point energies in reactions that involve isotopic exchange.

Example 12.14 Calculate the K_P for the reaction $H_2 + D_2 = 2HD$ as a function of temperature, from 200 to 700 K. The vibration frequency of H_2 is 12.95×10^{13} Hz.

Nuclei are so massive compared to electrons that one can assume that the nuclei are fixed in position when one calculates the interaction energy of electrons and nuclei in a molecule. Consequently, all three hydrogen molecules, H_2, D_2, and HD, have the same equilibrium internuclear distance R_e, the same force constant κ, and the same spectroscopic energy of dissociation D_e. The potential-energy curve of the molecules is shown in Fig. 12.7.

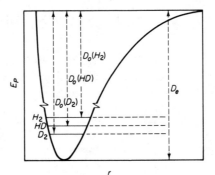

FIGURE 12.7 Potential-energy curves for H_2, HD, and D_2 are virtually identical, but zero-point energies (and other energy levels) differ.

$$K_P = \frac{m_{HD}^3}{(m_{H_2} m_{D_2})^{3/2}} \frac{4I_{HD}^2(1 - e^{-h\nu_{H_2}/kT})(1 - e^{-h\nu_{D_2}/kT})}{I_{H_2} I_{D_2}(1 - e^{-h\nu_{HD}/kT})^2}$$
$$\times\; e^{-(2h\nu_{HD} - h\nu_{H_2} - h\nu_{D_2})/2kT}$$

Note that for the isotopic exchange $\Delta\epsilon_0$ is simply the difference in zero-point energies, $\Delta\epsilon_0 = 2(\tfrac{1}{2}h\nu_{HD}) - \tfrac{1}{2}h\nu_{D_2} - \tfrac{1}{2}h\nu_{H_2}$. Since all the R_e are the same, the ratios of I are simply the ratios of reduced masses μ. Thus $I_{HD}/I_{D_2} = \tfrac{4}{3}$ and $I_{HD}/I_{H_2} = \tfrac{2}{3}$. At $T \leq 700$ K the vibrational terms $(1 - e^{-h\nu/kT})$ are all very close to unity. Since $\nu = (1/2\pi)\sqrt{\kappa/\mu}$, and all the κ are the same,

$$\nu_{HD} = [(\tfrac{1}{2})/(\tfrac{2}{3})]^{1/2}\nu_{H_2} = (\tfrac{3}{4})^{1/2}\nu_{H_2}$$
$$\nu_{D_2} = [(\tfrac{1}{2})/1]^{1/2}\nu_{H_2} = (\tfrac{1}{2})^{1/2}\nu_{H_2}$$

Thus

$$K_P = (\tfrac{9}{8})^{3/2}(4)(\tfrac{8}{9})e^{-[3^{1/2} - 1 - (1/2)^{1/2}][h\nu_{H_2}/2kT]}$$
$$= 4(1.06)e^{-77.7/T}$$

Note the preponderant effect of the symmetry factor σ, which gives the factor 4 in K_P. Calculated and experimental values of K_P are shown in Fig. 12.8.

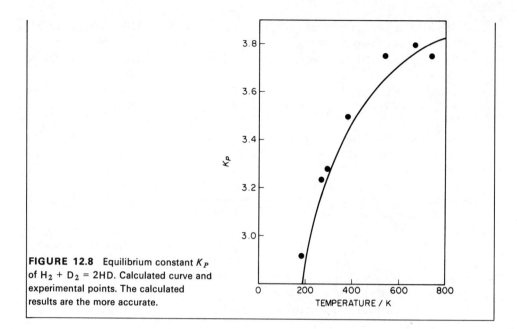

FIGURE 12.8 Equilibrium constant K_P of $H_2 + D_2 = 2HD$. Calculated curve and experimental points. The calculated results are the more accurate.

Problems

1. Calculate the molecular translational partition function for CH_4 in a volume of 20 dm³ at 250 K.

2. Calculate the molar translational partition function for argon at 300 K and $P = 100$ kPa. State any assumption made. How good is the assumption?

3. The internuclear distance in NaCl is 251 pm. Calculate the molecular partition function of $Na^{35}Cl(g)$ for (a) translation and (b) rotation, at 1000 K in a volume of 1 m³.

4. Calculate the translational and rotational contributions to the entropy S_m of $^1H_2$, $^2H_2$, $^1H^2H$, $^3H_2$ at 500 K and $P = 100$ kPa. The internuclear distance is 74.0 pm.

5. Consider a system that can exist only in either one of two states separated by an energy E. Derive expressions for U, S, A, and C_V for such a system and plot them vs. T.

6. For H_2, $\Theta_r = 85.3$ K. Calculate the rotational energy per mole of H_2 at 70 K from Eqs. (12.28) and (12.26). Why do these results differ?

7. Calculate the moment of inertia of SiF_4 from the Si—F bond distance 155 pm. What is the translational and rotational entropy of SiF_4 at 298 K and 101 kPa?

8. The bond energy of C_2 is 531 kJ mol⁻¹, $R_e = 131.2$ pm, and the fundamental vibration frequency is $\nu_0 = 4.92 \times 10^{13}$ Hz. Calculate $\Delta G°$ for $C_2(g) \longrightarrow 2C(g)$. In the spectrum of a flame $C_2 \longrightarrow 2C$ to the extent 20%. If the partial pressure of C_2 is 50 Pa, what is the flame temperature?

9. The bond length in NH_3 is 101 pm and the bond angle is 107.3°. Calculate the three principal moments of inertia of NH_3. Hence obtain the molecular partition function for rotation at 298 K.

10. There are six vibrational degrees of freedom in NH_3, at $\tilde{v} = 3336(2)$, $950(2)$, 3410, and 1625 cm^{-1}. Calculate the total entropy $S°$ (298.15 K) of NH_3. (See Problem 9.)

11. What fraction of HBr molecules are in the state $v = 3$, $J = 5$ at 800 K? $\Theta_v = 3700$ K, $\Theta_r = 12.1$ K.

12. The ionization energy of $K \longrightarrow K^+ + e$ is 4.33 eV. Calculate the degree of dissociation of K at 1600 K and 100 Pa. (Assume that electrons have the partition function of a dilute gas.)

13. Calculate K_P for $Na_2 \longrightarrow 2$ Na at 800, 1000, and 1200 K. The $D_0 = 0.730$ eV, $\tilde{v} = 159.2$ cm^{-1}, and the internuclear $R_e = 308$ pm. The ground state of Na_2 is a singlet and that of Na is a doublet.

14. NO has a doubly degenerate ground state and a doubly degenerate excited state only 121 cm^{-1} higher. Calculate the electronic partition function. At what temperature would the electronic heat capacity C_V be a maximum?

15. Calculate $(G° - U_0°)/T$ for 1 mol N_2 at intervals of 200 K from 200 to 1400 K given $R_e = 108.76$ pm, $\tilde{v} = 2357.55$ cm^{-1}; and $M = 28.0134$ g mol^{-1}. The ground state is a singlet.

16. At low temperatures the Debye theory of heat capacity of solids gives $C_V = 3nR \, (4\pi^4/5) \, T^3/\Theta_D^3$. Calculate the $S°$ of (a) Pb and (b) Be at 30 K. (See Table 12.6.)

17. The heat capacity of copper at 100 K is 15.5 J K^{-1} mol^{-1}. To what value of $\Theta_E = hv/k$ would this correspond? Calculate C_V at 20 K and at 500 K for Cu with this Θ_E (on the Einstein model).

18. What would be the zero-point energy of a crystal of gold with $\Theta_D = 180$ K?

13

Chemical Kinetics

Bonfires and explosions, hardboiled eggs and ale, antibiotics that cure and poisons that kill, the origin and existence of life on earth—all these raise questions concerning the rates of chemical reactions. In everyday life as in the most recondite syntheses of organic chemistry, the principles of chemical kinetics apply.

In this chapter we shall describe how reaction rates are measured and quantitatively recorded. In Chapter 15, we shall delve into the theoretical interpretation of rates in terms of structures and interactions of molecules. Chemical kinetics can be divided into two parts, homogeneous and heterogeneous. Homogeneous reactions occur entirely within a single phase, whereas heterogeneous reactions occur at an interface between two phases. Another way of dividing the subject of kinetics is based on the source of the energy that activates the reacting molecules. This can be thermal energy—the random motions of molecules—in which case we speak of *thermal* reactions. Absorption of electromagnetic radiation by molecules may cause *photochemical* reactions. Reactions caused by high-energy particles such as alpha and beta rays from radioactive substances are the subject of *radiation chemistry*. Electric energy can cause reactions at electrodes, the rates of which are studied in *electrochemical kinetics*. In this chapter we consider only thermal reactions, reserving the other varieties for later treatment.

13.1 The Rate of Chemical Change

The *rate of reaction* v_R is defined as

$$v_R \equiv d\xi/dt \qquad (13.1)$$

where ξ is the *extent of reaction* (Section 1.8). The dimensions of rate of reaction are (amount of substance)/(time), in SI, mol s^{-1}.

For the general reaction $aA + bB \rightarrow cC + dD$, a stoichiometric equation may be written as

$$cC + dD - aA - bB = 0$$

This equation is based on the usual convention: products positive, reactants negative. The rate of this reaction is

$$v_R \equiv \frac{d\xi}{dt} = -\frac{1}{a}\frac{dn_A}{dt} = -\frac{1}{b}\frac{dn_B}{dt} = \frac{1}{c}\frac{dn_C}{dt} = \frac{1}{d}\frac{dn_D}{dt} \tag{13.2}$$

where n_A, n_B, n_C, and n_D are the amounts of A, B, C, and D in the reacting system at any time.

If the volume V of the system is constant, the reaction rate is expressed in terms of concentrations, $c_j = n_j/V$. The Eq. (13.2) becomes

$$v_R = \frac{d\xi}{dt} = -\frac{V}{a}\frac{dc_A}{dt} = -\frac{V}{b}\frac{dc_B}{dt} = \frac{V}{c}\frac{dc_C}{dt} = \frac{V}{d}\frac{dc_D}{dt} \tag{13.3}$$

In such systems of constant volume, the rate of reaction per unit volume, v_R/V, is often called simply the "rate of reaction."

In 1850, Ludwig Wilhelmy published the first important research on the kinetics of a chemical reaction. He measured the hydrolysis of sucrose in aqueous solutions of various acids:

$$H_2O + C_{12}H_{22}O_{11} = C_6H_{12}O_6 + C_6H_{12}O_6$$

sucrose   glucose   fructose

Wilhelmy found that the rate of decrease in concentration of sucrose c with time t is proportional to the concentration of sucrose remaining unconverted. The rate law becomes

$$-\frac{dc}{dt} = k_1 c \tag{13.4}$$

The proportionality constant k_1 is called the *rate constant* of the reaction. The value of k_1 depends on the acid used and increases with the concentration of acid. Since the acid does not appear in the stoichiometric equation for the reaction, it is acting as a *catalyst*, increasing the rate of reaction without being consumed itself.

Wilhelmy rearranged Eq. (13.4) as $-(1/c)\,dc/dt = -d\ln c/dt = k_1$, and integrated it to obtain $\ln c = -k_1 t + \text{const}$. At $t = 0$, the concentration of sucrose has its initial value $c = c_0$, so that const. $= \ln c_0$. Hence

$$\ln c = \ln c_0 - k_1 t, \qquad \ln(c/c_0) = -k_1 t$$

$$c = c_0 e^{-k_1 t} \tag{13.5}$$

The experimental concentration of sucrose closely follows this exponential decrease with time.

To measure the rate of a chemical reaction, we must follow the concentration of a reactant or product as a function of time. The best methods of analysis are those based on measurements that do not require removal of successive samples from the reaction mixture. For example, Wilhelmy used the change in optical rotation of sugar solutions. Other physical methods include (1) absorption spectra; (2) measurement of dielectric constant; (3) measurement of refractive index; (4) dilatometric methods, based on the change in volume due to reaction; and (5) change in pressure in some gas reactions.

Flow systems are often used for study of rapid reactions. The *stopped-flow method* has been extensively applied to reactions in solution, especially enzymatic reactions. An example of a stopped-flow apparatus for a gas-phase reaction is shown in Fig. 13.1. It was designed to study $2NO_2 + O_3 = N_2O_5 + O_2$ under conditions in which reaction is complete within 0.1 s. A stream of $O_2 + NO_2$ is mixed with a stream of $O_3 + O_2$ in a chamber with tangential jets. After mixing, which is complete within 0.01 s, a magnetically operated steel gate traps a portion of the gas mixture. The disappearance of NO_2, a brown gas, is followed by the change of intensity of a beam of transmitted light.

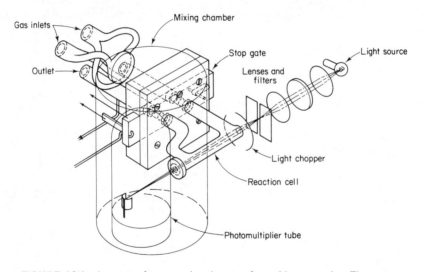

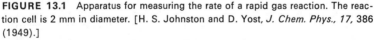

FIGURE 13.1   Apparatus for measuring the rate of a rapid gas reaction. The reaction cell is 2 mm in diameter. [H. S. Johnston and D. Yost, *J. Chem. Phys., 17,* 386 (1949).]

For reactions much faster than the millisecond range, various relaxation methods are available; they will be discussed in Section 13.14. These methods avoid the problem of initial mixing by starting the observations after a sharp initial disturbance of an existing equilibrium condition; for example, a jump in temperature might be the disturbance.

Let us consider the rate of a reaction as measured by the decrease in the concentration c_A of a particular reactant A. Concentrations are often denoted by square brackets, thus: $[A] \equiv c_A$. In many cases, as in the hydrolysis of sucrose, the rate $-dc_A/dt$ is found to be proportional to the concentration of A at any instant during the reaction. Then,

$$-dc_A/dt = k_1 c_A \qquad \text{or} \qquad -d[A]/dt = k_1[A]$$

Such an equation is called a *rate law*. The exponent of c_A in this rate law is unity. This exponent is called the *order of the reaction with respect to A*. In this instance, the reaction is *first order* with respect to A and k_1 is a *first order rate constant*. The hydrolysis of sucrose as studied by Wilhelmy is first order with respect to sucrose. First-order reactions are common. Another example is the gas-phase reaction $2N_2O_5 = 4NO_2 + O_2$. The rate law is $-d[N_2O_5]/dt = k_1[N_2O_5]$.

Another common case is a reaction rate that is proportional to the product of the concentrations of two different reactants. For example,

$$-d[B]/dt = k_2[B][C]$$

In this case we say that the reaction is first order with respect to B, first order with respect to C, and second order overall. Reaction rates of this type were first studied by Harcourt and Esson in 1865, in the oxidation of oxalic acid by permanganate:

$$2MnO_4^- + 5H_2C_2O_4 + 6H^+ = 2Mn^{2+} + 10CO_2 + 8H_2O$$

They found that $-d[MnO_4^-]/dt = k_2[MnO_4^-][H_2C_2O_4]$.

Second-order reactions may also follow a rate law of the type

$$-d[X]/dt = k_2[X]^2$$

Such a reaction is second order with respect to X. An example is the thermal decomposition of nitrogen dioxide, $2NO_2 = 2NO + O_2$. The rate law is found to be

$$-d[NO_2]/dt = k_2[NO_2]^2.$$

First- and second-order reactions are special cases of a more general form of reaction-rate law,

$$\frac{-d[A]}{dt} = k_r[A]^n[B]^m[C]^\ell \qquad (13.6)$$

The overall order of the reaction is the sum of the exponents in this equation, $n + m + \ell$. The order with respect to a given substance is the exponent of its particular concentration term. The exponents $n, m, \ell, \ldots$ need not be integers. For instance, the rate law for the decomposition of acetaldehyde, $CH_3CHO = CH_4 + CO$, is

$$\frac{-d[CH_3CHO]}{dt} = k_r[CH_3CHO]^{3/2}$$

The order of this reaction is $\frac{3}{2}$. The order of a reaction is determined solely by the best fit between a *rate equation* and the empirical data. In the decomposition of CH_3CHO, the best experimental order might be, for instance, 1.53 and not exactly 1.5.

Some rate laws do not have the form of Eq. (13.6) (see, for example, Section 13.16). In such cases, the concept of reaction order may not be applicable.

It is important to realize that there is no necessary connection between the form of the *stoichiometric equation* for the reaction and the kinetic order. For instance, the decompositions of N_2O_5 and of NO_2 have stoichiometric equations of similar forms, yet one follows first-order and the other, second-order kinetics.

The units of the rate constant depend on the order of the reaction. For first order, $-dc/dt = k_1c$; hence the usual units of k_1 are $(mol\ dm^{-3}\ s^{-1})/(mol\ dm^{-3}) = s^{-1}$. For second order, $-dc/dt = k_2c^2$; the units of k_2 are $(mol\ dm^{-3}\ s^{-1})/(mol\ dm^{-3})^2 = mol^{-1}\ dm^3\ s^{-1}$, often written $M^{-1}\ s^{-1}$. In general, for a reaction of the nth order, the dimensions of the constant k_n are $(time)^{-1}\ (concentration)^{1-n}$.

13.4   Reduced Rate Constants

Sometimes it may be convenient to express rate equations in concentrations relative to concentrations in a standard state. The standard state is usually at $c° = 1$ mol dm^{-3}. Equation (13.6) becomes

$$-d(c_A/c°)/dt = k_r'(c_A/c°)^n(c_B/c°)^m(c_C/c°)^l \tag{13.7}$$

Then k_r' has dimensions $(time)^{-1}[SI,\ s^{-1}]$, irrespective of the order of reaction.

Rate constants defined by Eq. (13.7) are called *reduced rate constants*. They can be related to equilibrium constants. Consider, for example, the reversible reaction, $A + B \rightleftharpoons C$. The forward rate is $-d(c_A/c°)/dt = k_f'(c_A/c°)(c_B/c°)$. The backward rate is $d(c_A/c°)/dt = k_b'(c_C/c°)$. At equilibrium, $dc_A/dt = 0$, and $k_f'(c_A/c°)(c_B/c°) = k_b'(c_C/c°)$. Thus

$$\frac{(c_A/c°)(c_B/c°)}{c_C/c°} = \frac{k_f'}{k_b'} = K_c$$

The equilibrium constant K_c is dimensionless, as it must be (page 159).

13.5   Reaction Molecularity and Reaction Order

Many chemical reactions are not kinetically simple; they proceed through a number of steps between initial reactants and final products. Each of the individual steps is called an *elementary reaction*. Complex reactions are made up of a sequence of elementary reactions.

In the earlier literature, the terms *unimolecular*, *bimolecular*, and *trimolecular* were used to denote reactions of the first, second, and third orders. We now apply the concept of *molecularity* only to elementary reactions. The molecularity indicates how many molecules of reactants are involved in the elementary reaction. For example, careful studies have been made of the reaction $NO + O_3 = NO_2 + O_2$. When an NO molecule strikes an O_3 with sufficient kinetic energy, it can capture an O atom, thus completing the reaction. This elementary reaction involves two molecules and it is therefore called a *bimolecular* reaction.

At the molecular level, a chemical reaction is a rearrangement of the chemical bonds of reactant molecules to form the chemical bonds of product molecules. It is necessary to change the energy state of a reactant molecule to allow the original bonds to change over to the new bonds. For example, consider the reaction $HCl + Br_2 = HBr + BrCl$. One H—Cl bond and one Br—Br bond are transformed into one H—Br bond and one Br—Cl bond.

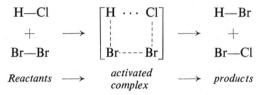

The transition state between reactants and products is called the *activated complex*.

The *molecularity* of a reaction can be defined as the number of molecules of reactants that are used to form the activated complex. The word "molecule" is used in its general sense to include atoms and ions also. In the examples of $NO + O_3$ and $HCl + Br_2$, the complex is formed from two molecules, and the reactions are bimolecular. Clearly, the molecularity of a reaction must be a whole number, and, in fact, is always found to be one, two, or occasionally three.

Experimental measurements show that the rate law for the reaction of NO with O_3 is $-d[NO]/dt = k_2[NO][O_3]$. This reaction is therefore second order. All bimolecular reactions are second order, but the converse is not true; many second-order reactions are not bimolecular.

Chemical reactions that are unimolecular are either isomerizations or decompositions. The isomerization of cyclopropane to propene in the gas phase is one of the most carefully studied unimolecular reactions:

$$\begin{matrix} & CH_2 & \\ & \diagup \quad \diagdown & \\ CH_2 & - & CH_2 \end{matrix} \longrightarrow CH_3—CH=CH_2$$

Experimental measurements show that $-d[\text{cyclopropane}]/dt = k_1[\text{cyclopropane}]$. This reaction is therefore first order. All unimolecular reactions are first order, but the converse is not true; some first-order reactions are not unimolecular.

13.6 Reaction Mechanisms

Two meanings for the term *reaction mechanism* are in common use. In one sense, *reaction mechanism* means the particular *sequence of elementary reactions* that gives the overall chemical change whose kinetics are under study. In the second sense, *reaction mechanism* means the detailed analysis of how chemical bonds (or nuclei and electrons) in the reactants rearrange to form the activated complex. For the present, we shall understand the mechanism of a reaction to be established when we have found a sequence of elementary reactions that explains the observed kinetic behavior. Each of these elementary reactions itself has a definite detailed mechanism, in terms of the motions of nuclei and electrons.

Consider, for example, the gas reaction, $2O_3 \rightarrow 3O_2$. We cannot predict the kinetic law which this reaction follows simply by looking at its stoichiometric equation. Experiment shows that the rate law is $-d[O_3]/dt = k_a[O_3]^2/[O_2]$. With this information, we can suggest a reasonable *mechanism:*

$$O_3 \underset{k'_{-1}}{\overset{k'_1}{\rightleftharpoons}} O_2 + O, \qquad O + O_3 \xrightarrow{k_2} 2O_2$$

The reversible dissociation is assumed to be rapid, leading to an equilibrium concentration of oxygen atoms,

$$\frac{[O_2][O]}{[O_3]} = K \qquad \text{or} \qquad [O] = \frac{K[O_3]}{[O_2]}$$

where $K = k'_1/k'_{-1}$. Then the slower second step gives the net rate of decomposition of O_3,

$$\frac{-d[O_3]}{dt} = k_2[O][O_3] = \frac{k_2 K[O_3]^2}{[O_2]}$$

Thus the proposed mechanism allows us to derive the observed rate law. This agreement does not prove that the mechanism is correct. It is a necessary but not a sufficient condition for its correctness.

Once a mechanism has been found that is consistent with the kinetics observed, it can be tested in various ways. One might measure independently the individual reaction rates and equilibrium constants involved, to see whether their predicted relation is confirmed. In the ozone decomposition, for instance, $k_a = k_2 K$. One could measure or calculate K and measure k_2 by introducing oxygen atoms of known concentration into ozone.

13.7 First-Order Rate Equations

Consider a reaction $A \rightarrow B + C$ in a system of constant volume V at temperature T. Let the initial concentration of A be a. If after a time t, a concentration x of A has decomposed, the remaining concentration of A is $a - x$, and a concentration x of B or C has been formed. The rate of formation of B or C is thus dx/dt. For a first-order reaction, this rate is proportional to the instantaneous concentration of A, so that

$$\frac{dx}{dt} = k_1(a - x) \tag{13.8}$$

Separating the variables and integrating, we obtain

$$-\ln(a - x) = k_1 t + C$$

where C is the constant of integration. The usual initial condition is that $x = 0$ at $t = 0$, whence $C = -\ln a$, and the integrated equation becomes

$$\ln \frac{a}{a - x} = k_1 t, \qquad x = a(1 - e^{-k_1 t}) \tag{13.9}$$

If $\ln [a/(a - x)]$ is plotted against t, a straight line passing through the origin should be obtained if the reaction kinetics is first order. The slope of the line is the first-order rate constant k_1.

[Integrate Eq. (13.8) between limits x_1 at t_1 and x_2 at t_2.]

Applications of these equations to the first-order decomposition of gaseous N_2O_5 are shown in Table 13.1 and Fig. 13.2. As this reaction proceeds, the pressure

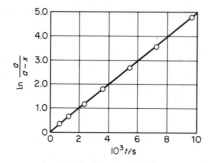

FIGURE 13.2 A first-order reaction, the thermal decomposition of nitrogen pentoxide (N_2O_5). The data are plotted in accord with Eq. (13.9).

in a closed system increases. From the increase in pressure, the amounts of undecomposed N_2O_5 can be calculated, and hence the partial pressure of N_2O_5 at various times. These are the $P(N_2O_5)$ given in Table 13.1. For an ideal gas $P = cRT$. Figure 13.2 is a plot of $\ln [a/(a - x)] = \ln [P_0/(P_0 - P)]$, where P_0 is the initial pressure of N_2O_5 and P is the partial pressure listed in Table 13.1. The data yield an excellent straight line, showing that this reaction follows the first-order rate law of Eq. (13.9). The slope of the line is $k_1 = 4.8 \times 10^{-4} \text{ s}^{-1}$.

TABLE 13.1
DECOMPOSITION OF DINITROGEN PENTOXIDE (T = 318.2 K)

Time, t (s)	$P_{N_2O_5}$ (kPa)	k_1 (s^{-1})	Time, t (s)	$P_{N_2O_5}$ (kPa)	k_1 (s^{-1})
0	46.4		4200	5.9	0.000478
600	32.9		4800	4.4	0.000475
1200	24.7	0.000481	5400	3.2	0.000501
1800	18.7	0.000462	6000	2.4	0.000451
2400	14.0	0.000478	7200	1.3	0.000515
3000	10.4	0.000493	8400	0.67	0.000590
3600	7.7	0.000484	9600	0.40	0.000467
			∞	0	

Another test of a first-order reaction is found in its *half-life* τ, the time required to reduce the concentration of reactant to half its initial value. In Eq. (13.9), when $x = a/2$, $t = \tau$, and

$$\tau = \frac{\ln 2}{k_1} \qquad (13.10)$$

Thus the half-life of a first-order reaction is independent of the initial concentration of reactant. In a first-order reaction, it would take just as long to reduce the reactant concentration from 0.1 to 0.05 M as it would to reduce it from 10 to 5 M.

Example 13.1 The decomposition, $SO_2Cl_2 = SO_2 + Cl_2$, is first order with a rate constant $k_1 = 2.20 \times 10^{-5}$ s^{-1} at 593 K. What percent of a sample of SO_2Cl_2 is decomposed by heating at 593 K for 1.50 h?

From Eq. (13.9), $x/a = 1 - e^{-k_1 t}$. With $k_1 t = (2.20 \times 10^{-5}$ s$^{-1})(1.50 \times 3600$ s$) = 0.118$, $x/a = 0.111$, so that 11.1% is decomposed.

13.8 Second-Order Rate Equations

Consider a reaction $A + B \longrightarrow C + D$. Let the initial concentrations at $t = 0$ be a of A and b of B. After a time t, concentrations x of A and x of B will have reacted, giving concentrations x of C and of D. If a second-order rate law is followed,

$$\frac{dx}{dt} = k_2(a - x)(b - x) \tag{13.11}$$

Separating the variables, we have

$$\frac{dx}{(a - x)(b - x)} = k_2\, dt$$

The expression on the left is integrated by breaking it into partial fractions, giving

$$\frac{\ln(a - x) - \ln(b - x)}{a - b} = k_2 t + C$$

where C is the constant of integration. When $t = 0$, $x = 0$, and hence $C = \ln(a/b)/(a - b)$. Therefore, the integrated second-order rate law is

$$\frac{1}{a - b} \ln \frac{b(a - x)}{a(b - x)} = k_2 t \tag{13.12}$$

An example of a reaction with second-order kinetics is that between ethylene bromide and potassium iodide in 99% methanol,

$$C_2H_4Br_2 + 3KI = C_2H_4 + 2KBr + KI_3$$

(Note that the stoichiometric equation in this case bears no relation to the order of reaction.) Sealed bulbs containing the reaction mixture were kept in a thermostat. At intervals of 2 or 3 min, a bulb was withdrawn, and its contents were analyzed for I_2 ($= KI_3$) by thiosulfate titration. If a is the initial concentration of $C_2H_4Br_2$ and b is that of KI, the second-order rate law is $d[I_2]/dt = dx/dt = k_2[C_2H_4Br_2][KI] = k_2(a - x)(b - 3x)$. This equation can be rewritten in the standard form of Eq. (13.11) as $dx/dt = 3k_2(a - x)[b/3 - x]$. The integrated equation from Eq. (13.12) is

$$\frac{1}{3a - b} \ln \frac{b(a - x)}{a(b - 3x)} = k_2 t$$

[Please prove this.]

Figure 13.3 is a plot of the left side of this equation against time. The excellent straight line confirms the second-order rate law. The slope of the line is the rate constant, $k_2 = 0.299 \text{ M}^{-1} \text{ min}^{-1}$ at 333 K.

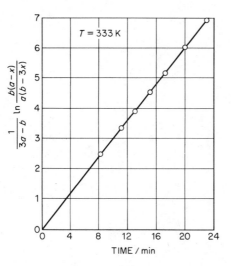

FIGURE 13.3 A second-order reaction, $C_2H_4Br_2 + 3KI = C_2H_4 + 2KBr + KI_3$, data at 333 K. [R. T. Dillon, *J. Am. Chem. Soc., 54*, 952 (1932)]

A special case of the general second-order equation (13.11) arises when the initial concentrations of both reactants are the same, $a = b$. This condition can be purposely arranged in any case, but it will necessarily be true whenever only one reactant is involved in a second-order reaction. An example is the decomposition of gaseous hydrogen iodide, $2HI = H_2 + I_2$, which follows the rate law, $-d[HI]/dt = k_2[HI]^2$.

In this case, the integrated equation (13.12) cannot be applied, since when $a = b$, it reduces to $k_2 t = 0/0$, which is indeterminate. It is best to return to the differential equation, which becomes $dx/dt = k_2(a - x)^2$. Integration then yields $1/(a - x) = k_2 t + C$. When $t = 0$, $x = 0$, so that $C = a^{-1}$. The integrated rate law is, therefore,

$$\frac{x}{a(a - x)} = k_2 t \qquad (13.13)$$

The half-life τ of the second-order decomposition of a single reactant is found from Eq. (13.13) by setting $x = a/2$ when $t = \tau$, so that

$$\tau = 1/k_2 a \qquad (13.14)$$

The half-life varies inversely as the initial concentration, whereas for a first-order process, the half-life is independent of initial concentration. [A first-order reaction $A \rightarrow B$ and a second-order reaction $2A \rightarrow C$ have the same half-life. Will they also have the same "quarter-life" when $A = [A]_0/4$?]

Example 13.2 The reaction $2HI = H_2 + I_2$ is second-order with a rate constant k_2 $= 1.20 \times 10^{-3}$ M^{-1} s^{-1} at 700 K. A sample of HI at a pressure of 100 kPa is maintained at 700 K. How long will it take to decompose 40.0% of the original HI?

From Eq. (13.13) $t = x/k_2a(a - x)$; $t = 0.4a/k_2a(0.6)a = 4/6k_2a$. We must calculate a in mol dm^{-3} from the pressure given. The gas will not deviate greatly from ideality. Hence, from $PV = nRT$, $c = a = n/V = P/RT = (100 \times 10^3$ Pa$)/$ $(8.314$ J K^{-1} mol$^{-1})(10^3$ dm^3 m$^{-3})(700$ K$) = 0.0172$ mol dm^{-3}, and

$$t = 4/6 (1.20 \times 10^{-3}) (0.0172) = 32.2 \times 10^3 \text{ s} = 8.97 \text{ h}.$$

13.9 Determination of Reaction Order

In simple reactions of first or second order, it is easy to establish the order and evaluate the rate constants. Graphical methods that may give linear plots are useful, but quantitative calculations of rate constants are best made by a simple least-squares program on a calculator.

In more complicated reactions, other methods can provide a survey of the kinetics. For a sufficiently slow reaction, the initial reaction rate dx/dt can be found with some precision before there has been any extensive chemical change in the system. It is then possible to assume that all the reactant concentrations are still effectively constant at their initial values. If $A + B + C \rightarrow$ products, and the initial concentrations are a, b, and c, the rate can be written quite generally as

$$\frac{dx}{dt} = k(a - x)^{n_1}(b - x)^{n_2}(c - x)^{n_3}$$

The initial rate, when x is very small, is $dx/dt = ka^{n_1}b^{n_2}c^{n_3}$. While we keep b and c constant, the initial concentration a can be varied, and the resultant change of the initial rate measured. In this way, the value of n_1 is estimated. Similarly, by keeping a and c constant while we vary b, a value of n_2 is found; and with a and b constant variation of c yields n_3. Biochemical reactions catalyzed by enzymes are usually studied by this initial-rate method.

If one reactant occurs in large excess, its concentration may be effectively constant during the measured course of the reaction. This situation often occurs for reactions in solution when one reactant is the solvent. For example, in the hydrolysis of ethyl acetate, $CH_3COOC_2H_5 + H_2O = CH_3COOH + C_2H_5OH$, the ester concentration is usually much lower than that of the solvent, water. The reaction follows a rate law that is first order with respect to ester, and the effectively constant concentration of water does not appear in the rate equation:

$$-d[CH_3COOC_2H_5]/dt = k_2[CH_3COOC_2H_5][H_2O] = k_1[CH_3COOC_2H_5]$$

Such a reaction is called a *pseudo-first-order reaction*. The hydrolysis of sucrose studied by Wilhelmy is a pseudo-first-order reaction of this kind.

In many reactions, the position of equilibrium is so far on the product side that for all practical purposes one can say that the reaction goes to completion. This is the case in the N_2O_5 decomposition and the oxidation of iodide ion that have been described. In other cases, however, a considerable concentration of reactants may remain when equilibrium is reached, for example, in the hydrolysis of ethyl acetate. In such instances, as the product concentrations increase, the rate of the reverse reaction becomes appreciable. The measured net rate of change is thereby decreased, and to deduce a rate equation to fit the empirical data, we must take into consideration the two *opposing reactions*.

For opposing first-order reactions, $A \rightleftharpoons B$, let the first-order rate constant in the forward direction be k_1, in the reverse, k_{-1}. Initially, at $t = 0$, the concentration of A is a, and that of B is b. If, after time t, a concentration x of A has been transformed into B, the concentration of A is then $a - x$, and that of B is $b + x$. The differential rate equation is, therefore,

$$\frac{dx}{dt} = k_1(a - x) - k_{-1}(b + x)$$

or

$$\frac{dx}{dt} = (k_1 + k_{-1})(r - x)$$

where $r = (k_1 a - k_{-1} b)/(k_1 + k_{-1})$. Integration yields

$$-\ln (r - x) = (k_1 + k_{-1})t + C$$

When $t = 0$, $x = 0$, so that $C = -\ln r$. Thus

$$\ln \frac{r}{r - x} = (k_1 + k_{-1})t \tag{13.15}$$

From the Guldberg and Waage Principle, the equilibrium constant $K = k_1/k_{-1}$. Thus equilibrium measurements can be combined with rate data to separate the forward and reverse rate constants in Eq. (13.15).

Example 13.3 In the reversible optical isomerization D-$CH_3CHBrCH_2CH_3 \rightleftharpoons$ L-$CH_3CHBrCH_2CH_3$, both forward and reverse reactions are first order with half-lives of 10.0 h at 400 K. If one starts with 1.000 mol of the D-bromide, how much will be left after 10.0 h at 400 K?

From Eq. (13.10), $k_1 = \ln 2/\tau = 0.0693$ h^{-1}. From Eq. (13.15), $\ln [r/(r - x)] = 1.386$. Now $r = 0.500$ mol and thus $x = 0.375$ mol. Thus 0.625 mol of D-bromide is left.

The case of opposing second-order reactions was first treated by Max Bodenstein in his classic study of the combination of hydrogen and iodine,

$$H_2 + I_2 \underset{k_{-2}}{\overset{k_2}{\rightleftharpoons}} 2HI$$

Between 523 and 773 K, the forward reaction can be conveniently studied, but at higher temperatures equilibrium lies too far on the reactant side. Even in the temperature range cited, we must consider the reverse reaction to obtain satisfactory rate constants. The rate constants are shown in Table 13.2, with values of $K = k_2/k_{-2}$.

TABLE 13.2
RATE CONSTANTS FOR THE REACTION $H_2 + I_2 \rightleftharpoons 2HI$

T (K)	$dm^3\ mol^{-1}\ s^{-1}$		$K = k_2/k_{-2}$
	k_2	k_{-2}	
300	2.04×10^{-19}	2.24×10^{-22}	912
400	6.61×10^{-12}	2.46×10^{-14}	371
500	2.14×10^{-7}	1.66×10^{-9}	129
600	2.14×10^{-4}	2.75×10^{-6}	77.8
700	3.02×10^{-2}	5.50×10^{-4}	54.9

13.11 Consecutive Reactions

It often happens that the product of one reaction becomes the reactant in a following reaction. There may be a series of consecutive steps. Such kinetics is especially important in polymerization and depolymerization processes. Only in the simplest cases has it been possible to obtain analytic solutions to the differential equations of these reaction systems. With modern computers, however, any of these sequential reaction schemes can be integrated numerically for the parameters and times of interest.

A simple consecutive-reaction scheme that can be treated exactly is one involving only irreversible first-order steps. The general case of n steps has been solved, but only the example of two steps will be discussed. This can be written

$$A \xrightarrow{k_1} B \xrightarrow{k_1'} C$$
$$\quad x \qquad y \qquad z$$

The simultaneous differential equations are

$$-\frac{dx}{dt} = k_1 x, \qquad -\frac{dy}{dt} = -k_1 x + k_1' y, \qquad \frac{dz}{dt} = k_1' y$$

The first equation can be integrated directly, giving $-\ln x = k_1 t + C$. When $t = 0$, let $x = a$, the initial concentration of A. Then $C = -\ln a$, and $x = ae^{-k_1 t}$. The concentration of A declines exponentially with time, as in any first-order reaction. Substitution of the value found for x into the second equation gives

$$\frac{dy}{dt} = -k_1' y + k_1 ae^{-k_1 t}$$

This is a linear differential equation of the first order, with the solution*

$$y = e^{-k_1't}\left[\frac{k_1a\,e^{(k_1'-k_1)t}}{k_1' - k_1} + C\right]$$

When $t = 0$, $y = 0$, so that $C = -k_1a/(k_1' - k_1)$.

We now have expressions for x and y. In the reaction sequence, there is no change in the total number of molecules, since every time an A disappears, a B appears, and every time a B disappears, a C appears. Thus $x + y + z = a$, and z is calculated to be

$$z = a\left(1 - \frac{k_1'e^{-k_1t}}{k_1' - k_1} + \frac{k_1e^{-k_1't}}{k_1' - k_1}\right) \tag{13.16}$$

In Fig. 13.4, the concentrations x, y, and z are plotted as functions of time for the case $k_1 = 2k_1'$. The intermediate concentration y rises to a maximum and then falls asymptotically to zero, while the final product rises gradually to the value of a.

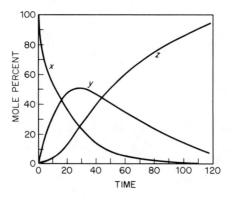

FIGURE 13.4 Concentration changes with time in a sequence of first-order reactions $x \to y \to z$.

Such a reaction sequence was found in the thermal decomposition (pyrolysis) of acetone,

$$(CH_3)_2CO \longrightarrow CH_2{=}CO + CH_4, \qquad CH_2{=}CO \longrightarrow \tfrac{1}{2}C_2H_4 + CO$$

The concentration of the intermediate, ketene $CH_2{=}CO$, rises to a maximum and then declines during the course of the reaction.

13.12 Parallel Reactions

Sometimes a given substance can react in more than one way. Then alternative parallel reactions must be considered in the analysis of the kinetic data. Consider a pair of parallel first-order reactions,

$$A \xrightarrow{k_1} B, \qquad A \xrightarrow{k_2} C$$

In the case of such parallel processes, the most rapid rate determines the predominant path of the overall reaction. If $k_1 \gg k_2$, the decomposition of A will yield mostly B.

*W. A. Granville, P. F. Smith, and W. R. Longley, *Elements of Calculus* (Boston: Ginn and Company, 1957), p. 380.

For example, primary alcohols can be either dehydrated to olefins or dehydrogenated to aldehydes,

$$C_2H_5OH \xrightarrow{k_1} C_2H_4 + H_2O, \qquad C_2H_5OH \xrightarrow{k_2} CH_3CHO + H_2$$

By suitable choice of catalyst and temperature, one rate can be made much faster than the other. In such a case, the composition of the mixture of reaction products depends on the relative rates of the parallel reactions and not on their equilibrium constants.

13.13 Chemical Relaxation

The application of relaxation methods to the study of rapid chemical reactions in solutions was initiated by Manfred Eigen about 1950. The basic idea is to subject a reaction system that is already at equilibrium to a sharp variation in some physical parameter on which the value of the equilibrium constant K is dependent. The system then must change to a new state of chemical equilibrium, and the rate of this change, called *relaxation*, can be followed. The usual parameters that have been applied in relaxation studies are temperature, pressure, and electric field. An apparatus for the temperature (T)-jump method is sketched in Fig. 13.5. The sample volume is usually about 1 cm³, and the T-jump is produced by discharge of a capacitor through the reaction mixture, with an input of about 50 kJ in 1 μs (a power of 5×10^7 W).

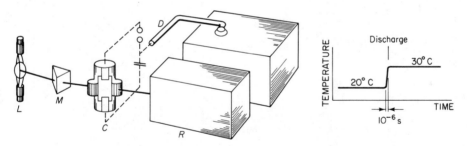

FIGURE 13.5 Apparatus for the T-jump method. An initial sharp displacement from equilibrium is followed by a relaxation to a new equilibrium. *L*, light source, *M*, monochromator; *C*, reaction cell; *D*, spark gap; *R*, detector and recorder. [Manfred Eigen, Max Planck Institut, Göttingen]

Provided that the displacement from equilibrium is small enough, the rate of restoration of equilibrium (rate of relaxation) will always follow first-order kinetics, irrespective of the kinetics of the forward and reverse reactions. Thus, if Δx_0 is the initial ($t = 0$) displacement of some quantity x that describes the composition of the mixture, and Δx is the value of this displacement at any time t after the initial disturbance,

$$d(\Delta x)/dt = \frac{\Delta x}{\tau} \qquad \text{and} \qquad \Delta x = \Delta x_0 e^{-t/\tau} \qquad (13.17)$$

where τ is the chemical *relaxation time* for the reaction system. The relations of τ to the rate constants of the system have been worked out for most cases of interest.* Two examples will be given.

Consider a reversible first-order reaction,

$$A \underset{k_{-1}}{\overset{k_1}{\rightleftharpoons}} B$$

Let a be the concentration of A + B, and x the concentration of B at any time. Then, $dx/dt = k_1(a - x) - k_{-1}x$. If x_e is the equilibrium concentration, $\Delta x = x - x_e$ or $x = \Delta x + x_e$. Since $d(\Delta x)/dt = dx/dt$, we obtain

$$\frac{d(\Delta x)}{dt} = k_1(a - \Delta x - x_e) - k_{-1}(\Delta x + x_e) \tag{13.18}$$

At equilibrium, however, $dx/dt = 0 = k_1(a - x_e) - k_{-1}x_e$. We use this equation to eliminate x_e from Eq. (13.18). Thus Eq. (13.18) becomes $d(\Delta x)/dt = -(k_1 + k_{-1})\Delta x$, so that, from Eq. (13.17) one obtains

$$\tau = (k_1 + k_{-1})^{-1} \tag{13.19}$$

A somewhat more complicated case is a first-order forward reaction with a second-order reverse,

$$A \underset{k_2}{\overset{k_1}{\rightleftharpoons}} B + C$$

This type includes the ionization of a weak acid in a large excess of solvent, for instance, $CH_3COOH + H_2O \rightleftharpoons CH_3COO^- + H_3O^+$. Let a be the concentration of CH_3-$COOH + CH_3COO^-$, and x the concentration of H_3O^+ (equal to that of CH_3COO^-). Including the constant water concentration in k_1, we can write $dx/dt = k_1(a - x) - k_2x^2$. Therefore, with $\Delta x = x - x_e$,

$$\frac{d(\Delta x)}{dt} = k_1(a - x_e - \Delta x) - k_2(x_e + \Delta x)^2 \tag{13.20}$$

At equilibrium,

$$\frac{dx}{dt} = 0 = k_1(a - x_e) - k_2x_e^2 \tag{13.21}$$

Where departure from equilibrium Δx is very small, the term in $(\Delta x)^2$ in Eq. (13.20) can be neglected compared to those in Δx. We use Eq. (13.21) to eliminate a from Eq. (13.20) to obtain

$$\frac{d(\Delta x)}{dt} = -(k_1 + 2k_2x_e)\,\Delta x$$

Then, from Eq. (13.17), we find that

$$\tau = (k_1 + 2k_2x_e)^{-1} \tag{13.22}$$

If we combine Eq. (13.22) with $K = k_1/k_2$, we can obtain the forward and reverse rate constants for the ionization reaction from τ and the equilibrium constant K.

*G. H. Czerlinski, *Chemical Relaxation* (New York: Marcel Dekker, Inc., 1966).

Example 13.4 A study by the T-jump method gave a relaxation time $\tau = 40$ μs at 25°C for $H_2O = H^+ + OH^-$. Calculate k_1 and k_2 for the reaction. At 25°C, $K_w = [H^+][OH^-] = 10^{-14}$ (mol dm^{-3})^2.

$K = [H^+][OH^-]/[H_2O] = 10^{-14}$ M^2/55.5 M $= 1.8 \times 10^{-16}$ M. From Eq. (13.21), $\tau = 40 \times 10^{-6}$ s $= (1.8 \times 10^{-16} k_2 + 2k_2 \times 10^{-7})^{-1}$. Thus $k_2 = 1.4 \times 10^{11}$ M^{-1} s^{-1} and $k_1 = Kk_2 = 2.5 \times 10^{-5}$ s^{-1}.

Table 13.3 lists a few of many interesting results that have been obtained by various relaxation methods. Experimental details can be found in the original papers. With present techniques, relaxation times from about 1 to 10^{-12} s can be measured. The shortest times are approaching the periods of molecular vibrations and rotations.

TABLE 13.3
EXPERIMENTAL RATE CONSTANTS FOR RAPID REACTIONS
IN AQUEOUS SOLUTIONS[a]

Reaction	T (K)	Method	Reference	k_2 (dm^3 mol^{-1} s^{-1})
$H^+ + HS^- \longrightarrow H_2S$	298	Electric field	(1)	7.5×10^{10}
$H^+ + N(CH_3)_3 \longrightarrow N(CH_3)_3H^+$	298	NMR	(2)	2.5×10^{10}
$H^+ + (NH_3)_5CoOH^{2+} \longrightarrow$ $H_2O + (NH_3)_5Co^{3+}$	285	T-jump	(3)	1.4×10^9
$H^+ + AlOH^{2+} \longrightarrow H_2O + Al^{3+}$	298	Electric field	(4)	3.8×10^9
$H^+ + OH^- \longrightarrow H_2O$	298	T-jump	(5)	1.5×10^{11}

[a]References are as follows: (1) M. Eigen and K. Kustin, *J. Am. Chem. Soc., 82,* 5952 (1960); (2) E. Grunwald et al., *J. Chem. Phys., 33,* 556 (1960); (3) M. Eigen and W. Kruse, *Z. Naturforsch., 186,* 857 (1963); (4) L. P. Holmes, D. L. Cole, and E. M. Eyring, *J. Phys. Chem., 72,* 301 (1968); (5) M. Eigen, *Discussions Faraday Soc., 17,* 194 (1954).

13.14 Reactions in Flow Systems

All the rate equations discussed so far apply to *static systems*, in which the reaction mixture is enclosed in a vessel at constant volume and temperature. Most industrial reactions, however, are carried out in *flow systems**, in which reactants enter continuously at the inlet of a reaction vessel, while the product mixture is withdrawn at the outlet. We shall describe two examples of flow systems: (1) a reactor in which there is no stirring; (2) a reactor in which complete mixing is effected at all times by vigorous stirring.†

*K. G. Denbigh, *Chemical Reactor Theory* (Cambridge, England: Cambridge University Press, 1965).

†W. C. Herndon, *J. Chem. Educ., 41,* 425 (1964).

Figure 13.6 shows a tubular reactor through which the reaction mixture passes at a volume rate of flow u (e.g., in $m^3 s^{-1}$). Let us consider an element of volume dV sliced out of this tube, and focus attention on one particular component κ, which enters this volume element at a concentration c_κ and leaves at $c_\kappa + dc_\kappa$. If there is no longitudinal mixing, the net change with time of the amount of κ within dV, (dn_κ/dt), will be the sum of two terms, one due to chemical reaction within dV and the other equal to the excess of κ entering dV over that leaving. Thus

$$\frac{dn_\kappa}{dt} = r_\kappa \, dV - u \, dc_\kappa \tag{13.23}$$

The chemical reaction rate *per unit volume* is denoted by r_κ. The explicit form of r_κ is determined by the rate law for the reaction: for a reaction of first order with respect to κ, $r_\kappa = -k_1 c_\kappa$; for second order, $r_\kappa = -k_2 c_\kappa^2$; and so on.

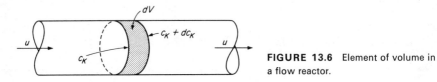

FIGURE 13.6 Element of volume in a flow reactor.

After reaction in the flow system has continued for some time, a *steady state* is attained, in which the number of moles of each component in any volume element no longer changes with time, the net flow into the element exactly balancing the reaction within it. Then $dn_\kappa/dt = 0$, and Eq. (13.23) becomes

$$r_\kappa \, dV - u \, dc_\kappa = 0 \tag{13.24}$$

After r_κ is introduced as a function of c_κ, the equation can be integrated. For example, with $r_\kappa = -k_1 c_\kappa$, $-k_1(dV/u) = dc_\kappa/c_\kappa$. The integration is carried out between the inlet and the outlet of the reactor.

$$\frac{-k_1}{u} \int_0^{V_0} dV = \int_{c_{\kappa1}}^{c_{\kappa2}} \frac{dc_\kappa}{c_\kappa}$$

$$-k_1 \frac{V_0}{u} = \ln \frac{c_{\kappa2}}{c_{\kappa1}} \tag{13.25}$$

The total volume of the reactor is V_0, and $c_{\kappa2}$ and $c_{\kappa1}$ are the concentrations of κ at the outlet and inlet, respectively.

Equation (13.25) reduces to the integrated rate law for a first-order reaction in a static system if the time t is substituted for V_0/u. The quantity V_0/u is called the *contact time* for the reaction; it is the average time that a molecule would take to pass through the reactor. Thus Eq. (13.25) allows us to evaluate the rate constant k_1 from a knowledge of the contact time and of the concentrations of any reactant species at the inlet and outlet of the reactor. For other reaction orders also, the correct equation for a flow reactor is obtained by substituting V_0/u for t in the equation for the static system. Many reactions that are too swift for convenient study in a static system can be followed readily in a flow system, in which the contact time is reduced by use of a high flow rate and a small volume.

The derivation of Eq. (13.24) tacitly assumed that there was no volume change

ΔV as a result of the reaction. If $\Delta V \neq 0$, the flow rate at constant pressure would not be constant. In liquid-flow systems, effects due to ΔV are generally negligible, but for gaseous systems, the form of the rate equations is considerably modified when $\Delta V \neq 0$.

Example 13.5 The reaction $NO + Cl_2 = NOCl + Cl$ is second order with $k_2 = 8.0 \times 10^5$ dm³ mol⁻¹ s⁻¹ at 1200 K. A stream of $NO + Cl_2$ in equimolar amounts in an inert carrier gas is passed through a tubular furnace 1.0 cm radius and 300 cm long at 1200 K, 101 kPa pressure, at a rate of 0.010 m³ s⁻¹. If Cl_2 is 1.0% of gas stream at entry to furnace, what will be the percent of Cl in the gas stream at exit?

The contact time is

$$\frac{V_0}{u} = \frac{\pi (0.010)^2 (0.30) \text{ m}^3}{0.010 \text{ m}^3 \text{ s}^{-1}} = 0.94 \times 10^{-2} \text{ s}$$

The initial concentration a of reactants, from $PV = nRT$ is

$$c = n/V = P/RT = \frac{(10^{-2})(101) \, 10^3 \text{ Pa}}{8.314 \text{ J K}^{-1} \text{ mol}^{-1} \times 1200 \text{ K}}$$

$$= 0.10 \text{ mol m}^{-3} = 1.0 \times 10^{-4} \text{ mol dm}^{-3}$$

From Eq. (13.13),

$$\frac{x}{1.0 \times 10^{-4}(1.0 \times 10^{-4} - x)} = (8.0 \times 10^5)(0.94 \times 10^{-2}) = 7.5 \times 10^3 \text{ dm}^3 \text{ mol}^{-1}$$

$$x = 0.43 \times 10^{-4} \text{ mol dm}^{-3}$$

Thus the percent of Cl at exit is 0.43%.

An example of a stirred-flow reactor is shown in Fig. 13.7. The reactants enter the vessel at A, and stirring at high speed (about 3000 rpm) effects mixing within about 1 s. The outflow of product mixture at B exactly balances the feed. After a steady state is attained, the composition of the mixture in the reactor remains unchanged as long as the composition and rate of supply of reactants is unchanged. An equation like Eq. (13.24) still applies, but in this case dV becomes V_0, the total reactor volume, and dc_K becomes $c_{K2} - c_{K1}$, where c_{K1} and c_{K2} are the initial and final

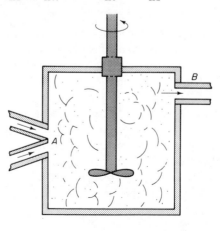

FIGURE 13.7 A stirred-flow reactor. Reactants enter at A and a reaction mixture having a steady-state composition is withdrawn at B.

concentrations of reactant κ. Thus

$$r_K = \frac{u}{V_0}(c_{K2} - c_{K1}) = \frac{dc_K}{dt} \tag{13.26}$$

With this method, there is no need to integrate the rate equation. One point on the rate curve is obtained from each steady-state measurement, and a number of runs with different feed rates and initial concentrations is required to determine the order of the reaction.

An important application of the stirred-flow reactor is the study of transient intermediates, the concentration of which in a static system might quickly reach a maximum value and then fall to zero (as was shown in Fig. 13.4). For example, when Fe^{3+} is added to $Na_2S_2O_3$, a violet color appears, which fades within 1 or 2 min. In a stirred-flow reactor, the conditions can be adjusted so that the color is maintained, and the violet intermediate, which appears to be $FeS_2O_3^+$, can be studied in a leisurely way by absorption spectrometry.

There is a basic analogy between a living cell and a continuous stirred-flow reactor, and the same theoretical analysis can often be applied to chemical kinetics in both. In the case of the cell, there is no obvious internal stirrer, but distances from one part to any other part are usually small so that mixing by diffusion should be adequate to maintain the "well-stirred" condition. For a cell diameter of 10^{-3} cm, the mean time for a small molecule to diffuse across the cell would be about 10^{-1} s. The cell does not have a definite inlet and outlet like a reactor, but the entire cell wall serves these functions.

13.15 Stationary States and Dissipative Processes

In a static (or batch) reaction vessel, the concentrations of the reactants and products change with time until an equilibrium condition is attained. At constant T and P, this equilibrium represents the minimum in Gibbs free energy G with respect to fractional conversion of reactants to products. In a static system, the change in composition occurs in the time coordinate, whereas in a flow system the change in composition is shifted to a space coordinate, for example along the axis of a tubular flow reactor. If the rate of feed to such a flow system is constant, the composition as a function of the appropriate space coordinate will generally approach a steady-state value. This steady-state conversion will not be at the minimum of G but at some other value determined by the flow rates and rate constants of the system. These flow systems are open systems (page 8) in the thermodynamic sense, and hence their time-invariant states are not equilibrium states but stationary states.

Chemical kinetics systems can sometimes give rise to remarkable phenomena, in which localized order arises from chaos. The system maintains itself in a steady state of high order at the expense of a *dissipation of the energy* provided by the chemical reaction. The reactor is similar in this respect to a living system—a localized region of order that maintains itself by feeding on the free-energy stores of its environment.

For example, an ordered structure in a dissipative system may occur when malonic acid reacts in an oxidizing solution made up from Ce^{3+} and BrO_3^- ions. The reacting solution first displays temporal oscillations, passing periodically from a red color, indicating an excess of Ce^{3+}, to a blue color, indicating an excess of Ce^{4+}. These oscillations do not occur at the same instant in all parts of the solution; they begin at one point and propagate in all directions at various rates. After a variable number of oscillations, a small region of inhomogeneous concentration appears, from which layers of coloration, alternately red and blue, proceed one by one until they fill the tube. The evolution of these structures is shown in Fig. 13.8. These dissipative structures appear only when the system is reacting far from equilibrium. As the reaction approaches equilibrium the colored layers disappear and the solution again becomes homogeneous.

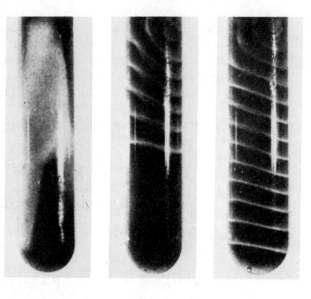

FIGURE 13.8 Development of a spatially ordered structure in an initially homogeneous chemical reaction mixture far from equilibrium. The lighter bands are blue regions with excess Ce^{4+} and the darker bands are red regions with excess Ce^{3+}. [Marcelle Herschkowitz, *C. R. Acad. Sci. (Paris)* 270, 1049 (1970)]

13.16 Chain Reactions: Formation of Hydrogen Bromide

After Bodenstein completed his study of the hydrogen–iodine reaction, he turned to $H_2 + Br_2 = 2HBr$, probably expecting to find another example of second-order kinetics. The results* were surprisingly different, for the reaction rate was found to fit a rather complicated expression,

$$\frac{d[\text{HBr}]}{dt} = \frac{k_a[\text{H}_2][\text{Br}_2]^{1/2}}{k_b + [\text{HBr}]/[\text{Br}_2]}$$

where k_a and k_b are constants. Notice that [HBr] occurs in the denominator of the rate expression, so that the rate is inhibited by the product HBr.

There was no interpretation of this curious rate law for 13 years. Then the prob-

*M. Bodenstein and S. C. Lind, *Z. Phys. Chem.*, **57**, 168 (1906).

lem was solved independently and almost simultaneously by Christiansen, Herzfeld, and Polanyi. They proposed a chain of reactions with the following steps:

$$\text{Chain initiation:} \qquad (1) \ Br_2 \xrightarrow{k_1} 2Br$$

$$\text{Chain propagation:} \ (2) \ Br + H_2 \xrightarrow{k_2} HBr + H$$

$$(3) \ H + Br_2 \xrightarrow{k_3} HBr + Br$$

$$\text{Chain inhibition:} \qquad (4) \ H + HBr \xrightarrow{k_4} H_2 + Br$$

$$\text{Chain breaking:} \qquad (5) \ 2Br \xrightarrow{k_5} Br_2$$

The reaction is initiated by bromine atoms from the thermal dissociation $Br_2 \rightarrow 2Br$. The chain propagating steps (2) and (3) form two molecules of HBr and regenerate the bromine atom, ready for another cycle. Thus a few bromine atoms suffice to cause an extensive reaction. Step (4) is introduced to account for the observed inhibition by HBr; since this inhibition is proportional to the ratio $[HBr]/[Br_2]$, it is evident that HBr and Br_2 compete, so that the atom being removed must be H rather than Br.

The atoms Br and H are always present in concentrations that are low compared to those of the stable species H_2, Br_2 and HBr. Except toward the beginning and end of reaction, we can assume that the atoms are in *steady-state* concentrations, in which their rate of formation equals their rate of disappearance. This assumption leads to the following equations:

$$\frac{d[Br]}{dt} = 0 = 2k_1[Br_2] - k_2[Br][H_2] + k_3[H][Br_2] + k_4[H][HBr] - 2k_5[Br]^2$$

$$\frac{d[H]}{dt} = 0 = k_2[Br][H_2] - k_3[H][Br_2] - k_4[H][HBr]$$

These two simultaneous equations are solved for the steady-state concentrations of the atoms, giving

$$[Br] = \left[\frac{k_1}{k_5}[Br_2]\right]^{1/2}, \qquad [H] = k_2 \frac{(k_1/k_5)^{1/2}[H_2][Br_2]^{1/2}}{k_3[Br_2] + k_4[HBr]}$$

The rate of formation of the product, HBr, is

$$\frac{d[HBr]}{dt} = k_2[Br][H_2] + k_3[H][Br_2] - k_4[H][HBr]$$

Introducing the expressions for [H] and [Br] and rearranging, we find that

$$\frac{d[HBr]}{dt} = 2\frac{k_3 k_2 k_4^{-1} k_1^{1/2} k_5^{-1/2}[H_2][Br_2]^{1/2}}{k_3 k_4^{-1} + [HBr][Br_2]^{-1}}$$

This agrees exactly with the empirical expression, but now the constants k_a and k_b are interpreted as composites of constants for step reactions in the chain.

For many years the reaction $H_2 + I_2 \rightarrow 2HI$ and the reverse reaction $2HI \rightarrow H_2 + I_2$ were considered to be purely bimolecular processes, but at higher temperatures part of the reaction has a chain mechanism. In 1967, Sullivan* showed that the

*J. H. Sullivan, *J. Chem. Phys.*, **46**, 73 (1967).

nonchain part of the $H_2 + I_2$ reaction also occurs with iodine atoms as intermediates:

$$I_2 \xrightleftharpoons{k} 2I, \qquad H_2 + 2I \xrightarrow{k_3} 2HI$$

[Show that this mechanism gives $d[HI]/dt = 2k_3 K[H_2][I_2]$.]

13.17 Free-radical Chains

In 1934, Frank Rice and Karl Herzfeld* showed that free-radical chain mechanisms could be devised for the thermal decomposition (pyrolysis) of many organic compounds. A typical example is the mechanism for decomposition of acetaldehyde, $CH_3CHO = CH_4 + CO$:

$$\text{(1)} \quad CH_3CHO \xrightarrow{k_1} CH_3 + CHO$$

$$\text{(2)} \quad CH_3CHO + CH_3 \xrightarrow{k_2} CH_4 + CO + CH_3$$

$$\text{(3)} \quad 2CH_3 \xrightarrow{k_3} C_2H_6$$

One primary split to give a methyl radical can result in the decomposition of many CH_3CHO molecules, since the chain carrier, CH_3, is regenerated in step (2). The steady-state treatment of the CH_3 concentration yields

$$\frac{d[CH_3]}{dt} = 0 = k_1[CH_3CHO] - 2k_3[CH_3]^2$$

so that $[CH_3] = [k_1/2k_3]^{1/2}[CH_3CHO]^{1/2}$. The reaction rate based on methane formation is then

$$\frac{d[CH_4]}{dt} = k_2[CH_3][CH_3CHO] = k_2\left(\frac{k_1}{2k_3}\right)^{1/2}[CH_3CHO]^{3/2}$$

Sometimes a test for a radical mechanism can be made by studying a mixture of isotopically substituted species. Suppose, for example, that we heat a mixture of CH_3CHO and CD_3CDO. If an *intramolecular* mechanism is followed, $CH_3CHO \rightarrow CH_4 + CO$, $CD_3CDO \rightarrow CD_4 + CO$. We should then obtain a mixture of CH_4 and CD_4 in the products. If the chain mechanism is followed, we should obtain also CH_3D and CD_3H from the steps $CH_3 + CD_3CDO \rightarrow CH_3D + CO + CD_3$ and $CD_3 + CH_3CHO \rightarrow CD_3H + CO + CH_3$. Actually, the isotopically mixed methanes are found, so that the radical mechanism is indicated.

Many industrial polymers, such as polyethylene and polystyrene, are produced by chain reactions with radical intermediates. The polymerization mechanism always includes three basic steps: (1) chain initiation, (2) chain propagation, and (3) chain termination. The initial radicals are usually produced from an initiator I, a rather unstable compound that decomposes thermally to yield a free radical R: $I \rightarrow R + M$. This R then adds to a double bond in the unsaturated monomer to yield a larger radical. For example, with C_2H_4,

*J. Am. Chem. Soc., 56, 284 (1934).

$$R + CH_2{=}CH_2 \longrightarrow R{-}CH_2{-}CH_2{-}$$

The new radical in turn can add to a new molecule of monomer,

$$R{-}CH_2{-}CH_2{-} + CH_2{=}CH_2 \longrightarrow R{-}CH_2{-}CH_2{-}CH_2{-}CH_2{-}$$

Chain propagation continues, building up a long linear polymer, until termination occurs, usually when two radical chain carriers react. [What products are formed when two $R{\cdot}CH_2{\cdot}CH_2{-}$ radicals react?]

13.18 Branching Chains—Explosive Reactions

The formation of H_2O from H_2 and O_2 displays the upper and lower pressure limits characteristic of many explosions, as shown in Fig. 13.9. If the pressure of a 2:1 mixture of H_2 and O_2 is kept below the lower line on the diagram, the thermal reaction proceeds slowly. At a temperature of 750 K, this lower pressure limit is shown at 240 Pa, but its value depends on the size of the reaction vessel and the material of its walls. If the pressure is raised above this value, the mixture explodes. As the pressure is raised still further, there is a limit of 3.3 kPa at 750 K above which there is no longer an explosion, but once again a comparatively slow reaction. This second explosion limit is strongly temperature dependent, but it does not vary with vessel size.

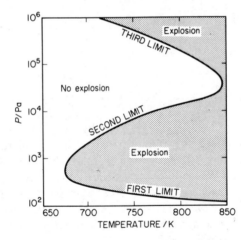

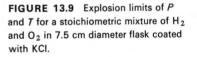

FIGURE 13.9 Explosion limits of P and T for a stoichiometric mixture of H_2 and O_2 in 7.5 cm diameter flask coated with KCl.

If an exothermic reaction is carried out in a confined space, the heat evolved often cannot be dissipated. The temperature therefore increases, so that the rate of reaction increases and there is a corresponding rise in the rate of heat production. The reaction velocity increases practically without bound and the result is called a *thermal explosion*. The third explosion limit in Fig. 13.9 arises in this way.

Chain reactions with reactive atomic and free-radical intermediates can lead to explosions if *branching chains* can occur, i.e., if one reactive chain carrier can yield more than one reaction carrier at some step in the reaction. For example, in the

$H_2 + O_2$ reaction, $H_2 + O \rightarrow OH + H$ would be such a step. The low-pressure explosion limit in chemical explosions is caused by diffusion of reactive chain carriers to the wall of the reaction vessel, where they are destroyed. The high-pressure limit is caused by destruction of chain carriers in the gas phase. (A nonchemical instance of explosion due to branching of chains is in a nuclear-fission bomb, in which a neutron is absorbed by a ^{235}U nucleus, causing it to undergo fission and releasing several neutrons in the process.)

13.19 How Reaction Rates Depend on Temperature

The effect of temperature on chemical reaction rates is of practical and theoretical importance. When an actual reaction is carried out, the choice of the proper temperature determines the success or failure of the operation. Here we shall consider how rates depend on temperature, and in Chapter 15 consider theories of reaction rates that explain this temperature dependence.

The van't Hoff equation for the temperature coefficient of the equilibrium constant in terms of concentrations K_c is

$$\frac{d \ln K_c}{dT} = \frac{\Delta U}{RT^2} \qquad (13.27)$$

where ΔU is the internal energy of the products minus that of the reactants. The Law of Mass Action of Guldberg and Waage indicates that $K_c = k'_f/k'_b$, where k'_f, k'_b are the reduced rate constants of forward and reverse reactions (Section 13.4). Based on these concepts, Svante Arrhenius in 1889 suggested that a reasonable equation for a rate constant k_r would be

$$\frac{d \ln k_r}{dT} = \frac{E_a}{RT^2} \qquad (13.28)$$

The parameter E_a is called the *activation energy* of the reaction.

An illustration of the relation between E_a and ΔU for an elementary reaction can be seen in Fig. 13.10. We see that ΔU is the difference between the activation energy

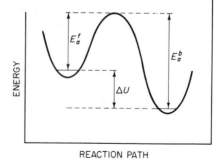

FIGURE 13.10 Reaction profile to illustrate the concept of energy of activation, E_a. The difference between E_a in foward and backward direction is the ΔU of reaction.

E_a^f of the forward reaction and the activation energy E_a^b of the backward reaction:

$$\Delta U = E_a^f - E_a^b \qquad (13.29)$$

The illustration in Fig. 13.10 is sometimes called a "reaction profile." It should be considered as a graphic device and not as the plot of a precise mathematical relationship. We shall see later (Chapter 15), however, that the concept of reaction coordinate can be defined more precisely.

If E_a does not depend on temperature, Eq. (13.28) gives on integration

$$\ln k_r = \frac{-E_a}{RT} - \ln A \qquad (13.30)$$

where $\ln A$ is the constant of integration. Integration between two limits gives

$$\ln [k_r(T_2)/k_r(T_1)] = \frac{E_a}{R} \left(\frac{1}{T_1} - \frac{1}{T_2} \right) \qquad (13.31)$$

Equation (13.30) is equivalent to

$$k_r = A \exp(-E_a/RT) \qquad (13.32)$$

Here A is called the *frequency factor* or *preexponential factor*. Equation (13.32) is the famous Arrhenius equation for the rate constant. The Arrhenius equation is not exact but it is usually a close approximation for the temperature dependence of a rate constant, provided that the range of temperature is not too large.

Equation (13.30) predicts that a plot of the logarithm of the rate constant $\ln k_r$ vs. the reciprocal of the absolute temperature T^{-1} should be a straight line. The slope of the line is $-E_a/R$ and the intercept of the line extrapolated to $T^{-1} = 0$ gives $\ln A$. Note that k_r and A must have the same units.

Experimental data from the thermal rearrangement of isopropenylallyl ether are reproduced in Fig. 13.11(a). The slopes of the lines yield the first-order rate constants k_1 for the reaction at the different temperatures. In Fig. 13.11(b), these k_1 are plotted in accord with the Arrhenius equation as $\ln k_1$ vs. T^{-1}. In this instance, the slope of the best straight line is -1.475×10^4 K, so that the activation energy $E_a = (8.314$ J K^{-1} mol^{-1}) $(14\,750$ K$) = 122\,600$ J mol^{-1}. The intercept is at $\ln A = 27.0$, so that $A = 5.4 \times 10^{11}$ s^{-1}. For this reaction, therefore, the Arrhenius equation (13.32) is $k_1 = 5.4 \times 10^{11} e^{-14\,750/T}$. The fact that the data points in Fig. 13.11(b) fall on a straight line implies that E_a is a constant independent of temperature. In some reactions, however, a curvature of the Arrhenius plot can be detected.

According to Arrhenius, Eq. (13.32) indicated that molecules must attain a certain critical energy E_a before they can react, and $e^{-E_a/RT}$ is the Boltzmann factor giving the fraction of the molecules that do achieve the critical energy. This interpretation is still essentially correct, but a more detailed theory will be discussed in Chapter 15. We shall see also that the pre-exponential factor A is related to the frequency of collisions between molecules in gas-phase bimolecular reactions, and to a molecular vibration frequency in unimolecular reactions. For these reasons A is called the *frequency factor*.

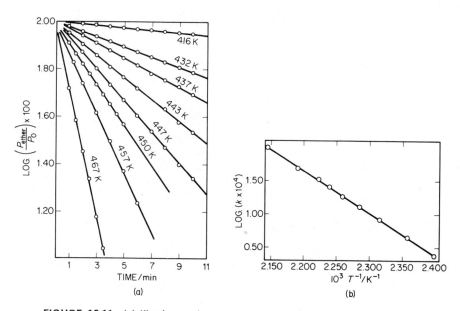

FIGURE 13.11 (a) Kinetic experiments over a range of temperatures on the rearrangement of isopropenyl allyl ether to allyl acetone.

$$CH_2{=}CH{-}CH_2{-}O{-}\underset{\underset{CH_2}{\|}}{C}{-}CH_3 \quad \rightarrow \quad CH_2{=}CH{-}CH_2{-}CH_2{-}C\underset{CH_3}{\overset{O}{\diagup\!\!\diagdown}}$$

(b) Plot of the rate constants as log k vs. T^{-1}. The straight line is in accord with the Arrhenius equation (13.30). [Reprinted with permission from L. Stein and G. W. Murphy, *J. Am. Chem. Soc.*, **74**, 1041. Copyright 1952 American Chemical Society.]

13.20 Examples of Temperature Dependence of Reaction Rates

Example 13.6 Many reactions that proceed at readily measurable rates at room temperature roughly double in rate for a 10-K rise in temperature. What value of E_a would correspond to such an effect?

From Eq. (13.31) for a doubled rate between 293 and 303 K:

$$\ln 2 = \frac{E_a}{R}\left(\frac{1}{293} - \frac{1}{303}\right)$$

$$E_a = \frac{8.314 \text{ J K}^{-1} \text{ mol}^{-1} \ln 2 \, (303 \text{ K})(293 \text{ K})}{10 \text{ K}} = 51\,200 \text{ J mol}^{-1} \approx 50 \text{ kJ mol}^{-1}$$

Even if one does not know the rate law for a chemical reaction, it is possible to use the Arrhenius expression to calculate the effect of temperature on reaction rate. Suppose that the times required to reach a certain extent of reaction at two different temperatures T_1 and T_2 are denoted by τ_1 and τ_2. Then,

$$\tau_1/\tau_2 = e^{-E_a/RT_2}/e^{-E_a/RT_1} \tag{13.33}$$

This result is based on the assumption that the kinetics of the reaction follow the same functional form at both temperatures and only the rate constants differ.

[Please derive Eq. (13.33).]

Example 13.7 The decomposition of cyclobutane to ethylene is one-third complete in 20 min at 717 K and in 40 min at 705 K. What is the activation energy of the reaction? How long would it take to go to one-third completion at 750 K?

From Eq. (13.33),

$$\ln(\tau_1/\tau_2) = \frac{E_a}{R}\left(\frac{1}{T_1} - \frac{1}{T_2}\right)$$

$$\ln 2 = \frac{E_a}{8.314 \text{ J K}^{-1} \text{ mol}^{-1}}\left(\frac{1}{705 \text{ K}} - \frac{1}{717 \text{ K}}\right); \qquad E_a = 240 \text{ kJ mol}^{-1}$$

At 717 K and 750 K,

$$\tau_2/\tau_3 = e^{-E_a/RT_3}/e^{-E_a/RT_2}$$

$$= e^{-38.97}/e^{-40.76} = 5.99$$

$$\tau_3 = 20 \text{ min}/5.99 = 3.3 \text{ min}$$

Problems

1. A first-order reaction and a second-order reaction have the same half-life τ and initial reactant concentration [A]. What are the reactant concentrations when $t = 3\tau$?

2. The reaction $C_2H_5NO_2 + OH^- = C_2H_4NO_2^- + H_2O$ was followed by measuring the OH^- concentration. The initial $[OH^-] = [C_2H_5NO_2] = 5.00 \times 10^{-3}$ M. After 5 min, $[OH^-] = 2.60 \times 10^{-3}$ M; 10 min, 1.70×10^{-3} M; 15 min, 1.30×10^{-3} M. Find the order of reaction and the rate constant by a graphical method.

3. The rate constant for decomposition of dimethyl ether is $k_1 = 4.00 \times 10^{-4}$ s^{-1} at 500°C. $(CH_3)_2O = CH_4 + H_2 + CO$. 1.40 g of $(CH_3)_2O$ is introduced into a 1000-cm^3 bulb at 500°C. Calculate the initial pressure and the pressure after 10 min.

4. The curie (Ci) is a unit of radioactivity equal to 3.700×10^{10} disintegrations per second. ^{60}Co has a half-life of 5.26 years. What would be the mass of ^{60}Co in a 10^5 Ci source?

5. If all reactants have initial concentration a and the reaction rate has order n, show that the half-life $\tau = (2^{n-1} - 1)/a^{n-1}(n-1)k_n$, where k_n is the rate constant.

6. The rate of reaction $2NO + 2H_2 = N_2 + 2H_2O$ was studied with equal amounts of NO and H_2 at different initial pressures.

P_0 (kPa)	50.0	47.2	38.4	32.4	26.9
$\tau_{1/2}$ (min)	95	81	140	176	224

Calculate the overall order of reaction. (See Problem 5.) Devise a mechanism that would give this order.

7. The reaction *cis*-stilbene $\longrightarrow$ *trans*-stilbene at 594 K follows Eq. (13.15).

Time (s)	0	630	1206	1806	∞
% *trans*	0	42.8	62.5	74.9	93.2

Calculate k_1, k_{-1}, and $K = k_1/k_{-1}$.

8. The dissociation constant of imidazole is 1.10×10^{-7} when concentrations are in mol dm^{-3}. After a sharp *T*-jump, the relaxation time is 6.50 μs. Calculate the rate constants for dissociation and for association. [Imidazole] $= 8.65 \times 10^{-5}$ mol dm^{-3}.

9. The thermal decomposition of cyclobutanone can take two alternative pathways: (a) to $C_2H_4 + H_2C{=}CO$; (b) to cyclopropane $+$ CO. In an experiment at 656 K and initial $[C_4H_6O] = 6.50 \times 10^{-5}$ M:

Time (min)	0.50	1.00	3.00	6.00
$10^5[C_2H_4]$ (M)	0.31	0.68	1.53	2.63
$10^7[C_3H_6]$ (M)	0.21	0.47	1.24	2.20

Calculate k_{1_a} and k_{1_b} for the two parallel first-order reactions. What is concentration of C_4H_6O after 12 min?

10. The decomposition of dimethyl ether has two sequential first-order steps: (a) $CH_3OCH_3 \longrightarrow CH_4 + HCHO$; (b) $HCHO \longrightarrow H_2 + CO$. At 770 K, $k_{1_a} = 8.5 \times 10^{-3}$ s^{-1}, and $k_{1_b} = 4.5 \times 10^{-2}$ s^{-1}. In an experiment with CH_3OCH_3 at initial concentration 1.0×10^{-4} mol/dm^3, what is maximum concentration of HCHO in the system, and at what time does it occur?

11. The isomerization of cyclopropane to propene is first order with $k_1 = 1.50 \times 10^{15}$ exp $(-32\,500/T)$ s^{-1}. Cyclopropane is passed through a heated tube (40 cm diameter) at 500°C at a rate of 1.0 dm^3 s^{-1}. How long must the tube be to yield 10% propene in the exit gases?

12. $NH_4OH \underset{k_{-1}}{\overset{k_1}{\rightleftharpoons}} NH_4^+ + OH^-$ was studied at 20°C by the *T*-jump method with initial $[NH_4OH] = 1.5 \times 10^{-4}$ M, and $[NH_4^+] = [OH^-] = 0.51 \times 10^{-4}$ M. The relaxation time $\tau = 3.7 \times 10^{-7}$ s. NH_4OH has dissociation constant $K_c = 1.8 \times 10^{-5}$. Calculate k_1 and k_{-1}.

13. A certain reaction is 25% complete in 18.0 min at 300 K and in 2.50 min at 350 K. Estimate its activation energy E_a.

14. For decomposition of N_2O_5:

T (K)	298	308	318	328	338
$10^5 k_1$ (s^{-1})	1.72	6.65	24.95	75	240

Calculate A and E_a in the expression $k_1 = Ae^{-E_a/RT}$.

15. The hydrolysis of mustard gas in water is a two-step reaction.

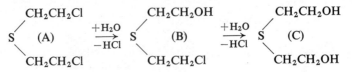

At 25°, the first-order rate constants are 1.96×10^{-3} s^{-1} and 2.77×10^{-3} s^{-1} respectively. If the initial concentration of $S(CH_2CH_2Cl)_2$ is 0.010 M, draw the concentrations of A, B, and C as functions of time.

16. For opposing second-order reactions,

$$A + B \underset{k_{-2}}{\overset{k_2}{\rightleftharpoons}} C + D$$

with $[A]_0 = [B]_0 = a$, show that

$$x/a = [1 + (k_{-2}/k_2)^{1/2} \coth{(at \sqrt{k_{-2}k_2})}]^{-1}$$

where $[C] = [D] = x$.

17. For $H_2 + I_2 \rightleftharpoons 2HI$ at 700 K, $k_2 = 30.2$ cm^3 mol^{-1} s^{-1} and $k_{-2} = 0.550$ cm^3 mol^{-1} s^{-1}. An equimolar mixture of H_2 and I_2 is heated for 10 min at 700 K and 10^2 kPa. Calculate the final concentration of [HI].

18. $C_6H_5N_2^+ + H_2O = C_6H_5OH + N_2 + H^+$

T (K)	288.16	292.99	297.80	303.07	308.13	313.20	317.99	322.98	328.20
$k_1 \times 10^5$ (s^{-1})	0.930	2.01	4.35	9.92	20.7	42.8	81.8	158	301

By the method of least squares, fit these data to an Arrhenius expression, $\ln k_1 = \ln A - E/RT$, and give the standard deviations in the values of A and E.

19. $Hg_2^{2+} + Tl^{3+} = 2Hg^{2+} + Tl^+$. The rate law is

$$-d[Tl^{3+}]/dt = \frac{k[Hg_2^{2-}][Tl^{3+}]}{[Hg^{2+}]}$$

Devise a mechanism that gives such a law. A rate law with an inverse dependence on a particular concentration occurs when the substance is the product of a reversible step prior to the rate-determining reaction.

20. A mechanism proposed for the first-order reaction $2N_2O_5 = 4NO_2 + O_2$ is [R. A. Ogg, *J. Chem. Phys.*, **18**, 572 (1950)]

$$N_2O_5 \xrightarrow{k_1} NO_2 + NO_3$$
$$NO_2 + NO_3 \xrightarrow{k_2} N_2O_5$$
$$NO_2 + NO_3 \xrightarrow{k_3} NO_2 + O_2 + NO$$
$$NO + N_2O_5 \xrightarrow{k_4} 3NO_2$$

Find the steady-state expressions for [NO] and [NO$_3$] and show that the overall first-order rate constant is $k_a = 2k_1k_3/(k_2 + k_3)$.

21. The reaction $2NO + O_2 \longrightarrow 2NO_2$ has a third-order rate constant $k_3 = 12.9 \times 10^{-39}$ (cm^3/molecule)^2 s^{-1} at 370 K. If the initial pressure of NO is 40 kPa, and of O_2,

20 kPa, what would be pressure in the reaction system after 10 min at 370 K? At 470 K, $k_3 = 9.04 \times 10^{-39}$. What sort of explanation would you suggest for this unusual temperature effect?

22. The kinetics of hydrolysis of $CH_3COOC_2H_5$ by OH^- was followed in a 500-cm^3 stirred-flow reactor at 25°C. Solutions of ethyl acetate [E] (0.0200 M) and KOH (0.0050 M) were introduced at flow rates of 1000 cm^3/h. The exit concentration of OH^- was 0.0011 M. The reaction follows the rate law $-d[E]/dt = k_2[E][OH^-]$. What was k_2 in this experiment?

23. The hydrolysis of methyl chloride was measured in aqueous solution at 313.3 K, $CH_3Cl + H_2O = CH_3OH + H^+ + Cl^-$.

$10^{-3}t$ (min)	0	6.80	12.62	19.75	24.10	31.30	37.06	41.34	47.10	54.21	∞
[Cl⁻](mmol L⁻¹)	0	4.22	7.02	10.40	12.44	15.46	18.00	19.44	21.64	23.40	42.92

Calculate the first-order rate constant k_1 and its standard deviation by a least-squares fit of the data to the theoretical equation.

24. $^{239}_{92}U$ decays to $^{239}_{93}Np$, which in turn decays to $^{239}_{94}Pu$. The half-lives are 23.5 min and 2.35 days, respectively. Plot the relative amounts of the three nuclides as functions of time in a sample of originally pure $^{239}_{92}U$.

25. Consider the polymerization of isoprene (M) initiated by benzoyl peroxide (I) Experiments show that the rate of polymerization is given by the expression

$$k[M][I]^{1/2}\left(\frac{k^*[M]}{1 + k^*[M]}\right)^{1/2}$$

where k and k^* are constants. Matheson postulated the following reaction mechanism:

$$(1) \quad I \xrightarrow{k_1} 2R\cdot$$

$$(2) \quad R\cdot + R\cdot \xrightarrow{k_2} I$$

$$(3) \quad R\cdot + M \xrightarrow{k_3} M\cdot$$

$$\begin{array}{ccc} \cdot & \cdot & \cdot \\ \cdot & \cdot & \cdot \\ \cdot & \cdot & \cdot \end{array}$$

$$(4) \quad M_n^{\cdot} + M \xrightarrow{k_4} M_{n+1}^{\cdot}$$

$$(5) \quad M_n^{\cdot} + M_r^{\cdot} \xrightarrow{k_5} \text{inactive products}$$

Step (2) represents the reaction between two "caged" free radicals, not the usual free-radical recombination process. As the presence of one caged free radical implies the presence of the other, Matheson wrote the rate as a *first-order* reaction:

$$\frac{-d[R\cdot]}{dt} = 2k_2[R\cdot]$$

Show that Matheson's scheme predicts the kinetics observed. Identify k and k^* in terms of this model. What order with respect to monomer is the rate of polymerization if $k_2 \gg k_3$?

26. Consider the solution polymerization of vinyl acetate in benzene at 60°C initiated by benzoyl peroxide. The initial concentrations of monomer and initiator are 0.50 M and 5.0×10^{-3} M, respectively. The initiator decomposition rate constant is $k_i = 2.0 \times$

10^{-6} s^{-1}. For vinyl acetate, $k_p = 2640$ dm^3 mol^{-1} s^{-1} and $k_t = 1.17 \times 10^8$ dm^3 mol^{-1} s^{-1} where k_p and k_t are the rate constants for propagation and termination. Termination is predominantly by disproportionation.

Calculate:

(a) The steady-state concentration of free radicals.

(b) The average number of monomer units added per second to each growing free radical initially.

(c) The average lifetime of a free radical in the early stages of polymerization.

(d) The average extent of polymerization,

$$\langle x \rangle = \frac{\text{rate of propogation}}{\text{rate of termination}}$$

Assume that all primary free radicals initiate polymerization.

14

Catalysis

The word *catalysis* was coined in 1835 by Berzelius, who wrote: "Catalysts are substances that by their mere presence evoke chemical reactions that would not otherwise take place." The idea of catalysis, however, extends far back into the early history of chemical processes. The philosopher's stone of the alchemists was a mythical catalyst that would heal the sick and turn base metals into gold.

In the laboratory and in chemical industry, catalysts are used to achieve a reasonable reaction rate for almost all reactions under practical conditions. The living cell depends on thousands of protein catalysts, the enzymes, to carry out metabolic processes at relatively low temperatures. Enzymes are remarkably specific in the reactions that they catalyze, but chemists are only beginning to understand how to design such highly specific catalysts.

14.1 Catalysts Influence Rate but not Equilibrium

A catalyst has been compared to a coin inserted in a slot machine that yields valuable products but also returns the coin. In a chemical reaction, a catalyst enters at one stage of the mechanism and leaves at another. The essence of catalysis is not the entering but the falling out.

Wilhelm Ostwald defined a catalyst as a "substance that changes the rate of a chemical reaction without itself appearing in the final products." He emphasized the fact that a catalyst changes the rate of a chemical reaction while it has no effect on the position of equilibrium. Ostwald gave a proof of this statement based on the First

Law of Thermodynamics. Consider a gas reaction that proceeds with a change in volume, for example $N_2 + 3H_2 = 2NH_3$. The gas mixture at equilibrium is confined in a cylinder fitted with a piston. Inside the cylinder the catalyst is kept in a small receptacle, which has a sliding lid so that the catalyst can be alternately exposed to the reacting gases or separated from them. If the equilibrium position of the reaction mixture were changed by its exposure to the catalyst, the pressure in the cylinder would change, and the piston would move. On covering the catalyst, the pressure would be restored to its previous level, and the piston would return to its original position. By alternately exposing and covering the catalyst, the piston would move up and down, and the system would function as a perpetual motion machine continually yielding work by a cyclic process, contrary to the First Law. Hence the catalyst cannot change the position of equilibrium.

Since a catalyst changes the rate of a reaction but not the equilibrium point, it must accelerate forward and reverse reactions by the same factor. Thus catalysts that accelerate hydrolysis of esters must also accelerate esterification of alcohols; dehydrogenation catalysts such as nickel and platinum are also good hydrogenation catalysts; enzymes, such as pepsin and papain, that catalyze splitting of the peptide bond must also catalyze its formation from the reaction products.

A catalyst usually provides an alternate reaction path having a lower activation energy E_a than the uncatalyzed reaction. This effect is shown schematically in Fig. 14.1. The new reaction path lowers the activation energies of forward and reverse reactions by the same amount ΔE_a. Although the usual catalytic effect is to lower E_a, catalysts can also alter the preexponential factor A in the expression for the rate constant, $k_2 = Ae^{-E_a/RT}$.

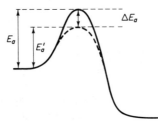

FIGURE 14.1 A catalyst can provide a new reaction path in which the activation energies of both forward and backward reactions are lowered by the same amount.

Three classes of catalysis can be distinguished. In *homogeneous catalysis* the entire reaction occurs within a single phase. In *heterogeneous catalysis* the catalyzed reaction occurs at the interface between two phases. Heterogeneous catalysis by solid catalysts is also called *contact catalysis*. In *phase-transfer catalysis*, the catalyst acts by transferring one reactant to the phase of a second reactant.

14.2 Homogeneous Catalysis in Gas Reactions

A catalyst works by first entering into a reaction with the reactant to form an intermediate complex. This complex then either reacts with a second reactant to yield the product, or decomposes directly to the product. In abstract form these mechanisms are as follows, where C denotes the catalyst:

(1) $A + C \longrightarrow (AC) \longrightarrow$ product $+ C$

(2) $A + C \longrightarrow (AC), \quad (AC) + B \longrightarrow$ product $+ C$

Various more complicated versions of these basic mechanisms also occur.

An example of the first type of mechanism is the catalysis by HBr of the gas-phase dehydration of tertiary butanol,

$$(CH_3)_3C\text{---}OH \longrightarrow (CH_3)_2C\text{=}CH_2 + H_2O$$

In the absence of catalyst the alcohol does not decompose at an appreciable rate below about 700 K. On addition of a small amount of HBr, rapid decomposition occurs at 500 K. The uncatalyzed reaction is first order, $-d[t\text{-BuOH}]/dt = k_1[t\text{-BuOH}]$, with a rate constant, $k_1 = 4.8 \times 10^{14} \exp(-32\,700/T)\,s^{-1}$. In the presence of HBr, the rate equation becomes $-d[t\text{-BuOH}]/dt = k_2[t\text{-BuOH}][\text{HBr}]$, with $k_2 = 9.2 \times 10^{12} \exp(-15\,200/T)\,s^{-1}\,dm^3\,mol^{-1}$. Note that the rate is proportional to the concentration of catalyst [HBr]. The catalyst in this case has markedly lowered the activation energy for the decomposition. A possible mechanism for this catalytic effect is based on the intermediate formation of t-BuBr as a rate-determining step, $t\text{-BuOH} + \text{HBr} \xrightarrow{k_2} t\text{-BuBr} + H_2O$. This is followed by rapid decomposition of t-BuBr, releasing the catalyst HBr for another reactive cycle, $t\text{-BuBr} \rightarrow (CH_3)_2C\text{=}CH_2 + \text{HBr}$.

A reaction of environmental importance is the catalysis by Cl atoms of the gas-phase decomposition of ozone. The reaction occurs as follows: $\text{Cl} + O_3 \rightarrow \text{ClO} + O_2$, $\text{ClO} + O \rightarrow \text{Cl} + O_2$. Oxygen atoms are formed by photochemical decomposition of O_3 or O_2 by ultraviolet light. In the stratosphere (10 to 50 km) the ozone concentration is only a few parts per million, but it plays an important role in absorbing ultraviolet radiation from the sun. The chlorofluorocarbons $CFCl_3$ and CF_2Cl_2 have been used as propellants in aerosol sprays; they are stable compounds and might diffuse into the stratosphere, where they would be photochemically dissociated to yield Cl atoms. Wide use of these sprays might eventually lead to a decrease in stratospheric ozone through the chlorine-catalyzed decomposition, and thus indirectly to a large increase in skin cancer due to unfiltered ultraviolet rays. As a precaution, the use of chlorinated propellants has been banned in some countries.

14.3   Acid–Base Catalysis

Most examples of homogeneous catalysis have been studied in liquid solutions. Catalysis by acids and bases governs the rates of many reactions in solution. The earliest studies in this field were by Kirchhoff in 1812 on the conversion of starch to glucose by the action of dilute acids, and by Thenard in 1818 on the decomposition of hydrogen peroxide in alkaline solutions. The classic investigation of Wilhelmy in 1850 dealt with the rate of inversion of cane sugar by acid catalysts.

The hydrolysis of esters, catalyzed by both acids and bases, was extensively studied in the latter half of the nineteenth century. The catalytic activity of an acid in these reactions became one of the accepted measures of acid strength. Table 14.1

TABLE 14.1
CATALYTIC CONSTANTS OF ACIDS

Acid	Relative conductivity	k_2'(ester)	k_2'(sugar)
HCl	100	100	100
HBr	101	98	111
HNO$_3$	99.6	92	100
H$_2$SO$_4$	65.1	73.9	73.2
CCl$_3$COOH	62.3	68.2	75.4
CHCl$_2$COOH	25.3	23.0	27.1
HCOOH	1.67	1.31	1.53
CH$_3$COOH	0.424	0.345	0.400

gives some of Ostwald's results on inversion of sucrose and hydrolysis of methyl acetate. If we write the acid as HA, these reactions are

$$C_{12}H_{22}O_{11} + H_2O + HA \longrightarrow C_6H_{12}O_6 + C_6H_{12}O_6 + HA$$

$$CH_3COOCH_3 + H_2O + HA \longrightarrow CH_3COOH + CH_3OH + HA$$

For the latter reaction, the rate can be written as

$$dx/dt = k'[CH_3COOCH_3][H_2O][HA]$$

Since water is present in large excess, its concentration is effectively constant. The rate therefore reduces to $dx/dt = k_2'[HA][CH_3COOCH_3]$. Now k_2' is called the *catalytic constant*. The values in Table 14.1 are all relative to 100 for k_2' with HCl. Ostwald and Arrhenius showed that the catalytic constant of an acid is proportional to its molar conductance. They concluded that the nature of the anion is unimportant, the only active catalyst being the hydrogen ion H^+.

In other reactions, however, it is necessary to consider the effect of the OH^- ion and also the rate of the uncatalyzed reaction. The result is a three-term equation for the observed rate constant, $k_2 = k_0 + k_{H^+}[H^+] + k_{OH^-}[OH^-]$. In aqueous solution $K_w = [H^+][OH^-]$, and k_2 becomes

$$k_2 = k_0 + k_{H^+}[H^+] + \frac{k_{OH^-}K_w}{[H^+]} \tag{14.1}$$

Since K_w is about 10^{-14} (mol/L)^2, in 0.1 M acid $[OH^-]$ is about 10^{-13} mol L^{-1} and in 0.1 M base $[OH^-]$ is 10^{-1} mol L^{-1}, there is a 10^{12}-fold change in $[OH^-]$ and $[H^+]$ in passing from dilute acid to dilute base. Therefore, OH^- catalysis is negligible in dilute acid and H^+ catalysis negligible in dilute base, except in the unusual event that the catalytic constants for H^+ and OH^- differ by as much as 10^{10}. By measurements in acid and basic solutions, it is therefore generally possible to evaluate k_{H^+} and k_{OH^-} separately.

Example 14.1 The decomposition of diacetone alcohol is catalyzed by OH^- ions.

$$(CH_3)_2 \cdot C(OH) \cdot CH_2 \cdot COCH_3 + OH^- = 2CH_3 \cdot CO \cdot CH_3 + OH^-$$

The rate constant is $k_2 = 1.31 \times 10^{11} \exp(-9070/T)$ dm^3 mol^{-1} s^{-1}. How long would it take to decompose half of the diacetone alcohol in a solution at 30°C and pH 10; at 20°C and pH 11?

At 30°C, $k_2 = 1.31 \times 10^{11} \exp(-29.9) = 1.35 \times 10^{-2}$ L mol^{-1} s^{-1}
At 20°C, $k_2 = 1.31 \times 10^{11} \exp(-31.0) = 4.51 \times 10^{-3}$ L mol^{-1} s^{-1}
At 30°C and pH 10, $k_2(OH^-) = k_2(10^{-4}) = 1.35 \times 10^{-6}$ s$^{-1} = k_1$
At 20°C and pH 11, $k_2(OH^-) = k_2(10^{-3}) = 4.51 \times 10^{-6}$ s$^{-1} = k_1$
(We have taken $K_w = [OH^-][H^+] = 10^{-14}$ and ignored its variation with T.) Then

$$\tau = \ln 2/k_1 = 0.513 \times 10^6 \text{ s} = 143 \text{ h at } 30°C, \text{ pH } 10$$

$$= 0.154 \times 10^6 \text{ s} = 42.7 \text{ h at } 20°C, \text{ pH } 11$$

The mechanism of specific catalysis by H_3O^+ or OH^- involves formation of an ionized intermediate, which then yields products. Ester hydrolysis in aqueous solution is catalyzed only by H_3O^+ and OH^-; no other acids or bases are effective.

14.4   General Acid–Base Catalysis

Brønsted and Lowry defined an acid as any molecule that can give up an H^+ ion, and a base on any molecule that can accept an H^+ ion. Thus other acids and bases, as well as H_3O^+ and OH^-, may be effective catalysts.

The essential feature of catalysis by an acid is the transfer of a proton from acid to *substrate* (substance whose reaction is being catalyzed). Catalysis by a base involves the acceptance of a proton from the substrate by the base. Thus in Brønsted–Lowry nomenclature, the substrate acts as a base in acid catalysis, or as an acid in base catalysis.

For example, the hydrolysis of nitramide is susceptible to base but not to acid catalysis, the mechanism being

$$NH_2NO_2 + OH^- = H_2O + NHNO_2^-, \qquad NHNO_2^- = N_2O + OH^-$$

Not only the OH^- ion but also other bases can act as catalysts, e.g., the acetate ion,

$$NH_2NO_2 + CH_3COO^- = CH_3COOH + NHNO_2^-$$

$$NHNO_2^- = N_2O + OH^-$$

$$OH^- + CH_3COOH = H_2O + CH_3COO^-$$

The reaction rate with different bases B is always $v_R = k_B(B)(NH_2NO_2)$.

Brønsted found that there is a relation between the catalytic constant k_B and the dissociation constant K_b of the base, namely

$$k_B = BK_b^\beta \quad \text{or} \quad \log k_B = \log B + \beta \log K_b \tag{14.2}$$

Here, B and β are constants for bases of a given charge type. Thus the stronger the base, the higher the catalytic constant. Figure 14.2 shows data on nitramide hydrolysis that are in good accord with the Brønsted relation (14.2).

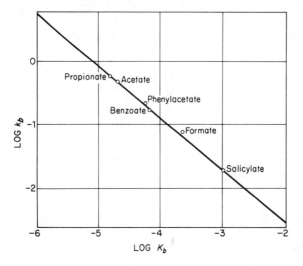

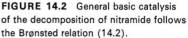

FIGURE 14.2 General basic catalysis of the decomposition of nitramide follows the Brønsted relation (14.2).

Nitramide hydrolysis displays general basic catalysis. Other reactions provide examples of general acid catalysis, with a relation like Eq. (14.2), $k_A = BK_a^\alpha$, where K_a is the dissociation constant of the acid. Some reactions occur with both general acid and general basic catalysis.

Since a solvent such as water can act as either an acid or a base, it is often itself a catalyst. What was formerly believed to be the uncatalyzed reaction, represented by k_0 in Eq. (14.1), is in most cases a reaction catalyzed by the solvent acting as acid or base.

14.5   Catalysis by Enzymes

Enzymes are specific catalysts that are produced by living cells. It is estimated that a typical mammalian cell contains 10^4 different enzymes. H. Büchner, in 1897, was the first to establish that intact cells are not necessary for many of the reactions of physiological chemistry, which can be carried out *in vitro* in cell-free filtrates. All known enzymes are proteins. They range in molar mass from about 10^4 to 2×10^6 g mol^{-1}. The diameters of equivalent spherical molecules would range from 3 to 30 nm, but the actual shapes are often not spherical.

Enzymes are specific in their catalytic actions. Urease catalyzes the hydrolysis of urea, $(NH_2)_2CO$, in dilutions as high as 1 part of enzyme in 10^7 of solution, yet it has no detectable effect on the hydrolysis rate of substituted ureas, such as methyl urea, $(NH_2)(CH_3NH)CO$. Pepsin catalyzes the hydrolysis of the peptide glycyl-L-glutamyl-L-tyrosine, but it is completely ineffective if one of the amino acids has the opposite optical configuration of the D form, or if the peptide is slightly different, e.g., L-glutamyl-L-tyrosine. Such specificity is not absolute, however, and many enzymes are effective to some extent with a variety of substrates having similar structures.

Mechanisms for enzyme kinetics usually begin with a first step in which enzyme E combines with substrate S to form a complex. This formulation was given originally by Henri in 1903 and extended by Michaelis and Menten in 1913, and Briggs and Haldane in 1925. The simplest mechanism of this type is

$$E + S \underset{k_{-1}}{\overset{k_1}{\rightleftharpoons}} ES \underset{k_{-2}}{\overset{k_2}{\rightleftharpoons}} E + P$$

The reverse reaction between E and P to reform ES is often slow enough to be neglected, and always becomes negligible in the early stages of reaction when [P] is very low.

Enzyme-catalyzed reactions are usually extremely fast, and the concentration of enzyme is always much lower than that of substrate. As a result, the rate of appearance of product is always almost equal to the rate of disappearance of substrate, $(d[P]/dt = -d[S]/dt)$. This condition implies that the concentration of enzyme–substrate complex [ES] is in a steady state. In the mechanism above,

$$\frac{d[ES]}{dt} = k_1[E][S] - k_{-1}[ES] - k_2[ES] = 0$$

The steady-state concentration [ES] is, therefore,

$$[ES] = \frac{k_1[E][S]}{k_{-1} + k_2} = \frac{[E][S]}{K_m} \tag{14.3}$$

The *Michaelis constant* is defined by

$$K_m = \frac{k_{-1} + k_2}{k_1} \tag{14.4}$$

The Michaelis constant K_m is not the dissociation constant K_d of the enzyme–substrate complex, which is $K_d = k_{-1}/k_1$. In many instances, however, $k_2 \ll k_{-1}$, and then K_m is approximately equal to K_d.

From Eq. (14.3), the rate of the enzyme-catalyzed reaction (per unit volume) is

$$v = -\frac{d[S]}{dt} = k_2[ES] = \frac{k_2[E][S]}{K_m}$$

In this form, the equation is not so useful for practical purposes, since it includes the concentration of free enzyme [E], whereas the experimentally known quantity is $[E_0]$, the total concentration of enzyme, both free and combined, in the reaction mixture. Since $[E_0] = [E] + [ES]$, we obtain, by substituting for [E] in Eq. (14.3),

$$[ES] = \frac{[E_0][S]}{K_m + [S]}$$

The rate is now given by the *Michaelis–Menten equation*,

$$v = \frac{-d[S]}{dt} = k_2[ES] = \frac{k_2[E_0][S]}{K_m + [S]} \tag{14.5}$$

It is often convenient to rewrite this equation in terms of the maximum rate v_m, which is reached when [S] becomes so large that $[S] \gg K_m$, and hence, from Eq. (14.5),

$v_m = k_2[E_0]$. One obtains

$$\frac{v}{v_m} = \frac{[S]}{[S] + K_m} \qquad (14.6)$$

From Eq. (14.6), when $[S] = K_m$, $v = v_m/2$. The Michaelis constant equals the substrate concentration at which the rate has fallen to one-half its maximum value. The variation of v/v_m with $[S]$ for the hydrolysis of sucrose by the enzyme invertase is shown in Fig. 14.3(a).

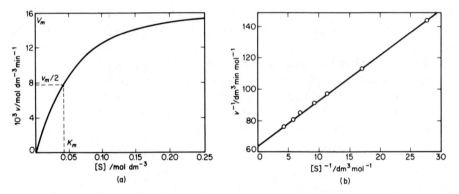

FIGURE 14.3 Hydrolysis of sucrose catalyzed by the enzyme invertase: (a) Rate vs. concentration of substrate; (b) Rate data of (a) plotted in accord with the Lineweaver–Burk equation (14.7).

It is convenient to transform Eq. (14.6) into a linear equation, for $1/v$ as a function of $1/[S]$:

$$\frac{1}{v} = \frac{K_m}{v_m}\frac{1}{[S]} + \frac{1}{v_m} \qquad (14.7)$$

This is called the *Lineweaver–Burk Equation*. The value of v as a function of $[S]$ can usually be obtained from the initial slopes of rate curves. Then, by plotting $1/v$ vs. $1/[S]$, one can obtain K_m and v_m from the slope and intercept of the linear graph [Fig. 14.3(b)]. Since $[E_0]$ is known, v_m gives $k_2 = v_m/[E_0]$. We cannot, however, determine k_1 and k_{-1} from this kind of analysis based on steady-state kinetics.

Example 14.2 The data plotted in accord with the Lineweaver–Burk equation in Fig. 14.3(b) were obtained with a concentration of invertase of 10^{-6} mol dm^{-3}. Calculate K_m and k_2 for this reaction.

The slope of the straight line in [14.3(b)] is $K_m/v_m = 2.90$ min^{-1}, and the intercept is $1/v_m = 64$ (dm^3 mol^{-1}) min^{-1}. Hence $K_m = 0.0453$ mol dm^{-3}. The maximum rate $v_m = 0.0156$ mol dm^{-3} min$^{-1} = k_2[E_0]$, so that $k_2 = 1.56 \times 10^4$ min^{-1}.

Many enzyme reactions do not follow simple Michaelis–Menten kinetics. With modern computational techniques, however, complicated mechanisms can be checked against the experimental data, even when graphical methods are not practical.

14.6 The Structure of an Enzyme—Carboxypeptidase A

Twenty years ago the dream of enzyme chemists was to decipher the complete three-dimensional structure of an enzyme. They thought that this information would unveil the secret of the catalytic power of the enzyme. X-ray crystallographers (see Chapter 28) have now worked out the structures of a number of enzymes and enzyme–substrate complexes *in the crystalline state* to a resolution of 200 pm. These structures allow considerable insights into the mechanisms of enzymatic action, but they have not solved the problem of exactly how enzymes exert their catalytic activity. The enzyme molecule in a crystal has a structure that is usually quite similar to that in solution, as far as the static equilibrium arrangement of its atoms is concerned. But an enzyme molecule acting as a catalyst is not a static structure, and internal motions of parts of the molecule at the active site where the substrate is held (and even motions distant from that site) may contribute to the dynamic activity of the enzyme catalyst.

A good example of the application of X-ray crystallography to enzyme structure is carboxypeptidase (CPA). This enzyme catalyzes the hydrolysis of a peptide bond adjacent to a free carboxyl group. CPA has a molar mass of 34 000 g mol^{-1} and contains 306 amino acid residues per molecule in a single folded polypeptide chain.

The crystal structure of CPA has been determined by W. N. Lipscomb and his colleagues. One view of a model of the structure is shown in Fig. 14.4. The atoms are denoted by circles and the bonds by lines. This is a computer drawing analogous to a ball-and-stick model. The molecule is a compact protein with a shape that is roughly an ellipsoid with axes 5000 × 4200 × 3800 pm. One side of the molecule contains a

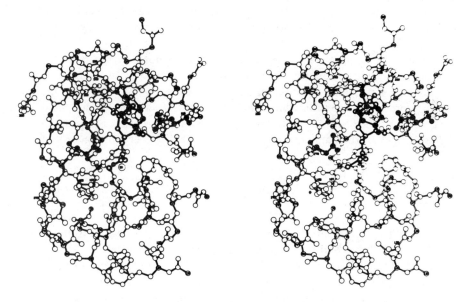

FIGURE 14.4 Stereoscopic drawing of the x-ray crystal structure of carboxypeptidase A, showing the active site of the enzyme and the binding of a substrate (heavy open circles). The black circles indicate certain amino-acid residues of enzyme especially important in catalytic action. [W. N. Lipscomb, Harvard University]

deep cleft, which includes the active site at which a substrate is bound. CPA contains one atom of zinc per molecule and the Zn is coordinated to three amino acid side chains of the protein, histidine 69, glutamic acid 72, and histidine 196, its fourth ligand being a water molecule.

A typical substrate of the enzyme is glycyl-L-tyrosine:

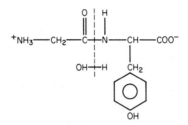

This substrate peptide is hydrolyzed slowly and it was possible to crystallize the enzyme–substrate complex and determine its crystal structure also. The conformation of the enzyme changes appreciably when the substrate is bound, and the cleft in the protein molecule changes from a hydrophilic to a hydrophobic region as four water molecules are expelled. The water ligand of the zinc atom is replaced by the carbonyl group of the substrate's peptide bond. The sidechain of tyrosine 248 snaps shut on the cleft, closing it off from the surrounding solvent water. While the CO group of the peptide bond is held by the Zn, the OH group of tyrosine 248 can donate a proton to the nearby —NH of the peptide bond. The —COO⁻ group of glutamic acid 270 either makes a nucleophilic attack on the —C=O of the peptide bond or facilitates an attack by H_2O (general base catalysis). All in all, the picture of the action of this enzyme is quite satisfactory, but the speed of the enzymatically catalyzed reaction is still amazing.

In general terms, the enzyme–substrate complex holds everything in exactly the right position, and the next step in the reaction can proceed with an activation energy that is much lower than that of the uncatalyzed reaction. The enzyme molecule is like a jig on a machine tool, which holds the work exactly placed for the operation to be done on it. If you set up the work by trial and error, a long time might elapse before you get all the alignments right, but with the jig everything goes swiftly.

14.7   Surface Catalysis

Many reactions that are slow in a homogeneous gas or liquid phase proceed rapidly if a catalytic solid surface is available. An interesting case is the bromination of ethylene: $H_2C=CH_2 + Br_2 \rightarrow BrH_2C—CH_2Br$. This reaction goes readily in a glass vessel at 470 K; it was at first thought to be an ordinary homogeneous combination, but a curious fact was noticed: the rate is higher in smaller reaction vessels. When a vessel is packed with glass beads, the rate is enhanced, but when a vessel is coated on the inside with paraffin wax, the rate is reduced. It was found that the reaction is mainly occurring not in the gas phase but on the glass surface of the reaction vessel. The glass is acting as a catalyst.

The decomposition of formic acid illustrates the specificity often displayed by surface reactions. If the acid vapor is passed through a heated glass tube, the reaction is about one-half dehydration and one-half dehydrogenation:

$$\text{(1) HCOOH} \longrightarrow H_2O + CO; \quad \text{(2) HCOOH} \longrightarrow H_2 + CO_2$$

If the tube is packed with Al_2O_3, only reaction (1) occurs; but if packed with ZnO, only reaction (2) occurs. Different surfaces can accelerate different parallel paths, and thus the nature of the products can be determined by the catalyst used.

A surface reaction can usually be divided into the following elementary steps: (1) diffusion of reactants to surface: (2) adsorption of reactants at surface; (3) chemical reaction on the surface; (4) desorption of products from surface; and (5) diffusion of products away from the surface. These are consecutive steps, and if any one has a much slower rate constant than all the others, it will become rate determining. Steps (1) and (5) are usually fast, but with active catalysts they can affect the overall rate, especially if the catalyst has a porous structure. Many industrial catalysts are in the form of porous pellets and pore diameters and pellet sizes must be carefully designed to ensure that all the catalyst surface is available for reaction. Steps (2) and (4) generally have higher specific rates than step (3), but reactions are known in which they may be rate determining. In some cases, instead of reaction occurring entirely on the surface, a molecule from the fluid phase may react with an adsorbed species.

The kinetics of many surface reactions can be treated successfully on the basis of the following assumptions: (1) The rate-determining step is a reaction of adsorbed molecules. (2) The reaction rate per unit surface area is proportional to θ, the fraction of surface covered by reactant. To proceed with our analysis of the kinetics of contact catalysis, we must find out how θ depends on pressure (for gases) or on concentration (for liquid solutions).

14.8 Langmuir Adsorption Isotherm

An *adsorption isotherm* is an expression that gives the fraction θ of a surface that is covered by adsorbed molecules in equilibrium at constant temperature as a function of pressure or concentration. Adsorption on a solid surface is classified as chemisorption or physisorption. *Chemisorption* involves the formation of chemical bonds between substrate molecule and solid surface, and often the breaking of pre-existing bonds in the adsorbed molecule. For example, when H_2 adsorbs on Pt, it dissociates, breaking its H—H bond and forming surface complexes Pt—H. In some cases the chemisorption step requires an activation energy. *Physisorption*, on the other hand, involves forces similar to the van der Waals forces that lead to condensation of vapors to liquids. Adsorption of a substrate on a catalyst site leading to chemical reaction generally is a chemisorption.

The first quantitative theory of the adsorption of gases on solid surfaces was given in 1916 by Irving Langmuir. He based his theory on the following assumptions:

1. The solid surface contains a fixed number of adsorption sites. At equilibrium at any temperature and gas pressure, a fraction θ of the sites is occupied by adsorbed molecules, and a fraction $1 - \theta$ is not occupied.

2. Each site can hold one adsorbed molecule.

3. The heat of adsorption is the same for all sites and does not depend on the fraction covered θ.

From these assumptions, Langmuir derived an adsorption isotherm by considering the kinetics of adsorption and desorption of gas molecules at the surface. If θ is the fraction of surface area covered by adsorbed molecules at any time, the rate of desorption of molecules from the surface is proportional to θ or equal to $k_d\theta$, where k_d is a constant at constant T. The rate of adsorption of molecules on the surface is proportional to the fraction of surface that is bare, $1 - \theta$, and to the rate at which molecules strike the surface, which varies directly with gas pressure. The rate of adsorption is therefore set equal to $k_aP(1 - \theta)$, where k_a is a rate constant for adsorption. At equilibrium, the rate of adsorption equals the rate of desorption, $k_d\theta = k_aP(1 - \theta)$. Solving for θ, we obtain

$$\theta = \frac{k_aP}{k_d + k_aP} = \frac{bP}{1 + bP} \tag{14.8}$$

where b is the ratio of rate constants, k_a/k_d, called the *adsorption coefficient*.

The Langmuir isotherm of Eq. (14.8) is plotted in Fig. 14.5(a). Sometimes it is more convenient to plot it in the form of a straight line,

$$\frac{1}{\theta} = 1 + \frac{1}{bP} \tag{14.9}$$

Figure 14.5(b) shows some data for the adsorption of gases on silica plotted in this

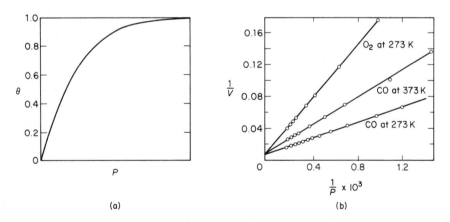

(a)

(b)

FIGURE 14.5 (a) Langmuir adsorption isotherm, fraction of surface covered by adsorbed gas vs. pressure of gas in equilibrium with adsorbate. (b) Adsorption of gases on silica plotted in accord with Eq. (14.9). The volume adsorbed V (proportional to θ) is in cm^3 at STP per gram silica and the pressure is in kPa.

form. The straight lines indicate that the adsorptions conform to the Langmuir isotherm.

Example 14.3 From the data on adsorption of CO on silica at 273 K in Fig. 14.5(b), calculate the adsorption coefficient b in the Langmuir isotherm, Eq. (14.9).

We must convert volumes of gas adsorbed V to fraction of surface covered θ. The intercept at $1/P = 0$ corresponds to $\theta = 1$ at $1/V = 0.010$. Thus $V = 100 \text{ cm}^3 \text{ g}^{-1}$ corresponds to $\theta = 1$. The slope is 0.090 Pa cm^{-3} or, in terms of θ, $9.0 \times 10^{-4} \text{ Pa}$, so that $b = (9.0 \times 10^{-4})^{-1} \text{ Pa}^{-1} = 1.1 \times 10^3 \text{ Pa}^{-1}$.

14.9 Adsorption on Nonuniform Surfaces

The Langmuir isotherm often fits experimental data on both chemisorption and physisorption quite well, but the assumptions on which it is based are too restrictive. Even specially prepared solid surfaces are nonuniform on a 1-nm scale and technical catalysts have rough polycrystalline surfaces. In Chapter 29 we shall discuss various types of defects in the solid state—all these imperfections contribute to the nonuniformity of catalytic surfaces. As John Updike wrote in his poem, "The Dance of the Solids,"*

> Solidity is an imperfect state.
> Within the cracked and dislocated Real
> Nonstoichiometric crystals dominate.
> Stray Atoms sully and precipitate;
> Strange holes, excitons wander loose; because
> of Dangling Bonds, a chemical substrate
> Corrodes and catalyzes—surface Flaws
> Help Epitaxial Growth to fix adsorptive Claws.

When chemisorption occurs on a nonuniform surface, the adsorbed atoms or molecules first cover sites having higher adsorption coefficients. Such sites are usually those with more negative enthalpies of adsorption, ΔH_{ad}. As adsorption proceeds to higher coverages θ, the $-\Delta H_{ad}$ declines.

Table 14.2 summarizes the adsorption isotherms most frequently used in work on contact catalysis. The Freundlich isotherm was originally empirical, but later it was shown to be consistent with a fraction of surface sites g that depends on ΔH_{ad} as $g = a \exp(-\Delta H_{ad}/Q_0)$ where a and Q_0 are constants. The Temkin isotherm arises from a linear dependence of g on ΔH_{ad}. The B.E.T. isotherm is based on a model of physical adsorption that leads to a buildup of successive layers of adsorbed molecules like a surface condensation of vapor to liquid. It has been widely used to determine surface areas of catalysts (see Problem 12).

*From *Midpoint and Other Poems* (New York: Alfred A. Knopf, 1968).

TABLE 14.2

ADSORPTION ISOTHERMS

Name	Equation[a]	Applicability
Langmuir	$V/v_m = \theta = \dfrac{bP}{1 + bP}$	Chemisorption and physisorption
Freundlich	$V = kP^{1/n} \quad (n > 1)$	Chemisorption and physisorption
Slygin–Frumkin (Temkin)	$V/v_m = \theta = \dfrac{1}{a} \ln C_0 P$	Chemisorption
Brunauer–Emmett–Teller (B.E.T.)	$\dfrac{P}{V(P_0 - P)} = \dfrac{1}{v_m c} + \dfrac{c - 1}{v_m c} \dfrac{P}{P_0}$	Multilayer physisorption
Harkins–Jura	$\ln \dfrac{P}{P_0} = B - \dfrac{cA^{1/2}}{\theta}$	Physisorption

[a] V, volume of gas adsorbed; v_m, volume adsorbed to form monolayer; θ, fraction of monolayer at equilibrium pressure P; P_0, vapor pressure of adsorbent in liquid state; A, surface area. Other symbols are constants.

14.10 Mechanism of Surface Reactions

It is virtually impossible to obtain a unique mechanism for a surface reaction simply from a study of its formal kinetics. Many additional experiments need to be done, such as adsorption studies, isotopic exchange, and direct spectroscopic examination of the reacting species at the surface.

For the case of a molecule A reacting on a uniform solid surface, it is often assumed that the reaction rate v_r is proportional to the fraction of surface θ_A covered by A molecules. If θ_A is given by the Langmuir isotherm and k_s is the surface rate constant,

$$v_R = k_s \theta_A = k_s \frac{b_A P_A}{1 + b_A P_A} \tag{14.10}$$

This expression is identical in form to the Michaelis–Menten equation (14.5), which was subsequently derived for enzymes.

The decomposition of PH_3 on a heated tungsten wire follows an expression like Eq. (14.10), $v_R = k_s[bP_{PH_3}/(1 + bP_{PH_3})]$. At P_{PH_3} from 1 to 10 Pa, $bP_{PH_3} \ll 1$ and the rate becomes first order, $v_R = k_s bP_{PH_3}$. From 10 to 10^3 Pa, the rate follows the full expression (14.10). Above 10^3 Pa, $bP_{PH_3} \gg 1$ and the rate becomes $v_R = k_s P_{PH_3}^0 = k_s$. The rate is now independent of the pressure of PH_3, i.e., proportional to P to the power zero. The reaction is *zero order*.

In many cases, a reaction *product* is appreciably adsorbed. An example is the decomposition of N_2O on Mn_3O_4: $2N_2O \rightarrow 2N_2 + O_2$. The rate fits the equation

$$v_R = \frac{k_s bP_{N_2O}}{1 + bP_{N_2O} + b'P_{O_2}^{1/2}}$$

The product O_2 is adsorbed as oxygen atoms (hence the exponent $\frac{1}{2}$ on the $P_{O_2}^{1/2}$ term).

This kind of inhibition by product is often observed in reaction of gases on solid catalysts.

Example 14.4 The decomposition of GeH_4 ($GeH_4 = Ge(s) + 2H_2$) on a germanium surface at 55 K is a zero-order reaction. In one experiment the initial pressure of GeH_4 was 41.0 kPa; after 200 min the pressure of GeH_4 was 11.6 kPa. Estimate the rate constant and the half-life τ of the reaction.

From Eq. (14.10), $-dP/dt = k_0$. On integration, $P = k_0 t + \text{const.}$ When $t = 0$, $P = P_0$, so that $P_0 - P = k_0 t$. When $P = P_0/2$, $P_0/2 = k_0 \tau$ and $\tau = P_0/2k_0$. For the GeH_4 decomposition, $(41.0 - 11.6) \text{ kPa} = k_0 (200 \text{ min})$, $k_0 = 0.147 \text{ kPa min}^{-1}$, and $\tau = 41.0 \text{ kPa}/2 \times 0.147 \text{ (kPa min}^{-1}) = 139 \text{ min}$. (From volume of reaction vessel, the rate constant k_0 could be calculated in units of $\text{mol dm}^{-3} \text{ s}^{-1}$.)

The development of catalysts for the polymerization of olefins at low pressures resulted in the creation of a new industry to produce products such as polypropylene and synthetic "natural rubber." In the 1950s Karl Ziegler and Giulio Natta discovered that a mixture of a transition-metal compound and an alkyl of a strongly electropositive metal, suspended in a hydrocarbon solvent, would catalyze the polymerization of ethylene at low pressures. These catalysts also can yield stereochemically regular polymers. The nomenclature of such polymers is illustrated in Fig. 14.6 for the example of polypropylene. The isotactic polypropylene has far better mechanical properties and is the only one of industrial importance.

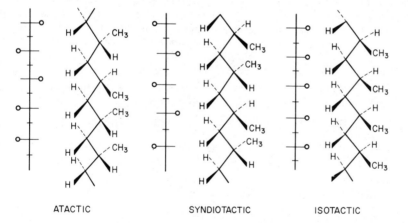

ATACTIC   SYNDIOTACTIC   ISOTACTIC

FIGURE 14.6 Atactic, syndiotactic, and isotactic polymers of propylene. [F. A. Bovey and F. H. Winslow, *Macromolecules, An Introduction to Polymer Science*, (New York: Academic Press, 1979)]

A typical catalyst for isotactic polymerization of propylene is a mixture of $TiCl_3$ and $Al(C_2H_5)_3$. When these compounds are brought together in solution, a brown precipitate is formed, and the polymerization proceeds on the surface of the suspended solid particles. A simplified picture of the mechanism is as follows. The Ti is in a state of octahedral coordination, with one position vacant and one bound to an alkyl radical. The propylene first forms a π-bond at the empty site on the Ti. The alkyl group adds to the propylene and a new vacant site appears. A second molecule of C_3H_6 binds to this site and the alkyl addition is repeated. These steps are summarized in Fig. 14.7(a), and Fig. 14.7(b) is an electron micrograph of the working catalyst.

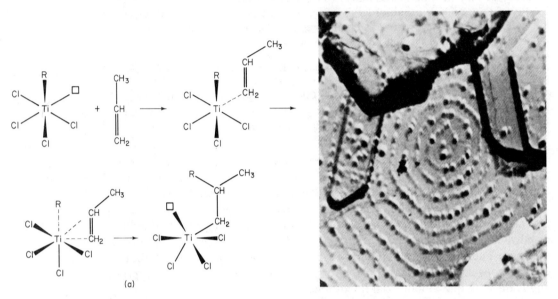

FIGURE 14.7 (a) Mechanism of Ziegler–Natta catalysis. (b) An electron micrograph of initial polymerization of propylene on hexagonal crystallites of $TiCl_3$ in Ziegler–Natta catalyst. The dots are traces of polymer on what appears to be a crystal growth spiral (see Fig. 29.13). [L. A. M. Rodriguez and J. A. Gabant, *J. Polym. Sci.* A1, 4, 1971 (1966)]

Many oxidation catalysts consist of platinum-group metals on ceramic supports, but the cost of these metals limits their applications. An example of another kind of specific oxidation catalyst is $BiMoO_4$, used to produce acrolein from propylene,

$$CH_3—CH{=}CH_2 + O_2 \rightarrow CH_2{=}CH—CHO + H_2O.$$

Experiments with oxygen-18 showed that oxygen appearing in the acrolein came from oxide ions in the crystalline catalyst. The MoO_4^{2-} transfers O to the propylene substrate, and O_2 from the gas oxidizes the Bi sites, from which oxide ions then move to replenish the oxygen in the MoO_4 groups.

Catalysis by acids in liquid solutions was discussed in Section 14.3. Many important catalysts function as solid acids, e.g., the aluminosilicates. If a Al^{3+} ion replaces a Si^{4+} ion, an acidic site is created in the crystal structure, either a Lewis acid in the absence of water or a Brønsted acid if water is present. The acidic site can catalyze cracking and isomerization of hydrocarbons through carbonium-ion intermediates. For example, the Al site can act as a strong Lewis acid and accept a pair of electrons,

$$
\begin{array}{ccc}
& \mathrm{O} & \mathrm{O} \\
& | & | \\
\mathrm{RH} + \mathrm{Al}\!-\!\mathrm{O}\!-\!\mathrm{Si} & \longrightarrow & \mathrm{R^+}\ \ \mathrm{H^-:Al}\!-\!\mathrm{O}\!-\!\mathrm{Si} \\
& | & | \\
& \mathrm{O}\ \ \mathrm{M^+} & \mathrm{O}\ \ \mathrm{M^+}
\end{array}
$$

The most important of the aluminosilicate acid catalysts are based on the crystal structures of the zeolites. A naturally occurring zeolite, faujasite, is drawn in Fig. 14.8. Zeolites can be designed so that only molecules of a certain size and shape can enter the pores, thereby yielding catalysts with shape and size specificity. One important process based on such catalysts is the conversion of methanol to gasoline. This process would be the final stage in production of high-octane gasoline from natural gas or biomass compounds.

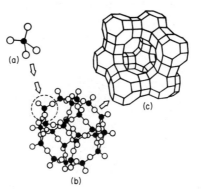

FIGURE 14.8 An SiO_2 unit polymerizes to yield faujasite. Zeolite catalysts with faujasite structure display specificity based on molecular size. [A. W. Sleight, *Science 208*, 895 (1980). Copyright 1980 by the American Association for the Advancement of Science.]

Problems

1. The decomposition of ozone, $2O_3 \longrightarrow 3O_2$, catalyzed by N_2O_5, follows the rate law,

$$-d[O_3]/dt = k_a[O_3]^{2/3}[N_2O_5]^{2/3},$$

   [R. Schumacher, *Z. Phys. Chem.*, *A140*, 281 (1929)]. Outline a mechanism that would explain this rate law. Recall that $N_2O_5 \rightleftharpoons NO_2 + NO_3$.

2. The decomposition of nitramide is catalyzed by OH^-. A possible mechanism is

   (1) $\quad NO_2NH_2 + OH^- \underset{}{\overset{k_1}{\rightleftharpoons}} H_2O + NO_2NH^-$

   (2) $\qquad\qquad NO_2NH^- \overset{k_2}{\longrightarrow} N_2O + OH^-$

   Two possible mechanisms have been considered: (a) Reaction (1) is fast in both directions with equilibrium constant K. Reaction (2) is slow and rate determining. (b) Reaction (1) is rate determining and (2) is much faster. What experiments might distinguish these two mechanisms?

3. $C_2H_4I_2 = C_2H_4 + I_2$ in CCl_4 solutions has a rate, $-d[C_2H_4I_2]/dt = k_a[C_2H_4I_2][I_2]^{1/2}$ What catalytic mechanism is likely?

4. For the mutarotation α-glucose $\rightleftharpoons$ β-glucose, the reaction at 25°C has $k_H = 9.98 \times 10^{-3}$ dm^3 mol^{-1} s^{-1}, $k_{OH} = 3.76 \times 10^2$ and $k_0 = 3.68 \times 10^{-4}$, for the reaction catalyzed by H^+, OH^-, and H_2O, respectively. Calculate the overall rate constant at pH 1, pH 7, and pH 10.

5. For acetylcholine esterase at 37°C the number of substrate (acetylcholine) molecules hydrolyzed in 1 min is about 10^6 (turnover number). If the acetylcholine released by a nerve impulse at a neuromuscular junction is 1 pmol, what would be the minimum number of enzyme molecules required to destroy 99.9% of the acetylcholine within 1 ms?

6. For propionic acid $pK_a = 4.89$ at $25°C$. For the acid-catalyzed hydrolysis of acetal the Brønsted constant $\alpha = 0.83$. In a solution $0.15\,M$ in propionic acid, $0.15\,M$ in propionate, calculate the fractions of reaction due to acid catalysis by propionic acid, H_2O, and H_3O^+. Assume for H_2O $pK_a = 15.74$ and for H_3O^+, $pK_a = -1.74$.

7. The initial rate of $ATP \longrightarrow ADP + P_i$ catalyzed by the enzyme myosin at $25°C$.

$[ATP]_0$ (μM)	7.1	11	23	40	77	100
$10^2 d[P_i]/dt$ ($\mu M\ s^{-1}$)	2.4	3.5	5.3	6.2	6.7	7.1

The $[E_0]$ was $0.040\,g\,L^{-1}$ ($M = 20\,000\,g\,mol^{-1}$). Calculate the Michaelis constant, v_m, and k_2 for this reaction.

8. The enzyme fumarase catalyzes the reaction fumarate $+ H_2O \longrightarrow$ L-malate. At $25°C$ with $[E_0] = 10^{-6}\,M$, initial rates at various substrate concentrations were:

$10^6[S]$ (M)	1.0	2.0	3.8	5.0	8.0	10	20	5
$10^4 v$ (M s^{-1})	2.6	4.3	6.3	7.2	8.7	9.3	10.8	12.0

Calculate the Michaelis constant and k_2 for this reaction.

9. Catalase catalyzes $H_2O_2 \longrightarrow H_2O + \frac{1}{2}O_2$ with a turnover number of $9 \times 10^6\,s^{-1}$. If H_2O_2 is being produced in a cell at the rate of $10^{-10}\,mol\,dm^{-3}\,s^{-1}$, what is the minimum concentration of catalase in the cell required to keep the $[H_2O_2]$ below $10^{-9}\,M$?

10. The adsorption of C_2H_4 on activated charcoal at $273\,K$ is as follows:

$P(C_2H_4)$ (MPa)	0.405	0.982	1.36	1.93	2.75
g C_2H_4 ads/g charcoal	0.163	0.189	0.198	0.206	0.206

Fit these data to a Langmuir isotherm and determine the adsorption coefficient b. If one molecule of C_2H_4 occupies $0.21\,nm^2$ of surface, calculate the specific area of the charcoal ($m^2\,g^{-1}$).

11. How would you explain the following kinetics of surface reactions on the basis of the Langmuir isotherm? (a) The decomposition of NH_3 on W is zero order. (b) Decomposition of N_2O on Au is first order. (c) Decomposition of NH_3 on Pt depends on P_{NH_3}/P_{H_2}.

12. The surface area of a catalyst was investigated by applying the B.E.T. isotherm to data on adsorption of N_2 at $77\,K$ where $P_0 = 100\,kPa$. (See Table 14.2.)

P (kPa)	10.0	15.0	20.0	25.0	30.0
V (cm^3 g^{-1} STP)	6.84	7.56	8.07	8.82	9.51

The cross-sectional area of an adsorbed N_2 molecule is $0.167\,nm^2$. Calculate the specific area. [*Hint:* Plot $P/V(P_0 - P)$ vs. P/P_0.]

13. Consider enzyme kinetics in the presence of an inhibitor I which competes with substrate S for the active site of an enzyme. If K_I is the dissociation constant of the enzyme–

inhibitor complex, show that the rate of the enzyme-catalyzed reaction with inhibitor concentration [I] is

$$v = \frac{v_m[S]}{[S] + K_m[1 + [I]/K_I]}$$

14. Succinic dehydrogenase catalyzes the dehydrogenation of succinic acid (S) to fumaric acid. It is competitively inhibited by malonic acid. At 25°C, $K_m = 10^{-5}$ and $K_I = 10^{-5}$ (standard state $c° = 1$ M). In the absence of inhibitor, $v = v_m/2$ when $[S] = K_m$. With $[I] = 2 \times 10^{-5}$ M, what must [S] be to reach the same velocity $v_m/2$?

15

Theory
of
Reaction Rates

The preceding two chapters described some of the facts about the rates of chemical reactions. We shall now consider some theories and models that have been used to calculate the values of rate constants from molecular properties. Considerable attention will be given to simple gas reactions since in this case chemical interaction between a pair of molecules can be investigated without complications due to surrounding dense matter.

15.1 Collision Theory of Gas Reactions—Collision Frequency

The kinetic-molecular theory of gases underwent an extensive development during the latter part of the nineteenth century, particularly in the work of Maxwell and Boltzmann. The first theories of rates of reactions between molecules in the gas phase were based on this kinetic theory. The papers of M. Trautz (1916) and W. C. McC. Lewis (1918) were especially significant. Their fundamental idea was that reactions occur during collisions between gas molecules when a rearrangement of chemical bonds forms new molecules from the old ones. The rate of reaction is set equal to the number of collisions in unit time (a frequency factor) multiplied by the fraction of these collisions that result in chemical reaction (an activation factor).

The simplest treatment of the frequency of collisions between molecules in a gas is based on the model that treats molecules as rigid spheres. Let us suppose that a gas contains two kinds of molecules A and B, with hard-sphere diameters d_A and d_B. The hard-sphere model requires that the only interactions between molecules are elastic collisions whenever the distance r between the centers of spheres equals $(d_A + d_B)/2$. In other words, when $r > (d_A + d_B)/2$, the intermolecular potential energy $U(r) = 0$, but when $r \leq (d_A + d_B)/2$, $U(r) = \infty$.

To calculate the frequency of collisions between molecules of different species A and B, suppose that the center of molecule A is the center of a sphere of radius $d_{AB} = (d_A + d_B)/2$. A collision between molecule A and molecule B occurs whenever the center of a B molecule comes within this sphere of radius d_{AB} [see Fig. 15.1(a)].

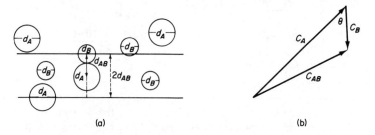

(a)   (b)

FIGURE 15.1 (a) Collision between hard-sphere molecules A and B. (b) Relative velocity c_{AB} is vector difference of c_A and c_B.

Suppose that all molecules B are stationary while molecule A zooms through the gas volume with a mean speed $\bar{c}_A$. In unit time the sphere of influence of moving molecule A sweeps out a cylindrical volume $\pi d_{AB}^2 \bar{c}_A$. The area $\pi d_{AB}^2 = \sigma_{AB}$ is called the *collision cross section* for molecules A and B on the rigid-sphere model. Other more detailed models for interaction between molecules would lead to different values for σ_{AB}. The concept of collision cross section is thus not restricted to collisions between hard spheres.

If the number of B molecules in unit volume of gas is N_B/V, the collision cross section in unit time will sweep through $(N_B/V)(\pi d_{AB}^2 \bar{c}_A)$ centers of molecules B. The collision frequency of a single molecule A with molecules B therefore is $z_{AB} = (N_B/V)$ $(\pi d_{AB}^2 \bar{c}_A)$. If there are N_A/V molecules of A per unit volume, the total frequency of collisions between molecules of A and of B is $Z_{AB} = (N_A/V)(N_B/V)\pi d_{AB}^2 \bar{c}_A$.

There is an error in this formula caused by the assumption that the B molecules do not move. Instead of $\bar{c}_A$, the mean speed of molecules A, we should use $\bar{c}_{AB}$, the mean *relative speed* of molecules A and B. The kinetic-theory expression for the mean speed $\bar{c}$ of a molecule of mass m is [Eq. (5.36)] $\bar{c} = (8kT/\pi m)^{1/2}$. The expression for the mean relative speed of molecules of masses m_A and m_B is

$$\bar{c}_{AB} = (8kT/\pi\mu)^{1/2} \tag{15.1}$$

The reduced mass is $\mu = m_A m_B/(m_A + m_B)$.

The concept of relative speed is illustrated in Fig. 15.1(b), where the velocities c_A and c_B are drawn as vectors, and the relative velocity c_{AB} is the vector difference between c_A and c_B. If θ is the angle between the two velocity vectors, the magnitude of the relative velocity (i.e., the relative speed) is $c_{AB} = (c_A^2 + c_B^2 - 2c_A c_B \cos\theta)^{1/2}$.

In terms of relative speed, the collision frequency of one molecule A with molecules B is

$$z_{AB} = (N_B/V)(\pi d_{AB}^2 \bar{c}_{AB}) \tag{15.2}$$

The total collision frequency between A and B molecules is

$$Z_{AB} = (N_A/V)(N_B/V)(\pi d_{AB}^2 \bar{c}_{AB}) \tag{15.3}$$

Example 15.1 What is the mean relative speed of H_2 and I_2 molecules in a gas at 1000 K?

The reduced mass of H_2 and I_2 molecules is

$$\mu = \frac{M_1 M_2}{M_1 + M_2} \frac{10^3}{L} = \frac{2.02 \times 254}{2.02 + 254} \frac{10^{-3}}{6.02 \times 10^{23}} = 3.33 \times 10^{-27} \text{ kg}$$

$$\bar{c}_{AB} = (8kT/\pi\mu)^{1/2} = \left(\frac{8 \times 1.38 \times 10^{-23} \text{ J K}^{-1} \times 10^3 \text{ K}}{\pi \times 3.33 \times 10^{-27} \text{ kg}} \right)^{1/2}$$

$$= 3.25 \times 10^3 \text{ m s}^{-1}$$

Example 15.2 At 1000 K and $P = 101.3$ kPa, what are the collision frequencies z_{AB} and Z_{AB} in an equimolar mixture of H_2 and I_2? The hard-sphere diameter of H_2 is 200 pm; of I_2, 400 pm.

In Eq. (15.2) the number density,

$$N_B/V = PL/RT = \frac{10^3(101.3/2)(\text{N/m}^2) \times 6.02 \times 10^{23} \text{ mol}^{-1}}{8.314 \text{ J K}^{-1} \text{ mol}^{-1} \times 1000 \text{ K}} = 36.7 \times 10^{23} \text{ m}^{-3}$$

$$z_{AB} = (N_B/V)(\pi d_{AB}^2 \bar{c}_{AB}) = (36.7 \times 10^{23} \text{ m}^{-3})(3.14)(300 \times 10^{-12} \text{ m})^2$$

$$\times (3.25 \times 10^3 \text{ m s}^{-1}) = 3.38 \times 10^9 \text{ s}^{-1}$$

$$Z_{AB} = (36.7 \times 10^{23} \text{ m}^{-3}) z_{AB} = 1.24 \times 10^{34} \text{ s}^{-1} \text{ m}^{-3}$$

If we consider a gas that contains only one kind of molecule, then $m_A = m_B = m$, and the relative speed from Eq. (15.1) becomes $\bar{c}_{AA} = \sqrt{2}(8kT/\pi m)^{1/2} = \sqrt{2}\,\bar{c}$. The number of collisions experienced by a single molecule in unit time is

$$z_{AA} = \sqrt{2}\,\pi d^2 N_A \bar{c}/V \tag{15.4}$$

The total number of collisions per unit time is

$$Z_{AA} = \tfrac{1}{2}\sqrt{2}\,\pi d^2 N_A^2 \bar{c}/V^2 \tag{15.5}$$

Note the factor $\frac{1}{2}$ introduced into this expression compared to Eq. (15.3) so that collisions A + A will not be counted twice. (A collision AB is not the same as BA, but there is only one AA.)

An important quantity in kinetic theory of gases is the *mean free path* λ, which is the mean distance traveled by a gas molecule between collisions. In unit time a molecule travels a mean distance $\bar{c}$ and in a gas containing only one kind of molecule it experiences z_{AA} collisions. Hence the mean free path is

$$\lambda = \bar{c}/z_{AA} = [\sqrt{2}\,\pi d^2(N/V)]^{-1} \tag{15.6}$$

[How does λ depend upon T at constant V? at constant P?]

Example 15.3 If the hard-sphere molecular diameter of O_2 is 290 pm, estimate the mean free path in O_2 at 500 K and 100 kPa; at 1 Pa.

At 100 kPa, since $N/V = PL/RT$,

$$\lambda = \left[\sqrt{2}\, \pi \frac{100 \times 10^3 \text{ N m}^{-2} \times 6.02 \times 10^{23} \text{ mol}^{-1}}{8.314 \text{ J K}^{-1} \text{ mol}^{-1} \times 500 \text{ K}} \times (290 \times 10^{-12})^2 \right]^{-1}$$

$$= 185 \times 10^{-9} \text{ m}$$

At 1 Pa, $\lambda = 18.5$ mm.

15.2   Collision Theory of Gas Reactions—The Rate Constant

Not all collisions between molecules lead to reaction. If they did, the reaction at ordinary pressures would be complete in a fraction of a second. The collision theory of gas reactions is based on the idea that collisions lead to reaction only when the energy E of the pair of colliding molecules exceeds a certain critical value E_a, called the *activation energy*. In the simplest form of the theory, reaction never occurs when the energy E is less than E_a, $E < E_a$, and reaction always occurs when the energy E is equal to or greater than E_a, $E \geq E_a$.

We must, however, ask what energy of the molecules are we talking about. Molecules can have energy in many degrees of freedom—translational kinetic energy as well as internal (rotational, vibrational, electronic) energy. In the simplest form of collision theory the only energy to be considered is the kinetic energy of translation along the line of centers between molecules as they approach each other on a collision course. The effective energy in the collision is the translational kinetic energy in two degrees of freedom, one for each molecule. The probability that the relative translational kinetic energy of a pair of molecules in a collision be equal to or greater than a critical value ϵ_a is $e^{-\epsilon_a/kT}$ or $e^{-E_a/RT}$, where $E_a = L\epsilon_a$ (see Problem 3). This is the activation factor that is multiplied into the collision frequency to give the number of reactive collisions in unit time.

Example 15.4   In a gas at 300 K, what fraction of the collisions between molecules will have a relative kinetic energy per mole along the line of centers greater than 10 kJ? 100 kJ? What fractions in a gas at 1000 K?

At 300 K:

$E_a = 10$ kJ: $e^{-E_a/RT} = \exp(-10\,000 \text{ J mol}^{-1}/8.314 \text{ J K}^{-1} \text{ mol}^{-1} \times 300 \text{ K})$
$$= 1.81 \times 10^{-2}$$

$E_a = 100$ kJ: $e^{-E_a/RT} = \exp(-100\,000/8.314 \times 300)$
$$= 3.87 \times 10^{-18}$$

At 1000 K:

$E_a = 10$ kJ: $e^{-E_a/RT} = \exp(-10\,000/8.314 \times 1000)$
$$= 3.00 \times 10^{-1}$$

$E_a = 100$ kJ: $e^{-E_a/RT} = \exp(-100\,000/8.314 \times 1000)$
$$= 5.98 \times 10^{-6}$$

10 kJ would be a very low activation energy, 100 kJ a fairly typical one.

The reaction rate per unit volume, from Eq. (15.3) and (15.1) is

$$v_R = Z_{AB}e^{-E_a/RT} = (N_A/V)(N_B/V)(\pi d_{AB}^2)\left(\frac{8kT}{\pi\mu}\right)^{1/2}e^{-E_a/RT}$$

For such a bimolecular reaction, $v_R = k_2'(N_A/V)(N_B/V)$, where k_2' is a second-order rate constant. Hence

$$k_2' = (\pi d_{AB}^2)\left(\frac{8kT}{\pi\mu}\right)^{1/2}e^{-E_a/RT} \qquad (15.7)$$

This is the simple hard-sphere collision theory expression for a bimolecular rate constant.

The units in this expression for the rate constant should be noted. The rate was calculated as (molecules/unit volume) per unit time. Hence k_2', in SI units, is $(m^3/\text{molecule})\,s^{-1}$. To convert into molar terms [SI units of $(m^3/\text{mol})\,s^{-1}$], we must multiply by the Avogadro constant $L: k_2 = Lk_2'$. Rate constants, however, are usually cited in $(dm^3/\text{mol})s^{-1}$, so that multiplication by 10^3 is then necessary to change k_2 to these units.

Example 15.5 Calculate the second-order rate constant for the reaction $2NOCl = 2NO + Cl_2$ at 600 K if the critical energy for activation is 103 kJ mol^{-1} and the hard-sphere collision diameter is 283 pm.

From Eq. (15.7), $k_2' = \frac{1}{2}(\pi d_{AA}^2)(8k/\pi\mu)^{1/2}T^{1/2}e^{-E_a/RT}$.

$$k_2' = \pi/2(283 \times 10^{-12}\text{ m})^2$$

$$\times \left[\frac{8(1.38 \times 10^{-23}\text{ J K}^{-1})}{\pi} \times \frac{(6.02 \times 10^{23}\text{ mol}^{-1})}{(32.8 \times 10^{-3}\text{ kg mol}^{-1})}\right]^{1/2} T^{1/2}e^{-12\,400/T}$$

$$= 3.20 \times 10^{-18}T^{1/2}e^{-12\,400/T}\text{ (molecules/m}^3)^{-1}\text{ s}^{-1}$$

$$k_2 = Lk_2' = 6.08 \times 10^6\,T^{1/2}e^{-12\,400/T}\text{ (mol/m}^3)^{-1}\text{ s}^{-1}$$

At 600 K:

$$k_2 = 0.158\text{ m}^3\text{ mol}^{-1}\text{ s}^{-1} = 1.58 \times 10^2\text{ (mol/dm}^3)^{-1}\text{ s}^{-1}$$

The experimental value is $k_2 = 0.60 \times 10^2\text{ (mol/dm}^3)^{-1}\text{ s}^{-1}$.

15.3 Molecular Diameters

Diameters of gas molecules can be obtained from a variety of experimental data. We should not be surprised if the values obtained from different sources do not agree closely. The hard-sphere model of a molecule is only a rough approximation to the actual intermolecular forces; consequently, hard-sphere diameters calculated from different kinds of measurements may have different values.

From Eq. (15.6) we see that any measurement that gives a value for the mean free path λ can be used to obtain an equivalent molecular diameter. The kinetic theory of gases gives theoretical equations that relate the *transport properties* of gases to the mean free path λ on a hard-sphere model. These transport properties are viscosity η, thermal conductivity κ, and diffusion coefficient D. They are called transport properties because they measure the rate of transport of some physical quantity down a gradient. Thus:

1. Thermal conductivity κ measures transport of kinetic energy $\frac{1}{2}mv^2$ by gas molecules down a gradient in temperature.
2. Viscosity η measures the transport of momentum mv by gas molecules down a gradient in velocity.
3. Diffusion coefficient D measures the transport of mass m by gas molecules down a gradient in concentration (or, more generally, in chemical potential).

The kinetic-theory expression for the three transport coefficients are, where ρ is the gas density and $\bar{c}$ the average molecular speed:

(1) $\kappa = 1.261 \rho \bar{c} \lambda c_V$, where c_V is the specific heat at constant volume (C_V per mole/M, the molar mass).

(2) $\eta = 0.499 \bar{c} \lambda \rho$ \hfill (15.8)

(3) $D = 0.599 \bar{c} \lambda$

Experimental methods for measuring transport properties of gases can be found in standard textbooks on the kinetic theory, with derivations of the hard-sphere equations for the coefficients. Of course, we are not restricted to a hard-sphere model in the theory of transport processes, and more realistic models can be used to derive information on intermolecular forces from the transport coefficients.

Example 15.6 The viscosity of ethylene at 273 K and 100 kPa is 9.33×10^{-6} kg m^{-1} s^{-1}. What is the corresponding hard-sphere diameter of the C_2H_4 molecule?

From Eqs. (15.6) and (15.8), $d^2 = 2(0.499)(mkT/\pi^3)^{1/2}/\eta$.

$$d^2 = 2(0.499)\left[\frac{28.1 \times 10^{-3} \text{ kg mol}^{-1} \times 1.38 \times 10^{-23} \text{ J K}^{-1} \times 273 \text{ K}}{6.02 \times 10^{23} \text{ mol}^{-1} \times \pi^3}\right]^{1/2} \Big/$$

$$9.33 \times 10^{-6} \text{ (kg m}^{-1} \text{ s}^{-1})$$

$d^2 = 0.254 \times 10^{-18}$ m^2, $d = 0.504 \times 10^{-9}$ m $= 0.504$ nm.

Another way to estimate molecular diameters is from the van der Waals constant b. We recall that b is the excluded volume occupied by molecules in a gas, which is four times the molar volume V_m of the hard-sphere molecules. Hence $b = 4V_m = 4L(\pi d^3/6)$

Molecular diameters for a selection of molecules from gas viscosity and van der Waals b data are given in Table 15.1. These two methods probably give the molecular diameters most suitable for the calculation of collision frequencies in chemical kinetics. When a molecule is not spherical, the diameter is that of a sphere having an equivalent volume.

TABLE 15.1

HARD-SPHERE MOLECULAR DIAMETERS

Molecule	d (pm), gas viscosity	d (pm), van der Waals b
Ar	286	286
CO	380	318
CO_2	460	324
Cl_2	370	330
He	200	246
H_2	218	275
Hg	360	238
Ne	234	264
N_2	316	314
O_2	296	290
H_2O	272	287

15.4 Collision Theory vs. Experiment

The simple hard-sphere collision theory is not too bad when combined with experimental energies of activation; it gives values for the rate constants of bimolecular reactions that are usually of the right order of magnitude, i.e., good to within a factor of 10.

An attempt was made to modify the simple rigid-sphere collision theory by including a term called the *steric factor p*. Then instead of Eq. (15.7), one writes

$$k_2' = p(\pi d_{AB}^2)(8kT/\pi\mu)^{1/2}e^{-E_a/RT} \tag{15.9}$$

The need for a steric factor is evident. For example, suppose that we are studying the rate of reaction of sodium atoms with *n*-butylbromide,

$$Na + CH_3 \cdot CH_2 \cdot CH_2 \cdot CH_2 \cdot Br \longrightarrow NaBr + CH_3 \cdot CH_2 \cdot CH_2 \cdot CH_2 \cdot$$

If a Na atom strikes a molecule of C_4H_9Br, the probability of reaction certainly depends on where it strikes. If Na hits at or near the methyl end, the chance of reaction must be small. On the other hand, a hit near the bromide moiety will probably lead to reaction.

The deficiencies of simple hard-sphere collision theory can be realized by the following analogy. A blue billiard ball hits a yellow billiard ball. They are instantaneously changed into two green billiard balls that fly away from each other. This is how simple collision theory simulates a reaction like $H_2 + D_2 = 2HD$, where D is the hydrogen isotope of mass number 2. Detailed quantum-mechanical calculations show that the reaction is a more complex process. As the H_2 molecule approaches D_2, weak attractive forces between the H and D atoms begin to be felt when the molecules are still separated by several hard-sphere diameters. The H—H and D—D bonds begin to stretch and weaken, while incipient H—D bonds begin to form. The kinetic energy of the colliding molecules is converted into intermolecular potential energy. If the relative kinetic energy of $H_2 + D_2$ is sufficiently large, the molecules will reach a transition state called the *activated complex*.

$$H_2 + D_2 \longrightarrow \begin{array}{c} H—H \\ | \quad | \\ D—D \end{array} \longrightarrow 2HD$$

In some instances, the reaction will then continue, the H—H and D—D bonds will break and two HD molecules will fly apart. The particular "four-center mechanism" shown here has, however, not been proven.

15.5 Potential-Energy Surfaces—Example of D + H$_2$

Let us consider D + H$_2$ → DH + H, which is one of the most simple examples of a chemical reaction. To describe the configuration of this reacting system at any point along the reaction path requires three spatial coordinates. As shown in Fig. 15.2, these coordinates can be taken to be the internuclear distance between H and H, the distance of D from the midpoint of the H—H bond, and the angle θ between the H—H bond and the vector from the midpoint of the bond to D.

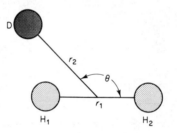

FIGURE 15.2 Coordinates to describe the relative positions of the atoms in the reaction D + H$_2$ → DH + H.

It can be shown, however, that one particular direction of approach of D to H—H is energetically more favorable than any other. This is the path in which D approaches H—H along the line of centers of the three-body system, i.e., the angle θ is always 0 or 180°. The reason for this result is that when D approaches H—H along the $\theta = 0$ direction, the D atom feels appreciable repulsion from only one of the H atoms, whereas in an approach from any other direction the D atom enters a repulsive field from both H atoms.

For the approach of D to H—H along the line of centers of the atoms, the potential energy E_p is a function of only two coordinates, namely, the distances D—H and H—H. We can plot one distance along the x axis and the other distance along the y axis in a plane, and plot the energy E_p along the z axis normal to the plane. We can then visualize the potential energy $E_p(x, y)$ of the reacting system as a surface in three dimensions.

The idea that a chemical reaction can be represented by such a potential-energy surface was suggested by Marcelin* in 1915, but the first surface was actually computed in 1931 in a paper by Eyring and Polanyi.† With the techniques and calculators available at that time, they were not able to perform a purely theoretical calculation, but relied on a semiempirical approach based on spectroscopic data. In any event, later

---

*Ann. Phys., 3, 158 (1915).

†H. Eyring and M. Polanyi, Z. Phys. Chem., B12, 279 (1931).

more exact calculations showed that the surface they constructed by a combination of calculation and intuition is essentially correct. It is shown in Fig. 15.3(a, b) in the form of drawings of a three-dimensional model.

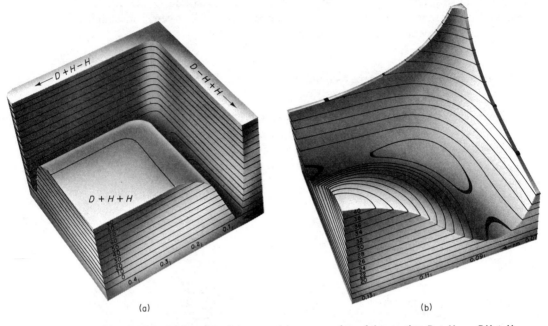

FIGURE 15.3 (a) Drawing of the potential-energy surface of the reaction $D + H_2 \rightarrow DH + H$ as constructed by Goodeve (1934) from the calculations of Eyring and Polanyi (1931). (b) Closeup view of the region near the saddle point between the two valleys of potential energy corresponding to H_2 and DH. [After F. H. Johnson, H. Eyring, and B. J. Stover, *The Theory of Rate Processes in Biology and Medicine* (New York; John Wiley & Sons, Inc., 1974)]

We can trace the path of the reaction over this surface; it is the path from reactant side to product side that follows the contour of minimal potential energy. If interconversion of translational and vibrational energy is considered, the path of reaction is more like that of a bobsled sliding on its run. The path traverses a deep valley $(D + H_2)$, rises over a hump to a saddle point at the activated complex $(D - H - H)$, and then passes down the other side of the pass into another valley $(DH + H)$. The reaction path is called the *reaction coordinate*. Figure 15.3(c) is a contour map of the region of the saddle point. The elevation of the saddle point is 38 kJ, which is the activation energy of the reaction. This is at the configuration of the activated complex, where $r_1 = r_2 = 93$ pm. This distance is considerably greater than the normal internuclear separation in H_2, which is 74 pm.

If the potential energy is drawn as a function of distance along the reaction path, Fig. 15.4 is obtained. This picture can be compared to the conventional *reaction profile* in Fig. 13.10 that is so often used to illustrate the concepts of activation energy and activated complex. We can now understand what is actually plotted there—the potential energy of the system vs. the distance along a reaction path that follows the route of lowest energy.

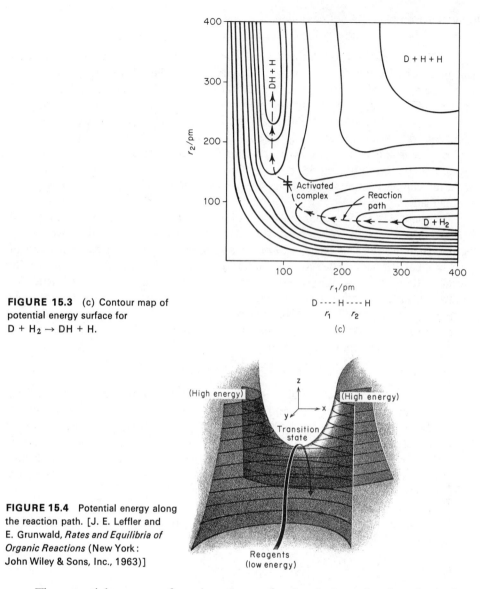

FIGURE 15.3 (c) Contour map of potential energy surface for $D + H_2 \rightarrow DH + H$.

$$D \cdots\cdots H \cdots\cdots H$$
$$r_1 \qquad r_2$$

(c)

FIGURE 15.4 Potential energy along the reaction path. [J. E. Leffler and E. Grunwald, *Rates and Equilibria of Organic Reactions* (New York: John Wiley & Sons, Inc., 1963)]

The potential-energy surface gives a map of a chemical reaction from beginning to end. In any reaction, there is always a particular configuration at the saddle point. In many respects this configuration of atoms, the activated complex, is like an ordinary molecule, except that it is not an equilibrium configuration. The activated complex and its properties are designated with the symbol $\neq$. We can now consider that any reaction takes place in two stages: (1) the reactants come together to form an activated complex; (2) the activated complex decomposes into the products. These stages are not sharply defined in any way, and from a dynamic point of view the reaction process is smooth and continuous. We can, however, designate a transition state as the highest region of the potential energy surface along the reaction path.

The quantitative formulation of rate constants in terms of activated complexes was first extensively used in the work of Henry Eyring. The theory has been applied to a wide variety of rate processes in addition to chemical reactions, such as flow of liquids, diffusion, dielectric loss, and internal friction in high polymers. It is possible to formulate the reaction rate entirely in terms of properties (1) of the reactants and (2) of the transition state at which they have formed the activated complex. The rate of reaction is the number of activated-complex species passing in unit time over the top of the potential-energy barrier. This rate is equal to the concentration of activated complexes times the average frequency with which a complex moves to the product side.

The calculation of the concentration of activated complexes is greatly simplified if we use the equations of ordinary statistical thermodynamics. In other words, we assume that the ratio of concentration of activated complexes to concentration of reactants has the same value as the equilibrium ratio. It is well to emphasize again, however, that activated complexes are not stable intermediate complexes (such as catalyst–substrate complexes, for instance). The activated complex is simply one stage in a continuous process leading from reactants to products.

We shall consider the calculation by activated-complex theory of the specific rate of a gas-phase bimolecular reaction,

$$A + B \longrightarrow (AB)^{\ddagger} \longrightarrow \text{products}$$

A and B represent reactant molecules (not necessarily atoms) and $(AB)^{\ddagger}$ is the activated complex. The reactants and the complex are assumed to behave as perfect gases. The reaction path is indicated in Fig. 15.4, where potential energy is plotted against reaction coordinate. A narrow region of the mountain pass or col, of arbitrary length δ, defines the existence of the activated complex. Energies ϵ_0^A, ϵ_0^B, and $\epsilon_0^{\ddagger}$ denote the zero-point energies of the three species, and $\Delta\epsilon_0^{\ddagger} = \epsilon_0^{\ddagger} - \epsilon_0^A - \epsilon_0^B$ is the difference in zero-point energies between activated complex and reactants.

We now assume that the concentration $C^{\ddagger}$ of activated complexes (SI units, molecules per m³) can be calculated from Eq. (12.50), in terms of the molecular partition functions z' per unit volume,

$$C^{\ddagger} = C_A C_B \frac{z'_{\ddagger}}{z'_A z'_B} e^{-\Delta\epsilon_0^{\ddagger}/kT} \tag{15.10}$$

where $z'_{\ddagger}$ is the partition function (per unit volume) of the activated complex. Note that $\Delta\epsilon_0^{\ddagger}$ is the height of the lowest energy level of the complex above the sum of the lowest energy levels of the reactants A + B.

According to activated-complex theory, the rate of the second-order reaction with rate constant k'_2 is

$$\frac{-dC_A}{dt} = k'_2 C_A C_B = v^{\ddagger} C^{\ddagger} \tag{15.11}$$

The frequency $v^{\ddagger}$ of passage of $(AB)^{\ddagger}$ over the barrier is equal to the frequency with which a complex flies apart into products. The complex flies apart when one of its vibrations becomes a translation, and what was formerly one of the bonds holding the complex together becomes the direction of translation of the fragments of the separated complex.

From Eqs. (15.10) and (15.11), we can write for the rate constant,

$$k_2' = \nu^{\ddagger} \frac{z_{\ddagger}'}{z_A' z_B'} e^{-\Delta\epsilon_0^{\ddagger}/kT} \tag{15.12}$$

Now, $z_{\ddagger}'$ is just like a partition function for a normal molecule, except that one of its vibrational degrees of freedom is in the act of changing into a translation along the reaction coordinate. The partition function in one vibrational degree of freedom from Eq. (12.35) is

$$z_v^{\ddagger} = (1 - e^{-h\nu^{\ddagger}/kT})^{-1} \tag{15.13}$$

[The zero-point energy term is already included in Eq. (15.12).] For the particular anomalous vibration along the reaction coordinate, we can be sure that $h\nu^{\ddagger}/kT \ll 1$, since, at any temperature at which reaction is detectable, this "decomposition vibration" of the complex must, by hypothesis, be completely excited. Hence, if we expand

$$e^{-h\nu^{\ddagger}/kT} = 1 - \frac{h\nu^{\ddagger}}{kT} + \frac{1}{2}\left(\frac{h\nu^{\ddagger}}{kT}\right)^2 - \cdots$$

we can discard terms beyond the first power in $(h\nu^{\ddagger}/kT)$, and Eq. (15.13) becomes

$$z_v^{\ddagger} = \left(\frac{h\nu^{\ddagger}}{kT}\right)^{-1} = \frac{kT}{h\nu^{\ddagger}}$$

Our next step is to factor this particular $z_v^{\ddagger}$ out of the complete $z_{\ddagger}'$, so that

$$z_{\ddagger}' = z_v^{\ddagger} z^{\ddagger\prime} = \left(\frac{kT}{h\nu^{\ddagger}}\right) z^{\ddagger\prime}$$

When we substitute this into Eq. (15.12), we obtain the Eyring equation for the rate constant,

$$k_2' = \frac{kT}{h} \frac{z^{\ddagger\prime}}{z_A' z_B'} e^{-\Delta\epsilon_0^{\ddagger}/kT} \tag{15.14}$$

This is the theoretical expression given by activated-complex theory for a bimolecular rate constant. One can see that it includes explicitly terms that depend on properties of the reactant molecules and of the activated complex. We can often deduce a reasonable structure for the activated complex from our general knowledge of chemical bonding, or in some cases we can obtain the structure of the complex from the calculated potential-energy surface for the reaction. Then we can use the general formulas for molecular partition functions (Table 12.3) to calculate k_2' from Eq. (15.14) and the measured or calculated activation energy.

The dimensions of k_2' are Vt^{-1} [SI units, (molecules/m³)$^{-1}$ s^{-1}]. In concentration units, $k_2 = Lk_2'$ [SI, (mole/m³)$^{-1}$ s^{-1}].

15.7 Activated-Complex Theory in Thermodynamic Terms

The properties of activated complexes can be expressed in terms of thermodynamic functions instead of partition functions. If we consider again the reaction $A + B \longrightarrow (AB)^{\ddagger} \longrightarrow$ products, with

$$K_c^{\ddagger} = \frac{c^{\ddagger}c^0}{c_A c_B} \qquad \text{and} \qquad v = \frac{kT}{h} c^{\ddagger}$$

we can write the rate constant from Eq. (15.11) as

$$k_2 = \frac{kT}{h} \left(\frac{1}{c^0}\right) K_c^{\ddagger} \tag{15.15}$$

Recall that c^0 is the concentration of the standard state, usually 1 mol dm^{-3}. We define $\Delta G^{0\ddagger} = -RT \ln K_c^{\ddagger}$, with $\Delta G^{0\ddagger} = \Delta H^{0\ddagger} - T \Delta S^{0\ddagger}$. Eq. (15.15) becomes

$$k_2 = \frac{kT}{h} \left(\frac{1}{c^0}\right) e^{-\Delta G^{0\ddagger}/RT} = \frac{kT}{h} \left(\frac{1}{c^0}\right) e^{\Delta S^{0\ddagger}/R} e^{-\Delta H^{0\ddagger}/RT} \tag{15.16}$$

The quantities $\Delta G^{0\ddagger}$, $\Delta H^{0\ddagger}$, and $\Delta S^{0\ddagger}$ are called the Gibbs free energy of activation, the enthalpy of activation, and the entropy of activation, respectively. The superscripts ^{0} are usually omitted from the symbols, but they must be understood.

The temperature coefficient of the rate constant is conveniently derived from Eq. (15.15) by taking logarithms and differentiating,

$$\frac{d \ln k_2}{dT} = \frac{1}{T} + \frac{d \ln K_c^{\ddagger}}{dT}$$

Since $K_c^{\ddagger}$ is an equilibrium constant in terms of concentration, $d \ln K_c^{\ddagger}/dT = \Delta U^{\ddagger}/RT^2$. [Please prove this from Eqs. (8.40) and (6.23).] Therefore, from Eq. (15.15),

$$\frac{d \ln k_2}{dT} = \frac{RT + \Delta U^{\ddagger}}{RT^2}$$

Thus the Arrhenius activation energy of Eq. (13.32) is

$$E_a = RT + \Delta U^{\ddagger} \tag{15.17}$$

From Eq. (6.14), $\Delta U^{\ddagger} = \Delta H^{\ddagger} - \Delta(PV)^{\ddagger}$. In liquid and solid systems, $\Delta(PV)^{\ddagger}$ is very small at ordinary pressures, and one can take

$$E_a \approx \Delta H^{\ddagger} + RT \qquad \text{(condensed systems)} \tag{15.18}$$

For reactions of ideal gases, from Eq. (6.25),

$$\Delta H^{\ddagger} = \Delta U^{\ddagger} + \Delta v^{\ddagger} RT \tag{15.19}$$

which $\Delta v^{\ddagger}$ is the stoichiometric number of complex, always equal to 1, minus the sum of the numbers for reactants. Thus we have, from Eqs. (15.17) and (15.19),

$$\Delta H^{\ddagger} = E_a + (\Delta v^{\ddagger} - 1)RT \tag{15.20}$$

The standard state to which $K^{\ddagger}$ and $\Delta G^{0\ddagger}$ are referred is usually taken to be 1 mol dm^{-3}, in which case the corresponding units of k_2 in Eq. (15.16) are (mol/dm^3)$^{-1}$ s^{-1}.

15.8 The Entropy of Activation

The experimental entropy of activation can be calculated from the rate constant at a given temperature and the experimental activation energy.

> **Example 15.7** For dimerization of butadiene, $2C_4H_6 \longrightarrow C_8H_{12}$ (3-vinylcyclohexene), from 440 to 660 K, the experimental rate constant is $k_2 = 9.20 \times 10^6 \exp\left(-99.12 \text{ kJ}/RT\right)$ dm³ mol⁻¹ s⁻¹. What are the $\Delta H^{\ddagger}$ and $\Delta S^{\ddagger}$ for the reaction at 600 K?
>
> From Eq. (15.16), with $\Delta H^{\ddagger} = E_a - 2RT$ from Eq. (15.20),
>
> $$k_2 = \frac{kT}{hc^{\circ}} e^{\Delta S^{\ddagger}/R} e^{-E_a/RT} e^{2RT/RT}$$
>
> At 600 K $\Delta H^{\ddagger} = 89.1$ kJ mol⁻¹, and with $c^{\circ} = 1$ mol dm⁻³,
>
> $$9.20 \times 10^6 \text{ dm}^3 \text{ mol}^{-1} \text{ s}^{-1} = (1.25 \times 10^{13} \text{ dm}^3 \text{ mol}^{-1} \text{ s}^{-1})\, e^{\Delta S^{\ddagger}/R}\, e^2$$
>
> $$e^{\Delta S^{\ddagger}/R} = \frac{9.20 \times 10^6}{(1.25 \times 10^{13})(7.360)} = 1.00 \times 10^{-7}$$
>
> $$\Delta S^{\ddagger} = -134 \text{ J K}^{-1} \text{ mol}^{-1}$$
>
> (Note that the value of $\Delta S^{\ddagger}$ depends on the units of the rate constant k_2, which are determined by the standard state chosen for the activated complex.)

The experimental activation entropy $\Delta S^{\ddagger}$ provides one of the best indications of the nature of the transition state. A positive activation entropy $\Delta S^{\ddagger}$ means that the entropy of the complex is greater than the entropy of the reactants. A loosely bound complex has a higher entropy than a tightly bound one. More often there is a decrease in entropy in passing to the activated state simply because the complex is more organized than the sum of its separate components. In bimolecular reactions, the complex is formed by association of two individual molecules, and there is a loss of translational and rotational freedom, so that $\Delta S^{\ddagger}$ is usually negative. In fact, sometimes $\Delta S^{\ddagger}$ is not notably different from ΔS° for the complete reaction. When this situation occurs in reactions of the type $A + B \longrightarrow AB$, it indicates that the activated complex $[AB]^{\ddagger}$ is similar in structure to the product molecule AB. Formerly, such reactions were considered to be abnormal, since they have unusually low steric factors. With the advent of transition-state theory, it became clear that the low steric factor is the result of the increase in order, and consequent decrease in entropy, when the complex is formed.

15.9 Chemical Dynamics

We have now discussed two important methods for theoretical calculation of reaction rates, collision theory and activated-complex theory. These methods have in common their reliance on a statistical treatment of the reacting system to give information about the average behavior of collections of large numbers of reacting molecules. But why don't we follow the individual molecule and see what happens when it meets another molecule and either reacts or bounces away? Physicists, with their accelerators, cloud or bubble chambers, and devices for particle detection, first looked at the dynamics of individual reactions of nuclei and particles. Only later did they realize that the interiors of stars are fiery furnaces in which nuclear reactions occur as a result of thermal activation. Chemists, on the other hand, spent many years studying thermal reactions in laboratory furnaces, before they began to look at individual molecular collisions.

To study the dynamics of molecular reactions, we must have a source of atoms or molecules that have defined directions and energies. The distribution of the energy among translational, rotational, vibrational, and electronic degrees of freedom should also be known. Most problems in the dynamics of molecular reactions are concerned with the internal motions of the molecules as well as with the velocity (speed and direction) of their translation. Therefore, two principal experimental techniques are employed in molecular reaction dynamics: the use of molecular beams of defined velocity and the use of sophisticated spectroscopic devices to detect changes in the intramolecular energy levels.

15.10   Reactions in Molecular Beams

An experimental apparatus for the study of chemical reactions in molecular beams is shown in Fig. 15.5. It includes two sources, one for each reactant species, and a detector of products, all mounted in a large high-vacuum chamber. This apparatus is used to study reactions in crossed beams. In some cases, however, only one reactant is in the form of a beam, which is allowed to intercept a region containing the second reactant in more diffuse form. The most exacting work requires that each beam be subjected to velocity analysis before it enters the reactive zone. The velocities of the reactant molecules are independently controlled, and the directions of intersection of the crossed beams are varied.

The most interesting data from molecular-beam experiments are the angular distributions of the reaction products. These data are plotted as contour maps that show the flux of products $P(\theta)$ as a function of the angle θ made by the trajectory of the product molecules from the line of centers of the colliding pair of reactant molecules.

Results for the reaction $K + I_2 \rightarrow KI + I$ are shown in Fig. 15.6(a). In this reaction, most of the KI molecules are scattered in the forward direction, i.e., in the same direction as the K beam or at small angles θ to it. This kind of reaction is called a *stripping reaction*. The translational kinetic energy of the KI molecules is quite small and hence the KI product will usually have considerable vibrational energy.

Results for $K + ICH_3 \rightarrow KI + CH_3$ are shown in Fig. 15.6(b). This is a typical backward or *rebound reaction*, with the KI product scattered backward. Most of the exothermic energy of reaction is transferred to the kinetic energy of translation of the rebounding KI.

Figure 15.6(c) shows the $P(\theta)$ contour map for the reaction $Cs + RbCl \rightarrow CsCl + Rb$. Here we see an approximate forward–backward symmetry in the CsCl flux distribution. Such a result indicates that a complex RbClCs has been formed with an appreciably long lifetime. The complex then dissociates along its symmetry axis.

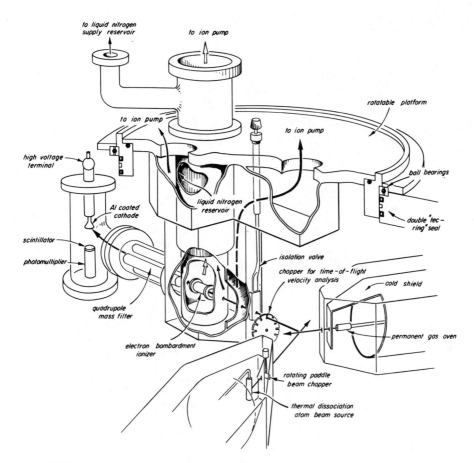

FIGURE 15.5 Apparatus for studying chemical reactions in crossed molecular beams. [Y. T. Lee, J. D. McDonald, P. R. LeBreton, and D. R. Herschbach, *Rev. Sci. Inst. 40,* 1402 (1969)]

15.11 Theory of Unimolecular Reactions

From 1918 to 1935, a number of gas reactions were found to be kinetically first order and apparently simple unimolecular decompositions. These reactions presented a paradox: The necessary activation energy must come from kinetic energy transferred during collisions, yet the reaction rate did not depend on collision frequency.

In 1922, F. A. Lindemann showed how a collisional mechanism for activation could lead to first-order kinetics. Consider a molecule A that decomposes according to $A \rightarrow B + C$, with a first-order rate law, $-d[A]/dt = k_{ex}[A]$. In a vessel full of A, the intermolecular collisions are continually causing a redistribution of molecular energies, producing molecules with energies both higher and lower than the average. Sometimes molecules acquire an energy above the critical value necessary for the

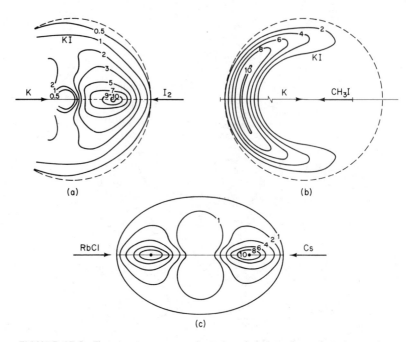

FIGURE 15.6 Flux contour maps to show the relative numbers of product molecules collected at different angles. (a) The product KI from the reaction $K + I_2 \rightarrow$ KI + I. [R. D. Levine and R .B. Bernstein, *Molecular Reaction Dynamics,* (New York: Oxford University Press, 1974)] (b) The product KI from the reaction $K + ICH_3 \rightarrow$ KI + CH_3. [A. M. Rulis and R. B. Bernstein, *J. Chem. Phys. 57,* 5497 (1972)] (c) The product CsCl from the reaction Cs + RbCl $\rightarrow$ CsCl + Rb. [R. D. Levine and R. B. Bernstein, *Molecular Reaction Dynamics,* (New York: Oxford University Press, 1974)]

activation that precedes decomposition. Let us suppose that a certain time lag exists between activation and decomposition; the activated molecule does not immediately fall to pieces, but moves for a while in its activated state. Sometimes it may meet an energy-poor molecule, and in the ensuing collision, it may be robbed of enough energy to become deactivated.

The situation can be represented as follows:

$$A + A \underset{k_{-2}}{\overset{k_2}{\rightleftharpoons}} A + A^*, \qquad A^* \overset{k_1}{\longrightarrow} B + C$$

Activated molecules are denoted by A^*. The bimolecular rate constant for activation is k_2, and for deactivation, k_{-2}. The decomposition of an activated molecule is a true unimolecular reaction with rate constant k_1.

The process called *activation* consists essentially in the transfer of translational kinetic energy into energy stored in internal degrees of freedom, especially vibrational degrees of freedom. The mere fact that a molecule is moving rapidly, i.e., has a high translational kinetic energy, does not make it unstable. To cause reaction, the energy must get into the chemical bonds, where vibrations of high amplitude will lead to ruptures and rearrangements. The transfer of energy from translation to vibration can occur only in collisions with other molecules or with the wall. The situation is like that of two rapidly moving cars; their kinetic energies will not wreck them unless they happen to collide and the kinetic energy of the whole is transformed into internal energy of the parts.

The point of the Lindemann theory is that there is a lag between the activation of internal degrees of freedom and the subsequent decomposition of the molecule. The reason is that a polyatomic molecule can take up collisional energy into a number of its $3N - 6$ vibrational degrees of freedom, and then some time may elapse before enough energy flows into the bond that breaks or rearranges.

The rate equations for the mechanism above are:

$$\frac{d[A^*]}{dt} = k_2[A]^2 - k_{-2}[A^*][A] - k_1[A^*]$$

$$\frac{-d[A]}{dt} = k_2[A]^2 - k_{-2}[A^*][A]$$

$$\frac{d[B]}{dt} = k_1[A^*]$$

The *steady-state approximation* is applied. After the reaction has been under way for a short time, the rate of formation of activated molecules can be set equal to their rate of disappearance, so that the net rate of change in $[A^*]$ is zero, $d[A^*]/dt = 0$. The first of the preceding equations then gives for the steady-state concentration $[A^*] = k_2[A]^2/(k_2[A] + k_1)$. The reaction velocity is the rate at which A^* decomposes into B and C, or

$$\frac{d[B]}{dt} = k_1[A^*] = \frac{k_1 k_2[A]^2}{k_{-2}[A] + k_1}$$

If the rate of decomposition of A^* is much greater than its rate of deactivation, $k_1 \gg k_{-2}[A]$, and the rate reduces to $d[B]/dt = k_2[A]^2$. This is a second-order law.

On the other hand, if the rate of deactivation of A^* is much greater than its rate of decomposition, $k_{-2}[A] \gg k_1$, and the overall rate becomes

$$\frac{d[B]}{dt} = \frac{k_1 k_2}{k_{-2}}[A] = k_{ex}[A] \tag{15.21}$$

This is a first-order law. It is thus evident that first-order kinetics can be obtained even with a collisional mechanism for activation. This will be the result whenever the activated molecule has so long a lifetime that it is usually deactivated by collision before it can break into fragments.

As pressure in the reacting system is decreased, the rate of deactivation, $k_{-2}[A^*]$ [A], must likewise decrease, and at low enough pressures the condition for first-order kinetics must always fail when $k_{-2}[A]$ is no longer much greater than k_1. The observed first-order rate constant should therefore decline at low pressures.

Figure 15.7 shows the rate constants for the first-order thermal isomerization of cylcopropane at various pressures,

$$\underset{\substack{| \quad | \\ CH_2-CH_2}}{\overset{CH_2}{\diagdown}} \longrightarrow CH_3-CH=CH_2$$

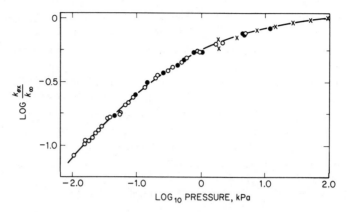

FIGURE 15.7 Dependence of the apparent first-order rate constant k_{ex} for isomerization of cyclopropane on pressure. The log of k_{ex}/k_∞ is plotted vs. log P, where k_∞ is the limiting high-pressure value of k_{ex}. [H. O. Pritchard, R. G. Sowden, and A. F. T. Trotman–Dickenson, *Proc. R. Soc. A* **217**, 563 (1953)]

The marked falloff in k_{ex} as the pressure is lowered confirms the theoretical prediction, in a qualitative way. If the decrease of k_{ex} at low pressures is merely the result of a lowered probability of deactivation of A*, it should be possible to restore the initial rate by adding a sufficient pressure of a completely inert gas. This effect of inert gas has been confirmed in a number of cases.

Table 15.2 lists some reactions now believed to be unimolecular processes. The quantitative theory of such reactions is an important field of contemporary research.

TABLE 15.2
UNIMOLECULAR GAS PHASE DECOMPOSITIONS[a]

Reactant	Products	log A (s^{-1})	E_a (kJ mol^{-1})
$CH_3 \cdot CH_2Cl$	$C_2H_4 + HCl$	14.6	254
$CCl_3 \cdot CH_3$	$CCl_2{=}CH_2 + HCl$	12.5	200
t-Butyl bromide	Isobutene + HBr	14.0	177
t-Butyl alcohol	Isobutene + H_2O	11.5	228
$ClCOOC_2H_5$	$C_2H_5Cl + CO_2$	10.7	123
$ClCOOCCl_3$	$COCl_2$	13.15	174
Cyclobutane	C_2H_4	15.6	262
Perfluorocyclobutane	C_2F_4	15.95	310
N_2O_4	NO_2	16	54

[a]From S. W. Benson, *The Foundations of Chemical Kinetics* (New York: McGraw-Hill Book Company, 1960).

15.12 Reactions in Solution

The statistical mechanics of liquid solutions is not nearly so advanced as that of gases. Consequently, the general theory of reaction rates in solution is less complete than the theory for gas reactions. There is, however, a vast amount of experimental data on reactions in solution, and many aspects of rate processes in solution are quite well understood.

Many first-order reactions, such as the decomposition of N_2O_5, Cl_2O, or CH_2I_2, and the isomerization of pinene, proceed at about the same rate in gas phase and in solution. It appears, therefore, that the rate is the same whether a molecule becomes activated by collision with solvent molecules or by collisions in the gas phase with others of its own kind.

It is more remarkable that many second-order, presumably bimolecular, reactions have rates close to those predicted from the gas-kinetic collision theory. Some examples are shown in the last column of Table 15.3. The explanation of such an agreement seems to be the following. Any given reactant solute molecule must diffuse

TABLE 15.3
EXAMPLES OF REACTIONS IN SOLUTION

Reaction	Solvent	E_a (kJ mol^{-1})	A [Eq. (15.7)] (dm^3 mol^{-1} s^{-1})	A_{calc}/A_{obs}
$C_2H_5ONa + CH_3I$	C_2H_5OH	81.6	2.42×10^{11}	0.8
$C_2H_5ONa + C_6H_5CH_2I$	C_2H_5OH	83.3	0.15×10^{11}	14.5
$NH_4CNO \longrightarrow (NH_2)_2CO$	H_2O	97.1	42.7×10^{11}	0.1
$CH_2ClCOOH + OH^-$	H_2O	108.4	4.55×10^{11}	0.6
$C_2H_5Br + OH^-$	C_2H_5OH	89.5	4.30×10^{11}	0.9
$(C_2H_5)_3N + C_2H_5Br$	C_6H_6	46.9	2.68×10^2	1.9×10^9
$CS(NH_2)_2 + CH_3I$	$(CH_3)_2CO$	56.9	3.04×10^6	1.2×10^5
$C_{12}H_{22}O_{11} + H_2O \longrightarrow 2C_6H_{12}O_6$ (sucrose)	$H_2O(H^+)$	107.9	1.5×10^{15}	1.9×10^{-4}

for some distance through the solution before it meets another reactant molecule. Thus the number of such encounters will be lower than in the gas phase. Having once met, however, the two reactant molecules will remain close to each other for a considerable time, being surrounded by a "cage" of solvent molecules. Thus repeated collisions may occur between the same pair of reactant molecules. The net result is that the effective collision number may not be much different from that in the gas phase.

A neat experimental demonstration of the existence of the solvent cage was given by Lyon and Levy in their study of the photochemical decomposition of azomethane, $CH_3-N{=}N-CH_3 \rightarrow C_2H_6 + N_2$. When an azomethane molecule absorbs a quantum of ultraviolet light $h\nu$, it dissociates to yield two methyl radicals, $CH_3-N{=}N-CH_3 \xrightarrow{h\nu} 2CH_3 + N_2$. The radicals subsequently combine to give ethane, $2CH_3 \rightarrow C_2H_6$. When an equimolar mixture of $CH_3-N{=}N-CH_3$ and the perdeuterated analog $CD_3-N{=}N-CD_3$ is irradiated in the gas phase, the ethane formed consists of CH_3-CH_3, CH_3-CD_3, and CD_3-CD_3 in about the statistically expected ratio of 1:2:1. When the photochemical decomposition of a mixture of the $CH_3-N{=}N-CH_3$ and $CD_3-N{=}N-CD_3$ is carried out in solution in isooctane, the product ethanes are almost exclusively CH_3-CH_3 and CD_3-CD_3. This experiment strongly supports the idea that the methyl radicals formed in the primary photochemical splitting of azomethane in solution are securely trapped in a solvent cage, so that the radicals from each azomethane molecule combine with each other before they have time to diffuse away and encounter radicals from other azomethane molecules.

Other bimolecular reactions in solution have rates that differ markedly from those of the corresponding reactions in the gas phase, by factors ranging from 10^9 to 10^{-9}. A high-frequency factor corresponds to a large positive ΔS^+, and a low-frequency factor to a negative ΔS^+. The remarks on the significance of ΔS^+ in gas reactions apply equally well here. Association reactions have low-frequency factors, owing to the decrease in entropy when the activated complex is formed. An example is the Menschutkin reaction: $(C_2H_5)_3N + C_2H_5I = (C_2H_5)_4NI$. The $\Delta S^+ = -160$ $J\,K^{-1}\,mol^{-1}$. [What would be the ratio of pre-exponential factors corresponding to $\Delta S^+ = -160\,J\,K^{-1}\,mol^{-1}$ and $\Delta S^+ = 0$ at 300 K?]

15.13 Diffusion-Controlled Reactions

In a gas, the upper limit for the rate of a bimolecular reaction is set by the collision frequency. If the activation energy is zero and the steric factor is unity, all collisions will lead to reaction. In a liquid, the upper limit to the reaction rate is set by the frequency of first encounters between reactant molecules moving randomly through the solution.

In 1917, M. Smoluchowski worked out a theory for the growth of a colloidal particle by accretion of smaller particles that diffuse toward it and become incorporated at its surface. Peter Debye applied this theory to calculate the rate of first encounters for reactions in solution. If every encounter leads to reaction, the limiting second-order rate constant under conditions of diffusion control is

$$k_2 = 4\pi d_{AB} L(D_A + D_B) \tag{15.22}$$

D_A and D_B are the diffusion coefficients (see Section 16.12) of reactants A and B and d_{AB} is the effective collision diameter. For typical values of $d_{AB} = 5 \times 10^{-10}$ m and $D_A = D_B = 10^{-9}$ m^2 s^{-1} for small molecules, $k_2 = 4 \times 10^9$ dm^3 mol^{-1} s^{-1}. For a gas reaction, the collision frequency would give a maximum rate constant of about $k_2 = 10^{11}$ dm^3 mol^{-1} s^{-1}. The fastest reactions in liquid solutions are thus subject to a limiting diffusion control and never attain a rate as high as the gas-phase maximum.

For spherical particles of radius a moving in a medium of viscosity η, the diffusion coefficient D is given by the Stokes–Einstein Relation,

$$D = kT/6\pi\eta a \tag{15.23}$$

It is often a good approximation to set $D_A \approx D_B$ and $a_A = a_B = d_{AB}/2$ in Eq. (15.22) and then to use Eq. (15.23) to obtain

$$k_2 = 8RT/3\eta \tag{15.24}$$

Note that the collision diameter has canceled out of this expression, so that to this approximation the rates of all diffusion-controlled reactions are the same, irrespective of the reaction partners.

Example 15.8 If the recombination of iodine atoms in hexane solution at 298 K is a diffusion-controlled reaction, estimate the second-order rate constant. The viscosity of hexane is 3.25×10^{-4} kg m^{-1} s^{-1} at 298 K.

When A = B, Eq. (15.24) becomes $k_2 = 4RT/3\eta$.

$$k_2 = (4)(8.314 \text{ J K}^{-1} \text{ mol}^{-1})(298 \text{ K})/3(3.25 \times 10^{-4} \text{ kg m}^{-1} \text{ s}^{-1})$$

$$= 1.01 \times 10^7 \text{ m}^3 \text{ mol}^{-1} \text{ s}^{-1} \text{ or } 1.01 \times 10^{10} \text{ dm}^3 \text{ mol}^{-1} \text{ s}^{-1}$$

(The experimental value is $k_2 = 1.3 \times 10^{10}$ dm^3 mol $^{-1}$ s^{-1}.)

Problems

1. Calculate the collision frequency between (a) H_2 molecules and (b) H_2 and D_2 molecules in an equimolar mixture of $H_2 + D_2$ at 100 kPa and 200 K.

2. At $T < 800$ K, $2C_2F_4 \rightarrow$ cyclo-C_4F_8 is second order with $k_2 = 10^{11.07}e^{-12\,870/T}$ cm^3 mol^{-1} s^{-1}. The molecular diameter of C_2F_4 is 5.12×10^{-10} m. Calculate k_2 from simple hard-sphere collision theory. Hence estimate the steric factor p at 725 K.

3. The Maxwell velocity distribution function $f(c)$ in two degrees of freedom is $f(c) = (m/kT) \exp(-mc^2/2kT)c$. Transform this into an expression in the kinetic energy $\epsilon = \frac{1}{2}mc^2$, $f(\epsilon) = (kT)^{-1}e^{-\epsilon/kT}$. Hence show that the fraction of molecules with $\epsilon > \epsilon_0$ is $e^{-\epsilon_0/kT}$.

4. The activation energy E_a of $2HI = H_2 + I_2$ is 186 kJ mol^{-1}. Calculate the second-order rate constant at 600 K and 800 K in L mol^{-1} s^{-1}, given the molecular diameter $d = 180$ pm.

5. The self-diffusion coefficient of CO is $D = 1.75 \times 10^{-5}$ m^2 s^{-1} at 273 K and 1 atm, and the density $\rho = 1.25$ kg m^{-3}. Calculate the molecular diameter.

6. In a high-vacuum chamber the pressure is 10^{-6} Pa. What is the mean free path at 300 K of a N_2 molecule with diameter 375 pm?

7. Calculate the rate constant of $S + O_2 = SO + O$ at 1000 K from activated-complex theory. Assume a linear activated complex S—O—O with S—O bond 135 pm, O—O bond 125 pm, and vibration frequencies at 1150, 520, and 1360 cm^{-1}. The experimental $E_a = 23.4$ kJ mol. The experimental $k_2 = 1.76 \times 10^7$ L mol^{-1} s^{-1}.

8. $CH_3I + (CH_3)_2 \cdot N \cdot C_6H_5 \rightarrow (CH_3)_3 \cdot N^+ \cdot C_6H_5)I^-$.

T (K)	298.0	313	333	353
$10^5 k_2$ (L mol^{-1} s^{-1})	8.39	21.0	77.2	238

Calculate $\Delta G^{\ddagger}$, $\Delta H^{\ddagger}$ and $\Delta S^{\ddagger}$ for this reaction with standard state 1 mol L^{-1}.

9. For internal rotations it is a good approximation to assume $\Delta S^{\ddagger} = 0$. The cis–trans isomerization of CH_3—O—N=O by rotation about the O—N bond has a rate constant $k_1 = 6.95 \times 10^5$ s^{-1} at 298 K. Assuming that $\Delta S^{\ddagger} = 0$, calculate the height of the barrier to internal rotation.

10. For the decomposition $N_2O_5 = N_2O_4 + \frac{1}{2}O_2$,

T (K)	298	308	318	328	338
$10^5 k_1$ (s^{-1})	1.72	6.65	24.95	75	240

Calculate $\Delta G^{\ddagger}$, $\Delta S^{\ddagger}$, $\Delta U^{\ddagger}$, and $\Delta H^{\ddagger}$ at 325 K.

11. From Eq. (15.16) one can calculate the effect of pressure on rate constants from the thermodynamic relation $(\partial \Delta G^{\ddagger}/\partial P)_T = \Delta V^{\ddagger}$. For the alkaline hydrolysis of propionamide, at 25°C, a hydrostatic pressure of 10^2 MPa increased the rate constant by a factor of 2. Estimate $\Delta V^{\ddagger}$, the activation volume for this reaction $[V^{\ddagger} - V(\text{reactants})]$.

12. At 360°C the rate constant for decomposition of isopropyl chloride is 150 times higher than that for ethyl chloride. The respective activation energies are 212.5 and 241.8 kJ mol^{-1}. Calculate the difference in the $\Delta S^{\ddagger}$ for the reactions. Is all the difference in rates due to the activation energy terms?

13. The hydrolysis of ATP catalyzed by myosin has a rate constant $k_2 = 1.6 \times 10^{22}$ $e^{-10\,500/T}$; catalyzed by H_3O^+, $k_2 = 2.4 \times 10^9$ $e^{-10\,600/T}$. What conclusions can you draw about the way the enzyme works in this reaction?

14. If the reaction

$$\underset{\text{O}}{\overset{\|}{CH_3 \cdot C \cdot CH_3}} + CH_3COOH \longrightarrow \underset{\text{OH}^+}{\overset{\|}{CH_3 \cdot C \cdot CH}} + CH_3COO^-$$

is diffusion controlled, estimate its rate in aqueous solutions at 25°C, where the viscosity of water is 8.937×10^{-4} kg m^{-1} s^{-1}.

15. Suppose that Eqs. (15.16) and (15.7) both gave the same theoretical rate constant for a reaction $A + B \longrightarrow AB$. Derive an expression for $\Delta S^{\ddagger}$. What would be the value of this $\Delta S^{\ddagger}$ at 1000 K if molar masses of A and B were 60 and 80 g mol^{-1}, respectively?

16. Typical values for partition functions are given in Table 12.3. Estimate the pre-exponential factor in Eq. (15.14) for the following types of reaction.
(a) $X + X \longrightarrow (XX)^{\ddagger}$
(b) $X + X_2 \longrightarrow (X\text{—}X\text{—}X)^{\ddagger}$
(c) $X + X_2 \longrightarrow \underset{\underset{X}{|}}{(X\text{—}X)^{\ddagger}}$
(d) $X_2 + X_2 \longrightarrow (X_2^{\ddagger})_2$, linear and nonlinear

17. Show that the probability that a molecule in a gas can travel a distance ℓ without a collision is $p = e^{-\ell/\lambda}$, where λ is the mean free path. Calculate p for $\ell = 1.0$ μm in N_2 gas at 300 K and pressures of 10^{-3}, 1.0, 10^3, and 10^5 Pa.

18. Show that the fraction of molecules in a gas having a kinetic energy greater than ϵ^* in three degrees of freedom is

$$N(\epsilon^*)/N_0 = 2(\epsilon^*/\pi kT)^{1/2} e^{-\epsilon^*/kT} + 1 - \text{erf}\,(\epsilon^*/kT)^{1/2}$$

For Br_2 at 500 K calculate this fraction for $\epsilon^* = 10$ kJ mol^{-1} and 100 kJ mol^{-1}. Compare the results with those for two degrees of freedom (Problem 3). What would be the result for helium?

19. The reaction cyclopropane $\longrightarrow$ propylene in homogeneous gas phase is catalyzed by HBr and by BCl_3. The second-order rate constants k_2 (dm^3 mol^{-1} s^{-1}) are

T (K)	650	685	715	740
HBr—k_2	0.0272	0.126	0.417	1.05
BCl_3—k_2	0.624	1.71	3.75	6.87

Calculate $\Delta H^{\ddagger}$ and $\Delta S^{\ddagger}$ for the reactions and comment on relevance to mechanisms of catalysis. [R. J. Johnson and V. R. Stimson, *Aust. J. Chem.* 28, 477 (1975)].

20. The Lindemann theory of unimolecular gas reactions in Section 15.11 predicts that the apparent first-order rate constant k_{ex} will decline at low pressures from its limiting high-pressure value k_∞. Show that the reactant concentration [A] at which $k_{ex} = k_\infty/2$ is $[A]_{1/2} = k_\infty/k_2$. The isomerization $CH_3NC \longrightarrow CH_3CN$ at 473 K has $k_{ex} = 63 \times 10^{-6} \text{ s}^{-1}$, and k_{ex} has declined to half this value at 2.67 kPa. Calculate k_2.

21. In the reaction of Problem 20, the temperature dependence gives an activation energy $E_a = 161 \text{ kJ mol}^{-1}$. According to the theory, this is the critical energy for collisional activation of A to A*. Calculate the rate constant k_2 on the basis of hard-sphere collision theory with a molecular diameter of 400 pm. Compare the result with that in Problem 20 and discuss any discrepancy.

16

Electrochemistry—
Ions in Solution

In 1813, Michael Faraday, a 22-year-old bookbinder's apprentice, went to the Royal Institution in London as the laboratory assistant of Humphry Davy. First with Davy and later independently, Faraday made detailed experiments on the decomposition of solutions of acids, bases, and salts by an electric current. With the help of William Whewell, an historian of science, he invented an elegant vocabulary for his electrochemical studies: electrode, electrolysis, ion, anion, and cation. The word *ion* comes from the Greek for *wanderer*. *Cations* were defined as positive ions: they move through the electrolyte toward the *cathode* (from the Greek, the *down road*). *Anions* are negatively charged ions: they move toward the *anode* (the *up road*).

16.1   Electrochemical Equivalence—the Faraday

Faraday discovered that in reactions at electrodes a constant amount of electric charge Q is always associated with a fixed equivalent amount of chemical change Δn. Thus

$$\frac{\Delta m}{M} = \Delta n = \frac{Q}{|z|F} = \frac{It}{|z|F} \tag{16.1}$$

The amount of electricity $Q = It$ is the product of the current I and the time t of its passage through the solution. The absolute value of the charge number of the ion that reacts at the electrode is $|z|$. The proportionality factor $F = 96\,485$ C mol^{-1} is now called the *faraday*. It is the amount of charge in coulombs carried by one mole of singly charged ions. The elementary unit of electric charge, the charge of a proton, is $e = 1.60 \times 10^{-19}$ C. Therefore, $F = Le$, where L is the Avogadro constant.

A measurement of the extent of chemical reaction in an electrolytic cell provides a measure of the amount of electric charge that has passed through the cell. A device for measuring electric charge $Q = It$ is called a *coulometer*. An example is the silver coulometer, which uses platinum electrodes in an aqueous solution of silver nitrate. The increase in mass of the cathode is measured after the passage of current. The reaction at the cathode is $Ag^+ + e = Ag$. From Eq. (16.1), one mole of silver, 0.107 870 kg, is deposited on the cathode for each faraday passed through the coulometer. Thus one coulomb of charge (1 C) is equivalent to $0.107\ 870/96\ 485 = 1.1180 \times 10^{-6}$ kg of silver. Note that a coulometer measures total charge passed. The current I might change during the time t of measurement, so that the coulometer measures $Q = \int_0^t I \, dt'$.

Example 16.1 A solution of $ZnSO_4$ was electrolyzed for 30.0 min at a constant current of 0.750 A with platinum electrodes. In this time 0.352 g Zn was deposited on the cathode. Bubbles of H_2 were observed to leave the cathode during electrolysis. What volume of H_2 at STP was formed?

From Eq. (16.1), if all the charge passed through the cell discharged Zn^{2+} ions,

$$\Delta m = M \, \Delta n = MIt/|z| F = \frac{(65.38 \times 10^{-3} \text{ kg mol}^{-1})(0.750 \text{ A})(1800 \text{ s})}{|2|(96\ 485 \text{ C mol}^{-1})}$$

$$= 4.57 \times 10^{-4} \text{ kg}$$

The yield of Zn was therefore $3.52/4.57 = 0.770$. The rest of the charge led to evolution of H_2 in the amount

$$\Delta n = \frac{(0.230)(0.750 \text{ A})(1800 \text{ s})}{96\ 485 \text{ C mol}^{-1}} = 0.00322 \text{ mol}$$

or

$$V = \Delta n \, RT/P = \frac{(0.00322 \text{ mol})(8.314 \text{ J K}^{-1} \text{ mol}^{-1})(273 \text{ K})}{101.3 \times 10^3 \text{ Pa}}$$

$$= 72.1 \times 10^{-6} \text{ m}^3$$

16.2 Conductivity of Solutions

One of the first theoretical problems in electrochemistry was to understand how solutions of electrolytes conduct an electric current.

Metallic conductors were known to obey Ohm's Law,

$$I = \frac{\Delta \Phi}{\mathcal{R}} \tag{16.2}$$

where I is the current (in SI, amperes), $\Delta \Phi$ is the difference in electric potential between terminals of the conductor (SI, volts), and the proportionality factor $\mathcal{R}$ is the resistance

(SI, ohms Ω). The resistance depends on the dimensions of the conductor. For a conductor of uniform cross section,

$$\Re = \frac{\rho \ell}{A} \tag{16.3}$$

Here ℓ is length, A cross-sectional area, and ρ is called the *resistivity*. The reciprocal of resistance is *conductance* and the reciprocal of resistivity is *conductivity* κ. Since $\Delta \Phi = E\ell$, where E is the magnitude of the electric field, from Eqs. (16.2) and (16.3) one has

$$I/A = \kappa E \tag{16.4}$$

The conductivity of solutions is measured with an alternating-current (a-c) bridge, such as that shown in Fig. 16.1. With a-c frequencies in the audio range, 1000 to 4000 Hz, the direction of the current changes so rapidly that buildup of charge at the electrodes (polarization) is eliminated. The balance point of the bridge is indicated on the cathode-ray oscilloscope. A typical conductivity cell X is also shown in Fig. 16.1.

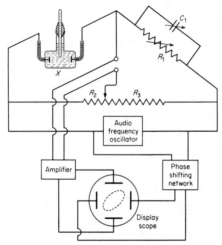

FIGURE 16.1 An a-c Wheatstone bridge for measurement of conductance of electrolytes.

As soon as reliable conductivity data were available, it became apparent that solutions of electrolytes follow Ohm's Law. Conductivity is independent of potential difference. Any conductivity theory must explain this fact: The electrolyte is always ready to conduct electricity, and this capability is not something produced by the applied electric field. On this score, the ingenious theory proposed in 1805 by Grotthuss was not acceptable. He supposed the molecules of electrolyte to be polar, with positive and negative ends. An applied field lines them up in a chain. Then the field causes the molecules at the end of the chain to dissociate, the free ions thus formed being discharged at the electrodes. Despite its shortcomings, the Grotthuss theory was valuable in emphasizing the necessity of having free ions in the solution to explain the observed conductivity. A mechanism with some features similar to that of Grotthuss actually occurs in some cases (see Fig. 16.5). In 1857, Clausius proposed that especially energetic collisions between undissociated molecules, in electrolytes, maintained at equilibrium a small number of ions. These ions were believed to be responsible for the observed conductivity.

16.3 Molar Conductivity

From 1869 to 1880, Friedrich Kohlrausch and his coworkers published a long series of careful conductivity investigations. The measurements were made over a range of temperatures, pressures, and concentrations. Typical of this painstaking work was the extensive purification of the water used as a solvent. After 42 successive distillations *in vacuo*, they obtained a conductivity water with $\kappa = 4.3 \times 10^{-6} \, \Omega^{-1}$ m^{-1} at 18°C. Ordinary distilled water in equilibrium with carbon dioxide of the air has a conductivity of about $70 \times 10^{-6} \, \Omega^{-1} \, m^{-1}$.

To reduce conductivities to a common concentration basis, a function called *molar conductivity* is defined by

$$\Lambda = \frac{\kappa}{c} \tag{16.5}$$

(Note that molar conductivity is not conductivity per mole but conductivity per unit concentration.) The SI unit of Λ is $\Omega^{-1} \, m^{-1}/mol \, m^{-3} = \Omega^{-1} \, m^2 \, mol^{-1}$. Some values for Λ are plotted in Fig. 16.2.

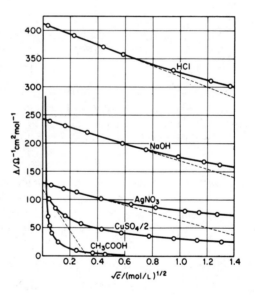

FIGURE 16.2 Molar conductivities Λ at 298.15 K of electrolytes in aqueous solution vs. square root of concentration.

On the basis of conductivities, two broad classes of electrolytes are distinguished. *Strong electrolytes,* such as most salts, and acids like hydrochloric, nitric, and sulfuric, have high molar conductivities that increase only moderately with increasing dilution. *Weak electrolytes,* such as acetic and other organic acids and aqueous ammonia, have much lower molar conductivities at high concentrations, but their values increase greatly with increasing dilution.

The value of Λ extrapolated to zero concentration is called *molar conductivity at infinite dilution*, Λ_0. The extrapolation is made readily for strong electrolytes but is impossible to make accurately for weak electrolytes because of their steep increase in Λ at high dilutions, where the experimental measurements become uncertain. Data for strong electrolytes are fairly well represented by an empirical equation,

$$\Lambda = \Lambda_0 - k_c c^{1/2} \tag{16.6}$$

where k_c is an experimental constant. The dashed lines in Fig. 16.2 illustrate this equation.

Example 16.2 What is the conductivity at 25°C of an aqueous solution containing 50.0 g L^{-1} $CuSO_4 \cdot 5H_2O$?

The concentration of the solution is $c = 50.0$ g $L^{-1}/249.7$ g $mol^{-1} = 0.200$ mol L^{-1}. From Fig. 16.2, for $CuSO_4/2$, $c = 0.400$ mol L^{-1}, $\sqrt{c} = 0.632$ (mol L^{-1})$^{1/2}$, $\Lambda = 0.0030$ Ω^{-1} m^2 mol^{-1}. From Eq. (16.5),

$$\kappa = \Lambda c = (0.00380 \ \Omega^{-1} \ m^2 \ mol^{-1})(0.400 \times 10^3 \ mol \ m^{-3})$$

$$= 1.52 \ \Omega^{-1} \ m^{-1}$$

Kohlrausch discovered that the difference in Λ_0 for pairs of salts having a common ion is always approximately constant. For example, at 298 K in units of Ω^{-1} m^2 mol^{-1},

	Λ_0		Λ_0		Λ_0
NaCl	0.01281	NaNO$_3$	0.01230	NaOH	0.02465
KCl	0.01498	KNO$_3$	0.01455	KOH	0.02710
	0.00217		0.00225		0.00245

No matter what the anion may be, there is an approximately constant difference between the Λ_0 values of potassium and sodium salts. This behavior can be explained if Λ_0 is the sum of two independent terms, one characteristic of the anion and one of the cation. For an electrolyte that yields v^+ cations and v^- anions,

$$\Lambda_0 = v^+ \Lambda_0^+ + v^- \Lambda_0^- \tag{16.7}$$

where Λ_0^+ and Λ_0^- are the *molar ionic conductivities* at infinite dilution. This is Kohlrausch's *Law of the Independent Migration of Ions*.

This rule makes it possible to estimate the Λ_0 for weak electrolytes, such as organic acids, from values for their salts, which are strong electrolytes. For example (at 298 K),

$$\Lambda_0(HAc) = \Lambda_0(NaAc) + \Lambda_0(HCl) - \Lambda_0(NaCl)$$

$$= 0.0091 + 0.0425 - 0.0128 = 0.0388 \ \Omega^{-1} \ m^2 \ mol^{-1}$$

16.4 Arrhenius Ionization Theory

From 1882 to 1886, Julius Thomsen published data on the enthalpies of neutralization of acids and bases. He found that ΔH of neutralization of a strong acid by a strong base in dilute solution is always nearly the same, about 57.7 kJ per equivalent at 298 K. The neutralization heats of weak acids and bases are lower, and indeed the strength of an acid appears to be proportional to its enthalpy of neutralization by a strong base.

These results and available conductivity data led Svante Arrhenius in 1887 to propose a new theory for the behavior of electrolyte solutions. He suggested that an equilibrium exists in solution between undissociated solute molecules and ions that arise from these by *electrolytic dissociation*. Strong acids and bases are almost completely dissociated, so that their interaction in every case is equivalent to $H^+ + OH^- = H_2O$, thus explaining the constant ΔH of neutralization.

While Arrhenius was working on this theory, van't Hoff's papers on the osmotic pressures of solutions appeared. His results provided strong support for the ideas of Arrhenius. Van't Hoff found that the osmotic pressures of dilute solutions of nonelectrolytes followed the equation $\Pi = cRT$. The osmotic pressures of electrolytes were always higher than predicted from this equation, often by a factor of 2, 3, or more, so that a modified equation was written as

$$\Pi = icRT \tag{16.8}$$

The van't Hoff "*i* factor" for strong electrolytes was close to the number of ions that would be formed if a solute molecule dissociated according to the Arrhenius theory. For NaCl, KCl, and other uniunivalent electrolytes, $i = 2$; for $BaCl_2$, K_2SO_4, and other unibivalent species, $i = 3$; for $LaCl_3$, $i = 4$.

On April 13, 1887, Arrhenius wrote to van't Hoff as follows:

It is true that Clausius had assumed that only a minute quantity of dissolved electrolyte is dissociated, and that all other physicists and chemists had followed him, but the only reason for this assumption, so far as I can understand, is a strong feeling of aversion to a dissociation at so low a temperature, without any actual facts against it being brought forward. . . . At extreme dilution all salt molecules are completely dissociated. The degree of dissociation can be simply found on this assumption by taking the ratio of the equivalent conductivity of the solution in question to the equivalent conductivity at the most extreme dilution.

Thus Arrhenius would write the degree of dissociation α as

$$\alpha = \frac{\Lambda}{\Lambda_0} \tag{16.9}$$

16.5 A High Dielectric Constant of Solvent Facilitates Separation of Ions

Since the time of Arrhenius and Ostwald we have learned that crystalline salts are formed of ions in regular arrays, so that there is no question of ionic dissociation of a molecule when they dissolve. The process of solution simply allows the ions to move apart from one another. The magnitude of the electrostatic force between two ions with charges Q_1 and Q_2 separated by a distance r in a medium of dielectric constant ϵ_r is given by Coulomb's Law as

$$F = \frac{Q_1 Q_2}{4\pi\epsilon_0\epsilon_r r^2} \tag{16.10}$$

In aqueous solutions the separation of oppositely charged ions is facilitated by the high dielectric constant of water, $\epsilon_r = 78.5$ at 298 K. Let us compare, for water and a vacuum, the energy necessary to separate two ions, say Na^+ and Cl^-, from a distance of 200 pm, as in a crystal, to 2000 pm, their average separation in a 0.1 molar solution. The element of work in separating the ions by a distance dr is $dw = -F\,dr$, and this equals the change in potential energy dE_p. Thus the ΔE_p in separating a pair of ions from r_1 to r_2 is

$$\Delta E_p = -\int_{r_1}^{r_2} F\,dr = \frac{Q_1 Q_2}{4\pi\epsilon_0\epsilon_r}\left(\frac{1}{r_2} - \frac{1}{r_1}\right) \tag{16.11}$$

With $Q_1 = -Q_2 = 1.602 \times 10^{-19}$ C, the permittivity of vacuum, $\epsilon_0 = 8.854 \times 10^{-12}$ $C^2\,J^{-1}\,m^{-1}$, and r_1, r_2 as given, $\Delta E_p = 1.038 \times 10^{-18}/\epsilon_r$, J. Calculated as $L\,\Delta E_p$ per mole of ion pairs, to separate the ions in vacuum ($\epsilon_r = 1$) thus requires 625 kJ mol^{-1}, and to separate them in water at 298 K ($\epsilon_r = 78.5$) requires only 8 kJ mol^{-1}. We can now see in quantitative terms how the high dielectric constant of water greatly facilitates the separation of ions.

16.6 Transport Numbers and Mobilities

The fraction of the current carried by a given kind of ion j in solution is called the *transport number* or *transference number* t_j of that ion. The average velocity of the ions j in the direction of an electric field of unit strength is the *mobility* u_j of the ions. If Λ_j is the molar ionic conductivity of ions j, from the Kohlrausch equation (16.7), the transport numbers t_0^+ and t_0^- of cation and anion at infinite dilution may be written

$$t_0^+ = \frac{\Lambda_0^+}{\Lambda_0}, \qquad t_0^- = \frac{\Lambda_0^-}{\Lambda_0}, \qquad t_j = \frac{\Lambda_j}{\Lambda_0} \tag{16.12}$$

It should be emphasized that Λ_j depends on the concentrations and on the identities of other ions present in solution. Only in the limit of infinite dilution does Λ_j become a property of the individual ion as in the Kohlrausch Law [Eq. (16.7)].

The current I can be carried by several different ions, the mobilities of which are denoted by u_j, the charges by $z_j e$ and the numbers in unit volume by $C_j = N_j/V$. The conductivity of the solution is then the sum of contributions from the different ions. From Eq. (16.4), one obtains

$$\kappa = \sum \kappa_j = \sum C_j |z_j e| u_j \tag{16.13}$$

The absolute value $|z_j e|$ of ionic charge appears in the expression for conductivity, since current I has the same direction whether carried by positive ions moving down a gradient of electric field or by negative ions moving up. We note in Eq. (16.13) that conductivity is determined by the concentrations of mobile electric charges and the mobilities of these charges.

From Eqs. (16.5) and (16.13), since $C_j = Lc_j$,

$$\Lambda_j = \frac{\kappa_j}{c_j} = \frac{C_j}{c_j}|z_j e| u_j = L|z_j e| u_j$$

$$\Lambda_j = t_j \Lambda = F|z_j| u_j \tag{16.14}$$

This relation applies to each ion in the solution. If we know the transport number t_j of the ion, we can therefore calculate its mobility u_j from the molar conductivity of the solution.

16.7 Transport Numbers—Hittorf Method

The measurement of transport numbers by the method of Hittorf is based on concentration changes in the neighborhoods of the electrodes, caused by passage of current through the electrolyte. A typical Hittorf apparatus is shown in Fig. 16.3(a). Figure 16.3(b) illustrates the principle of the method by means of a cell divided into three com-

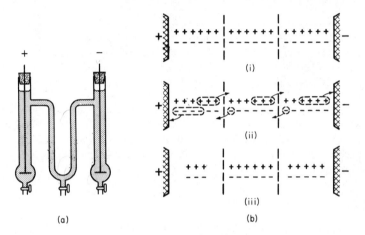

(a)   (b)

FIGURE 16.3 (a) A Hittorf cell. (b) Principle of Hittorf method for measuring transport numbers: (i) distribution of ions before passage of charge; (ii) effect of passage of three faradays of charge; (iii) final distribution of ions.

partments. The situation of the ions before the passage of any current is represented schematically as (i), each $+$ or $-$ sign indicating one faraday of the corresponding ion. We assume in this example that the ions have equal charge numbers, and that the mobility of the positive ion is three times that of the negative ion, $u_+ = 3u_-$. Let $4F$ of electric charge be passed through the cell. At the anode, therefore, four equivalents of negative ions are discharged, and at the cathode, four equivalents of positive ions. Four faradays must pass across any boundary plane drawn through the electrolyte normal to the path of the currents. Since the positive ions travel three times faster than the negative ions, $3F$ are carried across the plane from left to right by the positive ions with $1F$ is being carried from right to left by the negative ions. This transfer is depicted in part (ii) of the figure. The final situation, after discharge of ions at the electrodes, is shown in (iii). The change in number of equivalents in the anode compartment is $\Delta n_a = 6 - 3 = 3$; in the cathode compartment, it is $\Delta n_c = 6 - 5 = 1$. The ratio of these concentration changes is identical to the ratio of the ionic mobilities: $\Delta n_a / \Delta n_c = u_+ / u_- = 3$.

Suppose that the amount of electricity Q passed through the cell is measured by a coulometer in series. Provided that the electrodes are inert, Q/z_+F moles of cations have therefore been discharged at the cathode, and $Q/|z_-|F$ moles of anions at the anode. The net loss of solute from the cathode compartment is

$$\Delta n_c = \frac{Q}{z_+F} - t_+ \frac{Q}{z_+F} = \frac{Q}{z_+F}(1 - t_+) = \frac{Qt_-}{z_+F}$$

Thus

$$t_- = \frac{\Delta n_c}{Q|z_+|F}, \qquad t_+ = \frac{\Delta n_a}{Q|z_-|F}, \qquad \text{and} \qquad \frac{\Delta n_a}{\Delta n_c} = \frac{t_+z_+}{t_-|z_-|} \qquad (16.15)$$

where Δn_a is net loss of solute from the anode compartment. Since $t_+ + t_- = 1$, both transport numbers can be determined from measurements on either compartment, but it is useful to have both analyses as a check.

Example 16.3 A 4.000 molal solution of $FeCl_3$ was electrolyzed in a Hittorf apparatus with platinum electrodes. The reaction at the cathode was $Fe^{3+} + e = Fe^{2+}$. After electrolysis, the cathode compartment was 3.150 molal in $FeCl_3$ and 1.000 molal in $FeCl_2$. What was the transport number of Fe^{3+}?

Since the final cathode solution was 1.000 molal in Fe^{2+}, 1 faraday of electricity passed through the cell per 1 kg of water in the cathode chamber. The Fe^{3+} that migrated into the cathode compartment was $3.150 + 1.000 - 4.000 = 0.150$ mol per kg of water. Since $z_+ = 3$, the Fe^{3+} ions carried $3F$ per mol into the cathode chamber. Hence $t_+ = 0.45/1.00 = 0.45$.

16.8   Transport Numbers—Moving Boundary Method

The typical apparatus shown in Fig. 16.4 is used to follow the moving boundary between two liquid solutions. The electrolyte to be studied, CA, is introduced into the apparatus in a layer above a solution of a salt with a common anion, C′A, and a cation with a mobility considerably less than that of the ion C^+. As an example, a layer of KCl solution could be introduced above a layer of $CdCl_2$ solution. The mobility of Cd^{2+} is considerably lower than that of K^+. When a current is passed through the cell, A^- ions move downward toward the anode, while C^+ and $C′^+$ ions move upward toward the cathode. A sharp boundary is preserved between the two solutions, because the more slowly moving $C′^+$ ions never overtake the C^+ ions; nor do these following ions, $C′^+$, fall far behind, because if they began to lag, the solution behind the boundary would become more dilute, and its higher resistance and therefore steeper potential drop would increase the ionic velocity. Even with colorless solutions, a sharp boundary is visible owing to the different refractive indices of KCl and $CdCl_2$ solutions.

Suppose that the boundary moves a distance x for the passage of Q coulombs. The number of faradays transported is then Q/F, of which t_+Q/F are carried by positive ions. The faradays of positive charge per unit volume of solution are cz_+, where c

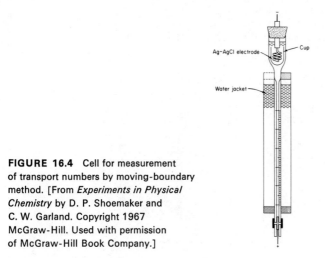

FIGURE 16.4 Cell for measurement of transport numbers by moving-boundary method. [From *Experiments in Physical Chemistry* by D. P. Shoemaker and C. W. Garland. Copyright 1967 McGraw-Hill. Used with permission of McGraw-Hill Book Company.]

is the concentration of electrolyte CA. Hence the volume of solution swept out by the boundary during the passage of Q coulombs is t_+Q/Fcz_+. If A is the cross-sectional area of the tube, $xA = t_+Q/Fcz_+$, or

$$t_+ = \frac{FxAcz_+}{Q} \tag{16.16}$$

16.9 Results of Transport Experiments

Some measured transport numbers are summarized in Table 16.1. With these values, it is possible to calculate from Eq. (16.12) the molar ionic conductivities at infinite dilution, some of which are given in Table 16.2. By use of the Kohlrausch rule, Eq. (16.7), they may be combined to yield values for the molar conductivities Λ_0 of a variety of electrolytes. For example, Λ_0 (BaCl$_2$) is $[127.28 + 2(76.34)] \times 10^{-4} = 276.96 \times 10^{-4}\ \Omega^{-1}\ m^2\ mol^{-1}$.

TABLE 16.1
TRANSPORT NUMBERS OF CATIONS IN WATER SOLUTIONS AT 298 K AT VARIOUS CONCENTRATIONS OF ELECTROLYTES

c (mol dm^{-3})	AgNO$_3$	$\frac{1}{2}$BaCl$_2$	LiCl	NaCl	KCl	KNO$_3$	$\frac{1}{3}$LaCl$_3$	HCl
0.01	0.4648	0.4400	0.3289	0.3918	0.4902	0.5084	0.4625	0.8251
0.05	0.4664	0.4317	0.3211	0.3876	0.4899	0.5093	0.4482	0.8292
0.10	0.4682	0.4253	0.1368	0.3854	0.4898	0.5103	0.4375	0.8314
0.50	—	0.3986	0.300	—	0.4888	—	0.3958	—
1.00	—	0.3792	0.287	—	0.4882	—	—	—

Cation	$10^4\Lambda_0^+$ $(\Omega^{-1}\ m^2\ mol^{-1})$	Anion	$10^4\Lambda_0^-$ $(\Omega^{-1}\ m^2\ mol^{-1})$
H^+	349.82	OH^-	198.0
Li^+	38.69	Cl^-	76.34
Na^+	50.11	Br^-	78.4
K^+	73.52	I^-	76.8
NH_4^+	73.4	NO_3^-	71.44
Ag^+	61.92	CH_3COO^-	40.9
Ca^{2+}	119.0	ClO_4^-	68.0
Ba^{2+}	127.28	SO_4^{2-}	159.6
Mg^{2+}	106.12		
La^{3+}	208.8		

The observed transport numbers are not those of bare ions but of hydrated ions. Mobilities of ions can be calculated from molar conductivities by means of Eq. (16.14). Some results are given in Table 16.3. The effect of hydration is seen in the set of values for Li^+, Na^+, and K^+. Although Li^+ is the smallest bare alkali-metal ion, it has the lowest mobility, i.e., the resistance to its motion through the solution is highest. This resistance must be partly due to a tightly held sheath of water molecules, bound by the intense electric field of the small ion.*

TABLE 16.3
MOBILITIES OF IONS IN WATER SOLUTIONS AT 298.15 K

Cation	Mobility $(m^2\ s^{-1}\ V^{-1})$	Anion	Mobility $(m^2\ s^{-1}\ V^{-1})$
H^+	36.30×10^{-8}	OH^-	20.52×10^{-8}
K^+	7.62×10^{-8}	SO_4^{2-}	8.27×10^{-8}
Ba^{2+}	6.59×10^{-8}	Cl^-	7.91×10^{-8}
Na^+	5.19×10^{-8}	NO_3^-	7.40×10^{-8}
Li^+	4.01×10^{-8}	HCO_3^-	4.61×10^{-8}

As a direct consequence of this difference in size of the hydrated ions, the membranes of living cells are generally much more permeable to K^+ ions than to Na^+ ions. Typically, the interior of the cell has a higher concentration of K^+ than the exterior, and the reverse is true for Na^+. Such ionic concentration gradients lead to differences in electric potential across the cell membranes. Many important physiological mechanisms are thus based on the hydration of ions and the way in which hydration influences the mobilities of ions.

*The important subject of hydration of ions has been treated in detail in a monograph by B. E. Conway, *Ionic Hydration* (New York: Academic Press, 1981).

Example 16.4 In a 0.1 M LiCl solution, the transport number of the Li^+ ion is 0.318 and the mobility of the Li^+ ion $u^+ = 4.01 \times 10^{-8}$ m^2 V^{-1} s^{-1}. Calculate the mobility of the Cl^- ion and the migration speed of the Cl^- ion in a field of strength $E = 100$ V m^{-1}.

$u_-/u_+ = t_-/t_+ = (1 - t_+)/t_+ = 0.682/0.318$. Thus $u(Cl^-) = (0.682/0.318)(4.01 \times 10^{-8}) = 8.60 \times 10^{-8}$ m^2 V^{-1} s^{-1}. Speed $v(Cl^-) = u(Cl^-) \times E = 8.60 \times 10^{-6}$ m s^{-1}.

16.10 Electrolytic Dissociation of Water

In pure water and aqueous solutions, water acts as an amphoteric substance, and dissociates as $2H_2O \rightarrow H_3O^+ + OH^-$. From Eq. (10.25) the dissociation constant for water is

$$K_a = \frac{a_{H^+} \cdot a_{OH^-}}{a_{H_2O}} = \frac{[H^+][OH^-]}{[H_2O]} \frac{\gamma_{H^+} \cdot \gamma_{OH^-}}{\gamma_{H_2O}} \tag{16.17}$$

Here $[H^+]$ etc., are really concentrations relative to $c°$ of the standard state (page 210), but we have omitted the $c°$. For convenience also we write the hydrogen ion as H^+, although we know it is almost always H_3O^+.

For a pure condensed phase at $P°$, $a = 1$, so that Eq. (16.17) becomes

$$K_w \equiv K_a a_{H_2O} = a_{H^+} \cdot a_{OH^-} = \gamma_{H^+}[H^+]\gamma_{OH^-}[OH^-] \tag{16.18}$$

Let c_w be the concentration of water $[H_2O]$, 55.3 mol dm^{-3} at 298 K. Then, if the fractional dissociation is α, $[H^+] = [OH^-] = \alpha c_w$, and Eq. (16.18) gives

$$K_w = \alpha^2 c_w^2 \gamma_{H^+} \gamma_{OH^-} \tag{16.19}$$

In pure water the concentration of ions H^+ and OH^- is very low so that γ_{H^+} and γ_{OH^-} equal unity, and Eq. (16.19) becomes

$$K_w = [H^+][OH^-] = \alpha^2 c_w^2 \tag{16.20}$$

The value of K_w can be calculated from the results of conductivity measurements on very pure water. At 298 K, $\kappa(H_2O) = 5.50 \times 10^{-6}$ Ω^{-1} m^{-1}. Thus

$$\Lambda = \frac{5.50 \times 10^{-6}(\Omega^{-1} m^{-1})}{55.3 \times 10^3(mol\ m^{-3})} = 0.995 \times 10^{-10}\ \Omega^{-1}\ m^2\ mol^{-1}$$

The hypothetical value of Λ_0 for water can be calculated by virtue of the principle of independent mobility of ions:

$$\Lambda_0(H_2O) = \Lambda_0(H^+) + \Lambda_0(OH^-) = (349.8 + 198.0)10^{-4}\Omega^{-1}\ m^2\ mol^{-1}$$

$$= 547.8 \times 10^{-4}\ \Omega^{-1}\ m^2\ mol^{-1}$$

Hence, from Eq. (16.9), $\alpha = \Lambda/\Lambda_0 = 0.995 \times 10^{-10}/547.8 \times 10^{-4} = 1.82 \times 10^{-9}$. Then,

$$K_w = \alpha^2 c_w^2 = (3.30 \times 10^{-18})(3.06 \times 10^9) = 1.01 \times 10^{-8}(mol\ m^{-3})^2$$

$$= 1.01 \times 10^{-14}(mol\ dm^{-3})^2$$

The ionic concentration in pure water at 298 K is thus

$$[H^+] = [OH^-] = K_w^{1/2} = 1.01 \times 10^{-7} \text{mol dm}^{-3}$$

16.11 Mobilities of Hydrogen and Hydroxyl Ions

Table 16.3 indicates that, with two exceptions, ionic mobilities in aqueous solutions do not differ in order magnitude, all being about 6×10^{-8} m^2 s^{-1} V^{-1}. The exceptions are the hydrogen and hydroxyl ions with the abnormally high mobilities of 36.3×10^{-8} and 20.5×10^{-8} m^2 s^{-1} V^{-1}, respectively.

The high mobility of H^+ and OH^- in hydroxylic solvents such as H_2O and CH_3OH is due to a Grotthuss type of mechanism, as shown in Fig. 16.5. Protons can shift rapidly from H_3O^+ to H_2O or from H_2O to OH^-. The effect of the change in partners is to transport the electric charge rapidly from one location to another.

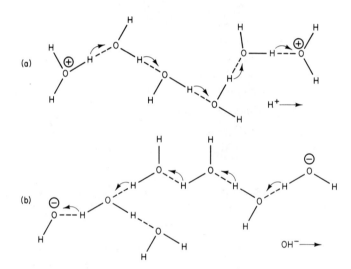

FIGURE 16.5 Grotthuss conductivity of (a) H^+ and (b) OH^- ions in water.

16.12 Diffusion and Ionic Mobility

Electrical conductivity depends on the motion of ions in an electric field, **E**. In the absence of a field, ions in solution (and solvent molecules as well) are moving about in random directions as a result of their kinetic energies. If gradients of concentration exist for a component, the system being maintained at constant T and P, there will be a drive toward an equilibrium condition in which the gradients of concentration are all smoothed out. The migration if a component in solution down a gradient of its concentration, i.e., from a region of higher to a region of lower concentration, is called *diffusion*. More precisely, diffusion occurs down a gradient of chemical potential μ_i of the component i. Chemical potential is the partial molar Gibbs free energy, $\mu_i = (\partial G/\partial n_i)_{T, p, n_i}$. Thus in a system at constant T and P, diffusive flow of a component from higher to lower μ_i lowers the Gibbs free energy.

The early development of diffusion theory was based on gradients of concentration. *Flux* is a technical term meaning flow through unit cross-sectional area. In 1855, Rudolf Fick stated his *First Law of Diffusion:* The flux of a component i by diffusion in the direction x is proportional to the gradient of concentration in that direction,

$$S_{ix} = -D_i(\partial c_i/\partial x) \tag{16.21}$$

The proportionality factor is the *diffusion coefficient* D_i.

Figure 16.6 shows one typical arrangement for measurement of diffusion. The cell is prepared, for example, with a solution on one side and pure solvent on the other. The barrier between the two sides is carefully retracted and the diffusion process begins.

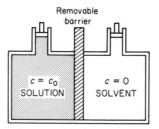

FIGURE 16.6 Diffusion cell—initial condition before start of experiment.

The differential equation for diffusion is derived as follows. Consider a thin slice of the cell of unit cross-sectional area between x and $x + dx$. The increase in concentration of i with time in this slice is $(\partial c_i/\partial t)_x$. This equals the excess of i molecules diffusing into the slice over those diffusing out, divided by the volume dx of the slice. Thus

$$(\partial c_i/\partial t)_x = \frac{1}{dx}[S_i(x) - S_i(x + dx)] = -(\partial S_i/\partial x)_t$$

since

$$S_i(x + dx) = S_i(x) + (\partial S_i/\partial x)_t\, dx$$

From Fick's First Law, Eq. (16.21), therefore,

$$(\partial c_i/\partial t)_x = \frac{\partial}{\partial x}\left(D_i \frac{\partial c_i}{\partial x}\right)_t$$

If D_i is independent of x,

$$(\partial c_i/\partial t)_x = D_i(\partial^2 c_i/\partial x^2)_t \tag{16.22}$$

This equation is called *Fick's Second Law of Diffusion*. Special treatises give solutions to the equation for all the situations of physicochemical interest.* A quick estimate

---

*J. Crank, *The Mathematics of Diffusion* (London: Oxford University Press, 1959).

of the mean square distance that a molecule diffuses in time t is given by

$$x^2 = 2Dt \qquad (16.23)$$

The diffusion coefficient D_i measures the mobility of molecules or ions i due to their thermal energy, which is proportional to kT. The electrical mobility u_i measures the mobility of ions i due to the kinetic energy imparted by an electric field E, an energy proportional to their charge Q_i. Nernst and Einstein showed that

$$\frac{D_i}{kT} = \frac{u_i}{Q_i} \qquad (16.24)$$

Example 16.5   The diffusion coefficient of Na^+ in a 0.1 M NaCl solution at 25°C was measured with a radioactive tracer Na* as $D = 1.30 \times 10^{-9}$ m² s⁻¹. Calculate the electric mobility of Na^+ in this solution, and compare it to the value in Table 16.4.

From Eq. (16.24), $u_i = D_i Q_i / kT = (1.30 \times 10^{-9}$ m² s⁻¹$)(1.60 \times 10^{-19}$ C$)/(1.38 \times 10^{-23}$ J K⁻¹$)(298$ K$) = 5.06 \times 10^{-8}$ m² s⁻¹ V⁻¹ (since J = V C). (Experimental value 5.19×10^{-8} m² s⁻¹ V⁻¹.)

16.13   Activities and Standard States

As shown in Section 10.7, the standard state for a component considered as a solute B is based on Henry's Law. In the limit as $X_B \rightarrow 0$, $a_B \rightarrow X_B$, and the activity coefficient $\gamma_B \rightarrow 1$. The departure of γ_B from unity is a measure of the departure of the behavior of the solute from that prescribed by Henry's Law. Henry's Law implies the absence of interaction between molecules of solute (component B sees only solvent A surrounding it). Therefore, deviation of the activity coefficient from unity measures the effect of interactions between solute molecules in solution.

Two changes must be made in the definition of standard state given in Section 10.7 before it can be used for solutions of electrolytes. The composition of electrolytic solutions is almost always expressed in molalities instead of mole fractions. Also, we need to consider the effect of dissociation of the electrolyte, by which a molecule of added solute yields two or more ions in the solution. Thus, for NaCl in water, the limiting form of Henry's Law is $f_B = k_H m_B^2$, where f_B and m_B are the fugacity and molality of NaCl. For an electrolyte yielding v particles on dissociation,

$$f_B = k_H m_B^v \qquad (16.25)$$

The usual standard state for an electrolyte in solution is illustrated in Fig. 16.7, for the case of NaCl. It is a hypothetical state in which the solute would exist at unit molality and pressure $P°$ but would still have the environment typical of an extremely dilute solution that follows Henry's Law.

The activity of B is $a_B = {}^m\gamma_B m_B / m°$, where $m°$ is the standard-state molality, 1 mol kg⁻¹.

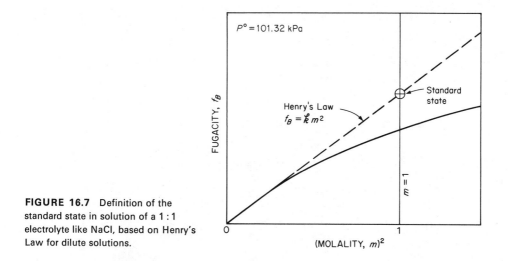

FIGURE 16.7 Definition of the standard state in solution of a 1 : 1 electrolyte like NaCl, based on Henry's Law for dilute solutions.

16.14 Ion Activities

In dealing with solutions of electrolytes, it would apparently be convenient to use activities of individual ionic species, but there are serious difficulties in the way of such a procedure. The requirement of overall electrical neutrality in the solution prevents any increase in the charge due to positive ions without an equal increase in the charge due to negative ions. For example, we can change the concentration of a solution of sodium chloride by adding equal numbers of sodium and chloride ions. It is not possible to add sodium ions alone or chloride ions alone. There is no experimental way to separate effects due to positive ions from those due to the accompanying negative ions in an uncharged solution. Thus there is no way to measure individual ion activities.

Nevertheless, it is convenient to define an expression for the activity of an electrolyte in terms of the ions into which it dissociates. Consider, for example, a solute like NaCl, dissociated in solution according to $NaCl \longrightarrow Na^+ + Cl^-$. If we denote the activity of the cation as a_+ and that of the anion as a_-, $a = a_+a_-$. (This result comes from $\mu = \mu_+ + \mu_-$ and $\mu = \mu° + RT \ln a$.) We then define the *mean ionic activity* $a_\pm$ by

$$a = a_+a_- = a_\pm^2 \qquad (16.26)$$

The quantity $a_\pm$ is the geometric mean of a_+ and a_-.

For more complex types of electrolytes, these definitions can be generalized. Consider an electrolyte that dissociates as

$$C_{\nu_+}A_{\nu_-} \longrightarrow \nu_+C^+ + \nu_-A^-$$

The total number of ions is $\nu = \nu_+ + \nu_-$. We then write

$$a = a_+^{\nu_+}a_-^{\nu_-} = a_\pm^\nu \qquad (16.27)$$

For example, $La_2(SO_4)_3 \longrightarrow 2La^{+3} + 3SO_4^{-2}$, $a = a_{La}^2 a_{SO_4}^3 = a_\pm^5$.

We can also define individual ionic activity coefficients γ_+ and γ_- by*

$$a_+ = \gamma_+ m_+ \quad \text{and} \quad a_- = \gamma_- m_- \tag{16.28}$$

The experimentally measured activity coefficient is $\gamma_\pm$, the geometric mean of the individual ionic coefficients, where

$$\gamma_\pm^\nu = \gamma_+^{\nu_+} \gamma_-^{\nu_-} \tag{16.29}$$

Equation (16.27) can then be written $a = m_+^{\nu_+} m_-^{\nu_-} \gamma_+^{\nu_+} \gamma_-^{\nu_-}$ or

$$a_\pm = a^{1/\nu} = (m_+^{\nu_+} m_-^{\nu_-} \gamma_+^{\nu_+} \gamma_-^{\nu_-})^{1/\nu} \tag{16.30}$$

Substituting Eq. (16.30) into Eq. (16.29), we obtain

$$\gamma_\pm = \frac{a_\pm}{(m_+^{\nu_+} m_-^{\nu_-})^{1/\nu}} \tag{16.31}$$

This equation applies whether the ions are added together as a single salt or added separately as a mixture of salts.

For a solution of a single salt, of molality m, $m_+ = \nu_+ m$ and $m_- = \nu_- m$. In this case, Eq. (16.31) becomes

$$\gamma_\pm = \frac{a_\pm}{m(\nu_+^{\nu_+} \nu_-^{\nu_-})^{1/\nu}} = \frac{a_\pm}{m_\pm} \tag{16.32}$$

In the case of $La_2(SO_4)_3$, for example, $\nu_+ = 2$ and $\nu_- = 3$, so that

$$\gamma_\pm = \frac{a_\pm}{m(2^2 3^3)^{1/5}} = \frac{a_\pm}{108^{1/5}}$$

The activity coefficient as defined in Eq. (16.32) becomes unity at infinite dilution.

Activities can be determined by several different methods. Among the most important are measurements of the colligative properties of solutions, such as freezing-point depression and osmotic pressure, measurements of the solubilities of sparingly soluble salts, and methods based on the emf of electrochemical cells. We shall describe the emf method in Chapter 17.

16.15 Experimental Activity Coefficients of Ionic Solutions

Mean activity coefficients of aqueous electrolytes are summarized in Table 16.4 and plotted in Fig. 16.8. For comparison, the activity coefficient in water of a typical nonelectrolyte, sucrose, is also shown. The activity coefficients for electrolytes decline markedly with increasing concentration in dilute solution, but then pass through minima and rise again in more concentrated solutions. The interpretation of this behavior constitutes one of the principal problems in the theory of strong electrolytes.

Why does the activity coefficient as a function of concentration pass through a minimum? Consider the chemical potential of an ionic solute, $\mu = \mu^\circ + RT \ln \gamma m$. Interionic attractions lower the chemical potential of the ions, and hence γ

*We use γ for $^m\gamma$ in the remainder of the chapter.

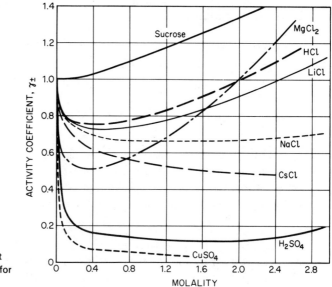

FIGURE 16.8 Mean molal activity coefficients of aqueous electrolytes at 25°C, with values for sucrose shown for comparison.

tends to decrease with increasing ionic concentration. On the other hand, the ions also exert attractive forces on the water molecules, so that the chemical potential of the water is lowered. For example, if the hydration number of NaCl is about 6, in a 1.0 molal solution of NaCl only about 90% of the water is free solvent. As shown by the Gibbs–Duhem relation (Section 10.8), this effect raises the chemical potential and hence the activity coefficient of the solute. The two opposite effects outlined may lead to a minimum in γ vs. m.

TABLE 16.4
MEAN MOLAL ACTIVITY COEFFICIENTS OF ELECTROLYTES
IN WATER AT 25°C

m	0.001	0.002	0.005	0.010	0.020	0.050	0.100	0.200	0.500	1.00	2.00	4.00
HCl	0.966	0.952	0.928	0.904	0.875	0.830	0.796	0.767	0.758	0.809	1.01	1.76
HNO$_3$	0.965	0.951	0.927	0.902	0.871	0.823	0.785	0.748	0.715	0.720	0.783	0.982
H$_2$SO$_4$	0.830	0.757	0.639	0.544	0.453	0.340	0.265	0.209	0.154	0.130	0.124	0.171
NaOH	—	—	—	—	—	0.82	—	0.73	0.69	0.68	0.70	0.89
AgNO$_3$	—	—	0.92	0.90	0.86	0.79	0.72	0.64	0.51	0.40	0.28	—
CaCl$_2$	0.89	0.85	0.785	0.725	0.66	0.57	0.515	0.48	0.52	0.71	—	—
CuSO$_4$	0.74	—	0.53	0.41	0.31	0.21	0.16	0.11	0.068	0.047	—	—
KCl	0.965	0.952	0.927	0.901		0.815	0.769	0.719	0.651	0.606	0.576	0.579
KBr	0.965	0.952	0.927	0.903	0.872	0.822	0.777	0.728	0.665	0.625	0.602	0.622
KI	0.965	0.951	0.927	0.905	0.88	0.84	0.80	0.76	0.71	0.68	0.69	0.75
LiCl	0.963	0.948	0.921	0.89	0.86	0.82	0.78	0.75	0.73	0.76	0.91	1.46
NaCl	0.966	0.953	0.929	0.904	0.875	0.823	0.780	0.730	0.68	0.66	0.67	0.78
ZnSO$_4$	0.734	0.610	0.477	0.387	—	0.202	0.148	0.110	0.063	0.043	0.035	—

> **Example 16.6** What is the mean activity of H_2SO_4 in a 0.100 molal solution at 25°C?
>
> From Table 16.4, $\gamma_\pm = 0.265$. From Eq. (16.32),
> $$a_\pm = \gamma_\pm (m_+^{\nu^+} m_-^{\nu^-})^{1/\nu} = (0.265)(0.2^2 \times 0.1)^{1/3} = 0.0421$$

16.16   The Ionic Strength

Many properties of ionic solutions depend on electrostatic interactions between ionic charges. The electrostatic force between a pair of doubly charged ions is four times the force between a pair of singly charged ions. A useful function of ionic concentration, devised to include such effects of ionic charge, is the *ionic strength I*, defined by

$$I = \tfrac{1}{2} \sum m_i z_i^2 \tag{16.33}$$

The summation is taken over all the different ions in a solution, multiplying the molality of each by the square of its charge. For example, a 1.00 molal solution of NaCl has an ionic strength $I = \tfrac{1}{2}(1.00) + \tfrac{1}{2}(1.00) = 1.00$. A 1.00 molal solution of $La_2(SO_4)_3$ has $I = \tfrac{1}{2}[2(3)^2 + 3(2)^2] = 15.0$. In dilute solutions, the activity coefficients of electrolytes, the solubilities of sparingly soluble salts, rates of ionic reactions, and other related properties become functions of ionic strength.

The molar concentration $c = m\rho/(1 + mM)$, where ρ is the density of the solution and M is the molar mass of solute. In dilute solution, $mM \ll 1$ and this relation approaches $c = \rho_0 m$, where ρ_0 is the density of the solvent. To this approximation, $I = \tfrac{1}{2} \sum m_i z_i^2 \approx (1/2\rho_0) \sum c_i z_i^2$.

16.17   Debye–Hückel Theory

The theory of Debye and Hückel (1923) is based on the assumption that strong electrolytes are completely dissociated into ions. Observed deviations from ideal behavior are then ascribed to electrical interactions between ions. To obtain theoretically the equilibrium properties of ionic solutions, it is necessary to calculate the extra Gibbs free energy arising from these electrostatic interactions.

If the ions were distributed completely at random, the chances of finding either a positive or a negative ion in the neighborhood of a given ion would be identical. Such a random distribution would have no electrostatic energy, since, on the average, attractive configurations would be exactly balanced by repulsive ones. It is evident that this cannot be the physical situation, since in the immediate neighborhood of a positive ion, a negative ion is more likely to be found than another positive ion. Indeed, were it not for the fact that ions are continually being batted about by molecular collisions, an ionic solution might acquire a well-ordered structure similar to that of an ionic crystal. The thermal motions effectively prevent any complete ordering, but the final situation is a dynamic compromise between electrostatic interactions tending to produce ordered configurations and kinetic collisions tending to destroy them. Our problem is to calculate the average electric potential Φ of a given ion in solution due to all the other ions. Knowing Φ, we can calculate the work that must be expended to charge the ions reversibly to this potential, and this work will be the energy due to electrostatic interactions. The extra electric energy is directly related to the ionic activity coefficient, since both are a measure of the deviation from ideality.

The deviation from ideal behavior of ions of a given species j in solution can be specified by their ionic activity coefficient γ_j. The electrostatic energy of the interaction between the ions j and other ions in the solution is then, per mole,

$$\Delta\mu_j = RT \ln a_j - RT \ln c_j,$$

or

$$\Delta\mu_j = RT \ln \gamma_j \tag{16.34}$$

Note that this electrical energy equals the Gibbs free energy of the ions in the real solution minus their free energy in an ideal solution. The basic hypothesis of the Debye–Hückel theory is that all the deviation from ideality is due to electrostatic interactions of ions.

Example 16.7 In a 0.5 M aqueous solution of KCl at 25°C, $\gamma_\pm = 0.901$. What is the extra Gibbs free energy due to electrostatic interactions of the ions in the solution?

From Eq. (16.34), since $\gamma_\pm^2 = \gamma_+\gamma_-$,

$$\Delta\mu = RT(\ln \gamma_+ + \ln \gamma_-) = RT \ln \gamma_+\gamma_- = 2RT \ln \gamma_\pm$$
$$= 2(8.314 \text{ J K}^{-1} \text{ mol}^{-1})(298 \text{ K}) \ln (0.901) = -517 \text{ J mol}^{-1}$$

(This electrical energy is much less than $RT = 2480 \text{ J mol}^{-1}$.)

16.18 The Ionic Atmosphere

On the average, a given ion is surrounded by a spherically symmetrical distribution of other ions, forming the ionic atmosphere. Figure 16.9 depicts a central ion with a section of a spherical shell at a distance r. The closest approach of any other ion to the center is designated by $2a$. We wish to obtain the average electrostatic potential $\Phi(r)$ due to the central ion and its surrounding atmosphere.

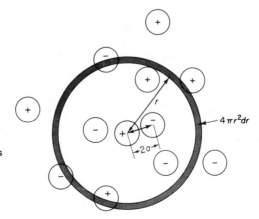

FIGURE 16.9 The ionic atmosphere. A central positive ion of radius a is surrounded by negative and positive ions of radius a. The problem is to calculate the electric potential at the central ion due to the excess concentration of negative ions in its surroundings.

Consider first the electric potential Φ at a distance r from a central ion of charge $z_j e$ in a medium of dielectric constant ϵ_r. This Φ equals the work of bringing a unit test charge from infinity to r. From Eq. (16.11), therefore, with $r_1 = \infty$ and $Q_1 = 1$, $Q_2 = z_j e$,

$$\Phi = \frac{z_j e}{4\pi\epsilon_0\epsilon_r r} \tag{16.35}$$

This expression is based on the questionable assumption that the bulk dielectric constant ϵ_r is the correct measure of the effect of solvent on the forces between ions in solution, even when they are quite near each other.

The effect of an ionic atmosphere, such as that shown in Fig. 16.9, is to shield the central ion from interactions with other ions in the solution at distances outside the radius of the inner sphere of counter ions. The thickness of the ionic atmosphere b is now introduced into the theory. This thickness corresponds roughly with the average distance of the innermost counterions from their central ion.

Instead of a potential that falls off as $1/r$, as in Eq. (16.35), Debye and Hückel found a potential due to the central ion that falls off much more steeply, in fact as $(1/r)e^{-r/b}$. This is called a shielded Coulomb potential or Debye–Hückel potential,

$$\Phi = \frac{z_j e}{4\pi\epsilon_0\epsilon_r r} e^{-r/b} \tag{16.36}$$

Figure 16.10 compares the unshielded potential and the shielded potential. At distances larger than b, the shielded potential falls off rapidly with r.

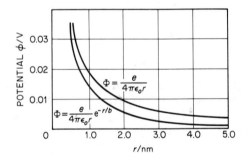

FIGURE 16.10 Coulombic potential and shielded Coulombic potential.

The thickness of the ionic atmosphere is an important parameter in the theory of ionic solutions. We can understand the factors that control b. As more ionic charge is added to the solution, the counterions crowd more closely about the central ion and thus reduce b. From the definition of ionic strength I in Eq. (16.33), the concentration of charge is proportional to $I^{1/2}$, so that b is inversely proportional to this factor. On the other hand, the higher the average thermal energy kT of the ions, the more freely can counterions move away from the ionic atmosphere. Since the average speed of an ion is proportional to $(kT)^{1/2}$, b is proportional to this factor. The exact equation of Debye and Hückel is

$$b = \left(\frac{\epsilon_0\epsilon_r kT}{4LIe^2\rho_0}\right)^{1/2} \tag{16.37}$$

where ρ_0 is the density of solvent. Table 16.5 summarizes some values of b for various combinations of ionic charge.

Example 16.8 Verify that b in Eq. (16.37) has the dimensions of length.

Consider SI units of the factors in Eq. (16.37).

$$\text{units of } b = \left[\frac{(C^2\, J^{-1}\, m^{-1})(\epsilon_r \text{ dimensionless})(J\, K^{-1})(K)}{(mol^{-1})(mol\, kg^{-1})(C)^2(kg\, m^{-3})} \right]^{1/2}$$

$$= [m^2]^{1/2} = m$$

TABLE 16.5
DEBYE LENGTH (nm) (EFFECTIVE RADIUS OF IONIC
ATMOSPHERE) IN AQUEOUS SOLUTIONS AT 298 K

Concentration (mol dm^{-3})	Salt type			
	1:1	1:2	2:2	1:3
10^{-1}	0.96	0.55	0.48	0.39
10^{-2}	3.04	1.76	1.52	1.24
10^{-3}	9.6	5.55	4.81	3.93
10^{-4}	30.4	17.6	15.2	12.4

16.19 Debye–Hückel Limiting Law

Using their shielded Coulomb potential, Debye and Hückel calculated the electrostatic energy due to the ionic atmosphere, and hence from Eq. (16.34) obtained for the activity coefficient,

$$\ln \gamma_i = \frac{-z_i^2 e^2}{8\pi\epsilon_0\epsilon_r kTb} \tag{16.38}$$

This result applies only in the limit of very dilute solutions and hence it is called the *Debye–Hückel limiting law*.

Example 16.9 Estimate γ_i for Na$^+$ ions at 25°C in an aqueous solution of NaCl of concentration 1.00×10^{-2} mol dm^{-3}.

From Table 16.5, $b = 3.04$ nm; $\epsilon_r = 78.5$, and from Eq. (16.38),

$$\ln \gamma_i = -(1.60 \times 10^{-19}\, C)^2/8\pi(8.85 \times 10^{-12}\, C^2\, J^{-1}\, m^{-1})$$

$$\times (78.5)(1.38 \times 10^{-23}\, J\, K^{-1})(298\, K)(3.04 \times 10^{-9}\, m)$$

$$= -0.117, \qquad \gamma_i = 0.890$$

Since the individual ion activity coefficients cannot be measured, the mean activity coefficient is calculated, to give an expression that can be compared with experimental data. From Eq. (16.29),

$$(\nu_+ + \nu_-) \ln \gamma_\pm = \nu_+ \ln \gamma_+ + \nu_- \ln \gamma_-$$

Therefore, from Eq. (16.38),

$$\ln \gamma_\pm = -\left(\frac{\nu_+ z_+^2 + \nu_- z_-^2}{\nu_+ + \nu_-}\right) \frac{e^2}{8\pi\epsilon_0\epsilon_r kTb}$$

Since $\nu_+ z_+ = |\nu_- z_-|$, $\nu_+ z_+^2 = \nu_- |z_+ z_-|$ and $\nu_- z_-^2 = \nu_+ |z_+ z_-|$, so that

$$\ln \gamma_\pm = -|z_+ z_-| \frac{e^2}{8\pi\epsilon_0\epsilon_r kTb} \tag{16.39}$$

Let us now transform Eq. (16.39) into base-10 logarithms, substitute b from Eq. (16.37), and introduce the values of the universal constants: $e = 1.602 \times 10^{-19}$ C, $\epsilon_0 = 8.854 \times 10^{-12}$ C^2 J^{-1} m^{-1}, $k = 1.381 \times 10^{-23}$ J K^{-1}, $L = 6.022 \times 10^{23}$ mol^{-1}. The result is the Debye–Hückel *Limiting Law* for the mean activity coefficient,

$$\log \gamma_\pm = -1.825 \times 10^6 |z_+ z_-| \left(\frac{I\rho_0}{\epsilon_r^3 T^3}\right)^{1/2} = -A|z_+ z_-|I^{1/2} \tag{16.40}$$

For water at 298 K, $\epsilon_r = 78.54$, $\rho_0 = 997$ kg m^{-3}, and the equation becomes

$$\log \gamma_\pm = -0.509 |z_+ z_-| I^{1/2} \tag{16.41}$$

In Fig. 16.11, some experimental activity coefficients are plotted against the square roots of the ionic strengths. These data were obtained from the solubilities of sparingly soluble complex salts in the presence of added salts, such as NaCl, BaCl$_2$, and KNO$_3$. The straight lines indicate the theoretical curves predicted by the limiting law, and it is evident that these limiting slopes are followed at low ionic strengths. The Debye–Hückel theory gives correct quantitative results only at extreme dilutions. Note that the highest concentration in Fig. 16.11 is about 0.01 M.

The theory can be improved by combining it with the general theory of nonideal solutions, as given by Macmillan and Mayer. Thus the theory of the thermodynamic properties of strong electrolytes is now quite well understood. No simple theoretical

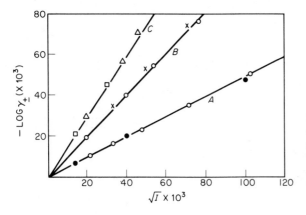

FIGURE 16.11 Activity coefficients of sparingly soluble salts as functions of square root of ionic strength follow the Debye–Hückel limiting law (Data of J. N. Brønsted and V. K. La Mer): *A*, uni-univalent salts; *B*, uni-bivalent; *C*, uni-trivalent.

equation for activity coefficients can be obtained, but semi-empirical equations are available that give $\gamma_\pm$ up to concentrations of about 1 molar.*

In concentrated ionic solutions, definite, through transient, pairs of oppositely charged ions are brought together by electrostatic attraction. The formation of pairs will be greater the lower the dielectric constant of the solvent and the smaller the ionic radii, both of these factors tending to increase the electrostatic attractions. The degree of association may become appreciable even in a solvent of high dielectric constant ϵ_r, such as water. In one-molar aqueous solution, univalent ions having a diameter of 282 pm are 14% associated; of 176 pm, 29% associated. Such association into ion pairs decreases the ionic activity coefficients.

Problems

1. We wish to introduce exactly 1% O_2 into a stream of N_2 gas flowing at 1.00 L s^{-1} at STP. What current should be passed through a KOH solution to liberate the required O_2 at the anode?

2. A conductivity cell filled with 0.100 M KCl (conductivity $\kappa = 1.1639$ Ω^{-1} m^{-1}) at 25°C has a resistance of 33.37 Ω, and water with $\kappa = 7.0 \times 10^{-6}$ Ω^{-1} m^{-1} is used to prepare the solutions. Filled with 0.0100 M acetic acid, the resistance of the cell is 2715 Ω. Calculate Λ of the acetic acid solution.

3. The solubility of $CaSO_4$ at 25°C is 0.667 g L^{-1}. Calculate the conductivity of a saturated solution of $CaSO_4$ in pure water.

4. The molar conductivity of $CNCH_2COOH$ in water at 25°C was measured at several concentrations c.

$10^3 c$ (mol dm^{-3})	0	0.466	1.856	7.335
$10^4 \Lambda$ (Ω^{-1} m^2 mol^{-1})	[386.1]	347.0	282.6	193.9

Calculate the ionization constant of the acid.

5. The conductivity of a saturated solution of AgCl in water at 293 K is $\kappa = 1.37 \times 10^{-4}$ Ω^{-1} m^{-1} and that of the water itself $\kappa = 1.13 \times 10^{-5}$ Ω^{-1} m^{-1}. Calculate the solubility of AgCl. (See Table 16.2.)

6. At 25°C, κ of 0.100 mM aqueous Na_2SO_4 is 2.60×10^{-2} Ω^{-1} m^{-1} and it rises to 7.00×10^{-2} Ω^{-1} m^{-1} when solution is saturated with $CaSO_4$. Calculate $K_{sp} = [Ca^{2+}][SO_4^{2-}]$ for $CaSO_4$.

7. The resistance $\mathfrak{R}$ of a conductivity cell filled with 0.0200 M KCl at 291 K is 17.60 Ω; filled with 0.100 acetic acid, its $\mathfrak{R}$ is 91.8 Ω. If κ of 0.200 M KCl at 291 K is 0.2399 Ω^{-1} m^{-1} and $\Lambda_0(H^+) = 0.0315$ Ω^{-1} m^2 mol^{-1} and $\Lambda_0(CH_3COO^-) = 0.00350$ Ω^{-1} m^2 mol^{-1}, calculate the degree of dissociation of acetic acid in 0.100 M solution.

8. If the purest water has $\kappa = 6.20 \times 10^{-6}$ Ω^{-1} m^{-1} at 25°C, calculate κ for a saturated solution of CO_2 in water if $P_{CO_2} = 2.67$ kPa and for $H_2O(\ell) + CO_2(aq) \rightleftharpoons HCO_3^- +$

*K. S. Pitzer, "Electrolyte Theory—Improvements since Debye and Hückel," *Acc. Chem. Res.*, **10**, 371 (1977).

H^+, $K_c = 4.16 \times 10^{-7}$. The solubility of CO_2 in water follows Henry's Law with $k_H = 2.86 \times 10^{-4}$ mol dm^{-3} kPa^{-1}. [D. MacInnes and R. Belek, *J. Am. Chem. Soc.*, **55**, 2630 (1933).]

9. Plot the conductivity of a solution in which 10.0 cm^3 of 0.100 M NaOH is titrated with 0.100 M HCl. Use the Λ_0 values (Table 16.2).

10. A Hittorf transport-number apparatus with silver electrodes is filled with 0.010 M $AgNO_3$. At the end of the run, 20.09 g of solution in the anode compartment contains 39.66 mg Ag and 27.12 g of solution in cathode compartment contains 11.14 mg Ag. In a silver coulometer in series with the Hittorf cell, 32.10 mg Ag has been deposited. Calculate $t(Ag^+)$.

11. L. G. Longsworth [*J. Am. Chem. Soc.*, **54**, 2741 (1932)] determined $t(Na^+)$ in 0.020 M NaCl solution at 25°C by the moving-boundary method. The boundary between NaCl and $CdCl_2$ solutions moved 6.00 cm in 34.5 min with a current 1.60 mA. The cell cross section is 0.120 cm^2. Calculate $t(Na^+)$.

12. From $\Delta H^\circ_{298} = 55.84$ kJ mol^{-1} for $H_2O \rightleftharpoons H^+ + OH^-$, estimate the pH of pure water at 80°C.

13. A solution initially 0.2000 molal $CuSO_4$ in a Hittorf apparatus: at the end of the electrolysis 100 g H_2O in anode compartment contained 0.0100 mol H_2SO_4 and 0.0160 mol $CuSO_4$. Calculate $t(Cu^{2+})$.

14. The solubility c_s of a cobalt complex salt $[Co(NH_3)_4(NO_2)(CNS)]^+[Co(NH_3)_2(NO_2)_2(C_2O_4)]^-$ was measured in water at 15°C as a function of added NaCl.

c(NaCl) mol dm^{-3}	0.0000	0.0003	0.0010	0.0020	0.0100	0.0200
$10^4 c_s$(complex) mol dm^{-3}	3.355	3.377	3.405	3.451	3.627	3.790

For AX(s) $\rightleftharpoons A^+ + X^-$, $K_{sp} = a^+ a^- = \gamma_+ \gamma_- c_+ c_-$ or $\gamma_\pm = K_{sp}^{1/2}/c_s$. Determine K_{sp} and $\gamma_\pm$ from data and compare the $\gamma_\pm$ with the result of the Debye–Hückel Limiting Law.

15. At 0°C, $\epsilon_r = 87.91$; at 100°C, $\epsilon_r = 56.23$ (water). Calculate the thicknesses of the ionic atmospheres of the Al^{3+} and Cl^- ions in 10^{-3} M $AlCl_3$.

16. The solubility of $AgIO_3$ in KNO_3 solutions was measured by I. M. Kolthoff and J. Lingane [*J. Phys. Chem.*, **42**, 133 (1938)] at 25°C.

c(KNO_3) (mol dm^{-3})	0	0.001301	0.003252	0.006503	0.01410
$10^4 c$($AgIO_3$) (mol L^{-1})	1.761	1.813	1.863	1.908	1.991

Calculate $\gamma_\pm$ for $AgIO_3$ in these solutions and compare results with the Debye–Hückel Limiting Law.

17. For H_2CO_3(aq) $\rightleftharpoons H^+ + HCO_3^-$, $\Delta G^\circ(298 \text{ K}) = 36.26$ kJ mol^{-1}. Calculate the concentration of H^+ ions in a 0.010 M solution of H_2CO_3 (a) assuming that $\gamma_\pm = 1$; (b) with $\gamma_\pm$ from the Debye–Hückel Limiting Law.

18. For $Mg^{2+} + SO_4^{2-} \rightleftharpoons MgSO_4$ at 298 K, $K_c = 10^{-2.141}$ at $c = 1.6196 \times 10^{-4}$ mol dm^{-3} and $K_c = 10^{-2.125}$ at $c = 3.2672 \times 10^{-4}$ mol dm^{-3}. Calculate K_a at these two concentrations from the Debye–Hückel theory.

19. From the Debye–Hückel limiting law, estimate $\gamma_{\pm}$ in the following solutions:
   (a) Aqueous NaCl, $CuSO_4$, $Al_2(SO_4)_3$ at 25°C, all 0.0050 M.
   (b) The same solutions at 75°C.
   (c) The same in a dioxane–water mixture of $\rho = 0.930$ g cm^{-3} and $\epsilon_r = 45$.

20. Activity coefficients can be calculated from data on freezing-point depression, ΔT_f. [G. N. Lewis and M. Randall, *J. Am. Chem. Soc.* **43**, 1112 (1921)]. Let $j = 1 - (\Delta T_f / vmK_F)$. Then

$$\ln \gamma_{\pm} = -j - \int_0^m (j/m')\, dm'$$

Calculate $\gamma_{\pm}$ of NaCl in 0.050 molal solution from data:

m (mol kg^{-1})	0.0100	0.0200	0.0500	0.100	0.200
ΔT_f (K)	0.0361	0.0714	0.1758	0.3770	0.6850

21. The mean-square displacement of a particle due to diffusion is $x^2 = 2Dt$. Consider an average Na$^+$ ion in a cell, where the electric field is 1.00 V cm^{-1}. Estimate its mean diffusion distance and its mean displacement by the electric field in 10 s.

22. Derive the Nernst-Einstein equation (16.24).

17

Electrochemical Cells

The preceding chapter discussed experiments and theories concerning ionic solutions. Now we take up the equilibrium properties of systems in which electrodes are immersed in ionic solutions and connected via an external conductor. The reaction at the surface of an electrode is a transfer of electric charge, usually in the form of electrons. The cathode acts as a source supplying electrons to the solution. The anode acts as a sink accepting electrons from the solution.

A pair of electrodes dipping into an ionic solution and connected by an external metallic conductor constitutes a typical electrochemical cell. If the cell is used to supply electrical energy—that is, if it converts the Gibbs free energy of a physical or chemical change into electrical energy—it is called a *galvanic cell*. A cell in which an external supply of electrical energy is used to bring about a physical or chemical change is called an *electrolytic cell*. These two basic kinds of electrochemical cells are illustrated in Fig. 17.1.

17.1 Metal Electrodes

An important type of electrode consists of a piece of metal dipping into a solution containing ions of that metal. An example is a silver rod dipping into a solution of silver nitrate. This system is conventionally written as $Ag | Ag^+(c)$, where c is the concentration of silver ion and the vertical bar denotes a separation between two phases. Such an electrode system is sometimes called a *half-cell*.

Silver metal consists of a regular array of silver ions Ag^+ permeated by a gaslike

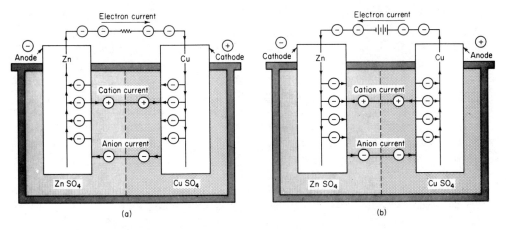

FIGURE 17.1 (a) Galvanic cell. (b) Electrolytic cell.

cloud of electrons (see Section 29.3). In the absence of an electric field the electrons are uniformly distributed through the piece of metal. If an electric field is applied, the electrons drift up the gradient of field until a new equilibrium is established. They move up rather than down because they are negatively charged. If Φ is the electric potential, the electric field is the negative gradient of the potential, $\mathbf{E} = -\partial\Phi/\partial x$, for a one-dimensional case.

Now suppose that we immerse a piece of metal into a solution of its ions, say Ag into $Ag^+NO_3^-$. An Ag^+ ion can leave the metal and enter the solution, leaving an electron behind. The reverse of this reaction can also occur: an Ag^+ ion from the solution can strike the metal surface and become part of the metal structure. Thus a reversible reaction can occur *at the interface* between Ag and $AgNO_3$ solution,

$$Ag(\text{metal}) \rightleftharpoons Ag^+(\text{solution}) + e(\text{in metal})$$

A model of the electrode is depicted in Fig. 17.2.

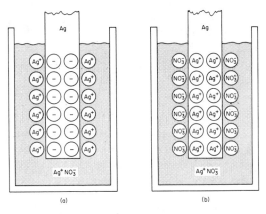

FIGURE 17.2 A silver electrode consists of a silver rod in a solution of Ag^+ ions: (a) low $[Ag^+]$; (b) high $[Ag^+]$.

If the concentration of Ag^+ ions in the solution is very low, at first more Ag^+ ions enter than leave the solution, causing the metal to become negatively charged. As the negative charge on the metal builds up, however, it becomes more difficult for positive Ag^+ ions to leave the metal. The Ag^+ ions that have left the metal cause a positive charge to accumulate in the solution. This positive charge in the solution tends to collect near the negative charge on the metal. Thus an *electric double layer* is built up as shown in Fig. 17.2(a). Since ions in solution and electrons in metals are both highly mobile (thermal motions) the double layer is actually not fixed in two parallel planes as shown, but is more diffuse. Nevertheless, the simple picture is adequate for the present. It shows that the double layer is like a capacitor with opposite charges on its plates, and in fact electrodes in solutions have a double-layer capacitance that can be measured quantitatively. We have considered the case in which $[Ag^+]$ in solution is initially very low so that ions leave the metal to enter the solution. If, however, $[Ag^+]$ is high, the initial net flow of ions is in the other direction, and the double layer at equilibrium has the opposite polarity, as shown in Fig. 17.2(b).

The equilibrium at the surface of an electrode, like all chemical equilibria, is dynamic and not static. Thus ions are continually passing back and forth across the interface. The equal current in each direction is called the *exchange current* I_0.

17.2   The Electrochemical Potential

In the absence of an electric field the equilibrium condition at constant T and P for a component j is governed by its chemical potential μ_j (see Section 8.19). In the presence of an electric field, we must add to μ_j a term that gives the electric potential energy of the component. For a particle of charge $z_j e$ at an electric potential Φ, this electric potential energy is $z_j e\Phi$ per particle or $z_j L e\Phi = z_j F\Phi$ per mole, where F is one faraday of charge. We therefore define the electrochemical potential as

$$\tilde{\mu}_j = \mu_j + z_j F\Phi \tag{17.1}$$

The electrochemical potential plays the role for charged components that the chemical potential plays for uncharged components. If no electric field is present, $\tilde{\mu}_j$ simply reduces to μ_j.

We cannot actually separate the two terms on the right of Eq. (17.1); i.e., we cannot measure μ_j and Φ separately. We must always use the combined function $\tilde{\mu}_j$ when we consider electrochemical problems. Therefore, instead of Eq. (8.53), we have, for electrochemical systems, the equilibrium condition

$$\sum v_j \tilde{\mu}_j = 0 \tag{17.2}$$

Let us apply this equilibrium equation to the surface of a silver electrode in a solution of silver ions, $Ag \mid Ag^+$. The electrochemical reaction is $Ag(\text{metal}) \rightleftharpoons Ag^+$ (solution) $+ e(\text{metal})$. Hence Eq. (17.2) becomes $\tilde{\mu}(Ag^+) + \tilde{\mu}(e) = \mu(Ag)$. ($\tilde{\mu} = \mu$ for the metal since it is an uncharged species.) Introducing Eq. (17.1), we find that

$$\mu(Ag^+) + F\Phi(solution) + \mu(e) - F\Phi(metal) = \mu(Ag)$$

which gives

$$F[\Phi(solution) - \Phi(metal)] = \mu(Ag) - \mu(Ag^+) - \mu(e)$$

or

$$F \Delta\Phi = \mu(Ag) - \mu(Ag^+) - \mu(e)$$

We cannot use this equation to calculate the $\Delta\Phi$ at a single electrode, since there is no way to calculate $\mu(e)$ and $\mu(Ag^+)$. We must use two electrodes to make a complete electrochemical cell before we can get measurable potential differences.

17.3   Contact Between Two Metals

At absolute zero, electrons in a metal would fill a range of energy levels that extends up to some highest level characteristic of the particular metal. In the theory of the metallic state this level is called the *Fermi level* ϵ_F. Figure 29.5 shows the probability p that an energy level is occupied, as a function of the energy. At 0 K, when $\epsilon < \epsilon_F$, $p = 1$, and when $\epsilon > \epsilon_F$, $p = 0$. At temperatures above 0 K some electrons acquire thermal energy and move into higher energy levels, as shown in Fig. 29.5. Then the Fermi level corresponds to the probability $p = \frac{1}{2}$. The Fermi level is identical to the electrochemical potential $\tilde{\mu}$ of the electrons, $\tilde{\mu}_e = \epsilon_F$.

Suppose that two different metals are brought into contact as shown in Fig. 17.3. Electrons flow from the metal of higher $\tilde{\mu}_e$ (higher Fermi level) into the levels of

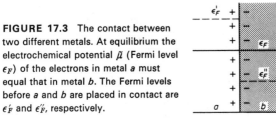

FIGURE 17.3   The contact between two different metals. At equilibrium the electrochemical potential $\tilde{\mu}$ (Fermi level ϵ_F) of the electrons in metal a must equal that in metal b. The Fermi levels before a and b are placed in contact are ϵ_F' and ϵ_F'', respectively.

the metal of lower $\tilde{\mu}_e$. The metal that gains net electrons acquires a negative charge and the metal that loses net electrons acquires a positive charge. Thus an electric potential difference $\Delta\Phi$ occurs at the contact between the two metals. The condition of equilibrium is the equality of electrochemical potential for electrons in the two metals. From Eq. (17.2), $\mu_e^a - F\Phi^a = \mu_e^b - F\Phi^b$, or

$$\Delta\Phi = \Phi^b - \Phi^a = \mu_e^b - \mu_e^a \tag{17.3}$$

In this way an equilibrium *contact potential* arises at the interface between any pair of different metals. Just as in the case of the $\Delta\Phi$ between a metal and a solution, however, we cannot measure this contact potential directly.

A metal electrode consists of a piece of metal in contact with a solution containing ions of that metal. The example of the silver electrode $Ag\,|\,Ag^+$ has already been discussed. It is sometimes convenient to form a metal electrode by using an amalgam instead of a pure metal. A liquid amalgam has the advantage of eliminating nonreproducible effects due to strains in the solid metal. In some instances, a dilute amalgam electrode can be successfully employed where the pure metal would react violently with the solution, for example, in the sodium amalgam half-cell, $NaHg(c_1)\,|\,Na^+(c_2)$.

Gas electrodes can be constructed by placing a strip of nonreactive metal, usually platinum or gold, in contact with both the solution and a gas stream. The hydrogen electrode consists of a platinum strip exposed to a current of hydrogen, and partly immersed in an acid solution. The hydrogen is probably dissociated into atoms at the catalytic surface of the platinum, the electrode reactions being

$$\tfrac{1}{2}H_2 \longrightarrow H, \qquad H \longrightarrow H^+(aq) + e, \qquad \text{overall:} \quad \tfrac{1}{2}H_2 \longrightarrow H^+(aq) + e$$

In an *oxidation–reduction electrode*, an inert metal dips into a solution containing ions in two different oxidation states, e.g., $Pt\,|\,Fe^{2+}, Fe^{3+}$. When electrons are supplied to the electrode, the reaction is $Fe^{3+} + e \longrightarrow Fe^{2+}$.

Metal–insoluble-salt electrodes consist of a metal in contact with one of its slightly soluble salts; this salt is in turn in contact with a solution containing a common anion. An example is the silver–silver-chloride half-cell: $Ag\,|\,AgCl\,|\,Cl^-(c_1)$. The electrode reaction can be considered in two steps:

$$(1)\ AgCl(s) \rightleftharpoons Ag^+ + Cl^-; \qquad (2)\ Ag^+ + e \rightleftharpoons Ag(s)$$

Or, overall, $AgCl(s) + e \rightleftharpoons Ag(s) + Cl^-$. Such an electrode is thermodynamically equivalent to a chlorine electrode $(Cl_2\,|\,Cl^-)$, in which the gas is at a pressure equal to the dissociation pressure of AgCl according to $AgCl \rightleftharpoons Ag + \tfrac{1}{2}Cl_2$.

17.5   Classification of Cells

When two suitable half-cells are connected, we have an *electrochemical cell*. The connection is made by bringing the solutions in the half-cells into contact, so that ions can pass between them. If these two solutions are the same, there is no liquid junction, and we have a *cell without transference*. If the solutions are different, the transport of ions across the junction causes irreversible changes in the two electrolytes, and we have a *cell with transference*.

The $-\Delta G$ that provides the driving force in a cell may come from a chemical reaction or from a physical change. Cells in which the driving force is a change in concentration (almost always a dilution process) are called *concentration cells*. The change in concentration can occur either in the electrolyte or in the electrodes. Examples of changes in concentration in electrodes are found in amalgams or alloy electrodes with different concentrations of solute metal and in gas electrodes with different pressures of gas.

The varieties of electrochemical cells can be classified as follows:

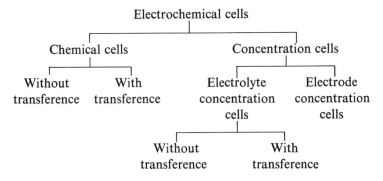

17.6 An Electrochemical Cell

Figure 17.1 shows a typical galvanic cell. A zinc electrode dips into a $1.0\,m$ solution of $ZnSO_4$ and a copper electrode dips into a $1.0\,m$ solution of $CuSO_4$. The two solutions are separated by a porous barrier, which allows electrical contact, but prevents excessive mixing of the solutions by interdiffusion. This cell is represented by the diagram

$$Zn\,|\,Zn^{2+}(1.0\,m)\,||\,Cu^{2+}(1.0\,m)\,|\,Cu \qquad (A)$$

The double bar represents the liquid junction between the two cells. It is, in fact, the locus of a small potential difference (about 10 mV) but to a first approximation we shall ignore this. A double bar implies this elimination of the liquid junction potential.

It is always possible to measure the difference in electric potential between two pieces of the same kind of metal. We attach to each electrode a length of copper wire and connect these copper leads to a voltmeter or some other device for measuring their difference in potential. The potential difference $\Delta\Phi$ of the cell is equal in sign and magnitude to the electric potential of a metallic conducting lead on the right minus that of an identical lead on the left.

$$\Delta\Phi = \Phi_R - \Phi_L \qquad (17.4)$$

Note that we cannot measure Φ_R or Φ_L separately, but their difference is a directly measurable quantity. The meaning of "left" and "right" refers to the cell as written in diagram (A); clearly it has nothing to do with how the actual cell is arranged on the bench.

The potential difference between the electrodes of the cell in Fig. 17.1 is measured as 1.10 V and the copper electrode is found to be positive, i.e., $\Phi_R > \Phi_L$. Therefore, cell (A) has $\Delta\Phi = \Phi_R - \Phi_L = +1.10$ V. The cell

$$Cu\,|\,Cu^{2+}(1.0\,m)\,||\,Zn^{2+}(1.0\,m)\,|\,Zn \qquad (B)$$

has $\Delta\Phi = \Phi_R - \Phi_L = -1.10$ V.

17.7 Cell Diagram and Cell Reaction

Conventional cell diagrams are always written so as to be read from left to right. Thus in the diagram (A) the reaction at the left-hand electrode is an oxidation process (loss of electrons), $Zn \rightarrow Zn^{2+} + 2e$. Electrons are supplied to the external circuit by the electrode reaction. The reaction at the right-hand electrode is a reduction process (gain of electrons), $Cu^{2+} + 2e \rightarrow Cu$. Electrons are accepted from the external circuit by the electrode reaction. In accord with this rule, the chemical reaction in cell (A) is $Zn + Cu^{2+} \rightarrow Zn^{2+} + Cu$. The reaction in cell (B) is $Cu + Zn^{2+} \rightarrow Cu^{2+} + Zn$.

The actual reaction that occurs in an experimental cell obviously has nothing to do with how the cell reaction is written. When the cell is actually made in the laboratory and the circuit completed through an external conductor, one side will be positive and the other negative, and electrons will flow through the metallic conductors from the negative to the positive pole.

17.8 Equilibrium Condition in an Electrochemical Cell

When equilibrium has been established across the phase boundaries in an electrochemical cell, we can apply the condition of equilibrium Eq. (17.2) to obtain the $\Delta\Phi_i$ at each boundary. We can then sum these $\Delta\Phi_i$ to obtain the $\Delta\Phi$ between the two copper terminals of the cell. The results are as follows:

$$Cu \,|\, Zn: \qquad F\Delta\Phi_1 = \mu_e(Zn) - \mu_e(Cu)$$

$$Zn \,|\, Zn^{2+}: \qquad F\Delta\Phi_2 = \tfrac{1}{2}\mu(Zn) - \tfrac{1}{2}\mu(Zn^{2+}) - \mu_e(Zn)$$

$$Zn^{2+} \,|\, Cu^{2+}: \qquad \Delta\Phi_3 = \text{liquid junction potential, neglected}$$

$$Cu^{2+} \,|\, Cu: \qquad F\Delta\Phi_4 = \tfrac{1}{2}\mu(Cu^{2+}) + \mu_e(Cu) - \tfrac{1}{2}\mu(Cu)$$

$$F(\Phi_R - \Phi_L) = F\Delta\Phi = F(\Delta\Phi_1 + \Delta\Phi_2 + \Delta\Phi_4) = \tfrac{1}{2}\mu(Zn) - \tfrac{1}{2}\mu(Zn^{2+}) + \tfrac{1}{2}\mu(Cu^{2+}) - \tfrac{1}{2}\mu(Cu)$$

The right side is simply $-\Delta\mu = -\Delta G$ for the cell reaction $\tfrac{1}{2}Zn + \tfrac{1}{2}Cu^{2+} = \tfrac{1}{2}Zn^{2+} + \tfrac{1}{2}Cu$. Therefore, at equilibrium,

$$F\,\Delta\Phi = F(\Phi_R - \Phi_L) = -\Delta G \qquad (17.5)$$

This important result shows that under equilibrium conditions the measurable difference in electric potential $\Delta\Phi$ between the two copper terminals of the cell is determined by $-\Delta G/F$, where ΔG is the change in Gibbs free energy of the cell reaction. The unmeasurable individual electric and chemical potential differences at the boundaries between unlike phases have been eliminated from the final result. During most of the nineteenth century there were fierce arguments among electrochemists about the locations of the potential differences in a cell. We can now understand that all the phase boundaries are sites of $\Delta\Phi$'s and the overall measurable $\Delta\Phi$ includes all these contributions.

17.9   Electromotive Force (EMF) of a Cell

We have calculated the potential difference $\Delta\Phi = \Phi_R - \Phi_L$ of a cell under equilibrium conditions. This potential difference is called the *electromotive force (emf)* E of the cell.* If an electric conductor is placed in the circuit between the two terminals of the cell, current flows through the cell. Suppose, for example, that an ammeter and a voltmeter are placed in the circuit. The voltmeter is a high-resistance (R) device, which limits the current I. As soon as any current flows between the terminals, the potential difference $\Delta\Phi = \Phi_R - \Phi_L$ becomes equal to IR, the voltage drop across the resistance between the terminals. This $\Delta\Phi$ will always be less than the equilibrium value E, the emf given by Eq. (17.5). The electromotive force (emf) E of the cell can be measured as the limiting value of the electric potential difference $\Delta\Phi$ as the current through the cell goes to zero.

$$E = \Delta\Phi_{(I=0)} \qquad (17.6)$$

It is possible to measure E under conditions in which the current drawn from the cell is so small as to be negligible. The method, devised by Poggendorf, uses a circuit known as the *potentiometer*. A basic potentiometer circuit is shown in Fig. 17.4. The slide wire is calibrated with a scale so that any setting of the contact cor-

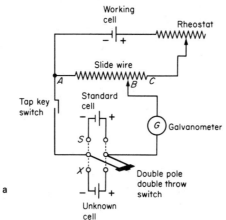

FIGURE 17.4 Basic circuit of the potentiometer by which the voltage of a cell can be measured while negligible current is drawn from the cell.

responds to a certain voltage. With the double-pole, double-throw switch in the standard cell position S, we set the slide wire to the voltage reading of the standard cell, and adjust the rheostat until no current flows through the galvanometer G. At this point, the potential difference between A and B, the IR along the section AB of the slide wire, just balances the emf of the standard cell. We then set the switch to the unknown cell position X, and readjust the slide-wire contact until no current flows through G. From the new setting, we can read directly from the scale of the slide wire the emf of the unknown cell. Since it is not difficult to balance such a circuit so that less than 10^{-12} A is drawn from the cell, we can satisfy for all practical purposes the condition specified by the definition of emf, i.e., measurement of $\Delta\Phi$ as $I \to 0$.

*Do not confuse E (emf) with E (magnitude of electric field) used in Chapter 16.

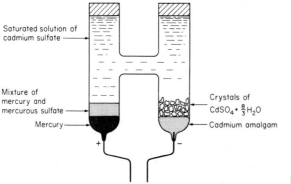

Saturated solution of cadmium sulfate

Mixture of mercury and mercurous sulfate

Mercury

Crystals of CdSO$_4 \cdot \frac{8}{3}$H$_2$O

Cadmium amalgam

+

−

FIGURE 17.5   The Weston standard cell.

17.10   A Standard Cell

The most widely used standard is the Weston cell, shown in Fig. 17.5, which is written

$$\text{Cd(Hg)} \,|\, \text{CdSO}_4 \cdot \tfrac{8}{3}\text{H}_2\text{O(s)} \,|\, \text{CdSO}_4\text{(sat.)} \,|\, \text{Hg}_2\text{SO}_4\text{(s)} \,|\, \text{Hg}$$

The cell reaction is

$$\text{Cd(Hg)} + \text{Hg}_2\text{SO}_4\text{(s)} + \tfrac{8}{3}\text{H}_2\text{O}(\ell) = \text{CdSO}_4 \cdot \tfrac{8}{3}\text{H}_2\text{O(s)} + 2\text{Hg}(\ell)$$

Its emf in volts is given by

$$E = 1.01845 - 4.5 \times 10^{-5}(T - 293) - 9.5 \times 10^{-7}(T - 293)^2$$

Thus, at 293 K, $E = 1.01845$ V; at 298 K, $E = 1.01832$ V. The small temperature coefficient of E is one advantage of this cell.

17.11   Reversible Cells

We shall be interested primarily in the class of cells called *reversible cells*. These can be recognized by the following criterion: The cell is connected with a potentiometer arrangement for emf measurement by the compensation method. The $\Delta\Phi$ of the cell is measured (1) with a small current flowing through the cell in one direction, (2) then with an imperceptible flow of current, and (3) finally with a small flow in the opposite direction. If a cell is reversible, its $\Delta\Phi$ changes only slightly during this sequence, and there is no discontinuity in the value of the $\Delta\Phi$ at the point of balance (2). Reversibility implies that any chemical reaction occurring in the cell can proceed in either direction, depending on the flow of current, and at the null point the driving force of the reaction is just balanced by the compensating voltage of the potentiometer. If a cell is reversible, the half-cells comprising it are both reversible.

One source of irreversibility in cells is the *liquid junction*, like that in the Daniell cell of Fig. 17.1. At the junction between ZnSO$_4$ and CuSO$_4$, we have the following situation:

$$\underset{(1.0\,m)}{Zn^{2+}} \Bigm|_{SO_4^{2-}} \underset{(1.0\,m)}{Cu^{2+}}$$

If we pass a small current through the cell from left to right, it is carried across the junction by Zn^{2+} ions and by SO_4^{2-} ions. But if we go through the balance point, and pass a small current in the opposite direction, it is carried across the junction from right to left by Cu^{2+} and SO_4^{2-} ions. Thus a cell with such a liquid junction is inherently not reversible.

Before we can apply reversible thermodynamics to such cells, we must eliminate the liquid junction. We can do this with considerable success by means of a *salt bridge*. This device consists of a connecting tube filled with a concentrated solution of a salt, usually KCl. The solution can be made up in a gel to decrease mixing with the solutions in the two half-cells. Now most of the current will be carried across the junction by K^+ and Cl^- ions. There will still be some irreversible effects where the bridge enters the two solutions, but these are regarded as minimal.

A better way to avoid irreversible effects is to avoid liquid junctions, by using a single electrolyte. The Weston cell does this with a solution of $CdSO_4$ that is also saturated with the sparingly soluble Hg_2SO_4. We shall discuss later other examples of cells without liquid junctions. Even in such cells, however, changes in electrolyte concentration around the electrodes, as a consequence of the cell reaction, introduce small irreversible effects.

17.12   Thermodynamics of Cell Reactions

We found in Eq. (17.5) that the equilibrium potential difference E(emf) for an electrochemical cell is related to the ΔG of the cell reaction by $\Delta G = -FE$, for the transfer of charge F through the cell. For a reaction in which the charge transferred is $|z|F$, the relation is

$$\Delta G = -|z|FE \qquad (17.7)$$

Example 17.1   In the Daniel cell (Fig. 17.1) the emf at 25°C is $E = 1.100$ V. What is the ΔG of the cell reaction?

$$Zn + CuSO_4(1\ m) \longrightarrow Cu + ZnSO_4(1\ m)$$

Since Cu^{2+} and Zn^{2+} are doubly charged ions, $2F$ coulombs of charge are transferred through the cell per mole of the reaction as written. Hence

$$\Delta G = -2(96\ 487\ \text{C mol}^{-1})(1.100\ \text{V}) = -212\ 300\ \text{V C mol}^{-1}\ \text{or J mol}^{-1}$$

A reaction can proceed spontaneously at constant T and P only if $\Delta G < 0$. Hence from Eq. (17.7) a cell reaction can proceed spontaneously only if $E > 0$. If $E < 0$ for a cell reaction, an external voltage must be applied to the cell to cause the reaction to occur. The cell then acts as an electrolytic cell.

Example 17.2 The Hall process for the production of aluminum is based on the electrolysis of a molten solution of Al_2O_3 in cryolite (Na_3AlF_6) with carbon electrodes. Calculate the minimum voltage that must be applied to the cell to produce Al at 1300 K, if the cell operated reversibly.

The cell reaction is taken to be $2Al_2O_3(\text{melt}) + 3C = 4Al(\ell) + 3CO_2$. We use thermodynamic data to calculate ΔG and hence E. The solubility of Al_2O_3 in cryolite at 1300 K is 12%. This corresponds to mole fraction $X(Al_2O_3) = (12/102)/[(12/102) + (88/210)] = 0.219$. For $Al_2O_3(s) = Al_2O_3(\text{melt})$, if we assume an ideal solution, $\Delta G_{1300} = -RT \ln X = 16.4$ kJ mol^{-1}. Thermodynamic data in the form of equations for $\Delta G°$ as a function of T are given by Kubachewski [*Metallurgical Thermochemistry*, 5th ed. (Oxford: Pergamon Press Ltd., 1979)]. The C, Al, and CO_2 can be taken to be in their standard states at 1300 K.

$$C + O_2 = CO_2 \qquad\qquad \Delta G°_{1300} = -395 \text{ kJ mol}^{-1}$$
$$2Al(\ell) + 1\tfrac{1}{2}O_2 = Al_2O_3(s) \qquad = -1260 \text{ kJ mol}^{-1}$$
$$\underline{Al_2O_3(s) = Al_2O_3(\text{melt}) \qquad\qquad = 16 \text{ kJ mol}^{-1}}$$
$$2Al_2O_3(\text{melt}) + 3C = 4Al(\ell) + 3CO_2$$
$$\Delta G = -2(16) - 2(-1260) + 3(-395) = 1303 \text{ kJ mol}^{-1}$$
$$E = -\Delta G/12F = -1\,303\,000/12(96\,485) = -1.13 \text{ V}$$

The calculated decomposition potential is 1.13 V. Experimentally, the decomposition potential is 1.70 V. We can conclude that the cell is far from reversible; the formation of O_2 and the oxidation of carbon at the anodes are major sources of irreversibility.

The application of the Gibbs–Helmholtz equation (8.38) to the relation $\Delta G = -|z|FE$ allows us to calculate the ΔH and ΔS of a cell reaction from the temperature coefficient of the reversible emf.

$$\Delta S = -(\partial \Delta G/\partial T)_P = |z| F(\partial E/\partial T)_P \qquad (17.8)$$

Since, at constant T, $\Delta H = \Delta G + T \Delta S$, Eqs. (17.7) and (17.8) give

$$\Delta H = -|z| FE + |z| FT(\partial E/\partial T)_P \qquad (17.9)$$

Example 17.3 Apply these relations to the Weston standard cell to calculate ΔG, ΔS, and ΔH at 25°C. For the cell as shown in Fig. 17.5, $E = 1.01832$ V at 298 K. The temperature coefficient of the emf is $dE/dT = -5.00 \times 10^{-5}$ V K^{-1}.

Since $|z| = 2$ per mole of the cell reaction, Eq. (17.7) gives

$$\Delta G = -2(96\,485 \text{ C mol}^{-1})(1.01832 \text{ V}) = -196\,505 \text{ J mol}^{-1}$$
$$\Delta S = 2(96\,485 \text{ C mol}^{-1})(-5.00 \times 10^{-5} \text{ V K}^{-1}) = -9.65 \text{ J K}^{-1} \text{ mol}^{-1}$$
$$\Delta H = \Delta G + T\Delta S = (-196\,505 \text{ J mol}^{-1}) + (298 \text{ K})(-9.65 \text{ J K}^{-1} \text{ mol}^{-1})$$
$$= -199\,390 \text{ J mol}^{-1}$$

17.13 The Standard emf of Cells

Let us consider the generalized cell reaction: $aA + bB = cC + dD$. By comparison with Eq. (8.33), the change in Gibbs free energy in terms of activities of the reactants is

$$\Delta G = \Delta G^\circ + RT \ln \frac{a_C^c a_D^d}{a_A^a a_B^b} \tag{17.10}$$

From $\Delta G = -|z| FE$, when the activities of all the products and reactants are unity, the value of the emf is $E^\circ = -\Delta G^\circ / |z| F$. This E° is called the *standard emf* of the cell. It is related to the equilibrium constant of the cell reaction, since

$$|z| FE^\circ = -\Delta G^\circ = RT \ln K_a \tag{17.11}$$

Division of Eq. (17.10) by $-|z| F$ yields

$$E = E^\circ - \frac{RT}{|z| F} \ln \frac{a_C^c a_D^d}{a_A^a a_B^b} \tag{17.12}$$

We shall call Eq. (17.12) the *Nernst equation*. A similar expression was given Walther Nernst (1890) in terms of concentrations instead of activities.

The determination of the standard emf of a cell is one of the most important procedures in electrochemistry. A useful method is shown in the following example. The cell in Fig. 17.6 consists of a hydrogen electrode and a silver–silver chloride elec-

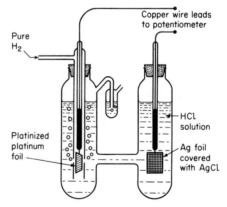

FIGURE 17.6 A cell consisting of a hydrogen electrode and a silver–silver chloride electrode.

trode immersed in a solution of hydrochloric acid: $Pt(H_2) | HCl(m) | AgCl | Ag$. This is a chemical cell without a liquid junction. The electrode reactions are

$$\tfrac{1}{2}H_2 = H^+ + e$$

$$AgCl + e = Ag + Cl^-$$

Overall reaction: $\quad AgCl + \tfrac{1}{2}H_2 = H^+ + Cl^- + Ag$

From Eq. (17.12) the emf of the cell is

$$E = E^\circ - \frac{RT}{F} \ln \frac{a_{Ag} a_{H^+} \cdot a_{Cl^-}}{a_{AgCl} a_{H_2}^{1/2}}$$

Setting the activities of the solid phases equal to unity, and choosing the hydrogen pressure so that $a_{H_2} = 1$ (for ideal gases, $P° = 101.3$ kPa) we obtain the equation

$$E = E° - \frac{RT}{F} \ln a_H \cdot a_{Cl^-}$$

Introducing the mean activity of the ions defined by Eq. (16.26) gives

$$E = E° - \frac{2RT}{F} \ln a_{\pm} = E° - \frac{2RT}{F} \ln \gamma_{\pm} m$$

On rearrangement,

$$E + \frac{2RT}{F} \ln m = E° - \frac{2RT}{F} \ln \gamma_{\pm} \qquad (17.13)$$

According to the Debye–Hückel theory, in dilute solutions, $\ln \gamma_{\pm} = Am^{1/2}$, where A is a constant. Hence for dilute solutions Eq. (17.13) becomes

$$E + \frac{2RT}{F} \ln m = E° - \frac{2RTA}{F} m^{1/2} \qquad (17.13a)$$

If the quantity on the left is plotted against $m^{1/2}$, and extrapolated back to $m = 0$, the intercept at $m = 0$ gives the value of $E°$. The plot of the data for this cell is shown in Fig. 17.7. The extrapolation gives $E° = 0.2225$ V at 25°C.

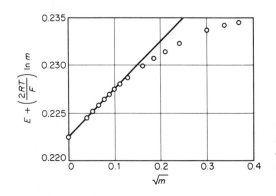

FIGURE 17.7 Graph of data on emf E of cell shown in Fig. 17.6 for determination of its standard emf $E°$. At low concentrations the straight line fits Eq. (17.13a).

Once the standard emf has been determined in this way, Eq. (17.13) can be used to calculate mean activity coefficients for HCl from the measured emf's E in solutions of different molalities m. This method has been the most important source of precise data on ionic activity coefficients.

17.14 Standard Electrode Potentials

Rather than tabulate data for all the numerous cells that have been measured, it would be much more convenient to make a list of *single-electrode potentials* of the various half-cells. Cell emf's could then be obtained simply by taking differences between these electrode potentials.

Although we cannot measure absolute single-electrode potentials, cell emf's can be reduced to a common basis by expressing all the values relative to the same reference electrode. The choice of a conventional reference state does not affect the values of differences between the electrode potentials, i.e., the cell emf's. The reference electrode is taken to be the standard hydrogen electrode, SHE, which is assigned by convention the value $E° = 0$. The SHE is a hydrogen electrode in which (1) the pressure of hydrogen is $P° = 101.3$ kPa (strictly, unit fugacity, but the gas may be taken to be ideal) and (2) the solution contains hydrogen ions at a mean ionic activity $a_\pm = 1$. Thus the SHE is $Pt \,|\, H_2(P°) \,|\, H^+(a_\pm = 1)$.

If a cell is formed by combining any electrode X with a SHE, the measured potential of the electrode X relative to that of the SHE taken as zero is called the *relative electrode potential*, or in short, the *electrode potential*, of X. Thus if the electrode X is positive with respect to the SHE, the electrode potential of X is positive. The electrode potential so defined has a definite sign, which in no way depends on how the electrode is written. The sign of the electrode potential always is the observed experimental sign of its polarity when it is coupled with a standard hydrogen electrode.

The electrode potential is the emf of the cell:

$$Pt \,|\, H_2(P°) \,|\, H^+(a_\pm = 1) \,|\, X^+ \,|\, X$$

The emf is $E = \Phi_R - \Phi_L$, so that

$$E(X^+ \,|\, X) - E(H_2 \,|\, H^+) \equiv E(X^+ \,|\, X)$$

If E is the standard $E°$, the electrode potential is the standard electrode potential. When we speak of an electrode potential, we usually mean this standard potential, which is the one given in tables. The electrode potential for any other choice of activities can be calculated from Eq. (17.12).

Table 17.1 is a collection of standard potentials. Electrodes are written as $X^+ \,|\, X$, to recall that in combination with the $H_2 \,|\, H^+$ electrode they will always be on right side of cell as written. Many electrode potentials that are not given in the table can be calculated from the tabulated values.

Example 17.4 Calculate the electrode potential at 25°C of the Cr^{3+}, $Cr_2O_7^{2-}$ electrode at pH 1 and pH 3 in a solution 0.01 M in both Cr^{3+} and $Cr_2O_7^{2-}$, on the assumption that all activity coefficients are unity.

From Table 17.1, the electrode reaction is $Cr_2O_7^{2-} + 14H^+ + 6e = 2Cr^{3+} + 7H_2O$ and $E° = 1.33$ V.

$$E = E° - \frac{RT}{6F} \ln \frac{[Cr^{3+}]^2}{[Cr_2O_7^{2-}][H^+]^{14}}$$

$$= 1.33 - \frac{(8.314 \text{ J K}^{-1} \text{ mol}^{-1})(298 \text{ K})}{6(96\ 485 \text{ C mol}^{-1})} \ln \frac{[0.01]^2}{[0.01][10^{-pH}]^{14}}$$

$$= 1.33 - 0.00985 \log \frac{[0.01]^2}{[0.01][10^{-pH}]^{14}}$$

$$= 1.33 - 0.00985[-2 + 14 \text{ pH}]$$

At pH $= 1$, $E = 1.21$ V; at pH $= 3$, $E = 0.94$ V.

TABLE 17.1

STANDARD ELECTRODE POTENTIALS AT 25°C[a]

Electrode	Electrode reaction	$E°$ (V)
Acid Solutions		
$Li^+\,\vert\,Li$	$Li^+ + e \rightleftharpoons Li$	-3.035
$K^+\,\vert\,K$	$K^+ + e \rightleftharpoons K$	-2.925
$Cs^+\,\vert\,Cs$	$Cs^+ + e \rightleftharpoons Cs$	-2.923
$Ra^{2+}\,\vert\,Ra$	$Ra^{2+} + 2e \rightleftharpoons Ra$	-2.92
$Ba^{2+}\,\vert\,Ba$	$Ba^{2+} + 2e \rightleftharpoons Ba$	-2.906
$Ca^{2+}\,\vert\,Ca$	$Ca^{2+} + 2e \rightleftharpoons Ca$	-2.866
$Na^+\,\vert\,Na$	$Na^+ + e \rightleftharpoons Na$	-2.713
$Mg^{2+}\,\vert\,Mg$	$Mg^{2+} + 2e \rightleftharpoons Mg$	-2.366
$Pu^{3+}\,\vert\,Pu$	$Pu^3 + 3e \rightleftharpoons Pu$	-2.03
$Al^{3+}\,\vert\,Al$	$Al^{3+} + 3e \rightleftharpoons Al$	-1.662
$Zn^{2+}\,\vert\,Zn$	$Zn^{2+} + 2e \rightleftharpoons Zn$	-0.7628
$Fe^{2+}\,\vert\,Fe$	$Fe^{2+} + 2e \rightleftharpoons Fe$	-0.440
$Cd^{2+}\,\vert\,Cd$	$Cd^{2+} + 2e \rightleftharpoons Cd$	-0.403
$Sn^{2+}\,\vert\,Sn$	$Sn^{2+} + 2e \rightleftharpoons Sn$	-0.136
$Pb^{2+}\,\vert\,Pb$	$Pb^{2+} + 2e \rightleftharpoons Pb$	-0.126
$D^+\,\vert\,D_2\,\vert\,Pt$	$2D^+ + 2e \rightleftharpoons D_2$	-0.0034
$H^+\,\vert\,H_2\,\vert\,Pt$	$2H^+ + 2e \rightleftharpoons H_2$	0
$S_2O_3^{2-}, S_4O_6^{2-}\,\vert\,Pt$	$S_4O_6^{2-} + 2e \rightleftharpoons 2S_2O_3^{2-}$	$+0.09$
$Sn^{4+}, Sn^{2+}\,\vert\,Pt$	$Sn^{4+} + 2e \rightleftharpoons Sn^{2+}$	$+0.14$
$Cu^{2+}, Cu^+\,\vert\,Pt$	$Cu^{2+} + e \rightleftharpoons Cu^+$	$+0.153$
$Cu^{2+}\,\vert\,Cu$	$Cu^{2+} + 2e \rightleftharpoons Cu$	$+0.337$
$Fe(CN)_6^{4-}, Fe(CN)_6^{3-}\,\vert\,Pt$	$Fe(CN)_6^{3-} + e \rightleftharpoons Fe(CN)_6^{4-}$	$+0.355$
$I^-\,\vert\,I_2\,\vert\,Pt$	$I_2 + 2e \rightleftharpoons 2I^-$	$+0.535$
$Fe^{2+}, Fe^{3+}\,\vert\,Pt$	$Fe^{3+} + e \rightleftharpoons Fe^{2+}$	$+0.771$
$Hg_2^{2+}\,\vert\,Hg$	$Hg_2^{2+} + 2e \rightleftharpoons 2Hg$	$+0.792$
$Ag^+\,\vert\,Ag$	$Ag^+ + e \rightleftharpoons Ag$	$+0.7994$
$Hg_2^{2+}, Hg^{2+}\,\vert\,Pt$	$2Hg^{2+} + 2e \rightleftharpoons Hg_2^{2+}$	$+0.907$
$Br^-\,\vert\,Br_2\,\vert\,Pt$	$Br_2 + 2e \rightleftharpoons 2Br^-$	$+1.075$
$Mn^{2+}, H^+\,\vert\,MnO_2\,\vert\,Pt$	$MnO_2 + 4H^+ + 2e \rightleftharpoons Mn^{2+} + 2H_2O$	$+1.23$
$Cr^{3+}, Cr_2O_7^{2-}, H^+\,\vert\,Pt$	$Cr_2O_7^{2-} + 14H^+ + 6e \rightleftharpoons 2Cr^{3+} + 7H_2O$	$+1.33$
$Cl^-\,\vert\,Cl_2\,\vert\,Pt$	$Cl_2 + 2e \rightleftharpoons 2Cl^-$	$+1.359$
$Ce^{3+}, Ce^{4+}\,\vert\,Pt$	$Ce^{4+} + e \rightleftharpoons Ce^{3+}$	$+1.70$
$Co^{2+}, Co^{3+}\,\vert\,Pt$	$Co^{3+} + e \rightleftharpoons Co^{2+}$	$+1.80$
$SO_4^{2-}, S_2O_8^{2-}\,\vert\,Pt$	$S_2O_8^{2-} + 2e \rightleftharpoons 2SO_4^{2-}$	$+2.01$
Basic Solutions		
$OH^-\,\vert\,Ca(OH)_2\,\vert\,Ca\,\vert\,Pt$	$Ca(OH)_2 + 2e \rightleftharpoons 2OH^- + Ca$	-3.02
$H_2PO_2^-, HPO_3^{2-}, OH^-\,\vert\,Pt$	$HPO_3^{2-} + 2e \rightleftharpoons H_2PO_2^- + 3OH^-$	-1.565
$ZnO_2^{2-}, OH^-\,\vert\,Zn$	$ZnO_2^{2-} + 2H_2O + 2e \rightleftharpoons Zn + 4OH^-$	-1.215
$SO_3^{2-}, SO_4^{2-}, OH^-\,\vert\,Pt$	$SO_4^{2-} + H_2O + 2e \rightleftharpoons SO_3^{2-} + 2OH^-$	-0.93
$OH^-\,\vert\,H_2\,\vert\,Pt$	$2H_2O + 2e \rightleftharpoons H_2 + 2OH^-$	-0.82806
$OH^-\,\vert\,Ni(OH)_2\,\vert\,Ni$	$Ni(OH)_2 + 2e \rightleftharpoons Ni + 2OH^-$	-0.72
$CO_3^{2-}\,\vert\,PbCO_3\,\vert\,Pb$	$PbCO_3 + 2e \rightleftharpoons Pb + CO_3^{2-}$	-0.509
$OH^-, HO_2^-\,\vert\,Pt$	$HO_2^- + H_2O + 2e \rightleftharpoons 3OH^-$	$+0.878$

[a]For a more complete critical survey, see International Union of Pure and Applied Chemistry, Electrochemical Commission, G. Charlot, ed., *Selected Constants: Oxidation–Reduction Potentials in Aqueous Solution* (London: Butterworth & Co. (Publishers) Ltd., 1971).

> **Example 17.5** What is the standard electrode potential of the half-cell $Cu^+|Cu$?
>
> From Table 17.1:
>
> | | | E° | $\Delta G^\circ = -zFE^\circ$ |
> |---|---|---|---|
> | | $Cu^{2+} + 2e = Cu$ | 0.337 V | $-2F(0.337)$ |
> | | $Cu^{2+} + e = Cu^+$ | 0.153 V | $-F(0.153)$ |
> | $Cu^+|Cu$: | $Cu^+ + e = Cu$ | 0.521 V | $-F(0.521)$ |
>
> Note that E° must be multiplied by the number of faradays transferred to give the ΔG° of the half-cell reactions and these ΔG° are combined to give ΔG° for the new half-cell reaction, which then gives $E^\circ = -\Delta G^\circ/|z|F$.

17.15 Calculation of the emf of a Cell

As a typical example, let us calculate the emf at 25°C of the cell,

$$Zn\,|\,ZnSO_4(1.0\ m)\,\|\,CuSO_4(0.1\ m)\,|\,Cu$$

The cell reaction is $Zn + CuSO_4 = ZnSO_4 + Cu$. From Eq. (17.4), the standard emf E°, when we take the electrode potentials from Table 17.1, is

$$E^\circ = E^\circ_R - E^\circ_L = +0.337 - (-0.763) = +1.100\ V$$

The Nernst equation becomes

$$E = E^\circ - \frac{RT}{2F}\ln\frac{a(ZnSO_4)a(Cu)}{a(CuSO_4)a(Zn)}$$

or, since $a(Cu) = a(Zn) = 1$ for the pure solids,

$$E = 1.100 - 0.0295\log\frac{a^2_\pm(ZnSO_4)}{a^2_\pm(CuSO_4)} = 1.100 - 0.059\log\frac{a_\pm(ZnSO_4)}{a_\pm(CuSO_4)}$$

From Eq. (16.32), $a_\pm = \gamma_\pm m$. From Table 16.4, for $CuSO_4$ at $m = 0.10$ mol kg^{-1}, $\gamma_\pm = 0.41$; $ZnSO_4$ at $m = 1.00$ mol kg^{-1}, $\gamma_\pm = 0.045$. Thus

$$E = 1.100 - 0.059\log\frac{0.045}{0.041} = 1.098\ V$$

This example shows that, provided activity coefficients are available for the electrolytes used, we can calculate the cell emf from the tabulated standard electrode potentials and the Nernst equation (17.12). In many cases, the use of molalities or concentrations instead of activities (i.e., the assumption that all activity coefficients are unity) provides an adequate estimate of E.

Example 17.6 What is the equilibrium constant K_a for the reaction $2Cu^+ = Cu^{2+} + Cu$?

From Example 17.5 and Table 17.1:

Electrode reactions	Electrode potentials
$2(Cu^+ + e = Cu)$	$E° = 0.521$ V
$Cu^{2+} + 2e = Cu$	$E° = 0.337$ V
$2Cu^+ \quad = Cu^{2+} + Cu$	$E° = 0.184$ V

Then, $\Delta G° = -RT \ln K_a = -2FE°$.

$\ln K_a = 2(96\,485$ C mol$^{-1})(0.184$ V$)/(8.314$ J K^{-1} mol$^{-1})(298$ K$) = 14.33$

$K_a = 1.68 \times 10^6$

Note that in a complete cell reaction (unlike a half-cell reaction) the electrons cancel out of the reaction equation. You can calculate $E°$ for the overall reactions simply by subtracting the $E°$ of the two half-cell reactions, since $E°$ measures the free energy *per electron transferred.* (You can always work with the ΔG values.) The large K_a indicates that Cu^+ is unstable in aqueous solution. If cuprous oxide (Cu_2O) is dissolved in dilute H_2SO_4, Cu^{2+} goes into solution and solid Cu precipitates. The $Cu^+|Cu$ half-cell thus cannot actually be made, but as shown here and in Example 17.5, it can be studied indirectly from other data.

17.16 Calculation of Solubility Products

Standard electrode potentials can be combined to yield the $E°$ and thus the $\Delta G°$ and equilibrium constant for the solution of salts. In this way, we can calculate the solubility of a salt even when an extremely low value makes direct measurement difficult.

As an example, consider silver iodide, which dissolves according to: $AgI = Ag^+ + I^-$. The *solubility product constant* is $K_{sp} = a(Ag^+)\, a(I^-)$. A cell whose net reaction corresponds to the solution of silver iodide can be formed by combining a $Ag|AgI$ electrode with a Ag electrode, $Ag|Ag^+, I^-|AgI(s)|Ag$.

Electrode reactions	Electrode potentials
$AgI(s) + e = Ag + I^-$	$E° = -0.1518$ V
$Ag^+ + e = Ag$	$E° = +0.7991$ V
Overall: $AgI(s) = Ag^+ + I^-$	$E° = -0.9509 \quad [E_R - E_L]$

Then, from $\Delta G° = -|z|\, FE° = -RT \ln K_{sp}$,

$$\log_{10} K_{sp} = \frac{-0.9509 \times 96\,485}{2.303 \times 8.314 \times 298.2} = -16.07$$

The activity coefficients are unity in the very dilute solution of AgI, so that K_{sp} corresponds to a solubility of 2.17×10^{-6} g dm^{-3} at 25°C.

17.17 Electrolyte-Concentration Cells

In the cell, $Pt \,|\, H_2 \,|\, HCl(c) \,|\, AgCl \,|\, Ag$, the cell reaction is

$$\tfrac{1}{2}H_2 + AgCl = Ag + HCl(c)$$

Measurements with two different concentrations c of HCl gave the following results at 25°C:

$c(\text{mol dm}^{-3})$	E (V)	$-\Delta G(\text{J})$
0.0010	0.5795	55 920
0.0539	0.3822	36 880

If two such cells with different values of electrolyte concentration c oppose each other, the combined cell can be written as

$$Ag \,|\, AgCl \,|\, HCl(c_2) \,|\, H_2 \,|\, HCl(c_1) \,|\, AgCl \,|\, Ag$$

The overall change in this cell is simply the difference between the reactions in the two separate cells: for the passage of each faraday, the transfer of one mole of HCl from concentration c_2 to c_1, $HCl(c_2) \rightarrow HCl(c_1)$. Note, however, that there is no direct transference of electrolyte from one side to the other. The HCl is removed from the left side by the reaction $HCl + Ag \rightarrow AgCl + \tfrac{1}{2}H_2$. It is added to the right side by the reverse of this reaction. From the preceding data, if $c_1 = 0.001$ and $c_2 = 0.0539$ mol dm^{-3}, the Gibbs free energy of dilution is $\Delta G = -19\,040$ and $E = 0.1973$ V. This cell is an example of a concentration cell without transference.

If the two HCl solutions of different concentrations contact each other directly, as through a sintered glass plug, we have a concentration cell with transference.

$$Ag \,|\, AgCl \,|\, HCl(c_2) \,||\, HCl(c_1) \,|\, AgCl \,|\, Ag$$

Ions can now pass across the liquid junction between the two solutions. As 1 F passes through the cell, it is carried across the liquid junction partly by the H$^+$ ion (t_+F) and partly by the Cl$^-$ ion (t_-F), where t_+ and t_- are transport numbers. In the case of the cell shown, the electrodes are reversible to the Cl$^-$ ion, so that 1 F of Cl$^-$ ions enters the electrolyte at the left and leaves at the right. There is, therefore, a net transfer of t_+ mole of HCl from left to right, and the emf of the concentration cell is

$$E = t_+ \frac{RT}{F} \ln \frac{a_2}{a_1} \qquad (17.14)$$

or just t_+ times the emf for the cell without transference. For a case in which the electrodes are reversible to the H$^+$ ion (hydrogen electrodes), the transference number t_- would appear in Eq. (17.14).

Actually, the argument given here is not rigorous, since the cell with transference is not a reversible cell. There is always a diffusion process taking place at the liquid junction, and consequently a *liquid-junction potential* arises, which cannot be treated by equilibrium thermodynamics. More exact treatments show, however, that Eq. (17.14) is a close approximation to the cell emf.

The concept of pH* was introduced in 1909 by the Danish chemist Sørensen in terms of the concentration of the hydrogen ion [H$^+$] as

$$\text{pH} = -\log[\text{H}^+] \tag{17.15}$$

Sørensen based the electrometric measurement of pH on the cell Pt|H$_2$|[H$^+$], solution X|sat. KCl salt bridge|0.1 M KCl|Hg$_2$Cl$_2$|Hg, where X is the solution whose [H$^+$] is to be measured. The emf of this cell, from Eq. (17.12), is

$$E = E^\circ - \frac{RT}{F}\ln\frac{a(\text{H}_2)a(\text{Hg}_2\text{Cl}_2)}{a(\text{H}^+ \text{ in } X)a(\text{Hg})}$$

If activities of pure substances are set equal to unity and if corrections due to non-ideality of the solutions are neglected, so that concentrations are used in place of activities, the emf of this cell would be

$$E = E^{\circ\prime} + \frac{RT}{F}\ln\frac{1}{[\text{H}^+]} \tag{17.16}$$

where $E^{\circ\prime}$ is the emf when [H$^+$] = 1 mol dm^{-3}. Equation (17.16) also implies that the two liquid-junction potentials at the salt bridge exactly cancel each other.

To avoid the problems of defining the pH scale in terms of a thermodynamically undefined single-ion activity, an international standard conventional pH scale has been adopted. A pH is assigned to a standard solution S and the pH of any other solution X is defined on the basis of this standard. The emf E of the cell is measured with S and with X, and

$$\text{pH}(X) = \text{pH}(S) + \frac{(E_X - E_S)F}{RT\ln 10} \tag{17.17}$$

For the standard buffers, and for dilute solutions of simple salts, the measured pH value is close to $-\log_{10} a_\text{H}$, where $a_\text{H} = \gamma_\text{H} m_\text{H}$, with m_H the molality of the hydrogen ion and γ_H as calculated from the Debye–Hückel equation (16.40).

Example 17.7 The pH of an aqueous solution of acetic acid in 0.0500 molal sodium acetate is 4.98 at 25°C. What is the molality m_H of hydrogen ion in the solution?

The ionic strength is $I = 0.0500 + m_\text{H} \approx 0.0500$ mol kg^{-1}. Hence $-\log m_\text{H} = -\log a_\text{H} + \log \gamma$. From Eq. (16.41), $-\log m_\text{H} = \text{pH} - 0.509\sqrt{0.0500} = 4.87$ or $m_\text{H} = 1.35 \times 10^{-5}$ mol kg^{-1}. If necessary, a second approximation can be obtained with $I = 0.0500 + m_\text{H}$.

The hydrogen electrode is the primary standard for electrometric measurement of pH but laboratory measurements are almost always made with a glass electrode. Glasses have been developed that exchange H$^+$ ions between the surface of the glass and the solution in which it is immersed, H$^+$(glass) $\rightleftarrows$ H$^+$(solution). The glass

---

*The standard reference work on all matters concerning pH is R. G. Bates, *Determination of pH: Theory and Practice* (New York: John Wiley & Sons Inc., 1973).

electrode consists of a thin bulb of the special glass containing a solution of HCl into which is fitted a Ag|AgCl or calomel electrode. When the bulb is immersed in a solution whose pH is to be determined, the difference in [H⁺] on the two surfaces of the glass membrane causes a difference in electric potential. The glass electrode is coupled with a standard electrode. The emf of the complete cell is measured with an electronic voltmeter of high input resistance, which draws negligible current from the cell.

The great convenience of measurements of hydrogen ion with the glass electrode has stimulated the development of many other ion selective electrodes. Thus one can measure a wide variety of ionic concentrations in solution with these special electrodes and a pH meter to provide the potentiometer circuits.

17.19 Biological Membrane Potentials

Differences in electric potential usually exist across the boundary membranes of living cells in both animals and plants. These potentials have a variety of causes and many important physiological functions. Basic electrochemistry is a necessary foundation for any study of the electrical properties of living cells and organs.

Figure 17.8 depicts a cell with several different ionic species, the concentrations

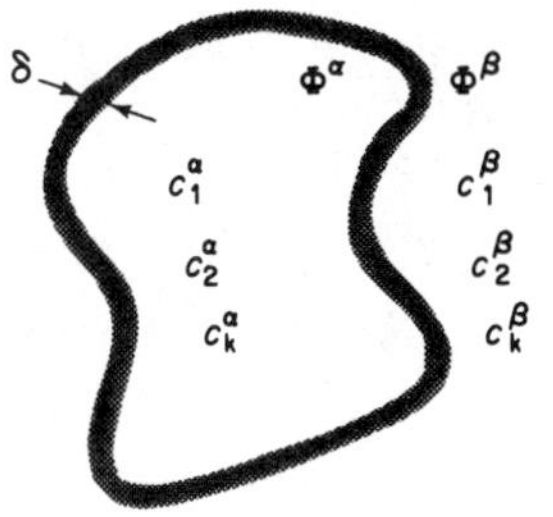

FIGURE 17.8 Two solutions α and β are separated by a membrane of thickness δ. The concentrations c and electric potentials Φ are uniform within each of phases α and β.

of which are denoted as c_i^α on the inside and c_i^β on the outside. If the membrane is permeable to all the ions, equilibrium will be reached only when the chemical potential of each ion is the same on both sides of the membrane. In this case the transmembrane potential at equilibrium will be zero, $\Delta\Phi = 0$. Suppose, however, that only one kind of ion, denoted by subscript k, can pass through the membrane. The equilibrium condition is that the electrochemical potential of this ion be the same on both sides of the membrane, $\tilde{\mu}_k^\alpha = \tilde{\mu}_k^\beta$. Equation (17.1), therefore, gives

$$\mu_k^\alpha + z_k F \Phi^\alpha = \mu_k^\beta + z_k F \Phi^\beta \qquad (17.18)$$

Since $\mu_k = \mu_k^\circ + RT \ln a_k$, where a_k is the activity of ion k, Eq. (17.18) becomes

$$\Delta\Phi = \Phi^\beta - \Phi^\alpha = \frac{RT}{z_k F} \ln \frac{a_k^\alpha}{a_k^\beta}$$

In dilute solutions, the ratio of activities approaches the ratio of concentrations $a_k^\alpha / a_k^\beta \approx c_k^\alpha / c_k^\beta$, and

$$\Delta\Phi = \frac{RT}{z_k F} \ln \frac{c_k^\alpha}{c_k^\beta} \qquad (17.19)$$

This potential difference is just that which is necessary to prevent the equalization of the concentration of ions k by diffusion across the membrane. It is possible to fix the ratio of concentrations in Eq. (17.19) whereupon an equilibrium $\Delta\Phi$ is established that corresponds to the ratio chosen. This $\Delta\Phi$ is sometimes called the *Nernst potential*.

Membranes of typical mammalian nerve cells are permeable to K^+ but, in their resting states, relatively impermeable to Na^+, Cl^-, and other ions. The concentration $[K^+]$ inside the cell is about 20 times the $[K^+]$ outside. From Eq. (17.19), the equilibrium potential across the membrane at 25°C is, therefore, about

$$\Delta\Phi = \frac{(8.31 \text{ J K}^{-1} \text{ mol}^{-1})(298 \text{ K})}{96\,485 \text{ C mol}^{-1}} \ln\left(\frac{1}{20}\right) = 25.7 \ln\left(\frac{1}{20}\right) = -78 \text{ mV}$$

(the inside being negative relative to the outside). This $\Delta\Phi$ is close to the experimentally observed value, and, to a first approximation, the nerve membrane potential can be regarded as an equilibrium K^+ potential. Closer consideration indicates, however, that since the membrane is permeable to some extent to ions other than K^+, the observed potential must be interpreted as a steady-state potential for a system in which the K^+ permeability is considerably greater than that of any other ion.

When several different ions can permeate the membrane at different rates, the equilibrium condition no longer fixes the value of the membrane potential. Instead, a steady-state is reached at which the total current due to all the ions is zero. For thin membranes, a good approximate solution was given by David Goldman in 1943, in which the electric field in the membrane becomes simply $(\Phi^\alpha - \Phi^\beta)/\delta$, where δ is the membrane thickness as shown in Fig. 17.8. The Goldman equation is

$$\Delta\Phi = \frac{RT}{F} \ln \frac{\sum D_p c_p(\delta) + \sum D_n c_n(o)}{\sum D_p c_p(o) + \sum D_n c_n(\delta)} \tag{17.20}$$

Here c_p and c_n refer to the concentrations of positive and negative ions, respectively, either outside the cell (o) or inside (δ), and D_p and D_n are the corresponding diffusion coefficients in the membrane.

Example 17.8 In cells of frog muscle, the ionic concentrations are as follows in mmol dm^{-3}.

	$c(K^+)$	$c(Na^+)$	$c(Cl^-)$
Inside	125	15	1.2
Outside	2.5	120	120

The diffusion coefficients are $D(K^+) = 6.0 \times 10^{-10}$, $D(Na^+) = 3.5 \times 10^{-10}$, $D(Cl^-) = 1.0 \times 10^{-9}$ cm^2 s^{-1}. Calculate the steady-state membrane potential at 25°C.

$$\Delta\Phi = 0.0257 \text{ V} \ln \frac{(6 \times 125) + (3.5 \times 15) + 10(120)}{(6 \times 2.5) + (3.5 \times 120) + 10(1.2)}$$

$$= 0.0385 \text{ V} = 38.5 \text{ mV}$$

Another kind of potential can arise at equilibrium across a membrane that is permeable to solvent and to some but not all ions. For example, protein molecules may carry a considerable number of positive or negative charges depending upon pH.

Suppose that a protein P^{z+} has z^+ net positive charges. The protein is inside the cell and cannot pass through the membrane. Negative ions, such as Cl^- ions, are required inside the cell to compensate for these positive charges. In other words, the concentration of small permeant ions on the two sides of the membrane is considerably altered by the presence of charged protein molecules in the intracellular fluid. An example of such a system is shown in Fig. 17.9.

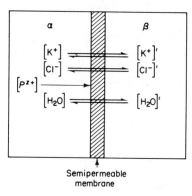

FIGURE 17.9 Donnan membrane equilibrium with high polymer cations P^{z+} and KCl as the neutral salt.

The theory of membrane equilibria of this type was first given by F. G. Donnan in 1911. Typical concentration differences and potentials are given in Table 17.2 for the case of uniunivalent salts (KCl, NaCl) with a protein of charge z^+ inside the cell. Donnan potentials are equilibrium values and in actual living cells steady-state rather than equilibrium conditions are more apt to prevail.

TABLE 17.2
EXAMPLES OF DONNAN MEMBRANE EQUILIBRIA AT 25°C[a]

	Concentrations (mol dm^{-3})			$\dfrac{c_+(0)}{c_+(\delta)} = \dfrac{c_-(\delta)}{c_-(0)}$	$\Delta\Phi$ (mV)
$z^+c(P^{z+})$	$c_+(0) = c_-(0)$	$c_+(\delta)$	$c_-(\delta)$		
0.002	0.0010	0.00041	0.00241	2.44	22.90
	0.0100	0.00905	0.01105	1.10	2.56
	0.100	0.0990	0.1010	1.01	2.58
0.02	0.0010	0.00005	0.02005	20.05	76.96
	0.0100	0.00414	0.02414	2.41	22.65
	0.100	0.0905	0.1105	1.10	2.56

[a]Adapted from *Physical Chemistry of Macromolecules*, by Charles Tanford (New York: John Wiley & Sons, Inc., 1961).

17.20   Nerve Conduction

Nerve cells communicate with one another and with muscles by sending out electrical pulses, called *action potentials*, that travel from the cell body down tubular fibers called *axons*. In 1850, Helmholtz measured for the first time the velocity of nerve conduction, obtaining a value of 30 m s⁻¹ in a frog nerve. This is the velocity at which a pulse of electric depolarization, of about 10^{-3} s duration, travels down the axon.

In the resting state the potential difference across the axonal membrane is about $\Delta\Phi = -70$ mV, with the inside negative. This value is close to the equilibrium Nernst potential for the K^+ ions. When the nerve "fires" an action potential a region of membrane where the axon joins the cell body (the axon hillock) experiences a sharp rise in conductance for Na^+ ions. The Na^+ ions rush through the membrane, driven by the high gradient of electrochemical potential. The permeability of the membrane to K^+ ions also increases. Consequently, the $\Delta\Phi$ across the membrane first falls to zero and then overshoots to reach about $\Delta\Phi = 50$ mV. This value is actually close to a Nernst Na^+ potential for the existing concentrations $[Na^+]$ inside and outside the axon. The form of the action potential is shown in Fig. 17.10.

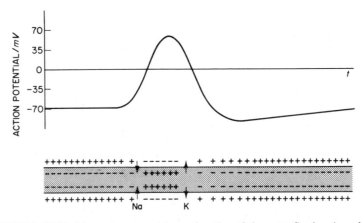

FIGURE 17.10 The action potential as a function of time at a fixed region of a nerve-cell membrane.

A potential difference of 70 mV across a membrane 7 nm thick gives an electric field strength of 10^5 V cm⁻¹. When this high electric field is abruptly reduced, channels in the membrane open to permit the passage of Na^+ ions. The molecular mechanism that controls the gates to these channels is not yet known, but it is likely to involve changes in the conformation of protein molecules caused by the large changes in electric field.

The interior of the axon is a good conductor of electricity so that the initial region of membrane depolarization spreads down the axon as a result of electric current. As new areas of membrane become depolarized, their sodium gates open in turn: hence the high conduction velocity of the nervous impulse.

The excess sodium ions are pumped back out of the interior of the axon by "sodium pumps," which use the free energy of adenosine triphosphate (ATP) hydrolysis to pump the Na^+ ions against their concentration gradient (*active transport*). Thus the action potential itself uses free energy already stored in the ionic gradients across the membrane, but this free energy must ultimately be provided by ATP, the fuel that operates the sodium pumps. The human brain requires a power of about 25 W, and almost all this power is expended to run the sodium pumps.

Problems

1. For the cell $Ag|Ag_2SO_4(s)|$ sat. soln. $Ag_2SO_4, Hg_2SO_4|Hg_2SO_4(s)|Hg$, the emf at 25°C is $E = 0.140$ V and $dE/dT = 0.00015$ V K^{-1}. (a) Write the cell reaction. (b) What are ΔG, ΔS, and ΔH for the cell reaction? (c) One mole each of Hg, $Ag_2SO_4(s)$, $Hg_2SO_4(s)$, and some saturated solution of the two salts are mixed at 25°C. What solid phases will be present at equilibrium and in what amounts?

2. For the cell, $H_2(P°)|HCl(0.01\ m)|AgCl(c)|Ag$, $E(V) = -0.0960 + 1.90 \times 10^{-3}T - 3.04 \times 10^{-6}T^2$. What is the cell reaction? Calculate ΔG, ΔH, ΔS, and ΔC_p for the reaction at 275, 300, and 325 K.

3. For the cell in Problem 2, calculate E if the P_{H_2} is $10P°$; $100P°$. ($P° = 101.32$ kPa at 300 K.)

4. A hydrogen electrode and a calomel electrode are used to measure the pH of a solution on a mountaintop where the pressure is 500 mmHg. The H_2 is allowed to bubble out of the electrode at this atmospheric pressure. The pH is found to be 4.00 by a visiting scientist from the lowlands, who forgets to correct for pressure. What is the correct pH?

5. The emf of the cell $H_2(P)|0.100\ m\ HCl|HgCl|Hg$ was studied as a function of P_{H_2} at 25°C.

P (MPa)	0.10132	3.840	5.228	11.17	29.04	74.15	104.9
E (mV)	399.0	445.6	449.6	459.6	473.4	489.3	497.5

Calculate the fugacity coefficient ($\gamma = f/P$) of H_2 and plot as $\gamma(P)$ over the given range of P.

6. The emfs of the cell $H_2(Pt)|HCl(m)|AgCl|Ag$ at 25°C were measured by H. S. Harned and G. Ehlers [J. Am. Chem. Soc., 54, 1350 (1932)]; see Fig. 17.6.

m (mol kg^{-1})	0.01002	0.01031	0.04986	0.09642	0.20300
E (V)	0.46376	0.46228	0.38582	0.35393	0.31774

Calculate the activity coefficients $\gamma_\pm$ of HCl at the concentrations given and compare with the values from the Debye–Hückel equation (16.41).

7. The $E_1°$ of the cell $H_2(Pt)|KOH(m_1), KCl(m_2)|AgCl(s)|Ag(s)$ can be used to determine $K_w = [H^+][OH^-]$. At 25°C:

m_1 (mol kg^{-1})	0.0100	0.0100	0.0100	0.0100
m_2 (mol kg^{-1})	0.0100	0.0200	0.0300	0.0400
E_1 (V)	1.05069	1.03295	1.02258	1.01521

For AgCl(s) $+ \frac{1}{2}H_2(g) \rightarrow$ Ag(s) $+ H^+ + \bar{Cl}^-$, $E_2^\circ = 0.2224$ V, at 25°C. (Determine E_1° as in Problem 6, with $m = m_1 + m_2$.) Then find K_w from $E_1^\circ - E_2^\circ$.

8. Calculate K_a at 298 K for $2H^+(aq) + D_2 = H_2 + 2D^+(aq)$. (See Table 17.1.) What is ΔG° for the reaction?

9. Devise cells in which the cell reactions are:
    (a) $H_2(g) + I_2(s) = 2HI(aq)$
    (b) $2I^- + S_2O_8^{2-} = I_2 + 2SO_4^{2-}$
    (c) $Sn^{2+} + 2Fe^{3+} = Sn^{4+} + 2Fe^{2+}$
    (d) $Pb^{2+} + CO_3^{2-} = PbCO_3(s)$
    Calculate E° for these cells. (See Table 17.1.)

10. The solubility of $I_2(s)$ in water at 25°C is 0.0013 mol L^{-1}. For $I_2(s) + 2e = 2I^-$, $E^\circ = 0.5355$ V, and for $I_3^- + 2e = 3I^-$, $E^\circ = 0.5365$ V. What is the equilibrium [I_3^-] in a saturated solution of I_2 in 1.0 M KI?

11. Calculate the emf of the cell Pt | Cu | CuSO$_4$(0.01 M) ‖ ZnSO$_4$(0.05 M) | Zn | Pt. ($\gamma_\pm$: Table 16.4)

12. The reaction $H_2 + \frac{1}{2}O_2 = H_2O(\ell)$ is carried out in a fuel cell. (a) Outline a cell that could be used. (b) What is the E° of the cell reaction at 25°C?

13. Cytochrome c is a protein containing iron for which the reaction Fe(III) $+ e =$ Fe(II) has $E^\circ = -0.25$ V at pH 7 and 25°C. The following cell is set up:

$$Pt \,|\, H_2 \,|\, H^+(pH\ 7),\ cyt\ c(III),\ cyt\ c(II) \,|\, Pt$$

What is the emf of the cell at 25°C? Hydrogen is bubbled over a platinized platinum electrode into a solution 0.10 M in cyt c at pH 7. What will be the ratio cyt c(III)/cyt c(II) at equilibrium?

14. The E° for the half-cell Pt | Sn^{2+}, Sn^{4+} is 0.140 V at 25°C. Hydrogen at $P = 1$ atm is bubbled through a solution initially 0.100 M in HCl and 0.100 M in Sn^{4+}. What will be the concentration of Sn^{4+} at equilibrium?

15. Hydrogen is bubbled through a solution initially 0.100 molar in Sn^{2+}. If the solution is maintained at pH 7, what is the equilibrium concentration of Sn^{2+}?

16. The cell: Cd–Hg(m_1) | CdSO$_4$ | Cd–Hg(m_2) can be used to measure the activity of cadmium in amalgams.

$10^4 m_1$ (mol kg^{-1})	889.6	88.96	8.896	0.8896
$10^4 m_2$ (mol kg^{-1})	88.96	8.896	0.8896	0.08896
E (V)	0.02966	0.02960	0.02956	0.02950

Calculate the activity coefficients γ for Cd in the amalgams for the molalities given.

17. What chemical reactions, if any, can be predicted in the following experiments?
    (a) 1 M Ce (SO$_4$)$_2$ is added to 1 M KI.
    (b) A Pb rod is immersed in 1 M AgNO$_3$.
    (c) A Zn rod is immersed in 1 M Pb(NO$_3$)$_2$.
    (d) A Pb rod is immersed in 1 M HCl.

18. For Zn | Zn^{2+} and Co | Co^{2+} the E° are -0.76 and -0.28 V, respectively. A solution that initially is 1 molar in Zn^{2+} and Co^{2+} is electrolyzed with platinum electrodes. What is the approximate concentration of Co^{2+} when Zn^{2+} begins to be deposited?

19. $Hg(NO_3)_2$ and $Fe(NO_3)_2$ each in 0.01 M solution are mixed in equal volumes at 25°C. What is the equilibrium concentration of Fe^{3+} in the solution?

20. The standard emf $E°$ of cell $Pt\,|\,H_2(P°)\,|\,NaOH,\ NaCl\,|\,AgCl\,|\,Ag$ was determined at different temperatures. What is the cell reaction? Calculate $\Delta G°$, $\Delta H°$, and $\Delta S°$ for the cell reaction at 298.15 K. [*J. Chem. Soc. Faraday Trans. I*, *69*, 949 (1973)]

T (K)	293.15	298.15	303.15
$E°$ (V)	1.04774	1.04864	1.04942

18

Surfaces and Colloids

Thomas Graham in 1861 introduced the word *colloid* (from the Greek *gluelike*) to describe suspensions of one material in another that did not separate on long standing. Colloids thus consist of a dispersed phase and a dispersion medium. Dispersed materials with a particle size less than about 0.2 μm are classed in the colloidal state. The limit of resolution of a microscope using ordinary light is about 0.2 μm, so that direct observation of colloidal particles usually requires the use of an electron microscope, but sometimes they can be visualized by scattered light in dark-field observation with a light microscope.

Colloidal suspensions are called *sols*. If the bulk dispersed phase spontaneously enters the dispersion medium, they are *lyophilic* (solvent loving) sols. Examples are high-polymer solutions, such as proteins in water or rubber in benzene. These solutions display many of the physical properties of colloidal suspensions owing to the high molecular mass of the solute. Colloidal suspensions of essentially insoluble materials are called *lyophobic* (solvent hating) sols. They may be prepared by condensation or dispersion methods. In simplest terms, they owe their stability to the fact that the particles bear electrical charges of the same sign and thus cannot approach one another closely enough to coagulate.

Surface properties are of primary importance in the study of colloids. For example, suppose that a sphere of material 10 mm in diameter (d) is reduced to colloidal dimensions in the form of 10^{15} little spheres 0.1 μm in diameter. The surface-to-volume ratio of a sphere, $\pi d^2 / \frac{1}{6}\pi d^3 = 6/d$. For the larger sphere the ratio is 0.6 mm^{-1}, whereas for the small colloidal spheres the ratio is 6×10^4 mm^{-1}. The area of the larger sphere is $10^2 \pi$ mm^2; the total area of the colloidal spheres is $10^7 \pi$ mm^2.

Matter that is at or near an interface is not in the same state as matter in the interior of a phase. Therefore, additional thermodynamic variables, such as surface tension γ, are required to describe the equilibrium properties of surfaces and colloids.

18.1 Surface Tension

Figure 18.1(a) represents a liquid in contact with its vapor. Molecules at the interface experience a net attractive force toward the interior. To extend the area of the interface, molecules must be moved from the interior into interfacial regions, and work must be done on the system against cohesive forces of the liquid. The interface is therefore at a higher chemical potential than the bulk liquid.

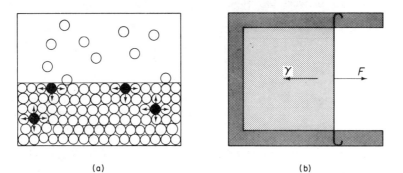

(a)   (b)

FIGURE 18.1 (a) Liquid–vapor interface—work must be done on the system to extend the interface by bringing more molecules from the interior. (b) Liquid–vapor interface viewed from above. The work done to extend the interface requires a force F that opposes the surface tension γ.

In 1805, Thomas Young showed that the mechanical properties of a surface could be related to those of a hypothetical membrane under tension stretched over the surface. A tension is a negative pressure and pressure is force per unit area, so that surface tension γ is force per unit length. The SI unit of surface tension is newton per meter (N m^{-1}). For a plane surface, surface tension is the force acting parallel to the surface and perpendicular to a line of unit length anywhere in the surface. Surface tension opposes any extension of surface area.

An interface separating two phases, α and β, is a region of small but finite thickness, of the order of 1 nm, in which there is a gradual change from properties of α to those of β. The important contribution of Young was to show that so far as its mechanical properties are concerned, such an interfacial region can be replaced by the conceptual model of a stretched membrane of infinitesimal thickness. The location of this dividing plane between the two phases is called the *surface of tension*. It can be proved rigorously that the properties of the surface layer suffice to establish completely (1) the position of this surface of tension, and (2) the value of the surface tension acting in it. Figure 18.1(b) shows a picture of a surface of tension. A soap film might be an example, but in that case there are two surfaces, one on either side.

The work done on the system by extending its surface area by $d\alpha$ is

$$dw = \gamma \, d\alpha \qquad (18.1)$$

The work done on the system at constant T and V increases its Helmholtz free energy by $dw = d\alpha$. Thus we can regard surface tension as $\gamma = (\partial A / \partial \alpha)_{T,V}$. Note that the units of γ, N m^{-1}, are equivalent to joules per square meter, J m^{-2}.

18.2 Equation of Young and Laplace

The idea of a surface tension was one of those great simplifying concepts that open up the development of a scientific field. By means of the concept, Young, and later Laplace, were able to derive explicitly the conditions for mechanical equilibrium at a curved surface between two phases. Consider in Fig. 18.2 a spherical surface of

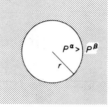

FIGURE 18.2 At equilibrium there is a difference in pressure across a curved phase boundary.

radius r separating two phases α and β. The equation of Young and Laplace for a spherical surface is

$$\Delta P = P^\alpha - P^\beta = 2\gamma/r \qquad (18.2)$$

This equation says that as a consequence of the existence of surface tension at a spherical surface of radius of curvature r, mechanical equilibrium is maintained between two fluids at different pressures P^α and P^β. (Note that the fluid on the concave side of a surface is at a pressure P^α, which is greater than P^β on the convex side.) In the case of a plane surface, of course, the radius of curvature goes to infinity and the condition of equilibrium is simply $P^\alpha = P^\beta$.

We can derive the Young–Laplace equation by supposing that the volume of phase α is increased in a reversible expansion by a slight amount dV, which requires work $dw = (P^\alpha - P^\beta)dV$. From (18.1) this work is also $\gamma \, d\alpha$. The spherical volume is $V = \frac{4}{3}\pi r^3$ and the area $\alpha = 4\pi r^2$. Thus $dV = 4\pi r^2 \, dr$ and $d\alpha = 8\pi r \, dr$, so that $(P^\alpha - P^\beta)4\pi r^2 \, dr = \gamma 8\pi r \, dr$, and $P^\alpha - P^\beta = 2\gamma/r$, the equation of Young and Laplace.

Example 18.1 What is ΔP at the surface of a droplet of mercury with $r = 1.00 \times 10^{-4}$ m?

For the mercury–air interface, $\gamma = 476 \times 10^{-3}$ N m^{-2} at 293 K. Hence

$$P^\alpha - P^\beta = \frac{2(476 \times 10^{-3})}{1.00 \times 10^{-4}} = 9.52 \times 10^3 \text{ N m}^{-2} \text{ (Pa)}$$

(Note that this is the difference in hydrostatic pressure across the interface and, of course, does not refer to the vapor pressure of the mercury. It is an appreciable pressure, about $0.1P^\circ$.)

18.3 Capillarity

The rise or fall of liquids in capillary tubes and its application to the measurement of surface tension may be treated quantitatively as a consequence of the fundamental equation (18.2). Whether a liquid rises in a glass capillary, like water, or is depressed, like mercury, depends on the relative magnitude of the forces of cohesion between the liquid molecules themselves, and the forces of adhesion between the liquid and the walls of the tube. These forces determine the contact angle θ, which the liquid makes with the tube walls (Fig. 18.3). If this angle is less than 90°, the liquid is said to *wet the surface* and a concave meniscus is formed. If the contact angle is greater than 90°, the liquid does not wet the surface and a convex meniscus is formed.

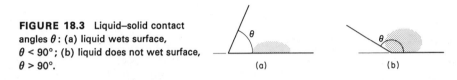

FIGURE 18.3 Liquid–solid contact angles θ: (a) liquid wets surface, $\theta < 90°$; (b) liquid does not wet surface, $\theta > 90°$.

(a)　　　　(b)

The occurrence of a concave meniscus leads to a capillary rise, whereas a convex meniscus leads to a capillary depression. As shown in Fig. 18.4(a), as soon as the concave meniscus is formed, the pressure P^β in the liquid under the curved surface is less than the pressure P^α under the plane surface. The liquid thus rises in the tube until the weight of the liquid column just balances the pressure difference $\Delta P = P^\alpha - P^\beta$ and restores the hydrostatic equilibrium. The liquid column acts as a manometer to register the pressure difference across the curved meniscus.

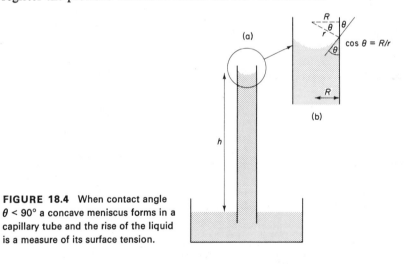

FIGURE 18.4 When contact angle $\theta < 90°$ a concave meniscus forms in a capillary tube and the rise of the liquid is a measure of its surface tension.

Consider in Fig. 18.4(b) a cylindrical tube whose radius R is sufficiently small that the surface of the meniscus can be taken as a spherical surface of radius r. Then, since $\cos \theta = R/r$, from Eq. (18.2) $\Delta P = (2\gamma \cos \theta)/R$. If the capillary rise is h, and if ρ and ρ_0 are the densities of liquid and gas, the weight of the cylindrical liquid column is $\pi R^2 gh(\rho - \rho_0)$, where g is the gravitational acceleration. The force per unit area balancing the pressure difference is $gh(\rho - \rho_0)$, so that

$$\frac{2\gamma \cos \theta}{R} = gh(\rho - \rho_0),$$

and

$$\gamma = \frac{1}{2}gh(\rho - \rho_0)\frac{R}{\cos \theta} \qquad (18.3)$$

Equation (18.3) should be corrected for the weight of liquid above the bottom of the meniscus. To a first approximation, the meniscus is a hemisphere of radius R or volume $\frac{2}{3}\pi R^3$, so that the volume of this liquid is $\pi R^3 - \frac{2}{3}\pi R^3 = \frac{1}{3}\pi R^3$, and Eq. (18.3) becomes

$$\gamma = \frac{g(\rho - \rho_0)R}{2 \cos \theta}\left(h + \frac{R}{3}\right) \qquad (18.4)$$

For large capillaries, it is no longer a good approximation to assume that the meniscus has a spherical surface, but correction factors have been computed. When all necessary corrections have been made, the capillary-rise method is probably the most accurate way to measure surface tension, good to about 2 parts in 10^4. Some other methods, such as maximum bubble pressure, weight of drops, and the ring method of du Noüy, are included in problems at the end of this chapter.

Some idea of the range of values of surface tensions is provided by Table 18.1. Liquids with exceptionally high surface tensions are those in which cohesive forces are large. Water has a higher γ than most ordinary liquids as a result of its hydrogen-bonded structure, but the γ's for molten metals are an order of magnitude higher.

TABLE 18.1
SURFACE TENSIONS OF LIQUIDS

A. Surface tensions of pure substances at 293 K (N m^{-1} × 10^3)

Isopentane	13.7	Carbon tetrachloride	26.7
Nickel carbonyl	14.6	Benzene	28.9
Diethyl ether	17.1	Ethyl iodide	29.9
n-Hexane	18.4	Carbon bisulfide	32.3
Ethyl mercaptan	21.8	Methylene iodide	50.8
Ethyl bromide	24.2	Water	72.8

B. Surface tensions of liquid metals and molten salts (N m^{-1} × 10^3)

	T (K)	γ		T (K)	γ
Hg	273	476	NaBr	1273	88
Ag	1243	800	NaCl	1273	98
Au	1343	1000	AgCl	725	126
Cu	1403	1100	NaF	1283	260

Example 18.2 From the surface tension of mercury at 273 K in Table 18.1 calculate the capillary depression of mercury in a glass tube 1.0 mm diameter, if the contact angle is 140°.

From Eq. (18.4), since $p \ll p_0$

$$h = \frac{2\gamma \cos \theta}{g \rho R} - \frac{R}{3} = \frac{2(0.476 \text{ N m}^{-1})(-0.766)}{(9.80 \text{ m s}^{-2})(13.59 \times 10^3 \text{ kg m}^{-3})(0.50 \times 10^{-3} \text{ m})}$$

$$- \frac{0.50 \times 10^{-3} \text{ m}}{3} = (-10.9 - 0.2)10^{-3}$$

$$= -11 \times 10^{-3} \text{ m} = -11 \text{ mm}$$

Example 18.3 What vacuum is necessary at 293 K to draw all the water out of a sintered glass filter if the minimum pore size is 4.0 μm in diameter?

As bubbles of air are formed at the orifices of capillaries, the maximum pressure is reached when a hemispherical bubble is formed with a radius just equal to that of the capillary. The pressure is then given by Eq. (18.2) as

$$\Delta P = \frac{2\gamma}{r} = \frac{2(72.8 \times 10^{-3} \text{ N m}^{-1})}{2.0 \times 10^{-6} \text{ m}} = 7.28 \times 10^4 \text{ N m}^{-2} \text{ (Pa)} = 73 \text{ kPa}$$

This would be the pressure below $P° = 101.3$ kPa needed to remove the water from the pore.

18.4 Enhanced Vapor Pressure of Small Droplets—Kelvin Equation

One of the most interesting consequences of surface tension is the fact that the vapor pressure of a liquid is greater when it is in the form of small droplets than when it has a plane surface. This result was first deduced by William Thomson (Kelvin) in 1871.

Consider a spherical drop of liquid with curvature $1/r$ in equilibrium with its vapor. From Eq. (18.2), the condition for mechanical equilibrium is

$$dP'' - dP' = d\left(\frac{2\gamma}{r}\right) \tag{18.5}$$

For physicochemical equilibrium between two phases, $\mu_i' = \mu_i''$ or $d\mu_i' = d\mu_i''$. From $d\mu_i = -S_i \, dT + V_i \, dP$, at constant temperature, $V_i' \, dP' = V_i'' \, dP''$. On combining this with Eq. (18.5), we have

$$d\left(\frac{2\gamma}{r}\right) = \left(\frac{V_i' - V_i''}{V_i''}\right) dP' \tag{18.6}$$

If we neglect the molar volume of the liquid V_i'' compared to V_i', and treat the latter as an ideal gas volume, $V_i' = RT/P'$, Eq. (18.6) becomes $d(2\gamma/r) = (RT/V_i'')dP'/P'$. This equation is integrated (taking V_i'' as constant) from zero curvature ($1/r = 0$), at which $P' = P^\bullet$, the normal vapor pressure, to a curvature $1/r$, at which the vapor pressure is P, thus giving the *Kelvin equation*,

$$\ln \frac{P}{P^\bullet} = \frac{2\gamma}{r} \frac{V_l''}{RT} = \frac{2M\gamma}{RT\,\rho r} \tag{18.7}$$

where $V_l'' = M/\rho$, M being molar mass and ρ the liquid density. A similar equation can be derived for the solubility of small particles, by means of the relation between vapor pressure and solubility developed in Section 9.5. The conclusions of the Kelvin equation have been experimentally verified. Small droplets of liquid have a higher vapor pressure than bulk liquid, and small particles of solid have a greater solubility than bulk solid.

A bubble of vapor forming within a liquid is the inverse of a droplet of liquid forming within a vapor. The curvature of the bubble interface is negative, and $1/r$ in the Kelvin equation (18.7) becomes $-1/r$. Hence the equilibrium vapor pressure within a bubble is less than that at a planar liquid surface. Figure 18.5 shows the effect of curvature of the surface on the vapor pressure of water, both for droplets and for bubbles.

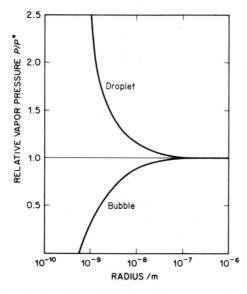

FIGURE 18.5 Effect of curvature on the vapor pressure of water as calculated from the Kelvin equation (18.7). The vapor pressures (relative to that of a planar surface) of spherical droplets and within spherical bubbles are plotted vs. the radius of curvature.

Example 18.4 Water boils at 373 K at $P^\circ = 101.3$ kPa at which temperature its surface tension $\gamma = 0.0580$ N/m. What would be the pressure of water vapor in bubbles of steam containing 50 molecules of H_2O?

50 molecules of H_2O would have a volume, at 373 K and P°, of $V = (50/L)(22.414 \times 10^{-3}\ \text{m}^3\ \text{mol}^{-1})(373/273) = 2.54 \times 10^{-24}\ \text{m}^3$. The radius of the bubble from $\frac{4}{3}\pi r^3 = 2.54 \times 10^{-24}$ is $r = 8.46 \times 10^{-9}$ m. From the Kelvin equation (18.7)

$$\ln(P/P^\bullet) = -2M\gamma/RT\,\rho r$$

$$= \frac{-2(18.0 \times 10^{-3}\ \text{kg mol}^{-1})(58.0 \times 10^{-3}\ \text{N m}^{-1})}{(8.314\ \text{J K}^{-1}\ \text{mol}^{-1})(373\ \text{K})(0.950 \times 10^3\ \text{kg m}^{-3})(8.46 \times 10^{-9}\ \text{m})}$$

$$\ln(P/P^\bullet) = -0.0838,\ P/P^\bullet = 0.920,\ P = 93.2\ \text{kPa}.$$

These results lead to the paradoxical problem of how new phases can ever arise from old ones. For example, if a container filled with water vapor at slightly below the saturation pressure is chilled suddenly, perhaps by adiabatic expansion as in the Wilson cloud chamber, the vapor may become supersaturated with respect to liquid water, and we may expect condensation to occur. A reasonable molecular model of condensation would be that several molecules of water come together to form a tiny droplet, and that this embryo of a new phase grows by accretion as additional molecules from the vapor happen to hit it. The Kelvin equation, however, indicates that this tiny droplet, being less than 10^{-6} mm in diameter, would have a vapor pressure many times that of bulk liquid. With respect to the embryos, the vapor would not be supersaturated at all. Such embryos should immediately reevaporate, and the emergence of a new phase at the equilibrium pressure, or even moderately above it, should be impossible.

There are two ways of escaping this dilemma. In the first place, the Second Law of Thermodynamics has a statistical basis. In any system at equilibrium there are always microscopic fluctuations about the average condition. There is always a chance that an appropriate fluctuation may lead to the formation of a stable embryo of a new phase. The chance of such a fluctuation is $e^{-\Delta S/k}$, where ΔS is the deviation of the entropy from the equilibrium value. This fluctuation mechanism is called *spontaneous nucleation*. In most cases, however, the chance $e^{-\Delta S/k}$ is very small. It is then more likely that dust particles act as nuclei for condensation in supersaturated vapors. For formation of bubbles in a liquid raised above its boiling point, gases adsorbed on the surface of the container may also act as nuclei. In the Wilson cloud chamber, condensation occurs along tracks of ionizing particles.

18.5   Surface Tension of Solutions

The addition of a solute to a pure solvent may either lower or raise the surface tension, depending on whether the solute is relatively more concentrated in the surface layers or in the bulk of the solution. Consider, for example, the surface tension of dilute solutions of n-butanoic acid, $CH_3CH_2CH_2COOH$, in water at 298 K shown in Fig. 18.6. The fatty acid contains a polar head group and a hydrocarbon tail. The

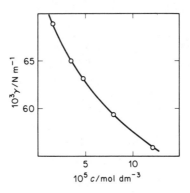

FIGURE 18.6 Surface tension of aqueous solutions of n-butanoic acid as a function of concentration at 298 K.

free energy of the system is minimized by a concentration of partially oriented fatty acid molecules in the surface of the solution. To extend the surface of such a solution is easier than for a pure liquid, since enlarging surface area permits more fatty acid molecules to enter the surface layers further lowering the free energy. The surface tension of pure water is markedly lowered by addition of fatty acid. The fatty-acid solute is said to be *positively adsorbed* at the surface.

Let n_{21} be the excess amount of the solute (2) in the surface layer compared to that in a solution of uniform composition. Then the lowering of free energy due to this adsorption of solute at the interface is $n_{21} \, d\mu_2$, where μ_2 is the chemical potential of solute. This lowering of free energy in the surface is equivalent to $-\mathcal{A} \, d\gamma$, where $\mathcal{A}$ is the surface area and $-d\gamma$ the lowering of the surface tension. Hence $n_{21} \, d\mu_2 = -\mathcal{A} \, d\gamma$. The surface excess of solute per unit area is $\Gamma_{21} = n_{21}/\mathcal{A}$, so that

$$\Gamma_{21} = -\left(\frac{\partial \gamma}{\partial \mu_2}\right)_T \tag{18.8}$$

where the partial derivative is used since the whole process occurs at a constant temperature T. This equation is called the *Gibbs adsorption isotherm*.

For dilute solutions, we can use an analog of Eq. (10.3) in terms of c instead of X, to find $d\mu_2 = RT \, d\ln c_2$, so that Eq. (18.8) becomes

$$\Gamma_{21} = -\frac{1}{RT}\left(\frac{\partial \gamma}{\partial \ln c}\right)_T = \frac{-c}{RT}\left(\frac{\partial \gamma}{\partial c}\right)_T \tag{18.9}$$

Example 18.5 From data in Fig. 18.6, calculate Γ_{21} for *n*-butanoic acid in water at 298 K at a concentration of 5.0×10^{-5} mol dm^{-3}.

The slope of the surface-tension curve at $c = 5.0 \times 10^{-5}$ mol dm^{-3} is $(\partial \gamma / \partial c) = -0.135$ N m^2 mol^{-1}. Then, from Eq. (18.9),

$$\Gamma_{21} = \frac{-5.0 \times 10^{-2} \text{ mol m}^{-3}}{(8.314 \text{ J K}^{-1} \text{ mol}^{-1})(298 \text{ K})}(-0.135 \text{ N m}^2 \text{ mol}^{-1}) = 2.7 \times 10^{-6} \text{ mol m}^{-2}$$

This figure would be a surface excess of fatty acid of about 16×10^{17} molecules/m^2 or an average surface area available to each molecule of fatty acid of about 0.6 nm^2.

18.6 Insoluble Surface Films

In 1917, Irving Langmuir devised a method for measuring the *surface pressure* exerted by surface films on liquids. The essential features of his *film balance* are shown in Fig. 18.7. The *fixed barrier*, which may be a strip of mica, floats on the surface of the water and is suspended from a torsion wire. At the ends of the floating barrier are attached strips of platinum foil or waxed threads, which lie on the water surface and connect the ends of the barrier to the sides of the trough, thus preventing leakage of surface film past the float. A *movable barrier* rests upon the sides of the trough and is in contact with the water surface. Other movable barriers are provided to sweep the surface clean.

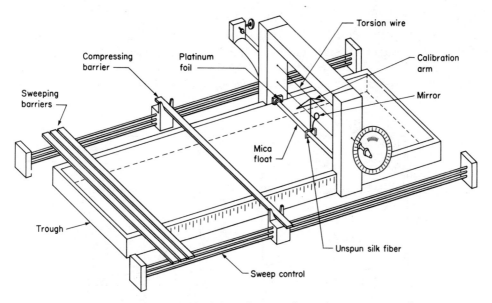

FIGURE 18.7 A Langmuir film balance for measuring surface pressure as a function of area.

In a typical experiment, a tiny amount of insoluble spreading substance is introduced onto a clean water surface. For example, a dilute solution of stearic acid in benzene might be used; the benzene evaporates rapidly, leaving a film of stearic acid. Then the movable barrier is advanced toward the floating barrier. The surface film exerts a surface pressure on the float, pushing it backward. The torsion wire, attached to a calibrated circular scale, is twisted until the float is returned to its original position. The required force divided by the length of the float is the force per unit length or surface pressure.

Surface pressure is simply another way of expressing the lowering of surface tension caused by a surface film. On one side of the float is a clean water surface with tension γ_0, and on the other side a surface covered to a certain extent with stearic acid molecules, with lowered surface tension γ. The surface pressure f is the negative of the change in surface tension.

$$f = -\Delta\gamma = \gamma_0 - \gamma \qquad (18.10)$$

The spreading of an insoluble substance on the surface of a liquid is an extreme case of the positive adsorption of a solute at the surface of the solution and the lowering of the surface tension in accord with the Gibbs adsorption isotherm [Eq. (18.9)].

Different substances in unimolecular films display a great variety of surface pressure vs. area ($f - \alpha$) isotherms. Sometimes, the film behaves like a two-dimensional gas, sometimes like a two-dimensional liquid or solid. There are also other types of monolayers having no exact analogs in the three-dimensional world. They can be recognized, however, to be definite surface phases by the discontinuities in the $f - \alpha$ diagram that signal their occurrence. If a surface film behaves as an ideal two-dimensional gas, it follows an equation similar to the three-dimensional $PV = nRT$,

$$f\alpha = n_{21}RT \qquad (18.11)$$

Figure 18.8(a) shows the $f - \alpha$ isotherms of different straight-chain aliphatic fatty acids spread on water. In the limit of large area per molecule, the lower members of series approach the $f - \alpha$ behavior of the two-dimensional gas of Eq. (18.11). At low areas, however, they form highly incompressible films that behave like two-dimensional solids.

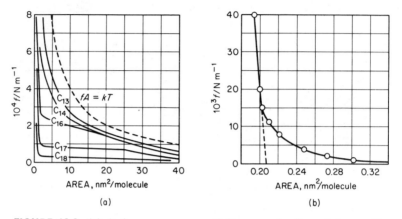

FIGURE 18.8 (a) Surface-pressure–area (f–α) curves for films of fatty acids $CH_3(CH_2)_n COOH$ on water. (b) The high-pressure region of the f–α curve for palmitic acid C_{16}.

18.7 Structure of Surface Films

The type of $f - \alpha$ isotherm observed when an organic compound is spread over water depends on the structure of the compound. It is not easy, however, to deduce the exact conformation and packing of the molecules in the surface layer. In the case of molecules with polar end groups and long hydrocarbon chains, the hydrophilic end group is in the water and the hydrophobic chain is in the air. This conclusion was reached by Langmuir as a result of his early observation that the surface area per molecule is the same, about $0.20\ \text{nm}^2$, for close packed surface films of the normal fatty acids from C_{14} to C_{18}.

The $f - \alpha$ isotherm for palmitic acid $C_{15}H_{31}COOH$ is drawn in Fig. 18.8(b). The steep linear portion of the $f - \alpha$ curve corresponds to compression of a closely packed surface layer. The critical area of $0.20\ \text{nm}^2$ is thus approximately equal to the cross-sectional area of the hydrocarbon chain. The molecular volume of solid palmitic acid is $M/L\rho = 0.556\ \text{nm}^3$. The length of the molecule is therefore $0.556\ \text{nm}^3/0.20\ \text{nm}^3 = 2.28\ \text{nm}$ or about $0.15\ \text{nm}$ per CH_2 group. This estimate agrees well with the value found by X-ray diffraction.

These monolayer structures are closely related to the structures of soap or detergent films. When a soap film thins, it ultimately reaches a *black film* state, where its thickness is below that required to give interference colors. The limiting thickness for sodium stearate black films is about 5.0 nm. This is about twice the length of the fully extended molecule, and the structure of the film is a bimolecular leaflet in which the polar end groups are in contact and the nonpolar chains are exposed to the exterior.

18.8 Surfactants and Micelles

Substances that drastically lower the surface tension of water even at low concentrations are called surface-active compounds or *surfactants*. As the name implies, they are strongly positively adsorbed at the air–water interface. The earliest surfactants were the soaps, such as sodium salts of long-chain fatty acids ($n = 14$ to 18): $CH_3(CH_2)_nCOO^-Na^+$. The anion has a polar end group and a long hydrocarbon tail, so that it tends to orient in the surface layer with the hydrophilic carboxyl in the water and the hydrophobic tail in the air. Synthetic surfactants also include anionic types, such as sodium dodecyl sulfate (SDS), $CH_3(CH_2)_{11}OSO_3^-Na^+$. Many household detergents are of this type. Cationic surfactants include cetyltrimethylammonium bromide, $C_{16}H_{33}(CH_3)_3N^+Br^-$. There are also many nonionic surfactants, for example, the Triton series,

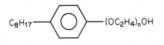

All these surfactant molecules are *amphipathic*; i.e., they have both polar and nonpolar moieties.

At concentrations c below about 10^{-3} mol dm^{-3}, most ionic surfactants in aqueous solution display conductance properties similar to other strong electrolytes, a slow decline in molar conductance Λ with $\sqrt{c}$. The surface tension γ declines steeply with c in this range. Between 10^{-3} and 10^{-1} mol dm^{-3}, however, sharp breaks occur in the Λ and γ curves vs. c. An example of such behavior is shown in Fig. 18.9. The concentration at which the break occurs is called the *critical micelle concentration (cmc)*. At this concentration, individual surfactant molecules aggregate to form

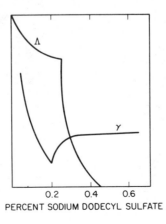

FIGURE 18.9 The critical micelle concentration can be detected by breaks in the Λ vs. c and γ vs. c curves. Data are for a surfactant in water.

PERCENT SODIUM DODECYL SULFATE

micelles containing well defined (but not absolutely fixed) numbers of molecules. Usually 50 to 100 molecules take part in a micelle. Data on the cmc for sodium dodecyl sulfate are summarized in Table 18.2.

TABLE 18.2
DATA ON MICELLES OF SODIUM DODECYL SULFATE AT 25°C

Medium	cmc (mol/dm³)	Aggregation number, n	Average charge per micelle, z
Water	0.0081	80	14.4
0.02 M NaCl	0.00382	94	13.2
0.10 M NaCl	0.00139	112	13.4
0.40 M NaCl	0.00052	126	16.4

The formation of micelles is another instance of the *hydrophobic interaction* discussed in Chapter 10. When hydrocarbon chains, or other nonpolar groups, of amphipathic molecules pack together, they can no longer modify the structure of water in their neighborhood. Thus a micellar arrangement lowers free energy by decreasing the modification of the water structure. The size of a micelle is in the range of typical colloids, about 0.1 mμ. A model of a micelle is shown in Fig. 18.10. The micelle presents a hydrophilic surface to the solution and has therefore little tendency to adsorb at the interface. This is why the decline in γ with c is arrested after the cmc is reached.

FIGURE 18.10 A model of a micelle of dodecyl sulfate with a molecule of pyrene segregated in its interior. [F. M. Menger, Emory University]

The micelle is a fascinating structure. It allows a small volume with an interior milieu like that of a lipid to exist suspended in an aqueous medium. The action of *detergents* is based on their ability to form micelles, which can sequester fatty materials, rendering them soluble in a surfactant solution.

Example 18.6 Suppose that a micelle contains 100 molecules of sodium stearate and that the dimensions of the hydrophobic tail of the molecule are $2.5 \times 0.45 \times 0.45$ nm. What area of contact of water with hydrophobic tails is eliminated when the micelle is formed? If the interfacial tension between water and hydrocarbon is $\gamma = 70 \times 10^{-3}$ N m^{-1}, what is the decrease in free energy due to hydrophobic interactions per mole of sodium stearate when the micelles are formed?

Assume that the hydrophobic tails have rectangular cross sections that pack together. The area of contact per pair of molecules is then $0.45 \times 2.5 = 1.13$ nm^2, or for 100 molecules, $50 \times 1.13 = 56.5$ nm^2. The interfacial free energy would be 70×10^{-3} N m$^{-1} \times 56.5 \times 10^{-18}$ m$^2 = 3.96 \times 10^{-18}$ J. Per mole, $\Delta G = 3.96 \times 10^{-18} \times 6.02 \times 10^{23}/100 = 23.8 \times 10^3$ J $= 23.8$ kJ mol^{-1}.

18.9 Cell Membranes

One of the most interesting applications of surface-film studies was made in 1925 by Gorter and Grendel. They found that the lipids extracted from red-cell membranes spread on water to a thickness just about half that of the membrane itself. Assuming, in accord with Langmuir, that the lipid layer is one molecule thick, they concluded that the membrane is essentially a double layer of lipid molecules, probably with some protein on its surfaces. In 1943, Davson and Danielli, on the basis of this work and data on the low surface tensions of membranes, suggested that a lipid bilayer coated with protein is a good model for the outer membranes of animal cells.

The lipid-bilayer model of the membrane is essentially correct as far as the location of the lipid molecules is concerned. The proteins, however, are not all situated on the outer surfaces, and some can penetrate the membrane. Protein molecules can diffuse laterally in lipid bilayer membranes, but only rarely can they perform a "flip-flop" from one side of the membrane to the other. The fluid-mosaic membrane model of Singer and Nicholson, as shown in Fig. 18.11, seems to correlate most properties

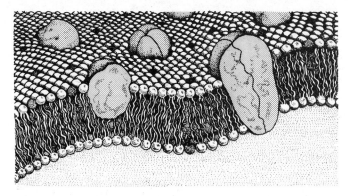

FIGURE 18.11 Fluid-mosaic model of a cell membrane. [S. J. Singer and G. L. Nicholson, *Science, 175,* 720 (1972)] Protein molecules are embedded in a lipid bilayer. [Drawing from *Cell Membranes*, ed. G. Weissmann and R. Claiborne, (New York: Hospital Practice Publishing Co. 1975)]

of membranes in a satisfactory fashion, but experimental determination of the exact arrangements of protein and lipid molecules is still a task for the future.

18.10   Colloidal Sols: Particle-Size Distribution

So far we have been discussing liquid–vapor interfaces. In colloidal sols, liquid–solid interfaces have the predominant role.

The dispersed phase in a colloidal sol or in a solution of a high polymer may have a distribution of particle sizes. An example is seen in Fig. 18.12 for a solution of brominated natural rubber in benzene. In a solution of styrene in benzene all the molecules of styrene are identical, and all methods of measuring the molar mass M of solute yield the same value. In a solution of a polymer such as polystyrene or rubber, however, the masses of the polymer molecules are distributed over a range of values. Consequently, different methods of measuring M may yield different values.

Colligative properties, such as osmotic pressure, depend on the *number* of particles in a volume of solution. Thus the molar mass calculated from a colligative

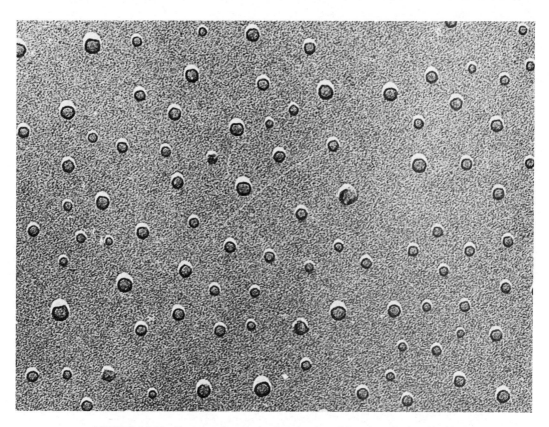

FIGURE 18.12 Electron micrograph of molecules of brominated natural rubber. [S. Nair, Rubber Research Institute of Malaya]

property is the *number average*,

$$\bar{M}_N = L \sum N_i m_i / \sum N_i \tag{18.12}$$

Other experimental measurements of molar mass, such as light scattering, depend on the mass of material in a volume of solution. These methods yield a *mass average*,

$$\bar{M}_m = L \sum N_i m_i \cdot m_i / \sum N_i m_i = L \sum N_i m_i^2 / \sum N_i m_i \tag{18.13}$$

Example 18.7 Consider a solution containing 10% by mass of polymer with $M_1 = 10\,000$ g mol^{-1} and 90% by mass of polymer with $M_2 = 100\,000$ g mol^{-1}. Calculate $\bar{M}_N$ and $\bar{M}_m$.

$$\bar{M}_N = \frac{0.1(10\,000) + 0.09(100\,000)}{0.19} = 52\,000 \text{ g mol}^{-1}$$

$$\bar{M}_m = \frac{0.1(10\,000 + 0.9(100\,000)}{1} = 91\,000 \text{ g mol}^{-1}$$

Usually, the particles in a lyophobic sol have a fairly wide distribution of size, but methods have been developed to produce *monodisperse sols* of uniform particle size. Such sols are convenient for theoretical studies. The basic idea in preparing a monodisperse sol is to adjust the condition of condensation or precipitation so that self-nucleation occurs only in one burst during a short period of time, after which the embryos grow from a solution of lower supersaturation.

Figure 18.13 shows the interaction of two monodisperse sols of widely different particle sizes.

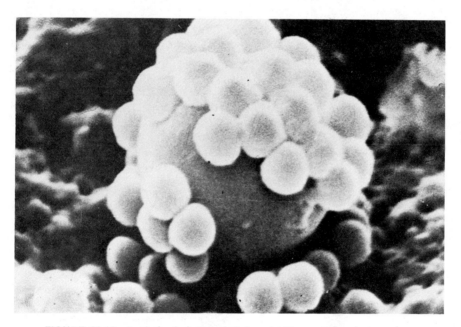

FIGURE 18.13 A sol of polyvinyl alcohol viewed with a scanning electron microscope. Two different uniform particle sizes were prepared and used to study the interactions of the smaller spheres with the larger ones. [R. H. Ottewill, University of Bristol]

The stability of colloidal sols of the lyophobic type is due to repulsion between like charges on the particles, which prevents coagulation (flocculation). The charges usually arise as a result of adsorption of ions from the solution, but in some cases they may be due to defect structures in the solid particle itself. For example, some clay minerals carry a charge due to substitution of Al^{3+} by Mg^{2+} or Li^+, etc. The surface charges on a particle are compensated by an array of ions of opposite sign, thus forming an *electrical double layer*.

The theory of the stability of lyophobic colloids was worked out about 1942 by Verwey and Overbeek and independently by Derjaguin and Landau. It is often called, therefore, the D.L.V.O. Theory. They calculated the free energy of interaction between particles and surfaces of various shapes. They took the sum of the electrical interaction between the double layers and the van der Waals interaction between the surfaces. Although van der Waals forces between individual molecules have a very short range, the situation is different for massive colloidal particles, which have an attractive interaction over a much longer range. [Why is this so?]

As in the Debye–Hückel theory of interionic forces, the presence of ions in colloidal sols tends to shield the charged particles from electrostatic interactions. The higher the ionic strength of the medium, therefore, the shorter the Debye length (or radius of the ionic atmosphere). Thus, as the ionic strength of the medium increases, ultimately a condition is reached in which van der Waals attraction between the particles overcomes the electrical repulsion and then coagulation of the sol occurs.

The earliest model of the double layer was given in 1879 by Helmholtz, who suggested the picture shown in Fig. 18.14(a), a layer of ions at a solid surface and a rigidly held layer of oppositely charged ions in the solution. The Helmholtz double layer is equivalent to a parallel-plate capacitor. If λ is the distance separating the oppositely charged plates and ϵ_r the dielectric constant of the medium, the capacitance per unit area of interface is $\epsilon_r\epsilon_0/\lambda$. If Q/α is the surface charge density, the potential difference $\Delta\Phi$ across the double layer is

$$\Delta\Phi = \frac{\lambda Q}{\epsilon_0 \epsilon_r \alpha} \tag{18.14}$$

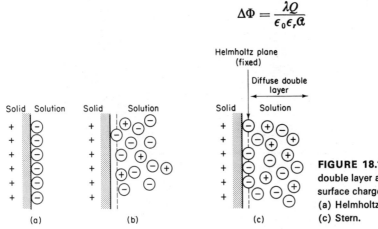

FIGURE 18.14   Models of the electric double layer at a solid surface: fixed surface charges and ions in solution. (a) Helmholtz; (b) Goüy–Chapman; (c) Stern.

Example 18.8 Calculate the magnitude of the electric field E in the double layer at the surface of colloidal silica in a 3.0 M NaCl solution if the excess surface charge is 10^5 μC m^{-2}.

In such a concentrated solution, E can be calculated from the Helmholtz formula (18.14) as $E = \Delta\Phi/\lambda = Q/\epsilon_0\epsilon_r\mathcal{C}$. The effective dielectric constant would be much less than that of bulk water, and we can estimate $\epsilon_r = 10$. Thus

$$E = \frac{10^5 \times 10^{-6} \text{ C m}^{-2}}{10 \times 8.854 \times 10^{-12} \text{ J}^{-1} \text{ C}^2 \text{ m}^{-1}} = 1.1 \times 10^9 \text{ V m}^{-1}$$

Note that this is a very high electric field.

The Helmholtz model of the double layer is inadequate, because the thermal motions of molecules and ions in solution could scarcely permit such a rigid array of charges at the interface. At high salt concentrations, however, it is not a bad approximation.

The theory of a diffuse double layer with a statistical distribution of ions in the electric field was given by Goüy in 1910 and Chapman in 1913. In the Debye–Hückel theory, the critical parameter is the thickness of ionic atmosphere b. In the Goüy–Chapman theory, the corresponding parameter is the thickness of double layer b'. This is given by the same formula as Eq. (16.37).

Values of b' computed at 25°C for different concentrations and valences are given in Table 18.3. Note the contraction of the double layer as ionic strength increases. The Goüy–Chapman model is shown in Fig. 18.14(b).

TABLE 18.3
THICKNESS OF THE DOUBLE LAYER CALCULATED FROM GOÜY–CHAPMAN THEORY

c (mol dm^{-3})	$z = 1$ b' (mm)	$z = 2$ b' (mm)
10^{-5}	10^{-4}	5×10^{-5}
10^{-3}	10^{-5}	5×10^{-6}
10^{-1}	10^{-6}	5×10^{-7}

A serious defect of the Goüy–Chapman theory is that it treats ions as point charges. It thus leads to absurdly high values for the charge concentration in the immediate neighborhood of the interface. In 1924, Stern provided a suitable correction in the form of an adsorbed layer of ions with a thickness δ about equal to the ionic diameters. This layer is assumed to be fixed at the surface. The Stern modification of the Goüy–Chapman model is shown in Fig. 18.14(c). There is a linear fall of potential in the rigid Stern layer, followed by a Goüy–Chapman behavior in the diffuse layer. [Sketch the potential Φ vs. x in the Stern model.]

The properties of double layers are important in the theory of colloidal stability, in electrodic rate processes, and in biophysical properties of cell membranes.

Problems

1. The surface tension γ of toluene is measured by the capillary rise method in an atmosphere of pure N_2 at $P°$. The diameter of the capillary tube is 1.014 mm, and the contact angle is $0°$. If the capillary rise is 5.44 mm, what is the value of γ from Eq. (18.3)? What would be the correction if Eq. (18.4) was used? ($T = 298$ K)

2. 20 cm^3 of a saturated solution of $PbSO_4$ (0.0040 g L^{-1}) containing as a tracer radioactive lead with an activity of 1600 counts/min (cpm) is shaken with 1.015 g of precipitated $PbSO_4$. The final count in the solution is 455 cpm. If only solid $PbSO_4$ in the surface layer exchanges with Pb^{2+} in solution, calculate the number of surface Pb^{2+} per gram of precipitate. If a $PbSO_4$ unit in the surface of solid has an effective area of 18.4×10^{-20} m^2, estimate the specific area (m^2 g^{-1}) of the precipitate.

3. If bubbles of air 1.00 μm in diameter but no other nuclei are present in water just below the boiling point, how far could the water be superheated before boiling begins?

4. A suspension of benzene and water is dispersed at $20°C$ with an ultrasonic probe. Estimate the work done on the system in producing a suspension of benzene in water with a mean droplet diameter of 0.100 μm.

5. At $20°C$ benzene has $\gamma = 28.86 \times 10^{-3}$ N m^{-1} and $\rho = 0.8788$ g cm^{-3}. At $40°C$, $\gamma = 26.21 \times 10^{-3}$ N m^{-1} and $\rho = 0.8573$ g cm^{-3}. Calculate the rises in a capillary with $R = 0.0205$ cm in air at $P°$. The contact angle is $13°$.

6. Consider a capillary with one end immersed in water that has risen to a height h. Describe what will happen when the capillary is lowered further into the water until the length remaining above the free liquid surface is only $h/2$.

7. For liquid mercury, $\gamma(N\ m^{-1}) = 0.4636 + 8.32 \times 10^{-5}T - 3.13 \times 10^{-7}T^2$. Calculate the surface free energy per unit area A^σ at 400 K, the surface energy $U^\sigma = \gamma - T(\partial\gamma/\partial T)_V$; and the surface entropy $S^\sigma = (-\partial A^\sigma/\partial T)_V$.

8. For mercury droplets of $R = 10$ nm at 400 K, calculate $P/P^\bullet$, where $P^\bullet$ is vapor pressure of bulk mercury. The density of Hg at 400 K is 13.29 g cm^{-3}.

9. An empirical equation due to Szyszkowski gives the surface tension of dilute aqueous solutions of organic compounds as

$$\gamma/\gamma_w = 1 - 0.411 \log_{10}(1 + X/a)$$

where γ_w is the surface tension of water, X is the mole fraction of solute, and a is a constant characteristic of the organic compound. For n-valeric acid, $a = 1.7 \times 10^{-4}$. Calculate the average area occupied by a molecule of n-valeric acid adsorbed from aqueous solution at $25°C$ onto the surface when $X = 0.01$.

10. When 0.0100 mg egg albumin spreads on the surface of water at $25°C$ in a Langmuir trough, an area of 0.0200 m^2 is covered and the surface tension is lowered by $\Delta\gamma = 2.86 \times 10^{-3}$ Nm^{-1}. Estimate the molar mass of the protein.

11. The surface tensions of aqueous solutions of sodium dodecyl sulfate (SDS) at $20°C$:

10^3c (mol dm^{-3})	0	2	4	6	8	10	12
$10^3\gamma$ (N m^{-1})	73.0	62.6	52.5	45.2	40.0	39.6	39.5

Calculate the area per molecule of SDS in the surface when adsorption is a maximum. Estimate the cmc.

12. The spreading of stearic acid on large bodies of water has been used to decrease evaporation in arid areas. What mass of stearic acid would cover an area of 1 km^2?

13. A glass rod of diameter 0.97 cm fits coaxially a glass tube of diameter 1.00 cm. What is the capillary rise in the annular space of water with $\gamma = 70 \times 10^{-3}$ N m^{-1} and zero contact angle?

14. If the attractive potential energy between two molecules is $U = -Ar^{-6}$, show that the interaction between a molecule in the gas and the surface of a solid composed of the same molecules, at a distance z is $U(z) = -\pi p'A/6z^3$, where p' is the number of molecules per unit volume in the solid.

15. Suppose that adsorbed H atoms are freely mobile on a solid surface of platinum. Calculate the translational entropy of the adsorbed gas per mole and compare it with that of a three-dimensional gas at the same temperature, 300 K.

16. Show that the ratio T/T_0 of equilibrium vaporization temperatures of a liquid in droplets of radius r and in bulk is given by

$$\ln (T/T_0) = -2V_\ell\gamma/\Delta H_v r$$

where V_ℓ is the molar volume and ΔH_v is the enthalpy of vaporization per mole.

17. Derive an equation for the capillary rise between two parallel plates of infinite length inclined at an angle ϕ to each other and meeting at the liquid surface. Assume zero contact angle and a circular cross section for the meniscus.

18. A monolayer of lipopolysaccharide on 0.2 M NaCl at 20°C gave the following surface pressures [*J. Colloid Interface Sci.*, *33*, 84 (1970)].

$m\,\mathcal{Q}^{-1}$ (mg m^{-2})	0.06	0.09	0.11	0.14	0.17	0.23
$10^6 f$(N m^{-1})	10.3	16.4	20.4	25.9	34.3	50.0

Calculate the molar mass M of the biopolymer. (See by analogy the limiting osmotic pressure method, page 193.)

19. The interfacial tension between mercury and a solution of stearic acid in hexane was measured at 30°C.

c (mol dm^{-3})	4.8×10^{-6}	8.5×10^{-6}	6.6×10^{-5}	2.7×10^{-4}	1.0×10^{-3}
$10^3\gamma$ (N m^{-1})	362	355	334	307	286

Evaluate the surface excess Γ of stearic acid at the mercury surface and draw the adsorption isotherm Γ vs. c.

20. In the du Noüy method of measuring surface tension, a fine platinum ring is suspended in the liquid surface from a torsion balance and the force necessary to pull the ring out of the surface (with its enclosed liquid layers) is measured. Calculate the force for a ring 10 mm in diameter for water at 20°C (see Table 18.1).

21. A protein with a monomeric molar mass of 20 000 g mol^{-1} associates ($2P = P_2$) into dimers with an association constant of 2.10×10^3 (Standard state 1 mol dm^{-3}) at 25°C. Calculate the number average and the mass average molar masses of the protein in solutions containing 10 g dm^{-3} and 50 g dm^{-3} protein.

22. Sketch the electric potential Φ vs. distance from surface x for the double layers shown in Fig. 18.14.

19

Electrochemical Rate Processes

Electrochemical processes are important in industrial chemistry and metallurgy, for example, in refining of copper and aluminum, in production of chlorine and caustic soda. Electrochemical methods are used in many industrial organic syntheses, such as production of the nylon intermediate adiponitrile from acrylonitrile, $2CH_2=CH-CN + 2H_2O + 2e^- \rightarrow NC-(CH_2)_4-CN + 2OH^-$. The design of an electrochemical process is based on the factors that control the rate of reaction at an electrode. Electrochemical kinetics is also basic to important analytical procedures such as electrophoresis and polarography. Biochemical preparations often depend on rates of migration of proteins in an electric field (e.g., in gel electrophoresis). Metallic structures, unless kept in the vacuum of outer space, are subject to corrosion, and corrosion is an electrochemical reaction.

The generation of electrical energy in fuel cells may allow a more efficient utilization of chemical fuels than is possible in thermal power stations. The theoretical efficiency of the fuel cell is unquestionable, being derived from the laws of thermodynamics. The practical problem is to achieve a high rate of reaction in the cell without an unacceptable loss in efficiency.

19.1   Electrode Kinetics

A metal electrode acts as a catalytic surface that facilitates transfer of electrons to and from chemical reactant molecules and ions. Thus an electrode reaction can be viewed as a succession of steps similar to those in heterogeneous catalysis, as listed in Section 14.7:

1. Diffusion of reactants to electrode
2. Adsorption of reactants on electrode
3. Transfer of electrons to or from adsorbed reactant species
4. Desorption of products from electrode
5. Diffusion of products away from electrode

In an electrode reaction, the energy of an electric field acts on charged species, such as ions and electrons, so as to help them surmount an activation-energy barrier. Since electrochemical reactions are always studied at some temperature $T > 0$, there will also be a thermal contribution to the energy of activation; consequently, electrochemical kinetics is based upon a combination of thermal and electrical activation.

At equilibrium, for each ionic species, the rate of electron transfer across an electrode in the cathodic direction is exactly balanced by an equal rate of electron transfer in the anodic direction, so that the *current density i* (current per unit area) is

$$i_c = i_a = i_0$$

The equilibrium difference in electric potential $\Delta\Phi_e$ is determined by this condition. As in any chemical reaction, the condition of equilibrium is not the cessation of all exchange, but the equality of forward and reverse reaction rates. The *current density* at equilibrium i_0 is called the *exchange current density*. Values of i_0 for some electrode reactions are given in Table 19.1. We see that these exchange reaction rates vary over many orders of magnitude.

TABLE 19.1
EXCHANGE-CURRENT DENSITIES i_0 AT 25°C FOR SOME
ELECTRODE REACTIONS[a]

Metal	System	Medium	Log i_0 (A cm^{-2})
Mercury	Cr^{3+}/Cr^{2+}	KCl	−6.0
Platinum	Ce^{4+}/Ce^{3+}	H_2SO_4	−4.4
Platinum	Fe^{3+}/Fe^{2+}	H_2SO_4	−2.6
Palladium	Fe^{3+}/Fe^{2+}	H_2SO_4	−2.2
Gold	H^+/H_2	H_2SO_4	−3.6
Platinum	H^+/H_2	H_2SO_4	−3.1
Mercury	H^+/H_2	H_2SO_4	−12.1
Nickel	H^+/H_2	H_2SO_4	−5.2

[a]After J. O'M. Bockris, *Modern Electrochemistry* (New York: Plenum Press, 1970).

The rate v of a chemical reaction at a surface is usually expressed in units of mol m^{-2} s^{-1}. The rate i of charge transfer at an electrode is expressed in units of A m^{-2}. If the charge number of the ionic species involved is $|z|$, $v = i/|z|F$, where F is the faraday, 96 485 C mol^{-1}. Thus, in electrochemical reactions a high current density means a high reaction rate. A typical rate would be 100 A m^{-2}, equivalent to about 10^{-3} mol m^{-2} s^{-1} for ions of unit charge.

Electrodes that have a high exchange-current density for a given reaction are said to be *nonpolarizable*. If the applied potential difference across such an electrode is increased, there is an increased flow of charge between electrode and solution, but the potential difference across the double layer is not altered. In other words, charge moves rapidly to and from the electrode and does not build up a charge density in the surface layers. An example of such an electrode is the calomel electrode.

Electrodes that have a low exchange-current density for a given reaction are said to be *polarizable*. If the applied potential difference across such an electrode is increased, there is little flow of charge into the solution. The charges remain in the double layer and increase the potential difference across it. An example is a mercury electrode in a solution of KCl.

19.2 Polarization

When an electrochemical cell is operating under nonequilibrium conditions, $i_c \neq i_a$ and there is a net current density $i = i_c - i_a$. The electric potential difference between the terminals of the cell then departs from the equilibrium value $\Delta\Phi = E$, the emf. If the cell is converting chemical free energy into electric energy, $\Delta\Phi < E$. If the cell is using an external source of electric energy to carry out a chemical reaction, $\Delta\Phi > E$. The actual value of $\Delta\Phi$ depends on the current density i at the electrodes. The difference,

$$\Delta\Phi(i) - \Delta\Phi(0) = \eta \qquad (19.1)$$

is called the *polarization* or, probably more commonly now, the *overpotential* of the cell. The value of η is determined in part by the potential difference (IR) necessary to overcome the resistance R in the electrolyte and leads. The corresponding electrical energy I^2Rt is dissipated as heat, being analogous to frictional losses in irreversible mechanical processes. The remaining part of η, which is the part of theoretical interest, is due to rate-limiting processes at the electrodes; the corresponding electrical energy is being used to provide part of the free energy of activation in one or more of the steps (previously outlined) in the electrode reaction.

In electrochemical kinetics, we usually wish to study reactions at a particular electrode. We measure the potential at this electrode by introducing into the cell an auxiliary reference electrode with a lead into the electrolyte very close to the experimental electrode. Such an arrangement is depicted in Fig. 19.1.

19.3 How the Electric Field Controls the Rate of an Electrode Reaction

The way in which an electric field can change the rate of an electrode reaction can be understood by study of Fig. 19.2. Let us suppose that a reactant ion M^{z+} has approached quite closely to an electrode. Reaction will occur if electrons are transferred from electrode to ion, $M^{z+} + ze \rightarrow M$.

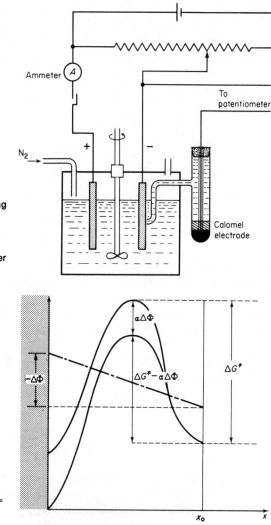

FIGURE 19.1 Apparatus for measuring the potential of an electrode relative to that of a reference electrode. The potential is measured as a function of the current density at the electrode under study.

FIGURE 19.2 Schematic Gibbs free-energy curves for the electrode reaction $M^{z+} + ze \rightarrow M$ in the neighborhood of the electrode surface, showing how the electric potential $\Delta\Phi$ lowers the free energy of activation $\Delta G^{\ne}$ by an amount $\alpha zF\,\Delta\Phi$.

We are going to apply the Eyring equation for the rate constant k_2 of a chemical reaction to this process, using the form in Eq. (15.16),

$$k_2 = (kT/hc^\circ) \exp\left(-\Delta G^{\circ\ne}/RT\right) \tag{19.2}$$

where $\Delta G^{\circ\ne}$ is the Gibbs free energy of activation. Figure 19.2 as drawn implies that the "reaction coordinate" in the Eyring formulation is normal to the electrode surface. The detailed mechanism of the reaction does not concern us here. (Like thermodynamics, the activated-complex theory of rate processes is not concerned with detailed molecular mechanisms.) Thus the curves in Fig. 19.2 represent reaction paths along the free-energy surfaces between reactant and product species. The maxima of the curves locate activated complexes for the reaction, and the higher $\Delta G^{\ne}$ indicates the free energy of activation for the thermal reaction.

The electrochemical reaction of electron transfer occurs in a region near the electrode that approximately coincides with the region of the electric double layer shown in Fig. 18.14. The strength of the electric field in a double layer can be very high, for example, a potential difference of 1 V across a typical double-layer thickness of 1 nm is equivalent to a field strength of 10^9 V m^{-1}. This is an enormous electric field, far greater than could be achieved over longer distances in laboratory apparatus. Such a high field can literally tear ions out of a solid metal and drag them into solution. If the potential difference across the double layer is changed by even a fraction of a volt, large effects on the rate of the electrode reaction can be anticipated.

The exact course of the potential Φ vs. distance x curve through the double layer is not known, but as an example a simple Helmholtz model can be assumed, with a linear dependence of Φ on x as shown in Fig. 19.2. The reactant molecule can be located at x_0, the position of the outer Helmholtz plane.

For an electrochemical reaction with ionic reactants, the potential difference $\Delta\Phi$ across the double layer at the electrode surface assists the transfer of an ion through the double layer in one direction but inhibits its transfer in the opposite direction. When the activated complex is reached, at the maximum in the free-energy barrier, only some fraction α of the electrical energy difference $zF\,\Delta\Phi$ has been used. This fraction α is called the *transfer coefficient*. Generally, α is in the neighborhood of 0.5.

As Fig. 19.2 shows, a change from the equilibrium potential increases the current in one direction and decreases it in the other. By convention, a net cathodic current is taken as positive and a net anodic current as negative. Thus α corresponds to the cathodic (reduction) process and $(1 - \alpha)$ to the anodic (oxidation) process.

From the general equation (19.2) for a rate process, given by the transition-state theory, we can write the anodic and cathodic current densities as follows:

$$\text{Anodic:} \quad i_a = zFk_a c_{0R} \exp\left[\frac{-\Delta G_a^\ddagger - (1 - \alpha)zF\,\Delta\Phi}{RT}\right]$$

$$\text{Cathodic:} \quad i_c = zFk_c c_{00} \exp\left(\frac{-\Delta G_c^\ddagger + \alpha zF\,\Delta\Phi}{RT}\right)$$

In these expressions, k_a and k_c are the pre-exponential parts of the rate constants for the forward (anodic) and reverse (cathodic) electron transfers, and c_{0R} and c_{00} represent the surface concentrations of reduced product and oxidized reactant of the electrochemical reaction, $M^{z+} + ze \longrightarrow M$. The $\Delta G_a^\ddagger$ and $\Delta G_c^\ddagger$ are the thermal Gibbs free-energy barriers for the anodic and cathodic electrode reactions, respectively. In a regime in which diffusion is sufficiently rapid to eliminate concentration gradients, the ionic concentrations adjacent to the electrode can be taken as constant, independent of i and also of time (when the extent of reaction is small).

At equilibrium, we can thus write for the exchange current per unit area,

$$i_0 = zFk_a c_{0R} \exp -\left[\frac{\Delta G_a^\ddagger + (1 - \alpha)zF\,\Delta\Phi_{\text{rev}}}{RT}\right]$$

$$= zFk_c c_{00} \exp -\left(\frac{\Delta G_c^\ddagger - \alpha zF\,\Delta\Phi_{\text{rev}}}{RT}\right)$$

In terms of i_0 and $\eta_t = \Delta\Phi - \Delta\Phi_{\text{rev}}$, the activation overpotential,

$$i = i_c - i_a = i_0\left[\exp\left(\frac{\alpha zF\eta_t}{RT}\right) - \exp\left(\frac{-(1 - \alpha)zF\eta_t}{RT}\right)\right] \qquad (19.3)$$

This is the important *Butler–Volmer equation*.

In Fig. 19.3 the current density i is plotted against the overpotential in accord with this equation. Two cases are shown. One, labeled A, is a case in which there are rather high exchange currents i_0 at both electrodes. (The individual electrode curves are labeled A' and A''.) In this case, even a small change in potential from the equilibrium value, i.e., a small overpotential, will produce an appreciable current flow through the cell. The other case, labeled B, corresponds to a very low exchange current i_0. (The individual electrode curves are not drawn.) In this case a large overpotential is required to cause appreciable current flow through the cell. We see that the governing factor that determines the activation overpotential, in accord with the Butler–Volmer equation, is the exchange-current density i_0. The transfer factor α influences the shape of the current density vs. overpotential curves. One way to measure α is to fit the experimental curve to the Bulter–Volmer equation (19.3).

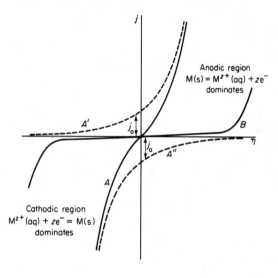

FIGURE 19.3 Variation of current density with overpotential for activation polarization. These experimental curves closely follow the Butler–Volmer equation (19.3).

19.4 The Tafel Equations

If the overpotential has large positive or negative values, $|\eta| \gg RT/zF$, one of the partial currents becomes much greater than the other, which is then negligible. In this case, either

$$\ln i_a = \ln i_0 - [(1 - \alpha)zF/RT]\eta \tag{19.4}$$

or

$$\ln i_c = \ln i_0 + (\alpha zF/RT)\eta \tag{19.5}$$

This type of logarithmic dependence of i on η was found empirically in 1905 by Tafel. The slope of linear $\ln i$ vs. η plots gives the transfer coefficient α and the intercept gives the exchange current density i_0 (Fig. 19.4).

Overpotentials can be caused by slow reactions in the solution adjacent to an electrode (reaction overpotential η_r) and in the process of deposition of a solid product on an electrode (crystallization overpotential η_c). In many cases, a diffusion

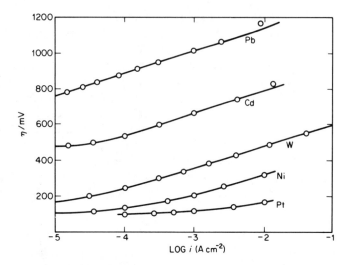

FIGURE 19.4 Variation of overpotential for discharge of H$^+$ on various metals with current density. The linear portions of the curves are in accord with the Tafel equation (19.5). [J. O'M. Bockris, Flinders University]

overpotential (concentration polarization) occurs together with an activation overpotential, and methods are available for separating these factors. Thus a great variety of interesting chemical kinetics can be studied by electrodic techniques. Practical applications of this branch of kinetics are found in such fields as fuel cells, storage batteries for vehicle propulsion, and electrochemical syntheses, where great selectivity can be achieved through the choice of electrodes and carefully controlled potentials.

Example 19.1 A solution of 1 mol dm^{-3} KOH is electrolyzed at 25°C with platinized platinum electrodes to produce oxygen O$_2$ at the anode. At an overpotential of 0.40 V, the current density $i = 10^{-3}$ A/cm^2. What does i become if the overpotential is increased to 0.60 V? Assume that $\alpha = 0.5$.

In this case $RT/F = 0.0257$ V, so that $\eta \gg RT/F$ and the Tafel equation (19.4) applies. The ratio of current densities is $i(\eta + 0.2)/i(\eta) = \exp\left[(1 - \alpha)F(0.2)/RT\right] \approx 10^2$. The current density and yield of O$_2$ is increased a hundredfold by increase of 0.2 V in overpotential.

Example 19.2 The exchange current density for the reaction H$^+ + e = \frac{1}{2}$H$_2$ on nickel at 25°C is 1.00×10^{-5} A cm^{-2}. What current density would be necessary to attain an overpotential of 0.100 V ($\alpha = 0.5$) (a) from the Butler–Volmer equation; (b) from the Tafel approximation?

(a) $i = i_0[e^{\alpha\eta F/RT} - e^{-(1-\alpha)\eta F/RT}]$
$= 10^{-5}[e^{(0.5)(0.1)(38.9)} - e^{-(0.5)(0.1)(38.9)}]$
$= 10^{-5}[e^{1.95} - e^{-1.95}] = 6.85 \times 10^{-5}$ A cm^{-2}
(b) $i = 6.99 \times 10^{-5}$ A cm^{-2}

> **Example 19.3** From the Tafel plot in Fig. 19.4 calculate the transfer coefficient α and the exchange current density i_0 for the discharge of H^+ on Pb.
>
> The slope in Eq. (19.5) $\alpha z F/RT = 17.8 \text{ V}^{-1}$ (remember that $\ln i = 2.303 \log i$). Hence
>
> $$\alpha = \frac{(8.314 \text{ J K}^{-1} \text{ mol}^{-1})(298 \text{ K})}{(1)(96485 \text{ C mol}^{-1})}(17.8 \text{ V}^{-1}) = 0.457$$
>
> The intercept at $\eta = 0$ is $\ln i_0 = -24.8$ or $i_0 = 1.27 \times 10^{-11} \text{ A cm}^{-2}$.

19.5 Kinetics of Discharge of Hydrogen Ions

Even since the pioneering work of Tafel, much effort has been devoted to study of the discharge of hydrogen ions (really H_3O^+), on metal electrodes. The overall reaction can be written $2H_3O^+ + 2e^- = H_2 + 2H_2O$. Some experimental overpotential vs. current density data are shown in Fig. 19.4.

With platinized platinum electrodes in acid solutions, it is possible to detect a diffusion control of the H_3O^+ discharge at extremely high current densities (about 100 A cm^{-2} at a concentration of 1.0 mol dm^{-3}), but under less extreme conditions diffusion is not rate determining. The occurrence of diffusion control is, nevertheless, important, since it proves that the species discharging is H_3O^+ and not H_2O ($2H_2O + 2e^- = H_2 + 2OH^-$).

> **Example 19.4** A 1.00 molar solution of $CdSO_4$ is electrolyzed at 25°C with a cadmium cathode and a platinum anode each of 50 cm² area at a constant current of 0.050 A. If H^+ ions are at unit activity, what will be the concentration of Cd^{2+} when evolution of H_2 begins at the cathode? Assume unit activity coefficient for the $CdSO_4$ solution.
>
> The current density is $0.050/50 = 1.00 \text{ mA cm}^{-2}$. From Fig. 19.4, the hydrogen overpotential is 0.650 V. When the potential of the Cd electrode reaches -0.650 V, evolution of H_2 will begin. The $E°$ of the Cd electrode is -0.403 V. Thus the limiting $[Cd^{2+}]$ is given by $E = E° + (RT/zF) \ln [Cd^{2+}]$
>
> $$-0.650 = -0.403 + 0.0296 \log [Cd^{2+}]$$
>
> $$\log [Cd^{2+}] = -8.35, \qquad [Cd^{2+}] = 4.5 \times 10^{-9} \text{ mol dm}^{-3}$$
>
> Thus the deposition of Cd is virtually quantitative.

19.6 Diffusion Overpotential

We have seen that when the various electrochemical reaction steps are sufficiently rapid, the overall rate of an electrode reaction can be controlled by a diffusion process. We recall that rapid reactions in solution and at catalytic surfaces can also become subject to diffusion control. The detailed theory of the diffusion process depends on the form of the electrode, whether it is moving or stationary, and on the extent of stirring of the electrolyte solution.

One of the earliest discussions was given by Nernst in 1904 for the case of steady-state diffusion toward a stationary planar electrode in a solution so vigorously stirred that only a thin layer adjacent to the electrode, about 10^{-2} to 10^{-1} mm thick, functioned as a diffusion barrier. In this simple model, the concentration of the ion being discharged at the electrode is constant in the bulk of the solution and then declines evenly through the Nernst diffusion layer (Fig. 19.5).

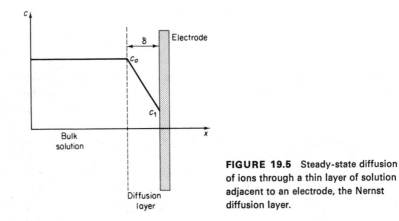

FIGURE 19.5 Steady-state diffusion of ions through a thin layer of solution adjacent to an electrode, the Nernst diffusion layer.

For concreteness, let us consider the discharge of Cu^{2+} and deposition of copper on the cathode. As Cu^{2+} is removed from the solution at the electrode, a layer of electrolyte solution of thickness δ is formed, in which concentration of Cu^{2+} is depleted. The rate of deposition of copper (amount n) will be the rate of diffusion of Cu^{2+} across the layer. By Fick's First Law,

$$-\frac{dn}{dt} = D\mathfrak{A}\frac{dc}{dx} = D\mathfrak{A}\frac{c_0 - c_1}{\delta}$$

where $\mathfrak{A}$ is the area of the electrode, and c_0 and c_1 are the concentrations of Cu^{2+} in the bulk electrolyte solution and at the electrode surface, respectively. The diffusion coefficient D of Cu^{2+} ion is assumed to be independent of its concentration.

The current to the cathode is

$$I_\delta = -zF\frac{dn}{dt} = \frac{zFD\mathfrak{A}(c_0 - c_1)}{\delta} \tag{19.6}$$

where z is the number of faradays transferred in the half-reaction, in this case 2 for the reduction of Cu^{2+}.

The difference in activity of Cu^{2+} across the diffusion layer leads to a potential difference of the form calculated for a concentration cell, namely

$$\eta_D = \frac{RT}{zF} \ln \frac{a_1}{a_0} \tag{19.7}$$

This η_D is called the *diffusion overpotential*. From Eq. (19.7),

$$c_1 = c_0 \left(\frac{\gamma_0}{\gamma_1}\right) \exp \left(\frac{zF\eta_D}{RT}\right) \tag{19.8}$$

where γ_0/γ_1 is the ratio of activity coefficients corresponding to the ratio of concentrations c_0/c_1.

From Eqs. (19.6) and (19.8),

$$\eta_D = \frac{RT}{zF} \ln\left[\frac{\gamma_1}{\gamma_0}\left(1 - \frac{\delta I_\delta}{\alpha c_0 D |z| F}\right)\right] \qquad (19.9)$$

The limiting value of I_δ corresponds to the discharge of every ion striking the electrode, so that $c_1 = 0$ and

$$I_{max} = \frac{zFD\alpha c_0}{\delta} \qquad (19.10)$$

As $I \to I_{max}$, $\eta \to -\infty$, but before this happens, some other ion will of course begin to be discharged. Since a typical value of δ is about 10^{-2} cm and $D \simeq 10^{-5}$ cm^2 s^{-1}, from Eq. (19.10) I_{max}/α is usually about $10^2\, c_0$ A cm^{-2} when c_0 is in mol cm^{-3}.

If $\gamma_1/\gamma_0 \approx 1$, Eqs. (19.9) and (19.10) give

$$\eta_D = \frac{RT}{zF} \ln\left(1 - \frac{I_\delta}{I_{max}}\right) \qquad (19.11)$$

19.7 Fuel Cells

Although the principle of the fuel cell had been known for over a hundred years, the first practical cell was invented by Bacon and Frost at Cambridge University in 1959. Fuel cells have been used in space vehicles, but are economic only in special situations on earth. For power generation at the 100-MW level, they are only marginally more efficient than steam turbines (and much more expensive), but they come into their own as small clean power sources at levels of 1 MW and below.

The fuel cell consists of a fuel electrode (the anode), an oxidant electrode (the cathode), and an electrolyte. A typical cell is shown in Fig. 19.6. The most common

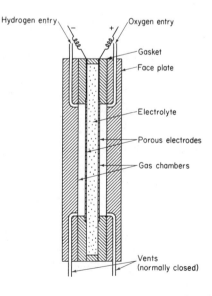

FIGURE 19.6 H$_2$–O$_2$ fuel cell with porous electrodes.

cell reaction is the oxidation of hydrogen, $2H_2 + O_2 = 2H_2O$. The electrolyte is concentrated NaOH. Hydrogen is fed into the anode, where it is catalytically dissociated and oxidized to H^+ ions, which react with OH^- to yield H_2O, $2H_2 + 4OH^-$ $\rightarrow 4H_2O + 4e$. The oxygen (usually as air) is fed into the cathode, where it is reduced, $O_2 + 2H_2O + 4e \rightarrow 4OH^-$. The emf of this cell is about 1 V.

The maximum efficiency of a fuel cell is the ratio of the electrical free energy produced $-\Delta G = |z|FE$ to the ΔH of the cell reaction,

$$\epsilon(\text{max}) = \frac{-\Delta G}{\Delta H} = \frac{|z|FE}{-\Delta H}$$

For example, in the H_2/O_2 fuel cell at 25°C, $\epsilon(\text{max}) = -229 \text{ kJ mol}^{-1}/-242 \text{ kJ mol}^{-1}$ $= 0.95$ (95 %). This thermodynamic efficiency cannot be achieved since overpotentials are required at both anode and cathode to make the electrode reactions go. Also, the output voltage must be reduced by the $I\mathcal{R}$ drop across the cell. Thus the working voltage is

$$\Delta\Phi_l = E - I\mathcal{R} - (\eta_{\text{anode}} + \eta_{\text{cathode}}) \tag{19.12}$$

The overvoltages η include those due both to diffusion and to electrode reactions. These terms can be calculated from Eqs. (19.4), (19.5), and (19.11).

Example 19.5 A H_2/O_2 fuel cell is based on platinized graphite electrodes. The exchange current densities are $i(\text{anodic}) = 1.00 \times 10^{-1} \text{ A m}^{-2}$, $i(\text{cathodic}) = 1.00 \times 10^{-3} \text{ A m}^{-2}$. Calculate the efficiency of the cell operated at $i = 0.300 \text{ A m}^{-2}$. Assume: $\alpha = 0.500$, $\mathcal{R} = 0.500 \ \Omega \text{ m}^{-2}$, and $I_\delta/I_{\text{max}} = 0.500$ at both electrodes, $T = 350$ K.

The Tafel equations can be used to calculate the activation overpotentials and Eq. (19.11) for the diffusion overpotential.
  Anodic overpotential:

$$\eta_a = \frac{RT}{\alpha F} \ln \frac{i}{i_0} = \frac{(8.314 \text{ J K}^{-1} \text{ mol}^{-1})(350 \text{ K})}{(0.5)(96\ 485 \text{ C mol}^{-1})} \ln \frac{i}{i_0}$$

$$= 0.0603 \ln \frac{i}{i_0} = 0.0603 \ln \frac{0.30}{10^{-1}} = 0.066 \text{ V}$$

Cathodic overpotential:

$$\eta_c = 0.0603 \ln \frac{0.30}{10^{-3}} = 0.344 \text{ V}$$

$I\mathcal{R}$ drop:

$$(0.30)(0.5) = 0.15 \text{ V}$$

Diffusion overpotential:

$$2(0.0603) \ln (1 - 0.5) = 0.084$$

Therefore,

$$V = 1.180 - 0.066 - 0.344 - 0.15 - 0.084 = 0.536$$

$$\epsilon = 0.536/1.180 = 45\% \text{ of thermodynamic efficiency}$$

The efficiency of fuel cells is markedly increased by the use of porous electrodes, an application of the basic theory of surface tension given in Section 18.2. The limiting current density in Eq. (19.10), $i = DzFc/\delta$, increases as the thickness δ of the diffusion layer decreases. At a planar electrode δ might be about 0.5 mm, but in a fine pore δ cannot exceed the pore radius, which may be about 10^{-4} mm. The gas pressure at a porous electrode is critical since it can neither be so large as to blow liquid out of the pores nor be so small as to allow liquid to flood them.

Storage batteries are quite similar to fuel cells with two special properties: (1) all the reactants are in the same container, and (2) the reaction can be reversed by putting electrical energy into the battery to recharge it. The development of light, efficient storage batteries for automotive applications is an important research field.

Problems

1. The cell $Ni|H_2|NaOH|O_2|Ni$ is run at $25°$ where $E° = 1.00$ V and at the cathode $\alpha = 0.58$. If $i_0 = 7.00 \times 10^{-6}$ A cm^{-2}, calculate the current through the cell at $\eta_t = 0.10, 0.50, 1.00, 2.00,$ and 5.00 V.

2. The exchange current density of the $Pt|Ce^{4+}, Ce^{3+}$ electrode is given in Table 19.1. The $E° = 1.61$ V. Calculate the current density i as a function of the applied voltage V, assuming only an activation overpotential and unit activity for the ions. ($\alpha = 0.50$)

3. Repeat the calculation in Problem 2 if the ions are at concentrations 0.05 M and $\gamma_\pm = 0.90$ and 0.80 for Ce^{3+} and Ce^{4+}, respectively.

4. One liter of a 1.00 M solution of $CuSO_4$ is electrolyzed with a platinum anode and a copper cathode of 100 cm^2 area at a constant current of 1.00 A. Estimate about how long electrolysis proceeds before hydrogen evolution begins and what fraction of the original copper then remains in solution. Assume deposition of copper is *not* limited by diffusion.

5. Overvoltage of cathode X is measured in the cell $Pt|X|H_2|H^+||KCl|HgCl|Hg|Pt$. At pH 7.00, the cell has $E = 1.325$ V at 25°C; what is the overvoltage on X?

6. Bowden and Rideal [*Proc. Roy. Soc., A120*, 59 (1928)] measured the following over-potentials η for H_2 evolution with a mercury cathode in dilute H_2SO_4 at 25°C:

10^7i (A cm^{-2})	2.9	6.3	28	100	250	630	1650	3300
η (V)	0.60	0.65	0.73	0.79	0.84	0.89	0.93	0.96

Calculate i_0 and α for this reaction.

7. The electrochemical oxidation of methanol in 1 M H_2SO_4 at 373 K and a current density of 200 A m^{-2} requires an overpotential of 0.44 V on platinum black, and 0.23 V on a Pt–Ru–Mo alloy catalyst. What are the relative reaction rates on the two catalysts at an overpotential of 0.30 V?

8. Would mild steel be a suitable container for a very dilute solution of acetic acid (pH 5), or would appreciable corrosion occur?

9. An impure sample of cadmium containing about 0.5% Zn and 0.5% Cu is refined electrolytically at a current density of 150 mA cm^{-2}. Assume that the cathode and anode have equal areas. What voltage is applied to the cell? The concentration of $CdSO_4$ in solution is 0.80 M, $\alpha = 0.50$, $i_0 = 1.0$ mA cm^{-2}. Assume no diffusion polarization and $T = 300$ K. Will pure Cd deposit on cathode?

20

Particles and Waves

In Chapter 4 the wave properties of material particles were used to define the allowed energy levels in molecules. The theoretical formulas were given there without any detailed discussion of the theory from which they were derived. In this chapter we shall consider the conceptual basis of the theory and some simple examples of its applications.

In 1926, Erwin Schrödinger and Werner Heisenberg independently discovered the basic principles for a new kind of mechanics, capable of dealing with the wave-particle duality of matter. The method of Schrödinger is called *wave mechanics* and that of Heisenberg, *matrix mechanics*. Despite their different mathematical formalisms, the two methods are essentially equivalent at the level of their basic physical concepts. They represent different forms of a fundamental theory called *quantum mechanics*. Schrödinger's mathematics is more familiar to the chemist, and it is usual to use his equations as the basis for chemical applications of quantum mechanics.

The heart of wave mechanics is an equation—the *Schrödinger wave equation*. The quantum mechanical descriptions of, say, the hydrogen atom or the vibrational motion of a diatomic molecule are begun by setting up the wave equation specifically tailored to each problem. Solutions to the equation are then obtained in the form of mathematical functions called *wavefunctions*. The wavefunctions contain all the information we can hope to learn about the hydrogen atom or the diatomic vibrator. That information can be uncovered by performing specific mathematical *operations* on the wavefunctions. These three steps—setting up the wave equation, solving it to obtain the wavefunction, and then using the wavefunction to learn about energies, momenta, spatial distributions and so forth—are the standard procedures for any quantum mechanical problem.

This chapter describes the foundations of Schrödinger's quantum mechanics. Four questions are considered. Where does the Schrödinger wave equation come from? What are the rules by which the equation is set up? How is the equation solved to yield the wavefunctions? What are the keys that unlock the physical information contained in the wavefunctions?

The answers to these questions are illustrated by application of quantum theory to the problem of translational energy. The results obtained are used to develop a theory, called *free electron molecular orbital theory*, which treats the behavior of electrons in molecules having conjugated π bonds.

20.1 Wave Motion

The Schrödinger wave equation of quantum mechanics is closely related to the treatment of waves found in Newtonian physics. This section is a review of aspects of wave motion that will be helpful later for quantum mechanics. We shall explain the mathematical description of plane sine waves and show how the differential wave equation of classical physics is developed.

Consider ocean waves approaching the shore on a gentle day. We might visualize these as ideal waves, being of uniform character everywhere. They travel with a velocity v that describes the motion of a particular wave crest in the direction of the shore, say the x direction. The wavelength λ is the distance between crests, and the number of crests passing a specified position per unit time is the frequency ν. These parameters are related by $\lambda\nu = v$. The period τ of the waves, the time required for successive crests to pass a given point, is $1/\nu$. The waves have a height or amplitude A, which is one half the vertical distance between crest and trough.

How is this information represented mathematically? First, we simplify the problem by ignoring the three-dimensional aspect of the ocean. Consider the waves running beside an imaginary wall that is parallel to the direction of propagation x. The coordinate z is the upward displacement of the water on the wall as the wave travels along it. It is convenient to describe the wave in terms of this displacement. The displacement z is a function of position x and time t, $z = f(x, t)$.

We wish to define the form of the function $f(x, t)$. Consider two snapshots of the water pattern along the wall, one at time $t = 0$ and one at later time t. During the interval t, a particular crest initially at x will have advanced to a position $x + vt$. To reproduce the original amplitude we must therefore subtract vt from the value of x. Thus

$$z = f(x - vt) \tag{20.1}$$

This is the common form in which variables x and t appear in the function describing the wave amplitude z.

What is the explicit function $f(x - vt)$? Wave trains of many shapes occur in nature. One of the simplest and most common is the *sine wave*. The electric and magnetic fields of light are sine waves. Violin strings sing with the motion of sine waves. Molecular vibrations are described to a close approximation by sine waves. The Schrödinger wave equation of quantum mechanics is closely connected with sine waves.

A sine wave traveling in the x direction with velocity v, wavelength λ, frequency v, and amplitude A is described mathematically by a sine function,

$$z = A \sin\left[\frac{2\pi}{\lambda}(x - vt)\right] = A \sin\left[2\pi\left(\frac{x}{\lambda} - vt\right)\right] \tag{20.2}$$

where z is called a *wavefunction*. The equation tells us the displacement z at any point x and at any time t.

The wave becomes easier to visualize if we follow the dependence of z upon only one variable at a time. Thus either t or x is frozen. First fix time at a constant value ($t = 0$ is simplest), and plot z vs. x [Fig 20.1(a)]. Alternatively, fix x at a constant value (say $x = 0$), and plot z vs. t [Fig. 20.1(b)]. In either case our choice of $t = 0$ or $x = 0$ is arbitrary. [Write the *same* wave as in Eq. (20.2) as a cosine function.]

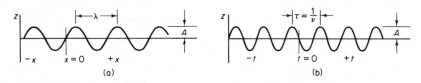

FIGURE 20.1 (a) A sine wave at a fixed time. Displacement z as a function of x. (b) A sine wave at a fixed position. Displacement z as a function of t.

20.2 The Classical Wave Equation

The sine (or cosine) function describes periodic motions arising from forces that occur in nature. The function can be derived directly from Newton's Second Law of Motion by considering the forces specific to the situation. In such a derivation, one obtains first a differential equation. This must then be solved to get the wave function $z = f(x, t)$. This differential equation occurs in many physical problems. It is the *general wave equation* or *classical wave equation*:

$$\frac{\partial^2 z(x, t)}{\partial x^2} = \frac{1}{v^2} \frac{\partial^2 z(x, t)}{\partial t^2} \tag{20.3}$$

The equation says that the wave function $z(x, t)$ has a remarkable property. The second derivative with respect to position x is directly proportional to the second derivative with respect to time t. The proportionality factor is simply $1/v^2$.

Equation (20.3) is important to us because this description of wave motion is a convenient starting point for finding a way to express the Schrödinger wave equation of quantum mechanics.

Example 20.1 Show that $z(x, t)$ in Eq. (20.1) satisfies the wave equation (20.3).

$z = A \sin (2\pi/\lambda)(x - vt)$.

$$\left(\frac{\partial z}{\partial x}\right)_t = A\left(\frac{2\pi}{\lambda}\right) \cos \frac{2\pi}{\lambda}(x - vt), \qquad \left(\frac{\partial z}{\partial t}\right)_x = -Av^2\left(\frac{2\pi}{\lambda}\right) \cos \frac{2\pi}{\lambda}(x - vt)$$

$$\left(\frac{\partial^2 z}{\partial x^2}\right)_t = -A\left(\frac{4\pi^2}{\lambda^2}\right) \sin \frac{2\pi}{\lambda}(x - vt), \qquad \left(\frac{\partial^2 z}{\partial t^2}\right)_x = -Av^2\left(\frac{4\pi^2}{\lambda^2}\right) \sin \frac{2\pi}{\lambda}(x - vt)$$

$$= -\left(\frac{4\pi^2}{\lambda^2}\right)z \qquad\qquad = -\left(\frac{4\pi^2}{\lambda^2}\right)v^2 z$$

Therefore,

$$(\partial^2 z/\partial x^2) = \frac{1}{v^2}(\partial^2 z/\partial t^2)$$

20.3 The Time-Independent Classical Wave Equation

Many problems in physics and chemistry are independent of time, and it is useful to have a form of the classical wave equation that does not contain time as a variable. A wave equation independent of time can be obtained from Eq. (20.3) if its general solution $z(x, t)$ can be written as a product of two functions, one of which depends only on x and the other only on t.

$$z(x, t) = \psi(x)\phi(t) \tag{20.4}$$

The independent variables x and t now occur in separate functions $\psi(x)$ and $\phi(t)$. The question is whether or not such a solution is possible. It turns out that it is, provided that certain conditions are satisfied. You can show that the solution $\psi(x)$ $\phi(t)$ is possible and see what the restrictions are by comparing the left and right sides of Eq. (20.3) when $z(x, t) = \psi(x)\phi(t)$. The left side becomes

$$\left(\frac{\partial^2 z}{\partial x^2}\right)_t = \phi(t)\frac{d^2\psi(x)}{dx^2} \tag{20.5}$$

The right side becomes

$$\frac{1}{v^2}\left(\frac{\partial^2 z}{\partial t^2}\right)_x = \psi(x)\frac{1}{v^2}\frac{d^2\phi(t)}{dt^2} \tag{20.6}$$

From Eq. (20.3) we can see that the right sides of Eqs. (20.5) and (20.6) must be equal if the solution $z = \psi(x)\phi(t)$ is valid, so that

$$\frac{1}{\psi(x)}\frac{d^2\psi(x)}{dx^2} = \frac{1}{v^2}\frac{1}{\phi(t)}\frac{d^2\phi(t)}{dt^2} \tag{20.7}$$

On the left we have functions only of x and on the right functions only of t. The equality in Eq. (20.7) can hold for all values of x and t only under a very special condition, namely if each side is equal to *the same constant*. For later convenience, we call this constant $-\beta^2$. Thus the condition has been found for which the solution $z(x, t)$ $= \psi(x)\phi(t)$ is valid. Equating each side of Eq. (20.7) to the constant $-\beta^2$ gives two equations, one in the variable x and one in t.

$$\frac{d^2\psi(x)}{dx^2} + \beta^2\psi(x) = 0 \qquad (20.8)$$

$$\frac{d^2\phi(t)}{dt^2} + v^2\beta^2\phi(t) = 0 \qquad (20.9)$$

Equation (20.8) is the *time-independent classical wave equation* that we seek. One of its solutions is

$$\psi(x) = A \sin \beta x \qquad (20.10)$$

where A is a constant that appears in the process of integration to obtain the solution.

Example 20.2 Verify that Eq. (20.10) is a solution to Eq. (20.8).

Substitute Eq. (20.10) into Eq. (20.8).

The solution in Eq. (20.10) is equivalent to that in Eq. (20.2), the solution to the classical wave equation, provided we set $t = 0$. Hence $\beta = 2\pi/\lambda$, and Eq. (20.8) thus becomes

$$\frac{d^2\psi(x)}{dx^2} + \frac{4\pi^2}{\lambda^2}\psi(x) = 0 \qquad (20.11)$$

Equation (20.11) is the *time-independent* description of a wave according to classical mechanics. When we are concerned with time-dependent processes, we shall need a related equation that gives $\Psi(x, t)$.

20.4   The Schrödinger Wave Equation

Strictly speaking, the Schrödinger wave equation cannot be derived from any more fundamental set of postulates. It is itself a postulate, occupying in quantum mechanics a position analogous to Newton's law $\mathbf{F} = m\mathbf{a}$ in classical mechanics. Nevertheless, the theory did not spring fully grown from the mind of its creator one sunny day on the beach near Kiel.

A more likely development would have followed from the new physics of small particles that had evolved by 1926. Recall the important elements of the new physics. First, Planck and Einstein demonstrated the surprising particle–wave duality of electromagnetic radiation. Bohr's successful description of the hydrogen-atom spectrum then forced revision of the classical picture of the atom by introducing the concepts of quantized energy and momentum. In 1924, Broglie showed how the Bohr orbits of the hydrogen atom could be derived by assigning wave properties to the motion of an electron about a proton. He made his idea of wave behavior applicable to all particles with his equation $\lambda = h/p$, where p is the magnitude of the momentum of the particle (see page 57). This equation was confirmed experimentally in 1926 by the results of electron diffraction.

Schrödinger's wave equation can be obtained by an arranged marriage of elements of Newton's classical physics and Broglie's idea of particle–wave duality. This path to Schrödinger's equation is particularly instructive since it shows so clearly the central idea of Schrödinger's quantum mechanics, namely that the properties of systems with atomic or molecular dimensions are manifestations of wave behavior. The development begins with the standard one-dimensional differential equation of Newtonian physics for wave behavior, the classical wave equation, in the special time-independent form of Eq. (20.11). Many interesting properties of atoms and molecules are independent of time, for example, structures and energy levels.

The key step is to assume that atomic and molecular particles obey Eq. (20.11), with the essential proviso that the wavelength of the particle conforms to the Broglie equation $\lambda = h/p$. First, we write the Broglie wavelength in a more useful form, using $E = E_k + U$, where E is the total energy of the particle, and E_k and U are the kinetic and potential energies, respectively. Now $p^2 = (mv)^2 = 2mE_k = 2m(E - U)$, so that $\lambda = h/mv$ yields

$$\lambda^2 = h^2/2m(E - U) \qquad (20.12)$$

Substituting Eq. (20.12) into Eq. (20.11) and rearranging, we get

$$\left(\frac{-h^2}{8\pi^2 m}\right)\frac{d^2\psi(x)}{dx^2} + U\psi(x) = E\psi(x) \qquad (20.13)$$

This is the Schrödinger wave equation. It is not the most general Schrödinger equation since it is (1) time-independent and (2) in one dimension (x) only. We shall see that this equation or its three-dimensional version is the starting point for all calculations of the time-independent behavior of atoms or molecules.

The Schrödinger equation is often written in an abbreviated notation,

$$\hat{H}\psi(x) = E\psi(x) \qquad (20.14)$$

The symbol $\hat{H}$ signifies the *Hamiltonian operator*,

$$\hat{H} = \frac{-h^2}{8\pi^2 m}\frac{d^2}{dx^2} + U(x) \qquad (20.15)$$

A special symbol $\hat{H}$ rather than an unadorned H is used to emphasize that $\hat{H}$ is a mathematical instruction rather than an ordinary function. Being an instruction to carry out a mathematical operation, it is called an *operator*. In this case the instruction calls for certain operations on the wavefunction $\psi(x)$. If they are carried out, the result will tell us the total energy E of the system.

20.5 Translational Energy

The Schrödinger wave equation will be set up for various systems in the following chapters. It will be used to describe the hydrogen atom, atoms of larger size, and electrons in molecules. It will be used to analyze molecular motions such as rotations and vibrations, and it will appear in connection with the magnetic properties of molecules. The simplest application, however, concerns the translational motion of a particle (electron, atom, or molecule) through space. This problem will illustrate how wave mechanics works.

It is unlikely that a real particle ever travels indefinitely through space free of interactions with matter or force fields. Thus a truly free particle is an artificial concept.

Interactions create boundaries to the free translation of a particle, just as the walls of a cylinder confine a gas. Thus a useful quantum mechanical treatment considers a particle confined within a volume.

The problem becomes simpler if reduced to one dimension, yet it retains its usefulness. The one-dimensional system is still called a *particle in a box*, although it is more realistically envisaged as a particle, say an electron, in a wire. The particle is constrained to move only within a restricted region of space, from $x = 0$ to $x = a$, and its potential energy U within that region is constant. For convenience we set $U = 0$. The constraints on the particle are established by infinite potential-energy walls at either end of the box. Thus the potential energy of a particle in this box becomes

$$U = 0 \text{ (a constant)} \quad \text{for } 0 < x < a$$

$$U = \infty \quad\quad\quad\quad \text{for } x \leq 0 \text{ and } x \geq a$$

Since the particle needs infinite energy to escape from its box, this condition confines it within the limits $x = 0$ and $x = a$, and the Schrödinger equation then considers only this region of space. The potential-energy function for the one-dimensional box is shown in Fig. 20.2.

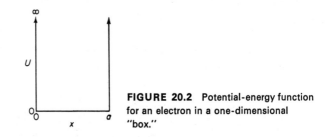

FIGURE 20.2 Potential-energy function for an electron in a one-dimensional "box."

If the Hamiltonian can be written, the Schrödinger wave equation follows immediately. Look at Eq. (20.15). You can see that the *Hamiltonian of one system is distinguished from that of another by its potential-energy term U(x)*. Thus to set up a problem, one first considers the potential energy. In this case $U(x) = 0$ everywhere that the particle can move, and the Hamiltonian operator (20.15) becomes $\hat{H} = (-h^2/8\pi^2 m)d^2/dx^2$. The Schrödinger wave equation $\hat{H}\psi = E\psi$ is then

$$\frac{-h^2}{8\pi^2 m} \frac{d^2\psi(x)}{dx^2} = E\psi(x) \tag{20.16}$$

The equation is rewritten as

$$\frac{d^2\psi(x)}{dx^2} + k^2\psi(x) = 0 \tag{20.17}$$

where

$$k^2 = \frac{8\pi^2 mE}{h^2} \tag{20.18}$$

Using this form, one can find that a general solution is

$$\psi(x) = c_1 \sin kx + c_2 \cos kx \tag{20.19}$$

where c_1 and c_2 are arbitrary constants.

Example 20.3 Show that Eq. (20.19) is a solution of Eq. (20.17) for all values of c_1 and c_2.

Calculate second derivative of $\psi(x)$ from Eq. (20.19) and substitute this and $\psi(x)$ into Eq. (20.17).

While Eq. (20.19) may seem to be the solution to our problem, there is still work to do before it is useful. The result in Eq. (20.19) shows a general property of all differential equations. Because of the constants of integration, c_1 and c_2, an infinite number of solutions exists; not all the solutions, however, are appropriate to the physical situation. The task now is to select those specific solutions that apply to the problem at hand. The constants in Eq. (20.19) are evaluated by using constraints that must apply to acceptable solutions $\psi(x)$. These constraints are known as *boundary conditions*. In order to learn what they are, we must now consider the significance and use of the wavefunctions $\psi(x)$.

20.6 Statistical Interpretation of Wavefunctions

Suppose that by some means the appropriate solutions $\psi(x)$ for a particular quantum mechanical problem are known. What do they mean? How can they tell us about the properties of the quantum mechanical system, be it a hydrogen atom, a Cl_2 molecule, or a particle in a box?

An answer was provided in 1926 by Max Born. He showed that a wavefunction can describe a particle in terms of probabilities. Specifically, *the probability $\mathcal{P}(x)dx$ of finding the particle in the region of space from x to $x + dx$ is given by $\psi^2(x)\,dx$*. In some cases $\psi(x)$ is a complex function, containing the imaginary number i. In such cases, the probability is $\psi^*\psi\,dx$, where ψ^* is the complex conjugate of ψ, formed by replacing i with $-i$ everywhere. For example, if $\psi = Ae^{-ix}$, $\psi^* = Ae^{ix}$

In the case of a particle in a box, Born's postulate tells us how to calculate the probability of finding the particle in any region of the box. It is a statistical answer to the question of position. According to wave mechanics, one can never expect to learn precisely where the particle is, but only the chance of finding it within a given region. Contrast this description with that of classical physics. To take an example from a classical wave, we can imagine a cork bobbing up and down at position x on an ocean wave. If the wave motion is described by Eq. (20.2) we can state precisely where the cork will be (up, down, or somewhere in between) by evaluating z for the given x and t. In a quantum mechanical system, such precision must be abandoned. One can specify only the *probability* of finding the particle in a given region. The experimental probabilities can be learned by making measurements on a large number of identical systems. The probability of finding the particle within a given range is then the number of systems having the particle within that range divided by the total number of systems.

Born's interpretation is consistent with the Heisenberg Uncertainty Principle (1927) which can be derived rigorously from the Schrödinger equation. The principle states that in a simultaneous measurement of momentum p and position x of a particle, the product of the uncertainties must always satisfy the relation,

$$\Delta p \cdot \Delta x \approx \frac{h}{2\pi} \tag{20.20}$$

This relation shows that if we were able to deduce the position of a particle *exactly* ($\Delta x \to 0$), then the uncertainty in its momentum would approach infinity ($\Delta p \to \infty$). This is an unacceptable physical situation. Hence quantum mechanics must use the language of probabilities.

Born's interpretation also emphasizes the wave nature of quantum mechanical descriptions of particles. The wavefunction can be considered as an amplitude function, just as in classical physics. In the case of a light wave, the intensity of light (or the energy of the electromagnetic field) is proportional to the square of the wave amplitude. In terms of light quanta or photons $h\nu$, the greater the amplitude of a light wave in any region, the greater is the probability of a photon being within that region. Born's interpretation of the quantum mechanical wavefunction is analogous. The highest probabilities of finding particles occur in regions of space where the wavefunction has its largest amplitude.

20.7 Further Characteristics of Wavefunctions

From Born's postulate we can immediately write the probability for finding a particle, described by $\psi(x)$, in a given interval of space, say between $x = c$ and $x = d$. This probability is the sum of the probabilities $\mathcal{P}(x)\,dx$ over all parts of that region. The sum is equivalent to the integral

$$\mathcal{P}(c \leq x \leq d) = \int_c^d \mathcal{P}(x)\,dx = \int_c^d \psi^2(x)\,dx \tag{20.21}$$

Borns' postulate in the form of Eq. (20.21) imposes other important requirements on the wavefunction $\psi(x)$. The first is called *normalization*. If the limits of Eq. (20.21) are extended to $\pm\infty$, the integral must equal unity since the particle must exist *somewhere* in that interval if it is to exist at all:

$$\int_{-\infty}^{\infty} \psi^2(x)\,dx = 1 \qquad \text{or} \qquad \int_{-\infty}^{\infty} \psi^*(x)\psi(x)\,dx = 1 \tag{20.22}$$

Wavefunctions meeting this requirement are said to be *normalized*. Other mathematical requirements must also be met by physically acceptable wavefunctions:

1. *$\psi(x)$ and $d\psi(x)/dx$ must be everywhere finite.*
2. *$\psi(x)$ and $d\psi(x)/dx$ must be everywhere single-valued.*
3. *$\psi(x)$ and $d\psi(x)/dx$ must be everywhere continuous.*

Figure 20.3 shows examples of wavefunctions that do and do not meet these requirements.

This function becomes infinite at a certain value of x.

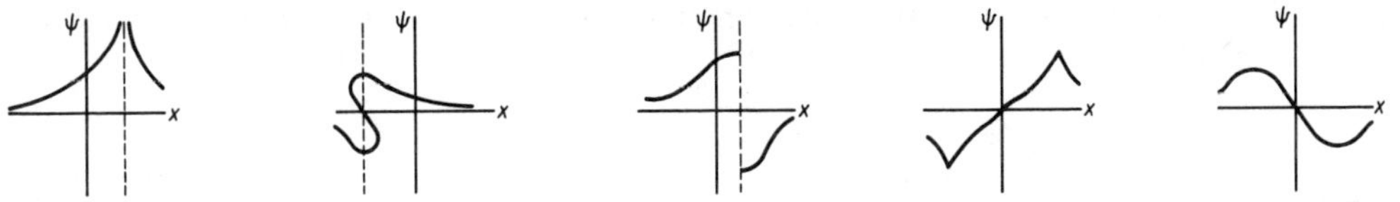

This function is not single valued at every position over the allowed range of x.

This function is not everywhere continuous.

This function has a discontinuous derivative at each cusp.

This meets all the requirements and hence is acceptable.

FIGURE 20.3 Examples of wavefunctions ψ that satisfy or fail to satisfy requirements for a physically acceptable ψ.

You can understand these conditions on $\psi(x)$ on the basis of Born's postulate. The wavefunction cannot become infinite since that would correspond to certainty of a particle's being at some definite point. If the wavefunction were double-valued, there would be two probabilities for the same position, which would be physically meaningless. In a similar spirit, a discontinuous wavefunction cannot correspond to a physically acceptable situation because it leads to a first derivative (and hence to a momentum) that is not finite (see Section 20.13). Analogous arguments concerning particle momentum establish the other requirements on the derivatives $d\psi(x)/dx$. It is essential to keep Born's postulate and the three requirements on wavefunctions in mind when working out solutions to Schrödinger's wave equation. They are the keys to the selection of physically correct solutions.

20.8 Orthogonality of Wavefunctions

Equation (20.22) expresses the condition of *normalization,* a property of all proper wavefunctions for quantized systems. There is another property, *orthogonality,* that is often useful. Two different wavefunctions ψ_n and ψ_m are said to be orthogonal if

$$\int_{-\infty}^{\infty} \psi_m \psi_n \, dx = 0 \quad \text{or} \quad \int_{-\infty}^{+\infty} \psi_m^* \psi_n \, dx = 0 \qquad (20.23)$$

Equation (20.23) is a general property of wavefunctions. Wavefunctions that are solutions of a given Schrödinger equation are usually orthogonal to one another.

Wavefunctions that are both orthogonal and normalized are called *ortho-normal.* They fulfill the condition

$$\int_{-\infty}^{\infty} \psi_m \psi_n \, dx = \delta_{mn} \quad \text{or} \quad \int_{-\infty}^{\infty} \psi_m^* \psi_n \, dx = \delta_{mn} \qquad (20.24)$$

where the *Kronecker delta* δ_{mn} is a function with two values, $\delta_{mn} = 1$ when $m = n$, $\delta_{mn} = 0$ when $m \neq n$.

An exception to the orthogonality rule occurs when two or more wavefunctions correspond to the same energy level. Such levels are said to be *degenerate.* Wavefunctions for degenerate levels are not always orthogonal to one another. However, they are orthogonal to all other wavefunctions that are solutions of the same Schrödinger equation.

20.9 Translational Wavefunctions

With Eq. (20.19) we were left with an infinite number of solutions to the wave equation. The task is now to select the physically acceptable solutions from that set. Born's postulate and the constraints on $\psi(x)$ as given in the preceding section enable the selection to be made.

First note that Born's postulate implies $\psi(x) = 0$ everywhere outside the box since there must be zero probability of finding the particle in those regions. This point may seem trivial considering that the problem was originally set up for only the region of the box. It becomes significant, however, when connected with the requirement that $\psi(x)$ be a continuous function. Since $\psi(x) = 0$ outside the box, $\psi(x)$ must go smoothly to zero at the walls. Thus the value of $\psi(x)$ is known at two useful positions:

$$\psi(x) = 0 \quad \text{at } x = 0 \quad \text{and} \quad \psi(x) = 0 \quad \text{at } x = a$$

These constraints on $\psi(x)$ are the *boundary conditions*.

The boundary conditions can be used to select the physically significant wavefunctions from the set in Eq. (20.19). Consider first the constraint $\psi = 0$ at $x = 0$. Equation (20.19) becomes $\psi = 0 = c_1 \sin 0 + c_2 \cos 0$. Since $\sin 0 = 0$ and $\cos 0 = 1$, $c_2 = 0$, and we are left with

$$\psi = c_1 \sin kx \tag{20.25}$$

as a reduced set of physically acceptable solutions.

The second constraint, $\psi = 0$ at $x = a$, now gives from Eq. (20.25)

$$\psi = 0 = c_1 \sin ka \tag{20.26}$$

Equation (20.26) can be satisfied only if

$$ka = n\pi \tag{20.27}$$

with $n = 0, 1, 2, 3$, etc., but $n = 0$ is excluded. [Why?] Substituting $k = n\pi/a$ into Eq. (20.25) gives the physically acceptable solutions,

$$\psi = c_1 \sin \left(\frac{n\pi}{a} x\right) \tag{20.28}$$

Finally, the constant c_1 needs to be evaluated. Evaluation is made with the normalization condition, Eq. (20.22),

$$1 = \int_{-\infty}^{\infty} c_1^2 \sin^2 \left(\frac{n\pi}{a} x\right) dx = c_1^2 \int_0^a \sin^2 \left(\frac{n\pi}{a} x\right) dx \tag{20.29}$$

The integration limits have been changed in Eq. (20.29). The particle can exist only between the values $0 < x < a$, so that the probability of finding the particle in this region must be unity. The integral is $a/2$, so that $c_1^2(a/2) = 1$, giving $c_1 = (2/a)^{1/2}$.

Thus the normalized wavefunctions for the particle in a one-dimensional box extending from $x = 0$ to $x = a$ are

$$\psi_n = \left(\frac{2}{a}\right)^{1/2} \sin\left(\frac{n\pi}{a}x\right) \qquad (20.30)$$

The boundary conditions, $\psi = 0$ at $x = 0$ and $x = a$, used to select solutions to the wave equation, give rise naturally to the integers $n = 1, 2, 3, \ldots$. The integer n is the *quantum number* of the particle, and it is customary to label specific solutions with the notation $\psi_n(x)$, indicating the quantum number. A particle represented by a solution ψ_n is said to be in the *state* ψ_n.

Wavefunctions for several states with low quantum numbers are shown in Fig. 20.4(a). The occurrence of nodes (positions where $\psi_n = 0$), the number of which increases with n, and regions of positive and negative values of ψ_n are common characteristics of quantum mechanical wavefunctions. Typically, the wavefunction with the lowest possible quantum number has no node.

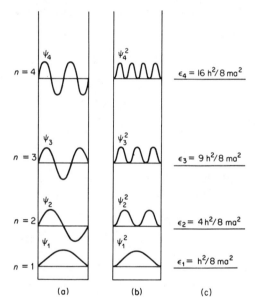

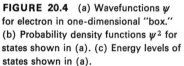

FIGURE 20.4 (a) Wavefunctions ψ for electron in one-dimensional "box." (b) Probability density functions ψ^2 for states shown in (a). (c) Energy levels of states shown in (a).

$n = 4$ ψ_4 ψ_4^2 $\epsilon_4 = 16\,h^2/8\,ma^2$

$n = 3$ ψ_3 ψ_3^2 $\epsilon_3 = 9\,h^2/8\,ma^2$

$n = 2$ ψ_2 ψ_2^2 $\epsilon_2 = 4\,h^2/8\,ma^2$

ψ_1 ψ_1^2

$n = 1$ $\epsilon_1 = h^2/8\,ma^2$

(a) (b) (c)

Particles that move through extended regions of constant potential energy have wavefunctions with clearly defined wavelengths such as those seen in Fig. 20.4(a). The wavelengths become shorter with larger values of n. It is more useful, however, to stress the correspondence between wavelength and the kinetic energy, which increases with n. As a general rule of quantum mechanics, *high kinetic energies are associated with short wavelengths*. These observations apply only to systems with constant or nearly constant potential energy.

Examples of the function $\psi^2(x)$ are plotted in Fig. 20.4(b). This function can be used to calculate the probability of finding the particle in any region of the one-dimensional box.

Example 20.4 Show that the wavefunctions in Eq. (20.30) are orthogonal for any pair of different values of n.

Suppose that $n = l, m$, two different integers.

$$\int_0^a \psi_l \psi_m \, dx = \frac{2}{a} \int_0^a \sin\left(\frac{l\pi x}{a}\right) \sin\left(\frac{m\pi x}{a}\right) dx$$

$$= \frac{2}{a} \int_0^a \frac{1}{2}\left[\cos(l-m)\frac{\pi x}{a} - \cos(l+m)\frac{\pi x}{a}\right] dx$$

$$= \frac{1}{\pi}\left[\frac{\sin(l-m)\frac{\pi x}{a}}{l-m} - \frac{\sin(l+m)\frac{\pi x}{a}}{l+m}\right]_0^a = 0$$

Example 20.5 (a) Calculate the wavelength for each quantum number in Fig. 20.4. (b) Show that the calculated wavelengths obey the Broglie relationship $\lambda = h/p$.

(a)

n	1	2	3	4
λ	$2a$	a	$2a/3$	$a/2$

(b) $\epsilon = p^2/2m$, $p = \sqrt{2m\epsilon}$, $n = 1$, $\epsilon_1 = h^2/8ma^2$, $p = \sqrt{h^2/4a^2} = h/2a$, $\lambda = h/p = 2a$, etc.

Example 20.6 Calculate the probability of finding the particle between $0.49a$ and $0.51a$ for the states ψ_1 and ψ_2.

From Eqs. (20.21) and (20.30),

$$\mathcal{P} = \int_{0.49a}^{0.51a} \psi^2 \, dx = \int_{0.49a}^{0.51a} \left(\frac{2}{a}\right) \sin^2 \frac{n\pi x}{a} \, dx$$

$$= \frac{2}{n\pi}\left[\frac{n\pi x}{2a} - \frac{1}{4}\sin\frac{2n\pi x}{a}\right]_{0.49a}^{0.51a}$$

$$n = 1, \mathcal{P}_1 = 0.0399; \quad n = 2, \mathcal{P}_2 = 0.0001.$$

20.10 Quantization of Energy

The restriction of the particle to discrete states $\psi_1, \psi_2, \ldots$ designated by the quantum number n leads to the quantization of energy. Combining Eq. (20.30), in which the quantum number appears, with Eq. (20.18), which defines k, gives

$$k^2 = 8\pi^2 mE/h^2 = n^2\pi^2/a^2 \tag{20.31}$$

Thus the allowed energy levels E_n of the particle in the box are

$$E_n = n^2h^2/8ma^2 \qquad (20.32)$$

Only certain energies are allowed for a particle of given mass m in a box of given length a. The energy is quantized. Some of the discrete energy levels as given by Eq. (20.32) are shown in Fig. 20.4(c). These translational energy levels were discussed in Section 4.4. This section should be reviewed, particularly with respect to the size of a translational energy quantum.

One should note especially how the separation of energy levels depends on the box length a. As space available to a particle increases, energy quanta become smaller and energy levels move closer together. As the box length becomes very large, quantization for all practical purposes disappears. There is a smooth transition from quantum behavior to classical behavior as box length increases. This result illustrates a general principle of quantum mechanics: Quantization results from restriction of the spatial regions that can be occupied by a particle.

20.11 Zero-point Energy and the Uncertainty Principle

The lowest energy state of a quantum mechanical system is called its *ground state*. Typically, the energy of this state, called the *zero-point energy*, is not zero. For example, the lowest allowed translational energy according to Eq. (20.32) is $E_1 = h^2/8ma^2$.

The existence of zero-point energy is a purely quantum-mechanical effect, detectable only for small masses. No experience in the everyday world of large particles would have led us to suspect such behavior—golf balls can be completely at rest with zero kinetic energy. Zero-point energy is a consequence of the uncertainty principle. Suppose that a particle could have $E = 0$. Its momentum $p = (2mE)^{1/2}$ would also be exactly zero, so that the uncertainty in momentum would be $\Delta p = 0$. But with $\Delta p = 0$, Eq. (20.20) indicates that $\Delta x = \infty$, so that the particle could not be located within its box. Thus a particle in a box must have some $E > 0$ even in its lowest allowed energy state.

20.12 The Free Particle

If a particle is truly free to move without boundaries, quantization disappears even though the particle motion is still described by a wave equation. For the free particle there are no constraints on ψ to establish boundary conditions. One of the constants c_1 or c_2 of Eq. (20.19) can be arbitrarily set to zero giving

$$\psi = c_1 \sin kx = c_1 \sin (8\pi^2 mE/h^2)^{1/2}x \qquad (20.33)$$

Further specification, however, is not possible. Equation (20.33) therefore gives the wavefunction for a free particle. Since there are no constraints on k, E may assume any nonzero positive value. Thus a continuous range of energies without quantization is available to such particles.

All the physical properties* of a particle moving in one dimension that do not depend upon time can be derived from the time independent (*stationary-state*) wavefunction $\psi(x)$ of the system. One example has already been discussed. Born's postulate $\mathcal{P}(x)\,dx = \psi^*(x)\psi(x)\,dx$ allows us to calculate the probability of finding the particle in any region of one-dimensional space. If we make an experiment to measure the position x of the particle, $\psi(x)$ does not permit us to calculate the result of that particular experiment. If, however, we make a large number of experiments, $\psi(x)$ permits us to calculate the mean (average) value of the position x as found in these replicate experiments. This mean value, denoted $\langle x \rangle$, is called the *expectation value* of the observable x. We can write

$$\langle x \rangle = \int \psi^* x \psi \, dx \qquad = \int x \psi^* \psi \, dx = \int x \mathcal{P}(x) \, dx$$

which is exactly the result stated by the Born postulate.

How do we calculate other physical quantities for the particle? Suppose we wish to calculate the expectation value $\langle G \rangle$ of some observable (such as momentum) which would be obtained experimentally as a result of repeated measurements of G for the system in the state $\psi(x)$. A basic postulate of quantum mechanics gives

$$\langle G \rangle = \int_{-\infty}^{\infty} \psi^* \hat{G} \, \psi \, dx \qquad (20.34)$$

Here $\hat{G}$ is the operator for the quantity G. Note that in general $\hat{G}$ is not a simple factor like x and thus Eq. (20.34) *cannot* be rearranged to give $\int \hat{G}\psi^*\psi \, dx$, but $\hat{G}$ must be allowed to operate on ψ as indicated.

In quantum mechanics, *observables* are represented by operators. The *operator* $\hat{G}$ corresponds to the observable G. The quantity of interest is first written as a function of position variables, x, y, z, and/or momentum components p_x, p_y, p_z. The operator is then formed according to the two specific rules in Table 20.1. Some examples show how it works.

TABLE 20.1
QUANTUM MECHANICAL OPERATORS

Classical variable	Quantum mechanical operator	Expression for operator	Operation
Position x	$\hat{x}$	x	Multiply by x
Momentum p_x	$\hat{p}_x$	$\dfrac{-ih}{2\pi}\dfrac{d}{dx}$	Take derivative with respect to x and multiply result by $-ih/2\pi$

*Effects due to spin are neglected here. See Section 21.9.

Example 20.7 What is the expectation value of the momentum p_x for a particle in the ground state of a one-dimensional box?

The wavefunction ψ is given in Eq. (20.30). According to Eq. (20.34) and Table 20.1, $\langle p_x \rangle$ is given by

$$\langle p_x \rangle = \int_0^a \left[\left(\frac{2}{a}\right)^{1/2} \sin\left(\frac{\pi x}{a}\right) \right]\left(\frac{-ih}{2\pi}\frac{d}{dx}\right)\left[\left(\frac{2}{a}\right)^{1/2} \sin\left(\frac{\pi x}{a}\right)\right] dx$$

$$= \frac{-ih}{2\pi}\left(\frac{2}{a}\right)\int_0^a \sin\left(\frac{\pi x}{a}\right)\left(\frac{\pi}{a}\right)\cos\left(\frac{\pi x}{a}\right) dx$$

$$= \frac{-ih}{2\pi}\left(\frac{1}{a}\right) \sin^2\left(\frac{\pi x}{a}\right)\Big|_0^a = 0$$

The result is not surprising since positive values of momentum, corresponding to motion in the $+x$ direction, are just as likely as negative values. (Note that the limits 0 and a are used instead of $\pm\infty$ since the wavefunction is known to be zero outside the box.)

Example 20.8 What is the expectation value of p_x^2, a quantity that is always positive, for the particle in a one-dimensional box in the state with $n = 1$?

The operator for p_x^2 is $(\hat{p}_x)(\hat{p}_x)$.

$$\hat{p}_x^2 = (\hat{p}_x)(\hat{p}_x) = \left(-\frac{ih}{2\pi}\frac{d}{dx}\right)\left(-\frac{ih}{2\pi}\frac{d}{dx}\right) = \frac{-h^2}{4\pi^2}\frac{d^2}{dx^2}$$

From Eqs. (20.30) and (20.34),

$$\langle p_x^2 \rangle = \frac{2}{a}\int_0^a \sin\left(\frac{\pi x}{a}\right)\frac{-h^2}{4\pi^2}\frac{d^2}{dx^2}\left[\sin\left(\frac{\pi x}{a}\right)\right] dx$$

$$= \frac{-h^2}{4\pi^2}\left(\frac{2}{a}\right)\int_0^a \sin\left(\frac{\pi x}{a}\right)\left[(-)\left(\frac{\pi}{a}\right)^2 \sin\left(\frac{\pi x}{a}\right)\right] dx$$

$$= \frac{h^2}{2a^3}\int_0^a \sin^2\left(\frac{\pi x}{a}\right) dx = \frac{h^2}{2a^3}\frac{a}{2} = \frac{h^2}{4a^2}$$

This result can be checked with the formula (20.32) for the energy levels

$$p_x^2 = 2mE = 2m\left(\frac{h^2}{8ma^2}\right) = \frac{h^2}{4a^2}$$

20.14   Operators

The concept of an *operator* is fundamental in quantum mechanics. As the name indicates, an operator is an instruction to carry out a mathematical operation on a function, which is called the *operand*. For example, in the expression $(d/dx)f(x)$ the operator is d/dx and the operand is $f(x)$. If $f(x) = x^2$,

$$\frac{d}{dx}f(x) = \frac{d}{dx}x^2 = 2x$$

In the expression $x \cdot f(x)$, we can consider x to be the operator that tells us to multiply $f(x)$ by x.

We can write the product of two operators, $\hat{O}_1$ and $\hat{O}_2$ as $\hat{O}' = \hat{O}_2\hat{O}_1$ or $\hat{O} = \hat{O}_1\hat{O}_2$. The product operator $\hat{O}_1\hat{O}_2$ tells us to perform first the operation $\hat{O}_2$ on the operand, and then to perform the operation $\hat{O}_1$ on the result. Consider, for instance, $\hat{O}_2 = d/dx$, $\hat{O}_1 = x$, $f(x) = x^2$. Then,

$$\hat{O}_1\hat{O}_2f(x) = x\frac{d}{dx}x^2 = 2x^2$$

It is important to note that

$$\hat{O}_2\hat{O}_1f(x) = \frac{d}{dx}x \cdot x^2 = 3x^2$$

Thus $\hat{O}_2\hat{O}_1 \neq \hat{O}_1\hat{O}_2$. The operators x and d/dx do not *commute*—the order in which they appear makes a difference in the product. Some pairs of operators commute and some do not.

An operator $\hat{O}$ is said to be *linear* when for any two functions f and g,

$$\hat{O}(\lambda f + \mu g) = \lambda(\hat{O}f) + \mu(\hat{O}g)$$

where λ and μ are arbitrary numbers, either complex or real. For example, d^2/dx^2 is a linear operator, but an operator SQ, which gives the command "take the square of following function," is not linear.

If, for a function f and an operator $\hat{O}$, we have

$$\hat{O}f = cf \tag{20.35}$$

where c is a number, then f is called an *eigenfunction* of the operator $\hat{O}$ and c is called an *eigenvalue* of the operator $\hat{O}$.

The terms "eigenfunction" and "eigenvalue" are commonly associated with solutions of differential equations with boundary-value conditions. If $\hat{O}$ is a differential operator, Eq. (20.35) is an expression for a differential equation in operator form, and the problem of finding the eigenfunctions and eigenvalues in Eq. (20.35) is mathematically equivalent to the solution of the differential equation and boundary-value problem. When the operator $\hat{O}$ associated with a certain observable O has an eigenvalue c, a measurement of the observable is predicted to give exactly the value c. [Does $\hat{x}$ have an eigenvalue c for the particle-in-a-box problem?]

20.15 The Hamiltonian Operator

The Hamiltonian operator $\hat{H}$ was introduced in Section 20.4 as part of the shorthand notation for the Schrödinger equation $\hat{H}\psi = E\psi$. We shall show that the rules of Table 20.1 make $\hat{H}$ *the time-independent quantum mechanical operator for energy.*

The Hamiltonian has its origin in classical mechanics where the equations describing the motion and energy of a particle can be derived from Newton's Second Law, $\mathbf{F} = m\mathbf{a}$. An alternative formulation of this law was developed by the

Irish mathematician William Hamilton in 1834 and is based on the total energy of a system rather than on forces. The total energy of a system is the sum of its kinetic energy E_k and potential energy U. The Hamiltonian H of classical physics is simply this sum: $H \equiv E_k + U$. For a one-dimensional system with one particle, $H = \frac{1}{2}mv^2 + U(x)$. The potential energy is a function only of position for all problems treated in this text. It is often more convenient to write H in terms of momentum $p_x = mv$ rather than velocity v, giving

$$H = \frac{p_x^2}{2m} + U(x) \tag{20.36}$$

Transformation of the classical Hamiltonian H of Eq. (20.36) into the quantum mechanical operator $\hat{H}$ now follows directly from Table 20.1.

$$\hat{H} = \left(\frac{1}{2m}\right)\left(\frac{-ih}{2\pi}\frac{d}{dx}\right)\left(\frac{-ih}{2\pi}\frac{d}{dx}\right) + U(x)$$

$$\hat{H} = \frac{-h^2}{8\pi^2 m}\frac{d^2}{dx^2} + U(x) \tag{20.37}$$

We now see that the operator $\hat{H}$, introduced in Eq. (20.15), has a simple origin. It is the energy prepared according to the recipe of quantum mechanics.

You can see in another way that $\hat{H}$ is the energy operator by using it to calculate the energy of a particle whose behavior is described by some wavefunction ψ_n. According to Eq. (20.34),

$$\langle H \rangle = \int_{-\infty}^{\infty} \psi_n \hat{H} \psi_n \, dx \tag{20.38}$$

The Schrödinger equation (20.14), $\hat{H}\psi_n = E_n\psi_n$, gives

$$\langle H \rangle = \int_{-\infty}^{\infty} \psi_n E_n \psi_n \, dx = E_n \int_{-\infty}^{\infty} \psi_n \psi_n \, dx \tag{20.39}$$

Since the integral in Eq. (20.39) is unity for normalized wavefunctions, $\langle H \rangle = E_n$. The energy E_n is the eigenvalue obtained when $\hat{H}$ operates on the eigenfunction ψ_n.

20.16 Free Electron Model for Conjugated Dyes

The simple particle-in-a-box theory can be applied to some interesting actual molecules with conjugated double bonds and hence mobile π electrons. Consider, for example, the polymethine dye,

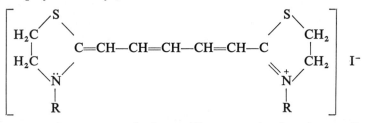

One of two resonating structures is shown. [Can you write the other one?] All bonds along the chain joining the two N atoms are equivalent. They are similar to C—C

bonds in benzene. The five π electrons on the carbons and the three extra electrons on the two nitrogens form a π electron system of eight electrons.

We can apply the particle-in-a-box theory to calculate the wavelength of the principal absorption band for these molecules in dilute solution. The energy levels are given by Eq. (20.32) as $E_n = n^2h^2/8ma^2$. Each level can hold two electrons of opposite spin in accord with the Pauli Principle (page 473). The pattern of levels is shown in Fig. 20.5. Thus the eight π electrons fill the four lowest levels. The absorp-

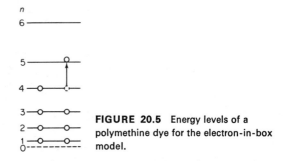

FIGURE 20.5 Energy levels of a polymethine dye for the electron-in-box model.

tion band at longest wavelength (lowest energy) arises from the excitation of an electron from the topmost filled level to the lowest empty level. For the general case of $N \pi$ electrons this transfer would be from level $N/2$ to $(N/2) + 1$. The energy difference is therefore

$$\Delta E = \frac{h^2}{8ma^2}\left[\left(\frac{N}{2} + 1\right)^2 - \left(\frac{N}{2}\right)^2\right] = \frac{h^2}{8ma^2}(N + 1) \qquad (20.40)$$

Since $hc/\lambda = \Delta E$, the wavelength of the light absorbed is

$$\lambda = \frac{8mc}{h}\frac{a^2}{N + 1} \qquad (20.41)$$

We can estimate a, the length of the "molecular box," from the average bond length along the chain, $\ell = 139$ pm. If we assume that the mobile electrons can move into the region beyond the N atoms to an extent of one extra bond length on each side, $a = 8\ell = 1112$ pm, and

$$\lambda = \frac{8(9.11 \times 10^{-31}\text{ kg})(3.00 \times 10^8\text{ m s}^{-1})}{6.63 \times 10^{-34}\text{ J s}}\frac{(1112 \times 10^{-12}\text{ m}^2)^2}{9} = 447\text{ nm}$$

The experimental value for this dye is $\lambda = 445$ nm. We see that electrons in molecules of this kind really do behave like waves in a box. We may have cheated a little to get such exact agreement between theory and experiment, since the length of the box beyond the N conjugated atoms could have been estimated in various ways. [What is the calculated λ if electrons cannot move beyond the N atoms?]

20.17 The Box in Three Dimensions

We can easily extend the results for one dimension to the case of a three-dimensional box in the form of a parallelepiped of sides a, b, c. The potential U equals zero everywhere within the box, and the Schrödinger equation is

$$\nabla^2 \psi \equiv \frac{\partial^2 \psi}{\partial x^2} + \frac{\partial^2 \psi}{\partial y^2} + \frac{\partial^2 \psi}{\partial z^2} = -\frac{8\pi^2 m}{h^2} E\psi \qquad (20.42)$$

The variables x, y, z can be separated by the substitution

$$\psi(x, y, z) = X(x)Y(y)Z(z) \qquad (20.43)$$

which gives

$$\frac{1}{X}\frac{\partial^2 X}{\partial x^2} + \frac{1}{Y}\frac{\partial^2 Y}{\partial y^2} + \frac{1}{Z}\frac{\partial^2 Z}{\partial z^2} = -\frac{8\pi^2 mE}{h^2} \qquad (20.44)$$

Since this equation must be true for all values of the independent variables x, y, z, we can conclude that each term on the left must equal some constant. Thus we can write

$$\frac{1}{X}\frac{d^2 X}{dx^2} = -k_x^2 \qquad \frac{1}{Y}\frac{d^2 Y}{dy^2} = -k_y^2 \qquad \frac{1}{Z}\frac{d^2 Z}{dz^2} = -k_z^2 \qquad (20.45)$$

with $k_x^2 + k_y^2 + k_z^2 = 8\pi^2 mE/h^2 = k^2$.

The equations (20.45) are similar to Eq. (20.17) previously solved for the one-dimensional case.

Instead of Eq. (20.30), the wavefunction becomes

$$\psi_{n_1 n_2 n_3}(x, y, z) = (8/abc)^{1/2} \sin\frac{n_1 \pi x}{a} \sin\frac{n_2 \pi y}{b} \sin\frac{n_3 \pi z}{c} \qquad (20.46)$$

specified by a set of three quantum numbers, n_1, n_2, n_3. The allowed energy levels are

$$E_{n_1, n_2, n_3} = \frac{h^2}{8m}\left(\frac{n_1^2}{a^2} + \frac{n_2^2}{b^2} + \frac{n_3^2}{c^2}\right) \qquad (20.47)$$

The eigenvalues E for this three-dimensional problem depend on three distinct integral quantum numbers, n_1, n_2, n_3. If the box is cubical with side a, Eq. (20.47) becomes

$$E_{n_1, n_2, n_3} = \frac{h^2}{8ma^2}(n_1^2 + n_2^2 + n_3^2) \qquad (20.48)$$

An important new feature now appears, namely the occurrence of more than one distinct wavefunction corresponding to the same value for the energy. For example, the three eigenfunctions, ψ_{121}, ψ_{211}, and ψ_{112}, correspond to different distributions in space, but they all have the same energy, $E = 6h^2/8ma^2$. This energy level is said to have a threefold *degeneracy*. In any statistical treatment of energy levels of the system, this level has a statistical weight of $g_k = 3$.

20.18 The Tunnel Effect

Take a baseball, place it in a strong box, and nail the lid down tightly. Any proper Newtonian will assure us that the ball is going to stay in the box until someone takes it out. There is no probability that the ball will be found on Monday inside the box and on Tuesday rolling along outside it. Yet if we transfer our attention from a baseball in a box to an electron in a box, quantum mechanics predicts exactly this unexpected behavior.

Consider in Fig. 20.6 a particle moving in a one-dimensional box with a kinetic energy E_k. It is confined by a potential-energy well; since the potential-energy barrier

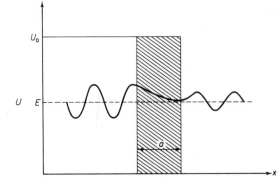

FIGURE 20.6 Model of electron in a "box" with a finite potential-energy wall of finite thickness. The tunnel effect.

is higher than the available kinetic energy, the classical probability of escape is nil. Quantum mechanics tells a different story. The wave equation (20.13) for the region of constant potential energy U is

$$(d^2\psi/dx^2) + (8\pi^2 m/h^2)(E - U)\psi = 0$$

This equation has the general solution

$$\psi = A \exp[(2\pi ix/h)\sqrt{2m(E - U)}]$$

[Please prove by substitution.]
In the region within the box $E > U$ and this solution is simply the familiar sine or cosine wave of Eq. (20.19) written in the complex exponential form: $e^{i\theta} = \cos\theta + i\sin\theta$. In the region within the potential-energy barrier, however, $U_0 > E$, so that the expression under the square-root sign is negative. One can therefore multiply out a $\sqrt{-1}$ term, obtaining

$$\psi = A \exp[-(2\pi x/h)\sqrt{2m(U_0 - E)}] \tag{20.49}$$

This exponential function describes the behavior of the wave function within the barrier. According to wave mechanics the probability of finding an electron in this region of negative energy is clearly not zero, but a certain positive number, which falls off exponentially with the distance of penetration within the barrier. The behavior of the wavefunction is shown in Fig. 20.6. As long as the barrier is neither infinitely high nor infinitely wide, there is always a certain probability that electrons (or particles in general) will leak through. This leakage is called the *tunnel effect*.

Many important phenomena involve the tunnel effect. An everyday example occurs when an electric circuit is completed by placing two metallic conductors in contact. The current of electrons flows freely across the contact even though the wires are covered by a thin insulating layer of oxide. The electrons readily tunnel through such a barrier; they do not need to have enough energy to surmount the barrier height. Many electrode processes also involve tunneling of electrons through potential barriers at an electrode surface. Massive particles usually cannot tunnel through barriers but protons can do so in some instances, leading to anomalous isotope effects for H^+ compared to D^+ reactions.*

*The theory and many fascinating applications are discussed by R. P. Bell in *The Tunnel Effect in Chemistry* (London: Chapman and Hall, 1980).

Problems

1. What is the wavelength of radiation absorbed when an electron in a one-dimensional "box" 100 pm long makes a transition from the ground state to the first excited state? What would be the wavelength for an α-particle in the same "box"?

2. The number of solar photons incident on 1 m^2 of the earth's surface rarely exceeds one mole per hour. If the light absorbed has $\lambda = 400$ nm, what is the maximum energy per hour that could be trapped for solar generation of power? If the conversion efficiency is 20%, what collection area would be required for a 1000-MW power station?

3. The potassium cathode surface of a photoelectric cell has a work function of 2.26 eV. What is the maximum speed of the electrons emitted when light of $\lambda = 350$ nm is incident on the cell?

4. Calculate the Broglie wavelength of:
   (a) Electrons accelerated to 1000 kV in an electron microscope.
   (b) Thermal neutrons emitted from a nuclear reactor at 300 K. (Take average energy $= \frac{1}{2}kT$.)
   (c) As in part (b) but at 10 K
   (d) An argon atom moving at 1.0 m s^{-1}
   (e) A snail of mass 1 g moving at 10^{-10} m s^{-1}

5. From the Heisenberg Principle [Eq. (20.20)] estimate the uncertainty in kinetic energy of an electron confined in a one-dimensional length of 10.0 nm and compare this value to its calculated energy in the lowest state.

6. An alternative form of the Heisenberg uncertainty principle is $\delta E \, \delta t \simeq h/2\pi$. When an electron makes a transition from a higher state to a lower state, a quantum of energy $h\nu$ is emitted. If the excited state has a lifetime of 10^{-9}s, what is the uncertainty in the frequency ν of the emitted quantum? This uncertainty causes a broadening of an observed spectral line. What is the line width (in cm^{-1}) in this case? How does it compare to the wave number (cm^{-1}) of a line in the visible region of the spectrum (where such electronic transitions often appear)?

7. What is (a) the momentum and (b) the mass of a photon of wavelength 100 nm; 100 μm; 100 mm?

8. For the particle in a one-dimensional box discussed in Section 20.9, show that the probability of its being found in the region $a/4 \leq x \leq a/2$ is

$$\frac{1}{4}\left[1 + \frac{2\sin(n\pi/2)}{n\pi}\right]$$

9. Which of the following functions are eigenfunctions of the operators d/dx and/or d^2/dx^2? (a) $\cos kx$; (b) e^{-bx}; (c) e^{ikx}; (d) e^{-kx^2}.

10. What is the degeneracy of the fourth energy level of a particle in a two-dimensional box?

11. Calculate the probability $\mathcal{P}$ that the particle be found in the middle third of the box in Problem 8 and discuss the behavior of $\mathcal{P}$ as n increases.

12. The absolute threshold of the dark-adapted human eye for perception of light at 510 nm was measured as 3.5×10^{-17} J at the surface of the cornea. To how many quanta does this correspond?

13. The diameter of a typical small nucleus is 10^{-15} m. Suppose that an electron was placed in a one-dimensional infinite potential well 10^{-15} m wide. What would be the lowest energy level? From this result, what would you conclude about the existence of electrons within the nucleus? What about the emission of β-rays from nuclei?

14. For a particle of mass m rotating in a circle of radius R, the kinetic energy is $L^2/2I$, where L is the magnitude of angular momentum and $I = mR^2$ is the moment of inertia. The operator for L^2 is $(h^2/4\pi^2)\partial^2/\partial\phi^2$, so that the Schrödinger equation $\hat{H}\psi = E\psi$ becomes $-(h^2/8\pi^2mR^2)\partial^2\psi/\partial\phi^2 = E\psi$. Solve this equation and determine the allowed energy levels. [Boundary condition $\psi(\phi) = \psi(\phi + 2\pi)$.]

15. Show that the ψ in Eq. (20.30) for the one-dimensional particle in a box is not an eigenfunction for the momentum p_x.

16. Consider an electron in a potential well of depth 10 eV approaching a barrier width that extends from $x = 0$ to $x = a$. If the kinetic energy of the electron is 9.5 eV and $a = 1.0$ nm, calculate the ratio of ψ^2 at $x = 0$ to ψ^2 at $x = a$. What if the width $a = 10$ nm?

17. For a particle in a one-dimensional box, does the operator for energy commute with the operator for momentum?

18. Consider a helium atom in a cubical box with side $a = 1$ mm. If the kinetic energy is $\frac{1}{2}mv^2 = (\frac{3}{2})kT$, find the value of n in a wavefunction that would correspond to this energy.

19. A proton with kinetic energy 1000 MeV makes an elastic collision with a stationary proton in such a way that after collision the two protons are moving at equal angles to the original direction of motion. Calculate the angle.

20. Write the Schrödinger equation for a one-dimensional harmonic oscillator (Section 3.6).

21. The wavefunction for a harmonic oscillator in its lowest energy state is $\psi_0 = (\alpha^2/\pi)^{1/4}e^{-\alpha^2x^2/2}$ where $\alpha = (4\pi^2\kappa\mu/h^2)^{1/4}$. Verify that this is a solution of the wave equation (Problem 20) and calculate the energy of the oscillator in its lowest state.

22. Sketch the wavefunction in Problem 21. At what value of x is ψ a maximum? Calculate the relative values of ψ^2 at the maximum, at the classical turning points of the vibrational motion, and at a distance 10% beyond the classical turning point.

21

Atomic Structure and Spectra

Chapter 20 introduced the basic ideas of wave mechanics and applied them to particles moving freely in space or confined to a box in which the potential energy is constant. We turn now to more difficult problems in which the particles, usually electrons, are components of atoms or molecules.

Spectroscopy is the most important experimental technique for obtaining data about the structures and energy levels of atoms and molecules. We shall see in this chapter how quantum mechanics has explained the energy levels of electrons in atoms and thereby elucidated the problems of atomic spectra. The same theory has also explained the periodic table of the elements.

The first correct application of quantum theory to the interpretation of spectra was not made in the field of atomic spectra but in connection with the absorption spectra of molecules. This advance was made by the Danish chemist, Niels Bjerrum, in 1912. He showed that absorption of infrared radiation by molecules could be explained by the uptake of rotational and vibrational energy in definite quanta.

21.1 Atomic Spectra

In 1885, J. J. Balmer discovered an empirical relationship between the wavelengths λ of atomic-hydrogen lines in the visible region of the spectrum. The wave numbers are given by $1/\lambda = \tilde{\nu} = \mathfrak{R}_H(1/2^2 - 1/n_1^2)$ with $n_1 = 3, 4, 5, \ldots$ The bright red H_α line at $\lambda = 656.28$ nm corresponds to $n_1 = 3$; the blue H_β line at 486.13 nm, to $n_1 = 4$, etc. The constant $\mathfrak{R}_H$, called the *Rydberg constant*, has the value 109 677.581 cm^{-1}. It is one of the most accurately known physical constants.

Other hydrogen series were observed later, which obey the more general formula,

$$\tilde{v} = \mathcal{R}_H \left(\frac{1}{n_2^2} - \frac{1}{n_1^2} \right) \tag{21.1}$$

Lyman found the series with $n_2 = 1$ in the far ultraviolet, and others were found in the infrared. Many similar series have been observed in the atomic spectra of other elements. In 1908, Walther Ritz showed that the observed frequencies are always differences between certain pairs of frequencies, called *spectral terms*.

In 1913, the interpretation of atomic spectra was greatly advanced by the work of a young Dane, Niels Bohr, who at that time was one of Rutherford's research students at Manchester. Bohr brought together two main streams of physics—the German school of theoretical physics exemplified by Planck and Einstein, and the English school of experimental physics of Thomson and Rutherford. The model of the nuclear atom proposed by Rutherford in 1971 consisted of a positively charged central nucleus with electrons revolving in orbits like planets around a sun. According to classical electromagnetic theory, the atom model of Rutherford would be unstable. The electrons revolving about the nucleus are accelerated charged particles; therefore, they should continuously emit radiation, lose energy, and execute descending spirals until they fall into the positive center. But the electrons were unaware of what was expected, and the facts of chemistry and physics pointed clearly to the Rutherford model. For Bohr there was only one conclusion: The old principles of theoretical physics do not apply to atoms.

Thus Bohr advanced the distinctly new principles:

1. Two different energy states of the electron in an atom define each spectral line. These are called *allowed stationary states*, and they correspond to the spectral terms of Ritz.

2. The Planck–Einstein equation $\Delta E = h\nu$ holds for the emission and absorption of radiation. Thus, if the electron makes a transition between two states with energies E_1 and E_2, the frequency ν of the spectral line is given by

$$h\nu = E_1 - E_2 \tag{21.2}$$

21.2 Bohr Orbits and Ionization Energies

To specify which orbits of electrons around the nucleus are allowed, Bohr postulated that the magnitude of the angular momentum L of an electron is quantized, with $L = n(h/2\pi)$, where n is integer. For a particle of mass m moving with a speed v in a circular path of radius r, $L = mvr$, and hence the Bohr condition becomes

$$m_e vr = \frac{nh}{2\pi} \tag{21.3}$$

where m_e is the mass of the electron. The integer n is called the *principal quantum number*.

An electron is held in its orbit by the electrostatic force that attracts it to the nucleus. If the nucleus has a charge Ze, this force is $Ze^2/4\pi\epsilon_0 r^2$, from Coulomb's Law. For a stable state, this is the centripetal force mv^2/r. Hence

$$\frac{Ze^2}{4\pi\epsilon_0 r^2} = \frac{m_e v^2}{r} \tag{21.4}$$

Then, from Eq. (21.3),

$$r = \frac{\epsilon_0 n^2 h^2}{\pi m_e e^2 Z} \tag{21.5}$$

In the case of hydrogen $Z = 1$, and the smallest orbit would be that with $n = 1$, which would have a radius

$$a_0 = \frac{\epsilon_0 h^2}{\pi m_e e^2} = 5.292 \times 10^{-11}\ \text{m} = 52.92\ \text{pm} \tag{21.6}$$

Bohr could calculate the energy of the electron in each allowed orbit, and then calculate the spectral line frequencies from Eq. (21.2). The energy levels so obtained are plotted in Fig. 21.1 and the transitions responsible for absorption or emission of radiation are shown as vertical lines. The Balmer series arises from transitions between an orbit with $n = 2$ and outer orbits; in the Lyman series, the lower term corresponds to the orbit with $n = 1$; the other series are explained similarly.

The energy levels are calculated as follows. The total energy E in any state is the sum of the kinetic and potential energies, $E = E_k + E_p = m_e v^2/2 - Ze^2/4\pi\epsilon_0 r$. From

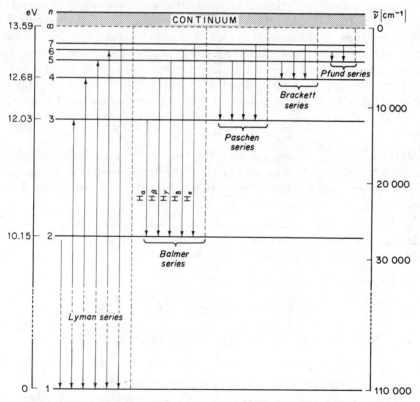

FIGURE 21.1 Diagram of energy levels of the hydrogen atom with transitions that occur in different spectral series.
[Why are only the Lyman-series arrows double headed?]

Eqs. (21.4) and (21.5),

$$E = \frac{Ze^2}{8\pi\epsilon_0 r} - \frac{Ze^2}{4\pi\epsilon_0 r} = \frac{-Ze^2}{8\pi\epsilon_0 r} = \frac{-m_e e^4}{8\epsilon_0^2 h^2}\frac{Z^2}{n^2}$$

Notice that the potential energy is twice the magnitude of the kinetic energy and opposite in sign. This result, called the *virial theorem*, is true at equilibrium for any system in which the forces are *central*, i.e., dependent only on the distance from a center. The closer the electron orbit is to the nucleus, the higher is the kinetic energy of the electron.

The frequency of a spectral line due to transition of the electron between levels with quantum numbers n_1 and n_2 is

$$\nu = \left(\frac{1}{h}\right)(E_{n_1} - E_{n_2}) = \frac{m_e e^4 Z^2}{8\epsilon_0^2 h^3}\left(\frac{1}{n_2^2} - \frac{1}{n_1^2}\right) \tag{21.7}$$

This theoretical expression has exactly the form of the experimental law found by Rydberg, and hence a theoretical value for the Rydberg constant can be obtained,

$$\mathcal{R}_H = m_e e^4/8\epsilon_0^2 ch^3 = 109\ 737\ \text{cm}^{-1} \tag{21.8}$$

The good agreement with the experimental value of $109\ 677.576 \pm 0.016\ \text{cm}^{-1}$ was a triumph for the Bohr theory.

Actually, a small correction makes the agreement between experiment and theory exact to within the experimental error of the constants and measurements. In Eq. (21.8), the mass m_e of the electron is used. Actually, however, the electron does not revolve about the proton, but about the center of mass of the proton–electron system. As shown in Section 3.5 we should therefore use the reduced mass μ of the two particles, where $\mu = m_e m_p/(m_e + m_p)$. Since m_e is 9.1095×10^{-31} kg and m_p, the proton mass, is 1.6727×10^{-27} kg, the calculated Rydberg constant for the H atom becomes

$$\mathcal{R}_H = \frac{m_e e^4}{8\epsilon_0^2 ch^3[1 + (m_e/m_p)]} = 109\ 677\ \text{cm}^{-1}$$

For other hydrogenlike atoms and ions (^{2}H, ^{4}He$^+$, ^{6}Li^{2+}, etc.) the Rydberg constant varies slightly with nuclear mass, as a result of the slight variation in μ from its value for the hydrogen atom, $109\ 737\ \text{cm}^{-1}$ being the value for infinite nuclear mass ($\mu = m_e$).

Example 21.1 Calculate the wavelength of the first line in the Balmer series of deuterium and compare it to that for hydrogen.

The Rydberg constant for the D atom is

$$\mathcal{R}_D = (\mu_D/\mu_H)\mathcal{R}_H = \left[\left(1 + \frac{m_e}{m_H}\right)\Big/\left(1 + \frac{m_e}{m_D}\right)\right]\mathcal{R}_H$$

$$\mathcal{R}_D = \left[\left(1 + \frac{9.1096 \times 10^{-31}\ \text{kg}}{1.6726 \times 10^{-27}\ \text{kg}}\right)\Big/\left(1 + \frac{9.1096 \times 10^{-31}\ \text{kg}}{3.3436 \times 10^{-27}\ \text{kg}}\right)\right]\mathcal{R}_H$$

$$= (1 + 0.000545)/(1 + 0.000272) = 1.000273\mathcal{R}_H = 109\ 708\ \text{cm}^{-1}$$

The H$_\alpha$ line of Balmer series is at 656.279 nm, so that the D$_\alpha$ line would be at 656.100 nm. Deuterium was discovered in 1931 by Harold Urey by detection of this faint line next to H$_\alpha$ in the spectrum of hydrogen prepared from residues of electrolysis of water.

In Fig. 21.1 the energy levels become more closely spaced as the height above the lowest state (*ground state*) increases. They finally converge to a limit whose height above ground level is the energy necessary to remove the electron completely from the field of the nucleus. In the observed spectrum, lines become more and more densely spaced and finally merge into a *continuum*, i.e., a region of continuous absorption or emission of radiation without any line structure. The reason for the continuum is that once an electron is completely free from the nucleus, it is no longer restricted to discrete quantized energy states, but may take up continuously the ordinary kinetic energy of translation $\frac{1}{2}m_e v^2$ corresponding to its speed in free space.

The difference in energy between the series limit and the ground level is the ionization energy, often called the *ionization potential*. From Eq. (21.7), the ionization energy of the H atom is obtained when $n_2 = 1$ and $n_1 = \infty$, as

$$I_1 = \Delta E = m_e e^4/8\epsilon_0^2 h^2 = \mathfrak{R}_H ch = 2.181 \times 10^{-18} \text{ J}$$

Example 21.2 The "atom" formed by a positive electron and a negative electron is called *positronium*. Calculate its ionization energy.

$I = \mu e^4/8\epsilon_0^2 h^2$. In this case,

$$\mu = m_e/2 = (9.11 \times 10^{-31} \text{ kg})/2 = 4.555 \times 10^{-31} \text{ kg}$$

$$I = \frac{(4.555 \times 10^{-31} \text{ kg})(1.602 \times 10^{-19} \text{ C})^4}{8(8.854 \times 10^{-12} \text{ J}^{-1} \text{ C}^2 \text{ m}^{-1})^2(6.63 \times 10^{-34} \text{ J s})^2}$$

$$= 1.09 \times 10^{-18} \text{ J} = 6.80 \text{ eV}$$

Note that positronium is stable relative to dissociation into an electron and a positron, but if the two particles collide, they can annihilate each other. The positron is an example of *antimatter*.

21.3 Schrödinger Equation for the Hydrogen Atom

The Bohr theory was developed intensively from 1913 to 1926, but its initial success with the H atom could not be equaled, and even the spectrum of He proved impossible to interpret. Then the work of Broglie, Heisenberg, and Schrödinger opened the way to a comprehensive theory of atomic structure and spectra based on quantum mechanics.

The theory of the hydrogen atom is concerned with the motion of a particle of charge $-e$ and mass μ in the electrostatic field of a charge $+e$. The situation is somewhat similar to that of a particle in a three-dimensional box. The "box" now has spherical symmetry with an infinitely deep potential well in the center and a rise in potential energy U with distance r from the center in accord with Coulomb's law,

$$U = -e^2/4\pi\epsilon_0 r \tag{21.9}$$

At $r = 0$, $U = -\infty$ and at $r = \infty$, $U = 0$. This potential-energy function is plotted in Fig. 21.2.

We obtain the Schrödinger equation $\hat{H}\Psi = E\Psi$ by following the rules of Section

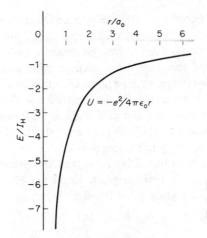

FIGURE 21.2 Potential energy of negative electron interacting with positive proton in accord with Coulomb's Law.

20.13 to write down the Hamiltonian operator $\hat{H}$. The classical Hamiltonian $H = E_k + U$. In Cartesian coordinates, the kinetic energy is $E_k = (p_x^2 + p_y^2 + p_z^2)/2\mu$, where the p's are the components of momentum. The operator for p_x, for example, is $\hat{p}_x = (h/2\pi i)\,\partial/\partial x$, so that $\hat{p}_x^2 = (-h^2/4\pi^2)\,\partial^2/\partial x^2$ and the operator for kinetic energy is thus $\hat{E}_k = (-h^2/8\pi^2\mu)(\partial^2/\partial x^2 + \partial^2/\partial y^2 + \partial^2/\partial z^2)$. The differential operator $\nabla^2 \equiv (\partial^2/\partial x^2 + \partial^2/\partial y^2 + \partial^2/\partial z^2)$ is called the *Laplacian* (read "del squared"). It occurs in many important problems in mathematical physics. The Schrödinger equation for the H atom, $\hat{H}\psi = E\psi$, now becomes

$$\nabla^2\psi + \frac{8\pi^2\mu}{h^2}\left(E + \frac{e^2}{4\pi\epsilon_0 r}\right)\psi = 0 \tag{21.10}$$

The spherical symmetry of the Coulombic potential energy (21.10) indicates that the equation may be solved most readily in spherical polar coordinates (r, θ, ϕ), which are shown in Fig. 21.3. The coordinate r measures the radial distance from the origin,

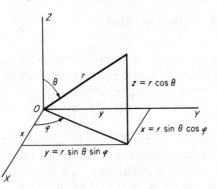

FIGURE 21.3 Spherical polar coordinates.

θ is a colatitude, and ϕ a longitude. Since the electron is moving in three dimensions, three coordinates suffice to describe its position at any time.

The variables r, θ, ϕ in the wave equation can be separated because the potential energy is a function of r alone. Hence

$$\psi(r, \theta, \phi) = R(r)\Theta(\theta)\Phi(\phi) \tag{21.11}$$

The wavefunction is a product of three functions, the first of which depends only on r, the second only on θ, and the third only on ϕ.

The wave patterns specified by the functions $R(r)$, $\Theta(\theta)$, and $\Phi(\phi)$ are not simple sine functions like those for the particle-in-a-box. Nevertheless, the waves that describe the electron in the hydrogen atom are subject to the requirements specified in Section 20.7. The boundary-value conditions imposed on wave function $\psi(r, \theta, \phi)$ then lead to three integral numbers, the quantum numbers n, ℓ, and m_ℓ, for the three-dimensional wave motion. Any allowed function for the electron in the hydrogen atom is specified by these three quantum numbers as $\psi_{n,\ell,m_\ell}(r, \theta, \phi)$.

21.4 The Radial Equation Gives the Energy Levels

Since the potential energy $U(r)$ is independent of θ and ϕ, the energy E of the H atom depends only on the function $R(r)$ and can depend in no way on the angular distributions $\Theta(\theta)$ and $\Phi(\phi)$. When the solutions $R(r)$ are used to calculate E, the quantum mechanical result is the same as that found from the Bohr theory (which is just as well since Bohr was exactly right in his calculated energies for the H atom). Thus the allowed energy levels are

$$E_n = \frac{-\mu e^4}{8\epsilon_0^2 h^2} \frac{1}{n^2} \tag{21.12}$$

These values of E_n are called the *eigenvalues* of the Schrödinger equation, since the eigenfunctions or wavefunctions ψ_n give

$$\hat{H}\psi_n(r, \theta, \phi) = E_n\psi_n(r, \theta, \phi)$$

The quantum number n, which can take values $1, 2, \ldots$, is called the *principal quantum number*.

The energy of the electron in the H atom remains constant as long as the wavefunction ψ_n that defines the state of the system is fixed. The energy E_n is one of the constants of motion for an electron under an inverse-square attractive force, just as it would be in a classical problem like the motion of two masses with mutual gravitational attraction.

21.5 The Angular Equation Gives the Angular Momenta

In classical mechanics the motion of a particle in a central field has two other constants of motion besides the energy, namely the total angular momentum and the component of angular momentum in the direction of one specified axis (conventionally chosen to be the z axis). In the quantum mechanics of the hydrogen atom the quantum numbers ℓ and m_ℓ specify the quantization of the total angular momentum and the z component of angular momentum, respectively.

The definition of angular momentum in classical mechanics is shown in Fig. 21.4. For a particle moving with a linear velocity **v** (or momentum **p**) at a position determined by a radius vector **r** from a fixed point O, the angular momentum vector **L** from O is defined by

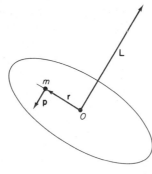

FIGURE 21.4 Angular momentum **L** of a particle of mass m with respect to point O is defined by $\mathbf{L} = \mathbf{r} \times \mathbf{p} = \mathbf{r} \times m\mathbf{v}$, where **p** is the linear momentum of the particle and **r** is the radius vector from O to the particle. The vector **L** is normal to the plane defined by **r** and **p**.

$$\mathbf{L} = \mathbf{r} \times \mathbf{p} = \mathbf{r} \times m\mathbf{v} \qquad (21.13)*$$

In Cartesian coordinates, the components of L are

$$L_x = yp_z - zp_y, \qquad L_y = zp_x - xp_z, \qquad L_z = xp_y - yp_x \qquad (21.14)$$

Application of the quantum mechanical rule from page 450 converts the angular-momentum components (21.14) to operators:†

$$\hat{L}_x = -(ih/2\pi)(y\partial/\partial z - z\partial/\partial y)$$
$$\hat{L}_y = -(ih/2\pi)(z\partial/\partial x - x\partial/\partial z) \qquad (21.15)$$
$$\hat{L}_z = -(ih/2\pi)(x\partial/\partial y - y\partial/\partial x)$$

The operator for the square of the magnitude of the total angular momentum is

$$\hat{L}^2 = \hat{L}_x^2 + \hat{L}_y^2 + \hat{L}_z^2 \qquad (21.16)$$

The hydrogenlike wavefunctions are eigenfunctions for the operators $\hat{L}_z$ and $\hat{L}^2$. Therefore, each eigenfunction corresponds to a definite measurable value of the total angular momentum and of the z component of angular momentum.‡

$$\hat{L}^2\Theta(\theta)\Phi(\phi) = (h^2/4\pi^2)\ell(\ell + 1)\Theta(\theta)\Phi(\phi) \qquad (21.17)$$

$$\hat{L}_z^2\Phi(\phi) = (h^2/4\pi^2)m_\ell^2\Phi(\phi) \qquad (21.18)$$

[Show the results above with the help of Eq. (21.15a). Note that $\hat{L}_z^2 = \hat{L}_z\hat{L}_z$, etc.] Thus $\ell(\ell + 1)h^2/4\pi^2$ gives the eigenvalues of the operator $\hat{L}^2$, and $m_\ell h^2/4\pi^2$ gives the eigenvalues of $\hat{L}_z^2$. This result means that a measurement of the magnitude of the total angular momentum of the hydrogen atom is predicted to give exactly one of the

---

*(The vector product **A** × **B** of two vectors is a vector of magnitude $AB \sin \theta$ and direction perpendicular to plane of **A** and **B** with the sense of a counterclockwise rotation of **A** into **B**.)

†Since the wavefunctions are usually given in spherical polar coordinates, it is convenient to transform the operators in Eq. (21.15) into this system:

$$\hat{L}_x = (ih/2\pi)\left(\cot \theta \cos \phi \frac{\partial}{\partial \phi} + \sin \phi \frac{\partial}{\partial \theta}\right)$$

$$\hat{L}_y = (ih/2\pi)\left(\cot \theta \sin \phi \frac{\partial}{\partial \phi} - \cos \phi \frac{\partial}{\partial \theta}\right) \qquad (21.15a)$$

$$\hat{L}_z = -(ih/2\pi)\left(\frac{\partial}{\partial \phi}\right)$$

‡The specification of the total angular momentum by $\ell(\ell + 1)$ is related to the uncertainty principle applied to the conjugate variables angle and angular momentum. An excellent discussion is given by R. S. Berry, S. A. Rice and J. Ross, *Physical Chemistry* (New York: John Wiley & Sons Inc., 1980), pp. 100–125.

values

$$L = \sqrt{\ell(\ell + 1)}h/2\pi \qquad (21.19)$$

and a measurement of the z component of angular momentum is predicted to give exactly one of the values

$$L_z = m_\ell h/2\pi \qquad (21.20)$$

Example 21.3 What is the angular momentum of an electron in the $2p$ orbital of atomic H? How does it compare with the angular momentum of rotation of a molecule of CO in its $J = 1$ rotational energy state?

The magnitude of the orbital angular momentum of an electron in an H atom is $L = \sqrt{\ell(\ell + 1)}(h/2\pi)$. For $2p$, $\ell = 1$,

$$L = \sqrt{2}(h/2\pi) = (\sqrt{2}/2\pi)(6.63 \times 10^{-34} \text{ J s})$$
$$= 1.49 \times 10^{-34} \text{ J s}$$

For a diatomic molecule, from Section 4.7, $p_\theta = \sqrt{J(J+1)}(h/2\pi)$. Thus for state with $J = 1$, $L = 1.49 \times 10^{-34}$ J s, the same as that for a $2p$ electron in the H atom. This is a strange result of quantum mechanics—only certain values of angular momentum are allowed in nature.

21.6 The Quantum Numbers

The wavefunctions ψ_{n,ℓ,m_ℓ} for the H atom are specified by three quantum numbers, as would be expected for a three-dimensional problem in wave mechanics. A *wavefunction for a single electron* is called an *orbital*. The old theory spoke of the *orbits* of electrons. In the new theory, there are no orbits and all the information about the position of an electron is summarized in its orbital ψ. The hydrogenlike orbitals are listed in Table 21.1 for lower values of the quantum numbers n and ℓ.

TABLE 21.1
NORMALIZED HYDROGENLIKE WAVEFUNCTIONS

	K shell
$n = 1, \ell = 0, m_\ell = 0$	$\psi(1s) = \dfrac{1}{\sqrt{\pi}}\left(\dfrac{Z}{a_0}\right)^{3/2} e^{-Zr/a_0}$
	L shell
$n = 2, \ell = 0, m_\ell = 0$	$\psi(2s) = \dfrac{1}{4\sqrt{2\pi}}\left(\dfrac{Z}{a_0}\right)^{3/2}\left(2 - \dfrac{Zr}{a_0}\right)e^{-Zr/2a_0}$
$n = 2, \ell = 1, m_\ell = 0$	$\psi(2p_z) = \dfrac{1}{4\sqrt{2\pi}}\left(\dfrac{Z}{a_0}\right)^{3/2}\dfrac{Zr}{a_0}e^{-Zr/2a_0}\cos\theta$
$n = 2, \ell = 1, m_\ell = \pm 1^a$	$\psi(2p_x) = \dfrac{1}{4\sqrt{2\pi}}\left(\dfrac{Z}{a_0}\right)^{3/2}\dfrac{Zr}{a_0}e^{-Zr/2a_0}\sin\theta\cos\phi$
	$\psi(2p_y) = \dfrac{1}{4\sqrt{2\pi}}\left(\dfrac{Z}{a_0}\right)^{3/2}\dfrac{Zr}{a_0}e^{-Zr/2a_0}\sin\theta\sin\phi$

ᵃThe functions here are real linear combinations of the $m_\ell = +1$ and $m_\ell = -1$ wavefunctions (see Table 21.2).

The *principal quantum number n* is the successor to the *n* introduced by Bohr in his theory of the hydrogen atom. The total number of nodes in the wavefunction is equal to $n - 1$. These nodes may be in the radial function $R(r)$, in the azimuthal function $\Theta(\theta)$, or in both.

The quantum number ℓ is called the *azimuthal* or *angular-momentum quantum number*. It is equal to the number of nodes in $\Theta(\theta)$, i.e., to the number of nodal surfaces passing through the origin. Since the total number of nodes in ψ is $n - 1$, the allowed values of ℓ run from 0 to $n - 1$. When $\ell = 0$, there are no nodes in the function $\Theta(\theta)$ and the wave function is spherically symmetric about the central nucleus. The electron angular momentum L is specified by the value of ℓ according to Eq. (21. 19). Orbitals with $\ell = 0$ therefore have zero angular momentum. Orbitals with $\ell = 0, 1, 2, 3$ are denoted as *s, p, d*, and *f* orbitals, respectively.

The quantum number m_ℓ is called the *magnetic quantum number*. If the hydrogen atom is placed in a magnetic field, a definite direction in space is physically established by the field, and the angular momentum vector precesses about this field direction like a spinning top precessing in the gravitational field of the earth. The solutions of the Schrödinger equation are such that not every orientation between the angular momentum vector and the field direction is allowed. The only allowed directions are those for which the components of angular momentum along the field direction (conventionally chosen to be the *z* axis) have certain quantized values given by Eq. (21.20) as $L_z = m_\ell h/2\pi$. This behavior is illustrated in Fig. 21.5 for the case in which

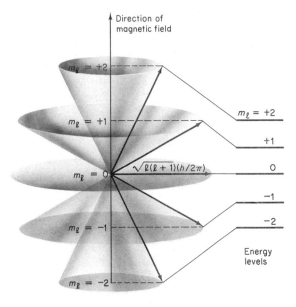

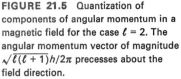

FIGURE 21.5 Quantization of components of angular momentum in a magnetic field for the case $\ell = 2$. The angular momentum vector of magnitude $\sqrt{\ell(\ell + 1)}h/2\pi$ precesses about the field direction.

the azimuthal number $\ell = 2$. The magnetic number m_ℓ can then have the values $-2, -1, 0, 1, 2$. For any value of ℓ, which specifies the total angular momentum, there are $2\ell + 1$ values of m_ℓ, which specify the allowed components of the angular momentum in the field direction. Consequently, the energy of the precessional motion is quantized. The allowed energy levels are spaced such that $\Delta E = m_\ell h\nu$, where ν is the frequency of precession of the angular momentum vector in the magnetic field. This ν is called the *Larmor frequency*.

21.7 The Radial Wavefunctions

Figure 21.6(a) is a plot of the radial parts of the wavefunctions for a few values of n and ℓ. The number of nodes in the radial function equals $n - \ell - 1$. The amplitude ψ of the electron wave can be positive or negative. The probability of finding the electron in the region between r and $r + dr$ is proportional to $\psi^*\psi = |\psi|^2$, the square of the absolute value of the amplitude.

Often we need to know the probability that the electron is at a given distance r from the nucleus, irrespective of direction. This is the probability that the electron lies between two spheres, of radii r and $r + dr$. The volume of this spherical shell is $4\pi r^2 \, dr$. Hence the probability of finding the electron somewhere within this shell is $4\pi r^2 \psi^*\psi \, dr$. The function $4\pi r^2 \psi^*\psi$ is called the *radial distribution function* $g(r)$. It has been plotted in Fig. 21.6(b) for the values of n and ℓ used in Fig. 21.6(a).

Example 21.4 Estimate the most probable value of the distance of a $1s$ electron in the uranium atom from the nucleus.

We assume that the wavefunction is a hydrogenlike $1s$ orbital, which is a rough approximation for an inner-shell $1s$ electron with $\psi_{1s} = (1/\sqrt{\pi})(Z/a_0)^{1/2}e^{-Zr/a_0}$ (see, however, Problem 6). The radial distribution function $g(r) = 4\pi r^2 \psi^2(r)$. At the radius of maximum probability $\partial g/\partial r = 0$.

$$0 = 4\pi r^2 \left(\frac{1}{\pi}\frac{Z}{a_0}\right)e^{-2Zr/a_0}\left(\frac{-2Z}{a_0}\right) + \left(\frac{1}{\pi}\frac{Z}{a_0}\right)e^{-2Zr/a_0}8\pi r$$

$$0 = \frac{-2Zr}{a_0} + 2, \qquad r(\text{max}) = a_0/Z$$

For $Z = 92$, $r(\text{max}) = 52.9 \text{ pm}/92 = 0.58 \text{ pm}$. This is an extraordinarily small distance and inner-shell electrons in heavy atoms spend most of their time very close to the nucleus and thus must have large kinetic energies.

As an example of the quantum mechanical model, consider the case $n = 1$, $\ell = 0$, the $1s$ state of the electron in the hydrogen atom. This is the lowest or ground state. In the old Bohr theory, the electron in this state revolved in a circular orbit of radius $a_0 = 52.9$ pm. The quantum mechanical result indicates that the electron has a certain probability of being in any region, from $r = 10^{-15}$ m, right in the center of the nucleus, to $r = 10^{20}$ m or beyond, out in the Milky Way. Nevertheless, the position of maximum probability for the electron does correspond to the value $r = a_0$, and the chance that the electron is far away from the nucleus is small. For example, what is the probability that the electron in its ground state in a H atom is at a distance from the nucleus of $10a_0$(529 pm) relative to probability of finding it at a_0? From Table 21.1, $\psi_{1s} = (\pi a_0^3)^{-1/2}e^{-r/a_0}$, so that the relative probability $(r^2\psi^2)$ at $10a_0$ compared to that at a_0 is $10^2(e^{-10}/e^{-1})^2 = 1.52 \times 10^{-6}$. This is a small but far from negligible number.

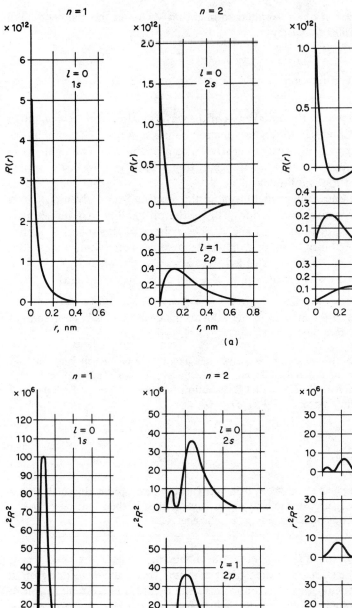

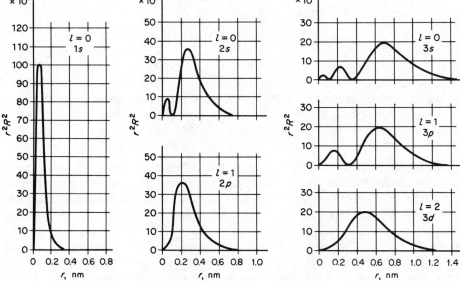

FIGURE 21.6 (a) Radial parts $R(r)$ of wave functions for hydrogen atom. (b) Radial distribution functions r^2R^2. The probability that an electron is within distances r and $r + dr$ from the nucleus (after averaging over angular variables) is $4\pi r^2R^2\, dr$. [G. Herzberg, *Atomic Spectra* (New York: Dover Publishing Co., 1944)]

Another way of expressing the result is to say that one out of about 7×10^5 hydrogen atoms at any instant might be expected to have its electron at a distance $10a_0$ from the nucleus.

In the $2s$ state, $n = 2$, $\ell = 0$, we find, in addition to the main peak in the radial distribution function at $r = 5.2a_0$, another little peak in the probability at a much smaller distance from the nucleus, $r = 0.8a_0$. A similar effect is seen in all states where $(n - 1) > \ell$. In heavier atoms, electrons in s orbitals are thus able to *penetrate* the outer electron distribution and sometimes to come close to the nucleus.

21.8   Angular Dependence of Hydrogen Orbitals

Orbitals with $\ell = 0$, s orbitals, are always spherically symmetric. In this case, ψ is a function of r alone, and does not depend on θ or ϕ. Orbitals with $\ell = 1$, p orbitals, have a marked directional character because the ψ function depends on θ and ϕ. The d orbitals, with $\ell = 2$, are also directional, with more complicated angular dependencies.

In the angular part of the wavefunction for the hydrogen atom,

$$\Phi_{m_\ell}(\phi) = (2\pi)^{-1/2} \, e^{im_\ell \phi} \tag{21.21}$$

where m_ℓ can take values $0, \pm 1, \pm 2, \pm 3, \ldots, \pm \ell$. For orbitals with $\ell = 1$, m_ℓ can be $-1, 0, +1$. For orbitals with $\ell = 2$, m_ℓ can be $-2, -1, 0, +1, +2$.

To display the angular dependence of these orbitals, it is helpful to form new real eigenfunctions by linear combinations of the complex ones in Eq. (21.21). (The general superposition property of solutions of linear differential equations ensures that such linear combinations of solutions will themselves be solutions.) Table 21.2 shows these linear combinations of the $\Theta(\theta)\Phi(\phi)$ functions, which are the basis for the usual discussion of their angular dependence. The subscripts on the real linear-combination orbitals denote the directional properties of the orbitals. Note, for example, that a p_x orbital does not designate only one value of m_ℓ, but a linear combination of orbitals with $m_\ell = 1$ and $m_\ell = -1$.

To see how the angular functions correspond to their designations in terms of x, y, and z, consider the example of p_x. From Tables 21.1 and 21.2, $\Theta(\theta)\Phi(\phi)$ for this orbital is

$$p_x = (1/\sqrt{2}) \sin \theta \, (1/\sqrt{2\pi})(e^{i\varphi} + e^{-i\varphi})$$

Since $(e^{i\varphi} + e^{-i\varphi})/2 = \cos \varphi$, and $x = r \sin \theta \cos \varphi$,

$$p_x = (1/\sqrt{\pi}) \sin \theta \cos \varphi = (1/\sqrt{\pi})(1/r)x$$

By similar transformations, we can verify the subscript designations of the other p and d orbitals.

A pictorial way to display the angular dependence of orbitals is to draw the surfaces $\Theta(\theta)\Phi(\theta)$, the spherical surface harmonics. Such surfaces are shown in Fig. 21.7 for orbitals with $\ell = 0$ (s), $\ell = 1$ (p), and $\ell = 2$ (d). To obtain the actual amplitudes of the wavefunctions, we must multiply the angular parts $\Theta(\theta)\Phi(\phi)$ shown in

TABLE 21.2

THE FUNCTIONS $\Theta(\theta)\,\Phi(\phi)$ IN THE HYDROGENLIKE ATOMIC ORBITALS

Complex forms	Real linear combinations
p orbitals	

$$p_{+1} = \frac{1}{\sqrt{2\pi}}\, e^{i\phi} \sin\theta \qquad\qquad p_x = \frac{1}{\sqrt{2}}(p_1 + p_{-1})$$

$$p_0 = \frac{1}{\sqrt{2\pi}}\cos\theta \qquad\qquad\qquad p_z = p_0$$

$$p_{-1} = \frac{1}{\sqrt{2\pi}}\, e^{-i\phi} \sin\theta \qquad\qquad p_y = \frac{-i}{\sqrt{2}}(p_1 - p_{-1})$$

d orbitals

$$d_{+2} = \frac{1}{\sqrt{2\pi}}\, e^{2i\phi}\sin^2\theta \qquad\qquad d_{z^2} = d_0$$

$$d_{+1} = \frac{1}{\sqrt{2\pi}}\, e^{i\phi}\sin\theta\cos\theta \qquad\quad d_{xz} = \frac{1}{\sqrt{2}}(d_{+1} + d_{-1})$$

$$d_0 = \frac{1}{\sqrt{2\pi}}(3\cos^2\theta - 1) \qquad\quad d_{yz} = -\frac{-i}{\sqrt{2}}(d_{+1} - d_{-1})$$

$$d_{-1} = \frac{1}{\sqrt{2}}\, e^{-i\phi}\sin\theta\cos\theta \qquad\quad d_{xy} = \frac{-i}{\sqrt{2}}(d_{+2} + d_{-2})$$

$$d_{-2} = \frac{1}{\sqrt{2\pi}}\, e^{-2i\phi}\sin^2\theta \qquad\qquad d_{x^2-y^2} = \frac{1}{\sqrt{2}}(d_{+2} - d_{-2})$$

Fig. 21.7 by the radial parts $R(r)$ shown in Fig. 21.6. The angular dependence is the same for all values of r. Thus the shapes in Fig. 21.7 do not imply that the orbitals are sharply defined in space. They simply depict the angular dependence of the orbitals for any value of r.

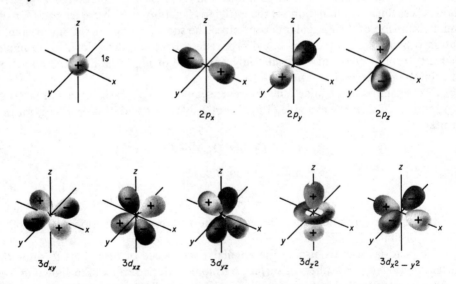

FIGURE 21.7 The angular dependence of the wavefunctions for atomic hydrogen. The contours show the shapes of the function $(\Theta(\theta)\Phi(\phi))^2$ at any value of r.

21.9 The Spinning Electron

The electron possesses an intrinsic angular momentum or spin, specified by a quantum number $s = \frac{1}{2}$. By analogy with the relation between the quantum number ℓ and orbital angular momentum, the spin angular momentum has a magnitude $S = \sqrt{s(s+1)}(h/2\pi) = (\sqrt{3}/2)(h/2\pi)$ since $s = \frac{1}{2}$. The components of S in the direction of the field are specified by a quantum number m_s, where $m_s = +\frac{1}{2}$ or $-\frac{1}{2}$. These relations are summarized in Fig. 21.8.

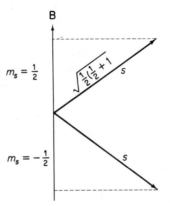

$$m_s = \frac{1}{2}$$

$$m_s = -\frac{1}{2}$$

FIGURE 21.8 The spin s of an electron in a magnetic field can be oriented in only two directions, corresponding to $m_s = \pm\frac{1}{2}$. In atoms of the alkali metals the motion of the valence electron in its orbit provides the magnetic field that splits the spin states, thus leading to doublets in the spectra.

The spinning electron acts like a little magnet. A beautiful demonstration of this fact was given by the Stern–Gerlach experiment in 1922. In atoms of the alkali metals (Li, Na, K, etc.) the outermost electron is in an s orbital and the inner electrons are in a closed shell with an inert-gas configuration. The outermost lone electron has spin-magnetic properties similar to those of an isolated electron. A beam of atoms of an alkali metal, when passed through a strong inhomogeneous magnetic field, is split into two separate beams. The field divides the atoms into two beams corresponding to $m_s = +\frac{1}{2}$ and $-\frac{1}{2}$ for the outermost electron.

The concept of electron spin first appeared in 1923 as a special new property of elementary particles, which needed to be tacked onto the rest of quantum theory. When Paul Dirac, an English theoretical physicist, worked out a relativistic form of wave mechanics for the electron in 1931, he found that the property of spin is a natural consequence of the theory, and no separate spin hypothesis is necessary.*

21.10 The Pauli Exclusion Principle

An exact solution of the Schrödinger wave equation for an atom has been obtained only in the case of the hydrogenlike atom—a single electron in the field of a positive charge. For more complex atoms it is often quite a good approximation to assume that the electrons move in a spherically symmetric field of a shielded nucleus.

*R. E. Powell, "Relativistic Quantum Chemistry," *J. Chem. Educ.*, **45**, 558 (1968).

In this *central field approximation*, the allowed stationary states of an electron in an atom can still be classified in terms of the four quantum numbers, n, ℓ, m_ℓ, and m_s.

An important rule governs the quantum numbers that can be assigned to electrons in an atom. This is the *Exclusion Principle*, first enunciated by Wolfgang Pauli in 1924. In the treatment of atomic structure by the central-field approximation, each electron is assigned to an orbital specified by its four quantum numbers. The Pauli Principle states that no two electrons in a given atom can have all four quantum numbers, n, ℓ, m_ℓ, and m_s, the same.

Consider in Fig. 21.9 the two electrons 1 and 2. Each electron can be denoted by a set of three space coordinates (x, y, z)(of which only x and y are shown) and a spin coordinate, which can have either of two values as shown by an arrow. We can exchange the spatial and/or spin coordinates of the two electrons, as shown in the diagram. The electrons are indistinguishable particles. Therefore when we make such an exchange, the wavefunction ψ of the system must either remain the same ($\psi \rightarrow \psi$) or simply change in sign ($\psi \rightarrow -\psi$). In the first case, we call ψ *symmetric* in the exchange; in the second case, we call ψ *antisymmetric* in the exchange. The total wavefunction for one electron can be written as a product of a spin part $\sigma [\alpha(+\frac{1}{2})$ or $\beta(-\frac{1}{2})]$ and a coordinate part ϕ; thus $\psi = \phi(x, y, z)\sigma$. The general statement of the Pauli Principle, independent of the central-field approximation, is as follows: A wavefunction for a system of electrons must be antisymmetric for exchange of the spatial and the spin coordinates of any pair of electrons. (If, therefore, $\phi \rightarrow -\phi$, $\sigma \rightarrow \sigma$; if $\phi \rightarrow \phi$, $\sigma \rightarrow -\sigma$, so that $\psi \rightarrow -\psi$, always.)

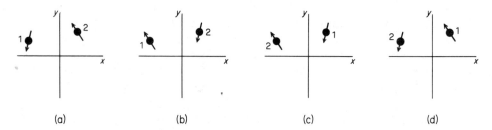

(a) (b) (c) (d)

FIGURE 21.9 Exchange of space and spin coordinates for two electrons: (a) original configuration; (b) spin coordinates exchanged; (c) space coordinates exchanged; (d) spin and space coordinates exchanged.

The statement of the Pauli Principle in terms of quantum numbers is a special case of the general statement. Consider two electrons 1 and 2, in states specified in the central-field model by n_1, ℓ_1, m_{ℓ_1}, m_{s_1}, and n_2, ℓ_2, m_{ℓ_2}, m_{s_2}. An antisymmetric function is

$$\psi = \psi_{n_1, \ell_1, m_{\ell_1}, m_{s_1}}(1)\psi_{n_2, \ell_2, m_{\ell_2}, m_{s_2}}(2) - \psi_{n_1, \ell_1, m_{\ell_1}, m_{s_1}}(2)\psi_{n_2, \ell_2, m_{\ell_2}, m_{s_2}}(1) \quad (21.22)$$

When (1) and (2) are exchanged, $\psi \rightarrow -\psi$, as is necessary. Suppose, however, that all four quantum numbers were the same for the two electrons. Then $\psi = 0$; i.e., no such state can exist.

Example 21.5 Write the antisymmetric ψ in Eq. (21.22) in the form of a determinant.

$$\psi = \begin{vmatrix} \psi_1(1) & \psi_1(2) \\ \psi_2(1) & \psi_2(2) \end{vmatrix} \qquad \text{where } \psi_1 = \psi_{n_1, \ell_1, m_{\ell_1}, m_{s_1}}, \text{ etc.}$$

When written as a determinant, ψ is always antisymmetric, since the interchange of any two columns of a determinant changes its sign.

21.11 Spectrum of Helium

The atomic spectrum of helium was more difficult to unravel than that of hydrogen. After much work, the various lines were sorted out and assigned to transitions between pairs of energy levels designated by their term symbols. The resulting term diagram is shown in Fig. 21.10. The significance of the symbols will be explained as

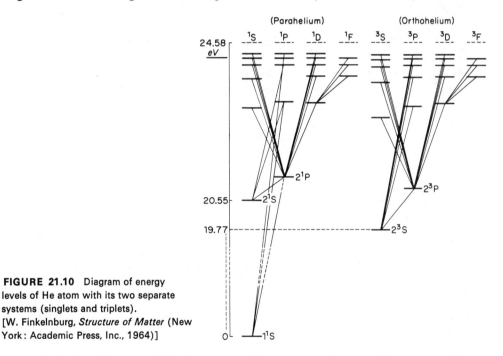

FIGURE 21.10 Diagram of energy levels of He atom with its two separate systems (singlets and triplets). [W. Finkelnburg, *Structure of Matter* (New York: Academic Press, Inc., 1964)]

we proceed. The terms are divided into two distinct sets, and spectral lines occur only by transition between terms in the same set. So definite is this separation between two sets of terms, that earlier workers believed they were dealing with two distinct kinds of helium, which they called *parahelium* and *orthohelium*. In modern notation, however, one set is said to consist of *singlets* and the other of *triplets*.

To a first approximation, we can specify a wavefunction ψ for each electron in the helium atom, i.e., an atomic orbital, by means of the same set of four quantum numbers that were found in the solution of the Schrödinger equation for the hydrogen atom, n, ℓ, m_ℓ, m_s. Of course, we no longer can use the solutions found for the H atom, and no exact solution of this type is available for the He atom itself. Nevertheless, we can imagine one electron in He to be gradually moved to infinity, forming the hydrogenlike ion He$^+$, and at no stage in such an imaginary process would we find an abrupt change in the wavefunction for the remaining electron. For this reason, a one-to-one correspondence can be drawn between the exact hydrogenlike orbitals and some approximate helium orbitals (one-electron wavefunctions). Thus one speaks freely of $1s$, $2s$, $2p$, etc, orbitals in helium and more complex atoms, even though the exact form of the wavefunction may not be known.

The ground state of the He atom has the electron configuration $1s^2$. The two electrons have quantum numbers as follows:

$$n = 1, \qquad \ell = 0, \qquad m_\ell = 0, \qquad m_s = \tfrac{1}{2}$$
$$n = 1, \qquad \ell = 0, \qquad m_\ell = 0, \qquad m_s = -\tfrac{1}{2}$$

In accord with the Pauli Principle, the two electrons do not have the same set of four quantum numbers.

The term symbol for the ground state is ^{1}S. The general notation for the term symbol is $^{2S+1}$L. The value of L, which specifies the total orbital angular momentum of all the electrons, is obtained from the vector sum of the ℓ_i, which specify the orbital angular momenta of individual electrons. According as $L = 0, 1, 2, 3$, etc., the term is called S, P, D, F, etc. The left-hand superscript gives the *multiplicity* of the term as $2S + 1$, where S is the total spin, specified by the sum of the individual m_s values. It is read as "singlet," "doublet," "triplet," and so on. In the case of the ground state of helium, $L = 0$ and $S = 0$, and hence the state is ^{1}S.

The lowest excited states of He are those in which one electron is in an orbital with principal quantum number $n = 2$. Two configurations are possible: $1s^1 2s^1$ and $1s^1 2p^1$. In the case of the H atom, the energy levels depend only on the value of n, and not on that of ℓ. For atoms with more than one electron, however, the one-electron energy levels depend strongly on the values of ℓ. The reason is that the electron is now subject to a Coulombic interaction not only with the nucleus but also with the other electrons. We can say that the positive nuclear charge is partly *shielded* by the other electrons. An orbital that permits an electron to penetrate the shielding charge and to spend more time close to the nucleus will therefore have a lower energy. From Fig. 21.6, we can see that for a given n, orbitals with lower ℓ, especially s orbitals with $\ell = 0$, permit the greater penetration. In the case of the states for He, the S terms therefore always lie below the P terms of the same principal quantum number.

The existence of singlet and triplet states of helium is due to the fact that the two electron spins can be either antiparallel ($S = 0$) or parallel ($S = 1$). Thus, except for the ground state, in which the Pauli Principle excludes the $S = 1$ state, each term is split into a singlet and a triplet. We note from the term diagram that the triplet states (for given values of n and L) always lie lower than the singlets. For example, for $n = 2$, ^{3}S is 6422 cm^{-1} lower than ^{1}S.

What is the reason for this large splitting of terms which have the same electron configuration and same L values, but which differ in their total spin? Let us state emphatically that this energy difference is *not* due to any magnetic interaction between the magnetic moments of the spins. Such a magnetic interaction does occur, but it is negligibly small compared to the observed difference in energies of the $1s^12s^1$, ^1S and $1s^12s^1$, ^3S states. The splitting of terms is due to a difference in electrostatic interactions in the system consisting of the $+2$ He nucleus and the two electrons. In the ^3S state, the two electrons have the same spin. Since they are in different orbitals, $1s$ and $2s$, no contradiction of the Pauli Principle is implied. Nevertheless, the two electrons with the same spin tend to stay away from each other. Consequently, the repulsive electrostatic energy in the ^3S state is less than in the ^1S state, where the two electrons have opposite spins. Thus the energy level of ^1S is higher than that of ^3S. The singlet–triplet splitting is an electrostatic interaction, but one that is caused by quantum mechanical rules. It is called the *exchange interaction*.

We have now explained the general structure of the term diagram of helium by means of a strong electrostatic Coulombic interaction which splits terms of different L, and a strong electrostatic exchange interaction which further splits terms of the same L but different S.

The total L and total S for a term can combine to yield a new *inner quantum number J*, and states of different J value are then split by the *spin–orbit interaction* between the orbital and spin magnetic fields. In the case of the ^1S states of He, since $L = 0$ and $S = 0$, J can only have the value 0. For the ^3S state with $L = 0$ and $S = 1$, J must be 1. Similarly, ^1P with $L = 1$ and $S = 0$ can have only $J = 1$. For ^3P, however, J can be 2, 1, or 0. (Figure 21.11 shows this result as a vector addition of L and S.)

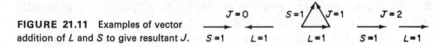

FIGURE 21.11 Examples of vector addition of L and S to give resultant J.

The correlation diagram for the first group of excited states in atomic He is summarized in Fig. 21.12. After all the internal interactions have been considered, the effect of an external magnetic field is shown. The total angular momentum, denoted by the value of J, can assume only those directions relative to the field that have components $M_J(h/2\pi)$ in the field direction. This relation of M_J to J is similar to the relation of m_ℓ to ℓ that was shown in Fig. 21.5.

21.12 Vector Model of the Atom

The physical picture for the various interactions between two electrons in the helium atom can be extended to atoms with any number of electrons. A systematic method of considering the interactions is provided by the *vector model* of the atom. We have seen how an angular momentum vector interacts with an external magnetic field and precesses around its direction. In a similar way, the angular momentum vectors of two different electrons within an atom can couple together to form a resultant total angular momentum, and each of the individual vectors then precesses around the resultant vector. We must consider the two different kinds of angular

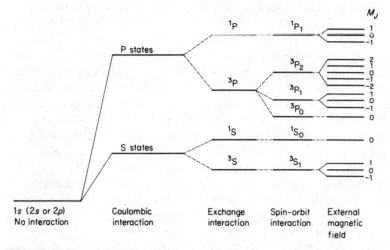

FIGURE 21.12 A diagram (not drawn to correct energy scale) to show the way in which a configuration of the He atom with one electron excited from $n = 1$ to $n = 2$ is split into distinct energy states by internal electrostatic and magnetic interactions, and finally by an external magnetic field. When all the levels have been separated, there are 16 different levels, as required by the 8 different orbitals for an excited $n = 2$ electron with spin.

momentum, the intrinsic angular momentum or electron spin and the orbital angular momentum due to motion of the electron about the nucleus.

The way in which these angular momenta interact is summarized in the scheme called *Russell–Saunders coupling*.

1. The individual spins s_i combine to form a resultant spin S: $\sum_i s_i = S$. The resultant must have integral or half-integral values. For example, three spins of $+\frac{1}{2}$ give $S = \frac{3}{2}$ or $\frac{1}{2}$. Two spins of $+\frac{1}{2}$ couple to give $S = 1$ or 0.

2. The individual orbital angular momenta combine to form a resultant L: $\sum \ell_i = L$. The quantum number L is restricted to integral values. The combination of the ℓ_i can be considered as a quantized vector addition of the corresponding angular momenta, which precess about the resultant L. This model is shown in Fig. 21.13.

3. The L and S represent the total orbital and the total spin angular momenta of all the electrons in the atom. They couple to form a resultant J called the *inner quantum number*. This coupling is called *spin-orbit interaction*. Magnetic fields are due to electric currents, i.e., to moving charges. Just as the sun moves from the viewpoint of Earth, so the nuclear charge Z may move from the viewpoint of an electron, and the resulting magnetic field due to the motion of this charge interacts with the spin magnetic moment of the electron. The spin-orbit interaction increases with nuclear charge Z.

The Russell–Saunders scheme applies to lighter atoms, but begins to break down in heavier atoms. In these, the large nuclear charge leads to a strong coupling between

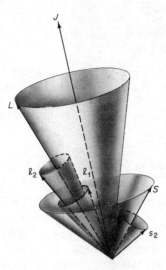

FIGURE 21.13 Russell–Saunders coupling. The ℓ_1 and ℓ_2 combine to form a resultant L. The s_1 and s_2 form a resultant S. L and S precess about their resultant J.

the s_i and ℓ_i *of each electron* due to spin–orbit interaction to give a resultant j_i. These j_i then couple to give J. This system is called *jj coupling*.

Example 21.6 What atomic terms can arise from a p and a d electron?

For a (pd) configuration, for p: $\ell_a = 1$, for d: $\ell_b = 2$. The quantum numbers ℓ are different and the Pauli principle will be obeyed for any assignments of m_ℓ and m_s. L may have value $|\ell_a + \ell_b| = 3, 2, 1$, so that F, D, P terms arise. S may have the value $|s_a + s_b| = 1, 0$. Thus we have singlet and triplet terms, ^3F, ^1F, ^3D, ^1D, ^3P, ^1D. J values $= |L + S|$ can occur for various terms as follows: $^3F_{4,3,2}$, $^3D_{3,2,1}$, $^3P_{2,1,0}$; $^1F_3$, $^1D_2$, $^1P_1$.

A set of rules due to Hund allows one to estimate the order of energies of the different terms that arise from a given electron configuration, but exceptions do occur:

1. The term with the highest spin multiplicity lies lowest in energy; i.e., electrons in atoms remain unpaired (parallel spins) whenever possible.
2. For terms with given spin multiplicity, the one with highest L lies lowest in energy.
3. For atoms with less than half-filled subshells, the term component having the lowest value of J lies lowest in energy. If subshells are more than half filled, the opposite holds. Recall that the value of ℓ defines the subshell.

Example 21.7 The ground configuration of the carbon atom $(1s^2 2s^2 2p^2)$ splits into the following terms: $^1S_0$, $^1D_2$, $^3P_2$, $^3P_2$, $^3P_0$. Arrange these in order of increasing energy.

By rule (1), the triplet terms are lowest, and since the $2p$ subshell is less than half filled, the order of increasing energy is $^3P_0$, $^3P_1$, $^3P_2$. By rule (2), $^1D_2$ is next, and $^1S_0$ is highest.

21.13 Atomic Orbitals and Energies—The Variation Method

Quantum mechanics gives an exact solution for the hydrogen atom. The calculated energy levels and electron distributions are both exact. Any experimental measurement of these quantities can hope only to be nearly as good as the theoretical values. For helium, with two electrons and a nuclear charge of $+2$, the situation is not so rosy. Already in this case we must face the hard truth that we can write down the Schrödinger equation for the system but we cannot solve it exactly by analytic methods.

The helium system is outlined in Fig. 21.14. The potential energy is

$$U = \frac{1}{4\pi\epsilon_0} \left(\frac{e^2}{r_{12}} - \frac{2e^2}{r_1} - \frac{2e^2}{r_2} \right) \tag{21.23}$$

The Schrödinger equation is

$$[\nabla_1^2 + \nabla_2^2 + 2(E - U)]\psi = 0 \tag{21.24}$$

Here, ∇_1^2 and ∇_2^2 are the Laplacians for coordinates of electrons (1) and (2).

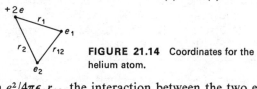

FIGURE 21.14 Coordinates for the helium atom.

The difficulty lies in the term $e^2/4\pi\epsilon_0 r_{12}$, the interaction between the two electrons. This term makes it impossible to separate the variables, i.e., the coordinates of electrons (1) and (2). Fortunately, powerful approximation methods allow us still to obtain accurate solutions in cases like this. We shall describe the *variation method*. Every student of chemistry should understand this method, because it lies at the heart of modern theoretical approaches to atomic and molecular structure.

The Schrödinger equation in operator form is $\hat{H}\psi = E\psi$. We multiply each side by ψ^*, and integrate over all space, to obtain

$$\int \psi^* \hat{H}\psi \, d\tau = \int \psi^* E\psi \, d\tau$$

Since E is a constant, it can be extracted from the integral as

$$E = \frac{\int \psi^* \hat{H}\psi \, d\tau}{\int \psi^* \psi \, d\tau} \tag{21.25}$$

This expression gives the energy of the system in terms of the correct wavefunction ψ, which is the solution to the Schrödinger equation. [Test this formula on the hydrogen atom by substituting ψ_{1s} from Table 21.1.]

But what use is Eq. (21.25) if we do not know the correct ψ? Suppose that we estimate an approximate $\psi^{(1)}$ based on our knowledge of what a reasonable electron distribution might be. Substitution of this trial $\psi^{(1)}$ into Eq. (21.25) yields an energy $E^{(1)}$. The variation theorem states that for any estimated $\psi^{(1)}$, $E^{(1)} \geq E$. The energy calculated from the trial $\psi^{(1)}$ can never be less than the true energy E.*

To apply the variation method, we take a trial expression for ψ that contains a number of variable parameters, $c_1, c_2, \ldots$. We obtain an expression for $E(c_1, c_2, \ldots)$ from Eq. (21.25) and minimize this E with respect to variations of each parameter. The variation method is a systematic effort to determine the electron distribution in a particular atom or molecule. The actual ground-state distribution is naturally the one with the lowest possible energy.

21.14 The Helium Atom

We now apply the variation method to the helium atom. If we simply neglect the effect of one electron on the motion of the other, we can assume that each electron moves in the field of an He^{2+} ion and has a hydrogenlike atomic orbital. Apart from constant factors, the wavefunction for electron (1) would be $1s(1) = e^{-Zr_1}$; for electron (2), $1s(2) = e^{-Zr_2}$. The function that expresses the probability for simultaneously finding electron (1) in $1s(1)$ and electron (2) in $1s(2)$ is the product.

$$\psi = e^{-Zr_1} e^{-Zr_2} \tag{21.26}$$

For $Z = 2$, the energy calculated from Eqs. (21.26) and (21.25) is $E^{(1)} = -74.81$ eV, compared to the experimental -79.99 eV. The discrepancy indicates that the effect of the interaction of the two electrons is important. The entire distribution of electron density is altered by the interaction between the two electrons.

We try next the wavefunction,

$$\psi^{(2)} = e^{-Z'r_1} e^{-Z'r_2} \tag{21.27}$$

This is like the first trial function except that Z' is now a variable parameter, which is adjusted until the minimum energy is found. The effect of changing Z' is to cause the electron distribution to expand (if $Z' < Z$) or to contract (if $Z' > Z$). We call this operation the *adjustment of the scale factor*. The minimum energy for He occurs when $Z' = 2 - 0.313 = 1.687$. The corresponding energy is $E^{(2)} = -77.47$ eV.

*A proof of the variation theorem: Take a complete set of eigenfunctions ψ_i of $\hat{H}$, with $\hat{H}\psi_i = E_i\psi_i$. Consider the expectation value for the energy of an arbitrary normalized function Φ in the space spanned by the eigenfunctions of $\hat{H}$. Then one can represent Φ as

$$\Phi = \sum_{i=0}^{\infty} c_i\psi_i \quad \text{with} \quad \int \Phi^*\Phi \, d\tau = 1$$

Compute

$$I = \int \Phi^* \hat{H}\Phi \, d\tau = \int (\sum c_i\psi_i^*)\hat{H}(\sum c_j\psi_j) \, d\tau = \sum c_i^2 E_i \tag{A}$$

because $\int \psi_i^*\psi_j \, d\tau = 0$ for $i \neq j$.

Now arrange the E_i in a monotonic nondecreasing sequence, $E_0 \leq E_1 \leq E_2 \leq \cdots$. Then we can replace E_i in each term of the sum (A) by E_0 with the assurance that we have never increased the value of the sum, but may have decreased it. Therefore, $I = \sum c_i^2 E_i \geq \sum c_i^2 E_0 = E_0 \sum c_i^2$. From the normalization condition of Φ, $\sum c_i^2 = 1$, so that $I \geq E_0$. Hence the Variation Principle is proved—that the expectation value of E as computed from Φ is an upper bound to the true ground-state energy E_0.

The wavefunction of Eq. (21.27) with $Z' < 2$ indicates that each electron partially screens the nucleus from the other one, reducing the effective nuclear charge from $+2$ to 1.687. The lower effective charge leads to an expansion of the electron density about the nucleus; i.e., the electrons have more space in which to move. As pointed out in Section 20.11 for the particle in a box (for which all the energy is kinetic), such delocalization causes a lowering of the kinetic energy. The expansion of the helium atom causes the potential energy to become less negative, but there is still a net decrease in energy, and hence greater stability.

Next we try a function that explicitly tells one electron to stay away from the other as much as possible:

$$\psi^{(3)} = (1 + br_{12})e^{-Z'r_1}e^{-Z'r_2} \tag{21.28}$$

This function becomes greater as r_{12} increases. The values of the parameters that minimize the energy are $b = 0.364$ and $Z' = 1.849$. The energy is -78.64 eV, close to the experimental -78.99 eV. By adding further terms to the trial function, exact agreement with the experimental energy can be achieved.

21.15 Heavier Atoms—The Self-Consistent Field

As the number of electrons increases, the application of quantum mechanics to the atom becomes more difficult. Most theoretical calculations on multielectron atoms are based on a method developed by Douglas Hartree, a type of variation treatment called the *method of the self-consistent field*. Hartree made the approximation that each electron moves in a spherically symmetric field, which is the sum of the field due to the nucleus and a spherically symmetric averaged field due to all the other electrons. This approximation has the great advantage that as long as each electron j has a potential energy with spherical symmetry, $U(r_j)$, we can separate the Schrödinger equation for all the N electrons in the atom into N equations, one for each separate electron. Thus it is possible to calculate one-electron wavefunctions (orbitals) and to describe them in terms of a set of four quantum numbers, n, ℓ, m_ℓ, m_s.

In the method as given originally by Hartree, the N-electron wavefunction was simply a product of one-electron wavefunctions. In 1930, however, Fock pointed out that most of the effects of electron spin could be taken into account by using, instead of products, antisymmetric linear combinations of wavefunctions as required by the Pauli Principle. The self-consistent field method calculated with antisymmetric wavefunctions is called the *Hartree–Fock method*. It can be applied to molecules as well as atoms.

The difference between the true energy of an atom or molecule and the calculated Hartree–Fock energy arises from two sources: (1) relativistic terms that are important for inner-shell electrons due to their high velocities, but which have little direct effect on chemical behavior; (2) the *correlation energy* due to interactions between electrons that cause the electrostatic field that they experience to differ from the averaged Hartree–Fock field. The correlation energy is important for the chemistry of atoms and molecules, since the energies involved (typically 1 or 2 eV per pair of valence electrons of opposite spin) are exactly in the range that govern chemical reactivities.

The explanation of the structure of the periodic table has been one of the greatest achievements in the history of chemistry. The periodic table is the result of two causes. First is the Pauli Exclusion Principle, which states that no two electrons in an atom can occupy the same orbital specified by quantum numbers n, ℓ, m_ℓ, and m_s. Second is the order of the energy levels of the orbitals, which can be predicted quantitatively by the central-field model. We arrange the different orbitals, each specified by its set of four quantum numbers, in the order of increasing energy and then feed the electrons one by one into the lowest open orbitals, until all the electrons, equal in number to the nuclear charge Z of the atom, have been accommodated. This process was called by Pauli the *Building Principle*.

Figure 21.15 shows the energy levels of the atomic orbitals calculated as func-

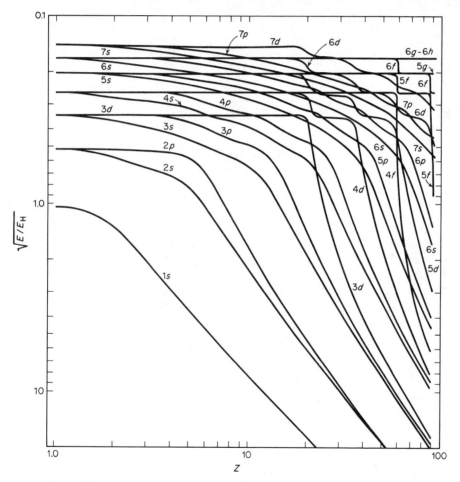

FIGURE 21.15 Energy levels E of atomic orbitals calculated as functions of nuclear charge number Z (atomic number). [Redrawn by M. Kasha from paper of R. Latter, *Phys. Rev. 99*, 510 (1955)] E_H is energy of hydrogen atom in its ground state, 13.6 eV.

tions of Z by the method of the self-consistent field.* In the limit of low Z, all orbitals having the same principal quantum number n fall together at the same energy, because there are too few electrons to cause splitting of the levels. In the limit of high Z, inner orbitals having the same principal quantum number n again fall together in energy. This convergence occurs because the nuclear attraction is now so great that interactions between electrons in the same shell are practically overwhelmed. These are the energy levels observed in X-ray spectra.

At intermediate Z, the sequence of energy levels can become tangled. This is the region in which interactions such as penetration effects can lead to departures from the sequence of principal quantum numbers. Consider, for example, the $3d$ orbital. Owing to the shielding of the nucleus by inner electrons, a $3d$ electron experiences an almost constant nuclear attraction until about $Z = 20$ (Ca). Its energy then begins to fall with Z. According to the calculation, the $3d$ crosses the $4s$ at $Z = 28$ (Ni). Actually, we know from chemical and spectroscopic evidence that the crossover occurs at about $Z = 21$ (Sc). Compared with experiment, all the calculated curves in Fig. 21.15 are displaced somewhat toward higher Z.

The electron configurations and ground-state terms for atoms of the chemical elements are tabulated inside the back cover of the text.

Problems

1. Calculate the ionization energies (in eV) of H, D, and T (^{3}H).

2. The ionization energy I_1 of the He atom, He $\longrightarrow$ He$^+$ + e, is 25.0 eV. Assume that the Bohr theory applies but that the effective nuclear charge is reduced to $Z - \sigma$ by the shielding effect of the second electron. Calculate σ. Calculate the second ionization energy, I_2 (He$^+$ $\longrightarrow$ He^{2+} + e).

3. Calculate the classical potential energy of an electron at a fixed distance a_0 from a proton. How does this energy compare to the energy of the H atom in its ground state? What is kinetic energy of the electron in its ground state?

4. What is the speed v of an electron in the Bohr hydrogen atom in its ground state? Does the speed increase in upper states? Calculate v for the state $n = 10$.

5. A new series in atomic hydrogen spectra was discovered by Humphreys. It begins at 12 638 nm. What are the transitions in this series? There is a line at 3281.4 nm; from which transition does it arise?

6. Calculate the v of an electron in the innermost (1s) shells of Ag and Au on basis of Bohr theory. Calculate the relativistic mass m'_e of the electrons at these speeds from $m'_e = m_e(1 - v^2/c^2)^{-1/2}$. What would be the changes in Bohr radii as a result of the increased values of m'_e? Such relativistic corrections are relevant to the question of why gold is yellow whereas silver is white. [K. S. Pitzer, *Acc. Chem. Res.*, *12*, 271 (1979)]

7. Compare the angular momentum of the electron in the ground state of the H atom as given by (a) the Bohr model; (b) the quantum mechanical model. How can the

*R. S. Berry, "Atomic Orbitals," *J. Chem. Educ.*, *43*, 283 (1966); R. Latter, *Phys. Rev.*, *99*, 510 (1955). Work on all the atoms has been reviewed by F. Herman, *Atomic Structure Calculations* (Englewood Cliffs, N.J.: Prentice-Hall, Inc., 1963).

energies be the same according to the two models, while the angular momenta are different?

8. For the H-atom problem, does the operator $\hat{L}_x$ commute with $\hat{L}_z$? What can you conclude about the possibility of simultaneously measuring two components of electronic angular momentum in the H atom?

9. If ψ_1 and ψ_2 are wavefunctions for a degenerate state of energy E, prove that $c_1\psi_1 + c_2\psi_2$ is also a wavefunction of energy E.

10. Show that ψ_{1s} and ψ_{2s} (Table 21.1) are each normalized. Show that they are orthogonal to each other.

11. For an electron in a $1s$ orbital of H, calculate $\langle 1/r \rangle$, and hence $\langle E_p \rangle$. Show that $\langle E_k \rangle = -(\langle E_k \rangle + \langle E_p \rangle)$ (the virial theorem).

12. Show that the average value of the distance between the nucleus and the electron in a hydrogen atom, when ℓ has its maximum value $n - 1$, is

$$\langle R \rangle = n(n + \tfrac{1}{2})a_0$$

Calculate $\langle R \rangle$ for the $1s$, $2p$, and $3d$ orbitals.

13. You wish to make an atomic model of a H atom on the scale 2 cm = 10^{-8} cm, and you wish the sphere to include 90% of the electron density of the atom. What diameter sphere will you take?

14. Calculate $\langle R \rangle$ for the $2s$, $3s$, and $3p$ states of the H atom and compare to value in Problem 12.

15. The $K_{\alpha 1}$ X-ray line is emitted when an electron falls from an L level ($n = 2$) to a hole in the K level. If Eq. (21.12) holds for the energy levels with an effective nuclear charge number Z' equal to the atomic number Z minus the number of electrons in shells between the given shell and the nucleus, estimate the wavelength of the $K_{\alpha 1}$ X-ray line in chromium. The experimental value is 228.5 pm. What is the main reason for discrepancy?

16. Consider an electron in a hydrogenlike $2p_x$ orbital. What is the probability that it will be around $\theta = 90°$ relative to that around $45°$ (i.e., on the X axis and $45°$ to it).

17. Apply the variation method to atomic hydrogen assuming that $\psi = e^{-\alpha r}$. Calculate E from Eq. (21.25). Minimize E with respect to α, solve for α, and calculate the minimum value of E. Compare to the actual value.

18. Determine the term symbols for the following atoms and the most likely ground terms: $N(p^3)$, $O(p^4)$, $F(p^5)$.

19. The first ionization potentials I_1 of Li, Na, and K are 5.39, 5.14, and 4.34 eV, respectively. To first approximation the outer electrons in these atoms are like $2s$, $3s$, and $4s$ electrons in hydrogen. What value of nuclear charge Z would then give the observed I_1?

20. The electron configurations of Fe^{3+}, Co^{3+}, and Ni^{3+} are $3d^5$, $3d^6$, and $3d^7$, respectively. Predict the ground-state terms.

21. Calculate and plot the radial distribution function $D(r)$ for a $4f$ H orbital.

22

The Chemical Bond

Most of structural and synthetic organic chemistry and a good part of inorganic chemistry depend on the arrangements in space of atoms bonded together to form molecules. The task of the physical chemist is to explain the nature of chemical bonds in these molecules, starting only from properties of atomic nuclei and electrons. The basic questions that need to be answered are the following:

1. Why do atoms unite to form molecules in some cases but not in others? For example, why is H_2 stable but not He_2?

2. Why do atoms combine only in certain definite proportions? For example, why do we have H_2O and H_2O_2 but not H_2O_3?

3. What determines the strengths of chemical bonds, i.e., the bond energies?

4. What determines the three-dimensional shapes of molecules, bond distances, and bond angles? For example, why is CO_2 linear and SO_2 triangular?

We can now give answers to all these questions. The answers are obtained by using quantum mechanics in the form of the Schrödinger equation to calculate the properties of systems of nuclei and electrons. Because of computational difficulties, we cannot calculate numerical values for the properties of large molecules from the basic theory. Nevertheless, the good agreement between theory and experiment for small molecules, like CH_4 and H_2O, makes us confident that the same theoretical principles underlie the structures of larger molecules.

Ever since the work of Berzelius early in the nineteenth century, chemists have believed that the affinity between atoms that causes them to combine into molecules must be due to attractions between positive and negative electric charges. Polar compounds, of which NaCl is a prime example, can be explained by an interaction between positive and negative ions. The nature of the chemical bond in nonpolar compounds, particularly homonuclear molecules like H_2 and N_2, was not understood until the advent of quantum mechanics.

In 1916, W. Kossel made an important contribution to the theory of the polar (electrovalent) bond, and in the same year G. N. Lewis proposed a theory for the nonpolar (covalent) bond. Kossel explained the formation of stable ions by a tendency of interacting atoms to gain or lose electrons until they achieve the stable inert-gas configuration. Potassium has $2 + 8 + 8 + 1$ electrons, and its ionization energy I (K $\rightarrow$ K$^+$ + e) is 4.35 eV. Chlorine has $2 + 8 + 7$ electrons, and its electron affinity A (Cl + e $\rightarrow$ Cl$^-$) is 3.60 eV. An atom of Cl can capture an electron from an atom of K because the electrostatic attractive energy in K$^+$Cl$^-$ is greater than $I - A = 0.75$ eV. Lewis proposed that bonds in nonpolar compounds result from the *sharing of pairs of electrons* between atoms in such a way as to form stable octets to the greatest possible extent. Thus carbon has an atomic number of 6, i.e., 6 electrons, or 4 less than the neon configuration. It can share four pairs of electrons with four hydrogen atoms. Each pair of shared electrons constitutes a single covalent bond. The number of pairs of shared electrons in a bond is called the *bond order*.

No static structure of electrical charges can be stable. The Bohr theory of the hydrogen atom emphasized the importance of the orbital motions of the electron, and showed that stable structures of nuclei and electrons must be dynamic and not static. From an energetic viewpoint, we must therefore always consider both the potential energy and the kinetic energy of the particles that comprise a molecule. The formation of a stable molecule from its constituent atoms simply means that the total energy of nuclei and electrons in the molecule is lower than their total energy in the separated atoms.

The theoretical calculation of the structure of a molecule is therefore the calculation of the dynamic pattern of nuclei and electrons that has minimum energy. To make such a calculation, however, is easier said than done, because the difference in energy between the molecule and its separated atoms is only a small fraction of the total energy of the system. We therefore must calculate a small difference between two large quantities. To calculate the difference accurately, we must calculate the total energies very accurately indeed. As an example, the various contributions to the energy of an H_2 molecule at 0 K are as follows:

Total electronic energy	3098 kJ mol^{-1}
Electronic energy of two H atoms	2643 kJ mol^{-1}
Electronic binding energy	455 kJ mol^{-1}
Zero-point vibrational energy	26 kJ mol^{-1}
Bond energy of H_2 at 0 K	429 kJ mol^{-1}

The binding energy is only about one-seventh the total electronic energy. For molecules more complex than H_2 the fraction is considerably less. We begin our discussion of the theory of chemical bonds, therefore, with the simplest case that we can find.

22.2 The Hydrogen-Molecule Ion H_2^+

The classic example of a covalent bond is found in the hydrogen molecule H_2, a system of two protons and two electrons. The system of two protons and only one electron also yields a stable molecule, H_2^+. Although we cannot isolate this charged species and it forms no stable salts $H_2^+X^-$, H_2^+ occurs in high concentrations in electric discharges through hydrogen gas, and it can be studied in the mass spectrometer. Its dissociation energy $(H_2^+ \rightarrow H^+ + H)$ is 2.78 eV (268 kJ mol^{-1}) and its bond distance H—H is 106 pm, almost exactly twice the Bohr radius a_0 of the hydrogen atom.

The two protons and one electron that comprise H_2^+ are indicated in Fig. 22.1 for one particular internuclear distance R and one particular position of the electron.

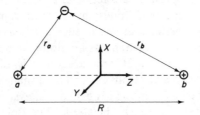

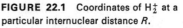

FIGURE 22.1 Coordinates of H_2^+ at a particular internuclear distance R.

The electrostatic potential energies U due to the charges on these particles are given by Coulomb's Law. For the two protons at a separation R, $U_1 = e^2/4\pi\epsilon_0 R$. For the electron–proton interactions, $U_2 = -e^2/4\pi\epsilon_0 r_a$ and $U_3 = -e^2/4\pi\epsilon_0 r_b$. The total potential energy of the system for a given configuration of the charges is, therefore,

$$U = \frac{e^2}{4\pi\epsilon_0}\left(\frac{1}{R} - \frac{1}{r_a} - \frac{1}{r_b}\right) \tag{22.1}$$

The theoretical problem for the hydrogen-molecule ion, H_2^+, is to solve the Schrödinger equation with the potential-energy expression given in Eq. (22.1) and to calculate the total energy of the system.

22.3 Born–Oppenheimer Approximation

The system of two protons and one electron presents a typical three-body problem, and no exact (analytic) solution is possible for the motions of three interacting bodies, either in classical or quantum mechanics.

An important approximate treatment, however, can be applied to this and other molecular problems that involve systems of nuclei and electrons. The masses of nuclei are some thousands of times the mass of an electron, and nuclear motions are slow and ponderous compared to those of electrons. The motions of nuclei as they vibrate in molecules are so slow compared to motions of the electrons that we can calculate the electronic states (wavefunctions and energy levels) *on the assumption that the nuclei are held in fixed positions.* This method was first used by Max Born and J. Robert Oppenheimer in 1927. It has been basic to most quantum mechanical calculations of the properties of molecules.

As a consequence of the Born–Oppenheimer approximation, we can fix the positions of the nuclei in a molecule and then calculate the stationary-state wavefunctions for the electrons in this system of fixed nuclear charges. In the case of the H_2^+ system in Fig. 22.1, the distance R is held constant, and only r_a and r_b are varied. Once this problem is solved, a new fixed value of R is taken, and the problem of the electronic motion solved for this new value. The internuclear distance R is thus a constant parameter in any one calculation.

When the calculations have been made for a number of values of R, the energy $E(R)$ of the system can be plotted against R. The curve that results from the calculations for H_2^+ is shown in Fig. 22.2. Such a plot is usually called the *potential-energy*

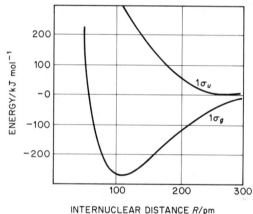

FIGURE 22.2 Energy of H_2^+ as function of internuclear distance R.

INTERNUCLEAR DISTANCE R/pm

curve for the diatomic molecule. The force between the nuclei is given by the derivative of this effective potential-energy function $E(R)$ at any separation R between the nuclei, $F = -(\partial E/\partial R)$. Note that the energy $E(R)$ contains the potential energy of internuclear repulsion and both the potential and kinetic energies of the rapidly moving electrons. The $E(R)$ is the effective potential energy for the vibrating motion of the nuclei. It is identical with the vibrational potential energy discussed in Section 3.6.

22.4 The Chemical Bond in H_2^+

From Eqs. (22.1), (19.17), and (19.18), the Schrödinger equation $\hat{H}\psi = E\psi$ for the H_2^+ system is

$$\left[\frac{-h^2}{8\pi^2 m_e}\nabla^2 + \frac{e^2}{4\pi\epsilon_0}\left(-\frac{1}{r_a} - \frac{1}{r_b} + \frac{1}{R}\right)\right]\psi = E\psi \tag{22.2}$$

By an appropriate choice of a system of coordinates, this equation can be solved exactly to give the electronic wavefunctions ψ and the allowed energy levels for H_2^+. We have already shown results of the energy calculations in Fig. 22.2. The minimum in the plot of $E(R)$ vs. R for the ground state indicates that H_2^+ is a stable molecule. The single electron is remarkably effective in bonding together the two protons.

What can we say about the physical interpretation of this chemical bond? Let us look at the wavefunction ψ to see how the electronic charge is distributed in the H_2^+ system. Recall that $\psi^2 \, d\tau$ gives the probability of finding the electron in any given volume element $d\tau$. Figure 22.3 depicts the wavefunction ψ in two different sections,

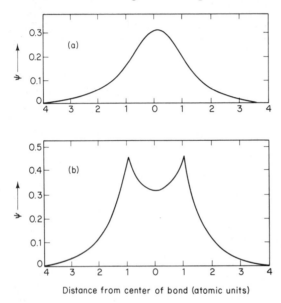

FIGURE 22.3 The exact wavefunction ψ of the electron in the ground state of H_2^+: (a) along a line normal to the H—H bond at its midpoint; (b) along a line passing through the two protons. The two peaks are at the positions of the protons.

Distance from center of bond (atomic units)

one along a line passing through the two protons and the other along a line perpendicular to this internuclear axis at its midpoint. The maxima in ψ occur at the positions of the protons but there is a considerable amplitude of ψ concentrated in the region between them. Since the density of electronic charge is proportional to ψ^2, there is a concentration of negative charge in the region between the protons. The electron cloud is drawn more closely around the positive nuclei. This effect lowers the electrical potential energy and makes an important contribution to the stability of H_2^+ as compared to $H^+ + H$.

We cannot, however, discuss the bond solely in terms of the potential energy. As the internuclear distance R decreases, the electron is confined in a smaller space and its kinetic energy must increase. This effect was discussed in Section 19.13 for the system of an electron in a box. An analogous effect occurs in molecules, and the kinetic energy of an electron rises as it is forced into smaller confines of space. The delicate balance between decreased electron–proton potential energy, increased proton–proton potential energy, and increased electron kinetic energy is such that there is a net lowering of $E(R)$ to give a minimum at $R_e = 106$ pm, the bond distance of the stable H_2^+ molecule. Only the methods of quantum mechanics can provide a calculation of the total energy, potential plus kinetic, of the electron as a function of R. Thus the theory of the chemical bond is based entirely on the quantum mechanical solutions for motions of electrons in a system of fixed positive nuclei.

22.5 Angular Momentum of H_2^+

When we discussed the wave mechanics of the hydrogen atom we found that, in addition to the total energy E, the angular momentum L, and the component of the angular momentum along any one fixed axis L_z are also constants of motion. They are quantized with quantum numbers ℓ and m_ℓ. In the case of H_2^+, we no longer have the spherical symmetry of H, but instead we have a system with cylindrical symmetry. The symmetry axis is the internuclear axis and components of angular momentum about this axis are quantized. The quantum number is denoted by λ, and the angular-momentum components are restricted to values $\lambda(h/2\pi)$.

As in the case of electrons in atoms, a wavefunction for a single electron is called an *orbital*. Thus in H_2^+ we have a set of *molecular orbitals*, which are wavefunctions for the motion of the single electron in the field of the two positive nuclei. The quantum number λ provides a basis for the classification of the molecular orbitals of H_2^+, and, by extension, of other diatomic molecules. The notation is similar to that for atomic orbitals based on quantum number ℓ, except that for the molecular case, Greek letters are used as follows: $\lambda = 0, 1, 2 \ldots$; orbital; $\sigma, \pi, \delta \ldots$.

A second important designation of the molecular orbitals in H_2^+ (and in other homonuclear diatomic molecules) is their symmetry with respect to inversion through the midpoint of the two identical nuclei. As Fig. 22.3 shows, the ground-state orbital is symmetrical (German, *gerade*) with respect to such inversion. It is therefore designated as the $1\sigma_g$ orbital. The first excited state is unsymmetrical with respect to this inversion (German, *ungerade*), i.e., it changes sign ($\psi \longrightarrow -\psi, -\psi \longrightarrow \psi$). This orbital is therefore called the $1\sigma_u$ orbital.

22.6 Simple Variation Theory of H_2^+

The variation theory described in Section 21.15 is the basis for most quantum mechanical calculations of molecular properties. Therefore, it is interesting to apply this method to H_2^+ and to compare the results with the exact solution. We take as variation function a linear combination of the two normalized hydrogen $1s$ atomic orbitals centered on protons a and b.

$$\psi = c_1 \psi_{1sa} + c_2 \psi_{1sb} \tag{22.3}$$

The molecular orbital (MO) in Eq. (22.3) is constructed from a linear combination of atomic orbitals (L.C.A.O.). Thus Eq. (22.3) is an example of a *MO–LCAO approximation* to the exact wavefunction. From Eq. (21.25),

$$E = \int \psi^* \hat{H} \psi \, d\tau \Big/ \int \psi^* \psi \, d\tau \tag{22.4}$$

where $\hat{H}$ is given in Eq. (22.2). We introduce the notation

$$H_{aa} = H_{bb} = \int \psi_a^* \hat{H} \psi_a \, d\tau = \int \psi_b^* \hat{H} \psi_b \, d\tau$$

$$H_{ab} = H_{ba} = \int \psi_a^* \hat{H} \psi_b \, d\tau = \int \psi_b^* \hat{H} \psi_a \, d\tau \qquad (22.5)$$

$$S = \int \psi_a^* \psi_b \, d\tau$$

The integral S is called the *overlap integral*.

From Eqs. (22.3) and (22.4), we obtain

$$E = \frac{c_1^2 H_{aa} + 2c_1 c_2 H_{ab} + c_2^2 H_{bb}}{c_1^2 + 2c_1 c_2 S + c_2^2} \qquad (22.6)$$

To minimize the energy with respect to c_1 and c_2, we set partial derivatives of E with respect to these coefficients equal to zero:

$$\frac{\partial E}{\partial c_1} = 0 = c_1(H_{aa} - E) + c_2(H_{ab} - SE)$$

$$\frac{\partial E}{\partial c_2} = 0 = c_1(H_{ab} - SE) + c_2(H_{bb} - E) \qquad (22.7)$$

These are simultaneous linear homogeneous equations. If we try to solve them in the usual way, by setting up the determinant of the coefficients and dividing it into the same determinant in which a given column is replaced by the constant terms (Cramer's rule), we get only the trivial solutions, $c_1 = c_2 = 0$. Only in the case where the determinant of the coefficients itself equals zero can we obtain nontrivial solutions, and then only for certain values of E, which are the *eigenvalues* of this problem. We therefore write the condition for obtaining nontrivial solutions as the vanishing of the determinant of the coefficients of the set of linear homogeneous equations.

$$\begin{vmatrix} H_{aa} - E & H_{ab} - SE \\ H_{ab} - SE & H_{aa} - E \end{vmatrix} = 0 \qquad (22.8)$$

In this case, the resulting equation is a quadratic in E. In the general case of N simultaneous equations, it is an equation of Nth degree in E. An equation of this type is called a *secular equation*. The solutions of Eq. (22.8) are

$$E(1\sigma_g) = \frac{H_{aa} + H_{ab}}{1 + S}$$

$$E(1\sigma_u) = \frac{H_{aa} - H_{ab}}{1 - S} \qquad (22.9)$$

When these eigenvalues are substituted back into Eq. (22.7), the equations can be solved for the ratio c_1/c_2, giving (as, indeed, is evident from symmetry) $c_1/c_2 = \pm 1$. Thus

$$\psi_g = c_1(\psi_{1sa} + \psi_{1sb}), \qquad \psi_u = c_1(\psi_{1sa} - \psi_{1sb})$$

The constant c_1 is eliminated by the normalization conditions,

$$\int \psi_g^2 \, d\tau = 1, \qquad \int \psi_u^2 \, d\tau = 1$$

$$c_1^2 \left[\int \psi_{1sa}^2 \, d\tau \pm 2 \int \psi_{1sa} \psi_{1sb} \, d\tau + \int \psi_{1sb}^2 \, d\tau \right] = 1$$

$$c_1^2[1 \pm 2S + 1] = 1, \qquad c_1 = 1/\sqrt{2 \pm 2S}$$

The two wavefunctions are, therefore,

$$\psi_g = \frac{1}{\sqrt{2+2S}}(\psi_{1sa} + \psi_{1sb}), \qquad \psi_u = \frac{1}{\sqrt{2-2S}}(\psi_{1sa} - \psi_{1sb}) \qquad (22.10)$$

To evaluate the energies from Eqs. (22.9), the wavefunctions $\psi_{1s} = (\pi a_0^3)^{-1/2} e^{-r/a_0}$ are used to calculate the integrals H_{aa}, H_{ab}, and S for various values of the internuclear distance R. For the $1\sigma_g$ orbital a minimum in energy of -171 kJ mol^{-1} occurs at $R_e = 132$ pm (exact values: -268 kJ mol^{-1} and 106 pm). No minimum in energy is found for the $1\sigma_u$ orbital.

22.7 The Covalent Bond in H_2

If two H atoms are brought together, the system consists of two protons and two electrons. If the atoms are far apart, their mutual interaction is effectively nil. In other words, the energy of interaction $E \longrightarrow 0$ as the internuclear distance $R \longrightarrow \infty$. At the other extreme, if the two atoms are forced closely together, there is a large repulsive force between the positive nuclei, so that as $R \longrightarrow 0$, $E \longrightarrow \infty$. Experimentally, we know that two hydrogen atoms can unite to form a stable hydrogen molecule, whose dissociation energy D_e is 458.1 kJ mol^{-1}. The equilibrium internuclear separation in the molecule is 74.1 pm. These facts about the interaction of two H atoms are summarized in the potential-energy curve of Fig. 22.4.

The system of two protons and two electrons is shown in Fig. 22.5, with coor-

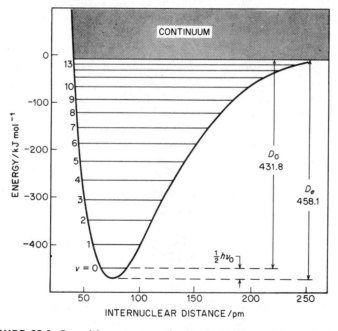

FIGURE 22.4 Potential energy curve for H_2. The vibrational energy levels are shown.

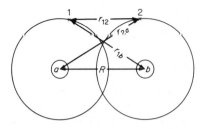

FIGURE 22.5 Coordinates for two protons and two electrons in interaction of two H atoms.

dinates appropriately labeled. This system is quite similar to that for the helium atom, shown in Fig. 21.13. The difference is that we have two nuclei with charges of $+1$ each instead of one nucleus with charge $+2$. Thus, instead of the expression in Eq. (21.24), the potential energy is now

$$U = (r_{12}^{-1} - r_{1a}^{-1} - r_{2b}^{-1} - r_{1b}^{-1} - r_{2a}^{-1} + R^{-1})\frac{e^2}{4\pi\epsilon_0} \qquad (22.11)$$

The Schrödinger equation is

$$\left[\nabla_1^2 + \nabla_2^2 + \frac{8\pi^2 m_e}{h^2}(E - U)\right]\psi = 0 \qquad (22.12)$$

where ∇_1^2 and ∇_2^2 refer to coordinates of electrons 1 and 2, respectively. The main difficulty in solving this equation is caused by the term r_{12}^{-1}. If this term were not present, the equation could be solved exactly, as was the H_2^+ problem.

Most of what we have learned about the covalent bond in H_2^+ also applies in the case of H_2. In general terms, the bond occurs as a consequence of the concentration of negative charge closer to the nuclei. The bond energy is 458.1 kJ mol^{-1} in H_2 as against 268.2 kJ mol^{-1} in H_2^+, and the bond distance is much shorter in H_2, 74.0 pm as compared to 106 pm in H_2^+.

To compare the theoretical energy with the experimental curve of Fig. 22.4, we calculate the energy of the system for a number of different values of the internuclear distance R. The repulsion between the nuclei always contributes a term $e^2/4\pi\epsilon_0 R$. The energy of the electrons is calculated by the variation method, as in the cases of the He atom and the H_2^+ molecule.

To begin the calculation, we must make some reasonable choice for the wavefunction for each electron in the molecule, the *molecular orbital*. An orbital is a one-electron wavefunction, i.e., a function of the coordinates of only one electron, for example $\psi(x_1, y_1, z_1)$. If a molecule contains more than one electron, the orbital treatment is only a first approximation to the exact wavefunction, which for a molecule with N electrons is a function of the coordinates of all the electrons, or $\psi(x_1 y_1 z_1, x_2 y_2 z_2, \ldots, x_N y_N z_N)$.

What shall we take as the first-order approximation for a molecular orbital in the H_2 molecule? If we pull the nuclei far apart, one electron will go with each nucleus, and the system can be represented as the sum of two H atoms. The molecular orbital accordingly becomes the sum of two $1s$ atomic orbitals for H atoms, one centered on nucleus a and the other on nucleus b.

$$\psi_{\text{MO}} = 1s_a(1) + 1s_b(1) \qquad (22.13)$$

The same MO–LCAO was used in the variation treatment of H_2^+. A wavefunction that expresses the combined probability of having electron (1) in ψ_1 and electron (2) in ψ_2 is $\psi_1(1)\psi_2(2)$. Our first trial wavefunction for the H_2 molecule can thus be written as

$$\Psi_{MO}^{(1)} = [1s_a(1) + 1s_b(1)][1s_a(2) + 1s_b(2)] \qquad (22.14)$$

For convenience, we now use the notation $\psi_{1s_a} = 1s_a$. Both electrons have been placed in the same molecular orbital, which is formed as the sum of two $1s$ atomic orbitals. Two electrons can go into such an orbital, in accord with the Pauli Exclusion Principle, provided they have antiparallel spins.

To calculate the energy, we substitute $\Psi_{MO}^{(1)}$ from Eq. (22.14) into Eq. (22.4). The calculation is similar to that given for H_2^+. The calculated dissociation energy $D_e(H_2 \rightarrow 2H)$ is 258.6 kJ mol^{-1}, and the calculated equilibrium internuclear distance R_e is 85.0 pm. The experimental values are 458.1 kJ mol^{-1} and 74.1 pm. The quantitative agreement is poor, but the fact that the calculation yields a stable molecule indicates that the model is reasonable.

Example 22.1 From the molecular orbital ψ_g in Eq. (22.10), calculate the relative probability of finding an electron in H_2 at one of the proton centers and at the midpoint between the two protons.

The probabilities are proportional to

$$\psi^2 = [1/2(1 + S)][\psi_a^2 + 2\psi_a\psi_b + \psi_b^2] \qquad \text{with } \psi = (\pi a_0^3)^{-1/2}e^{-r/a_0}$$

At $r_a = 0$, $r_b = 74$ pm: $\psi^2 \propto [1 + 2e^{-74/53} + e^{-148/53}]$

$$= [1 + 0.495 + 0.061] = 1.556$$

At $r_a = r_b = 37$ pm: $\psi^2 \propto [4e^{-74/53}] = 4(0.248) = 0.990$

The relative probability is 1.572.

The next step, as in the treatment of the He atom, is to introduce a scale factor to adjust the nuclear charge, giving $e^{-Z'r/a_0}$ for ψ_a and ψ_b. The lowest energy is found when $Z' = 1.197$, and gives $D_e = 334.7$ kJ mol^{-1} with $R_e = 73.2$ pm. In this case, unlike that of the He atom, the effective charge is greater than the charge on the individual nucleus. The electrons are thus squeezed into a smaller volume, closer to the nuclei, and their potential energy is lowered. True, the kinetic energy must be raised, but the total energy is lowered by pulling the electrons closer to the nuclei. The resulting improvement in D_e, however, is rather disappointing. The source of the trouble is obvious. We have not included the interaction between the two electrons. In fact, the MO of Eq. (22.14) places both electrons on the same nucleus for a considerable proportion of the time, and thus greatly overestimates the interelectronic repulsion energy.

One way to write a wavefunction that keeps electrons away from each other is to include what is called a *configuration interaction*. The MO of Eq. (22.14) is an L.C.A.O. formed entirely of $1s$ orbitals. If terms from $2s$ and other higher states are also included, the electrons can find additional regions out of each other's way. This treatment at best improves D_e to 386.2 kJ.

So far we have not used wavefunctions that include explicitly the interelectronic distance r_{12}. Even the most complicated functions, if they neglect this factor, will not yield a value of D_e greater than 410 kJ mol^{-1}, about 10% below the experimental value. Once r_{12} terms are included, however, the energy again begins to improve. Kolos and Roothaan in 1960 used a wavefunction with 50 terms and achieved exact agreement with experiment.

The formation of the H_2 molecule as two H atoms come together can be followed in terms of the kinetic energy E_k and the potential energy E_p of the system. Figure 22.6 shows how these energies and their sum, the total energy $E = E_k + E_p$,

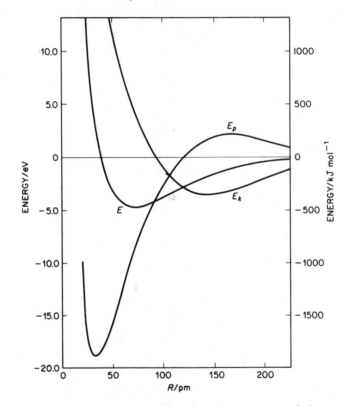

FIGURE 22.6 Calculated kinetic, potential, and total energies of electrons and protons in H_2 as functions of internuclear distance.

vary as the internuclear distance decreases. At first, E_p increases as electronic charge is drawn away from the H$^+$ nuclei into a region between the two nuclei, and E_k decreases as the electrons become more delocalized. E_k is a minimum and E_p a maximum around

140 pm, about twice the equilibrium internuclear distance R_e. As R decreases below 140 pm, E_p begins to fall drastically while E_k rises. The electrons are often close to both nuclei but their kinetic energy is rising as they become more localized in the internuclear region. At the equilibrium value of $R_e = 74$ pm, the total energy is a minimum, with $E_p = -472$ kJ mol^{-1} and $E_k = 236$ kJ mol^{-1}. Note that $E_p = -2E_k$. This result is an example of the *virial theorem*, which holds at equilibrium for any group of particles that interact by coulombic forces.

One of the most graphic ways to present the results of quantum mechanical calculations on simple molecules is to plot the calculated electron densities in defined planes through the molecules. These densities can be presented either as contour maps or as perspective drawings.

A contour map of the electron density in a plane containing the internuclear axis in H_2 is shown in Fig. 22.7. Figure 22.8(a) shows perspective drawings of the electron density ρ. The ρ is drawn in (i) on a linear vertical (electron density) scale and in (ii) on a logarithmic scale. The buildup in electron density between the nuclei is evident, as well as the maxima in density at the nuclei.

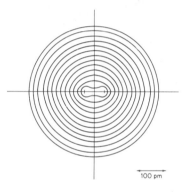

FIGURE 22.7 Contour map of electron density in H_2 in a plane containing the internuclear axis.

100 pm

A particularly instructive kind of drawing for study of the chemical bond is an *electron-density difference plot*. In this case, we subtract from the total electron density the density that would occur if unbonded atoms were centered at the nuclear positions. In the case of H_2, the resultant atomic difference plots are shown in Fig. 22.8(b). The linear plot in (i) shows that, compared to the unbonded atoms, electron density is increased at the nuclei and between them. Electronic charge is withdrawn from regions beyond the internuclear region, but the resultant decrease in density in these regions is barely discernible on a linear plot. In the logarithmic atomic-difference plot of (ii), this effect becomes very evident, however, and it gives a striking picture of the shift of electronic charge density that occurs when two hydrogen atoms form a hydrogen molecule.

The results just described are obtained from what theoretical chemists call *ab initio calculations*. No experimental data are used to evaluate any of the integrals or parameters that occur in the calculations. The results are obtained purely by means of the quantum mechanical theory *ab initio*, from the beginning.

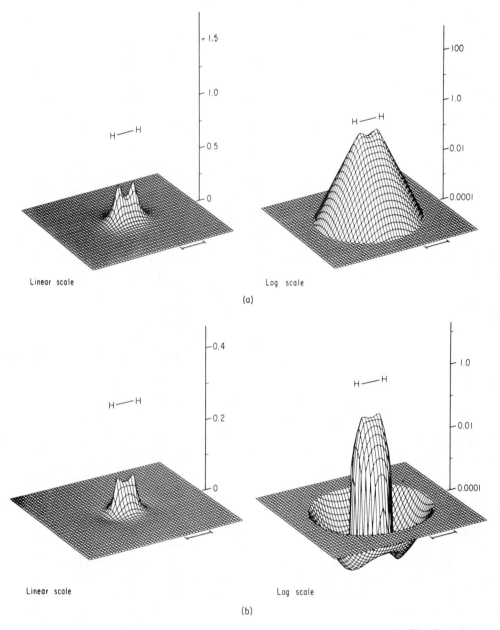

FIGURE 22.8 (a) Electron density in H_2 in the plane of the internuclear axis. The electron density is drawn on a vertical axis, in one case on a linear scale, and in the other case on a logarithmic scale. (b) Electron density difference plot. Here the electron density of the two H atoms has been subtracted from the electron density in H_2. These plots therefore show how the electron density changes as the result of formation of the H—H bond. [Reprinted with permission of Macmillan Publishing Co., Inc., from *Orbital and Electron Density Diagrams* by A. Streitwieser and P. H. Owens. © 1973 Macmillan Publishing Co., Inc.]

22.8 The Valence-Bond Method

The first quantitative theory of the covalent bond was given by Heitler and London in 1927. The method they used to calculate the energy of H_2 was different from the molecular-orbital theory that we have been describing. It was the first example of what is now called the *valence-bond* method in molecular quantum mechanics. This method is closely related to the classical structure theory of organic chemistry, which considers molecules to be constructed of atoms held together by chemical bonds. More precisely, the atoms contribute some of their outer or valence electrons to form bonds with other atoms, so that the molecule consists of atomic cores (atoms that may have provided valence electrons) and bonds between these cores. In the case of H_2, each atom provides one valence electron, and the cores are protons.

The approximate wavefunction chosen by Heitler and London was

$$\psi_{VB}^{(1)} = 1s_a(1)1s_b(2) + 1s_a(2)1s_b(1) \qquad (22.15)$$

The argument by which they arrived at this choice is interesting. They started by considering two H atoms, each with its own electron. A wavefunction that places electron 1 on nucleus a and electron 2 on nucleus b is the product $\psi = 1s_a(1)1s_b(2)$. They then pointed out that the electrons are *indistinguishable particles* and the wavefunction must express this fact. The function $\psi_{VB}^{(1)}$ is *symmetric* in the coordinates of the electrons since it does not change if the indices (1) and (2) are interchanged.

In the H_2 molecule, two electrons with opposite spins, in accord with the Pauli Principle, can be placed in the wavefunction $\psi_{VB}^{(1)}$. When the energy is calculated from $\psi_{VB}^{(1)}$ there is a deep minimum in the potential-energy curve, corresponding to the electron-pair valence bond in H_2. In the wavefunction $\psi_{VB}^{(1)}$ the two electrons do not pile up on the same nucleus as they did in the $\psi_{MO}^{(1)}$ of Eq. (22.14). On the other hand, $\psi_{VB}^{(1)}$ is less effective than $\psi_{MO}^{(1)}$ in delocalizing the electrons.

22.9 Molecular Orbitals for Homonuclear Diatomic Molecules

The VB method is based on the chemical concept that, in a sense, atoms exist within molecules, and that the structure of a molecule can be interpreted in terms of its constituent atoms and the bonds between them. The MO method discards the idea of atoms within molecules, and starts with the bare positive nuclei arrayed in definite positions in space. The total number of electrons is fed one by one into the electrostatic field of the nuclei. The MO theory is more physical than chemical in its view of a molecular structure, which it sees not as atoms connected by bonds, but as an electronic cloud of varying density, interspersed with some positive nuclear stars.

Just as the electrons in an atom can be assigned to definite atomic orbitals characterized by quantum numbers, n, ℓ, m_ℓ, and occupy the lowest levels consistent with the Pauli Principle, so the electrons in a molecule can be assigned to definite molecular orbitals, and at most two electrons with antiparallel spins can occupy any particular molecular orbital.

A molecular orbital can be constructed from a linear combination of atomic orbitals (LCAO*), as in Eq. (22.13),

$$\psi = c_1(1s_a) + c_2(1s_b) \tag{22.16}$$

Since the molecules are cylindrically symmetrical, c_1 must equal $\pm c_2$. Thus the two possible molecular orbitals from the $1s$ atomic orbitals are (except for normalization factors)

$$1\sigma_g = 1s_a + 1s_b$$
$$1\sigma_u = 1s_a - 1s_b \tag{22.17}$$

These molecular orbitals are shown schematically in Fig. 22.9. Both are symmetric *about the internuclear axis;* the angular momentum about the axis is zero,

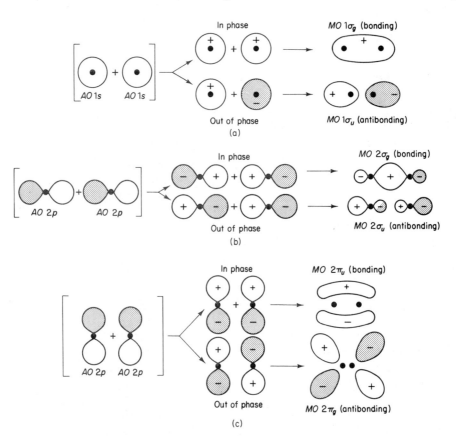

FIGURE 22.9 Formation of molecular orbitals by linear combination of atomic orbitals (MO–LCAO). The + and − signs indicate the signs (phases) of the orbitals, i.e., the one-electron wavefunctions.

*As emphasized in *Coulson's Valence* (R. McWeeny, Oxford Univ. Press, 1979, p. 90), the atomic orbitals in an MO formed by LCAO have no objective significance. We use them simply as convenient building blocks to construct approximate one-electron wavefunctions (orbitals) for molecules. They do allow us to visualize the MO by breaking it down into the more familiar atomic orbitals.

$\lambda = 0$, so that they are σ *orbitals*. The first one is designated a $1\sigma_g$ orbital. It is called a *bonding orbital*, because the concentration of charge between the nuclei bonds them together. The second one is designated $1\sigma_u$, and it is called an *antibonding orbital*, corresponding to a net repulsion because there is much less shielding between the positively charged nuclei. In the case of H_2, the two electrons enter the $1\sigma_g$ orbital. The configuration is $(1\sigma_g)^2$, and it corresponds to a single electron pair bond between the H atoms.

The molecular orbitals we have described are those for H_2, but we can use the same description for other homonuclear diatomic molecules. The next possible molecule is one with three electrons, and the configuration $(1\sigma_g)^2(1\sigma_u)^1$. This is He_2^+. There are two bonding electrons and one antibonding electron, so that a net bonding is to be expected. The molecule has, in fact, been observed spectroscopically; it has a dissociation energy of 290 kJ mol^{-1}. If two helium atoms are brought together, the configuration is $(1\sigma_g)^2(1\sigma_u)^2$. With two bonding and two antibonding electrons, no stable He_2 molecule exists, since the extra $1\sigma_u$ antibonding electron more than offsets the bonding energy of the He_2^+ molecule.

The next higher atomic orbitals are the $2s$, and their L.C.A.O. gives $2\sigma_g$ and $2\sigma_u$ molecular orbitals with accommodations for four more electrons. If we bring together two lithium atoms with three electrons each, the molecule Li_2 is formed. Thus

$$Li[1s^2 2s^1] + Li[1s^2 2s^1] \longrightarrow Li_2[(1\sigma_g)^2(1\sigma_u)^2(2\sigma_g)^2]$$

Actually, the molecular orbitals of inner K-shell electrons need not be explicitly designated since they no longer overlap with one another. The Li_2 configuration is therefore written as $KK(2\sigma_g)^2$. The molecule has a dissociation energy of 110 kJ mol^{-1}. Electron density maps for Li_2 are drawn in Fig. 22.10. The bond in Li_2 is quite weak and the electron distribution in the $2\sigma_g$ bonding orbital still resembles that of two separate atoms. In contrast, the electron distribution in the $1\sigma_g$ orbital of H_2, with its strong bond, is almost spherical (Fig. 22.7).

The hypothetical molecule Be_2, with eight electrons, does not occur, since the configuration would be $KK(2\sigma_g)^2(2\sigma_u)^2$ with no net bonding electrons.

The next atomic orbitals are the $2p$ shown in Fig. 22.9. There are three of these, p_x, p_y, and p_z, mutually perpendicular and with nodes at their origins. The most stable MO that can be formed from the atomic p orbitals is one with the maximum overlap along the internuclear axis. This MO is shown in Fig. 22.9(b). This bonding orbital and the corresponding antibonding orbital can be written

$$3\sigma_g = 2p_{za} + 2p_{zb}; \qquad 3\sigma_u = 2p_{za} - 2p_{zb}$$

These orbitals have the same symmetry about the internuclear axis as the orbitals formed from atomic s orbitals; thus they also have zero angular momentum about the axis, so that $\lambda = 0$ and they are σ orbitals.

The molecular orbitals formed from the p_x and p_y atomic orbitals have a distinctly different form, as shown in Fig. 22.9(c). As the nuclei are brought together, the sides of the p_x or p_y orbitals coalesce, and lead to two streamers of charge density, one above and one below the internuclear axis. These orbitals have an angular momentum of one unit, so that $\lambda = 1$ and they are π orbitals. There are two bonding orbitals π_u and two antibonding π_g.

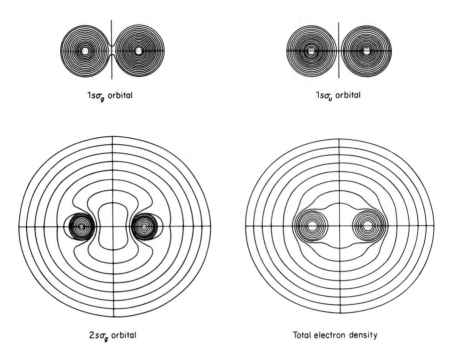

1sσ_g orbital   1sσ_u orbital

2sσ_g orbital   Total electron density

FIGURE 22.10 Contour maps of electron density in individual orbitals and of total electron density for Li_2 molecule. Adjacent contour lines differ by a factor of 2. The outermost contour is an electron density of 4.12×10^{-10} electrons pm^{-3}. The internuclear distance is the experimental value 267.2 pm. [A. C. Wahl, *Science 151*, 961 (1966). Copyright 1966 by the American Association for the Advancement of Science.]

22.10 Correlation Diagram

We can obtain a good understanding of the relative energy levels of molecular orbitals by means of the model of the *united atom*. Imagine that we start with two H atoms both in the same quantum state (1s, for example) and squeeze them gradually together until they coalesce to form the united atom of helium. For this imaginary process, we must assume that internuclear repulsion is ignored. In this way, we can correlate the initial atomic orbitals in the hydrogen atoms with the final atomic orbitals in the helium atom. The molecular orbitals of the H_2 molecule lie somewhere on a curve joining these two extremes, at the correct internuclear distance for H_2.

Such a correlation diagram is shown in Fig. 22.11. We can usually see how orbitals of the isolated atoms correlate with those of the united atom by looking at the symmetry properties of the orbitals. For example, suppose that nuclei A and B in the $1\sigma_u$ orbital shown in Fig. 22.9(a) are squeezed together. The result is an orbital with the typical *p* shape, and the lowest *p* orbital of the united atom is the 2p of helium. The *noncrossing rule* is also useful in establishing the correlation diagram: As the

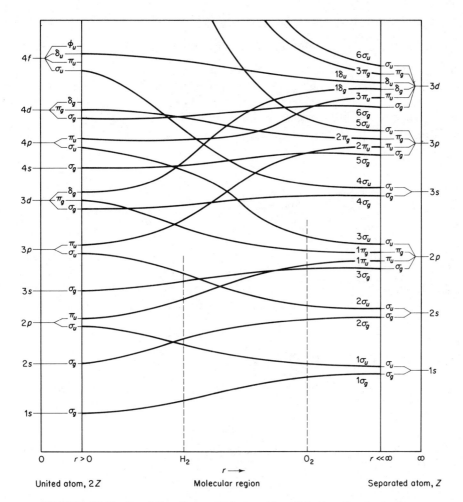

FIGURE 22.11 Correlation diagram for homonuclear diatomic molecules to show how molecular orbitals are made from atomic orbitals of separated atoms and correlate with atomic orbitals of the united atoms formed as the nuclei merge.

internuclear distance is varied, two curves for orbitals of the same symmetry cannot cross. Thus, for example, a σ_g can never cross a σ_g, but a σ_g can cross a σ_u. We can tell whether a molecular orbital is bonding or antibonding by examining the correlation diagram to see whether the energy rises or falls as the atoms are brought together to form this orbital.

The configurations of other homonuclear diatomic molecules can now be described simply by adding electrons to the available orbitals of lowest energy. The formation of N_2 proceeds as follows:

$$N[1s^2 2s^2 2p^3] + N[1s^2 2s^2 2p^3] \longrightarrow N_2[KK(2\sigma_g)^2(2\sigma_u)^2(1\pi_u)^4(3\sigma_g)^2]$$

Since there are six net bonding electrons, it can be said that there is a triple bond between the two N's. One of these bonds is a σ bond; the other two are π bonds.

Molecular oxygen is an interesting case:

$$O[1s^2 2s^2 2p^4] + O[1s^2 2s^2 2p^4] \longrightarrow O_2[KK(2\sigma_g)^2(2\sigma_u)^2(3\sigma_g)^2(1\pi_u)^4(1\pi_g)^2]$$

The contours of the MO's for O_2 are drawn in Fig. 22.12. Four of the six π electrons are assigned to the bonding $1\pi_u$ orbital and two to the antibonding $1\pi_g$. Four of the six σ electrons are assigned to the bonding $2\sigma_g$ and $3\sigma_g$ and two to the antibonding $2\sigma_u$. Thus the net number of bonding electrons is four, which constitute a double bond consisting of a σ bond and a π bond. [Why is order of $1\pi_u$ and $3\sigma_g$ reversed between N_2 and O_2?]

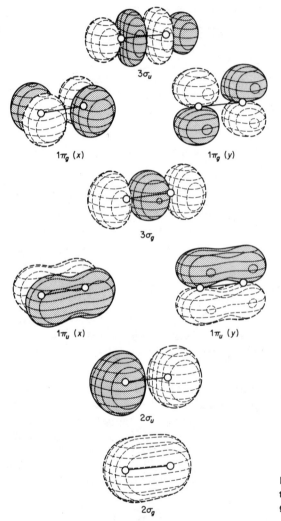

$3\sigma_u$

$1\pi_g\ (x)$ $1\pi_g\ (y)$

$3\sigma_g$

$1\pi_u\ (x)$ $1\pi_u\ (y)$

$2\sigma_u$

$2\sigma_g$

FIGURE 22.12 Contour surfaces of the valence shell molecular orbitals of the ground state of O_2.

In O_2, the $1\pi_g$ orbitals, which can hold four electrons, are only half filled. Because of electrostatic repulsion between the electrons, the most stable state is that in which the electrons are assigned as $(2\pi_{xg})^1(2\pi_{yg})^1$. The total spin of O_2 is therefore $S = 1$, and its spin mutiplicity is $2S + 1 = 3$. The ground state of oxygen is $^3\Sigma$. Because of its unpaired electron spins, O_2 is paramagnetic (Section 23.8). This simple explanation of the paramagnetism of O_2 was one of the first successes of molecular-orbital theory.

In the molecular-orbital method, all electrons outside closed shells make a contribution to the binding energy of the molecule. The shared electron-pair bond is not particularly emphasized. The way in which the excess number of bonding over antibonding electrons determines the tightness of bonding may be seen by reference to the molecules in Table 22.1.

TABLE 22.1
GROUND STATES OF HOMONUCLEAR DIATOMIC MOLECULES

Molecule	Electron configuration	Term symbol	Bond order	D_e (eV)	R_e (pm)
H_2	$(1\sigma_g)^2$	$^1\Sigma_g^+$	1	4.75	74.12
He_2	$(1\sigma_g)^2(1\sigma_u)^2$	$^1\Sigma_g^+$	0	—	—
Li_2	$KK(2\sigma_g)^2$	$^1\Sigma_g^+$	1	1.14	267.3
Be_2	$KK(2\sigma_g)^2(2\sigma_u)^2$	$^1\Sigma_g^+$	0	—	—
B_2	$KK(2\sigma_g)^2(2\sigma_u)^2(1\pi_u)^2$	$^3\Sigma_g^-$	1	3.0	159
C_2	$KK(2\sigma_g)^2(2\sigma_u)^2(1\pi_u)^4$	$^1\Sigma_g^+$	2	6.24	124.3
N_2	$KK(2\sigma_g)^2(2\sigma_u)^2(1\pi_u)^4(3\sigma_g)^2$	$^1\Sigma_g^+$	3	9.76	109.4
O_2	$KK(2\sigma_g)^2(2\sigma_u)^2(3\sigma_g)^2(1\pi_u)^4(1\pi_g)^2$	$^3\Sigma_g^-$	2	5.12	120.8
F_2	$KK(2\sigma_g)^2(2\sigma_u)^2(3\sigma_g)^2(1\pi_u)^4(1\pi_g)^4$	$^1\Sigma_g^+$	1	1.60	140.9

Example 22.2 N_2^+ is less stable than N_2, but O_2^+ is more stable than O_2 (D_e values are 8.86, 9.90, 6.77, and 5.21 eV, respectively). Explain on basis of MO theory.

The ionization $N_2 \longrightarrow N_2^+ + e$ removes an electron from the bonding $3\sigma_g$ orbital, but $O_2 \longrightarrow O_2^+ + e$ removes an electron from the antibonding $1\pi_g$ orbital.

22.11 Heteronuclear Diatomic Molecules

When the two nuclei of a diatomic molecule are not the same, there is no longer a center of symmetry in the molecule and thus the *g-u* designation of the molecular orbitals does not apply. The cylindrical symmetry about the internuclear axis remains, so that λ is still a good quantum number, and the molecular orbitals are still designated σ, π, δ. A correlation diagram like that for homonuclear diatomics is still used. Now, however, the corresponding atomic orbitals of the separated atoms have different energy levels.

The simplest stable uncharged heteronuclear diatomic molecule is LiH. There are now four electrons to be accommodated in the two lowest molecular orbitals. The lowest orbital 1σ is similar to the low-lying $1s$ orbital of Li. The $2s$ and $2p$ atomic orbitals of Li, however, can both participate in the bonding. In a case like this, the molecular orbital can be represented as a 2σ orbital that is a linear combination of the $1s$ orbital of H with both the $2s$ and $2p$ of Li.

$$2\sigma = c_1(1s_H) + c_2(2s_{Li}) + c_3(2p_{z,Li})$$

The electron configuration of LiH is $(1\sigma)^2(2\sigma)^2$.

The calculated electron density in LiH is shown in Fig. 22.13. There is a net positive charge in the region of the Li nucleus and net negative charge in the region of the H nucleus. The charge distribution is quite close to a Li^+H^- structure, but we can see some buildup of electron density between the nuclei.

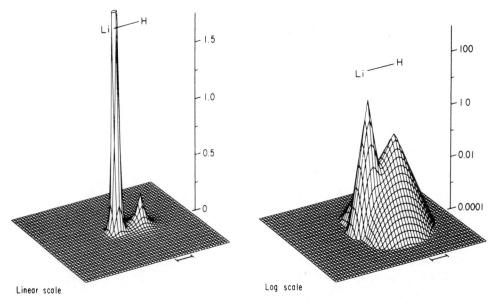

FIGURE 22.13 Electron density in LiH in the plane of the nuclei. [Reprinted with permission of Macmillan Publishing Co., Inc. from *Orbital and Electron Density Diagrams* by A. Streitwieser and P. H. Owens. © 1973 Macmillan Publishing Co., Inc.]

22.12 Electronegativity

Pauling defined electronegativity as "the power of an atom in a molecule to attract electrons to itself." The emphasis on atoms in molecules indicates that electronegativity is a bond property and not a property of isolated atoms. Thus when Pauling devised a numerical electronegativity scale, it was based on bond energies. He defined $\Delta(A\text{—}B)$ as the difference in energy between a bond A—B and the geometric mean of the energies of bonds A—A and B—B,

$$\Delta(A\text{—}B) = D(A\text{—}B) - [D(A\text{—}A)\,D(B\text{—}B)]^{1/2}$$

The quantity $\Delta(A\text{—}B)$ was observed to increase with the electronegativity difference between A and B. The data were fitted to an empirical expression,

$$\Delta(A—B) = 125(X_A - X_B)^2 \qquad (22.18)$$

where $X_A - X_B$ is the difference in electronegativity, and $\Delta(A—B)$ is in kilojoules. The 125 is an arbitrary factor to bring the electronegativity scale to a convenient numerical range. Note that Pauling electronegativities have the dimension of (energy mol^{-1})$^{1/2}$.

Mulliken defined an electronegativity X_j^* as an *atomic* property, the mean of the ionization energy I_j and the electron affinity A_j of the atom,

$$X_j^* = (I_j + A_j)/2 \qquad (22.19)$$

For example, for Cl, $I_j = 1259$ kJ mol^{-1} and $A_j = 349$ kJ mol^{-1}, $X_j^* = 804$ kJ mol^{-1}. Since X_j^* has the dimensions of energy per mole, it cannot be converted to the Pauling scale of X_j.

A summary of electronegativities on the Pauling scale is given in Table 22.2. These values have been found to correlate with a great variety of systematic data on chemical bonds—for example, nuclear quadrupole coupling constants, diamagnetic proton shielding in NMR studies, and frequencies of charge-transfer spectra in molecules with metal ligands. An empirical equation due to Hannay and Smyth relates ionic character of a bond to electronegativity difference:

$$\% \text{ ionic character} = 16(X_A - X_B) + 3.5(X_A - X_B)^2 \qquad (22.20)$$

TABLE 22.2
AVERAGE ELECTRONEGATIVITIES ON THE PAULING
SCALE DETERMINED FROM THERMOCHEMICAL DATA

H 2.1						
Li	Be	B	C	N	O	F
1.0	1.6	2.0	2.5	3.0	3.4	4.0
Na	Mg	Al	Si	P	S	Cl
0.9	1.3	1.6	1.9	2.2	2.6	3.2
K	Ca	—	Ge	As	Se	Br
0.8	1.0	—	2.0	2.2	2.5	3.0
Rb	Sr	—	Sn	Sb	Te	I
0.8	1.0	—	2.0	2.1	2.2	2.7

Example 22.3 Estimate the fractional ionic character FIC of the bond in ICl. If the MO–L.C.A.O. is represented as $\psi = \psi_I + \lambda\psi_{Cl}$, estimate λ.

From Eq. (22.20), $FIC = 0.16(3.2 - 2.7) + 0.035(3.2 - 2.7)^2 = 0.089$. The normalized $\psi = N(\psi_{Cl} + \lambda\psi_I)$. The electron probability density integrated over all space is $\int \psi^2 \, d\tau = 1 = N^2(1 + 2\lambda S + \lambda^2)$, where S is the overlap integral. The transfer of charge from I to Cl is proportional to $\lambda^2 - 1$, so that

$$FIC = \frac{\lambda^2 - 1}{1 + 2\lambda S + \lambda^2}$$

We do not have the exact value for S, but assume that $S = 0.5$, whence $FIC = (\lambda^2 - 1)/(1 + \lambda + \lambda^2) = 0.089$ or $\lambda = 1.14$, and $\psi = \psi_I + 1.14\psi_{Cl}$.

The simplest type of molecule to understand is that formed from two atoms, one of which is strongly electropositive (low ionization potential) and the other, strongly electronegative (high electron affinity), for example, sodium and chlorine. In crystalline sodium chloride, one should not speak of an NaCl molecule since the stable arrangement is a three-dimensional crystal structure of Na^+ and Cl^- ions. In the vapor, however, an NaCl molecule can be studied, in which bonding is due mainly to electrostatic attraction between Na^+ and Cl^- ions.

Such ionic molecules can be treated by the general MO–L.C.A.O. methods, but the ionic bond can be more readily described by a simple electrical model. The attractive force between two ions with charges Q_1 and Q_2 at an internuclear distance R can be represented at moderate distances of separation by a Coulombic force $Q_1 Q_2 / 4\pi\epsilon_0 R^2$ or by a potential $U = -Q_1 Q_2 / 4\pi\epsilon_0 R$. If the ions are brought so close together that their electron clouds begin to overlap, a mutual repulsion between the net positive charges of the shielded nuclei becomes evident. Born and Mayer suggested a repulsive potential having the form be^{-aR}, where a and b are constants. The net potential for the pair of ions is therefore

$$U = \frac{-Q_1 Q_2}{4\pi\epsilon_0 R} + be^{-aR} \qquad (22.21)$$

This function is plotted in Fig. 22.14 for NaCl. The zero of potential energy is for Na^+ and Cl^- at an infinite distance of separation. The minimum in the curve represents the stable internuclear separation for the molecule. Note, however, that at large separations, Na + Cl is a more stable system than $Na^+ + Cl^-$, and accordingly the molecule NaCl dissociates into atoms rather than ions.

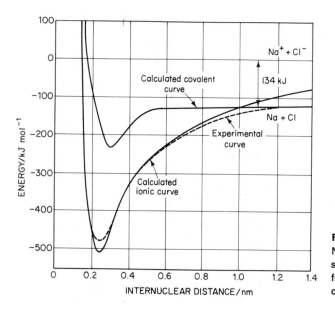

FIGURE 22.14 Potential energy of NaCl as a function of internuclear separation. The ionic curve was calculated from Eq. (22.21). Note the long range of the ionic interactions.

Some experimental properties of alkali-halide molecules are collected in Table 22.3. The chemical bonds in these molecules are never purely ionic. In particular the smaller positive ions tend to distort the electronic charge distribution of the larger negative ions. The effect of this *polarization* is to increase electron density in the region between the two nuclei.

TABLE 22.3
EXPERIMENTAL PROPERTIES
OF GAS-PHASE ALKALI-HALIDE MOLECULES

Molecule	Equilibrium internuclear distance, R_e (pm)	Fundamental vibration (cm^{-1})	Dipole moment, (10^{-30} C m)	Dissociation energy, D_e (kJ mol^{-1})
LiF	156.4	910.34	21.11	577
LiCl	202.1	641.	23.80	469
LiBr	217.0	563.	24.26	423
LiI	239.2	498.	20.9	351
NaF	192.6	536.1	27.22	477
Na^{35}Cl	236.1	364.6	30.05	406
Na^{79}Br	250.2	298.5	30.44	360
NaI	271.1	259.2	30.83	331
KF	217.2	426.0	28.68	490
K^{35}Cl	266.7	279.8	34.28	423
KI	304.8	186.53	36.88	335
RbF	227.0	373.3	28.53	485
CsI	331.5	119.20	40.4	343

Example 22.4 Calculate the dissociation energy D_e of Na^{35}Cl $\rightarrow$ Na $+$ ^{35}Cl on the assumption that the molecule has a purely ionic bond. The fundamental vibration frequency $\nu = 1.093 \times 10^{13}$ Hz and $R_e = 236.1$ pm. The ionization energy of Na is $I = 5.138$ eV and the electron affinity of Cl is $A = 3.613$ eV.

We must evaluate the constants b and a in Eq. (22.21). We use the conditions that $(dU/dR) = 0$ at $R = R_e$, and that the force constant for the vibration at the in potential-energy curve is $\kappa = d^2U/dR^2$. Thus

$$dU/dR = -abe^{-aR_e} + e^2/4\pi\epsilon_0 R_e^2 = 0 \qquad \text{(A)}$$

$$d^2U/dR^2 = a^2be^{-aR_e} - 2e^2/4\pi\epsilon_0 R_e^3 = \kappa \qquad \text{(B)}$$

From (A),

$$b = (e^2/4\pi\epsilon_0\, aR_e^2)e^{aR_e} \qquad \text{(C)}$$

From (C) and (B),

$$(e^2/4\pi\epsilon_0 R_e^3)(aR_e - 2) = \kappa \qquad \text{(D)}$$

We find κ from $\nu_e = (1/2\pi)(\kappa/\mu)^{1/2}$, where μ is the reduced mass, 2.303×10^{-26} kg, and thus $\kappa = 1.086 \times 10^2$ J m^{-2}. From (D), $a = 3.47 \times 10^{10}$ m^{-1}. Then from (A), $b = 4.325 \times 10^{-16}$ J. Now,

$$D_e = -be^{-aR_e} + e^2/4\pi\epsilon_0 R_e - I + A$$
$$= (-1.19 \times 10^{-19}\text{ J}) + (9.77 \times 10^{-19}\text{ J}) - (2.44 \times 10^{-19}\text{ J})$$
$$= 6.14 \times 10^{-19}\text{ J per molecule}$$

or

$$D_e = 370\text{ kJ mol}^{-1}$$

The experimental value is 406 kJ mol^{-1}. Most of the difference is due to a small covalent contribution to the Na—Cl bond.

22.14 Polyatomic Molecules—H$_2$O for Example

In principle, it is possible to describe a polyatomic molecule in terms of non-localized molecular orbitals, by placing the nuclei in fixed positions and pouring the electrons into the array of positive charges. Actually, however, such an approach ignores our chemical knowledge about the ways in which pairs of electrons make localized bonds in molecules. For example, for the water molecule the ΔH°_{298} for $2H + O = H_2O$ is 916 kJ mol^{-1}, which is only 10% greater than twice the value for the hydroxyl radical, $H + O = OH$, $\Delta H^\circ_{298} = 416$ kJ mol^{-1}. It is evident that the structure of H$_2$O can be described by two O—H bonds that are similar to the single bond in O—H. Unless the electrons are known from chemical evidence to be delocalized, as in aromatic carbon compounds, the structure and properties of polyatomic molecules can be discussed most effectively in terms of chemical bonds. Even electrons that are not paired in covalent bonds often are localized as "lone pairs," which also play an important role in determining the structures of molecules.

The advantages of chemical bonds can be maintained by introducing *bond orbitals* or localized molecular orbitals. For example, in the water molecule, the atomic orbitals that take part in bond formation in a first approximation are the 1s orbitals of the two hydrogens and the 2p_x and 2p_y of oxygen. The 2p_z orbital of oxygen cannot contribute since it has a node in the xy plane; the oxygen 2s lies much lower in energy and, in this model, is not used in bonding. The stable structure will be that in which there is maximum overlap of the atomic orbitals from the O and the H. Instead of making the molecular orbital by a linear combination of all four atomic orbitals, we take them in pairs to form two localized molecular orbitals corresponding to the two O—H bonds.

$$\psi_1 = 1s(H_a) + 2p_x(O)$$
$$\psi_2 = 1s(H_b) + 2p_y(O)$$

The formation of these MO's is shown schematically in Fig. 22.15.

An important property of the orbitals ψ_1 and ψ_2 is that they are orthogonal, i.e., $\int \psi_1\psi_2\,d\tau = 0$. Because ψ_1 and ψ_2 are orthogonal, each one makes its own contribution to the electron density of the molecule. Thus molecular orbitals that are orthogonal bond orbitals, such as ψ_1 and ψ_2, are mathematically appropriate to represent the description of molecular structures in terms of chemical bonds. There are two electrons from H and four from O to be accommodated in the molecular orbitals. A pair of electrons of opposite spin is placed in each of the orbitals, ψ_1 and

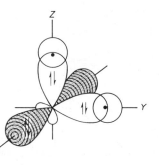

FIGURE 22.15 Formation of molecular orbitals for H_2O by overlap of $2p$ orbitals of O and $1s$ orbitals of H.

ψ_2, and the third pair is placed in the empty $2p_z$ orbital of oxygen, as a typical lone pair.

The observed valence angle in H_2O is not the 90° predicted by the orbitals above but actually 104.5°. The difference may be ascribed in part to the polar nature of the bond; the electrons are drawn toward the oxygen, and the residual positive charge on the hydrogens causes their mutual repulsion. In H_2S, the bond is less polar and the angle is 92°. Note the straightforward fashion in which directed valence has been explained in terms of the shapes of the molecular orbitals. Although the explanation is based on our prior knowledge of the experimental shape of the molecule, detailed quantum mechanical calculations confirm the conclusions of the simple model. Much more elaborate molecular orbitals are used, instead of ψ_1 and ψ_2, and their adjustable parameters are derived by the variation method.

Figure 22.16 shows results of such an *ab initio* quantum mechanical calculation of electron density in H_2O, in the xy plane of the molecule. From such quantum mechanical calculations, it is now possible to obtain reliable values for the energies, electron densities and other physical properties of small light molecules, such as H_2O.

Note in Fig. 22.16 the considerable density of electron charge that appears at the back of the oxygen nucleus (away from the bonds). This distribution indicates that the molecular orbital holding the "2s" lone pair is not spherically symmetric as in the rough model discussed above. This electron density is important in the formation of hydrogen bonds, as the electron-rich region can interact, for example, with an electron-deficient hydrogen atom in another water molecule.

22.15 Calculation of Molecular Geometries

Advances in computational techniques and in computer speeds have greatly extended the range of molecular-orbital calculations on small molecules. The geometries of many molecules, including all internuclear distances and bond angles, can often be calculated with an accuracy equal to that of the experimental measurements. Some examples are given in Table 22.4.

It is even possible to calculate molecular structures that have not yet been determined experimentally. For instance, many different positive ions occur in the ionic beams of mass spectrometers. Determination of their structures by electron-diffraction methods is a difficult procedure. One example is the structure of CH_5^+, which

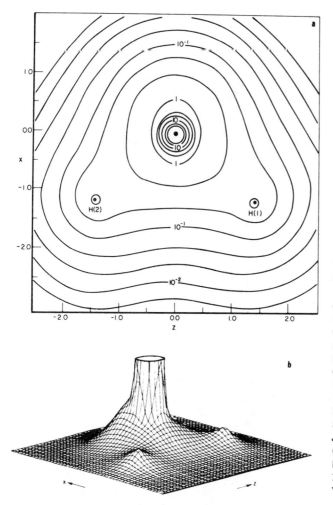

FIGURE 22.16 (a) The total electron density ρ of the ground state of water in the molecular (xz) plane in units of electrons per a_0^3 (a_0 = 52.9 pm). The large solid dot in the center represents the oxygen nucleus ($\rho \simeq 300$) at $(x, y, z) = (0, 0, 0)$ and the smaller circles the protons ($\rho \simeq 0.5$). The density gradient is $(0.1/a_0^3)$ per contour. Distances along the axes are in units of a_0. (b) A three-dimensional plot of the electron density shown in (a). The density is plotted along the axis perpendicular to the molecular (xz) plane. The viewer is oriented 45° clockwise from the line bisecting the HOH angle and 15° above the plane. [These plots were prepared by T. H. Dunning based on the calculations of S. Aung, R. M. Pitzer, and S. I. Chan, *J. Chem. Phys. 49*, 2071 (1968). From C. W. Kern and M. Karplus, "The Water Molecule," in *Water, A Comprehensive Treatise*, vol. 1, ed. Felix Franks (New York: Plenum Press, 1972)]

has been calculated to show two electrons forming an unusual three-center bond:

We can be confident that if and when this compound is studied experimentally it will have the predicted structure.

22.16 Nonlocalized Molecular Orbitals—Benzene

Electrons in molecules are not always essentially localized between two nuclei. Important examples of *delocalization* are found in conjugated and aromatic hydrocarbons. In the case of benzene, the carbon atomic orbitals are first prepared as

TABLE 22.4

COMPARISON OF THEORETICAL CALCULATIONS
AND EXPERIMENTAL MEASUREMENTS OF MOLECULAR
STRUCTURES OF SMALL MOLECULES[a]

Molecule	Point group[b]	Parameter	R_e(pm) and <(°)	
			MO theory[c]	Spectro-scopic[d]
LiH	$C_{\infty v}$	r(LiH)	164.3	159.6
CH	$C_{\infty v}$	r(CH)	112.6	112.0
CH_2	C_{2v}	r(CH)	108.0	107.8
		<(HCH)	131.8	136
CH_3	D_{3h}	r(CH)	108.1	107.9
CH_4	T_d	r(CH)	109.1	109.2
NH	$C_{\infty v}$	r(NH)	104.4	103.6
NH_2	C_{2v}	r(NH)	103.1	102.4
		<(HNH)	103.2	103.3
NH_3	C_{3v}	r(NH)	101.7	101.2
		<(HNH)	106.2	106.7
OH	$C_{\infty v}$	r(OH)	98.1	97.0
OH_2	C_{2v}	r(OH)	96.7	95.8
		<(HOH)	104.3	104.5
FH	$C_{\infty v}$	r(FH)	93.2	91.7
C_2H_2	$D_{\infty h}$	r(CC)	120.6	120.3
		r(CH)	106.6	106.1
C_2H_4	D_{2h}	r(CC)	133.4	133.9
		r(CH)	108.6	108.5
		<(HCH)	116.4	117.8
HCN	$C_{\infty v}$	r(CN)	115.8	115.3
		r(CH)	106.7	106.5
HNC	$C_{\infty v}$	r(CN)	117.4	117.2
		r(NH)	100.0	98.6
CH_3NH_2	C_s	r(CN)	146.6	147.1
		<(HNO)	109.5	110.3
H_2CO	C_{2v}	r(CO)	121.0	120.8
		r(CH)	110.4	111.6
		<(HCH)	116.0	116.5
CH_3OH	C_s	r(CO)	142.1	142.1
		r(OH)	96.7	96.3
N_2H_4	C_2	r(NN)	144.0	144.7
		$\omega(H_aNNH_b)$	90.9	88.9
O_2	$D_{\infty h}$	r(OO)	122.1	120.8

[a]D. J. DeFrees, K. Raghavachari, H. B. Schlegel, and J. A. Pople
[Carnegie–Mellon University, 1982].

[b]See Chapter 25; the point group specifies the shape of the molecule.

[c]Results are typical of contemporary MO calculations that include electron-correlation effects. They are all obtained by the same method.

[d]The accuracy of the spectroscopic results is such that most of the theoretical values lie within the experimental range.

trigonal sp^2 hybrids and then brought together with the hydrogens. These localized σ orbitals lie in a plane, as shown in Fig. 22.17(a). The atomic p orbitals extend their lobes above and below the plane [Fig. 22.17(b)], and when they overlap, they form delocalized molecular orbitals, above and below the plane of the ring. These orbitals hold six mobile electrons. The shapes of the three π orbitals with lowest energies are shown in Fig. 22.17(c).

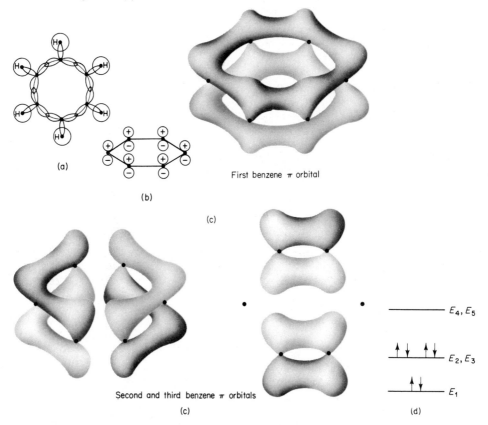

(a)

(b)

First benzene π orbital

(c)

Second and third benzene π orbitals

(c)

(d)

E_4, E_5

E_2, E_3

E_1

FIGURE 22.17 Molecular orbitals of benzene: (a) overlap of sp^2 orbitals; (b) the p_z orbitals which overlap to give π orbitals; (c) the three lowest π orbitals.

We can write a π molecular orbital in benzene as a linear combination of six atomic p orbitals:

$$\psi = c_1\psi_1 + c_2\psi_2 + c_3\psi_3 + c_4\psi_4 + c_5\psi_5 + c_6\psi_6 \tag{22.22}$$

This wavefunction ψ expresses the fact that a π electron can move around the benzene ring. We calculate the ground state energy by variation of the coefficients c_1, c_2, etc., until we find the ψ functions that give the lowest energy. The method is just like that used in Section 22.6 for H_2^+, except that now there are six terms in ψ instead of two. Thus the equation corresponding to Eq. (22.8) is

$$\begin{vmatrix} H_{11} - S_{11}E & H_{21} - S_{21}E & \cdots & H_{61} - S_{61}E \\ H_{12} - S_{12}E & H_{22} - S_{22}E & \cdots & H_{62} - S_{62}E \\ H_{13} - S_{13}E & H_{23} - S_{23}E & \cdots & H_{63} - S_{63}E \\ H_{14} - S_{14}E & H_{24} - S_{24}E & \cdots & H_{64} - S_{64}E \\ H_{15} - S_{15}E & H_{25} - S_{25}E & \cdots & H_{65} - S_{65}E \\ H_{16} - S_{16}E & H_{26} - S_{26}E & \cdots & H_{66} - S_{66}E \end{vmatrix} = 0 \qquad (22.23)$$

The secular equation (22.23), when multiplied out, is an equation of the sixth degree in E, and therefore has six roots.

Because of the difficulties in evaluating the integrals in the secular equation, an approximate treatment, developed by E. Hückel, has been used by organic chemists.[*] The approximations are the following:

1. $H_{jj} = \alpha$, the *Coulombic integral* for all j
2. $H_{jk} = \beta$, the *resonance integral* for $j \neq k$ and atoms that are bonded together
3. $H_{jk} = 0$ for $j \neq k$ atoms that are not bonded together
4. $S_{jj} = 1$
5. $S_{jk} = 0$ for $j \neq k$

With these approximations, the secular equation is greatly simplified, and it becomes

$$\begin{vmatrix} \alpha - E & \beta & 0 & 0 & 0 & \beta \\ 0 & \alpha - E & \beta & \beta & 0 & 0 \\ 0 & \beta & \alpha - E & \beta & 0 & 0 \\ 0 & 0 & \beta & \alpha - E & \beta & 0 \\ 0 & 0 & 0 & \beta & \alpha - E & \beta \\ \beta & 0 & 0 & 0 & \beta & \alpha - E \end{vmatrix} = 0$$

The roots of this equation of the sixth degree are

$$E_1 = \alpha + 2\beta, \qquad E_{2,3} = \alpha + \beta \text{ (twice)}, \qquad E_{4,5} = \alpha - \beta \text{ (twice)}, \qquad E_6 = \alpha - 2\beta$$

Since α and β are negative, the roots are given in order of increasing energy. When the values of E are put back into the system of linear equations, they can be solved for the coefficients c_j, and thus explicit expressions for the molecular orbitals are obtained.

The MO of lowest energy is

$$\psi_A = \psi_1 + \psi_2 + \psi_3 + \psi_5 + \psi_6$$

This orbital, which can hold two electrons with antiparallel spins, is shown in Fig. 22.17(c). There are two next lowest molecular orbitals, ψ_B and ψ'_B, which together hold four electrons. The six π electrons of benzene occupy these three orbitals of low energy, so that the exceptional stability of the structure is explained by the theory.

[*]See W. M. Flygare, *Molecular Structure and Dynamics* (Englewood Cliffs, N.J.: Prentice-Hall, Inc., 1978), p. 372.

The total energy is $E = 2(\alpha + 2\beta) + 4(\alpha + \beta) = 6\alpha + 8\beta$. If the structure of benzene consisted of three localized single bonds and three localized double bonds (i.e., a single Kekulé structure), the energy of the six π electrons in the ground state would be $6(\alpha + \beta) = 6\alpha + 6\beta$. Therefore, the lowering of energy due to delocalization of electrons is -2β, and this is often called the *resonance energy*.

In applying Hückel theory to organic compounds, one does not attempt to evaluate the integrals by *ab initio* calculations. Instead, the theoretical expressions are fitted to experimental data to obtain empirical values for the integrals. For example, from thermochemical data the resonance energy of benzene is 150 kJ. (This is the difference between the experimental ΔU_0 of formation of benzene and the ΔU_0 calculated from six C—H, six C—C, and three C=C bonds). Thus the empirical $\beta = -75$ kJ mol^{-1}. Given β, we can assign energy levels to the different π electron orbitals in benzene, as shown in Fig. 22.17(c).

22.17 Photoelectron Spectroscopy

The technique of photoelectron spectroscopy provides direct experimental information about the energy levels of electrons in molecular orbitals. Short-wavelength ultraviolet radiation or long-wavelength X rays are used to eject electrons from molecules. This process is called *photoionization:* $A + h\nu \rightarrow A^+ + e$. The energy of the photon $h\nu$ absorbed by the molecule A provides the ionization energy I_J, and any excess energy appears as kinetic energy ϵ_k of the electron:

$$I_J = h\nu - \epsilon_k \tag{22.24}$$

Since $h\nu$ is known from the frequency of the radiation used, a measurement of ϵ_k tells us the energy I_J necessary to release an electron from a particular quantum state in the molecule. The value of ϵ_k can be measured by various devices that analyze the velocities of the emitted electrons. [Why is ϵ_k of the ion missing from Eq. (22.24)?]

To the extent that the molecular-orbital approximation is valid, we can consider that the emitted electron comes from some definite molecular orbital in the molecule. By a theorem due to T. Koopmans (1934), to a good approximation, the ionization energy I_J equals the energy of the molecular orbital $-\epsilon_J$.

$$I_J = -\epsilon_J \tag{22.25}$$

Thus measurements of the I_J of a molecule in a photoelectron spectrometer give values for the energy levels of electrons in the molecule that can be compared directly with the results of quantum mechanical molecular-orbital calculations.

A diagram of a photoelectron spectrometer is shown in Fig. 22.18. A similar method can also be applied to solids. A convenient light source is a helium discharge lamp, the main output of which is a sharp line at wavelength $\lambda = 58.4$ nm from the $P(1s\,2p) \rightarrow S(1s^2)$ transition in the atom (photon energy 21.1 eV). The line from He$^+$ at 30.3 nm (40.8 eV) is also useful. The resolution of the analyzer of photoelectron energies is about 0.1 eV for work in the ultraviolet, and about 1 eV for X rays.

An example of a photoelectron spectrum is that for N_2 in Fig. 22.19. The bands in the photoelectron spectrum correspond with the molecular orbitals of N_2 shown in

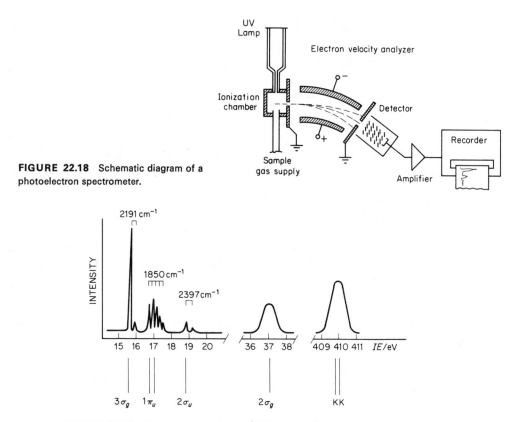

FIGURE 22.18 Schematic diagram of a photoelectron spectrometer.

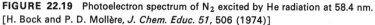

FIGURE 22.19 Photoelectron spectrum of N_2 excited by He radiation at 58.4 nm. [H. Bock and P. D. Mollère, *J. Chem. Educ. 51*, 506 (1974)]

Table 22.1. The structure of the bands is caused by the different vibrational levels of the N_2^+ ion that are excited when the electron is ejected: $N_2 \rightarrow N_2^+ + e$. The bond distance in the positive ion in general will be different from that in the neutral molecule, and the resultant abrupt change in bond length on ionization leads to excited vibrational states in the ion. The ionization energy will be somewhat different for ions with different vibrational energy.

We recall that the MO configuration of N_2 is $KK\, 2\sigma_g^2\, 2\sigma_u^2\, 1\pi_u^4\, 3\sigma_g^2\, 1\sigma_g^2$. Ionization from the orbital of highest energy is not seen in this spectrum. Ionizations from the filled orbitals $1\pi_u^4$ and $2\sigma_g^2$ give two bands for each orbital, since the spin of the ionized electron may be either parallel or antiparallel to that of the unpaired electron in the orbital.

Problems

1. The ionization energy I of Na is 5.08 eV, the electron affinity A of Cl is 3.82 eV. At what distance would the potential-energy curve of the ionic state of NaCl cross that of the covalent state? What distance for KBr, with $I = 4.34$ and $A = 3.37$ eV?

2. The dissociation energy of H_2 is $D_0 = 4.46$ eV and the zero-point energy $\frac{1}{2}h\nu_0 = 0.26$ eV. Calculate D_0 and D_e for H_2, D_2, HD, and T_2.

3. Show that the wavefunctions ψ_g and ψ_u in Eq. (22.10) are orthonormal.

4. Indicate which of the following molecules would be (a) less stable than their positive ion $(AB)^+$; (b) less stable than their negative ion $(AB)^-$: N_2, O_2, NO, CO, CN, C_2, and F_2.

5. Draw the correlation diagram for heteronuclear diatomic molecules between atoms in first row of periodic table, corresponding to Fig. 22.11 for homonuclears. Then write the probable electron assignments to the molecular orbitals in BeO, OH, NH, CO, CN, and LiN.

6. On the basis of Fig. 22.11, discuss the bonding in Na_2, P_2, S_2. How would you expect the strengths of bonding in these third row molecules to compare to these of the second row molecules in Table 22.1? Why? The actual D_e values are 0.73, 5.03, 4.4 eV, respectively.

7. The overlap integral for the ψ_g and ψ_u functions in Eq. (22.10) can be evaluated as $S = e^{-R/a_0}(1 + R/a_0 + R^2/3a_0^2)$ where $a_o = 52.9$ pm. Sketch S vs. R and note the values of S at R_e for H_2^+ and H_2.

8. From the "potential-energy curve" for H_2^+ in Fig. 22.2, calculate the kinetic energy of the electron in the molecule at the minimum in energy at $R_e = 106$ pm. [Hint: Remember the virial theorem.] Refer to Fig. 22.6 for the H_2 molecule. How does the average kinetic energy of an electron in H_2 at $R = 106$ pm compare to that in H_2^+ at this distance? Comment?

9. Mulliken defined an electronegativity scale as $(EN) = (I + A)/2$, where I is the ionization potential and A the electron affinity of an atom. For F, Cl, Br, I, the values of I are 17.42, 12.97, 11.81, and 10.44 eV, and of A, 3.48, 3.61, 3.40, and 3.11 eV, respectively. Calculate the Mulliken (EN) of these elements and compare them with the values on the Pauling scale.

10. The Morse function for potential energy of a diatomic molecule is $U = D_e[1 - e^{-\beta(r-r_0)}]^2$. Derive $\beta = \pi\nu(2\mu/D_e)^{1/2}$ from the fact that the force constant κ is the curvature of the potential-energy curve at $r = r_0$, $(\partial^2 U/\partial r^2)_{r_0}$.

11. Apply the Hückel theory of Section 22.16 to π bonding in the allyl radical, $CH_2 = CH—CH_2—$, starting with a MO, $\psi = c_1\psi_1 + c_2\psi_2 + c_3\psi_3$, where 2 is the C atom at center of molecule. Show that $E = [(c_1^2 + c_2^2 + c_3^2)\alpha + 2(c_1c_2 + c_2c_3)\beta][c_1^2 + c_2^2 + c_3^2 + 2(c_1c_2 + c_2c_3)S]^{-1}$, where $\alpha = H_{jj}$ and $\beta = H_{jk}$.

12. From the bond enthalpies in Table 6.4, calculate the difference in electronegativities of atoms in HF, HCl, HBr, and HI, and compare with the values in Table 22.2 based on electronegativity (H) $= 2.1$.

13. The term symbols of diatomic molecules are constructed from the molecular-orbital assignments of their electrons in a way that is analogous to the method for atoms given in Section 21.12. Thus the vector sum of λ_i give a Λ, and states with $\Lambda = 0, 1, 2, \ldots$ are Σ, Π, Δ, $\ldots$. The multiplicity of the state $2S + 1$ is a left superscript, where S is the number of unpaired spins. The g, u, symmetry for homonuclear diatomics is a right subscript. On this basis explain the ground-state term symbols in Table 22.1.

14. Consider the benzene molecule as a two-dimensional square box of side 350 pm. Assign the six π electrons to the lowest accessible energy levels. Calculate the wavelength of the radiation that would raise an electron to the first excited state. Compare this result with that from the Hückel theory in Section 22.16.

15. The molecules Na_2 and B_2 dissociate into atoms when the respective vapors are heated. If spin angular momentum must be conserved, what are the allowable spin states of the Na and B atoms that are formed? Will all the atoms be in their ground electronic states?

16. Carry out a Hückel MO analysis of the orbitals in butadiene and calculate the four lowest states in terms of α and β. Sketch the energy levels and orbitals.

17. The first two photoelectron bands in butadiene are at 9.03 and 11.46 eV. Calculate the resonance energy β.

18. The photoelectron spectrum of H_2S was studied with the helium resonance line at 58.4 nm. The ejected photoelectron energy was 10.74 eV. What is the energy of the orbital from which this electron is ejected?

23

Electric and Magnetic Properties of Molecules

Electric charges and magnetic moments in the interior of molecules respond in various ways to externally applied electromagnetic fields. The study of such responses can provide information about the structure of molecules. The purpose of this chapter is to describe the interpretation of some of these interactions of external fields with molecules. The electric and magnetic properties of materials have many technical applications in electronics and electrical engineering, and the theoretical foundation of these technologies begins with the molecular theories that we shall outline.

23.1 Relative Permittivity

Consider a parallel-plate capacitor with a vacuum between the plates. Let the charges on the plates be $+\sigma$ and $-\sigma$ per unit area. From basic electrostatic theory,* the electric field between the plates then has the magnitude

$$E_0 = \sigma/\epsilon_0 \tag{23.1}$$

where ϵ_0 is the vacuum permittivity, 8.854×10^{-12} F m^{-1} or C V^{-1} m^{-1}.

---

*A derivation can be found in *Currents, Fields and Particles* by F. Bitter (New York: John Wiley & Sons, Inc., 1956).

Example 23.1 The plates of an evacuated capacitor hold one electronic charge per square micrometer. What is the electric field between the plates?

From Eq. (23.1),

$$E_0 = \frac{1.602 \times 10^{-19} \times (10^6)^2 \text{ m}^{-2}}{8.854 \times 10^{-12} \text{ C m}^{-1} \text{ V}^{-1}} = 1.81 \times 10^4 \text{ V m}^{-1}$$

Note that even a low surface charge density suffices to cause a high electric field.

Now consider that the vacuum between the capacitor plates is filled with a nonconducting substance or *dielectric*, while the charge on the plates is not altered. The electric field then falls to a new value

$$E = \frac{E_0}{\epsilon_r} = \frac{\sigma}{\epsilon_r \epsilon_0} \tag{23.2}$$

The factor ϵ_r, which is always greater than unity, is called the *relative permittivity* or *dielectric constant* of the medium. ($\epsilon_r = \epsilon/\epsilon_0$, where ϵ is the *permittivity*.)

The *capacitance* is the ratio of the charge on the plates to the difference in electric potential between them,

$$C = \frac{Q}{\Delta\Phi} = \frac{\sigma A}{Ed} = \frac{\epsilon_r \epsilon_0 A}{d} \tag{23.3}$$

where A is the area of the plates and d is the distance between them. The relative permittivity ϵ_r is usually measured with a capacitance bridge as the ratio of capacitance with a dielectric between the plates to capacitance with a vacuum, $\epsilon_r = C/C_0$.

The values of ϵ_r for gases are close to unity; for various liquids ϵ_r ranges from about 2 to 100, and for a few special solids such as barium titanate up to about 1000. Some data on ϵ_r which depend on temperature, and on the frequency of the electric field, are given in Table 23.1.

TABLE 23.1
RELATIVE ELECTRIC PERMITTIVITIES ϵ_r AT ZERO FREQUENCY

Gases at STP		Liquids at 25°C	
Air	1.000583	Benzene	2.2725
He	1.000074	$CHCl_3$	4.724
CH_4	1.000886	CH_3OH	32.60
SO_2	1.009930	Nitrobenzene	34.89
		Water	79.45
		HCN	107

23.2   Polarization of Dielectrics—Dipole Moments

Why does the introduction of a dielectric between the plates of a capacitor decrease the electric field? The answer to this question was given by Faraday in 1837. He pointed out that an electric field *polarizes* a dielectric, i.e., causes a separation between the centers of positive and negative charges. An internal field is thus produced in the dielectric, which is opposite in direction to the external field. Polarization of the dielectric decreases the net field between the plates of a charged capacitor.

A pair of electric charges of opposite sign $\pm Q$ separated by a distance r, as shown in Fig. 23.1(a), is called an *electric dipole*. An electric dipole is characterized by its *dipole moment* $\boldsymbol{\mu}$, which is a vector having the magnitude Qr and the direction of the line from the negative to the positive charge. A dipole consisting of charges $\pm e$ separated by 100 pm (the order of magnitude of a chemical bond) would have a dipole moment of magnitude $\mu = (1.602 \times 10^{-19} \text{ C})(100 \times 10^{-12} \text{ m}) = 1.602 \times 10^{-29} \text{ C m}$. In honor of Peter Debye, the Dutch physical chemist who developed the theory of dipole moments, the unit 10^{-18} esu cm $= 3.338 \times 10^{-30}$ C m is called the *debye* (D).

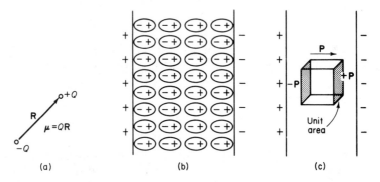

FIGURE 23.1 (a) An electric dipole and its dipole moment $\boldsymbol{\mu}$. (b) Polarization of a dielectric. (c) Definition of the polarization vector **P**.

The *polarization* of a dielectric [Fig 23.1(b)] is defined quantitatively as the *dipole moment per unit volume*. The polarization vector **P** shown in Fig. 23.1(c) is parallel in direction to the electric field vector. The polarization **P** is equivalent to the dipole moment that would be produced in a unit cube of dielectric by charges $\pm P$ on the unit faces of the cube that are parallel to the capacitor plates. Thus the field in the dielectric due to the polarization is P/ϵ_0. The resultant field in the dielectric is

$$E = \frac{E_0}{\epsilon_r} = \frac{\sigma}{\epsilon_0} - \frac{P}{\epsilon_0} = E_0 - \frac{P}{\epsilon_0}$$

The ratio $P/\epsilon_0 E$ is the *electric susceptibility* χ_e(khi) of the dielectric. Thus

$$\epsilon_r - 1 = \chi_e = P/\epsilon_0 E \tag{23.4}$$

The polarization of a dielectric by an applied field can occur in two ways—by induction of dipoles and by orientation of dipoles. An electric field always induces a dipole in a molecule by pulling apart to some extent its positive and negative charges. This inductive effect is independent of temperature. If the molecule has a permanent dipole (in the absence of a field) the external field tends to align this dipole parallel to the field direction. The random thermal motions of the molecules (intermolecular collisions) try to destroy this orienting effect of the field, but a net polarization remains. The orientation effect decreases with increasing temperature.

The polarization is thus the sum of two terms, $P = P_d + P_o$. The *induced* or *distortion polarization* P_d is due to the partial separation of charges caused by the field. The *orientation polarization* P_o is due to the preferential alignment of permanent dipoles by the field. We can further write $P_d = P_e + P_n$ where P_e is due to displacement of electrons and P_n to displacement of nuclei.

23.3 Polarizability

Let us consider first the dipole moment induced in an atom placed in an electric field. The field displaces the electrons and to a lesser extent the nucleus and thus induces a dipole moment $\mathbf{\mu}$ that is directly proportional to the field $\mathbf{E}^*$ (Fig. 23.2):

$$\mathbf{\mu} = \alpha \mathbf{E}^* \qquad (23.5)$$

The proportionality factor α is called the *polarizability* of the atom. (Be careful not to confuse *polarizability* and *polarization*—they are entirely different quantities.) The field is written $\mathbf{E}^*$ to indicate that it is the local field acting on the atom and not necessarily the same as the overall field $\mathbf{E}$.

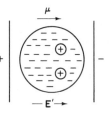

FIGURE 23.2 Induction of a molecular dipole by the electric field. The effect of field is exaggerated.

In SI, the units of α [from Eq. (23.5)] are $C\ m/V\ m^{-1} = C\ m^2\ V^{-1}$. The units of ϵ_0 are $C\ V^{-1}\ m^{-1}$ so that $\alpha/4\pi\epsilon_0$ has the dimensions of volume. For example, for the hydrogen atom $\alpha/4\pi\epsilon_0 = 4.5a_0^3$, whereas the volume of a sphere of radius equal to that of the Bohr orbit a_0 is $\frac{4}{3}\pi a_0^3 = 4.19a_0^3$. Table 23.2 lists values of $\alpha/4\pi\epsilon_0$ for different atoms.

TABLE 23.2
VALUES OF $\alpha/4\pi\epsilon_0$ FOR SOME ATOMS

Atom	H	He	Li	Be	C	Ne	Na	A	K
$(\alpha/4\pi\epsilon_0)$ $(10^{-30}\ m^3)$	0.66	0.21	12	9.3	1.5	0.4	27	1.6	34

Note the high polarizabilities of alkali-metal atoms. The electron density due to the valence electron in these atoms is easily distorted by an electric field.

Let us consider next the polarizability of methane, a molecule that has an almost spherical shape (an *isotropic* molecule). If we simply add the atomic polarizabilities, $(\alpha_C + 4\alpha_H)/4\pi\epsilon_0 = 4.1 \times 10^{-30}\ m^3$, but the experimental value for CH_4 is $\alpha/4\pi\epsilon_0 = 3.3 \times 10^{-30}\ m^3$. We can see here how the strong covalent C—H bonds inhibit distortion of the electron distribution by an external field.

When we consider a molecule that is distinctly *anisotropic* and not at all spherical in shape, the polarizability becomes a more complicated property. In CO_2, for example, the polarizability $\alpha_{\parallel}$ along the molecular axis is almost twice the polarizability $\alpha_{\perp}$ perpendicular to the axis. It is no longer possible to describe the polarizability of such a molecule by a single scalar quantity α. In such cases the induced electric dipole vector μ no longer is parallel to the electric field vector E^*. Each component of the dipole vector depends on each component of the field vector. Thus we must write the general relation as

$$\mu_x = \alpha_{xx}E_x + \alpha_{xy}E_y + \alpha_{xz}E_z$$
$$\mu_y = \alpha_{yx}E_x + \alpha_{yy}E_y + \alpha_{yz}E_z \qquad (23.6)$$
$$\mu_z = \alpha_{zx}E_x + \alpha_{zy}E_y + \alpha_{zz}E_z$$

The set of nine α_{ij} together comprise the *polarizability tensor*, which can be represented by a matrix:

$$\hat{\alpha} = \begin{bmatrix} \alpha_{xx} & \alpha_{xy} & \alpha_{xz} \\ \alpha_{yx} & \alpha_{yy} & \alpha_{yz} \\ \alpha_{zx} & \alpha_{zy} & \alpha_{zz} \end{bmatrix}$$

Only six of the components of the polarizability tensor are independent, since $\alpha_{xy} = \alpha_{yx}$, $\alpha_{xz} = \alpha_{zx}$, $\alpha_{yz} = \alpha_{zy}$. As a result of this symmetry, it is always possible to find a set of reference axes X, Y, Z for a molecule that will cause the off-diagonal components of the tensor to vanish, so that

$$\hat{\alpha} = \begin{bmatrix} \alpha_{XX} & 0 & 0 \\ 0 & \alpha_{YY} & 0 \\ 0 & 0 & \alpha_{ZZ} \end{bmatrix}$$

The mean polarizability is $\bar{\alpha} = (\alpha_{XX} + \alpha_{YY} + \alpha_{ZZ})/3$. We can determine values of $\bar{\alpha}$ from measurements of ϵ_r or of refractive index n_0 (Section 23.6). To find the polarizabilities $\alpha_{XX}, \alpha_{YY}, \alpha_{ZZ}$ for anisotropic molecules, however, requires special experimental measurements such as the effect of electric field on the refractive index (Kerr effect) and depolarization of scattered light. When the molecule has cylindrical symmetry $\alpha_{YY} = \alpha_{ZZ} = \alpha_{\perp}$ is the polarizability perpendicular to the axis of symmetry and $\alpha_{XX} = \alpha_{\parallel}$ is the polarizability parallel to that axis. Polarizabilities of a number of molecules are given in Table 23.3. [Is $\alpha_{\parallel}$ of benzene parallel to plane of ring?]

23.4 The Local Field

In a gas at low pressure, uncharged molecules are so far apart that they do not exert appreciable electric forces on one another. In this case, the field that polarizes a molecule is simply the external field E, and the induced dipole moment $\mu = \alpha E$. The distortion polarization P_d, the induced dipole moment per unit volume, is then the number of molecules in unit volume (N/V) times the induced dipole moment of a molecule. If V_m is the molar volume, $N/V = L/V_m$. Hence $P_d = L\,\alpha E/V_m$. From Eq. (23.4), the dielectric constant of a pure dilute gas of nonpolar molecules is

POLARIZABILITIES OF CYLINDRICALLY SYMMETRIC
MOLECULES AS MEASURED WITH VISIBLE LIGHT:
$\alpha/4\pi\epsilon_0$ IN UNITS OF 10^{-30} m³

Molecule	$\alpha/4\pi\epsilon_0$	$\alpha_\parallel/4\pi\epsilon_0$	$\alpha_\perp/4\pi\epsilon_0$
H_2	0.79	1.00	0.69
N_2	1.77	2.24	1.53
O_2	1.60	2.83	1.23
Cl_2	4.61	6.60	3.62
HCl	2.63	3.13	2.39
N_2O	3.01	5.03	2.00
CO_2	2.64	4.06	1.94
CH_3Cl	4.62	5.66	4.08
$CHCl_3$	8.56	6.74	9.46
NH_3	2.26	2.41	2.18
C_6H_6	10.50	6.68	12.36

$$\epsilon_r = 1 + \frac{L\alpha}{\epsilon_0 V_m}, \quad \text{so that} \quad \chi_e = \frac{L\alpha}{\epsilon_0 V_m} \tag{23.7}$$

If the dielectric is not a dilute gas, the local field $\mathbf{E}^*$ that acts to polarize a given molecule is partly due to the charges on the capacitor plates and partly due to charges on neighboring molecules. Imagine a small spherical cavity in the dielectric around the given central molecule (Fig. 23.3). The effective field $\mathbf{E}^*$ is then the overall field $\mathbf{E}$

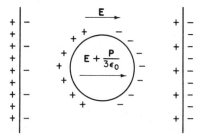

FIGURE 23.3 A spherical cavity in a dielectric. The charges on surface of cavity lead to an effective field $\mathbf{E}^* = \mathbf{E} + (\mathbf{P}/3\epsilon_0)$ within the cavity.

plus the field due to charges at the boundaries of the cavity. Lorentz showed† that this extra field is $\mathbf{P}/3\epsilon_0$, so that

$$\mathbf{E}^* = \mathbf{E} + (\mathbf{P}/3\epsilon_0) \tag{23.8}$$

It follows from Eq. (23.5) that $\boldsymbol{\mu} = \alpha[\mathbf{E} + (\mathbf{P}/3\epsilon_0)]$. Since $\boldsymbol{\mu} = (V_m/L)\mathbf{P}$, we solve these two equations for P/E and substitute this into Eq. (23.4) to obtain for the nonpolar condensed phase,

$$\frac{\epsilon_r - 1}{\epsilon_r + 2} V_m = \frac{L\alpha}{3\epsilon_0} \equiv P_m \tag{23.9}$$

This is called the *Clausius–Mossotti equation*. The *molar polarization* P_m, has dimensions of volume per mole. So far, P_m contains only the contribution from induced

†Derivation: D. A. McQuarrie, *Statistical Thermodynamics* (New York: Harper & Row, Publishers, 1973), p. 250.

dipoles. The contribution from permanent dipoles must next be considered, in order to complete the theory that relates the measured property ϵ_r to the molecular properties α and μ. [**P** is a vector, but we commonly call its magnitude *P polarization*.]

Example 23.2 The relative permittivity of methane at STP is $\epsilon_r = 1.000886$. Estimate the volume of the CH_4 molecule from this result and compare it to the value from van der Waals b.

For an ideal gas, $V_m = RT/P$ so that Eq. (23.7) becomes

$$\epsilon_r - 1 = LP\alpha/RT\epsilon_0, \qquad \alpha/\epsilon_0 = \frac{RT}{LP}(\epsilon_r - 1) = \frac{kT}{P}(\epsilon_r - 1)$$

$$\alpha/4\pi\epsilon_0 = \frac{(1.381 \times 10^{-23} \text{ J K}^{-1})(273 \text{ K})(0.000886)}{4\pi \times 101.32 \times 10^3 \text{ N m}^{-2}} = 2.63 \times 10^{-30} \text{ m}^3$$

Van der Waals $b = 4.278 \times 10^{-5} \text{ m}^3/\text{mol} = 4Lv_m$.

$$v_m = \frac{4.278 \times 10^{-5}}{4 \times 6.02 \times 10^{23}} = 1.78 \times 10^{-29} \text{ m}^3/\text{molecule}$$

23.5 Orientation of Dipoles in an Electric Field

When a molecule is placed in an electric field, an induced dipole moment is evoked almost instantaneously in the direction of the field. If the position of the molecule is disturbed by thermal collisions, a new dipole is immediately induced again by the field. Averaged over all the molecules in the sample, the induced dipole moment is independent of temperature. The contribution to the polarization caused by permanent dipoles, however, is less at higher temperatures, because the random thermal collisions of the molecules continually destroy the alignment of the dipoles in the electric field.

To obtain the contribution of permanent dipoles to the molar polarization, it is necessary to calculate the average component of a permanent dipole in the field direction as a function of temperature. Consider a dipole with random orientation. If there is no field, all orientations are equally probable. This fact can be expressed by saying that the number of dipole moments directed within a solid angle $d\omega$ is $A\, d\omega$, where A is a factor depending on the number of molecules under observation.

If a dipole moment μ is oriented at angle θ to a field of strength E^*, its potential energy is $U = -\mu E^* \cos \theta$. According to the Boltzmann equation, the number of molecules oriented within the solid angle $d\omega$ is then

$$dN = A \exp\left(\frac{-U}{kT}\right) d\omega = A \exp\left(\frac{\mu E^* \cos \theta}{kT}\right) d\omega$$

The average value of the dipole moment in the direction of the field, from Eq. (5.33), can be written

$$\bar{\mu} = \frac{\int A \exp\left(\mu E^* \cos \theta/kT\right)\mu \cos \theta\, d\omega}{\int A \exp\left(\mu E^* \cos \theta/kT\right) d\omega}$$

where the integrals are taken over all possible orientations. To evaluate this expression, let $\mu E^*/kT = x$, $\cos \theta = y$; then $d\omega = 2\pi \sin \theta \, d\theta = 2\pi \, dy$. Thus,

$$\bar{\mu}/\mu = \int_{-1}^{+1} e^{xy}y \, dy \Big/ \int_{-1}^{+1} e^{xy} \, dy$$

Since

$$\int_{-1}^{+1} e^{xy} \, dy = \frac{e^x - e^{-x}}{x} \qquad \text{and} \qquad \int_{-1}^{+1} e^{xy}y \, dy = \frac{e^x + e^{-x}}{x} - \frac{e^x - e^{-x}}{x^2}$$

then

$$\frac{\bar{\mu}}{\mu} = \frac{e^x + e^{-x}}{e^x - e^{-x}} - \frac{1}{x} = \coth x - \frac{1}{x} \equiv \mathcal{L}(x)$$

This theory was devised by Paul Langevin. The Langevin function $\mathcal{L}(x)$ is shown in Fig. 23.4.

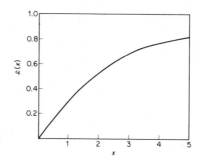

FIGURE 23.4 The Langevin function $\mathcal{L}(x) = \coth x - 1/x$.

In most cases, $x = \mu E^*/kT$ is a small fraction, so that on expanding $\mathcal{L}(x)$ in a power series, only the first term need be retained, leaving $\mathcal{L}(x) = x/3$, or

$$\bar{\mu} = \frac{\mu^2}{3kT}E^* \tag{23.10}$$

The orientation polarization due to permanent dipoles is now added to the induced polarization. Instead of Eq. (23.9), the total molar polarization is therefore

$$\frac{\epsilon_r - 1}{\epsilon_r + 2}V_m = P_m = \frac{L}{3\epsilon_0}\left(\alpha + \frac{\mu^2}{3kT}\right) \tag{23.11}$$

This equation was first derived by Debye.

The Debye equation makes it possible to evaluate both α and μ from the intercept and slope of P_m vs. $1/T$ plots, as shown in Fig. 23.5. The experimental data are values over a range of temperatures of the dielectric constant ϵ_r. The ϵ_r can be calculated from the measured capacitance of a parallel-plate capacitor in which the vapor or solution under investigation is the dielectric between the plates. The Debye equation is applicable to polar gases and to dilute solutions of polar molecules in nonpolar solvents. It is not satisfactory, however, for polar liquids (water, methanol, etc.), since the local field that gives the Clausius–Mossotti equation no longer applies. The theory of the local field in polar liquids has been developed by Lars Onsager and by John Kirkwood.

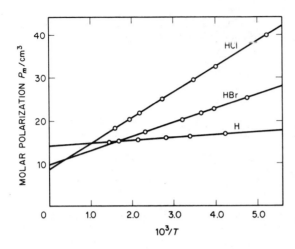

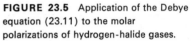

FIGURE 23.5 Application of the Debye equation (23.11) to the molar polarizations of hydrogen-halide gases.

Example 23.3 Consider molecules of nitrobenzene which have a permanent dipole moment $\mu = 13.35 \times 10^{-30}$ C m. In an electric field $E^* = 2.0 \times 10^6$ V m^{-1}, what fraction of the molecules at 298 K are oriented with their dipole moments within 0° and 10° of the field direction? What is the average moment in the field direction?

The fraction of molecules between angles θ_1 and θ_2 would be

$$\Delta = \int_{\theta_1}^{\theta_2} \exp\left(\mu E^* \cos \theta / kT\right) \sin \theta \, d\theta \Big/ \int_0^{\pi} \exp\left(\mu E^* \cos \theta / kT\right) \sin \theta \, d\theta$$

Let $\mu E^* \cos \theta / kT = x$, so that

$$\Delta = \frac{\int_{x_1}^{x_2} e^x \, dx}{\int_{\mu E^*/kT}^{-\mu E^*/kT} e^x \, dx} = \frac{e^{\mu E^* \cos \theta_2/kT} - e^{\mu E^* \cos \theta_1/kT}}{e^{-\mu E^*/kT} - e^{\mu E^*/kT}}$$

$\cos \theta_1 = \cos 0° = 1$, $\cos \theta_2 = \cos 10° = 0.9848$, $\mu E^*/kT = 6.50 \times 10^{-3}$. Hence

$$\Delta = \frac{1.00642 - 1.00652}{0.99352 - 1.00652} = \frac{-0.00010}{-0.013} = 0.0077$$

(Note that less than 1% of the molecules are in this range of angles.) The average moment is

$$\bar{\mu}/\mu = \mathcal{L}(x) \simeq \frac{x}{3} \qquad \text{with} \qquad x = \frac{\mu E^*}{kT} = 6.50 \times 10^{-3}$$

Hence $\bar{\mu}/\mu = 0.0022$ and $\bar{\mu} = 0.029 \times 10^{-30}$ C m.

23.6 How Relative Permittivity Depends on Frequency

Relative permittivities are usually measured with capacitance bridges operated in the radiofrequency range, say 50 to 500 kHz. As the electric field in the sample cell changes direction, the dipoles change their orientations in response. At higher frequencies, however, the dipoles can no longer follow the alternations in the field, and hence the contribution of permanent dipoles to the polarization of a dielectric (orientation polarization P_o) declines until only the induced polarization P_d remains. As a result, the relative permittivity ϵ_r depends on frequency ν. An example is shown in Fig. 23.6.

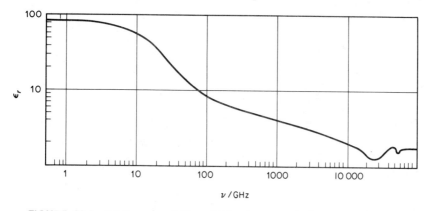

FIGURE 23.6 Relative permeability ϵ_r (dielectric constant) of water as a function of the frequency of the electric field.

At still higher frequencies, the part of the induced polarization due to shifts of the nuclei is lost and only the electronic part remains. With frequencies in the infrared range, both nuclei and electrons still follow the field, but with frequencies in the range of visible light, the contribution of nuclear motions to P_d is negligible.

Light always travels more slowly in a material medium than in a vacuum. The ratio of the speed in vacuum to that in the medium c/c_m is the refractive index n_0 of the medium. A light wave is a rapidly alternating electric and magnetic field. The electric field of the light wave acts to polarize the medium through which it passes, pulling the electrons back and forth in rapid alternation. One of the important results of Maxwell's electromagnetic theory of light is that $\epsilon_r = n_0^2$. This result, of course, holds only when the effects of permanent dipoles have been excluded from the relative permittivity ϵ_r.

When the Maxwell relation does hold, Eq. (23.9) becomes

$$P_m = R_m = \frac{n_0^2 - 1}{n_0^2 + 1} V_m = \frac{L\alpha}{3\epsilon_0} \qquad (23.12)$$

where R_m is called the *molar refraction*. This is the *Lorentz–Lorenz equation*. It enables us to calculate induced polarizations from measurements of refractive index. By combining these results with dielectric data, the dipole moment can then be calculated from the Debye equation.*

Example 23.4 The refractive index of benzene at 298 K for light of wavelength 600 nm is 1.498. The density of benzene is $\rho = 0.874 \times 10^3$ kg m^{-3}. Estimate the average polarizability of benzene at 600 nm.

---

*I. F. Halverstadt and W. D. Kumler, *J. Am. Chem. Soc.*, **64**, 2988 (1942).

From Eq. (23.12) the molar refraction of benzene is

$$R_m = \left(\frac{1.498^2 - 1}{1.498^2 + 2}\right)\left(\frac{78.1 \times 10^{-3} \text{ kg mol}^{-1}}{0.874 \times 10^3 \text{ kg m}^{-3}}\right) = 26.2 \times 10^{-6} \text{ m}^3 \text{ mol}^{-1}$$

$$R_m = L\bar{\alpha}/3\epsilon_0, \qquad \bar{\alpha} = \frac{3(26.2 \times 10^{-6} \text{ m}^3 \text{ mol}^{-1})(8.854 \times 10^{-12} \text{ C}^2 \text{ J}^{-1} \text{ m}^{-1})}{6.02 \times 10^{23} \text{ mol}^{-1}}$$

$$\alpha = 1.16 \times 10^{-39} \text{ C m}^2/\text{V}$$

23.7 Dipole Moment and Molecular Structure

In addition to results from dielectric data, dipole moments can also be obtained from the effect of electric fields on molecular spectra (Stark effect) and from an electric resonance method applied to molecular beams. A representative selection of dipole moment values is given in Table 23.4.

TABLE 23.4
MAGNITUDES OF DIPOLE MOMENTS

Compound	Dipole moment $\text{C m} \times 10^{30}$	Compound	Dipole moment $\text{C m} \times 10^{30}$
HCN	9.78	CH_3F	6.04
HCl	3.44	CH_3Cl	6.24
HBr	2.61	CH_3Br	6.00
HI	1.27	CH_3I	5.47
H_2O	6.17	C_2H_5Cl	6.84
H_2S	3.17	CH_2Cl_2	5.27
NH_3	4.97	CH_3COCH_3	9.47
SO_2	5.37	CH_3OH	5.64
CO	0.40	C_6H_5OH	5.67
NO	0.53	$C_6H_5NO_2$	13.4
LiH	19.6	CH_3NO_2	11.7
H_2O_2	7.34	$C_6H_5CH_3$	1.23

Dipole moments provide two kinds of information about molecular structure: charge distribution and molecular geometry. Consider a collection of positive and negative charges that is electrically neutral as a whole. To specify the positions of each charge, we can assign a set of Cartesian coordinates x, y, z, with respect to any convenient axes. Thus a charge Q_i is located at $x_i\ y_i\ z_i$ in this coordinate system. The coordinates $x_i y_i z_i$ specify a vector $\mathbf{r}_i$ from the origin to the position of the given charge. The dipole moment of this collection of charges is then defined by the vector sum

$$\boldsymbol{\mu} = \sum Q_i \mathbf{r}_i \tag{23.13}$$

The individual components of the dipole-moment vector are

$$\mu_x = \sum Q_i x_i, \qquad \mu_y = \sum Q_i y_i, \qquad \mu_z = \sum Q_i z_i \qquad (23.14)$$

We can illustrate this definition of the dipole moment by applying it to a simple model of a molecule. Consider in Fig. 23.7 the H_2O molecule, which has a planar

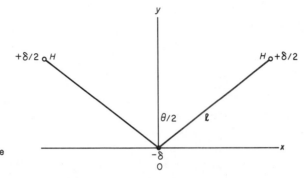

FIGURE 23.7 Calculation of the dipole moment of H_2O.

structure with O—H bond lengths $\ell = 95.7\,pm$ and a bond angle $\theta = 104.6°$. We draw X, Y axes as shown in the figure. We can ignore the Z axis since the molecule is planar. Because of the greater electronegativity of the oxygen atom we locate an effective net charge of $-\delta$ at its center, and then assign a charge of $+\delta/2$ to each of the H atoms. We now apply Eq. (23.14) to calculate the x and y components of the dipole moment for this model.

$$\mu_x = -\delta(0) + \frac{\delta}{2}x_1 + \frac{\delta}{2}x_2 = \frac{\delta}{2}\ell \sin\frac{\theta}{2} - \frac{\delta}{2}\ell \sin\frac{\theta}{2} = 0$$

Thus the x component of the dipole moment vanishes, as it must from the symmetry of the figure.

$$\mu_y = -\delta(0) + \frac{\delta}{2}y_1 + \frac{\delta}{2}y_2$$

$$= -\delta\ell \cos\frac{\theta}{2} = \delta(95.7 \times 10^{-12}\,m)(0.611) = 58.5\delta \times 10^{-12}\,C\,m$$

Thus the dipole moment of H_2O is directed along the line bisecting the H—O—H angle and has a magnitude $(58.5 \times 10^{-12}\,\delta)\,C\,m$. The observed $\mu = 6.14 \times 10^{-30}$ C m; hence, $\delta = 1.05 \times 10^{-19}$ C or $0.65e$, where e is the protonic charge. Of course, this is a crude model for the water molecule, and it is used only to show how the general formula [Eq. (23.13)] gives the dipole moment.

Electrons in molecules are not fixed in position and the methods of wave mechanics must be used to calculate effective charge distributions. We can, however, use Eq. (23.13) to obtain the quantum mechanical operator for the dipole moment of a molecule, and then we can calculate μ by the general methods of molecular quantum mechanics, provided adequate wavefunctions ψ are known for the molecule.

In applying dipole moments to problems of molecular geometry a consistent set of bond moments is useful. The bond moments are based on dipole moments of molecules of known structures. They can be used to draw conclusions about the bond angles in molecules of unknown structure.

The absence of any permanent dipole moment in a molecule can be due to the fact either that all its bonds are nonpolar (e.g., H_2, S_8) or that the bond moments add vectorially to zero because of the symmetry of the molecule (e.g., CO_2, CH_4). A molecule with a center of symmetry cannot have a dipole moment. The absence of a dipole moment in CO_2 indicates that the molecule is linear.

Consider the substituted benzene derivatives:

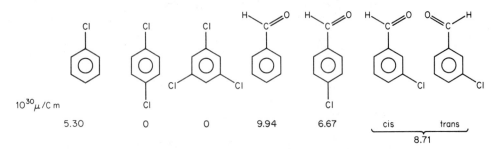

$10^{30}\mu$/C m						
5.30	0	0	9.94	6.67	cis	trans
					8.71	

The zero moments of *p*-dichloro- and *sym*-trichlorobenzene indicate that benzene is planar and that the C—Cl bond moments are directed in the plane of the ring, thereby adding to zero. The moments of the *p*- and *m*-chlorobenzaldehydes can be calculated quite accurately as vector sums of the moments of chlorobenzene and benzaldehyde. [For *o*-chlorobenzaldehyde this calculation is less accurate. Why?]

Example 23.6 From the dipole moments given above, calculate (a) the angle of the benzaldehyde dipole moment to the 1,4 axis of the benzene ring; (b) the equilibrium mole fractions of the *cis* and *trans* conformers of *m*-chlorobenzaldehyde.

(a) We assume that the μ of *p*-chlorobenzaldehyde is the vector sum of μ_1, for benzaldehyde, and μ_2, for chlorobenzene. From trigonometry

$$\cos \theta = (\mu_1^2 + \mu_2^2 - \mu^2)/2\mu_1\mu_2$$
$$= (9.94^2 + 5.30^2 - 6.67^2)/2(9.94)(5.30)$$
$$= 0.782 \quad \text{or} \quad \theta = 38.5°$$

(b) For the *cis* conformer,

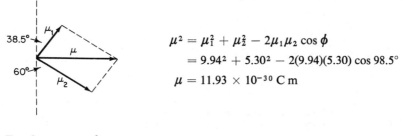

$$\mu^2 = \mu_1^2 + \mu_2^2 - 2\mu_1\mu_2 \cos \phi$$
$$= 9.94^2 + 5.30^2 - 2(9.94)(5.30) \cos 98.5°$$
$$\mu = 11.93 \times 10^{-30} \text{ C m}$$

For the *trans* conformer,

$$\mu^2 = 9.94^2 + 5.30^2 - 2(9.94)(5.30) \cos 21.5°$$
$$\mu = 5.37 \times 10^{-30} \text{ C m}$$

The observed μ^2 is the weighted average of values for cis and trans [since, from Eq. (23.11), the measured ϵ_r gives μ^2]. Thus $8.71^2 = X_1(11.93)^2 + (1 - X_1)(5.37)^2$ and $X_1 = 0.415$ or about 42% cis, 58% trans. [E. Bruce, G. L. D. Ritchie, and A. J. Williams, *Aust. J. Chem.*, **27**, 1809 (1974)]

23.8 Magnetic Properties of Molecules

Whereas electric properties are due to the static distribution of electric charges, magnetic properties are due to electric currents.

There are two magnetic field vectors: **H** the *magnetic field strength* and **B** the *magnetic induction* (or *flux density*). In a vacuum $B_0 = \mu_0 H$, where μ_0 is the vacuum permeability, $4\pi \times 10^{-7} \text{ N A}^{-2}$ (exactly). We may note that $\epsilon_0\mu_0 = c^{-2}$, where c is the speed of light in a vacuum. The SI unit of B is kg s^{-2} A^{-1}, called the tesla (T). It is equivalent to 10^4 gauss in electromagnetic units.

A molecule can have a magnetic moment μ_m, either a permanent moment or one induced by a magnetic field. The magnetic analog of the electric polarization P is the magnetization M, the magnetic moment per unit volume. The induction B in an isotropic medium is

$$B = B_0 + \mu_0 M \tag{23.15}$$

The ratio B/H is the permeability μ, and μ/μ_0 is the relative permeability μ_r. The *magnetic susceptibility* is

$$\chi = M/B_0 = (\mu_r - 1)/\mu_0 \tag{23.16}$$

Note that in SI, χ has the dimensions of μ_0^{-1} ($\mu_0\chi$ is dimensionless).

Whereas in the electrical case P/E is always positive, in the magnetic case M/B may be either positive or negative; i.e., the magnetization may be either in the direction of B or in the opposite direction. When $\chi < 0$, the substance is *diamagnetic*. When $\chi > 0$, the substance is *paramagnetic*. Paramagnetic substances are drawn into magnetic fields, whereas diamagnetic substances are pushed out of magnetic fields. In terms of the lines of force of the flux density, the two situations are shown schematically in Fig. 23.8. In a diamagnetic substance the flux density is less than in vacuum; in a paramagnetic substance, it is greater.

Vacuum   Paramagnetic substance   Diamagnetic substance

FIGURE 23.8 Magnetic lines of force are drawn into a paramagnetic substance so that the field inside is *higher* than in free space. The lines of force are pushed out of a diamagnetic substance, so that the field inside is *lower* than in free space.

One can obtain χ from a measurement of the force on the sample in an inhomogeneous magnetic field. This method was invented by Faraday and later used by Pierre Curie. A typical experimental apparatus is shown in Fig. 23.9.

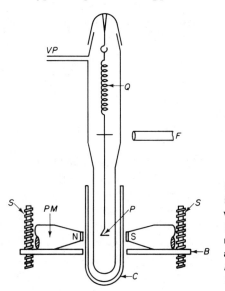

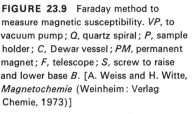

FIGURE 23.9 Faraday method to measure magnetic susceptibility. *VP*, to vacuum pump; *Q*, quartz spiral; *P*, sample holder; *C*, Dewar vessel; *PM*, permanent magnet; *F*, telescope; *S*, screw to raise and lower base *B*. [A. Weiss and H. Witte, *Magnetochemie* (Weinheim: Verlag Chemie, 1973)]

23.9   Diamagnetism and Temperature-Independent Paramagnetism

The counterpart of the distortion polarization in the electrical case is the magnetic moment induced in a molecule by an applied magnetic field. This effect is exhibited by all substances and is independent of temperature. Part of the induced magnetism is diamagnetic and the other part is paramagnetic. The net contribution to χ is thus the difference between these two terms.

A simple interpretation of the diamagnetic term is obtained if one imagines electrons moving about the nuclei to be like currents in a wire. If a magnetic field is applied, the velocity of the electrons is changed, inducing a magnetic field, which in accord with Lenz's Law is opposed in direction to the applied field. The diamagnetic contribution to the susceptibility χ is therefore always negative. In the case of an electric current in a wire, if the applied field is kept constant, the induced field quickly dies out owing to the ohmic resistance. Inside an atom or molecule, however, there is no resistance to the electronic current, so that the opposing field persists as long as an external magnetic field is maintained. The diamagnetic effect is small because of the small orbits of electrons in atoms.

The temperature-independent paramagnetic response of electrons in a molecule to an applied magnetic field was first discussed by van Vleck. It involves the interaction of excited states of the molecule with the ground state under the influence of the field. Its magnitude is usually less than that of the diamagnetic response.

23.10 Temperature-Dependent Paramagnetism

Temperature-dependent paramagnetism is the magnetic analog of the orientation polarization caused by an electric field. It is caused by the permanent magnetic moments μ_m of ions or molecules. As in the analogous case of permanent electric dipole moments in an electric field, the magnetic field tends to align the moments μ_m while the molecular collisions tend to knock them about and destroy the alignment. The theory, given originally by Langevin, leads to a result similar to Eq. (23.10), $\bar{\mu}_m = (\mu_m^2/3kT)B$. The paramagnetic susceptibility is thus

$$\chi = L\mu_m^2/3kTV_m \tag{23.17}$$

This is a theoretical expression for the law found experimentally by Curie to govern the dependence of χ on T. It was derived on the assumption that interactions between the magnetic moments are negligible and that $kT \gg \mu_m B$. When this type of paramagnetism occurs, the paramagnetic susceptibility is usually 10^3 to 10^4 times the diamagnetic susceptibility, so that the small diamagnetic effect is overwhelmed.

Paramagnetism is related to the orbital angular momenta and the spins of electrons. For ions and molecules, the contributions due to spin are predominant.

There is a natural unit of magnetic moment, $\mu_B = eh/4\pi m_e$, called the *Bohr magneton*. In SI units, its value is 9.2741×10^{-24} J T^{-1}. The magnetic moment in the field direction of one unpaired electron spin is one Bohr magneton. A measurement of the permanent magnetic moment of a molecule can tell us how many unpaired spins are in its structure. The magnetic moment for n unpaired spins is $\sqrt{n(n+2)}\,\mu_B$. [What is the maximum component in the field direction?]

Problems

1. The dielectric constant ϵ_r of $SO_2(g)$ is 1.00993 at 273 K and 1.00569 at 373 K ($P =$ 1 atm). Estimate the dipole moment of SO_2. If the bond dipole of S—O is 2.80 debye, estimate the OSO angle.

2. Solutions of mole fraction X of isopropyl cyanide in benzene had the following densities ρ and dielectric constant ϵ_r at 25°C [*J. Am. Chem. Soc.*, *69*, 457 (1947)]

X	0.00301	0.00523	0.00956	0.01301	0.01834	0.02517
ϵ_r	2.326	2.366	2.411	2.502	2.598	2.718
ρ (g cm^{-3})	0.87326	0.87301	0.87260	0.87226	0.87121	0.87108

For pure C_3H_7CN, $\rho = 0.76572$ g cm^{-3}, refractive index $n_0 = 1.3712$; pure benzene $\rho = 0.87345$, $\epsilon_r = 1.5016$. Calculate the dipole moment μ of C_3H_7CN.

3. The dipole moment of HBr is 2.604×10^{-30} C m and its relative permittivity is 1.00313 at 273 K and $P°$. Calculate the polarizability of HBr.

4. Calculate the polarization of water at 25°C in an electric field of 10^5 V m^{-1}. (See Table 23.1.)

5. The C—O—C bond angle in dimethyl ether is 111° and the dipole moment is 4.31×10^{-30} C m. Estimate the dipole moment of ethylene oxide, in which the C—O—C bond angle is 61°. (Ignore the effects of CH bond moments.)

6. From data in Example 23.5 and Problem 5, estimate μ for CH_3OH as function of bond angle COH. Experimentally, $\mu = 5.64 \times 10^{-30}$ C m. Estimate bond angle.

7. The molar polarization P_m of C_2H_5Br was determined as a function of T.

T (K)	203	223	243	263	283
P_m (10^{-6} m^3 mol^{-1})	104.2	99.1	94.5	90.5	80.8

Calculate the dipole moment and the polarizability.

8. The magnetic susceptibility of gold at 300 K is $\chi \mu_0 = -3.7 \times 10^5$. Calculate the magnetization M of gold in a magnetic flux density $B = 1$ tesla. Give SI units of M.

9. The paramagnetic susceptibility of O_2 (g) follows Curie's Law. At 300 K and $P° = 101.32$ K Pa, $\chi \mu_0 = 3.45 \times 10^{-3}$. What is the permanent magnetic moment μ_m of O_2? What is the value in Bohr magnetons and how many unpaired electron spins does it indicate? What is the magnetic moment per m^3 O_2 at the specified T and P in a magnetic flux density of 1 tesla?

10. The magnetic susceptibility of $CuSO_4 \cdot 5H_2O$ at 25°C is $\chi \mu_0 = 1.46 \times 10^{-3}$. What is the electronic configuration of the Cu^{2+} ion?

24

Magnetic Resonance

In 1946 Purcell, Torrey, and Pound at Harvard, and Bloch, Hansen, and Packard at Stanford published their independent discoveries of *nuclear magnetic resonance* (nmr). Nmr is a powerful technique for investigating the structures, motions, and reactions of molecules. The usefulness of nmr is based on the fact that nuclei are not unfeeling point charges. They have individual properties which make them sensitive to the internal electromagnetic fields in molecules.

Purcell wrote later: "I remember, in the winter of our first experiments... looking on snow with new eyes. There the snow lay around my doorstep—great heaps of protons quietly precessing in the earth's magnetic field. To see the world for a moment as something rich and strange is the private reward of many a discovery."

24.1 Electric and Magnetic Properties of Nuclei

A nucleus cannot have an intrinsic electric dipole moment, but it can have an electric quadrupole moment Q. A quadrupole moment arises from a distribution of four equal charges, two positive and two negative, arranged, for instance, like the signs on the d orbitals in Fig. 21.7. If a nucleus has a quadrupole moment, the distribution of its charge is not perfectly spherical; such a nucleus can be represented as an ellipsoid.

Some nuclei possess intrinsic nuclear spins and thus act as little magnets with magnetic dipole moments. The nuclear spin is specified by a quantum number I. For nuclei with odd mass numbers the values of I are odd multiples of $\frac{1}{2}$; for nuclei with even mass numbers the values of I are zero or even multiples of $\frac{1}{2}$.

We thus classify nuclei as shown in Fig. 24.1. Nuclei with electric quadrupole moments and/or magnetic dipole moments can act as delicate probes that report on the electromagnetic fields in their surroundings.

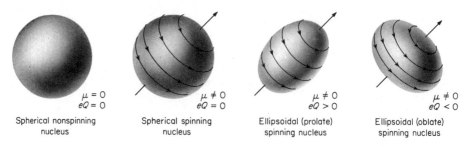

$\mu = 0$ $eQ = 0$	$\mu \neq 0$ $eQ = 0$	$\mu \neq 0$ $eQ > 0$	$\mu \neq 0$ $eQ < 0$
Spherical nonspinning nucleus	Spherical spinning nucleus	Ellipsoidal (prolate) spinning nucleus	Ellipsoidal (oblate) spinning nucleus

FIGURE 24.1 Nuclei classified according to their magnetic dipole moments and electric quadrupole moments.

The spin angular momentum of a nucleus is restricted to quantized values, $L = \sqrt{I(I+1)}(h/2\pi)$. In a magnetic field in the z direction, the z component of this angular momentum is also quantized, with a quantum number M_I, which can have values $I, I-1, I-2, \ldots, -I$. The z components of angular momentum are thus $L_z = M_I(h/2\pi)$, the maximum value being $I(h/2\pi)$. I and M_I for nuclear spin are analogous to s and m_s for electron spin (page 473).

A nucleus with nonzero spin has a magnetic dipole moment $\boldsymbol{\mu}_m$. The maximum value of the nuclear magnetic dipole moment in the field direction z is

$$\mu_z = \gamma I(h/2\pi) = g_N \mu_N I \tag{24.1}$$

This μ_z is usually called the *nuclear magnetic moment*. In Eq. (24.1), the ratio of magnetic moment to angular momentum is γ, the *magnetogyric ratio*. The nuclear g factor g_N is an experimental quantity that depends on the structure of the nucleus. The *nuclear magneton* is

$$\mu_N = \frac{eh}{4\pi m_p} = \frac{(1.602 \times 10^{-19}\ \text{C})(6.626 \times 10^{-34}\ \text{J s})}{4\pi(1.673 \times 10^{-27}\ \text{kg})} = 5.051 \times 10^{-27}\ \text{C J s/kg} \tag{24.2}$$

Since $\text{J} = \text{kg m}^2\ \text{s}^{-2}$ and $\text{C s}^{-1} = \text{A}$, $\mu_N = 5.051 \times 10^{-27}\ \text{A m}^2$ or J T^{-1}. The SI unit of magnetic flux density $\mathbf{B}$ is V s m^{-2}, called the *tesla* T. Since the mass of the proton m_p is almost 2000 times m_e, the mass of the electron, μ_N is about 2000 times less than the electronic or Bohr magneton $\mu_B = eh/4\pi m_e$.

Some nuclear properties are summarized in Table 24.1. Note that the nuclear g factor can sometimes be negative. The magnetic moment of the neutron is -1.9130 μ_N and the magnetic moment of a nucleus is the resultant of the orbital and intrinsic moments of the protons and neutrons of which it is composed. A negative g_N means that the nuclear magnetic moment is antiparallel to the nuclear spin angular momentum.

TABLE 24.1
PROPERTIES OF REPRESENTATIVE NUCLEI

Nuclide	Natural abundance (%)	Spin I $(h/2\pi)$	Magnetic moment, μ_z (nuclear magnetons)	Quadrupole moment $(e \times 10^{-30} \text{ m}^2)$	NMR frequency (MHz at field of 1 T)
^{1}H	99.9844	$\frac{1}{2}$	2.79270		42.577
^{2}H	0.0156	1	0.85738	0.277	6.536
^{10}B	18.83	3	1.8006	1.22	4.575
^{11}B	81.17	$\frac{3}{2}$	2.6880	3.55	13.660
^{13}C	1.108	$\frac{1}{2}$	0.70216		10.705
^{14}N	99.635	1	0.40357	2.0	3.076
^{15}N	0.365	$\frac{1}{2}$	-0.28304		4.315
^{17}O	0.07	$\frac{5}{2}$	-1.8930	-0.40	5.772
^{19}F	100	$\frac{1}{2}$	2.6273		40.055
^{31}P	100	$\frac{1}{2}$	1.1305		17.235
^{33}S	0.74	$\frac{3}{2}$	0.64272	-6.4	3.266
^{39}K	93.08	$\frac{3}{2}$	0.39094		1.987

Example 24.1   What are the magnetogyric ratio γ and the g_N factor of the ^{31}P nucleus, the magnetic moment of which is 1.1305 μ_N.

From Eq. (24.1),

$$\gamma = \mu_z/I(h/2\pi) = (1.1305)(5.05 \times 10^{-27} \text{ J T}^{-1})/(\tfrac{1}{2})(6.626 \times 10^{-34} \text{ J s}/2\pi)$$

$$= 1.083 \times 10^8 \text{ T}^{-1} \text{ s}^{-1}$$

$$g_N = \mu_z/\mu_N I = 1.1305/I = 1.1305/(\tfrac{1}{2}) = 2.261$$

24.2   Nuclear Magnetic Resonance

When a nucleus with spin I is placed in a magnetic field with flux density $\mathbf{B}_0$, the magnetic moment vector $\boldsymbol{\mu}_m$ precesses about the field direction z as shown in Fig. 24.2.* The components of $\boldsymbol{\mu}_m$ in the direction of the field are restricted to values $\mu_z = M_I g_N \mu_N$. Figure 24.3(a) shows an example for a nucleus with $I = \frac{1}{2}$ (e.g., ^{1}H). In the magnetic field, states with different values of M_I have slightly different energies. The potential energy of the nucleus when its magnetic moment has a component μ_z in the direction of the magnetic flux density $\mathbf{B}_0$ is

$$\epsilon_p = -\mu_z B_o = -M_I g_N \mu_N B_0 \tag{24.3}$$

The energy spacing between two adjacent levels ($\Delta M_I = \pm 1$) is $\Delta\epsilon = g_N \mu_N B_0$. The frequency of radiation absorbed or emitted in a transition between the energy levels is

$$\nu = \Delta\epsilon/h = g_N \mu_N B_0/h \tag{24.4}$$

*The equation of motion is $d\boldsymbol{\mu}_m/dt = -\gamma(\boldsymbol{\mu}_m \times \mathbf{B})$.

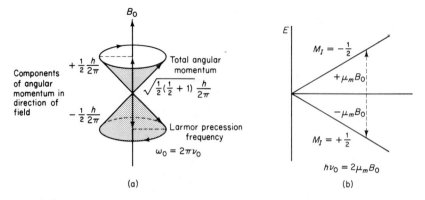

(a)   (b)

FIGURE 24.2 (a) A nucleus with spin $I = \frac{1}{2}$ has a total angular momentum $\sqrt{\frac{1}{2}(\frac{1}{2}+1)}(h/2\pi)$ which precesses about the field direction at angles such that its components in the field direction are $\pm\frac{1}{2}(h/2\pi)$. (b) The energy levels of two states with $M_I = \pm\frac{1}{2}$ are split by the magnetic field. The upper state has $M_I = -\frac{1}{2}$ and the lower state $M_I = +\frac{1}{2}$. The energy difference between the states is $2\mu_m B_0$, where μ_m is the magnetic moment of the nucleus.

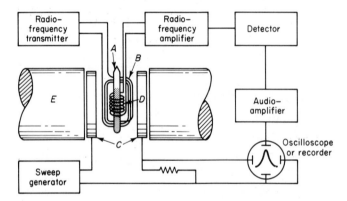

FIGURE 24.3 Block diagram of a nuclear magnetic resonance spectrometer: A, sample tube; B, transmitter coil; C, sweep magnet; D, receiver coil; E, main magnet. [F. Bovey, *Nuclear Magnetic Resonance Spectroscopy* (New York: Academic Press, Inc., 1971)]

When an external electromagnetic field of frequency ν interacts with the spin system, the absorption or emission of energy is called *nuclear magnetic resonance*. Note that the frequency ν varies directly as the magnitude of the flux density $\mathbf{B}_0$.

In a classical picture, ν is the frequency at which the magnetic moment of the nucleus precesses about the magnetic-field direction. The angular frequency (units, radians per second) would be

$$\omega = 2\pi\nu = 2\pi g_N \mu_N B_0/h = \gamma B_0 \tag{24.5}$$

This is called the *Larmor frequency*. In Fig. 24.2(b) the direction of the magnetic field is along the z axis. The component of the magnetic moment in the xy plane thus rotates about the z axis with the Larmor frequency ω.

Example 24.2 Consider the nucleus ^{19}F for which $g_N = 5.256$. What is its resonant frequency in a flux density of exactly 1 T?

From Eq. (24.4),

$$\Delta\epsilon = (5.256)(5.051 \times 10^{-27} \text{ J T}^{-1})(1 \text{ T})$$
$$= 2.655 \times 10^{-26} \text{ J}$$

The frequency of radiation for this $\Delta\epsilon$ would be in the radiofrequency shortwave region at

$$v = \frac{\Delta\epsilon}{h} = \frac{2.655 \times 10^{-26} \text{ J}}{6.626 \times 10^{-34} \text{ J s}} = 40.07 \times 10^6 \text{ s}^{-1} = 40.07 \text{ MHz}$$

Example 24.3 The magnetic field of the earth is about 5.0×10^{-5} T. What is the precession frequency of the protons that Purcell could see in the snow piles?

From Eq. (24.5) and Table 24.1,

$$\omega = (2\pi)(42.577 \times 10^6 \text{ s}^{-1})(5 \times 10^{-5}) = 13\ 400 \text{ rad s}^{-1}$$

24.3 Experimental Apparatus for NMR

The principle of a spectrometer to measure nuclear magnetic resonance is shown in Fig. 24.3. The field that aligns the nuclear magnetic moments can be provided by an electromagnet for B_0 up to about 2.3 T (corresponding to $v = 100$ MHz for ^1H). The upper limit for the field is caused by magnetic saturation of the iron core of the magnet. Higher fields can be achieved with superconducting solenoids of alloys such as niobium–titanium, run at liquid-helium temperatures. The limit of present technology is about $B_0 = 14$ T. A diagram of a typical superconducting magnet is shown in Fig. 24.4.

The nmr spectrum is excited by a magnetic field with flux density $\mathbf{B}_1$ produced by a transmitter coil with its axis normal to the sample tube. This coil is excited by a radiofrequency power supply operating close to the resonant frequency for the nucleus under observation. Radiation is picked up in a receiver coil whose axis is perpendicular to that of the transmitting coil. When the resonant condition is reached, the frequency of the transmitter just matches the Larmor frequency of the nucleus being observed, and energy is absorbed by the sample. The resultant fall in intensity of the transmitted signal is detected by the receiver coil. The spectrum can be scanned either by changing the magnetic field $\mathbf{B}_0$ or by changing the frequency of the transmitting coil. The field can be varied with a pair of small coils attached to the pole pieces of the large magnet, which yield a "sweep field" across the range of the spectrum. For example, a sweep field of about 25 μT gives a frequency range of about 1000 Hz, which suffices to scan a typical proton spectrum centered at 100 MHz. The line width of a resonance peak may be 1 Hz or less, so that extreme control and stability of $\mathbf{B}_0$ are required.

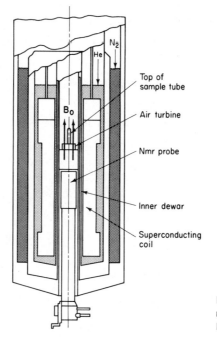

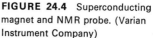

Top of
sample tube

Air turbine

Nmr probe

Inner dewar

Superconducting
coil

FIGURE 24.4 Superconducting
magnet and NMR probe. (Varian
Instrument Company)

24.4 Spin-Lattice Relaxation

The difference in energy between the ground state and the excited state in nmr transitions is usually small compared to kT. At equilibrium, the ratio of the number of nuclei in the lower state ($M_I = +\frac{1}{2}$) to the number in the upper state ($M_I = -\frac{1}{2}$) is given by the Boltzmann factor,

$$N^+/N^- = \exp\left(2\mu_z B_0/kT\right) \tag{24.6}$$

Example 24.4 Calculate the ratio of protons in ground state to protons in upper state at equilibrium at 300 K in a field of $B_0 = 2.349$ T.

At this B_0, the resonant frequency from Eq. (24.4) is 100 MHz, and $\Delta\epsilon = (6.63 \times 10^{-34}$ J s$)(10^8$ s$^{-1}) = 6.63 \times 10^{-26}$ J; whereas at 300 K, $kT = (1.38 \times 10^{-23}$ J K$^{-1})$ (300 K) $= 4.14 \times 10^{-21}$ J. Hence, from Eq. (24.6),

$$N^+/N^- = \exp\left(6.63 \times 10^{-26}/4.14 \times 10^{-21}\right)$$
$$= \exp\left(1.60 \times 10^{-5}\right)$$

Since $\Delta\epsilon \ll kT$, the first two terms in the power series for the exponential suffice: $N^+/N^- = 1 + (\Delta\epsilon/kT) = 1.000016$. At equilibrium the excess population in the ground state is only 16 protons in 10^6.

When the sample whose nmr is to be measured is subjected to the radiofrequency field of B_1, transitions occur from the ground to the excited state. The ratio

N^+/N^- then departs from its equilibrium value and may no longer exceed unity. If this happens, no further energy can be absorbed from the field, and no nmr measurement is possible. To record an nmr spectrum there must be some process of rapid *relaxation* to remove the excitation energy from the upper state and to restore the excess population in the ground state, so that energy can be continuously absorbed from the B_1 field. The transition from lower to upper state corresponds to flipping a nuclear magnetic moment vector from one allowed orientation to the other in the large B_0 field. In the ground state the protons are aligned so that the z components of their magnetic moments μ_z are in the direction of B_0. The B_1 field flips the moments to the opposite direction as energy is absorbed. The relaxation process flips them back to the original orientation. The spin system returns to the ground state by transferring its energy of excitation to the random thermal motions in its surroundings. We could say that the nuclear-spin system becomes cooler and its surroundings become warmer as a result of this transfer.

If the excess number of spins in the ground state is $N^+ - N^- = \Delta N$, the rate of change of ΔN after the B_1 field has caused a departure from the equilibrium value ΔN_e is

$$\frac{d\,\Delta N}{dt} = k_1(\Delta N_e - \Delta N) = \frac{\Delta N_e - \Delta N}{T_1} \tag{24.7}$$

Equation (24.7) has the typical form of a first-order rate law with $1/T_1 = k_1$ the first-order rate constant, and T_1 is called the *spin-lattice relaxation time*. On integration, one obtains

$$\Delta N = \Delta N_{eq}\,(1 - e^{-t/T_1}) \tag{24.8}$$

The name *spin-lattice relaxation* for the T_1 process originates from the fact that the first experiments were made on solid samples, in which the relaxation is caused by a transfer of spin excitation energy to lattice vibrations. The term *spin-lattice relaxation* has been retained, but now it refers more generally to relaxation due to fluctuating magnetic fields arising from random motions in the surroundings of the spin system. For example, a given nucleus in a molecule will sense the fluctuating magnetic moments of neighbor nuclei, those in the same molecule and those in nearby molecules. This "magnetic noise" has a frequency component at the Larmor value ω_0, which stimulates transitions between the spin states.

The spin-lattice relaxation time T_1 is also called the *longitudinal relaxation time*, since it measures the change in the z component of net magnetic moment of all the spins, i.e., in the longitudinal component, which is parallel to the direction of the magnetic field. The most common cause of the T_1 process is a magnetic dipole–dipole interaction. Electric quadrupole moments (as in ^2H) and paramagnetic substances, if present in the sample, increase the efficiency of nuclear spin relaxation and markedly decrease T_1.

24.5 Spin-Spin Relaxation

Another kind of relaxation process is due to the interaction of the spin of one nucleus with the spins of neighboring nuclei. Figure 24.5 shows two neighboring spins in a magnetic field of flux density $\mathbf{B}_0$. The pair of spins have the same or nearly the

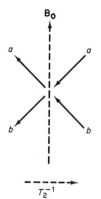

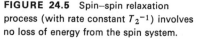

FIGURE 24.5 Spin–spin relaxation process (with rate constant T_2^{-1}) involves no loss of energy from the spin system.

same Larmor frequency and therefore they can sometimes interact to give an interchange in their orientations in the field. Unlike the T_1 process, such a switch of spins does not change the energy of the spin system. The spin-spin interaction is triggered by the fields in the xy plane (we have already seen how the externally applied B_1 field in this plane causes flippings of spin). The spin-spin relaxation time is denoted by T_2. The T_2 process is called *transverse relaxation* since it occurs in the xy plane perpendicular to the B_0 field direction. The value of T_2 can be estimated from the line width δv (at half height) of the resonance line by $1/T_2 = \pi\, \delta v$.

24.6 Chemical Shifts

If all the nuclei of a given species in a molecule, say all the protons, displayed resonance at one identical frequency, nmr would be of little interest to chemists. In fact, however, the environment of a nucleus has a small but measurable effect on the magnetic field sensed by the nucleus. The electrons near a nucleus are acted on by the external B_0 field to produce an induced magnetic field that modifies the local magnetic flux density at the nucleus. Thus

$$B(\text{local}) = B_0(1 - \sigma) \qquad (24.9)$$

Here σ is called the *shielding constant.*

When $\sigma > 0$, we have a *shielding effect* and when $\sigma < 0$, a *deshielding effect.* When a shielding effect occurs, the applied field must be increased to bring the nucleus into resonance, and there is an "upfield shift" in the field. When there is a deshielding effect, the applied field at resonance will be lowered, and there is a "downfield shift" in the field.

The shielding effect may amount to only one part per million of the external field, but the precision of nmr measurements is so great that the effect can be measured to within 1%. The resulting change in resonance frequency is called the *chemical shift.* Figure 24.6 shows an example in which the frequency of proton resonance in CH_3CH_2OH is somewhat different for each of the three different kinds of H atom in the structure. Because the chemical shift is due to magnetism induced by the external field its *absolute magnitude* depends on the strength of the field.

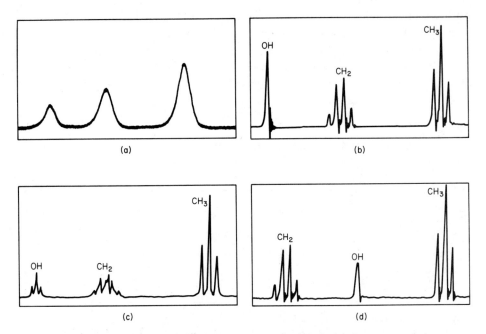

FIGURE 24.6 Aspects of the nmr spectrum of ethanol. (a) Spectrum at low resolution at 40 MHz. Peaks occur for the three nonequivalent protons in $CH_3 \cdot CH_2 \cdot OH$ and the peak areas are in the ratio $3:2:1$. (b) Spectrum of ethanol in presence of a trace of acid. The spin–spin splitting due to interaction between CH_2 and CH_3 protons is observed, but the OH proton exchanges so rapidly between molecules that no splitting of its resonance peak is observable. (c) Spectrum at high resolution of carefully purified dry ethanol. The OH proton peak is now split into a triplet by interaction with the two adjacent CH_2 protons, and the quartet of resonances for the CH_2 protons has become a rather poorly resolved octet. (d) Spectrum of 6% C_2H_5OH in $CDCl_3$. The solvent is not an acceptor of H bonds and hence the H bonds that occur in pure ethanol are mostly destroyed. Consequently, the chemical shift of the OH proton is greatly decreased. The H bond acts to deshield the proton from the external field. The OH peak now actually occurs upfield from the CH_2 peaks.

The chemical shift is expressed relative to that of some standard substance. In the case of proton resonance the standard in nonaqueous solvents is usually tetramethylsilane (TMS), which gives a single sharp nmr line. We define the relative chemical shift δ in terms of the measured resonance frequencies ν as

$$\delta = \frac{\nu(\text{reference}) - \nu(\text{specific peak})}{\nu(\text{reference})} \qquad (24.10)$$

The *relative* chemical shift in Eq. (24.10) is independent of the magnetic field at which the spectrometer is operating. The value of δ is expressed in parts per million, ppm. The larger the δ, the greater the downfield shift of the resonance. Chemical shifts provide a sort of catalog of the different environments of nuclei in molecular structures.

The carbon isotope ^{13}C occurs to the extent of 1.1% in natural abundance and, even without isotopic enrichment, ^{13}C nmr spectra are readily obtained. Chemical shifts for ^{13}C nuclei are much larger than those for protons because electron clouds around ^{13}C nuclei are much more dense and hence shielding effects are enhanced. In some cases ^{13}C shifts extend over a considerable range even for a specified chemical group, so that it is often possible to determine precisely the particular situation of a group in a molecular structure. Examples of chemical shifts in ^{13}C nmr spectra are shown in Table 24.2.

TABLE 24.2
^{13}C CHEMICAL SHIFTS IN ORGANIC COMPOUNDS[a]

Carbon nucleus		ppm	Carbon nucleus		ppm
$\ce{>C=O}$	Ketone	203–228	$\ce{-C#C-}$	Alkyne	74–92
$\ce{>C=O}$	Aldehyde	185–208	$\ce{>C-C<}$	C (Quaternary)	28–51
$\ce{>C=O}$	Acid	171–183	$\ce{>C-O}$		72–83
$\ce{>C=O}$	Ester, amide	160–174	$\ce{>C-N<}$		61–75
$\ce{-C#N}$	Nitrile	117–123	$\ce{>CH-C<}$	C (Tertiary)	32–60
$\ce{>C=N}$	Heteroaromatic	145–155			
$\ce{>C=C<}$	Alkene	105–143	$\ce{-CH2-C<}$	C (Secondary)	22–44
			$\ce{-CH3-C<}$	C (Primary)	5–29
$\ce{>C=C<}$	Aromatic	112–136	$\ce{CH3-Hal}$		2–28

[a] Relative to Tetramethylsilane

Electronic currents in aromatic rings may cause relatively large chemical shifts in nmr. Figure 24.7 depicts the currents and the magnetic lines of force for a benzene ring. Their effect is to produce shielding above and below the plane of the ring, and deshielding in the plane of the ring outside the inner hexagon of carbon atoms. Thus the resonant frequency of the ring protons in benzene is shifted to a lower value (downfield shift). These ring-current shifts have been used to elucidate the exact location and conformation of certain aromatic sidechains in protein structures.

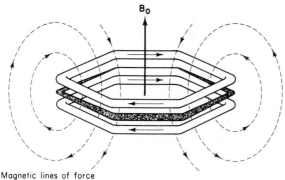

B_0

Magnetic lines of force

FIGURE 24.7 Origin of ring-current shifts due to magnetic fields produced by electrons of aromatic rings.

When the nmr spectrum in Fig. 24.6(a) is taken at higher resolution, the peak for CH_2 is split into four lines and that for CH_3 into three lines, as shown in Fig. 24.6(b). The splitting of the CH_2 and CH_3 peaks into multiplets is not a chemical shift. This conclusion is proved by the fact that the observed splitting does not depend on the strength of the applied field. The effect is caused by the interaction of the nuclear spins (magnetic moments) of one set of equivalent protons with those of another set. It is therefore called *spin–spin coupling*. The splitting of an nmr line due to interaction of the spins is measured by the *spin-spin coupling constant J* (units: Hz).

Consider one of the equivalent protons of the CH_3 group. How can the spin of a given proton in the CH_3 group be arranged in relation to the spins of two protons *a* and *b* of CH_2? There are four possible ways:

1. Spin parallel to both *a* and *b*
2. Spin antiparallel to both *a* and *b*
3. Spin parallel to *a*, antiparallel to *b*
4. Spin antiparallel to *a*, parallel to *b*

The last two arrangements clearly have the same energy. The result of the spin-spin interaction is that each proton in the CH_3 group can feel three slightly different magnetic fields, depending on its relation to the spins of the CH_2 protons. The energy levels in the magnetic field become a closely spaced triplet and the resonance signal thus splits into three components with intensity ratio 1:2:1, as shown in Fig. 24.8. In the same way, a proton in the CH_2 group feels slightly different fields depending on how its spin is related to those of the CH_3 group. There are four energetically different arrangements possible, as shown in Fig. 24.8, and the spectrum is a quartet with intensity ratio 1:3:3:1.

What are the effects of the interaction of the CH_3 and CH_2 protons with the proton in the OH group? Each of the lines shown in Fig. 24.8 should be split into a doublet, and the CH_2 should be split more than the CH_3 since it is closer to the OH group. These predictions are confirmed by the experimental spectrum in Fig. 24.6(c). We also see the spectrum of the OH proton, which is split into a 1:2:1 triplet by interaction with the CH_2.

If a trace of acid is added to the alcohol, a dramatic change occurs in the spectrum [Fig. 24.6(b)]. The OH triplet becomes a singlet and the splitting of CH_2 and CH_3 by the OH disappears. The H^+ of the acid acts as a catalyst for the rapid exchange of protons between the OH groups of different molecules. Thus the lifetime of an OH in any given conformation becomes too short to permit its individual observation and one sees only the average effect of all the OH groups.

The case of CH_3CH_2OH belongs to a simple type of spectrum in which $J \ll \Delta\nu$, where $\Delta\nu$ is the difference in chemical shifts of the two kinds of protons that are spin coupled by J. This is called a *first-order spectrum*. In a first-order spectrum the nuclei are weakly spin coupled. Nuclei that are weakly coupled are denoted by letters that are alphabetically far apart, such as A_3X_2 for the CH_3CH_2 group.

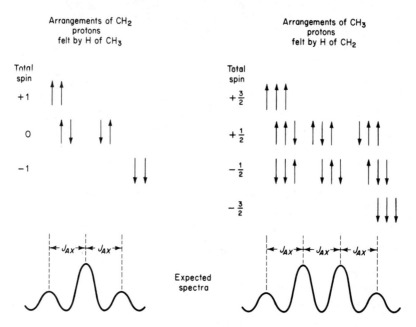

FIGURE 24.8 Spin–spin splitting of NMR spectrum of CH_3CH_2OH. Each proton in CH_2 senses four different spin arrangements in CH_3, and each proton in CH_3 senses three different spin arrangements in CH_2.

When δ/J is less than about 10 the nuclei are strongly spin coupled, and the spectra become more difficult to interpret. Nuclei that are strongly coupled are denoted by letters that are close together, for example, methanol CH_3OH is an AB_3 system.

An example of how an nmr spectrum changes as the spin coupling becomes stronger is shown in Fig. 24.9, which is the calculated spectrum of an AB system for $J = 10$ Hz and six different values of δ/J from 0 to 10.

If nuclei have exactly the same chemical shift they are called *chemical-shift equivalent* and are designated by the same letter. Nuclei are chemical-shift equivalent

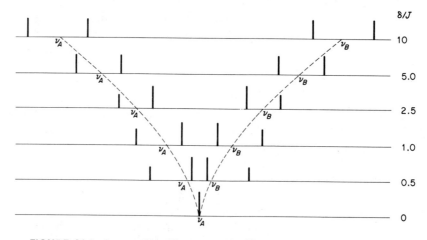

FIGURE 24.9 Spectra of an *AB* system with different values of ratio of spin–spin coupling constant J to difference in chemical shift $\Delta\nu$.

if they can be interchanged through one of the symmetry operations of the molecule (Chapter 25). Thus all the H in benzene are chemical-shift equivalent since they are related by the sixfold symmetry axis. Nuclei that can exchange very rapidly compared to their relaxation time T_1 are also chemical-shift equivalent.

Nuclei that are chemical-shift equivalent are not necessarily magnetically equivalent. For magnetic equivalence the nuclei must be coupled in exactly the same way to every other nucleus in the molecule. Consider, for example, p-fluoronitrobenzene.

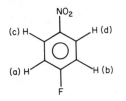

Protons (a) and (b) are chemical-shift equivalent since they are related by the mirror plane that bisects the molecule. Protons (a) and (b) are coupled to F ($I = \frac{1}{2}$) with identical bond distances and angles. However, protons (a) and (b) are not identically coupled to protons (d) and (c) and hence they fail the test for magnetic equivalence. Similarly, (c) and (d) are not magnetically equivalent. This spin system is thus of the type AA′BB′X (X being F). (The notation uses primed letters for nuclei that are chemical-shift equivalent but not magnetically equivalent. Of course, nuclei in this category must be strongly coupled since $\Delta v = 0$.)

Spin coupling is caused by an interaction between the magnetic dipole of one nucleus and that of the other. The energy of interaction through space between a pair of dipoles $\boldsymbol{\mu}_1$, $\boldsymbol{\mu}_2$ is

$$U \propto \frac{1}{r^3} \mu_1 \mu_2 (3 \cos^2 \theta - 1) \qquad (24.11)$$

where r is the distance and θ the angle between them. In a crystal this interaction energy is an important part of the spin coupling. In a fluid, however, the random tumbling of the molecules leads to an average value $3 \overline{\cos^2 \theta} = 1$, so that the direct dipole interaction of Eq. (24.11) vanishes.

There is also an indirect interaction, which is transmitted through the chemical bonds that intervene between the two magnetic dipoles (nuclear spins). This effect is called *Fermi contact interaction*. It arises from the fact that the wavefunction ψ of an electron in an s orbital has a finite value at the center of the nucleus (see page 470) and thus electrons in such orbitals can feel the nuclear dipoles quite intensely. Two nuclear spins separated by a chemical bond interact through the pair of electron spins in the bond: ↑↓↑↓. Coupling constants J depend on the torsional angles about chemical bonds and thus can yield information about the conformation of a molecule.

24.8 Dynamic NMR—Measurement of Reaction Rates

NMR spectroscopy can be used to measure the rates of certain fast reactions. The method can be applied only when the mean lifetime τ of the reactant is comparable in magnitude to one of the relaxation times T_1 or T_2 of the nucleus being monitored by nmr. At practical temperatures the rates of many reactions, including conformation changes, internal rotations, and proton transfers, fall into such a range. To make kinetic observations possible there must be a measurable difference between the chemical shifts of the resonant nucleus in the reactant and product molecules.

As an example, consider the spectra of 4,4-dimethylaminopyrimidine shown in Fig. 24.10 at a series of different temperatures. Unless rapid rotation occurs about the N—C bond to the ring, the two methyl groups are nonequivalent. At low temperatures, therefore, where rotation is inhibited, two distinct peaks occur, separated by 17.8 Hz, the difference in chemical shifts. As the temperature is raised from 228 K to 240 K, the internal rotation becomes faster, and the peaks first broaden and then merge into a broad single peak. As the temperature is further raised from 240 K to 265 K, the broad peak sharpens progressively. In this case the reaction that exchanges the nonequivalent sites is an internal rotation.

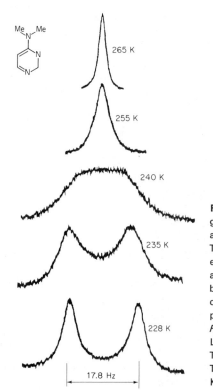

FIGURE 24.10 Spectra of the methyl groups of 4,4-dimethylaminopyrimidine at 100 MHz and various temperatures. This series shows a sequence of spectra essentially of the form of Fig. 24.6(a), as there is no appreciable coupling between the methyl protons and any other nuclei. [Reproduced with permission from *Nuclear Magnetic Resonance Spectroscopy* by R. M. Lynden-Bell and R. K. Harris (London: Thomas Nelson & Sons Ltd., 1970). The spectrum was produced by A. R. Katritzky and G. J. T. Tiddy.]

For a reaction $A \rightleftharpoons B$, the mean lifetime of species A is $\tau_R = k_1^{-1}$, where k_1 is the first-order rate constant for the exchange. Thus the width of the NMR peak at half-height is $\delta v = k_1/\pi$. Before using this relation to calculate k_1, we must subtract the natural line width in the absence of exchange to obtain the residual broadening due only to exchange. In the other limit, at high exchange rates when the individual peaks have fused, the line width is $\delta v = \pi(v_A - v_B)^2 k_1^{-1}$, where v_A and v_B are the individual frequencies of peaks A and B, as determined from the measurements at low temperatures.

Example 24.5 From the spectra in Fig. 24.13 calculate the rate constant k_1 for rotation about the N—C bond at 228 K and 255 K. The natural line width can be taken to be 0.5 Hz. Estimate the energy barrier for the internal rotation.

At 228 K, the peak width at half-height is $\delta v = 4.29$ Hz. At 255 K, $\delta v = 4.61$ Hz. At 255 K, $k_1 = \pi(17.8\ s^{-1})^2/4.11\ s^{-1} = 24\ s^{-1}$. At 228 K, $k_1 = \pi(3.79\ s^{-1}) = 11.9$ s^{-1}. From Eq. (13.31), $\ln(24/11.9) = (E_a/R)(1/228 - 1/255)$, $E_a = 53\ 900$ J mol^{-1}. (This treatment is only approximate.)

24.9 Fourier Transform NMR

The technique of nmr spectroscopy that was described in Section 24.3 is called *continuous wave* (CW) nmr. A newer technique, called *Fourier transform* (FT) nmr, gives the same spectra with great savings of time for a given signal-to-noise ratio. Spectra are generally presented in the form of intensity of radiation at the detector as a function of frequency, $I(v)$. This function can be transformed by a mathematical operation called the *Fourier transform* into a function of time $F(t)$, given by

$$F(t) = \int_{-\infty}^{\infty} e^{2\pi i v t} I(v)\, dv \qquad (24.12)$$

Conversely, $F(t)$ can be transformed back into $I(v)$ by the inverse Fourier transform,

$$I(v) = \int_{-\infty}^{\infty} e^{-2\pi i v t} F(t)\, dt \qquad (24.13)$$

$F(t)$ contains all the information about the spectrum that $I(v)$ contains, and *vice versa*. One spectrum is in the time domain and the other is in the frequency domain. With a rather small computer it is possible to carry out the Fourier transform of a spectrum in a few seconds. Thus it is natural to ask whether it would be useful to take a spectrum $F(t)$ and then transform it into the usual $I(v)$. What would be the advantage of this procedure? To measure a spectrum in the frequency domain takes a good deal of time, since one must gradually scan the range of frequency desired. To measure a spectrum in the time domain, however, takes only a large number of frequencies, and it is possible to apply all the frequencies at once to the sample within a very short interval of time. Instead of taking, say, 15 minutes to scan a spectrum, it can be done in a few seconds. The pioneering paper on Fourier transform nmr was published by R. R. Ernst and W. A. Anderson in 1966. Since then it has become a standard technique in nmr and other kinds of spectroscopy.

The band of frequencies needed to excite the nmr spectrum in the time domain is obtained by application of a short, approximately square-wave pulse of radio-frequency power in the xy plane. Figure 24.11 shows the Fourier synthesis of a square-wave pulse. The more terms that are added, the better the approximation to the square wave. Conversely, when a square wave $F(t)$ is analyzed into its Fourier conponents,

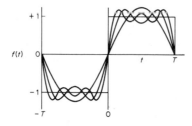

FIGURE 24.11 Fourier synthesis of a square-wave pulse. Superposition of sine waves of many different frequencies yields the square wave.

we find that it contains a whole spectrum of different superimposed harmonic waves. Applying a square-wave pulse to a system is equivalent to applying a whole spectrum of harmonic waves of different frequencies. In the pulsed Fourier transform nmr spectrometer, the pulse B_1 is from a radiofrequency magnetic field that has a frequency v_0, say 100 MHz in a typical case for protons. The Fourier components add to or subtract from this carrier frequency. If t_p is the length of the pulse (say 100 μs), the sample receives a range of frequencies about $v_0 \pm 1/t_p$, in this example, 100 MHz $\pm$ 10 kHz. The pulse length is chosen to be much less than the longitudinal relaxation time T_1 of the nucleus to be observed. [Why?]

When the B_1 pulse terminates, the magnetization in the xy plane begins to decay. This process is called *free induction decay* (FID) because the nuclei are freely precessing about B_0 in the absence of an applied radiofrequency field. The detector is along a definite axis, say y, and it is tuned to detect a fixed frequency v_0, which is continuously applied to it as a reference frequency. The free-induction signal rotates in the xy plane and comes alternately in and out of phase with the reference signal. The form of the resulting FID is shown in Fig. 24.12 for a typical nmr spectrum in which a number of ^{13}C nuclei at different chemical shifts are being simultaneously recorded.

To improve the ratio of signal to noise (S/N) many consecutive pulses can be applied to the sample. The FID's are collected in the computer and averaged before being Fourier transformed to give the $I(v)$ spectrum. In the case of ^{13}C nmr of proteins as many as 100 000 pulses may be averaged.

24.10 Electron Spin Resonance

Electron spin resonance (esr) is also called *electron paramagnetic resonance* (epr). The theory is similar to that of nmr, except that unpaired electron spins are studied instead of nuclear spins. Unpaired spins occur in transition metal complexes, in defect sites and color centers in crystals, in free radicals, and in triplet states.

The electron spin quantum number m_s gives the component of spin in the direction of the magnetic field. The energy spacing between two adjacent levels ($\Delta m_s = \pm 1$) by analogy with Eq. (24.4) is

$$\Delta \epsilon = hv = g\mu_B B_0 \tag{24.14}$$

where μ_B is the Bohr magneton and g is the g *factor*. For a free electron $g = 2.00232$. The resonant frequency v of the system is directly proportional to the magnetic flux

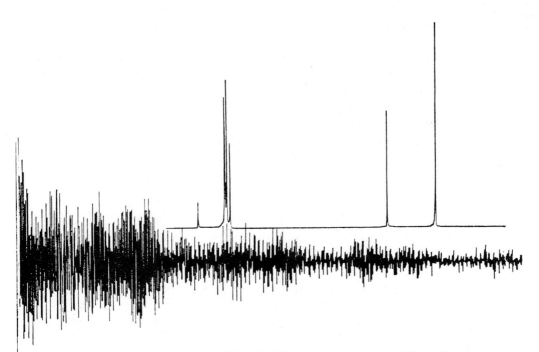

FIGURE 24.12 Free induction decay (FID): The ^{13}C time-domain spectrum of N-dimethyl-benzylamine and the spectrum transformed to the frequency domain. [With permission from K. Müllen and P. S. Pregosin, *Fourier Transform NMR Techniques*. Copyright Academic Press Inc. (London) Ltd., 1976]

density B_0. For a free electron at $B_0 = 0.3354$ T, from Eq. (24.14),

$$\nu = \frac{\Delta\epsilon}{h} = \frac{g\mu_B B_0}{h} = \frac{(2.00232)(9.2741 \times 10^{-24} \text{ J T}^{-1})(0.3354 \text{ T})}{6.626 \times 10^{-34} \text{ J s}}$$

$$= 9.40 \times 10^9 \text{ Hz} = 9.40 \text{ GHz}$$

This frequency is in the *X-band* of the microwave region.

Example 24.6 An irradiated specimen of MgO has a strong esr line at 0.16290 T at a frequency of 9.41756 GHz. What is the g value of the line?

From Eq. (24.14), $g = h\nu/\mu_B B_0$.

$$g = \frac{(6.6262 \times 10^{-34} \text{ J s})(9.41756 \times 10^9 \text{ Hz})}{(9.27408 \times 10^{-24} \text{ A m}^2)(0.16290 \text{ T})} = 4.1306$$

(Recall that the tesla, $\text{T} = \text{kg s}^{-2} \text{ A}^{-1}$. The resonance is due to the Fe^{+3} ion.)

The simple equation (24.14) would apply to an isotropic sample, i.e., one without directional properties, such as a liquid solution. For an anisotropic system, such as a crystal, the resonant frequency depends on the orientation of the crystal axes to the flux density vector $\mathbf{B_0}$. Any anisotropic system can be described by reference to three mutually perpendicular axes X, Y, and Z. The g values when $\mathbf{B_0}$ is along these directions are described as g_{xx}, g_{yy}, and g_{zz}. If a system has axial symmetry, the g value along the axis is $g_{\parallel} = g_{zz}$, and $g_{xx} = g_{yy} = g_{\perp}$.

A block diagram of an esr spectrometer is shown in Fig. 24.13. Klystron tubes for radar were developed for the X-band at 9.4 GHz, the K-band at 24 GHz, and the Q-band at 35 GHz. Esr spectrometers are based on these available sources of radiation, and the flux density of the large magnet is varied (by auxiliary coils) about the corresponding B_0 values to bring the sample with its unpaired spins into resonance. Modern spectrometers often use backward-wave oscillators as microwave sources; they allow the frequency to be varied over a considerable range.

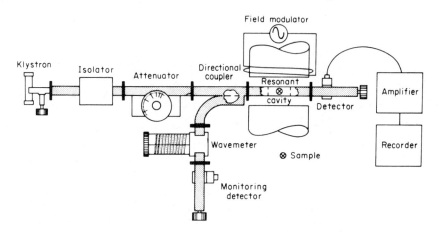

FIGURE 24.13 Block diagram of an X-band EPR spectrometer. [From *Electron Spin Resonance* by J. E. Wertz and J. R. Bolton. Copyright © 1972 McGraw-Hill Book Company. Used with permission of McGraw-Hill Book Company.]

The microwave radiation is channeled through waveguides. The key component of the spectrometer is the *magic-T* waveguide shown in Fig. 24.13, which acts as a microwave bridge. When the adjustable load is balanced against the sample cavity in an off-resonance condition, no radiation passes through the detector arm of the bridge. When the magnetic sweep coils bring the sample into resonance, the bridge is thrown off balance and a signal is transmitted to the detector. Actually, the microwave power is given a sine-wave modulation at, say, 100 kHz, so that the detected signal varies at this modulation frequency. Often, the detector responds to the rate of change of absorption, i.e., to the first derivative of the absorption line with respect to time. Typical spectral lines of this type are shown in Fig. 24.14.

The energy difference between lower and upper states of electrons in the field of flux density B_0 is much greater than that for nuclei in nmr, so that the Boltzmann factor in Eq. (24.6) corresponds to an excess population in the ground state of about 1 in 10^3 (instead of the 1 in 10^5 in nmr). Hence esr is a far more sensitive technique for spin detection and in fact 10^{-11} mol of a free radical or other paramagnetic substance can give a measurable signal. Because electrons are much more exposed to fluctuating fields than are nuclei, the spin-spin relaxation time T_2 is much shorter for esr than for nmr. Thus an absorption peak in an esr spectrum is much broader than in nmr, typically about 500 kHz as compared to a few hertz.

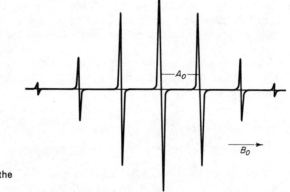

FIGURE 24.14 EPR spectrum of the benzene radical ion, $C_6H_6^-$.

24.11 Nuclear Hyperfine Interaction

If the only magnetic field acting on an unpaired electron were that of the external magnet (B_0), the esr spectrum would always consist of a single line, with a frequency specified by its g value. In fact, however, the electron spin also interacts with the magnetic moments of any nuclear spins in its neighborhood. As a result, the single esr line of the lone electron is split into a set of lines by this *nuclear hyperfine interaction* (Fig. 24.15).

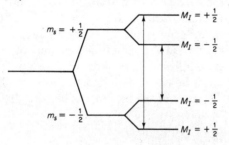

FIGURE 24.15

Consider, for example, a single electron (spin $m_s = \pm\frac{1}{2}$) interacting with a single proton (spin $M_I = \pm\frac{1}{2}$) in an external field of flux density B_0. The four energy levels that result are shown in Fig. 24.15, with the allowed transitions. The selection rules* are $\Delta m_s = \pm 1$, $\Delta M_I = 0$. The energies of the two allowed transitions are

$$\Delta\epsilon = h\nu = g\mu_B B_0 \pm \tfrac{1}{2}hA_0 \qquad (24.15)$$

where A_0 is the isotropic hyperfine coupling constant. If the electron spin interacts with a nucleus of spin I, there are $(2I + 1)$ esr lines of equal intensity. This treatment applies to systems in liquid or gas phases; in crystals the interaction becomes anisotropic and its dependence on orientation of the crystal must be considered.

---

*The photon has angular momentum $h/2\pi$. If a photon having the resonant esr frequency is absorbed, $\Delta m_s = +1$ and the spin angular momentum increases by $h/2\pi$, while the angular momentum $h/2\pi$ of the photon disappears. Hence ΔM_I must equal zero to conserve the total angular momentum of the system.

Example 24.7 For ^{23}Na, $A_0 = 886$ MHz. The ground state of the Na atom is ^2S and $I - 3/2$. Calculate the values of B_0 for esr resonance of the single unpaired electron of the Na atom at a fixed microwave frequency of 9.300 GHz if $g = 2.0022$.

The esr line is split into four lines by the isotropic hyperfine interaction. From Eq. (24.15) the resonant condition is $h\nu_0 = g\mu_B B_0 + hA_0 M_I$ with $M_I = \frac{3}{2}, \frac{1}{2}, -\frac{1}{2}, -\frac{3}{2}$. Hence $B_0 = h(g\mu_B)^{-1}[\nu_0 - A_0 M_I]$.

$$B_0 = \frac{6.626 \times 10^{-34} \text{ J s}}{(2.0022)(9.274 \times 10^{-24} \text{ A m}^2)} [9.300 \times 10^9 \text{ s}^{-1} - 8.86 \times 10^8 \text{ s}^{-1} M_I]$$

$$= 3.568 \times 10^{-11}[9.300 \times 10^9 - 8.86 \times 10^8 M_I] \text{ T}$$

For $M_I = \frac{3}{2}, \frac{1}{2}, -\frac{1}{2}, -\frac{3}{2}$, $B_0 = 0.2844, 0.3160, 0.3476,$ and 0.3792 T. (Note that one must specify both g and A_0 to calculate the resonant field at fixed frequency.)

24.12 Spectra of Free Radicals

When the electron spin interacts with more than one equivalent nuclear magnetic moment, the hyperfine splitting gives a pattern of lines that is similar to those seen in first-order spin-spin interactions in nmr (Section 24.7). Interesting examples are found in the esr spectra of organic free radicals. Free radicals are produced as intermediates in many reactions, particularly oxidations and reductions, and of course they often are formed in photochemical reactions.

For example, reduction of benzene yields the benzene radical anion, $C_6H_6^-$. The spectrum of this species is shown in Fig. 24.14. The g value of $C_6H_6^-$ is 2.0025. The spectrum consists of seven equidistant lines with relative intensities $1:6:13:20:15:6:1$. The splitting is due to the magnetic interaction between the spin of the unpaired electron and the nuclear spins of the six protons in the $C_6H_6^-$ ring. The possible arrangements of the proton spins are:

Statistical weight $(N!/N_1!N_2!)$

↓↓↓↓↓↓	1	↑↑↑↑↑↑
↑↓↓↓↓↓	6	↑↑↑↑↑↓
↑↑↓↓↓↓	15	↑↑↑↑↑↓↓
↑↑↑↓↓↓	20	

The number of arrangements of the spins with given numbers up and down determines the relative intensity of each line.

We can see at once from the spectrum that the single unpaired electron in a π orbital of $C_6H_6^-$ has an equal probability density (or *spin density*) at the sites of all six ring protons. The magnitude of the hyperfine splitting (i.e., of the spin-spin interaction) is proportional to the *spin density* at the positions of the protons.

Example 24.8 The esr spectrum of the ethyl radical C_2H_5 appears as follows:

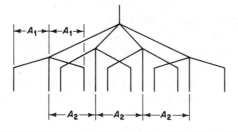

Explain the pattern of lines observed.

The hyperfine splitting (A_1) due to the two CH_2 protons yields a triplet of intensity ratio $1:2:1$. Each component is further split (A_2) by the CH_3 protons into four components of intensity ratio $1:2:2:1$. The resultant spectrum is:

Problems

1. What are the magnetogyric ratio and g_N factor for the ^{13}C nucleus, which has a magnetic moment of 0.70216 nuclear magneton? What is the magnitude of the maximum value of the angular momentum of the nucleus and its maximum component in the direction of a magnetic field? How do these values compare to those for electron spin?

2. Consider a nucleus of ^{10}B in a magnetic field of 10 T. What are the allowed energy levels? What will be the equilibrium difference in population of the lowest and highest energy level at 100 K?

3. Sketch the nmr spectra you would expect to find for (a) the protons in isopropyl bromide; (b) the protons in dimethylethylamine, $(CH_3)_2C_2H_5N$; (c) the ^{19}F nuclei in PF_3; (d) the ^{31}P nucleus in PF_3. In each case show qualitatively the line positions and relative intensities and indicate with suitable labels the origin of the various features shown in the spectra. Assume a 7.0-T spectrometer and room temperature; if you need to make additional assumptions, mention them explicitly.

4. The proton nmr spectrum of dimethylformamide at 40 MHz and room temperature is as follows:

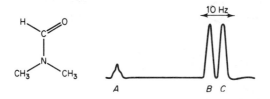

The doublet B, C might arise either from chemically different groups or from spin-spin coupling. (a) What feature of the spectrum immediately rules out one of these explanations in this case? (b) Explain the origin of the lines A, B, C and how you expect the spectrum to appear at a higher temperature.

5. Sketch the NMR spectra you would expect for (a) isopropyl alcohol and (b) *tert*-butyl alcohol with a superconducting NMR spectrometer operating at 7.2 T. Indicate the estimated chemical shifts, spin-spin splitting, and relative intensities of lines.

6. An esr spectrometer operates in the K-band at 24 GHz. What would be the strength of magnetic field required (B_0 value)? In detection of free radicals at 300 K, what would be the relative number of radicals with spin quantum numbers $m_s = \pm\frac{1}{2}$? If 10^{-11} mol dm^{-3} of radicals can be detected, what would be the *numbers* of these per dm^3 in upper and lower states?

7. The benzene radical anion has $g = 2.0025$. At what magnetic induction B_0 would its esr be centered at a frequency 9.350 GHz?

8. The electron in atomic hydrogen H has $g = 2.0032$. In a spectrometer at 9.250 GHz, the two lines from H appeared at 357.3 and 306.6 mT. Calculate the hyperfine coupling constant for H.

9. Sketch the appearance of the esr spectra of the radicals CH_3 and CD_3.

10. The fluorine nmr spectrum of F_2BrC—$CBrCl_2$ is shown at 40 MHz. Explain qualita-

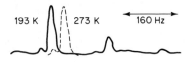

tively the change in spectra with T. Estimate the barrier to rotation about C—C bond in this compound.

11. In the nmr spectrum of HD, how many lines occur in (a) the H spectrum? (b) the D spectrum? The coupling constant is $J = 43.0$ Hz.

12. The coupling constant between H and D in CH_3D is 1.90 Hz. What would be the coupling constant between the two H in CH_4? Sketch the proton nmr spectra of CH_3D and CH_4.

13. In 1, 4 dibromobutane is the pair of protons on C_2 magnetically equivalent to the pair on C_3? Are the pair on C_2 magnetically equivalent?

14. An electron with magnetic moment $\mathbf{\mu}_B$ in a magnetic field of flux density $\mathbf{B}$ has two allowed energy levels $\pm\mu_B B$. Calculate the average energy of the spin system at 4 K and at 200 K. Draw the partition function as a function of T between 0 and 500 K.

15. Suppose that the intensity of a proton nmr peak is proportional to the difference in populations of the lower and upper levels in the magnetic field. A spectrum is observed at 100 MHz and 290 K. With no change in sample or time, the spectrum is rerun at 400 MHz. What is the difference in intensity of a typical line?

25

Symmetry

The word *symmetry* comes from a Greek word meaning "in the right measure." The idea of symmetry, explored by classic Greek sculptors and architects, is also applicable to the structures of molecules and crystals. It is particularly useful in the study of molecular spectra. This chapter provides an introduction to the elementary theory of groups of symmetry operations. It is not intended to be a systematic treatment of this important subject. We shall not give general proofs but instead take a simple example of a symmetry group and follow some important concepts as illustrated by this example.

25.1 Symmetry Operations

A symmetry operation transforms a spatial arrangement, such as a molecule, into an arrangement that is indistinguishable from the original one. For example, consider the ammonia molecule in Fig. 25.1. There are six different operations that will produce an NH_3 structure that coincides with the original one. These operations are shown in Fig. 25.1, where the H atoms have been labeled A, B, C, for convenience in description. The symmetry operations on NH_3 are as follows:

1. *E* is the identity operation, which leaves each point unchanged.
2. C_3 is a threefold rotation about an axis passing perpendicularly through the N atom. This operation rotates a representative point through an angle $2\pi/3$ rad (120°) in a positive (counterclockwise) direction. C_3 is called a *threefold*

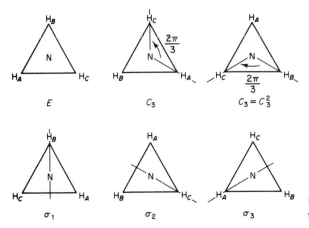

FIGURE 25.1 Symmetry operations of the NH_3 molecule—point group C_{3v}.

rotation because when the operation is repeated three times, the figure is returned to its original position.

3. $\bar{C}_3$ is a threefold rotation about an axis passing perpendicularly through the N atom. This operation rotates a representative point through an angle of $2\pi/3$ rad (120°) in a negative (clockwise) direction. (This operation could also be written as C_3^2, two successive applications of C_3 leading to a positive rotation of $4\pi/3$.)

4. σ_1 is a reflection in a *mirror plane* that is perpendicular to the plane of the H atoms and passes through the N atom and H_C. This operation takes a representative point to an equal distance the other side of the line bisecting the H_C vertex of triangle $H_A H_B H_C$.

5. σ_2 is an operation like σ_1, but involves reflection in the plane bisecting the vertex H_B.

6. σ_3 is an operation again like σ_1, but it involves reflection in the plane bisecting the vertex H_A.

We shall encounter additional symmetry elements in other structures, but first let us examine some properties of the set of elements just listed for NH_3. The product AB of two symmetry operations is defined to mean that operation B is first performed, and then operation A is performed on the result. If we examine all products AB between elements in the set of six symmetry operations for NH_3, we find that the product of any two operations always gives the same effect as one of the operations in the original set. This rule can be seen in the multiplication table for the set of operations, given in Table 25.1.

The table follows the convention that the intersection of a row with a column gives the product RC of the row element R and the column element C. As can be seen from the table, RC does not necessarily equal CR, although it may. When $RC = CR$, we say that R and C *commute*. [Do C_3 and C_3^2 commute? Do σ_1 and C_3?]

TABLE 25.1

MULTIPLICATION TABLE OF THE SYMMETRY GROUP C_{3v}

Operation R	Operation C					
	E	C_3	$\bar{C}_3 = C_3^2$	σ_1	σ_2	σ_3
E	E	C_3	$\bar{C}_3$	σ_1	σ_2	σ_3
C_3	C_3	$\bar{C}_3$	E	σ_3	σ_1	σ_2
$\bar{C}_3 = C_3^2$	$\bar{C}_3$	E	C_3	σ_2	σ_3	σ_1
σ_1	σ_1	σ_2	σ_3	E	C_3	$\bar{C}_3$
σ_2	σ_2	σ_3	σ_1	$\bar{C}_3$	E	C_3
σ_3	σ_3	σ_1	σ_2	C_3	$\bar{C}_3$	E

25.2 Definition of a Group

A set of elements constitutes a group if its members satisfy the following conditions:

1. The operation of taking the product of two elements having been defined, the product AB of any two elements of the set is itself a member C of the set, $AB = C$.
2. The set includes an identity element E such that for every member A, $EA = AE = A$.
3. Each element A has an inverse A^{-1}, which is also a member of the set, where $A^{-1}A = AA^{-1} = E$.
4. The associative law of multiplication holds, $A(BC) = (AB)C$.

Examination of the multiplication table of the symmetry operations of NH_3 show that they form a group. This particular group is designated $\mathbf{C_{3v}}$ in accord with a notation to be described later.

Any collection of the elements of a group that by themselves satisfy all the group postulates is called a *subgroup* of the original group. [From the multiplication Table 25.1, select a subgroup of $\mathbf{C_{3v}}$.]

Each symmetry operation of NH_3 leaves the position of the N atom invariant. A group of symmetry operations in which a point is left invariant under each operation is called a *point group*. All the symmetry groups of molecular structures are point groups. Other molecules belonging to the point group $\mathbf{C_{3v}}$ are methyl chloride and chloroform.

If A and X are members of a group, the element $Y = X^{-1}AX$ is also a member of the group. We say that Y is the result of a *similarity transformation* performed on element A. If we allow X to be each member of the group in turn, the set of elements Y that results is called a *class* of the group.

Example 25.1 Divide the elements of the group C_{3v} into classes.

The operations are all based on the multiplication Table 25.1. When $A = E$, the similarity transformation yields $X^{-1}EX = E$ for every element X of the group. Thus E forms a class of which it is the only member. When $A = C_3$ we obtain the following results:

$$
\begin{array}{ll}
X = E & Y = E^{-1}C_3E = C_3 \\
X = C_3 & Y = C_3^{-1}C_3C_3 = C_3 \\
X = \bar{C}_3 & Y = \bar{C}_3^{-1}C_3C_3 = \bar{C}_3E = \bar{C}_3 \\
X = \sigma_1 & Y = \sigma_1^{-1}C_3\sigma_1 = \sigma_1^{-1}\sigma_3 = \sigma_1\sigma_3 = \bar{C}_3 \\
X = \sigma_2 & Y = \sigma_2^{-1}C_3\sigma_2 = \sigma_2^{-1}\sigma_1 = \sigma_2\sigma_1 = \bar{C}_3 \\
X = \sigma_3 & Y = \sigma_3^{-1}C_3\sigma_3 = \sigma_3^{-1}\sigma_2 = \sigma_3\sigma_2 = \bar{C}_3
\end{array}
$$

Hence C_3 and $\bar{C}_3$ form a class of group C_{3v}. Similarly, σ_1, σ_2, and σ_3 form a class. Thus there are three classes: (E), $(C_3, \bar{C}_3)$, $(\sigma_1, \sigma_2, \sigma_3)$.

25.3 Further Symmetry Operations

In addition to the identity operation E, n-fold rotation axes C_n, and mirror planes σ, two other types of symmetry operation can occur in molecules.

1. In a *rotary reflection*, a representative point is rotated through an angle of $2\pi/n$ about an axis and then reflected in a mirror plane σ_h perpendicular to that axis. The order of these operations makes no difference. The operation of rotary reflection can thus be written symbolically as $S_n = \sigma_h C_n$. An example of a molecule with an S_4 axis is methane, as shown in Fig. 25.2.

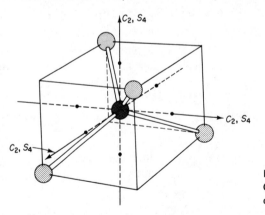

FIGURE 25.2 Model of methane CH_4, showing the C_2 axes, which coincide with S_4 axes.

2. Inversion in a center of symmetry O is denoted by i. The operation takes the representative point P into a point P', equidistant from O, and the direction of OP' is an extension of the line PO. An inversion center is equivalent to a twofold rotary reflection axis, $i = S_2$. Figure 25.3 shows an example of a molecule with a center of symmetry.

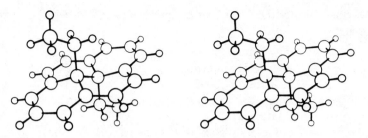

FIGURE 25.3 The molecule *trans*-15,16-diethyldihydropyrene has a center of symmetry and no other symmetry elements. Point group **S_2**. [From *Symmetry: A Stereoscopic Guide for Chemists* by I. Bernal, W. C. Hamilton, and J. S. Ricci. W. H. Freeman and Company. Copyright © 1972.]

We have now enumerated all the symmetry elements that can occur in molecules. They are summarized in Table 25.2. Many different groups of these elements can be formed; these are the *molecular point groups*. If we consider the equilibrium positions of the nuclei in a molecule, we can assign every molecule to a definite point group.

TABLE 25.2
SYMMETRY ELEMENTS AND SYMMETRY OPERATIONS

Symbol	Symmetry element	Symmetry operation
E	Identity	No change (identity operation)
C_n	n-fold rotation axis	Rotation of $2\pi/n$*
σ	Mirror plane	Reflection
i	Inversion center (center of symmetry)	Inversion
S_n	n-fold rotary reflection axis	Rotation by $2\pi/n$ and reflection in plane ⊥ axis

*Note that for cylindrical symmetry $n = \infty$.

25.4 Notation for Point Groups

In the Schoenflies notation, each point group is identified by a symbol that consists of a capital letter and a subscript. The capitals identify the principal symmetry elements as follows:

C   simple rotation axis
D n rotation axes ⊥ to a principal rotation axis
S   rotary reflection axis
T   symmetry based on tetrahedron
O   symmetry based on octahedron
I   symmetry based on icosahedron

The subscript indicates the order n of the principal axis and whether planes of symmetry occur.

s	only plane of symmetry
i	only center of symmetry
n	only rotation axis
nv	symmetry planes σ_v that contain the principal rotation axis
nh	symmetry planes $\sigma_h \perp$ principal rotation axis
nd	symmetry planes that contain the major symmetry axis and that bisect the angles between the twofold axes $\perp$ principal axis

Examples of molecules belonging to $\mathbf{D_5}$ and $\mathbf{D_{5h}}$ are shown in Fig. 25.4. The rare point group $\mathbf{I_h}$ is exemplified by the $B_{12}H_{12}^{2-}$ anion in Fig. 25.5.

The systematic assignment of the point group of a molecule is facilitated by the flowchart in Fig. 25.6.

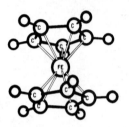
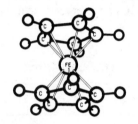

(a)

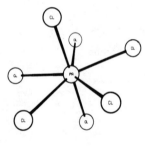
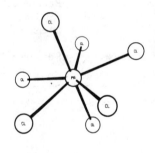

(b)

FIGURE 25.4 (a) Biscyclopentadienyl iron (II), $(C_5H_5)_2Fe$, belongs to point group $\mathbf{D_5}$. The fivefold axis C_5 passes through the Fe atom and the midpoints of the rings, and five twofold axes C_2 are perpendicular to C_5 and pass through the Fe. (b) Protactinium heptachloride belongs to $\mathbf{D_{5h}}$. In addition to the elements of $\mathbf{D_5}$, $\mathbf{D_{5h}}$ has a mirror plane perpendicular to C_5 and five vertical mirror planes, one bisecting each pair of Cl atoms in the plane of the Pa. [From *Symmetry: A Stereoscopic Guide for Chemists* by I. Bernal, W. C. Hamilton, and J. S. Ricci. W. H. Freeman and Company. Copyright © 1972.]

25.5 Point-groups and Molecular Properties

The symmetry elements of a molecule can restrict its properties. An example is the dipole moment $\boldsymbol{\mu}$. Only molecules belonging to one of the point groups $\mathbf{C_s}$, $\mathbf{C_n}$ or $\mathbf{C_{nv}}$ can have permanent dipole moments. (In all other molecular point groups, there will be a symmetry element that corresponds to turning the molecule upside down. The dipole-moment vector obviously cannot coincide with its original direction after the molecule is turned upside down.) For $\mathbf{C_n}$ or $\mathbf{C_{nv}}$ ($n > 1$), $\boldsymbol{\mu}$ must be in the direction of the symmetry axis; for $\mathbf{C_s}$, $\boldsymbol{\mu}$ must be in the plane of symmetry.

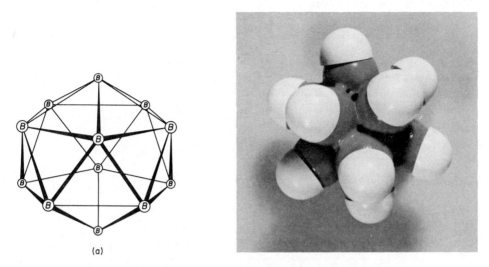

FIGURE 25.5 (a) Arrangement of the boron atoms in the icosahedral [$\mathbf{I_h}$] $B_{12}H_{12}^{2-}$ ion. (b) Model of the $B_{12}H_{12}^{2-}$ ion.

    The occurrence of optical activity is closely related to the molecular symmetry. A molecule can display optical activity only if it does not have a rotary reflection axis $S_n(n > 2)$, an inversion center i, or a symmetry plane σ_h. Since an S_2 axis corresponds to an inversion center and an S_1 axis corresponds to a mirror plane σ_h, the rule can be reduced to: A molecule is optically active if and only if it does not have any rotary reflection axis S_n ($n \geq 1$).

    Thus optically active molecules belong to the following point groups: $\mathbf{C_n}$, $\mathbf{D_n}$ and $\mathbf{T}$. Note that the absence of a center of symmetry does not necessarily imply optical activity. Molecules belonging to point group $\mathbf{S_4}$ lack the symmetry center but still are not optically active; an example is shown in Fig. 25.7.

25.6 Transformations of Vectors by Symmetry Operations

    The physical representation of symmetry elements as rotations, reflections, and so on, introduces many of the properties of the molecular point groups, but further development of the theory requires a mathematical formulation of the transformations that are produced by the symmetry operations. With reference to a set of Cartesian axes X, Y, Z, the coordinates of each atom i in the molecule can be specified by x_i, y_i, z_i. If these coordinates are given for each atom, the position and orientation of the molecule is exactly specified. The set of three coordinates $x_i y_i z_i$ defines the position of a vector from the origin to the particular atom i. This vector can be written as

$$\begin{pmatrix} x_i \\ y_i \\ z_i \end{pmatrix}$$

HOW TO SPECIFY A POINT GROUP

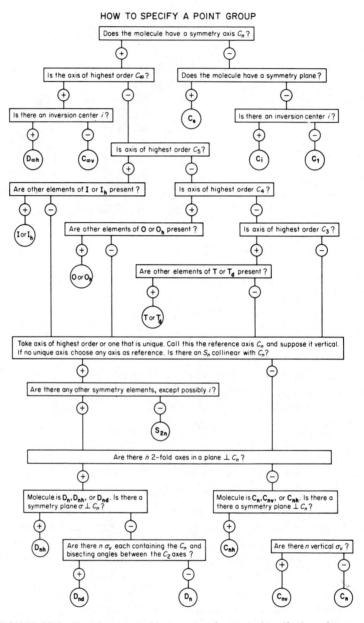

FIGURE 25.6 Flowchart to provide a systematic way to identify the point group of a molecule.

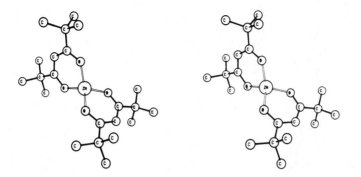

FIGURE 25.7 The molecule $Zn[(CH_3)_3C \cdot CO \cdot CH_2 \cdot COC(CH_3)_3]_2$ belongs to point group S_4. Although there is no center of symmetry, the molecule cannot have optical activity. [From *Symmetry: A Stereoscopic Guide for Chemists* by I. Bernal, W. C. Hamilton, and J. S. Ricci. W. H. Freeman and Company. Copyright © 1972.]

If the molecule is subjected to a symmetry operation, the coordinates x_i y_i z_i will be transformed to some new values x_i' y_i' z_i'. The transformation of coordinates can always be written as a set of linear equations of the general form

$$x_i' = a_{11}x_i + a_{12}y_i + a_{13}z_i$$
$$y_i' = a_{21}x_i + a_{22}y_i + a_{23}z_i$$
$$z_i' = a_{31}x_i + a_{32}y_i + a_{33}z_i$$

The usual notation for such a linear transformation is

$$\begin{pmatrix} x_i' \\ y_i' \\ z_i' \end{pmatrix} = \begin{pmatrix} a_{11} & a_{12} & a_{13} \\ a_{21} & a_{22} & a_{23} \\ a_{31} & a_{32} & a_{33} \end{pmatrix} \begin{pmatrix} x_i \\ y_i \\ z_i \end{pmatrix}$$

The new vector is obtained by multiplying the original vector by the transformation matrix. The transformation matrix is the same for all the vectors of a given molecule and would apply to the general vector (x, y, z). If we can represent all the symmetry operations by transformation matrices, we can represent the point group in terms of these matrices.

The identity operation E is a trivial case. The corresponding matrix is

$$E \longrightarrow \begin{pmatrix} 1 & 0 & 0 \\ 0 & 1 & 0 \\ 0 & 0 & 1 \end{pmatrix} \tag{25.1}$$

Consider next the operation of reflection in a mirror plane σ. Figure 25.8 shows a set of Cartesian axes with unit vectors **i, j, k** directed along these axes. The **k** vector (z axis) lies in this mirror plane which is at an angle β to the **i** vector (x axis). We now consider how the symmetry operation σ acts on each of the unit vectors. The transformed vectors are denoted by σ**i**, σ**j**, σ**k**. By inspection of the figure we find

$$\sigma\mathbf{i} = (\cos 2\beta)\mathbf{i} + (\sin 2\beta)\mathbf{j} + (0)\mathbf{k}$$
$$\sigma\mathbf{j} = (\sin 2\beta)\mathbf{i} - (\cos 2\beta)\mathbf{j} + (0)\mathbf{k}$$
$$\sigma\mathbf{k} = (0)\mathbf{i} + (0)\mathbf{j} + (1)\mathbf{k}$$

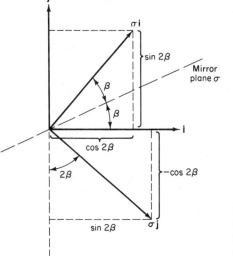

FIGURE 25.8 Transformation of unit vectors **i** and **j** by a mirror plane σ. The **k** axis is normal to the plane of the paper and lies in the mirror plane.

The transformation matrix that represents σ is thus

$$\sigma \longrightarrow \begin{pmatrix} \cos 2\beta & \sin 2\beta & 0 \\ \sin 2\beta & -\cos 2\beta & 0 \\ 0 & 0 & 1 \end{pmatrix} \tag{25.2}$$

Consider next the operation of rotation about an axis that is directed along the **k** vector. Figure 25.9 shows this operation $C(\alpha)$ for a rotation by an angle α. The unit vectors are thereby transformed to new vectors $C\mathbf{i}, C\mathbf{j}, C\mathbf{k}$, as follows

$$C\mathbf{i} = (\cos \alpha)\mathbf{i} - (\sin \alpha)\mathbf{j} + (0)\mathbf{k}$$
$$C\mathbf{j} = (\sin \alpha)\mathbf{i} + (\cos \alpha)\mathbf{j} + (0)\mathbf{k}$$
$$C\mathbf{k} = (0)\mathbf{i} + (0)\mathbf{j} + 1(\mathbf{k})$$

The matrix representing $C(\alpha)$ is thus

$$C(\alpha) \longrightarrow \begin{pmatrix} \cos \alpha & -\sin \alpha & 0 \\ \sin \alpha & \cos \alpha & 0 \\ 0 & 0 & 1 \end{pmatrix} \tag{25.3}$$

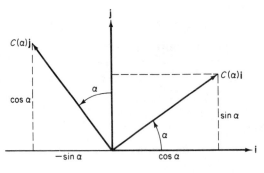

FIGURE 25.9 Transformation of unit vectors **i** and **j** by a symmetry axis $C(\alpha)$ directed along the **k** axis which is normal to **i** and **j**. The operation of rotation is $C(\alpha)$.

[Write the matrix corresponding to the operation of inversion at a symmetry center located at the origin of the $X Y Z$ axes.]

25.7 Matrix Representation of Group C_{3v}

We can now write down a set of matrices that correspond with the set of symmetry operations of the point group C_{3v}. Figure 25.10 shows the unit vectors superimposed on the structure of NH_3 that we used in Fig. 25.1 to illustrate the symmetry operations of C_{3v}. For the mirror plane σ_a, $\beta = 0$ and the transformation matrix of Eq. (25.2) becomes

$$R(\sigma_a) = \begin{pmatrix} 1 & 0 & 0 \\ 0 & -1 & 0 \\ 0 & 0 & 1 \end{pmatrix}$$

For σ_b and σ_c, $\beta = -60°$ and $+60°$, respectively. The corresponding matrices are listed in Table 25.3. For the operation C_3, $\alpha = 120°$, and for C_3^2, $\alpha = -120°$. Hence Eq. (25.3) gives the corresponding matrices in Table 25.3.

We now have a matrix corresponding to each of the symmetry operations of C_{3v}. These matrices obey the same multiplication table as that shown in Table 25.1

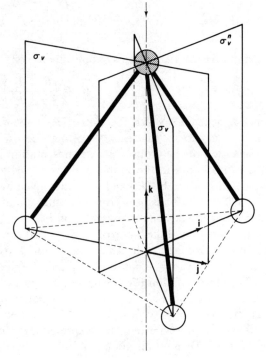

FIGURE 25.10 Unit vectors **i, j, k** superimposed on a figure with symmetry of point group C_{3v}. [D. S. Schonland, *Molecular Symmetry* (London: D. van Nostrand Company, 1965)].

TABLE 25.3

$$E = \begin{pmatrix} 1 & 0 & 0 \\ 0 & 1 & 0 \\ 0 & 0 & 1 \end{pmatrix} \quad C_3 = \begin{pmatrix} -1/2 & -\sqrt{3}/2 & 0 \\ \sqrt{3}/2 & -1/2 & 0 \\ 0 & 0 & 1 \end{pmatrix} \quad \bar{C}_3 = \begin{pmatrix} -1/2 & \sqrt{3}/2 & 0 \\ -\sqrt{3}/2 & -1/2 & 0 \\ 0 & 0 & 1 \end{pmatrix}$$

$$\sigma_1 = \begin{pmatrix} 1 & 0 & 0 \\ 0 & -1 & 0 \\ 0 & 0 & 1 \end{pmatrix} \quad \sigma_2 = \begin{pmatrix} -1/2 & -\sqrt{3}/2 & 0 \\ -\sqrt{3}/2 & 1/2 & 0 \\ 0 & 0 & 1 \end{pmatrix} \quad \sigma_3 = \begin{pmatrix} -1/2 & \sqrt{3}/2 & 0 \\ \sqrt{3}/2 & 1/2 & 0 \\ 0 & 0 & 1 \end{pmatrix}$$

for the symmetry operations themselves. This set of matrices is called a *representation* of the group C_{3v}. [Show by matrix multiplication that $R(\sigma_b)\, R(\sigma_c) = R(C_3)$.]

25.8 Irreducible Representations

There are many ways to derive groups of matrices that are representations of symmetry groups. A particular type of representation is fundamentally important. This is the *irreducible representation*.

If we inspect the particular representation of C_{3v} shown in Table 25.3, we will note that all the matrices have the same general form,

$$(R) = \begin{pmatrix} a_{11} & a_{12} & 0 \\ a_{21} & a_{22} & 0 \\ 0 & 0 & 1 \end{pmatrix}$$

Such a matrix can be written symbolically as

$$(R) = \left(\begin{array}{c|c} R' & 0 \\ \hline 0 & R'' \end{array} \right)$$

In such a case, we say that the matrix (R) is the *direct sum* of the matrices (R') and (R''). It is easy to see that each set of matrices (R') and (R'') separately furnishes a representation of the group C_{3v}. A representation is often denoted by the symbol Γ, and the direct sum in the example above is written $\Gamma = \Gamma_3 + \Gamma_1$, where Γ_3 is the representation given by (R') and Γ_1 is that given by (R''). We say that the representation Γ in Table 25.3 has been *reduced* to the sum of a two-dimensional representation Γ_3 and a one-dimensional representation Γ_1.

Examination of the group of matrices that comprise Γ_3 indicates that it cannot be reduced any further; i.e., it cannot be set equal to the direct sum of two one-dimensional representations. Thus both Γ_3 and Γ_1 are *irreducible representations* (I.R.).

The representation Γ_1, in which each element is represented by "1," is called the *identical representation*. It is trivial to write down, and always occurs, but nevertheless it has important applications.

There is another irreducible representation of C_{3v}, which can be obtained by inspection of the multiplication Table 25.1. In this I.R., (Γ_2), E, C_3, and C_3^2 are represented by $+1$ and $\sigma_a, \sigma_b, \sigma_c$ by -1.

Table 25.4 lists the three irreducible representations that we have obtained. Is there any other I.R. of C_{3v}? There is a theorem of group theory: The number of I.R.'s is equal to the number of classes in the group. Since there are three classes of elements in C_{3v}, there are three I.R.'s and we have the complete set in Table 25.4.

TABLE 25.4
THE IRREDUCIBLE REPRESENTATIONS OF C_{3v}

$$\Gamma_3 \quad E = \begin{pmatrix} 1 & 0 \\ 0 & 1 \end{pmatrix} \quad C_3 = \begin{pmatrix} -1/2 & -\sqrt{3}/2 \\ \sqrt{3}/2 & -1/2 \end{pmatrix} \quad \bar{C}_3 = \begin{pmatrix} -1/2 & \sqrt{3}/2 \\ \sqrt{3}/2 & -1/2 \end{pmatrix}$$
$$\chi(E) = 2 \qquad \chi(C_3) = -1 \qquad\qquad \chi(C_3) = -1$$

$$\sigma_1 = \begin{pmatrix} 1 & 0 \\ 0 & -1 \end{pmatrix} \quad \sigma_2 = \begin{pmatrix} -1/2 & -\sqrt{3}/2 \\ \sqrt{3}/2 & 1/2 \end{pmatrix} \quad \sigma_3 = \begin{pmatrix} -1/2 & \sqrt{3}/2 \\ \sqrt{3}/2 & 1/2 \end{pmatrix}$$
$$\chi(\sigma_1) = 0 \qquad \chi(\sigma_2) = 0 \qquad\qquad \chi(\sigma_3) = 0$$

$\Gamma_1 \quad E = (1) \quad C_3 = (1) \quad \bar{C}_3 = (1) \quad \sigma_1 = (1) \quad \sigma_2 = (1) \quad \sigma_3 = (1)$

$\Gamma_2 \quad E = (1) \quad C_3 = (1) \quad \bar{C}_3 = (1) \quad \sigma_1 = (-1) \quad \sigma_2 = (-1) \quad \sigma_3 = (-1)$

25.9 Characters of Irreducible Representations

All the symmetry groups have been studied and their irreducible representations can be obtained by methods similar to those we have used for C_{3v}. In some cases more powerful group theoretic methods must be used but you do not need these to apply the results of the theory.

For many applications of group theory to molecular spectra and quantum mechanics, we do not use the I.R.'s directly but only their *characters* χ. The *character* of an element in the matrix representation of a group is the *trace* of the matrix for that element. The trace of a matrix is the sum of its diagonal terms. The characters χ of the elements in the representation Γ_3 of C_{3v} are included in Table 25.4. For the one-dimensional representations Γ_1 and Γ_2, the characters are simply the elements themselves.

For each group, we can now write a *character table*. This table summarizes those essential properties of the symmetry group that are important in spectroscopic or theoretical investigations of any molecule belonging to that group. The character table of the group C_{3v} is shown in Table 25.5 in a standard format.

TABLE 25.5
CHARACTER TABLE OF GROUP C_{3v}

C_{3v}	E	$2C_3$	$3\sigma_v$		
A_1	1	1	1	z	$x^2 + y^2, 2z^2 - x^2 - y^2$
A_2	1	1	-1	R_z	
E	2	-1	0	$(x, y), (R_x, R_y)$	$(xz, yz), (xy, x^2 - y^2)$

This format can be summarized as follows:

Schoenflies symbol	Symbols for each class in group and number of elements in the class		
Mulliken symbols for each I.R.	Characters of the irreducible representations of each class	Symmetry properties of translations and rotations	Symmetry properties of functions of coordinates

The Mulliken symbol A or B denotes a one-dimensional I.R. E denotes a two-dimensional I.R. (not to be confused with the identity operation E), and T a three-dimensional I.R. A is used when the character for rotation about the principal axis (C_3 in this case) is $+1$ and B when this character is -1. The subscripts 1 and 2 designate symmetry or antisymmetry (change of sign) with respect to a C_2 normal to the principal axis, or if such a C_2 is lacking, with respect to a vertical plane of symmetry σ_v. A subscript g or u is used to denote a representation for which the character for an inversion through a center of symmetry is $+1$ (g) or -1 (u), if this operation occurs in the group.

The next columns list the characters of the I.R.'s arranged according to classes of symmetry operations. All operations of the same class always have the same character.

In the next column of the table, six symbols occur, x, y, z, R_x, R_y, and R_z. The molecule is always oriented so that its principal axis coincides with the Z axis of a system of Cartesian axes. Then x, y, and z denote the transformation properties of the Cartesian coordinates of any point in the molecule, and R_x, R_y, and R_z refer to rotations about the respective axes. Thus we say that, in the $\mathbf{C}_{3v}$ group, z transforms in accord with the irreducible representation A_1. Inspection of Fig. 25.10 shows that z is unchanged by any of the operations of the group in accord with the characters $+1$ for all the operations. The *rotation* about the Z axis is unchanged by E, C_3, or C_3^2, but reflection in a mirror planes changes the sense of the rotation; thus R_z transforms in accord with A_2.

The coordinates x and y transform according to the two-dimensional I.R., E, and indeed we can see that the result of the symmetry operations on x or y is to produce a linear combination of x and y coordinates. The result for R_x and R_y is similar.

The last column denotes some symmetry properties of functions of x, y, and z. These functions are particularly useful in quantum mechanical problems since they have the transformation properties of certain atomic orbitals.

The character tables of a number of groups that frequently occur in chemical problems are given in an appendix to this chapter. Complete tables with many applications and a much more detailed treatment of the theory are readily available.*

*F. A. Cotton, *Chemical Applications of Group Theory*, 2nd ed. (New York: John Wiley & Sons, Inc., 1971).

We shall mention two simple applications of group theory to atomic orbitals, and in Chapter 26 we shall discuss several applications to molecular spectra. In coordination compounds and in crystals a transition-metal ion experiences an electric field due to a regular array of surrounding ligands or neighboring ions. Typical examples would be $[Co(H_2O)_6]^{3+}$, $[FeF_6]^{3-}$, or in ruby the Cr^{3+} ion that substitutes for Al^{3+} in the structure of corundum (Al_2O_3). In the absence of a field, the five $3d$ orbitals in any one of these ions are equal in energy. The crystal field (or ligand) splits the energy level in a way that is determined by the symmetry of the field.

In the case of a field with octahedral symmetry, the five $3d$ orbitals are split into an upper set of two (e_g) and a lower set of three (t_{2g}). The d orbitals transform as functions of coordinates: xy, xz, yz, z^2, and $x^2 - y^2$. The symmetry group of the regular octahedron is $\mathbf{O_h}$. From the character table of this group, we find that xy, xz and yz transform as the I.R. T_{2g}, and $x^2 - y^2$ and z^2 transform as the I.R. E_g. [Note that the function listed in the Table, $2z^2 - x^2 - y^2 = 3z^2 - (x^2 + y^2 + z^2) = 3z^2 - r^2$, which transforms as z^2 since r^2 is spherically symmetrical.] This inspection of the character table shows immediately how an octahedral field must split the d orbitals. The notation for the orbitals is simply that for the I.R.'s written in lower-case letters.

Example 25.2   How are the d orbitals split in a field of tetrahedral symmetry?

The tetrahedral symmetry group is $\mathbf{T_d}$. Inspection of its character table indicates that the d orbitals will split into three t_2 and two e orbitals. Group theory does not explain the relative energies of the orbitals, but the e orbitals lie lower in energy since they are directed away from the ligands.

Another kind of question easily answered by group theory is how to choose a set of atomic orbitals that can give hybrid orbitals of specified geometry. For example, consider $Fe(CN)_6^{3-}$, a regular octahedral structure (Fig. 25.11). To determine the hybridization of the Fe orbitals take six vectors each directed along one of the Fe—CN bonds. Consider these vectors to be the bases for a representation of the group. The characters of the representations of the symmetry operations will be equal to the

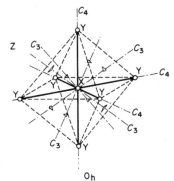

FIGURE 25.11 Symmetry elements of a regular octahedron.

number of vectors that are not shifted by the operation. We then get a reducible representation Γ_6 for the $\mathbf{O_h}$ group with the following characters:

$\mathbf{O_h}$	E	$8C_3$	$6C_2$	$6C_4$	$3C_4^2$	i	$8S_6$	$3\sigma_h$	$6\sigma_d$
Γ_6	6	0	0	2	2	0	0	4	2

The character table of $\mathbf{O_h}$ shows that this reducible representation is the sum of I.R.'s: $\Gamma_6 = A_{1g} + E_g + T_{1u}$. The transformation properties of the s, p, and d orbitals as listed in the character table indicate which orbitals transform in accord with each I.R.

$$A_{1g} \qquad s$$
$$E_g \qquad d_{z^2}, d_{x^2-y^2}$$
$$T_{1u} \qquad p_x, p_y, p_z$$

Hence the hybridization is d^2sp^3, and the only d orbitals that can be used are d_{z^2} and $d_{x^2-y^2}$.

Chapter Appendix: Character Tables for Some Point Groups

$\mathbf{C_{nv}}$ GROUPS

$\mathbf{C_{2v}}$	E	C_2	$\sigma_v(xz)$	$\sigma_v'(yz)$		
A_1	1	1	1	1	z	x^2, y^2, z^2
A_2	1	1	-1	-1	R_z	xy
B_1	1	-1	1	-1	x, R_y	xz
B_2	1	-1	-1	1	y, R_x	yz

$\mathbf{C_{3v}}$	E	$2C_3$	$3\sigma_v$		
A_1	1	1	1	z	$x^2 + y^2, z^2$
A_2	1	1	-1	R_z	
E	2	-1	0	$(x, y)(R_x, R_y)$	$(x^2 - y^2, xy)(xz, yz)$

$\mathbf{C_{nh}}$ GROUPS

$\mathbf{C_{2h}}$	E	C_2	i	σ_h		
A_g	1	1	1	1	R_z	x^2, y^2, z^2, xy
B_g	1	-1	1	-1	R_x, Y_y	xz, yz
A_u	1	1	-1	-1	z	
B_u	1	-1	-1	1	x, y	

D_{nh} GROUPS

D_{2h}	E	$C_2(z)$	$C_2(y)$	$C_2(x)$	i	$\sigma(xy)$	$\sigma(xz)$	$\sigma(yz)$		
A_g	1	1	1	1	1	1	1	1		x^2, y^2, z^2
B_{1g}	1	1	-1	-1	1	1	-1	-1	R_z	xy
B_{2g}	1	-1	1	-1	1	-1	1	-1	R_y	xz
B_{3g}	1	-1	-1	1	1	-1	-1	1	R_x	yz
A_u	1	1	1	1	-1	-1	-1	-1		
B_{1u}	1	1	-1	-1	-1	-1	1	1	z	
B_{2u}	1	-1	1	-1	-1	1	-1	1	y	
B_{3u}	1	-1	-1	1	-1	1	1	-1	x	

D_{3h}	E	$2C_3$	$3C_2$	σ_h	$2S_3$	$3\sigma_v$		
A_1'	1	1	1	1	1	1		$x^2 + y^2, z^2$
A_2'	1	1	-1	1	1	-1	R_z	
E'	2	-1	0	2	-1	0	(x, y)	$(x^2 - y^2, xy)$
A_1''	1	1	1	-1	-1	-1		
A_2''	1	1	-1	-1	-1	1	z	
E''	2	-1	0	-2	1	0	(R_x, R_y)	(xz, yz)

CUBIC GROUPS

T_d	E	$8C_3$	$3C_2$	$6S_4$	$6\sigma_d$		
A_1	1	1	1	1	1		$x^2 + y^2 + z^2$
A_2	1	1	1	-1	-1		
E	2	-1	2	0	0		$(2z^2 - x^2 - y^2, x^2 - y^2)$
T_1	3	0	-1	1	-1	(R_x, R_y, R_z)	
T_2	3	0	-1	-1	1	(x, y, z)	(xy, xz, yz)

O_h	E	$8C_3$	$6C_2'$	$6C_4$	$3C_2$ $(=C_4^2)$	i	$6S_4$	$8S_6$	$3\sigma_h$	$6\sigma_d$		
A_{1g}	1	1	1	1	1	1	1	1	1	1		$x^2 + y^2 + z^2$
A_{2g}	1	1	-1	-1	1	1	-1	1	1	-1		
E_g	2	-1	0	0	2	2	0	-1	2	0		$(2z^2 - x^2 - y^2, x^2 - y^2)$
T_{1g}	3	0	-1	1	-1	3	1	0	-1	-1	(R_x, R_y, R_z)	
T_{2g}	3	0	1	-1	-1	3	-1	0	-1	1		(xz, yz, xy)
A_{1u}	1	1	1	1	1	-1	-1	-1	-1	-1		
A_{2u}	1	1	-1	-1	1	-1	1	-1	-1	1		
E_u	2	-1	0	0	2	-2	0	1	-2	0		
T_{1u}	3	0	-1	1	-1	-3	-1	0	1	1	(x, y, z)	
T_{2u}	3	0	1	-1	-1	-3	1	0	1	-1		

D_{2d}	E	$2S_4$	C_2	$2C_2'$	$2\sigma_d$		
A_1	1	1	1	1	1		$x^2 + y^2, z^2$
A_2	1	1	1	-1	-1	R_z	
B_1	1	-1	1	1	-1		$x^2 - y^2$
B_2	1	-1	1	-1	1	z	xy
E	2	0	-2	0	0	(x, y); (R_x, R_y)	(xz, yz)

D_{3d}	E	$2C_3$	$3C_2$	i	$2S_6$	$3\sigma_d$		
A_{1g}	1	1	1	1	1	1		$x^2 + y^2, z^2$
A_{2g}	1	1	-1	1	1	-1	R_z	
E_g	2	-1	0	2	-1	0	(R_x, R_y)	$(x^2 - y^2, xy)$, (xz, yz)
A_{1u}	1	1	1	-1	-1	-1		
A_{2u}	1	1	-1	-1	-1	1	z	
E_u	2	-1	0	-2	1	0	(x, y)	

Problems

1. The symmetry operations of the point group C_{2v} are E, C_2, σ_v, and $\sigma_{v'}$. Write the multiplication table for the group. Identify the classes of the group.

2. Write the matrix representations of the group C_{2v} based on the transformation matrices in Section 25.6.

3. To what irreducible representation do the following belong: (a) $d_{x^2-y^2}$ and d_z^2 orbitals in O_h; (b) p_y orbital in D_{2h}?

4. Assign each of the following molecules to an appropriate point group: (a) BF_3; (b) C_2H_4; (c) benzene; (d) acetylene; (e) SO_2; (f) SiFClBr; (g) anthracene.

5. Consider all the isomeric n-bromobenzenes, where n can be from 1 to 6. Assign each molecule to its point group. Which molecules can have dipole moments?

6. Groups for which the commutative law $AB = BA$ holds for all elements are called commutative or Abelian groups. Which of the following point groups are Abelian groups: C_{2v}, C_{2h}, C_3, S_4, D_{2h}, O_h, C_{3v}?

7. Show that in an Abelian group each class possesses only one element.

8. Indicate on a regular octahedron the directions of (a) C_3 axes, (b) σ_d planes, (c) S_6 axes, and (d) C_4 axes, and specify the total number of each.

9. Write a complete list of symmetry operations for the following and state the point group for each molecule: (a) anthracene; (b) $(PtCl_4)^{2-}$; (c) m-dichlorobenzene; (d) naphthalene.

10. Under which symmetry species do the five d-orbitals transform in (a) an octahedral complex; (b) a tetrahedral complex.

11. To what irreducible representations do the following belong? (a) R_x and R_y in D_{5h}; (b) components of the dipole moment in D_{6h}; (c) components of the polarizability tensor in D_{2h}; (d) a p_y orbital in C_{4v}; (e) $d_{x^2-y^2}$ and d_z^2 in O_h?

26

Rotational and Vibrational Spectra— Microwave, Infrared, and Raman

Molecular spectroscopy is the study of the interaction of electromagnetic radiation with molecules. From each frequency v of radiation absorbed or emitted by a molecule, we obtain a spacing between two energy levels in the molecule from $\Delta\epsilon = hv$. Quantum mechanical theory is used to interpret the pattern of energy levels obtained from a spectrum. The theoretical analysis gives detailed information about the structure of the molecule, not only in its lowest energy level, the *ground state*, but also in various higher energy levels, the *excited states*.

For investigation of isolated molecules, samples in the gas phase are most suitable, but liquid solutions, pure liquids, crystals, and solid surfaces can also be studied by spectroscopic techniques. Spectroscopic methods are widely used in analytical chemistry, in astrophysics, and in many branches of biology and medicine.

26.1 Survey of Molecular Spectra

Table 26.1 summarizes the types of molecular spectra of major interest to the chemist. The human eye can see only a narrow range of electromagnetic rays, from about 14 000 cm^{-1} in the red to about 24 000 cm^{-1} in the violet but physical instruments and chemical detectors (such as photographic plates) can scan a vastly enlarged range of frequencies. The spectra from the near infrared to the ultraviolet are sometimes called *optical spectra*.

The spectra of atoms consist of sharp lines, and those of molecules appear to consist of bands, in which a densely packed line structure is often revealed with spectrographs of high resolving power, i.e., capable of separating small differences

TABLE 26.1

THE ELECTROMAGNETIC SPECTRUM

Energy, E (J mol^{-1})	Frequency, $\log_{10} \nu$ (Hz)	Region	Origin
	20—		Nuclear transitions
10^{10}—	19—	γ-rays	
10^9 —	18—	X-rays	Transitions of core electrons
10^8 —	17—	Vacuum	Loss of outer-shell electrons
10^7 —		ultraviolet	
10^6 —	16—		Transitions of valence
	15—	Ultraviolet	electrons
10^5 —		Visible	
10^4 —	14—	Infrared	Molecular vibrations
10^3 —	13—	Far infrared	Molecular rotations
10^2 —	12—	Microwave	
10 —	11—		Electron-spin resonance
1 —	10—		
10^{-1}—	9—	Radiofrequency	Nuclear magnetic resonance
10^{-2}—	8—		
10^{-3}—	7—		Nuclear quadrupole resonance
	6—		
	5—		

in frequency. The energy levels responsible for atomic spectra correspond to different allowed quantum states for the electrons in the atoms. These levels have been discussed in Chapter 21. In a molecule also, absorption or emission of energy can occur when an electron makes a transition between two stationary molecular quantum states, specified by the wave functions ψ_i and ψ_j. In addition, however, a molecule can absorb or emit energy in the vibrational and rotational motions of its nuclei. These internal energies, like the electronic energy, are quantized, so that the molecule can exist only in certain discrete levels of vibrational and rotational energy.

In the theory of molecular spectra it is customary, as a good first approximation, to consider that the energy of a molecule can be expressed simply as the sum of electronic, vibrational, and rotational contributions,

$$\epsilon = \epsilon_{\text{elec}} + \epsilon_{\text{vib}} + \epsilon_{\text{rot}} \tag{26.1}$$

This separation into three distinct categories is not strictly correct. The separation of electronic energy as an independent term is essentially the Born–Oppenheimer approximation, discussed in Section 22.3. Rotational and vibrational energies cannot be strictly separated from each other because (1) the atoms in a rapidly rotating molecule are pushed apart by centrifugal forces, which thereby affect the character of the vibrations, and (2) vibrations are not strictly harmonic, so that average bond distances in molecules increase with vibrational excitation, thus increasing the moments of inertia. Nevertheless, the approximation of Eq. (26.1) explains many of the observed characteristics of molecular spectra.

The separations between electronic energy levels are usually much larger than those between vibrational energy levels, which, in turn, are much larger than those between rotational levels. The type of energy-level diagram that results is shown in Fig. 26.1, which depicts several of the electronic energy levels of the CO molecule. For the ground electronic state, the vibrational energy levels are shown and, on an enlarged scale, the rotational levels for the lowest ($v = 0$) vibrational level. Associated with each electronic level is a similar set of vibrational levels, each of which is in turn associated with a set of rotational levels. The very small energy differences between successive rotational levels are responsible for the band structure of molecular spectra.

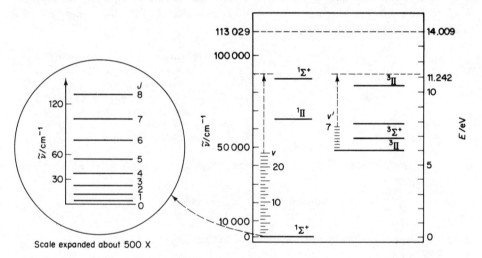

Scale expanded about 500 X

FIGURE 26.1 Energy-level diagram of the CO molecule. Several singlet and several triplet electronic states are shown. The vibrational levels associated with the lowest singlet and the lowest triplet are shown. Actually, these levels continue toward higher energies until the molecule dissociates into a ^3P carbon atom and a ^3P oxygen atom, at an energy level of 11.242 eV. The energy of ionization, CO → CO$^+$ + e, is 14.009 eV. The left of the diagram, on an expanded scale, shows the rotational levels associated with the lowest ($v = 0$) vibrational level.

Transitions between different electronic levels give rise to spectra in the visible or ultraviolet region, called *electronic spectra*. Transitions between vibrational levels within the same electronic state are responsible for spectra in the near infrared (<20 μm), called *vibration–rotation spectra*. Finally, spectra are observed in the far infrared (>20 μm), arising from transitions between rotational levels belonging to the same vibrational level; these are called *pure rotation* or *microwave spectra*.

As shown in Fig. 4.1, light is an electromagnetic radiation that traverses space as a traveling wave with an electric field **E** at right angles to a magnetic field **B**. Optical spectra arise mainly from interaction of the negative electrons and positive nuclei of molecules and crystals with the electric field of the light wave.

Consider in Fig. 26.2 two states m and n of a molecule, with energy levels ϵ_m and ϵ_n. Suppose that at a certain time, say $t = 0$, the molecule exists in a stationary state denoted by the wavefunction ψ_m, and that it is in the path of a light wave that has a frequency in a range including $\nu_{nm} = (\epsilon_n - \epsilon_m)/h$. The molecule may then absorb a quantum of energy $h\nu_{nm}$ from the electromagnetic field and make a transition from its initial state m (ψ_m) to a new stationary state n (ψ_n). The disturbance or *perturbation* of the molecule that causes such a transition to occur is due to the force exerted by the electric field of the light wave on the internal electric charges in the molecule. If the electric field at frequency ν_{nm} traverses a molecule that is already in the state n (ψ_n), it can cause that molecule to revert to m (ψ_m) by emission of a quantum $h\nu_{nm}$. This process is called *stimulated emission*. A molecule in state n (ψ_n) can also revert to m (ψ_m) by emission of a quantum $h\nu_{nm}$ without the stimulus of the alternating electric field. Such a process is called *spontaneous emission*.

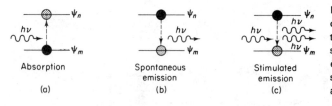

Absorption
(a)

Spontaneous
emission
(b)

Stimulated
emission
(c)

FIGURE 26.2 (a) Absorption of a quantum of energy $h\nu_{nm}$ causes molecule to make a transition from state ψ_m to state ψ_n. The molecule in state ψ_n can emit a quantum by either (b) spontaneous or (c) stimulated emission and thereby return to state ψ_m.

The relative rates of spontaneous and stimulated emission are given by a formula due to Einstein as

$$\frac{[\text{spontaneous}]}{[\text{stimulated}]} = \frac{A_{mn}}{B_{mn}I} = \frac{8\pi h(\nu/c)^3}{I} \tag{26.2}$$

where I is the energy density (or intensity) of the radiation at the frequency ν. Equation (26.2) shows that spontaneous emission becomes relatively more important at high frequencies (visible or UV), whereas stimulated emission will be relatively more important at low frequencies (microwave or radiofrequency). These processes are summarized in Fig. 26.2. In stimulated emission the emitted radiation is in phase with the radiation that stimulates the emission.

A necessary condition for the absorption of a quantum $h\nu_{nm}$ by a molecule is that the molecule has two states ψ_m and ψ_n with an energy separation $\epsilon_m - \epsilon_m = h\nu_{nm}$. This is not, however, a sufficient condition for the absorption of the quantum $h\nu_{mn}$ to occur. The quantum mechanical theory shows that once the condition $\epsilon_n - \epsilon_m = h\nu_{nm}$ is met, the probability of absorption depends on

$$\boldsymbol{\mu}_{nm} = \int \psi_m^* \hat{\boldsymbol{\mu}} \psi_n \, d\tau \tag{26.3}$$

Here $\hat{\mu}$ is the operator for the electric dipole moment,

$$\hat{\mu} = \sum_i e(x_i + y_i + z_i) \tag{26.4}$$

where the sum is over all the charged particles, electrons and nuclei, in the molecule. The quantity μ_{nm} is called the *transition dipole moment*. If $n = m$, μ_{mm} is simply the permanent dipole moment of the molecule in the state m. Thus μ_{nm} measures a sort of average molecular dipole that the electric field of the light wave acts upon during the actual process of transition of the molecule from state m (ψ_m) to state n (ψ_n). If $\mu_{nm} = 0$ for any pair of states, the electric field cannot perturb the molecule and thus no transition by this mechanism can occur. The transition is said to be *forbidden*. In such a case absorption or emission of a quantum $h\nu_{nm}$ is much less probable even if the incident radiation has a frequency component at ν_{nm}.

The electric-dipole mechanism just described is by far the most important way in which a light wave interacts with a molecule. Sometimes, however, when $\mu_{nm} = 0$ in Eq. (26.3), other less effective mechanisms (magnetic dipole, electric quadrupole) can lead to transitions between the states ψ_m and ψ_n, giving rise to weak absorption spectra.

26.3 Pure Rotation Spectra—Rigid Rotors

The rotational energy levels of diatomic and other linear molecules with moment of inertia I were given in Eq. (4.10) as

$$\epsilon = \frac{h^2 J(J + 1)}{8\pi^2 I} = BJ(J + 1) \tag{26.5}$$

These are the energy levels of a *rigid rotor* in which all the internuclear distances are assumed to be constant. The bonds in rotating molecules will actually stretch to some extent, owing to centrifugal forces, as the rotational energy increases, but the rigid-rotor model ignores this effect. Spacings between rotational energy levels correspond to transitions in the far infrared or microwave region of the spectrum.

To display a pure rotation spectrum, a molecule must have a permanent dipole moment. If a molecule does not already have a dipole moment μ in its ground state $J = 0$, rotation of the molecule in a higher state $J > 0$ can never create a dipole moment. Hence the transition moment μ_{nm} in Eq. (26.3) will always be zero in the absence of a permanent dipole moment. Thus, for example, HCl displays a pure rotational spectrum, but N_2 does not. When the rotational wave functions are substituted into Eq. (26.3), the selection rule for rotational transition is found to be $\Delta J = \pm 1$ for a diatomic molecule having no electronic angular momentum.*

From Eq. (26.5) the difference in energy between two levels J' and J ($J' > J$) is

$$\Delta \epsilon = h\nu = B[J'(J' + 1) - J(J + 1)]$$

*The photon is a particle with spin = 1. To conserve angular momentum in a rigid rotor, therefore, the rotational angular momentum must change by one unit ($\Delta J = \pm 1$) when a photon is absorbed or emitted by a molecule.

The selection rule, $J' - J = 1$, gives

$$\Delta\epsilon = 2BJ' = 2\tilde{B}\,hc\,J' \tag{26.6}$$

The rotational constant $\tilde{B}$ is always given in units of cm^{-1}.

Figure 26.3 shows the energy levels of a rigid linear rotor, and the allowed transitions responsible for its absorption spectrum. The predicted spacing between successive energy levels increases linearly with J'. The absorption spectrum is a series of lines with an equidistant spacing of $\Delta\tilde{v} = 2\tilde{B}$.

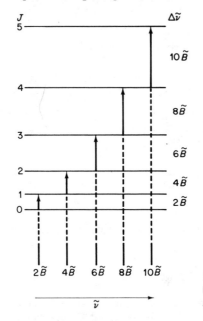

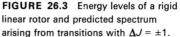

FIGURE 26.3 Energy levels of a rigid linear rotor and predicted spectrum arising from transitions with $\Delta J = \pm 1$.

Example 26.1 Absorption by $H^{35}Cl$ has been observed in the far infrared with a spacing between successive lines $\Delta\tilde{v}$ of about 20 cm^{-1}. The transition $J = 0$ to $J' = 1$ occurs at $\tilde{v} = 20.60$ cm^{-1}. Calculate the internuclear distance, R_e.

From Eqs. (26.5) and (26.6) with $\tilde{v} = \Delta\epsilon/hc = 2BJ'/hc$, $B = hc\tilde{v}/2J' = h^2/8\pi^2 I$, where $I = \mu R_e^2$, and μ is the reduced mass. Hence

$$I = h/8\pi^2 c\tilde{B} = \frac{6.626 \times 10^{-34}\ \text{J s}}{8\pi^2(2.998 \times 10^8\ \text{m s}^{-1})(10.301 \times 10^2\ \text{m}^{-1})}$$

$$= 0.2718 \times 10^{-46}\ \text{kg m}^2$$

$$\mu = \frac{(34.97)(1.008)}{34.97 + 1.008} \frac{10^{-3}}{6.022 \times 10^{23}} = 1.627 \times 10^{-27}\ \text{kg}$$

Hence

$$R_e = \left(\frac{0.2718 \times 10^{-46}\ \text{kg m}^2}{1.627 \times 10^{-27}\ \text{kg}}\right)^{1/2} = 1.293 \times 10^{-10}\ \text{m} = 129.3\ \text{pm}$$

26.4 Microwave Spectroscopy

The wavelengths of microwaves are in a range of about 1 to 10 mm. In ordinary absorption spectroscopy, the source of radiation is usually a hot filament or a high-pressure gaseous-discharge tube, giving in either case a wide distribution of wavelengths. This radiation is passed through the absorber and the intensity of the transmitted portion is measured at different wavelengths, after separation by means of a grating or prism. In microwave spectroscopy, the source is monochromatic, at a well defined single wavelength, which can, however, be rapidly varied. The frequency of an electronically controlled oscillator can be swept over the frequency range of the waveguide cell. After passage through the cell, which contains the substance under investigation, the microwave beam is picked up by a receiver, and after amplification is fed to a cathode-ray oscillograph or other recorder. The resolving power of this arrangement is 10^5 times that of the best infrared grating spectrometer, so that frequency measurements can be made to seven significant figures. The setup of a typical microwave spectrometer is shown in Fig. 26.4.

One of the most useful extensions of the microwave technique is the inclusion in the cell of a metallic septum by means of which an electric field can be applied to

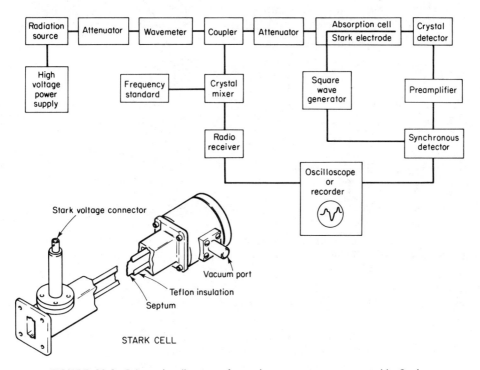

FIGURE 26.4 Schematic diagram of a microwave spectrometer with Stark modulation. A typical microwave absorption cell would have a cross section of 4 mm × 10 mm and a length of 2 m. The frequency range is from 8 to 40 GHz.

the gas while the spectrum is being scanned. The splitting of quantized energy levels and consequent splitting of spectral lines by an electric field is called the *Stark effect*. The rotational energy level of a molecule in an electric field depends on the dipole moment of the molecule, and the Stark effect in microwave spectra is one of the best methods for measuring dipole moments.

26.5 Rotational Spectra of Polyatomic Molecules

With moments of inertia obtained from microwave spectra, we can calculate internuclear distances to at least ± 0.2 pm. The rotational energy levels of linear polyatomic molecules are given by Eq. (26.5). The moment of inertia I, however, may depend on two or more different internuclear distances. For example, in the molecule OCS, I depends on the distances C—O and C—S. We cannot calculate both these distances from a single moment of inertia. In such cases, the method of isotopic substitution can be used. Internuclear distances are determined by the minimum energy of the system of electronic and nuclear charges in the molecule. Internuclear distances do not depend appreciably on the masses of isotopic nuclei. Thus isotopic substitution provides a molecule that has new moments of inertia but practically unchanged internuclear distances. For example, the microwave spectra of isotopically substituted OCS give the following rotational constants:

$$^{16}O—^{12}C—^{32}S \qquad \tilde{B} = 0.202864 \text{ cm}^{-1}$$

$$^{16}O—^{12}C—^{34}S \qquad \tilde{B} = 0.197910 \text{ cm}^{-1}$$

The two moments of inertia give for the C—O distance 116.5 pm and for C—S, 155.8 pm. (see Problem 7.)

26.6 Inversions and Internal Rotations

In addition to pure rotations, certain other molecular motions have frequencies in the microwave region.

The inversion of ammonia NH_3 was the basis of the first maser, which was invented by Charles Townes in 1954. (*Maser* is an acronym for "microwave amplification by stimulated emission of radiation.") NH_3 has a pyramidal structure with the N atom out of the plane of the three H atoms. Considered in terms of classical mechanics, the motion of the H atoms relative to the N atom somewhat resembles the motion of an umbrella turning inside out.

The potential energy of NH_3 is shown in Fig. 26.5 as a function of the distance z between the N and the plane of the H atoms. There are two potential energy wells corresponding to positive and negative values of z. In classical mechanics, the inversion of NH_3 would require an energy equal at least to the height of the energy barrier between the two wells.

The quantum mechanical picture is quite different. If ψ_a and ψ_b are the wave-

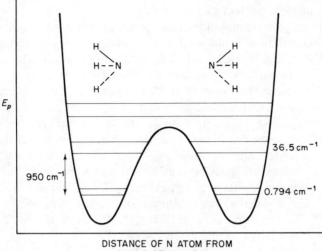

FIGURE 26.5 Potential energy of NH_3 molecule for the motion of "umbrella inversion." The lower energy levels are shown.

functions for the molecule in the wells a and b, the correct wavefunction for the molecule must be a linear combination:

$$\text{either } \psi_S = \psi_a + \psi_b \quad \text{or} \quad \psi_A = \psi_a - \psi_b$$

Each energy level for the vibration of NH_3 in the umbrella mode is split into a doublet, with energies ϵ_S and ϵ_A. The ψ_S state has a slightly lower energy than the ψ_A. The lowest level is split by 0.794 cm^{-1} (9.50 J mol^{-1}). The energy difference corresponds to a frequency of 23 786 MHz. The nitrogen atom cannot be said to be either above or below the plane of the hydrogen atoms. We must say that the ground state ψ_S of NH_3 is a symmetric superposition of a state of well a (ψ_a) and a state of well b (ψ_b). At any time the N atom has an equal probability of being above or below the plane of the H atoms. When NH_3 absorbs microwave energy at 23 786 MHz it undergoes a transition to ψ_A, which is the antisymmetric superposition of ψ_a and ψ_b. Note that the energy difference $\epsilon_A - \epsilon_S$ is much less than the height of the energy barrier. The inversion of NH_3 is thus an example of the tunnel effect (Section 20.18). The frequency 2.3786×10^{10} Hz can be envisaged as the frequency of tunneling. It is much lower than a typical vibration frequency. For example, the symmetric N—H stretching frequency of NH_3 is $\sim 3 \times 10^{13}$ Hz.

Molecules in the higher energy state ψ_A can be separated from those in the lower energy state ψ_S by a simple electrostatic focusing device. Thus it is possible to collect a selected population of the higher-energy molecules in a resonant cavity. If now the molecules in the cavity are irradiated with 23 786 MHz microwaves, stimulated emission will occur, as shown in Fig. 26.2. The input signal is amplified and a coherent beam of 23 786 MHz radiation is emitted. Actually, the NH_3 maser is not a practical amplifier. Its microwave signal, however, is of almost incredible purity, departing from a perfect sine wave by less than 1 part in 10^{11}. It provides a standard for clocks that would gain or lose less than 1 s per century.

For certain polyatomic molecules, the separation of internal degrees of freedom into vibrations and rotations is not valid. Let us compare, for example, ethylene and ethane, $CH_2=CH_2$ and $CH_3—CH_3$. The orientation of the two methylene groups in C_2H_4 is fixed by the double bond, so that there is a torsional or twisting vibration about the bond but no complete rotation. In C_2H_6, however, there is a hindered internal rotation of the methyl groups about the single bond. Thus one of the vibrational degrees of freedom is lost, becoming a hindered internal rotation. Such a rotation would not be difficult to treat if it were completely free and unrestricted, but usually there are potential-energy barriers, which must be overcome before rotation occurs. Consider ethane, shown in Fig. 26.6(a), as viewed along the C—C bond axis. The position shown is that of minimum potential energy, $U = 0$; when the CH_3 group is rotated through 60°, we have the H atoms aligned and a position of maximum potential energy, $U = U^*$. The variation of U with angle ϕ is shown in Fig. 26.6(b). This potential-energy curve is represented by

$$U = \tfrac{1}{2}U^*(1 - \cos \sigma\phi) \qquad (26.7)$$

where σ is the number of identical conformations produced by rotation: $\sigma = 3$ for ethane. The barrier to internal rotation in ethane is 11.5 kJ mol^{-1}. At 300 K, $kT = 2.5$ kJ mol^{-1}, so that at 300 K most ethane molecules at any instant are caught in one of the lower torsional states and are not internally rotating.

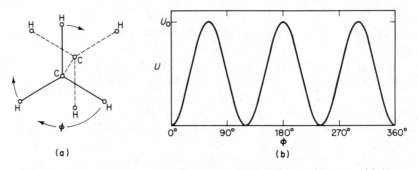

FIGURE 26.6 (a) Orientation of CH_3 groups in C_2H_6. (b) Potential energy of C_2H_6 as a function of CH_3 orientations.

26.7 The Harmonic Oscillator

Whereas molecular spectra due to transitions between rotational energy levels occur in the far infrared or microwave region, spectra due to transitions between vibrational energy levels occur in the near infrared. Since each vibrational level is associated with a set of rotational levels, these spectra appear as bands, which at high resolution reveal a fine structure of closely packed lines corresponding to the separate rotational levels.

The simplest model for the vibrational motion of a diatomic molecule is the harmonic oscillator in one dimension. The Schrödinger equation for this system is obtained by introducing the appropriate potential energy $U = \frac{1}{2}\kappa x^2$ into Eq. (20.13) to give

$$\frac{d^2\psi}{dx^2} + \frac{8\pi^2\mu}{h^2}\left(\epsilon - \frac{1}{2}\kappa x^2\right) = 0 \tag{26.8}$$

The allowed energy levels ϵ are the eigenvalues of the Schrödinger equation $\hat{H}\psi = \epsilon\psi$, specified by integral values of the vibrational quantum number v, as

$$\epsilon_v = (v + \frac{1}{2})h\nu \tag{26.9}$$

where ν is the fundamental vibration frequency. The wavefunctions corresponding with the energy levels in Eq. (26.9) are the eigenfunctions of the harmonic-oscillator problem, given by

$$\psi_v = N_v\, e^{-a^2x^2/2}\, \mathcal{H}_v(ax)$$

where

$$a = (4\pi^2\kappa\mu/h^2)^{1/4} \tag{26.10}$$

and $\mathcal{H}_v(ax)$ are called *Hermite polynomials*. The first few of these polynomials are given in Table 26.2. The N_v in Eq. (26.10) is a normalization factor to ensure that

TABLE 26.2
HERMITE POLYNOMIALS

$\mathcal{H}_0 = 1$	$\mathcal{H}_3 = 8y^3 - 12y$
$\mathcal{H}_1 = 2y$	$\mathcal{H}_4 = 10y^4 - 48y^2 + 12$
$\mathcal{H}_2 = 4y^2 - 2$	$\mathcal{H}_5 = 32y^5 - 160y^3 + 120y$

$\int_{-\infty}^{\infty} \psi^2\, dx = 1$, so that the probability of finding the oscillating mass somewhere between $-\infty$ and $+\infty$ is unity.

Figure 26.7 shows the first few wavefunctions for a harmonic oscillator super-

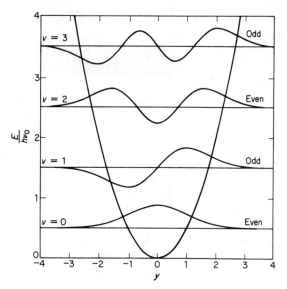

FIGURE 26.7 Some wavefunctions of the harmonic oscillator.

imposed on its potential-energy curve. When the wavefunctions ψ_v are substituted into the expression for the transition moment, Eq. (26.3), the selection rule for the harmonic oscillator is obtained as $\Delta v = \pm 1$.

Example 26.2   Calculate the normalization factor for the the lowest state $v = 0$ of the harmonic oscillator. Calculate the value of x for which ψ is a maximum in this lowest state.

From Eq. (26.10) and Table 26.2, $\psi_0 = N_0 e^{-a^2 x^2/2}$, and the normalization condition becomes $N_0^2 \int_{-\infty}^{\infty} e^{-a^2 x^2} \, dx = 1$.

$$N_0^2(\pi^{1/2}/a) = 1 \qquad \text{or} \qquad N_0 = (a^2/\pi)^{1/4}$$

Thus $\psi_0 = (a^2/\pi)^{1/4} e^{-a^2 x^2/2}$. The maximum value occurs at $d\psi_0/dx = 0$, or

$$e^{-a^2 x^2/2}(-2a^2 x/2) = 0 \qquad \text{or} \qquad x = 0$$

Thus the position of maximum probability for an oscillator in the state $v = 0$ is at the midpoint of its classical range of oscillation. This is the opposite of the result of classical mechanics, which would put the maximum probability at the turning points $x = \pm A$ of any oscillation, and the minimum probability at $x = 0$.

26.8   The Anharmonic Oscillator

The harmonic oscillator is not a good model for molecular vibrations except at low energy levels, near the bottom of the potential-energy curve. The restoring force in harmonic vibrations is directly proportional to the displacement x. The potential-energy curve is therefore a parabola and dissociation of a harmonic oscillator can never take place, no matter how great the amplitude of vibration. We know, however, that the restoring force must actually become weaker as the displacement increases, and for large enough amplitude of vibration, the molecule must fly apart. Potential-energy curves for real molecules therefore look like the one in Fig. 22.4, which shows the vibrational energy levels of the H_2 molecule.

A potential curve like that in Fig. 22.4 corresponds to the model of an *anharmonic oscillator*. The energy levels corresponding to an anharmonic potential-energy curve can be expressed as

$$\epsilon_v = h\nu_0[(v + \tfrac{1}{2}) - x_e(v + \tfrac{1}{2})^2] \tag{26.11}$$

where x_e is the *anharmonicity constant*. The energy levels are not evenly spaced, but fall more closely together as the quantum number increases. This behavior is illustrated in the levels superimposed on the curve in Fig. 22.4.

The selection rule governing vibrational transition is $\Delta v = \pm 1$ for the harmonic oscillator, but for a real (anharmonic) oscillator there will be *overtone* transitions with $\Delta v = \pm 2, \pm 3$, etc., although these will have much lower intensities than the *fundamentals* with $\Delta v = \pm 1$.

Example 26.3 The infrared absorption spectrum of HCl shows an intense band at 2886 cm^{-1} and a weak band at 5668 cm^{-1}. Calculate $\tilde{v}_0$, the fundamental absorption wave number, and x_e, the anharmonicity constant.

The band frequencies are roughly in the ratio 1:2 and we can conclude that they arise from the fundamental transition ($v = 0$ to $v' = 1$), and the first overtone ($v = 0$ to $v' = 2$). From Eq. (26.11),

$$\Delta \epsilon_v = h v_0 [(1 - x_e)(v' - v) - x_e(v'^2 - v^2)^2]$$

For $v = 0, v' = 1$: $\Delta \epsilon_v = h v_0 (1 - 2x_e)$

For $v = 0, v' = 2$: $\Delta \epsilon_v = 2h v_0 (1 - 3x_e)$

Thus,

$$2886 \text{ cm}^{-1} = \tilde{v}_0 (1 - 2x_e)$$

$$5668 \text{ cm}^{-1} = 2\tilde{v}_0 (1 - 3x_e)$$

and $x_e = 0.0174$, $\tilde{v}_0 = 2886 \text{ cm}^{-1}/(1 - 0.0348) = 2990 \text{ cm}^{-1}$.

Potential-energy curves of the type shown in Fig. 22.4 are so useful in chemical discussions that much effort has been devoted to obtaining convenient mathematical expressions for them. An empirical function that fits quite well was suggested by P. M. Morse:

$$U(x) = D_e[1 - e^{-\beta x^2}] \tag{26.12}$$

Here $x = R - R_e$ and β is given in terms of molecular parameters as $\beta = \pi v_0 (2\mu/D_e)^{1/2}$ or β is related to the anharmonicity constant as $\beta^2 = 8\pi^2 \mu v_0 x_e/h$. When the Morse function is used as the potential energy in the Schrödinger equation, the energy levels that are obtained for the oscillator correspond with those in Eq. (26.11).

Example 26.4 Estimate the spectroscopic energy of dissociation from the results in Example 26.3 on the basis of the Morse model of the potential-energy curve.

From the two expressions above for the constant β of the Morse function $D_e = h v_0/4x_e$,

$$D_e = \frac{(6.63 \times 10^{-34} \text{ J s})(2990 \text{ cm}^{-1})(2.998 \times 10^{10} \text{ cm s}^{-1})}{4(0.0174)}$$

$$= 8539 \times 10^{-22} \text{ J}$$

$$= 514\,200 \text{ J mol}^{-1} = 514 \text{ kJ mol}^{-1}$$

The experimental value is 445 kJ mol^{-1}. The Morse function is only a fair approximation for the potential-energy curve near the dissociation limit.

26.9 Vibration–Rotation Spectra of Diatomic Molecules

A diatomic molecule has only one degree of vibrational freedom and thus only one fundamental vibration frequency v_0. To absorb or emit quanta $h v_0$ of vibrational energy, the molecule must possess a permanent dipole moment, since otherwise the transition probability from Eq. (26.3) vanishes. Thus molecules such as CO and HCl display a spectrum in the near infrared due to transitions between vibrational energy levels, but molecules such as H_2 and Cl_2 display no infrared spectra in the gaseous state.

The vibrational spectrum, however, arises from transitions between definite rotational-energy levels belonging to definite vibrational levels. Therefore, it is a *vibration–rotation spectrum*. The expression for an energy level in the approximation corresponding to harmonic oscillator and rigid rotor is

$$\epsilon_{vr} = (v + \tfrac{1}{2})hv_0 + BJ(J + 1) \tag{26.13}$$

For a transition between an upper level v', J', and a lower level v'', J''.

$$\Delta\epsilon_{vr} = (v' - v'')hv_0 + B'J'(J' + 1) - B''J''(J'' + 1) \tag{26.14}$$

We must use different rotational constants B' and B'' for the upper and lower states, because the moment of inertia of the molecule is not exactly the same in different vibrational states.

The selection rules for transitions between the vibration–rotation levels of Eq. (26.14) are $\Delta v = \pm 1$, $\Delta J = \pm 1$. In the exceptional case that the diatomic molecule has an electronic angular momentum about the internuclear axis (i.e., quantum number $\Lambda \neq 0$), we can also have $\Delta J = 0$. The best known molecules of this type are NO and OH, which have $^2\Pi$ ground states.

A vibration–rotation band may display three branches corresponding to the three cases:

$$J' - J'' = \Delta J = +1, \quad \text{R branch}$$
$$J' - J'' = \Delta J = -1, \quad \text{P branch}$$
$$J' - J'' = \Delta J = 0, \quad \text{Q branch}$$

As an example, the fundamental infrared absorption spectrum of CO at 2168 cm^{-1} is shown in Fig. 4.2. This band arises from transitions between $v'' = 0$ and $v' = 1$. The appearance of the band at low resolution is shown in Fig. 4.2(a). The rotational fine structure is not resolved, but we can see the P and R branches. The band is shown in Fig. 4.2(b) at high resolution. The arrangement of energy levels is shown in Fig. 4.4. Each line corresponds to certain values of J' and J''. Note that the lines of greatest intensity do not correspond to $J'' = 0$ but to J'' about 4. [Why?]

Example 26.5   The origin of the fundamental infrared absorption band of HI is at 2230 cm^{-1}. The first two lines in the P branch are at 2216.7 and 2203.6 cm^{-1}, and the first two lines in the R branch are at 2242.2 and 2254.3 cm^{-1}. Calculate the moment of inertia and the internuclear distance R_e of HI in the states $v = 0$ and $v = 1$.

The lines given arise from the following values of v and J.

	v''	J''	v'	J'
2216.7	0	1	1	0
2203.6	0	2	1	1
2242.2	0	0	1	1
2254.3	0	1	1	2

From Eq. (26.14), to calculate $\tilde{B}''$ for $v = 0$, one must take the difference between transitions that end at same J' but start from different J', and to calculate $\tilde{B}'$, *vice versa*. Therefore,

$$2242.2 - 2203.6 = 39.6 = 6\tilde{B}''$$

$$2254.3 - 2216.7 = 37.6 = 6\tilde{B}'$$

and $\tilde{B}'' = 6.43 \text{ cm}^{-1}$, $\tilde{B}' = 6.27 \text{ cm}^{-1}$.

$$I = h/8\pi^2 c\tilde{B} = (6.626 \times 10^{-34} \text{ J s})/8\pi^2(2.998 \times 10^{10} \text{ cm s}^{-1})\tilde{B}$$

$$= 2.80 \times 10^{-46}/\tilde{B}$$

Then for $v = 0$, $I = 4.35 \times 10^{-47} \text{ kg m}^2$; $v = 1$, $I = 4.47 \times 10^{-47} \text{ kg m}^2$. From $I = \mu R_e^2$ with $\mu = 1.66 \times 10^{-27} \text{ kg}$, $v = 0$, $R_e = 162 \text{ pm}$; $v = 1$, $R_e = 164 \text{ pm}$. As a result of the anharmonicity, the equilibrium internuclear distance is somewhat larger in the upper vibrational level.

26.10 Infrared Spectrum of Carbon Dioxide

A polyatomic molecule need not have a permanent dipole moment to have a vibrational spectrum in the infrared, but any vibration that emits or absorbs radiation must cause a changing dipole moment. For example, CO_2 is a linear molecule with no permanent dipole. There are $3n - 5 = 4$ normal modes of vibration, as shown in Fig. 26.8. The symmetrical stretching vibration ν_1 cannot cause a changing dipole moment, and therefore this vibration is said to be *inactive* in the infrared. The twofold degenerate bending vibrations ν_2 cause a changing dipole moment, and are therefore *active* in the infrared, giving rise to a fundamental absorption band at 667 cm^{-1}. The

FIGURE 26.8 Normal vibrations of CO_2. The carbon dioxide molecule (a) is linear and symmetric. It has four degrees of vibrational freedom. In the symmetric stretching mode of vibration (b) the atoms of the molecule vibrate along the internuclear axis in a symmetric manner. In the bending mode (c) the vibration of atoms is perpendicular to the internuclear axis. This mode is doubly degenerate and there is a similar bending vibration normal to the plane of the paper. In the asymmetric stretching mode (d) the atoms vibrate along the internuclear axis in an asymmetric manner. The vibrational state of the molecule is described by three quantum numbers, v_1, v_2, and v_3, written as $(v_1 v_2 v_3)$, where v_1 denotes the number of vibrational quanta in the symmetric stretching mode, v_2 the number of vibrational quanta in the bending mode, and v_3 the number of vibrational quanta in the asymmetric stretching mode.

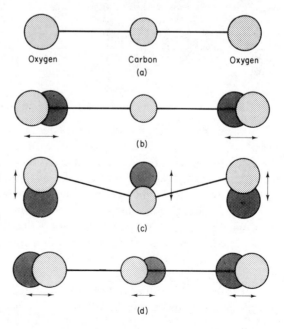

Oxygen   Carbon   Oxygen

(a)

(b)

(c)

(d)

antisymmetric stretching mode v_3 also causes a changing dipole and is observed in the fundamental absorption band at 2349 cm^{-1}. Notice that the stretching frequency is much higher than the bending one. This is a general result because it is easier to distort a molecule by bending than by stretching, and hence the force constant κ is lower for a bending mode. [Recall that $v = (1/2\pi)\sqrt{\kappa/\mu}$.]

In addition to the fundamental absorption bands, many combination and overtone bands occur in the infrared spectrum of CO_2, but these have lower intensities than the fundamental. (Their occurrence shows that the harmonic-oscillator selection rules are not rigorously obeyed.) Analysis of an infrared spectrum consists in sorting out the bands and correctly assigning the frequencies to their particular transitions. As we shall see, some vibrations inactive in the infrared spectra may be active in the Raman spectra, so that data should be available from both these sources. The assignment of some of the bands in the CO_2 spectrum is shown in Table 26.3.

TABLE 26.3
INFRARED VIBRATION BANDS OF CO_2

Wave number (cm^{-1})	Vibrational quantum numbers (v_1, v_2, v_3)						Assignment of bands
	Initial state			Final state			
667	0	0	0	0	1	0	Fundamental v_2
945	1	0	0	0	0	1	Combination $v_3 - v_1$
1044	0	2	0	0	0	1	Combination $v_3 - 2v_2$
1933	0	0	0	0	3	0	Overtone $3v_2$
2076	0	0	0	1	1	0	Combination $v_1 + v_2$
2140	0	1	0	2	0	0	Combination $2v_1 - v_2$
2349	0	0	0	0	0	1	Fundamental v_3
3684	0	0	0	0	2	1	Combination $2v_2 + v_3$
4985	0	0	0	1	2	1	Combination $v_1 + 2v_2 + v_3$

26.11   Lasers

The word *laser* is an acronym for "light amplification by stimulated emission of radiation." The basis of laser action is the stimulated emission of radiation which was discussed in Section 26.2. Stimulated emission is the exact reverse of absorption. In a classical picture, when a light wave passes through a medium and absorption takes place, the transmitted light wave has lower amplitude but an unchanged frequency and phase. If stimulated emission occurs, the transmitted wave has a higher amplitude but unchanged frequency and phase. To obtain an amplification of light through stimulated emission, there must be a population inversion among the molecules in the medium, so that there are more molecules in the upper energy state of the transition than in the lower state.

The principles of the laser can be illustrated by a particular example. The carbon dioxide laser is based on the vibrational energy levels of the CO_2 molecule as shown in

Fig. 26.9. The three quantum numbers v_1, v_2, v_3 refer to the symmetric stretching mode, the bending mode, and the asymmetric stretching mode, respectively. These normal modes are shown in Fig. 26.8. The rotational fine structure is not included in Fig. 26.9, but each of the vibrational levels is actually associated with a set of closely spaced rotational levels. Figure 26.9 also includes the first vibrational level of N_2, which is at nearly the same energy as the (001) level of CO_2.

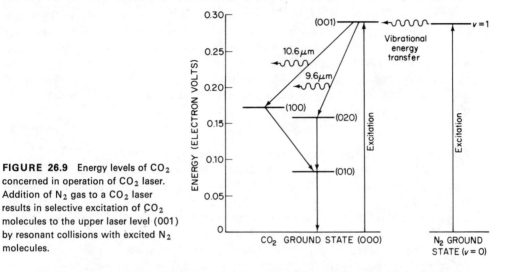

FIGURE 26.9 Energy levels of CO_2 concerned in operation of CO_2 laser. Addition of N_2 gas to a CO_2 laser results in selective excitation of CO_2 molecules to the upper laser level (001) by resonant collisions with excited N_2 molecules.

The (001) level of CO_2 is ideal as the upper level for a laser operation. It can be excited directly by passing an electric discharge through CO_2, but even more effectively if N_2 is mixed with CO_2. The N_2 is excited by the electric discharge and transfers its energy to the (001) excitation of CO_2. Since N_2 is a homonuclear diatomic molecule (no dipole moment), it cannot undergo vibrational transitions by absorption or emission of radiation. Hence, the N_2 excitation is not lost through radiation but rather transferred in collisions with CO_2 molecules. Because of the close match of energy levels, these are called *resonant collisions*. The first condition for laser operation is therefore achieved when an electric discharge is passed through a mixture of CO_2 and N_2, i.e., a population inversion that produces a high concentration of CO_2 molecules in the (001) vibrational state.

Two radiative transitions are possible from (001) as shown in Fig. 26.9, either to (100) with emission of infrared radiation at 10.6 μm (945 cm^{-1}), or to (020) with emission at 9.6 μm (1044 cm^{-1}). The 10.6-μm transitions are stronger than the 9.6 μm by a factor of about 10. The CO_2 molecules in the lower levels become deexcited by nonresonant collisions, which convert vibrational energy into translational kinetic energy.

The power output of the CO_2 laser can be made to occur almost exclusively in a single P-branch transition of the 10.6-μm band, usually in the P($J = 20$) transition at 10.59 μm. Typical CO_2 lasers are tubes about 2 m long with a continuous power output up to 150 W. Much larger CO_2 lasers have been constructed, with power up to 30 kW. These are awesome devices; the beam can drill a hole through 1 cm of stainless-steel plate in a few seconds.

In 1921, Chandrasekhara Raman, professor of physics at Calcutta, saw for the first time the beautiful blue opalescence of the Mediterranean Sea. He resolved to devote all his future research to the scattering of light by liquids. After Compton in 1923 observed the inelastic scattering of X rays, Raman began a search for an optical analog of the Compton effect. In 1928, he discovered it in the scattering of the indigo blue mercury line at 435.8 nm by liquid benzene. When the scattered light at right angles to the direction of incidence was analyzed in a spectrograph, in addition to the intense Rayleigh scattering at 435.8 nm, he observed faint lines at both longer and shorter wavelengths. The effect was interpreted as the addition or subtraction of quanta of vibrational energy of the liquid to or from the quantum of incident radiation $h\nu$. Thus the scattered quantum $h\nu'$ is given by

$$h\nu' = h\nu \pm (\epsilon_m - \epsilon_n) \tag{26.15}$$

where ϵ_m and ϵ_n are two energy levels in the scattering substance.

One should understand the distinction between Raman scattering and fluorescence. In fluorescence, the system first absorbs a quantum $h\nu$ and then re-emits a quantum $h\nu'$; the incident light must therefore be at an absorption frequency. In Raman scattering, the incident light can be at any frequency.

The Raman shift $\Delta\nu = \nu' - \nu$ is independent of the frequency of the incident light ν. We can observe pure rotational and vibration–rotational Raman spectra, which are the counterparts of the absorption spectra in the far and near infrared. The Raman spectra, however, are studied with light sources in the visible or ultraviolet. In many cases, the Raman and infrared spectra of a molecule complement each other, since vibrations and rotations that are not observable in the infrared may be active in the Raman. The application of intense monochromatic laser light sources has revolutionized Raman spectroscopy.

Raman scattering occurs as a result of the oscillations of a dipole moment **μ** *induced* in a molecule by the electric field of the incident light wave. The induced dipole moment depends on the strength of the electric field **E** as **μ** = α**E**, where α is the polarizability. In general, α is a tensor, as shown in Section 23.3, but we shall consider a one-dimensional case to illustrate the theory.

If α varies with an internal motion of the molecule (vibration of rotation), we can write to a first approximation

$$\alpha = \alpha_0 + \left(\frac{\partial\alpha}{\partial x}\right)x \quad \text{or} \quad \mu = \alpha_0 E + \left(\frac{\partial\alpha}{\partial x}\right)xE \tag{26.16}$$

The exciting light wave provides an alternating electric field,

$$E = E_0 \cos 2\pi\nu_0 t \tag{26.17}$$

The coordinate x within the molecule varies during the internal motion in accord with its own characteristic frequency ν_1,

$$x = x_0 \cos 2\pi\nu_1 t \tag{26.18}$$

From Eqs. (26.16, 17.18) we find that

$$\mu = \alpha_0 E_0 \cos 2\pi\nu_0 t + \left(\frac{\partial\alpha}{\partial x}\right) x_0 E_0 \cos 2\pi\nu_0 t \cos 2\pi\nu_1 t \qquad (26.19)$$

From the trigonometric identity,

$$\cos(\alpha \pm \beta) = \cos\alpha\cos\beta \mp \sin\alpha\sin\beta,$$

Eq. (26.19) gives

$$\mu = \alpha_0 E_0 \cos 2\pi\nu_0 t + \frac{1}{2}\left(\frac{\partial\alpha}{\partial x}\right) x_0 E_0 [\cos 2\pi(\nu_0 - \nu_1)t + \cos 2\pi(\nu_0 + \nu_1)t] \qquad (26.20)$$

The first term gives scattering without change in wavelength (Rayleigh scattering), and the second term gives the Raman scattering. The induced dipole within the molecule oscillates with frequencies $\nu_0 + \nu_1$ and $\nu_0 - \nu_1$ and hence emits radiation at these frequencies.

Quantum mechanical theory gives the transition moment for the Raman effect from Eq. (26.3) as

$$\boldsymbol{\mu}_{nm} = E \int \psi_n^* \alpha \psi_m \, d\tau \qquad (26.21)$$

If α is constant, this integral vanishes because of the orthogonality of the wavefunctions. In order that a vibration or rotation be active in the Raman, therefore, the polarizability must change during the rotation or vibration.

When any nonspherical molecule rotates, the polarizability changes. Thus rotational Raman spectra can be obtained from most molecules, giving detailed data on rotational energy levels to supplement those from microwave spectra. Figure 26.10 shows the rotational Raman spectrum of cyanogen C_2N_2 excited by an argon-ion laser at 488 nm.

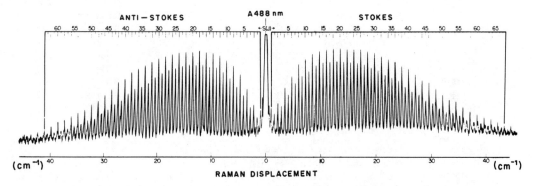

FIGURE 26.10 Rotational Raman spectrum of C_2N_2 excited by argon-ion laser at 488 nm. [I.-Y. Wang and A. Weber, *J. Chem. Phys.* **67**, 3084 (1977). © 1977 American Inst. Physics]

The selection rule for the pure rotational Raman spectrum of a rigid rotor is $\Delta J = \pm 2$. As the molecule rotates, its polarizability changes with a frequency *twice* the rotation frequency ν_1. [Please show why.] Thus in Eq. (26.18), $\nu_1 = 2\nu_r$. For vibration–rotation spectra the Raman selection rules are $\Delta v = \pm 1$, $\Delta J = 0, \pm 2$. A vibration–rotation band in the Raman is resolved into three branches O, Q, and S, corresponding to $\Delta J = -2, 0, +2$, respectively.

Spectroscopic measurements provide detailed data about the bond lengths, bond angles, vibration frequencies, and dissociation energies of molecules. Examples of such results are given in Tables 26.4 and 26.5.

TABLE 26.4
SPECTROSCOPIC DATA ON PROPERTIES OF DIATOMIC MOLECULES*

Molecule	Ground state	D_0/eV	$\tilde{\nu}_e$/cm^{-1}	$\tilde{\nu}_e x_e$/cm^{-1}	$\tilde{B}_e$/cm^{-1}	α_e/cm^{-1}	R_e/pm
$^{11}B_2$	$^3\Sigma_g^-$	3.02	1051.3	9.35	1.212	0.014	159.0
$^{79}Br_2$	$^1\Sigma_g^+$	1.9707	325.321	1.0774	0.08211	0.000319	228.11
$^{12}C_2$	$^1\Sigma_g^+$	6.21	1854.71	13.34	1.8198	0.0177	124.25
$^{40}Ca_2$	$^1\Sigma_g^+$	0.13	64.93	1.065	0.046113	0.000703	427.73
$^{35}Cl_2$	$^1\Sigma_g^+$	2.4794	559.7	2.68	0.2440	0.0015	198.8
$^{133}Cs_2$	$^1\Sigma_g^+$	0.394	42.022	0.0823	0.0127	0.0000264	447
$^{63}Cu_2$	$^1\Sigma_g^+$	2.03	264.55	1.025	0.10874	0.000614	221.97
$^{19}F_2$	$^1\Sigma_g^+$	1.602	916.64	11.236	0.89019	0.01385	141.193
$^1H_2$	$^1\Sigma_g^+$	4.4781	4401.21	121.34	60.853	3.062	74.144
$^1H^2H$	$^1\Sigma_g^+$	4.5138	3813.2	91.65	45.655	1.986	74.142
$^1H^3H$	$^1\Sigma_g^+$	4.5269	3597.1	81.68	40.595	1.664	74.142
$^2H^2$	$^1\Sigma_g^+$	4.5563	3115.5	61.82	30.444	1.0786	74.142
$^4He_2^+$	$^2\Sigma_u^+$	2.365	1699	35	7.211	0.224	108.1
$^{127}I_2$	$^1\Sigma_g^+$	1.5424	214.50	0.615	0.03737	0.000114	266.6
$^{39}K_2$	$^1\Sigma_g^+$	0.514	92.021	0.2829	0.05674	0.000165	390.51
$^7Li_2$	$^1\Sigma_g^+$	1.046	357.43	2.610	0.67264	0.00704	267.29
$^{24}Mg_2$	$^1\Sigma_g^+$	0.0501	51.12	1.645	0.09287	0.00378	389.1
$^{14}N_2$	$^1\Sigma_g^+$	9.759	2358.57	14.324	1.99824	0.01732	109.769
$^{14}N_2^+$	$^2\Sigma_g^+$	8.713	2207.00	16.10	1.93176	0.01881	111.642
$^{23}Na_2$	$^1\Sigma_g^+$	0.720	159.125	0.7255	0.15471	0.000874	307.89
$^{16}O_2$	$^3\Sigma_g^-$	5.116	1580.19	11.98	1.44563	0.0159	120.752
$^{16}O_2^+$	$^2\Pi_g$	6.663	1904.8	16.26	1.6913	0.01976	111.64
$^{31}P_2$	$^1\Sigma_g^+$	5.033	780.77	2.84	0.30362	0.00149	189.34
$^{32}S_2$	$^3\Sigma_g^-$	4.3693	725.65	2.844	0.2955	0.00157	188.92
$^{28}Si_2$	$^3\Sigma_g^-$	3.21	510.98	2.02	0.2390	0.0014	224.6
$^{129}Xe_2$	$^1\Sigma_g^+$	0.0230	21.12	0.65	—	—	436.1

The rotational constant in vibrational state v is $B_0 = B_e - \alpha_e(v + \frac{1}{2}) + \cdots$
The vibrational term $G(v) = \nu_e(v + \frac{1}{2}) - \nu_e x_e(v + \frac{1}{2})^2 + \cdots$
*[After K. P. Huber and G. Herzberg, *Molecular Spectra and Molecular Structure* (D. van Nostrand, Princeton, NJ. 1979). Many of these results are derived from electronic spectra (Chapter 27).]

26.14   Normal Modes of Vibration

Consider a molecule with N nuclei. The location of each nucleus in space can be described by a set of three coordinates x_i', y_i', z_i', so that $3N$ coordinates are required. The positions of the nuclei in the molecule oscillate about certain equilibrium sites.

TABLE 26.5
SPECTROSCOPIC PROPERTIES OF TRIATOMIC MOLECULES

Molecule A—B—C	R_{AB} (pm)	R_{BC} (pm)	Bond angle	$\tilde{\nu}_1$ (cm^{-1})	$\tilde{\nu}_2$ (cm^{-1})	$\tilde{\nu}_3$ (cm^{-1})
H—O—H	95.72	95.72	104.52	3657	1595	3776
H—O—D	95.72	95.72	104.52	2724	1403	3708
D—O—D	95.72	95.72	104.52	2666	1179	2787
H—S—H	133.4	133.4	92.27	2611	1183	2626
H—C—N	106.4	115.6	180	2096	712	3312
O—C—O	116.2	116.2	180	1388	667	2349
O—C—S	116.4	155.8	180	859	522	2050
N—N—O	112.6	119.1	180	1285	589	2224
O—N—O	119.7	119.7	134.25	1306	755	1621
O—O—O	127.8	127.8	116.82	1110	705	1043
O—S—O	143.3	143.3	119.55	1151	518	1362

The oscillations of each nucleus can be described in terms of a set of displacements from equilibrium, x_i, y_i, z_i. When each nucleus undergoes a simple harmonic vibration with the same frequency v and always in phase with every other nucleus, we have a *normal mode* of vibration. For example, the displacement from its equilibrium site of nucleus i would be

$$x_i = x_{i0} \cos (2\pi v t + \phi) \qquad (26.22)$$

where x_{i0} is the amplitude and ϕ the phase.

Given the frequencies of a set of normal vibrations from spectroscopic measurements, the practical problem is to assign each frequency to its correct normal vibration and to derive a set of force constants. The force constants outnumber the vibration frequencies, but isotopic substitution may provide additional data. For moderately complicated molecules ($N > 4$), some drastic reduction in the number of force constants is also necessary. Several models are available. One of the most convenient is to divide the displacements of the nuclei into those caused by (1) stretching of chemical bonds, (2) bending of bonds (alteration of bond angles), (3) twisting of bonds (torsional deformation), and (4) a limited number of nonbonded (van der Waals) interactions between atoms that come quite close together in the molecule.

26.15 Symmetry and Normal Vibrations

The type of the normal vibrations can often be described by simple symmetry considerations, based on the character table for the point group to which the molecule belongs. If a normal vibration is nondegenerate, i.e., if its particular frequency v belongs to only one normal vibration, the symmetry requirement can be seen as follows. The total energy of the vibrating molecule cannot change as the result of any symmetry operation performed on it. The energy can be expressed in terms of the *normal vibration coordinates* q_i as the sum of kinetic and potential energies,

$$\epsilon = \sum \tfrac{1}{2} m_i \ddot{q}_i^2 + \sum \tfrac{1}{2} \kappa_i q_i^2$$

To maintain constant energy, each q_i must be either symmetric ($q_i \longrightarrow q_i$) or antisymmetric ($q_i \longrightarrow -q_i$) with respect to each symmetry operation. As an example, Fig. 26.11 shows the effect of a symmetry plane on the normal vibrations of an ABC_2 molecule, such as formaldehyde, H_2CO. If the normal vibration ν_i is degenerate, the symmetry problem is somewhat more involved, because now a linear combination of normal modes must be taken, and the effect of the symmetry operation is usually more than just a change of sign.

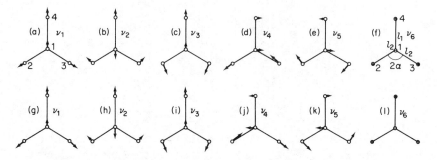

FIGURE 26.11 Normal vibrations of an ABC_2 molecule and their behavior for a reflection at the plane of symmetry through AB perpendicular to the plane of the molecule. Motions perpendicular to the plane of the paper are indicated by + or − signs in the circles representing nuclei. [From G. Herzberg, *Molecular Spectra and Molecular Structure* (New York: D. van Nostrand & Co., 1945)]

From the discussion of group representations in Chapter 25, it is evident that each normal mode can serve as the basis for a representation of the symmetry group of the molecule. The representation can be reduced to a set of irreducible representations. Then the normal mode (or, in a degenerate case, a linear combination of normal modes) transforms under the symmetry operations exactly as required by the characters of the irreducible representation (I. R.) to which it belongs.

As an example, consider H_2CO, whose $(3N - 6)$ normal vibrations are shown in Fig. 26.11. The point group is C_{2v} with the character table given on p. 574. All the I. R.'s are one dimensional; therefore, all the normal vibrations must be nondegenerate. The three normal vibrations ν_1, ν_2, ν_3 are totally symmetric and belong to species A_1. If we take the plane of the molecule to be the xz plane, vibrations ν_4 and ν_5 belong to species B_1, whereas ν_6 belongs to species B_2, being antisymmetric with respect to reflection $\sigma_v(xz)$ in the plane of the molecule. The following basic principle thus holds: The character table of the I. R.'s of the point group leads to the classification of the normal vibrations.

A fundamental vibrational transition occurs when a particular normal mode i is excited from its ground state $v_i = 0$ to its first excited state $v_i' = 1$, while all other normal nodes remain unexcited. The vibration that is excited must result in an oscillating dipole moment. For this to occur, the symmetry of the vibration, i.e., of the normal coordinate that is changing in the vibration, must correspond with one of the components of the dipole moment (x, y, or z). The character table of the symmetry group lists the irreducible representations to which x, y, and z belong. Therefore we can tell if a particular normal mode of vibration is active in the infrared simply by inspecting the character table to see if x, y, or z belongs to the same I. R. as the particular vibration.

For the example of the C_{2v} molecule, H_2CO, shown in Fig. 26.11, which of the normal vibrations will be active? From the character table, we can assign the normal vibrations and the Cartesian coordinates as follows:

$$\nu_1 \quad A_1 \quad \nu_4 \quad B_1 \quad x \quad B_1$$
$$\nu_2 \quad A_1 \quad \nu_5 \quad B_1 \quad y \quad B_2$$
$$\nu_3 \quad A_1 \quad \nu_6 \quad B_2 \quad z \quad A_1$$

In this case, each of the normal-mode representations (A_1, B_1, B_2) matches one of the dipole-moment representations (x, y, z); therefore, all the normal modes of formaldehyde are active as fundamentals in the infrared spectrum.

The character table for the symmetry group of a molecule also indicates which normal modes are active in the Raman spectrum. In accord with Eq. (26.21), a normal mode can be Raman active only if the electric field of the incident radiation can induce a changing dipole moment in the molecule. The nine contributions to the induced dipole moment are shown in Eq. (23.6). The product of each of these by the changing component of the normal-mode coordinate x, y, or z has one of the quadratic forms x^2, y^2, z^2, xz, xy, or yz. To ascertain whether a normal mode is Raman active, therefore, we simply inspect the character table to see whether its irreducible representation corresponds to one of the quadratic forms.

Example 26.6 The 12 normal vibrations of *trans*-$C_2H_2Cl_2$ are shown as follows:

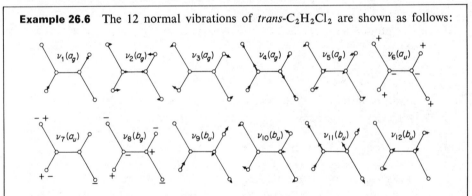

Which normal modes are active in the infrared and which in the Raman?

The molecule belongs to the point group C_{2h}, with the character table on page 574. The modes ν_1, ν_2, ν_3, ν_4, and ν_5 are totally symmetric and belong to the representation A_g. Modes ν_6 and ν_7 are antisymmetric with respect to the horizontal mirror plane σ_h and the inversion center, and therefore belong to A_u. Mode ν_8 is antisymmetric with respect to the mirror plane σ_h and the twofold axis C_2, and hence belongs to B_g. Modes ν_9, ν_{10}, ν_{11}, and ν_{12} are symmetric with respect to σ_h but antisymmetric with respect to C_2 and i, and hence belong to B_u. Since A_u and B_u include the coordinates x, y, or z, and A_g and B_g include the quadratic term, our conclusion is:

Infrared: ν_6, ν_7, ν_9, ν_{10}, ν_{11}, ν_{12}

Raman: ν_1, ν_2, ν_3, ν_4, ν_5, ν_8

Problems

1. The moment of inertia of the NH radical is 1.68×10^{-47} kg m². At what frequency, wavelength, and wave number (cm⁻¹) would one find the transition $J = 2 \rightarrow J = 3$?

2. The rotational transition $J = 0 \rightarrow J = 1$ has been measured in the microwave spectrum of $^{12}C^{16}O$ at 115 271.204 MHz. Calculate the internuclear separation R_e. What factors, if any, would limit its accuracy to less than the nine figures of the spectral data?

3. The internuclear distance of $^{79}Br^{19}F$ is 175.5 pm. Calculate the moment of inertia I of the molecule and predict the spacing in its pure rotational spectrum (in cm⁻¹). In what region of the electromagnetic spectrum would these lines be observed? The classical formula for the energy of rotation is $E = \frac{1}{2}I\omega^2$, where ω is the rotational frequency in radians per second. What is ω for the $J = 10$ state of $^{79}Br^{19}F$?

4. An unknown molecule XY has a vibration frequency at $\tilde{v} = 2331$ cm⁻¹ and a force constant $\kappa = 2245$ N m⁻¹. The diatomic oxide XO has $\tilde{v} = 1876$ cm⁻¹ and $\kappa = 1550$ N m. Identify the molecule XY.

5. The isotopes ^{35}Cl and ^{37}Cl have relative abundances 0.764 and 0.246, respectively. Sketch the appearance of the $v = 0$ to $v = 1$ band in the absorption spectrum of HCl at high resolution. Show only the first four lines in the P and R branches of ^{35}ClH and ^{37}ClH. Indicate where the ^{35}ClD band origin would be on same scale.

6. The moment of inertia of $Na^{35}Cl$ is 129.0×10^{-47} kg m². Draw to scale the first five rotational levels in $Na^{35}Cl$. Sketch the pure rotational absorption spectrum by drawing vertical lines at indicated wave numbers with the heights of lines proportional to intensity of absorption. In NaCl vapor at 1200 K, what would be the relative number of molecules in states $J = 0, 1, 2, 3, 4, 5$?

7. The far infrared spectrum of OCS was investigated at two different isotopic compositions:

$$^{16}O^{12}C^{32}S, \qquad B_1 = 0.202864 \text{ cm}^{-1}$$

$$^{16}O^{12}C^{34}S, \qquad B_2 = 0.197910 \text{ cm}^{-1}$$

Calculate the internuclear distances R_{co} and R_{cs}.

8. The following lines are observed in the microwave spectrum of HF:

J	0	1	2	3	4	5	6
$\tilde{v}$ (cm⁻¹)	41.08	82.19	123.15	164.00	204.62	244.93	285.01

For a nonrigid rotor, $v_J = 2\tilde{B}(J + 1) - 4\tilde{D}(J + 1)^3$. Calculate the best values of $\tilde{B}$ and $\tilde{D}$ from a plot of $v_J/(J + 1)$ vs. $(J + 1)^2$. Hence calculate the force constant κ of HF from $\tilde{D} = 16\tilde{B}^3\pi^2\mu c^2/\kappa$.

9. The relative populations of rotational energy levels are governed by the multiplicity factor $2J + 1$ and by the Boltzmann factor. For a rotor with $\tilde{B} = 10$ cm⁻¹, plot the relative population vs. J at 300 K.

10. The three normal modes of H_2O are at 1595, 3652, and 3756 cm⁻¹. Calculate the fraction of H_2O molecules in each vibrational state (v_1, v_2, v_3) in a sample of water vapor at 1300 K.

11. The pure rotation spectrum of $H^{35}Cl$ has lines at the following $\tilde{v}$ (cm⁻¹): 21, 42, 64, 85, 106, 127, 148, 170, 191, 212, 233, 254, 275, 297, 318, 339, 360, 381, 403, 424, 445, 466, 487, 509. Calculate the rotational partition function at 80 K (a) by direct summa-

tion; (b) from Eq. (12.26); (c) from Eq. (12.28). What is the average rotational energy per mole of HCl by each method?

12. (a) Write the Hamiltonian $\hat{H}$ for the problem of the one-dimensional harmonic oscillator with $U = \frac{1}{2}\kappa a^2$.
    (b) Show that $\psi_0 = \exp(-ax^2)$ and $\psi_1 = x\exp(-ax^2)$ are eigenfunctions of $\hat{H}$.
    (c) Normalize ψ_0 and ψ_1. Show that ψ_0 and ψ_1 are orthogonal.
    (d) If the eigenvalues are $\epsilon_v = (v + \frac{1}{2})hv$, determine a.

13. The infrared spectrum of N_2O shows three fundamental vibrational frequencies in the infrared, at 589, 1285, and 2224 cm^{-1}. What can you conclude about the structure of N_2O? Assign the normal modes.

14. The following microwave absorption lines were observed in the vapor of NaCl at 800°C.

v (MHz)	26 051	25 848	25 667	25 494	25 474	25 308	25 120
Rel. intensity	1.0	0.6	0.35	0.30	0.23	0.18	0.10

These lines arise from a given rotational transition in molecules that are in different vibrational states. There are two isotopic species ^{35}Cl and ^{37}Cl to be considered (see Problem 6). Determine R_e for each state and each isotopic species.

15. In the near infrared spectrum of CO there is an intense band at 2144 cm^{-1}. Calculate (a) the fundamental vibration frequency of CO; (b) the force constant; (c) the zero-point energy.

16. A C—H stretching frequency is observed for acetylene in the infrared at 3287 cm^{-1} and another one in the Raman at 3374 cm^{-1}. Sketch the normal modes of vibration corresponding to these frequencies.

17. For H_2, $R_e = 74.0$ pm, $D_e = 458.1$ kJ mol^{-1}, and $\tilde{v}_0 = 4159$ cm^{-1}. Draw the potential-energy curve of H_2 according to the Morse function and compare the result with the experimental curve in Fig. 22.4.

18. The fundamental and overtone vibrational bands of HCl occur at the following wave numbers: 2885.9, 5668.0, 8346.9, 10 923, and 13 397 cm^{-1}. Fit these data to the model of the anharmonic oscillator of Section 26.8 and calculate the anharmonicity constant x_e.

19. The Raman spectrum of N_2 is studied with excitation by an argon-ion laser at 540.80 nm. What wavelengths would be predicted for the first three Stokes ($v_0 - v$) and first three anti-Stokes ($v_0 + v$) lines in the pure rotational Raman spectrum?

20. The CH_3Cl molecule belongs to the point group C_{3v}, whose character table was derived in Section 25.9. The normal vibrations are shown on p. 602. Which vibrations will be active in the infrared and which in the Raman?

21. Calculate ΔU_0 for the reaction $H_2 + D_2 = 2HD$ from spectroscopic data.

22. The normal vibrations of acetylene, HC≡CH are:

$$\leftarrow H-C\equiv C-H\rightarrow \quad \leftarrow H-C\equiv C-H\rightarrow \quad H-C\equiv C-H \quad H-C\equiv C-H \quad H-C\equiv C-H$$
$$v_1 \qquad\qquad v_2 \qquad\qquad v_3 \qquad\qquad v_4 \qquad\qquad v_5$$

What is the point group of the molecule? Which vibrations are active in (a) the infrared; (b) the Raman?

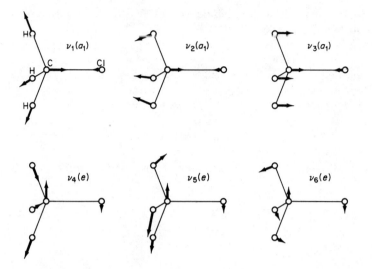

23. CH_4 is a spherical top with C—H bond distance 109 pm. Suppose that a methane molecule has $\frac{3}{2}kT$ of rotational energy. What is the J value of its rotational state at 300 K?

24. From the wavefunction in Eq. (26.10), show that the transition $\Delta v = \pm 2$ is forbidden for a harmonic oscillator.

27

Electronic Spectra
and Photochemistry

Spectra in the visible and ultraviolet regions arise from transitions between different energy levels of electrons in atoms or molecules. The energy of dark red light at the limit of visibility to the human eye (~ 750 nm) is about 160 kJ/mol, and that of violet light at the other limit (~ 400 nm) is about 300 kJ/mol. The colors that we see result when photons are caught by specific visual pigments in the cone cells of the retina. These pigments selectively absorb red, green, or blue light, and the resultant photochemical reactions cause electrical impulses to be relayed to the brain, where they are translated by the conscious mind into perceptions of the various colors. We cannot imagine what a new color would look like if our eyes were sensitive to slightly longer wavelengths in the infrared or slightly shorter ones in the ultraviolet. Honeybees can perceive ultraviolet radiation as a distinct color, but, on the other hand, cannot distinguish red from green.

The energies of quanta of visible and ultraviolet light are comparable to the energies of chemical bonds (100 to 500 kJ/mol). If a molecule absorbs a quantum of such energy, definite chemical effects may be expected. Photochemistry, the science of the chemical effects of light, is thus closely related to electronic spectroscopy.

27.1 Absorption of Light

Consider a beam of monochromatic light passing through a sample that absorbs some of the light. Let I be the intensity of light incident upon a thin layer of sample that is perpendicular to the beam (Fig. 27.1). The fractional change in intensity

603

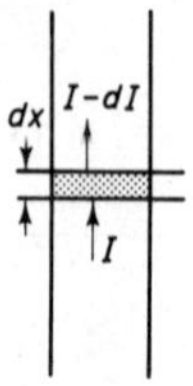

FIGURE 27.1 Absorption of light in a slice dx of absorbing medium.

$-dI/I$ as the light beam traverses a thickness dx of sample is

$$-dI/I = b\,dx \tag{27.1}$$

If all the incident light is either transmitted or absorbed (no scattering), b is called the *absorption coefficient*. This law of light absorption was originally stated in 1729 by Pierre Bouguer, and later rediscovered by J. H. Lambert. To find the light intensity I transmitted through a sample of finite thickness ℓ, Eq. (27.1) is integrated,

$$\int_{I_0}^{I} -dI'/I' = \int_0^{\ell} b\,dx$$

$$-\ln I/I_0 = b\ell \tag{27.2}$$

or $\qquad\qquad -\log I/I_0 = -\log \mathfrak{I} = A = b\ell/2.303 = a\ell \tag{27.3}$

In the equations above, I_0 is the light intensity incident on the sample, $\mathfrak{I} = I/I_0$ is the *transmittance*, and $-\log \mathfrak{I} = A$ is the *absorbance*, formerly called the *optical density*.

For solutions in which the chemical form of the absorbing species does not change with concentration (i.e., association, complex formation, or dissociation does not occur) the absorption coefficient is proportional to the concentration of absorbing species, $a = \epsilon c$ and Eq. (27.3) becomes

$$-\log (I/I_0) = \epsilon c\ell \tag{27.4}$$

where ϵ is the *molar absorption coefficient* or *molar absorptivity* (formerly called the *extinction coefficient*). Equation (27.4) is called the *Lambert–Beer Law*. Note that ϵ is an absorption coefficient divided by concentration. Its SI unit is $m^2\,mol^{-1}$, but it is usually cited in units of $(dm^3/mol)\,cm^{-1}$ [$M^{-1}\,cm^{-1}$], when concentrations are in $mol\,dm^{-3}$ and path lengths are in centimeters.

Example 27.1 The concentration of proteins in solution can be estimated by measuring the absorbance of the solution at $\lambda = 280$ nm; the light absorption is due to the amino acid tryptophan with $\epsilon = 540\ m^2\,mol^{-1}$. A protein has an absorbance $A = 0.54$ in 0.05 mM solution with 1 cm path length. How many tryptophan residues occur in the protein molecule?

From $A = \epsilon c\ell$, $c = A/\epsilon\ell = 0.54/(540\ m^2\,mol^{-1})(10^{-2}\ m)$

$$c = 10^{-1}\ mol\,m^{-3} = 10^{-4}\ mol\,dm^{-3}$$

Therefore, $n = 10^{-4}/5 \times 10^{-5} = 2$ tryptophan/molecule.

The molar absorption coefficient ϵ is a function of the wavelength λ or frequency ν of the absorbed radiation, $\epsilon(\nu)$. A plot of $\epsilon(\nu)$ vs. ν gives the absorption of a substance

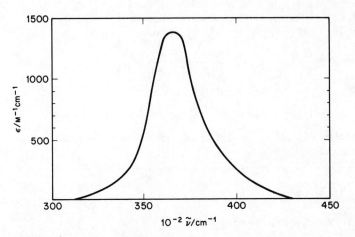

FIGURE 27.2 Absorption band of tyrosine in water at pH 7 and 25°C. Molar absorption coefficient ϵ is plotted vs. wave number $\tilde{\nu}$.

over a range of frequencies. An example is shown in Fig. 27.2, the absorption band of tyrosine. The integral of ϵ over the entire absorption band, i.e., the area under the curve in Fig. 27.2, is the integrated absorption coefficient, $E = \int \epsilon(\nu)\, d\nu$.

The integrated absorption coefficient is related to a dimensionless quantity called the *oscillator strength f* of the absorption band,

$$f = \left(\frac{4m_e c\epsilon_0}{Le^2}\right) 2.303E = (1.441 \times 10^{-18} \text{ mol s m}^{-2})E \qquad (27.5)$$

The oscillator strength of an absorption band is related directly to the transition moment described in Section 26.3, by

$$\mu_{nm}^2 = \frac{3}{8\pi^2} \frac{he^2}{m_e\nu} f_{nm} \qquad (27.6)$$

where ν is the frequency of the absorption maximum. The range of f is from about 10^{-3} for very weak absorption bands due to "forbidden" transitions to a maximum of $f \sim 2$ for "allowed" transitions.

Example 27.2 Calculate the oscillator strength f and the transition moment μ_{nm} for the absorption band in Fig. 27.2.

The area of the absorption band is measured as $4.83 \times 10^6 \text{ cm}^{-1} \text{ m}^2 \text{ mol}^{-1}$, and

$$E = (4.83 \times 10^6 \text{ cm}^{-1} \text{ m}^2 \text{ mol}^{-1})(3.00 \times 10^{10} \text{ cm s}^{-1})$$

$$= 14.5 \times 10^{16} \text{ s}^{-1} \text{ m}^2 \text{ mol}^{-1}$$

From Eq. (27.5),

$$f = (1.441 \times 10^{-18} \text{ mol s m}^{-2})(14.5 \times 10^{16} \text{ m}^2 \text{ mol}^{-1} \text{ s}^{-1})$$

$$= 0.209$$

We take ν at the peak of the absorption curve, $\nu = 36\,500\ \text{cm}^{-1} \times 3.00 \times 10^{10}$ $\text{cm s}^{-1} = 1.095 \times 10^{15}\ \text{s}^{-1}$. From Eq. (27.6)

$$\mu_{nm}^2 = \frac{3}{8\pi^2} \frac{(6.626 \times 10^{-34}\ \text{J s})(1.602 \times 10^{-19}\ \text{C})^2}{(9.110 \times 10^{-31}\ \text{kg})(1.095 \times 10^{15}\ \text{s}^{-1})}(0.209)$$

$$= 6.48 \times 10^{-58}\ \text{C}^2\ \text{m}^2$$

$$|\mu_{nm}| = 2.55 \times 10^{-29}\ \text{C}^2\ \text{m}^2\ (\text{or 7.64 debye units})$$

(The values of f and μ_{nm} are typical of a strong absorption band, not forbidden by selection rules).

27.2 Electronic Transitions and Band Spectra

The structure of a molecule in an excited electronic state may be quite different from that in the ground state. For example, acetylene in the ground state is linear with a C—C bond distance of 120.8 pm. In its first excited state, acetylene has a bent *trans* conformation with a C—C bond of 138.5 pm. Thus, in general, the force constants κ, anharmonicity constants x_e (Section 26.8), and moments of inertia of a molecule are different in its various electronic states. Each vibrational transition that accompanies the electronic transition gives an absorption or emission band in the spectrum. The wave number of the radiation absorbed or emitted is the sum of electronic, vibrational, and rotational terms:

$$\tilde{\nu} = \tilde{\nu}_{el} + \tilde{\nu}_v + \tilde{\nu}_J \tag{27.7}$$

An example is shown in Fig. 27.3, which depicts the Swan bands of C_2 and the vibrational levels in the upper and lower electronic states that lead to these bands. The Swan bands are important in the study of hydrocarbon flames. The bands appear in well defined *sequences*, groups that are designated by the pair of vibrational quantum numbers that determine the first band in each sequence. The position of each band is given by electronic and vibrational terms:

$$\tilde{\nu} = \tilde{\nu}_{el} + [G'(v') - G''(v'')]$$

where $$G(v) = (v + \tfrac{1}{2})\tilde{\nu}_e - (v + \tfrac{1}{2})^2 x_e \tilde{\nu}_e$$

The vibrational wave number is $\tilde{\nu}_e$ for a harmonic oscillator and x_e is the anharmonicity constant. The selection rule is that $v' - v'' = 0, 1, 2$, etc.

At a higher resolution, we would see that each vibrational transition in Fig. 27.3 gives a band of closely spaced lines arising from different rotational transitions. An example of such a high-resolution spectrum is Fig. 27.4, in which a single vibrational band ($v' = 0$, $v'' = 0$) in the electronic spectrum of SiN is shown. The rotational structure of such a band is similar to that described in Section 26.9 for vibration-rotation bands in infrared spectra.

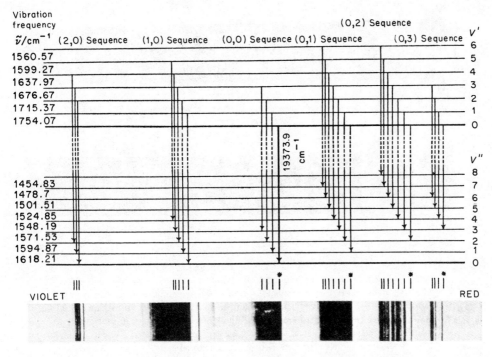

FIGURE 27.3 Swan bands of C_2 and the vibrational states in which they originate.

27.3 The Franck–Condon Principle

Figure 27.5 represents the potential-energy curves for the ground state of a diatomic molecule and for an electronically excited state. These curves can be plotted from the Morse function of Eq. (26.12) or constructed directly from the vibrational levels determined from the spectra.

A rule given by James Franck and Edward Condon helps us to understand the transitions that can occur between two such electronic states. An electron transition takes place in a time ($\sim 10^{-16}$s) that is very short compared to the period of vibration of atomic nuclei ($\sim 10^{-13}$s), which are heavy and sluggish compared to electrons. (Remember the Born–Oppenheimer approximation.) The positions and velocities of nuclei are therefore almost unchanged during an electronic transition. Thus we can represent such a transition by a vertical line drawn between two potential-energy curves. If we know the initial state of the molecule on one potential-energy curve, the vertical line will intersect the other curve at the final state. An arrow may be drawn upward for absorption of a quantum of energy or downward for emission.

In Fig. 27.5 the ranges of internuclear distance of possible transitions are indicated for absorption and emission. The square of the vibrational wavefunction ψ (Fig. 26.7) determines the most probable initial states of the vertical transitions. In

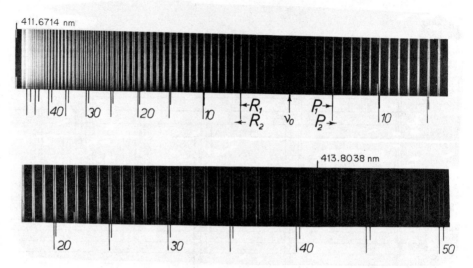

FIGURE 27.4 An example of an electronic band emission spectrum at high resolution. The spectrum shows the 0, 0 band ($v = 0$ in both upper and lower states) of the $^2\Sigma \rightarrow {}^2\Sigma$ transition of SiN in the violet region of the spectrum. The numbers denote the total angular momentum quantum number J. As explained in Section 26.9, $\Delta J = +1$ gives the R branch and $\Delta J = -1$ the P branch. In this spectrum, however, each line is split into a doublet as a result of what is called *spin doubling*. The molecule has a single unpaired electron outside of closed shells, so that $S = \frac{1}{2}$ and $2S + 1 = 2$. The rotational quantum number K combines with the spin quantum number S to yield $J = K \pm S$. The case $J = K + \frac{1}{2}$ gives the P_1 and R_1 branches and the case $J = K - \frac{1}{2}$ gives the P_2 and R_2 branches. The *band gap* of approximately $4B$ at the origin $\tilde{v}_0$ is an obvious feature of the spectrum and confirms the fact that transitions between $K' = 0$ and $K'' = 0$ are forbidden. The wavelengths at the top of spectrum refer to standard thorium lines used for calibration. [This beautiful spectrum was provided by Thomas Dunn of the University of Michigan.]

the lowest vibrational level $v = 0$, the maximum in ψ^2 occurs at the midpoint of the vibration, but in higher vibrational levels, the maximum in ψ^2 lies nearer the extremes of the vibration.

Example 27.3 From the spectral data in Fig. 27.3 calculate $\tilde{v}_e$ and $x_e\tilde{v}_e$ for the $A^3\Pi_u$ and $X^3\Pi_g$ states of C_2.

For ground state, $X^3\Pi_g$

$$\left.\begin{array}{l} 1618.21 = \tilde{v}_e - 2x_e\tilde{v}_e \\ 1594.87 = \tilde{v}_e - 4x_e\tilde{v}_e \end{array}\right\} \qquad \begin{array}{l} \tilde{v}_e \;= 1641.55 \text{ cm}^{-1} \\ X_e\tilde{v}_e = 11.67 \text{ cm}^{-1} \end{array}$$

For excited state, $A^3\Pi_u$

$$\left.\begin{array}{l} 1754.07 = \tilde{v}_e - 2x_e\tilde{v}_e \\ 1715.37 = \tilde{v}_e - 4x_e\tilde{v}_e \end{array}\right\} \qquad \begin{array}{l} \tilde{v}_e \;= 1792.77 \text{ cm}^{-1} \\ x_e\tilde{v}_e = 19.35 \text{ cm}^{-1} \end{array}$$

Unlike Fig. 27.5, the excited state has a slightly higher $\tilde{v}_e$ (thus higher force constant) and a higher anharmonicity.

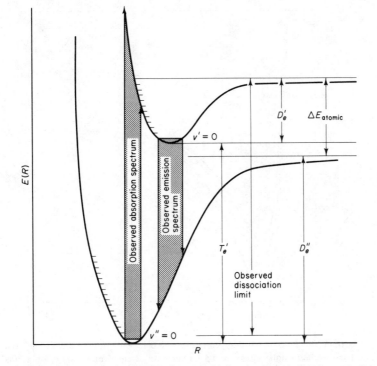

FIGURE 27.5 "Vertical transitions" permitted by the Franck–Condon principle between two electronic states. The dissociation energies of ground and excited electronic states are D''_e and D'_e respectively. ΔE (atomic) is the difference in energy between the separated atoms arising from the two states and T'_0, is the difference in energy between the two electronic states. [J. I. Steinfeld, *Molecules and Radiation*. (Cambridge, Mass.: The MIT Press, 1974)]

27.4 Excited States of Oxygen

Potential-energy curves for the ground state and several electronically excited states of the oxygen molecule are shown in Fig. 27.6. The ground state of O_2 is $^3\sum_g^-$; its molecular orbital representation was discussed in Section 22.10. There are two low-lying singlet states, $^1\Delta_g$ and $^1\sum_g^+$. Dissociation of O_2 from any of these states yields two oxygen atoms in their ground state, ^3P.

The lowest singlet states of molecular oxygen are especially interesting because they have such long half-lives (7 s and 2700 s) for radiative decay, as a consequence of the fact that transition from the singlet back to the triplet with emission of radiation is forbidden. For many years it has been known that a weak red chemiluminescence at about 633 nm accompanies the reaction, $H_2O_2 + OCl^- = O_2 + Cl^- + H_2O$. Michael Kasha and coworkers showed that this chemiluminescence involves a bimolecular collision of two molecules of singlet O_2, with the emission of a quantum

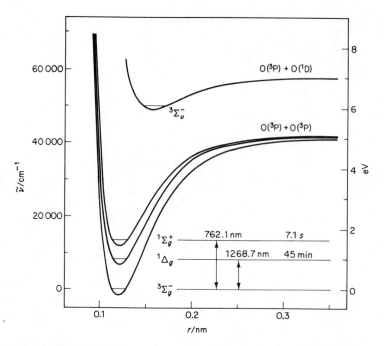

FIGURE 27.6 Potential-energy curves for the electronic ground state of O_2, two low-lying excited singlet states, and a higher stable excited triplet state. Vibrational dissociation of the three lower states yields O atoms in their ^3P ground states, but dissociation from the $^3\Sigma_u^-$ state yields one ^3P oxygen atom and one oxygen atom in an excited state ^1D.

or red light,

$$(^1\Delta_g) + (^1\Delta_g) \longrightarrow (^3\textstyle\sum_g^-) + (^3\textstyle\sum_g^-) + h\nu(633.4 \text{ nm})$$

Singlet oxygen may occur in various biological oxidations, in radiation effects on tissues, and in the formation of smog due to photo-oxidations of organic compounds in the atmosphere.

If a molecule dissociates from an excited electronic state, the fragments (atoms in the case of a diatomic molecule) are not always in their ground states. To obtain the energy of dissociation into atoms in the ground states, we must therefore subtract any excitation energies of the atoms.

Example 27.4 The UV absorption spectrum of O_2 includes a series of bands arising from transitions between the ground state and the $^3\sum_u^-$ state shown in Fig. 27.6. These *Schumann–Runge bands* converge to a continuum with onset at 175.9 nm. The dissociation yields one O atom in its ground state ^3P and one O atom in an excited state ^1D. Calculate the ΔH_{298}° for the reaction $O_2 \longrightarrow 2O$.

The energy corresponding to $\lambda = 175.9$ nm is hcL/λ per mole.

$$\Delta E = \frac{(6.626 \times 10^{-34} \text{ J s})(2.998 \times 10^{-8} \text{ m s}^{-1})(6.022 \times 10^{23} \text{ mol}^{-1})}{175.9 \times 10^{-9} \text{ m}}$$

$$= 680.2 \text{ kJ mol}^{-1}$$

The standard reference work (C. A. Moore, *Atomic Energy Levels*, Natl. Bur. Stand. Circ. 467, 1943), reveals that the ^1D state of O is 1.970 eV or 190.1 kJ mol^{-1} above the ground state ^3P. Hence for $O_2 = 2O$, $\Delta H_0 = 680.2 - 190.1 = 490.1$ kJ mol^{-1}. From Table A2, for O, $H^\circ_{298} - H^\circ_0 = C_p \Delta T = 6.20$ kJ mol^{-1} and for O_2, 8.66 kJ mol^{-1}. Then,

$$\Delta H^\circ_{298} = 490.1 + 8.66 - 2(6.20) = 486.3 \text{ kJ mol}^{-1}$$

27.5 Excited States of Polyatomic Molecules

The complete analysis of the electronic spectrum of a polyatomic molecule and the description of its excited states is a difficult problem. We can, however, give a simplified description of the types of molecular orbitals that are involved in the electronic transitions. The orbitals for the ground states of molecules are usually either σ or π type. Excited antibonding orbitals of σ and π types are denoted σ^* and π^* orbitals. There are also orbitals that hold lone pairs of electrons on hetero-atoms such as N or O. These are nonbonding orbitals, designated as n orbitals. Usually, they are the occupied orbitals of highest electron energy and hence are particularly important in electronic spectra. Groups such as —CO, —NO$_2$, —CONH, in molecules are called *chromophores* (color carriers) since they give absorption bands in the visible or near ultra-violet in structures that otherwise would not absorb in these regions.

The σ, π, and n orbitals and their excited states are illustrated in Fig. 27.7 for the case of a simple carbonyl compound like formaldehyde, HCHO. Schematic outlines of the electron-density contours of the orbitals and their approximate energy levels are shown. If a single electron in an excited-state orbital has a spin opposite to that of the single electron left in the lower orbital, the total spin of the excited state is $S = 0$, the multiplicity $2S + 1 = 1$, and the state is a *singlet*. If the two electrons in

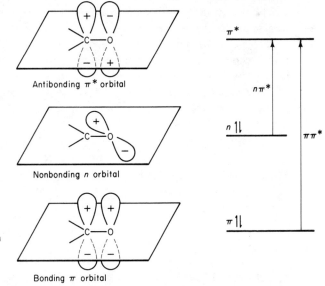

FIGURE 27.7 Molecular orbitals in carbonyl compounds. The $n\pi^*$ transition typically would be at $\lambda = 280$ nm with $\epsilon = 10$ and the $\pi\pi^*$ at $\lambda = 180$ nm with $\epsilon = 1000$.

Antibonding π^* orbital

Nonbonding n orbital

Bonding π orbital

the upper and lower orbitals have the same spin, $S = \frac{1}{2} + \frac{1}{2} = 1$, $2S + 1 = 3$, and the excited state is a triplet. Just as in the case of the helium atom in Section 21.11, the excited states are therefore classified as singlets or triplets, and in accord with Hund's Rule, the triplet is lower in energy than the corresponding singlet.

Important examples of $n \longrightarrow \pi^*$ and $\pi \longrightarrow \pi^*$ transitions are the ultraviolet absorptions by protein molecules due to the peptide bonds of the polypeptide chain: —C(O)⋯NH. The band at 210 nm is due to the $n \longrightarrow \pi^*$ transition and that at 190 nm to the $\pi \longrightarrow \pi^*$. Proteins also absorb in the UV at 240 to 280 nm, the chromophores being the aromatic sidechain rings of tryptophan, tyrosine, and phenylalanine.

27.6 Photochemical Principles

In 1818, Grotthuss and Draper stated the Principle of Photochemical Activation: Only light that is absorbed by a substance is effective in producing a photochemical change. At the time, the distinction between scattering processes and quantum transitions was not understood, so that this activation principle was helpful. It now appears to be an almost self-evident starting point for any discussion of photochemical reactions.

Stark in 1908 and Einstein in 1912 applied the concept of the quantum of energy to photochemical reactions of molecules. They stated the Principle of Quantum Activation: In the primary step of a photochemical process, one molecule is activated by one absorbed quantum of radiation. It is essential to distinguish clearly between the primary step of light absorption and the subsequent processes of chemical reaction. An activated molecule does not necessarily undergo reaction; on the other hand, in some cases one activated molecule may cause the reaction of many other molecules through a chain mechanism. Thus the principle of quantum activation should never be interpreted to mean that one molecule reacts for each quantum absorbed. The validity of the Stark–Einstein Principle depends upon the facts that the lifetimes of excited states are usually short and the intensity of illumination is usually quite low. With the development of laser sources of coherent high-intensity light, it has become possible to observe primary photochemical processes in which more than one quantum is absorbed by a given molecule, and other more complex interaction processes.

The energy $E = Lh\nu$, where L is the Avogadro constant, is called *one einstein*. The value of the einstein depends on the wavelength. For orange light with $\lambda = 600$ nm,

$$E = \frac{(6.02 \times 10^{23} \text{ mol}^{-1}) \times (6.63 \times 10^{-34} \text{ J s}) \times (3.00 \times 10^8 \text{ m s}^{-1})}{(600 \times 10^{-9} \text{ m})(10^{-3} \text{ kJ J}^{-1})} = 199 \text{ kJ}$$

This energy is enough to break certain rather weak covalent bonds, but for C—C bonds and others with energies higher than 300 kJ mol^{-1}, radiation in the ultraviolet region is required.

Example 27.5 A ruby laser with frequency doubling to 347.2 nm has an output of 100 J and a pulse time of 20 ns. If all the light is absorbed in 10 cm^3 of a 0.10 M solution of perylene, what fraction of the perylene molecules is activated?

Energy per light quantum $= hc/\lambda = (6.63 \times 10^{-34}$ J s$)(3.00 \times 10^8$ m s$^{-1})/347.2 \times 10^{-9}$ m $= 5.73 \times 10^{-19}$ J. Number of quanta per pulse $= 100$ J$/5.73 \times 10^{-19}$ J $= 1.75 \times 10^{20}$.

$$\text{Fraction of molecules activated} = \frac{1.75 \times 10^{20}}{(10 \times 10^{-3}\ dm^3)(10^{-1}\ mol\ dm^{-3})(6.02 \times 10^{23}\ mol^{-1})}$$

$$= 0.29.$$

The *quantum yield* Φ of a photochemical reaction is the number of molecules of reactant consumed or product formed per quantum of light absorbed. When we consider in more detail the mechanisms of photochemical activation, we shall need to define other more specialized quantum yields.

An *actinometer* is a device that measures the total amount of incident radiation. A basic type of actinometer is the *thermopile*, which consists of a number of thermocouples connected in series, with their hot junctions imbedded in a black surface that absorbs all the incident radiation and converts it into heat. Calibrated lamps of known energy output are available from the National Bureau of Standards. The voltage developed by the thermopile is measured first with the standard lamp, and then with the source of radiation of unknown intensity. In a photochemical study the reaction vessel is mounted between the thermopile and the source of light, and the radiation absorbed by the reacting system is measured by the difference between readings with the vessel filled and empty.

It is often more convenient to use various chemical actinometers based on quantum yields Φ that have been previously measured with high accuracy. An excellent reaction for use as a chemical actinometer is the decomposition of ferrioxalate,[*]

$$(C_2O_4)^{2-} + 2Fe^{3+} \xrightarrow{h\nu} 2Fe^{2+} + 2CO_2(g)$$

The quantum yield at 334 nm is $\Phi = 1.25$.

An experimental arrangement for a photochemical study is shown in Fig. 27.8. The light from the source passes through a monochromator, which yields a narrow band of wavelengths in the desired region. The monochromatic light passes through the reaction cell, and the part that is not absorbed is measured by an actinometer.

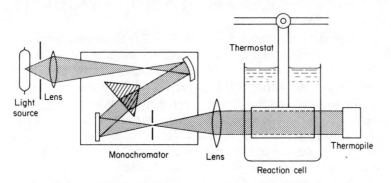

FIGURE 27.8 An apparatus for photochemical experiments.

[*]C. G. Hatchard and C. A. Parker, *Proc. R. Soc. (Lond.), A235* (1956).

Example 27.6 In the photochemical decomposition (photolysis) $2HI \longrightarrow H_2 + I_2$ with light at $\lambda - 253.7$ nm, absorption of 3070 J of energy decomposed 1.30×10^{-2} mol HI. What is the quantum yield Φ of the reaction?

The energy of the 253.7-nm quantum is $\epsilon = h\nu = (6.63 \times 10^{-34}$ J s$)(3.00 \times 10^8$ m s$^{-1})/(253.7 \times 10^{-9}$ m$) = 7.83 \times 10^{-19}$ J. The HI has absorbed $N = 3070/7.83 \times 10^{-19} = 3.92 \times 10^{21}$ quanta or $3.92 \times 10^{21}/6.02 \times 10^{23} = 6.51 \times 10^{-3}$ einstein. The quantum yield

$$\Phi = \frac{\text{mol reacted}}{\text{einsteins absorbed}} = \frac{1.30 \times 10^{-2}}{6.51 \times 10^{-3}} = 1.99$$

[How can you explain this value?]

Photosensitization occurs when a molecule absorbs a quantum and then transfers its excitation energy to a molecule of a different species. Photosensitization has important applications since it can lead to photochemical activation of a molecule that does not itself absorb light at the wavelength of interest. For example, one of the most reproducible photochemical reactions is the decomposition of oxalic acid photosensitized by uranyl salts. The uranyl ion UO_2^{2+} absorbs radiation from 250 to 450 nm, becoming an excited ion $(UO_2^{2+})^*$, which transfers its energy to a molecule of oxalic acid, causing it to decompose.

$$UO_2^{2+} + h\nu \longrightarrow (UO_2^{2+})^*;$$

$$(UO_2^{2+})^* + (COOH)_2 \longrightarrow UO_2^{2+} + CO_2 + CO + H_2O$$

This reaction has a quantum yield of 0.49 at $\lambda = 365$ nm.

27.7 Pathways of Molecular Excitation

In most molecules, the absorption of a quantum leads to a transition from a singlet ground state to a singlet excited state. There is usually a triplet state somewhat below this excited singlet. In the electronic excitation, one electron from an electron pair bond is excited to a higher state. If the excited electron has a spin antiparallel to that of its former mate, the excited state is a singlet, but if the spin of the excited electron is parallel, the state is a triplet.

A typical pattern of states is shown in Fig. 27.9. Immediately after the primary quantum jump, a series of extremely rapid events takes place, before any photochemical reaction or emission of luminescent radiation can occur. First, there is internal conversion: No matter which upper singlet state has been reached in the primary quantum jump, there is often a rapid radiationless transfer of electronic excitation energy into thermal energy of the environment, either unexcited molecules or molecules of solvent. Thus the excited molecules rapidly reach the lowest excited singlet state S_1. Second, there is intersystem crossing: In a competition between the lowest excited singlet and triplet states, S_1 and T_1, some of the energy is transferred to the triplet by a radiationless process. The excitation energy in a typical polyatomic molecule has a certain probability of reaching either the lowest excited singlet or the lowest excited triplet state. The basic model for excitation therefore involves three

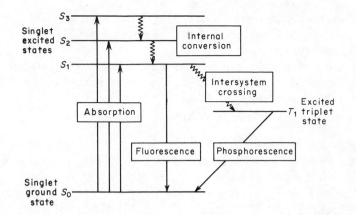

FIGURE 27.9 The molecular energy levels involved in photochemical processes are represented by a *Jablonski diagram.*

important states: the ground singlet S_0, the first excited singlet S_1, and the first excited triplet T_1.

As shown in Fig. 27.9, the excited molecule formed in the primary step can re-emit a quantum $h\nu$ of either the same or a different frequency. This emission is either fluorescence or phosphorescence. A radiative transition between two states of the same multiplicity (usually two singlets) is called *fluorescence*, and a radiative transition between two states of different multiplicity (usually triplet to singlet) is called *phosphorescence*. Processes in which the mechanism has not been decided are referred to as *luminescence*.

27.8 Fluorescence

Fluorescence is an emission of electromagnetic radiation from a real excited state. It should not be confused with the scattering of radiation, i.e., Rayleigh scattering without change in wavelength and Raman scattering with change in wavelength. In the scattering process, no stationary excited state occurs.

Fluorescence spectra are measured with a *spectrofluorimeter*. This instrument is based on two monochromators. The excitation monochromator provides a variable range of wavelengths from the light source to the sample cell. The emission monochromator analyzes the fluorescent light emitted from the sample. Two kinds of fluorescent spectra can be obtained. In the emission spectrum, the sample is illuminated with light of a fixed wavelength, and the spectrum of emitted light is recorded. In the excitation spectrum, the sample is illuminated with light over a range of wavelengths and the fluorescent light intensity recorded at some fixed wavelength.

The natural lifetime of an excited state in a molecule undisturbed by collisions is usually about 10^{-8} s, but it can lie anywhere in the range from 10^{-9} s to a few seconds. At a pressure of 100 kPa, a molecule experiences about 100 collisions in 10^{-8} s. Consequently, excited molecules in most gaseous systems at atmospheric pressures usually transfer their energies by collision before they have a chance to fluoresce. If the excitation energy is converted to thermal energy, the fluorescence is said to be *quenched*. In some such systems, fluorescence can be observed if the pressure is sufficiently reduced.

The kinetics of quenching of fluorescence is based on two parallel reactions of an excited molecule M*. It may emit a photon in fluorescence, or it may be deactivated by collision with a quenching agent Q.

$$\text{Fluorescence:} \quad M^* \xrightarrow{k_1} M + h\nu$$

$$\text{Quenching:} \quad M^* + Q \xrightarrow{k_2} M + Q$$

The deactivation of M* by collision with unexcited M molecules can be neglected if $[M] \ll [Q]$. The fraction of molecules that fluoresce is the fluorescent yield Φ_F, given by

$$\frac{1}{\Phi_F} = \frac{k_1[M^*] + k_2[M^*][Q]}{k_1[M^*]} = 1 + (k_2/k_1)[Q] \tag{27.8}$$

This is called the *Stern–Volmer equation*. Figure 27.10 shows data on the quenching of fluorescence of NO_2 plotted in accord with Eq. (27.8). The constant k_1 can be measured by irradiating a sample of NO_2 by a nanosecond light pulse from a laser and following the intensity of emission of fluorescent light. The measured rates of decay of fluorescence are extrapolated to zero pressure to give k_1.

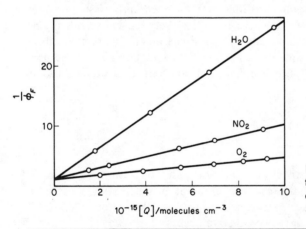

FIGURE 27.10 Quenching of fluorescence of NO_2 as a function of concentrations [Q] of various gases. Stern–Volmer plot.

Example 27.7 From fluorescent decay measurements for NO_2, $k_1 = 2.0 \times 10^4 \text{ s}^{-1}$. From the data in Fig. 27.10 at 300 K, calculate k_2 and the quenching cross section σ_Q for $H_2O + NO_2$ [$\sigma_Q = (\pi d_{AB}^2)$ from Eq. (15.7)].

The slope of the Stern–Volmer plot is $k_2/k_1 = 27.6 \times 10^{-16} \text{ cm}^3 \text{ molecule}^{-1}$ and $k_2 = 55.2 \times 10^{-18} \text{ m}^3 \text{ molecule}^{-1} \text{ s}^{-1} = \sigma_Q (8kT/\pi\mu)^{1/2}$.

$$\mu = (18 \times 46/64 \times 6.02 \times 10^{23})10^{-3} = 2.15 \times 10^{-26} \text{ kg}$$

$$\sigma_Q = (55.2 \times 10^{-18} \text{ m}^3 \text{ molecule}^{-1} \text{ s}^{-1})$$
$$(3.14 \times 2.15 \times 10^{-26} \text{ kg}/8 \times 1.38 \times 10^{-23} \times 300 \text{ J})^{1/2}$$

$$= 7.9 \times 10^{-20} \text{ m}^2$$

This is about half the hard-sphere collision cross section, so that two or three collisions suffice to quench the fluorescence.

27.9 Dissociation and Predissociation

If a molecule absorbs a photon with an energy greater than that of one of its bonds, dissociation to atoms or free radicals may ensue. Several different pathways to dissociation are possible.

If a transition occurs from the ground state to an unstable excited state, the energy of the upper state is not quantized and the transition is seen in the spectrum as a continuous absorption band without vibrational or rotational fine structure. In Fig. 27.5, the excited state is a stable state with quantized energy levels. Electronic transitions from the ground state to upper states can be defined by the Franck–Condon Principle by vertical lines between potential-energy curves. When the final states have definite quantized energy levels, an absorption band with vibration-rotational fine structure will be seen, similar to those in Figs. 27.3 and 27.4. If, however, the final state reached in a Franck–Condon transition lies above the dissociation energy of the excited state, the molecule will dissociate within the period of a single vibration and the structure of the absorption band will disappear at the onset of a continuous absorption.

Absorption spectra are sometimes observed in which the onset of a continuum is not sharply defined, but preceded by a series of rotational lines that are more broad and diffuse then the usual lines. An example is shown in Fig. 27.11(a). This effect is called *predissociation*. Its explanation in terms of potential-energy curves of excited states is shown in Fig. 27.11(b). In addition to the stable excited state A, there is a nonbonding excited state B, whose potential-energy curve "crosses" that of the stable state. The vibrating molecule in A can sometimes make a radiationless transition to B, which is followed by dissociation. Such transitions reduce the average lifetime τ of molecules in the excited state, and in accord with the Heisenberg relation $\Delta\epsilon$ $\Delta\tau = h/2\pi$, increase the energy uncertainty in the level of the excited state. The vibrational levels of the upper states are not much affected, since the molecule can still

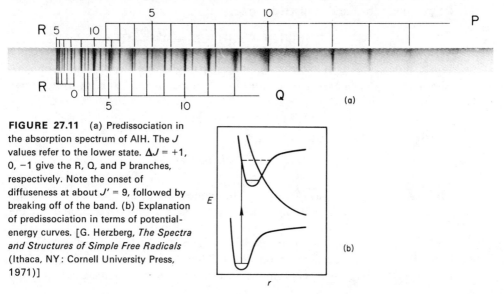

FIGURE 27.11 (a) Predissociation in the absorption spectrum of AlH. The J values refer to the lower state. $\Delta J = +1$, 0, −1 give the R, Q, and P branches, respectively. Note the onset of diffuseness at about $J' = 9$, followed by breaking off of the band. (b) Explanation of predissociation in terms of potential-energy curves. [G. Herzberg, *The Spectra and Structures of Simple Free Radicals* (Ithaca, NY: Cornell University Press, 1971)]

undergo several vibrations before it crosses to the nonbonding curve. The rotational period τ_r, however, is usually about 100 times longer than the vibrational period τ_v, and the molecule does not have time to establish a definite rotational state before it dissociates. As a result, the rotational lines in the absorption spectrum broaden markedly at the onset of predissociation. In the emission spectrum, predissociation is marked by the broadening and then complete disappearance of the rotational fine structure at certain J values.

27.10 Secondary Photochemical Processes

If a molecule is dissociated as a consequence of absorbing a quantum of radiation, extensive secondary reactions may occur, since the fragments are often highly reactive atoms or radicals. Sometimes, also, the products of the primary fission process are still in electronically or vibrationally excited states.

When a mixture of chlorine and hydrogen is exposed to light in the continuous region of the absorption spectrum of Cl_2 ($\lambda < 480$ nm), a rapid formation of hydrogen chloride ensues. The quantum yield Φ is 10^4 to 10^6. In 1918, Nernst explained the high value of Φ in terms of a long reaction chain. The first step is dissociation of a chlorine molecule,

$$(1) \quad Cl_2 + h\nu \longrightarrow 2Cl \qquad \Phi I_a$$

where I_a is the intensity of light absorbed. This is followed by

$$(2) \quad Cl + H_2 \longrightarrow HCl + H, \quad k_2$$
$$(3) \quad H + Cl_2 \longrightarrow HCl + Cl, \quad k_3$$
$$(4) \quad 2Cl \longrightarrow Cl_2 \text{ (on wall)} \, k_4$$

If we set up the steady-state expression for [Cl] and [H] in the usual way, (Section 13.16), we obtain for the rate of HCl production,

$$\frac{d[HCl]}{dt} = k_2[Cl][H_2] + k_3[H][Cl_2] = \frac{2k_2 \Phi I_a}{k_4}[H_2]$$

Instead of reaction (4), the chain-ending step might be a recombination of chlorine atoms in the gas phase with cooperation of a third body M to carry away excess energy.

$$(5) \quad Cl + Cl(+M) \longrightarrow Cl_2(+M) \qquad k_5$$

In this case, the calculated rate expression is

$$\frac{d[HCl]}{dt} = k_2 \left[\frac{\Phi I_a}{k_5[M]} \right]^{1/2} [H_2]$$

It is likely that reactions (4) and (5) both contribute to the ending of the chain under most experimental conditions, since in experiments with pure H_2 and Cl_2 the rate is proportional to I_a^n, with n somewhere between $\frac{1}{2}$ and 1.

In contrast with the high quantum yield of the $H_2 + Cl_2$ reaction are the low yields in the photochemical decompositions (photolyses) of alkyl iodides. These compounds have a region of continuous absorption in the near ultraviolet, which leads to a break into alkyl radicals and iodine atoms. For example, $CH_3I + h\nu \rightarrow$ $CH_3 + I$. The quantum yield of the photolysis is only about 10^{-2}. The reason for the low Φ is that the most likely secondary reaction is a recombination, $CH_3 + I \rightarrow CH_3I$. Only a few radicals react with another molecule of alkyl iodide, $CH_3 + CH_3I \rightarrow CH_4 + CH_2I$.

27.11 Flash Photolysis

The technique called *flash photolysis* is especially useful in the study of atoms and radicals that have only a short lifetime before reacting. A powerful flash of light, with energy up to 10^5 J and a duration of about 10^{-4} s, is obtained by discharging a bank of capacitors through an inert gas, such as argon or krypton. The reactants are in a vessel aligned parallel with the lamp, and at the instant of the flash, an extensive photolysis occurs. The primary products of the photolysis, usually radicals and atoms, are produced at concentrations much higher than those in an experiment with continuous illumination at low intensities. A good method for following the subsequent reactions of the radicals is to make a continuous record of their absorption spectra.

In the pioneering work of Ronald Norrish and George Porter, the spectra were recorded photographically at intervals following the flash. The first flash was followed by a second flash—the *specflash*—triggered electronically at precise short intervals after the photoflash. The specflash was arranged to photograph the absorption spectra of the products at successive times measured in the range of microseconds to milliseconds. Each point in a reaction sequence thus required a separate experiment. An example of the data obtained is shown in Fig. 27.12 on the formation and decay of ClO in mixtures of Cl_2 and O_2. This system is completely reversible; only the chloromonoxy radical ClO is formed in the photoflash, and it reacts completely to $Cl_2 +$

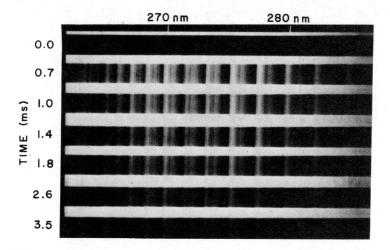

FIGURE 27.12 Time sequence of spectra of ClO in flash photolysis of a $Cl_2 + O_2$ mixture. The decay of [ClO] is bimolecular.

O_2 after the flash. The mechanism of formation of ClO is most likely through an intermediate complex ClOO, as

$$Cl_2 + h\nu \longrightarrow 2Cl, \quad Cl + O_2 \longrightarrow ClOO, \quad ClOO + Cl \longrightarrow ClO + ClO$$

The decay of ClO follows ideal bimolecular kinetics, $-d[ClO]/dt = k_2[ClO]^2$, with $k_2 = 4.8 \times 10^7 \ dm^3 \ mol^{-1} \ s^{-1}$ at 298 K.

27.12 Energy Transfer in Condensed Systems

In a condensed phase, energy can be transferred between molecules over considerable distances. Such long-range energy transfer was discovered in 1924 in exploratory work by Jean Perrin on the depolarization of fluorescence. When a dye in solution is irradiated with polarized light, the fluorescent light emitted is also polarized provided the dye is present in low concentration, but as the concentration of dye increases, the extent of polarization of the fluorescent light decreases. Perrin concluded that the molecule of dye that emits the quantum of depolarized fluorescent light is not the same as the molecule that absorbs the exciting quantum of polarized light. A transfer of electronic excitation energy can occur from one molecule to another over a considerable distance, up to 10 nm, in the solution.

In later work, two different solutes were included in the solution, and one was found to absorb light while the other emitted light. This phenomenon, called *sensitized fluorescence*, provided definite proof of the intermolecular transfer of energy of excitation. A clear-cut example is observed in a solution of 1-chloroanthracene and perylene, where most of the energy is absorbed by 1-chloroanthracene, whereas almost all the fluorescent emission is characteristic of perylene.

The theory of such long-range intermolecular transfers of energy was worked out, mainly by T. Förster, from 1948 onward. The transfer depends upon overlap between an emission band of the donor and an absorption band of the acceptor. The excited donor interacts with the unexcited receptor through a dipole–induced-dipole mechanism similar to that responsible for the London dispersion forces, to be discussed in Section 30.6. The interaction potential thus varies as R^{-6}, where R is the intermolecular distance.

When donor and receptor molecules are close together and can interact strongly, the Förster mechanism of energy transfer may be supplanted by the exciton mechanism of Davydov. An exciton is an electronic excitation in a crystal or in a regular array of molecules. An exciton can jump from one site to another (and back again) in a time that is short compared to the period of a molecular vibration. The difference between the Förster and Davydov mechanisms is summarized in Fig. 27.13.

Transfer of excitation energy is an essential part of the mechanism of photosynthesis in green plants. Each photosynthetic reaction center is associated with about 300 molecules of chlorophyll and other pigments, which act as harvesters of light, collecting quanta of light energy and rapidly transferring the excitation to the reaction center in the membrane of the chloroplast. Both the Förster and the Davydov mechanisms probably contribute to the energy transfer.

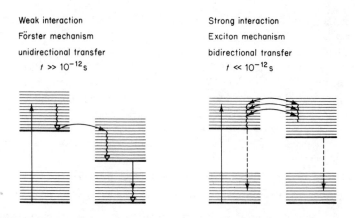

Weak interaction
Förster mechanism
unidirectional transfer
$t \gg 10^{-12}\,s$

Strong interaction
Exciton mechanism
bidirectional transfer
$t \ll 10^{-12}\,s$

FIGURE 27.13 Förster and Davydov mechanisms for energy transfer in condensed media.

Problems

1. The absorbance of a 5.3×10^{-5} M solution of Fe(III) cytochrome c at 530 nm and pH 6.8 was 0.54 in a cell with 1 cm path length. Calculate the molar absorptivity in $m^2\,mol^{-1}$ and $dm^3\,mol^{-1}\,cm^{-1}$.

2. If a solution contains two solutes x and y which each follow the Lambert–Beer Law, one can analyze the mixture by measuring the absorbance A_1, A_2 at two different wavelengths. Show that the concentrations of x and y are

$$[x] = D(\epsilon_{2y}A_1 - \epsilon_{1y}A_2)$$
$$[y] = D(\epsilon_{1x}A_2 - \epsilon_{2x}A_1)$$

where $D = (\epsilon_{1x}\epsilon_{2y} - \epsilon_{2x}\epsilon_{1y})^{-1}$.

3. The amino acids tryptophan (W) and tyrosine (Y) have the following molar absorptivities at pH 12.

W: $\epsilon(240\,nm) = 1960$, $\epsilon(280\,nm) = 5380$ M^{-1} cm^{-1}

Y: $\epsilon(240\,nm) = 11\,300$, $\epsilon(280\,nm) = 1500$ M^{-1} cm^{-1}

A solution of the two amino acids, measured in a 1-cm spectrophotometer cell, had $A(240) = 0.680$, $A(280) = 0.252$. Calculate the concentrations of W and Y in the solution. (See Problem 2.)

4. Suppose that in a given molecule there is a UV absorption band and an IR absorption band and they both have about the same oscillator strength f. How will the absorption intensities compare? What are the physical reasons for this difference?

5. Figure 27.4 shows the (0, 0) band of Si^{14}N. If the spectrum is taken with Si^{15}N, the band origin is observed to shift toward the violet. What can you conclude about the relative vibration frequencies in the upper and lower electronic states of SiN?

6. On the basis of the Franck–Condon Principle, draw potential-energy curves for ground and excited electronic states that would explain the following observation: A molecule has a broad structureless absorption band but its emission spectrum is quite sharp with a definite line structure.

7. What is the energy in kJ mol^{-1} of 1 einstein of ultraviolet light from a mercury lamp at 253.7 nm?

8. Propionaldehyde, $CH_3 \cdot CH_2 \cdot CHO$, is irradiated at 27.0 kPa and 300 K with radiation at 302 nm. The quantum yield for CO production is 0.54. If light absorption is 3.00 mW, calculate the rate of CO formation. What is the light absorption in einsteins per second?

9. The binding energy of O_2 is 5.16 eV per molecule. What is the maximum wavelength of radiation that could possibly dissociate an O_2 molecule into its atoms in their ground states? If O_2 is dissociated by UV light at $\lambda = 150$ nm, what is the average kinetic energy of the O atoms formed? At what temperature would O atoms have this average kinetic energy?

10. The reaction of photosynthesis in the presence of chlorophyll can be schematically written as $CO_2 + H_2O + xh\nu \longrightarrow \rangle CHOH + O_2$. The ΔU of reaction is 502 kJ mol^{-1}. Chlorophyll has an absorption maximum for light of $\lambda = 594$ nm. How many quanta x of this light would be required for the photosynthesis reaction?

11. The maximum intensity in sunlight is in the green region of the spectrum around 600 nm, but chlorophyll, being green, does not absorb appreciably in this region. How would you explain this situation from an evolutionary (or teleological) standpoint?

12. At 200°C, N_2 at $P = 210$ Pa reduces the fluorescence intensity of excited Na by 50%. If the natural lifetime of the excited Na atom is 10^{-7} s, calculate the quenching cross section for N_2.

13. The quantum yield for inactivation of the T1 bacteriophage is 3×10^{-4} at 260 nm. If the phage has a molar mass $M = 10^7$ g mol^{-1} and a specific volume 0.65 cm^3 g^{-1}, estimate the target cross section of the segment of phage at which absorption of a quantum has a lethal effect.

14. The photochemical chlorination of chloroform, $CHCl_3 + Cl_2 = HCl + CCl_4$, follows the rate equation $d[CCl_4]/dt = k'I_a^{1/2}[Cl_2]^{1/2}[CHCl_3]$, where I_a is the intensity of light absorbed. Devise a reaction scheme to explain these kinetics. If the reaction is studied with an intermittent (flashing) light source, the quantum yield at very slow flashing rate is one half that at very fast flashing rate. Explain.

15. What is the selection rule for electric dipole transitions of electrons in a one-dimensional box?

16. The effective cross section of H_2 for quenching fluorescence of mercury "resonance radiation" ($^3P \longrightarrow {}^1S$) is 6.07×10^{-20} m^2. If the fluorescent intensity in the absence of H_2 is 100, what will it be at a pressure of $H_2 = 10^2$ Pa at 300 K? The natural lifetime of the Hg ($^3P_1$) is 10^{-8} s.

17. The quenching reaction in Problem 16 is $Hg + H_2 = HgH + H$. What is the rate of production of H atoms in a vessel in which H_2 at 1 kPa and 300 K containing mercury vapor is irradiated with a 100-W Hg discharge lamp which emits the Hg resonance line (253.7 nm) with 10% efficiency if 1% of the UV radiation is absorbed in the reaction vessel?

18. A photographic plate is exposed for 10^{-3} s to a 10^3 W flash (600 nm effective wavelength) at a distance of 100 m. If 50% of the power is emitted as light to which the plate is sensitive, estimate the number of silver atoms that will be produced in a AgBr grain 10 μm in diameter.

28

Crystallography

In 1669, Niels Stensen, Professor of Anatomy at Copenhagen and Vicar Apostolic of the North, compared the interfacial angles in a collection of quartz crystals. He found that the corresponding angles in different crystals are always the same. After the invention of the contact goniometer in 1780, this conclusion was extended to other substances, and the constancy of interfacial angles has been called the *First Law of Crystallography*.

A crystal grows by the deposition of molecules or ions onto its faces. The faces with the largest area are those on which added particles deposit most slowly. An altered rate of deposition can change the form, or *habit*, of a crystal. Sodium chloride grows from water solution as cubes, but from 15% aqueous urea as octahedra. Urea is preferentially adsorbed on the octahedral faces, preventing deposition of sodium and chloride ions, and thereby causing these faces to extend in area as the crystals grow.

In 1665, Robert Hooke speculated on the reason for the regular forms of crystals, and decided that they were the consequence of a regular packing of small spherical particles.

> So I think, had I time and opportunity, I could make probable that all these regular Figures that are so conspicuously various and curious, and do so adorn and beautify such multitudes of bodies . . . arise only from three or four several positions or postures of Globular particles. . . .

The discovery of X-rays in 1895 was followed in 1912 by Max von Laue's inspiration to examine their diffraction by the regular array of atoms believed to exist in crystals. The subsequent progress of X-ray crystallography has been one of the most exciting scientific developments of the past 70 years.

28.1 Crystal Planes and Faces

The faces of crystals, and also planes within crystals, can be referred to a set of three noncoplanar axes. Consider in Fig. 28.1(a) three axes having lengths a, b, and c, which are cut by the plane ABC, making intercepts OA, OB, and OC. If a, b, and c are chosen as unit lengths, the length ratios of the intercepts can be expressed as OA/a, OB/b, OC/c. The reciprocals of these length ratios will then be a/OA, b/OB, c/OC. It is always possible to find a set of axes on which the reciprocal intercepts of crystal faces are small whole numbers. Thus, if h, k, and ℓ are small integers,

$$\frac{a}{OA} = h, \qquad \frac{b}{OB} = k, \qquad \frac{c}{OC} = \ell$$

This is equivalent to the Law of Rational Intercepts, first enunciated by R. J. Haüy in 1783. The use of reciprocal intercepts $(hk\ell)$ as indices to specify crystal faces was proposed by W. H. Miller in 1839. If a face is parallel to an axis, the intercept is at ∞, and the *Miller index* becomes $1/\infty$ or 0. The notation is also applicable to planes drawn within a crystal. As an illustration of the Miller indices, some of the planes in a cubic crystal are shown in Fig. 28.1(b).

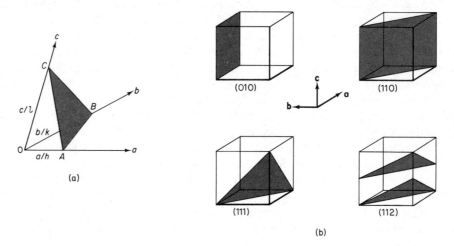

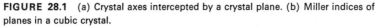

FIGURE 28.1 (a) Crystal axes intercepted by a crystal plane. (b) Miller indices of planes in a cubic crystal.

Example 28.1 A crystal face intercepts the three axes of a crystal at the following multiples of the axial lengths: 3/2, 2, and 1. What are the Miller indices of the face?

The reciprocal intercepts are 2/3, 1/2, and 1, which are in the integral ratio 4: 3: 6, so that it is a (436) face.

In the Miller index for a crystal face, only the ratio $h:k:\ell$ is significant. Thus (420) is the same face as (210). For planes within crystals, multiplication of the Miller index by an integer changes the interplanar spacing. Thus the planes 420 include all the planes 210, and in addition a set of planes midway between them. In crystallographic notation, $(hk\ell)$ refers to a crystal face and $hk\ell$ (without parentheses) to a set of planes. A negative index is written above the numeral, as in $(\bar{1}10)$. Curly brackets are used to designate all the equivalent faces of a crystal, i.e., a *form* of a crystal. For example, we say that cubic NaCl has the {100} form, and octahedral NaCl has the {111} form.

28.2   Crystal Systems

According to the set of axes used to represent their faces, crystals are assigned to seven *systems*, summarized in Table 28.1. They range from the completely general set

TABLE 28.1
THE SEVEN CRYSTAL SYSTEMS

System	Axes	Angles	Example
Triclinic	$a; b; c$	$\alpha; \beta; \gamma$	Potassium dichromate
Monoclinic	$a; b; c$	$\alpha = \gamma = 90°; \beta$	Monoclinic sulfur
Orthorhombic	$a; b; c$	$\alpha = \beta = \gamma = 90°$	Orthohombic sulfur
Rhombohedral	$a = b = c$	$\alpha = \beta = \gamma$	Calcite
Hexagonal	$a = b; c$	$\alpha = \beta = 90°; \gamma = 120°$	Graphite
Tetragonal	$a = b; c$	$\alpha = \beta = \gamma = 90°$	White tin
Cubic	$a = b = c$	$\alpha = \beta = \gamma = 90°$	Rocksalt

of three unequal axes (a, b, c) at three unequal angles (α, β, γ) of the triclinic system, to the highly symmetrical set of three equal axes at right angles of the cubic system. Examples of crystals belonging to each of the systems are shown in Fig. 28.2.

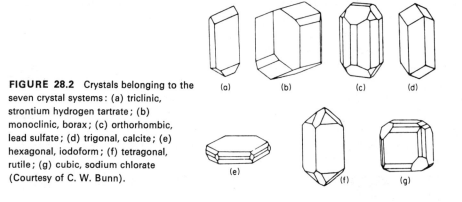

FIGURE 28.2 Crystals belonging to the seven crystal systems: (a) triclinic, strontium hydrogen tartrate; (b) monoclinic, borax; (c) orthorhombic, lead sulfate; (d) trigonal, calcite; (e) hexagonal, iodoform; (f) tetragonal, rutile; (g) cubic, sodium chlorate (Courtesy of C. W. Bunn).

An essential geometric concept in crystallography is the *lattice*. A lattice is an infinite regular array of *points* in space; the surroundings as viewed from any point are identical. A three-dimensional lattice can be described by three primitive translation vectors **a**, **b**, and **c**, such that the translation vector **r** from one lattice point to any other can be written in terms of the primitive vectors as

$$\mathbf{r} = u\mathbf{a} + v\mathbf{b} + w\mathbf{c} \tag{28.1}$$

where u, v, and w are integers or zero.

The three primitive translations can be used to define the edges of a parallelepiped which is called a *primitive unit cell* of the lattice. This unit cell can be repeated so as to cover the entire lattice. The same lattice can be broken up into unit cells in an infinite number of ways. For example, two different primitive unit cells are shown for the two-dimensional lattice in Fig. 28.3(a). Each primitive unit cell contains one lattice point, since each of the four corners of the cell is shared among four cells. Thus the number of points in each cell is $(\frac{1}{4})(4) = 1$.

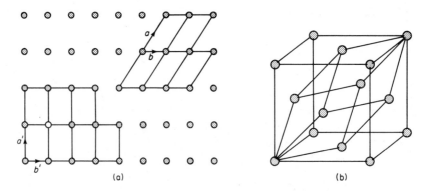

FIGURE 28.3 (a) Different primitive unit cells of a plane lattice. (b) Points on a face-centered cubic lattice can also be located on a primitive rhombohedral lattice.

A lattice can also be used to define nonprimitive unit cells. An example is shown in Fig. 28.3(b) in which the same lattice is represented by a face-centered cubic unit cell and by a primitive rhombohedral unit cell. The nonprimitive cell contains more than one lattice point. For example, the face-centered cubic lattice in Fig. 28.3(b) contains one point at its vertices [8 vertices each shared between 8 cells: $(8)(1/8) = 1$] and three on its faces [6 faces each shared between 2 cells: $(6)(\frac{1}{2}) = 3$], so that the number of points in this unit cell is 4. Nonprimitive unit cells are used in crystallography because they may have higher symmetries than the primitive unit cells.

In 1848, A. Bravais showed that all possible space lattices can be assigned among only 14 types. The 14 *Bravais lattices* are shown in Fig. 28.4. The choice of the 14 lattices is somwhat arbitrary, since in certain cases alternative descriptions are possible, but at least 14 different lattices are needed.

The location of any point within a unit cell is described by giving its coordinates

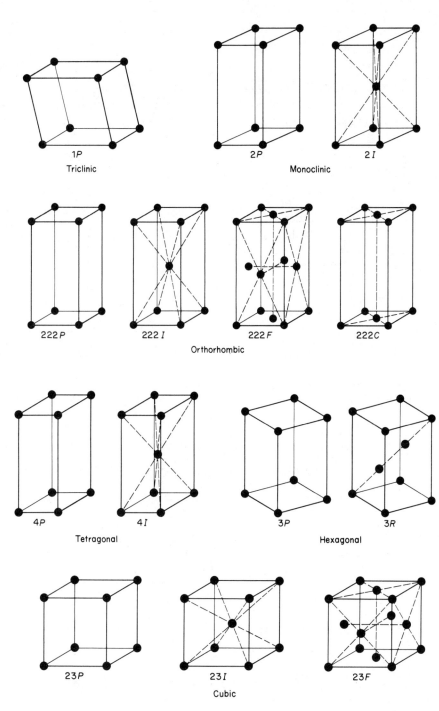

FIGURE 28.4 The 14 Bravais lattices.

$(x/a, y/b, z/c)$ as fractions of the lengths of the vectors that define the unit cell. For example, the face-centered positions in a fcc unit cell are $(\frac{1}{2}\frac{1}{2}0)$, $(\frac{1}{2}0\frac{1}{2})$ and $(0\frac{1}{2}\frac{1}{2})$; the body-centered position in a bcc unit cell is $(\frac{1}{2}\frac{1}{2}\frac{1}{2})$.

The volume of a unit cell is given by

$$V = abc(1 - \cos^2 \alpha - \cos^2 \beta - \cos^2 \gamma + 2 \cos \alpha \cos \beta \cos \gamma)^{1/2} \qquad (28.2)$$

The distance between planes $hk\ell$ in a unit cell is readily calculated for cells with orthogonal axes (cubic, tetragonal, orthorhombic). The calculation for two dimensions is shown in Fig. 28.5. Extension to three dimensions gives

$$d_{hk\ell} = \sqrt{\frac{h^2}{a^2} + \frac{k^2}{b^2} + \frac{\ell^2}{c^2}} \qquad (28.3)$$

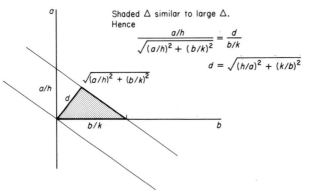

FIGURE 28.5 Derivation of formula for interplanar spacing in crystal systems with orthogonal axes.

Example 28.2 Iron crystallizes in a body-centered cubic structure with the length of the unit cell $a_0 = 286.1$ pm at 25°C. What is the nearest distance between iron atoms?

The nearest distance is between the corner atom at (0 0 0) and the body-centered atom at $(\frac{1}{2}\frac{1}{2}\frac{1}{2})$. The distance between any two points in a system of rectangular coordinates is

$$d = \sqrt{(x_1 - x_2)^2 + (y_1 - y_2)^2 + (z_1 - z_2)^2}$$

In this example,

$$d = (\sqrt{(\tfrac{1}{2} - 0)^2 + (\tfrac{1}{2} - 0)^2 + (\tfrac{1}{2} - 0)^2})\, a_0$$
$$= (\sqrt{3}/2)(286.1) = 247.8 \text{ pm}$$

28.4 Symmetry Properties—Crystal Classes

Closer consideration of the crystal systems or Bravais lattices reveals a curious and important fact. The only kinds of axes of symmetry that occur are C_2, C_3, C_4, or C_6. We never see a crystal with a C_5, C_7, C_8, or indeed any other axial symmetry except the four mentioned. As we saw in Chapter 25, C_5, C_7, and other axes do occur in isolated molecules. Ferrocene, for example, is a well known molecule with a C_5 symmetry axis. Why do C_5 axes occur in molecules but not in crystals? The reason is that it is impossible to fill all of space with figures of fivefold symmetry. We can see this result in Fig. 28.6. in two dimensions. It is possible to tile a floor with parallel-ograms (C_2), equilateral triangles (C_3), squares (C_4), or regular hexagons (C_6). It is impossible to tile a floor with regular pentagons, heptagons, and so on, without leaving gaps in the tiling. The fact that actual crystals never display axial symmetry elements other than C_2, C_3, C_4 and C_6, therefore leads to the following important conclusion: Crystals must be constructed of regular subunits that *fill all space* in a definite geometric array. The regular crystal forms observed in nature are outward manifestations of inner regularities in structure. The significance of the 14 Bravais lattices is now more evident—they are the 14 possible ways of arranging points in a regular array *so as to fill all space.*

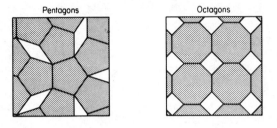

Pentagons   Octagons

FIGURE 28.6 It is not possible to cover an area completely with regular pentagons or octagons.

Symmetry elements in crystals are therefore restricted to the following: m, C_2, C_3, C_4, C_6, i, E. The crystallographic point groups are restricted to groups that can be formed from these symmetry elements. There are exactly 32 crystallographic point groups, which specify the 32 *crystal classes*. All crystals necessarily fall into one of the seven systems, but there are several classes in each system. Only one of these, called the *holohedral class*, possesses the complete symmetry of the system.

28.5   Crystal Structures

A crystal structure is a regular array of identical repeating units consisting of atoms or groups of atoms. A repeating unit of this kind is called the *basis* of the crystal structure. The *crystal structure* arises if the basis is assigned to each point of the lattice to which the structure belongs. We can write

$$\text{lattice} + \text{basis} \longrightarrow \text{crystal structure}$$

It is desirable to learn the correct definitions of *lattice* and *crystal structure*, and not to use the term *lattice* incorrectly when you mean an array of atoms, ions or molecules in a crystal *structure*.

The basis of a crystal structure can be a single atom. For example, the structure of iron is based upon a body-centered cubic (bcc) lattice in which each point is occupied by an iron atom, [Fig. 28.7(a)]. The basis consists of one Fe atom. The lattice then places Fe atoms at $(0\ 0\ 0, \frac{1}{2}\frac{1}{2}\frac{1}{2})$ in the unit cell. The structure of α-manganese is also based upon a bcc lattice, but here the basis contains 29 Mn atoms, as shown in Fig. 28.7(b).

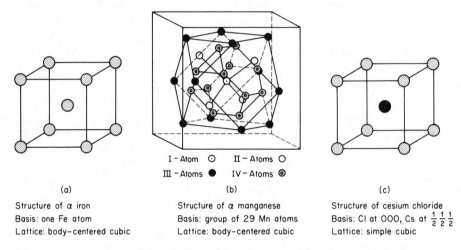

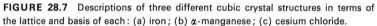

(a)

Structure of α iron
Basis: one Fe atom
Lattice: body–centered cubic

(b)

Structure of α manganese
Basis: group of 29 Mn atoms
Lattice: body–centered cubic

(c)

Structure of cesium chloride
Basis: Cl at OOO, Cs at $\frac{1}{2}\frac{1}{2}\frac{1}{2}$
Lattice: simple cubic

FIGURE 28.7 Descriptions of three different cubic crystal structures in terms of the lattice and basis of each: (a) iron; (b) α-manganese; (c) cesium chloride.

A basis can also consist of a group of atoms of more than one kind. For example, the crystal structure of CsCl is based on a simple cubic lattice and the basis is a CsCl unit, as shown in Fig. 28.7(c). Note that CsCl is *not* based upon a body-centered cubic lattice, and it should not be referred to as a bcc structure. The basis of the CsCl structure is [Cs (0 0 0), Cl($\frac{1}{2}\frac{1}{2}\frac{1}{2}$)]. The basis is described by giving the coordinates and identities of the atoms that comprise it.

The full symmetry of a lattice is modified in the crystal structure, as a result of replacing geometrical points by groups of atoms. Since these groups need not have so high a symmetry as the original lattice, classes of lower than holohedral symmetry can arise within each crystal system.

Example 28.3 The density ρ of α-Mn at 298 K is 7.40×10^3 kg m^{-3}. What is the length a of the unit cell in Fig. 28.7(b)?

The unit cell contains $Z = 2 \times 29$ Mn atoms with molar masses $M = 54.94 \times 10^{-3}$ kg mol^{-1}. Thus

$$a^3 = \frac{ZM}{L\rho} = \frac{(58)(54.94 \times 10^{-3} \text{ kg mol}^{-1})}{(6.02 \times 10^{23} \text{ mol}^{-1})(7.40 \times 10^{-3} \text{ kg m}^{-3})}$$

$$= 7.153 \times 10^{-22} \text{ m}^3$$

$$a = 8.94 \times 10^{-8} \text{ m} = 89.4 \text{ nm}$$

28.6 Space Groups

The crystal classes are the possible groups of symmetry operations of finite figures, e.g., actual crystals. They are made up of operations by symmetry elements that leave at least one point in the crystal invariant. This is why they are called *point groups*.

In a crystal structure, considered as an infinitely extended pattern in space, new types of symmetry operation are admissible, which leave no point invariant. These are called *space operations*. The new symmetry operations involve translations in addition to rotations and reflections. Only an infinitely extended pattern can have a space operation (translation) as a symmetry operation. The new symmetry elements that result from combination of translation with the symmetry operations of the point groups are *glide planes* and *screw axes*. A glide plane causes a reflection across a plane followed by a specified translation parallel to the plane. A screw axis combines rotation about an axis with a translation in the direction of the axis for a distance equal to some fraction of the lattice spacing in that direction.

The possible groups of symmetry operations of infinite figures are called *space groups*. They arise from combinations of the 14 Bravais lattices with the 32 point groups. There are exactly 230 possible crystallographic space groups. A space group may be visualized as a crystallographic kaleidoscope. If one structural unit is introduced into the group, the operations of the space group immediately generate the entire crystal structure, just as the mirrors of a kaleidoscope produce a symmetrical pattern from a few bits of colored paper. The space group expresses the totality of the symmetry properties of a crystal structure, and mere external form or bulk properties do not suffice for its determination. One must make a determination of the inner structure of the crystal, and this is made possible by the methods of X-ray diffraction.

28.7 X-Ray Crystallography

Physicists at the University of Munich in 1912 were interested in both crystallography and X rays. P. P. Ewald and Arnold Sommerfeld were studying the passage of light waves through crystals. In a discussion of this work, Max von Laue, a theoretical physicist, pointed out that if the wavelength of the radiation became about as small as the distance between atoms in crystals, a diffraction pattern should result. There was some evidence that X rays might have a wavelength in this range, and W. Friedrich and P. Knipping agreed to make an experimental test. An X-ray beam passed through a crystal of copper sulfate gave a definite diffraction pattern. Figure 28.8 shows a modern example of an X-ray diffraction pattern obtained by the Laue method. The wave properties of X rays were thus definitely established and the new science of X-ray crystallography began.

The Laue method uses a continuous spectrum of X radiation over a wide range of wavelengths, called *white radiation*, conveniently obtained from a tungsten target

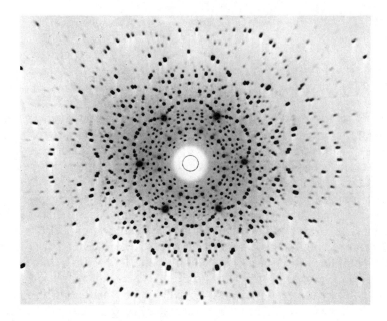

FIGURE 28.8 X-ray diffraction pattern from a crystal of beryl by the Laue method. [Eastman Kodak Research Laboratories]

at high voltages. In this case, at least some of the radiation should be at the proper wavelength to experience interference effects, no matter what the crystal-to-beam orientation.

28.8 The Bragg Analysis of X-ray Diffraction

When the news of the Munich work reached England, it was taken up by William Bragg and his son Lawrence, who had been working with X rays. Lawrence Bragg, using photographs of the Laue patterns, analyzed the structures of NaCl, KCl, and ZnS (1912, 1913). In the meantime (1913) the elder Bragg devised a spectrometer that measured the intensity of an X-ray beam by the amount of ionization it produced, and he found that the X-ray line spectrum characteristic of the element of the target could be used for crystallographic work. Thus the Bragg method uses a monochromatic (single wavelength) beam of X rays.

Example 28.4 The wavelength of PdK_α X rays is 58.6 pm. What is the minimum voltage $\Delta\Phi$ that must be applied between cathode and anode (the target) of an X-ray tube to produce these X rays?

The energy of an electron (charge, $-e$) accelerated through a voltage drop $-\Delta\Phi$ is $e\Delta\Phi$. If we neglect small energy losses in the target, the X-ray photon will have

---

energy $h\nu = hc/\lambda = e\Delta\Phi$. Thus

$$\Delta\Phi = \frac{hc}{\lambda e} = \frac{(6.63 \times 10^{-34} \text{ J s})(3.00 \times 10^8 \text{ m s}^{-1})}{(58.6 \times 10^{-12} \text{ m})(1.60 \times 10^{-19} \text{ C})}$$

$$= 2.12 \times 10^4 \text{ V}$$

The Braggs showed that the scattering of X rays could be represented as a "reflection" by successive planes of atoms in the crystal. Consider, in Fig. 28.9, a set of parallel planes in the crystal structure and a beam of X rays incident at an angle θ. When the incident X-ray beam is inclined at a certain angle θ to the set of planes, the diffracted X rays form a beam for which the angle of "reflection" equals the angle of incidence. The condition for a maximum in the intensity of diffracted X rays is that the difference δ in length of path between waves scattered from successive planes must be an integral number of wavelengths, $n\lambda$. If we consider the "reflected" waves at the point P, this path difference for the first two planes is $\delta = \overline{AB} - \overline{BC}$. Since $\overline{AB} = \overline{BD}$, $\delta = \overline{CD} = \overline{AD} \sin \theta = 2d \sin \theta$. The condition for reinforcement or Bragg "reflection" is thus

$$n\lambda = 2d \sin \theta \qquad (28.4)$$

According to this equation, there are different *orders* of "reflection" specified by the values $n = 1, 2, 3, \ldots$. The second-order diffraction from 100 planes may also be ascribed to a set of planes 200 with half the spacing of the 100 planes.

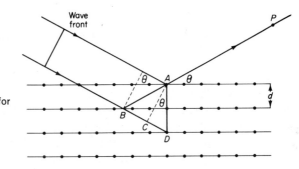

FIGURE 28.9 The Bragg derivation for the diffraction condition by monochromatic X-rays. In X-ray diffraction, a million planes may contribute to a diffraction maximum.

The Bragg equation indicates that for any given wavelength of X rays there is a lower limit to the spacings that can give observable diffraction spectra. Since the maximum value of $\sin \theta$ is 1, this limit is given by $d_{\min} = n\lambda/2 \sin \theta_{\max} = \lambda/2$.

28.9 Structures of Sodium and Potassium Chlorides

In the work of Bragg, a single crystal of NaCl or KCl was mounted on the diffractometer, as shown in Fig. 28.10, so that the X-ray beam was incident on one of the important crystal faces, (100), (110), or (111). The scattered beam entered the ionization chamber, which was filled with methyl bromide vapor.

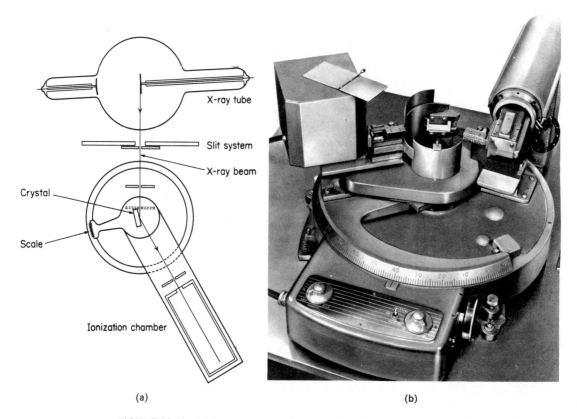

(a)   (b)

FIGURE 28.10 (a) Bragg X-ray spectrometer. (b) A modern version of the Bragg X-ray diffractometer. The head of the X-ray tube is at the right. The diffracted beam is measured by the counter at the left.

The experimental data are plotted in Fig. 28.11 as "intensity of reflected beam" vs. "twice the angle of incidence of beam to crystal." As the crystal is rotated, successive maxima "flash out" as angles are passed that conform to the Bragg condition, Eq. (28.4). In these first experiments the monochromatic X-radiation was obtained from a palladium target, but the wavelength λ was not known. From the external form of the crystals it is evident that both NaCl and KCl belong to the cubic system, and thus are based on one of the three cubic Bravais lattices, primitive, face-centered (fcc), or body-centered (bcc). By comparing spacings calculated from the X-ray data with those expected for these lattices, the proper assignment can be made.

From Eq. (28.3), the distance between the planes $hk\ell$ in a cubic lattice is

$$d_{hk\ell} = \frac{a_0}{(h^2 + k^2 + \ell^2)^{1/2}} \tag{28.5}$$

When combined with the Bragg equation (28.4), Eq. (28.5) gives

$$\sin^2 \theta = (\lambda^2/4a^2)(h^2 + k^2 + \ell^2) \tag{28.6}$$

Thus each observed value of $\sin \theta$ can be indexed by assigning to it the value of $hk\ell$ for the set of planes responsible for the diffraction. For a cubic lattice, interplanar spacings occur as follows:

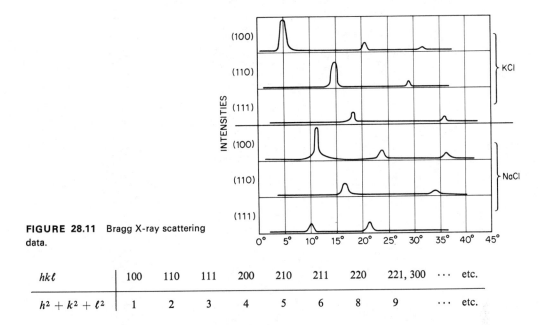

$hk\ell$	100	110	111	200	210	211	220	221, 300	⋯	etc.
$h^2 + k^2 + \ell^2$	1	2	3	4	5	6	8	9	⋯	etc.

Figure 28.12(a) shows the 100, 110, and 111 planes in a simple cubic lattice. A structure may be based on this lattice by replacing each lattice point by an atom. If an X-ray beam strikes such a structure at the Bragg angle, the rays scattered from one 100 planes are exactly in phase with the rays from successive 100 planes. A strong scattered beam is obtained for the first-order "reflection" from the 100 planes. A similar result is obtained for the 110 and 111 planes. A structure based on a simple cubic lattice gives a diffraction maximum from each set of planes $hk\ell$ since for any given $hk\ell$ all the atoms are included in these planes. The X-ray diffraction from a simple cubic crystal, plotted as intensity vs. $\sin^2 \theta$, gives a series of six equidistant maxima. Then the seventh is missing, since there is no set of integers $hk\ell$ such that $h^2 + k^2 + \ell^2 = 7$. Then follow seven equidistant maxima, with the fifteenth missing and so on (see Table 28.2).

Figure 28.12(b) shows a structure based on a body-centered cubic lattice. Exactly midway between any two 100 planes, there lies another layer of atoms. When X rays

TABLE 28.2
CALCULATED AND OBSERVED DIFFRACTION MAXIMA FOR $hk\ell$ PLANES

$hk\ell$ $h^2 + k^2 + \ell^2$	100 1	110 2	111 3	200 4	210 5	211 6	220 8	300, 221 9
Simple cubic	|	|	|	|	|	|	|	|
Body-centered cubic	⋮	|	⋮	|	⋮	|	|	⋮
Face-centered cubic	⋮	⋮	|	|	⋮	⋮	|	⋮
Sodium chloride	⋮	⋮	|	|	⋮	⋮	|	⋮
Potassium chloride	⋮	⋮	⋮	|	⋮	⋮	|	⋮

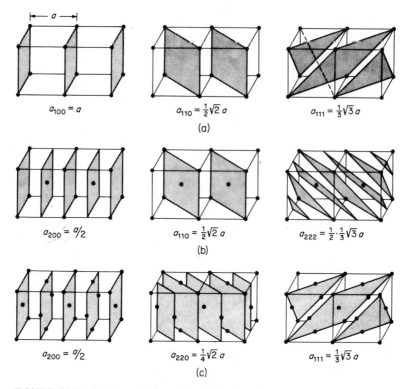

$a_{100} = a$ $a_{110} = \frac{1}{2}\sqrt{2}\,a$ $a_{111} = \frac{1}{3}\sqrt{3}\,a$

(a)

$a_{200} = a/2$ $a_{110} = \frac{1}{2}\sqrt{2}\,a$ $a_{222} = \frac{1}{2}\cdot\frac{1}{3}\sqrt{3}\,a$

(b)

$a_{200} = a/2$ $a_{220} = \frac{1}{4}\sqrt{2}\,a$ $a_{111} = \frac{1}{3}\sqrt{3}\,a$

(c)

FIGURE 28.12 Spacings of planes in cubic lattices: (a) simple cubic; (b) body-centered cubic; (c) face-centered cubic.

scattered from the 100 planes are in phase and reinforce one another, the rays scattered by the interleaved atomic planes are retarded by half a wavelength, and hence are exactly out of phase with the others. The observed intensity is therefore the difference between the scattering from the two sets of planes and the resultant intensity is reduced to zero by the destructive interference, so that no first-order 100 reflection appears.

The *second-order* diffraction from the 100 planes, occurring at the Bragg angle with $n = 2$ in Eq. (28.4), can equally well be expressed as the scattering by a set of planes, called the 200 planes, with just half the spacing of the 100 planes. In the body-centered cubic structure, all the atoms lie in these 200 planes, so that all the scattering is in phase, and a strong scattered beam is obtained. The same situation holds for the 111 planes: The first order 111 are extinguished, but the second-order 111, i.e., 222 planes, give strong scattering. The 110 planes pass through all the atoms in the bcc structure and hence a strong first-order 110 diffraction is observed. If we examine successive planes $hk\ell$ we find for the bcc structure the results shown in Table 28.2, in which planes that do not give diffraction maxima are indicated by dotted lines.

Table 28.2 also shows the expected lines for a face-centered cubic crystal and the observed results for NaCl and KCl. The Braggs found the explanation of these diffraction spectra from NaCl and KCl in an 1897 paper by William Barlow, who had been curator of crystals at the London Museum. Barlow showed different ways in which spheres of equal or unequal sizes can be packed together and one of his structures for an AB-type compound was equivalent to the NaCl structure shown in Fig. 28.13. The NaCl structure is based upon a face-centered cubic lattice. The basis of the structure is (Na$^+$ at 000, Cl$^-$ at $\frac{1}{2}$00).

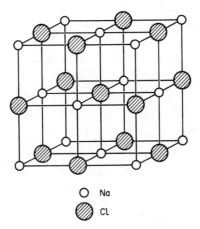

FIGURE 28.13 Rock salt structure. Both NaCl and KCl have this structure. It is based on a face-centered cubic lattice. The basis is an NaCl unit, Na(0, 0, 0), Cl($\frac{1}{2}$, 0, 0).

○ Na

◍ Cl

In this structure, the 100 and 110 planes contain equal numbers of Na$^+$ and Cl$^-$ ions, but the 111 planes contain either all Na$^+$ or all Cl$^-$ ions. When X rays scattered from the planes of Cl$^-$ ions are exactly in phase and meet the Bragg condition, scattering from the interleaved planes of Na$^+$ ions is exactly out of phase. The Cl$^-$ ions scatter X rays more strongly than the Na$^+$, and the observed intensity is proportional to the difference between the scattering intensities. Thus the lowered intensity of the first order diffraction from the 111 planes of NaCl is explained by the proposed structure.

The KCl structure was known to be the same as that of NaCl since NaCl and KCl form a continuous series of solid solutions, and thus Na$^+$ can replace K$^+$ in the structure in any proportion. K$^+$ and Cl$^-$ contain the same number of electrons, and hence their scattering intensities for X rays are almost identical. Thus the first, third, and all odd-order spectra from the 111 planes are entirely extinguished.

Example 28.5 Calculate the wavelength of the X rays used to obtain the data on NaCl in Fig. 28.10. The density of crystalline NaCl, $\rho = 2163$ kg m^{-3} at 25°C.

The molar volume is

$$M/\rho = \frac{58.45 \times 10^{-3} \text{ kg mol}^{-1}}{2163 \text{ kg m}^{-3}} = 27.02 \times 10^{-6} \text{ m}^3 \text{ mol}^{-1}$$

Then the volume occupied by each NaCl unit is

$$\frac{27.02 \times 10^{-6} \text{ m}^3 \text{ mol}^{-1}}{6.022 \times 10^{23} \text{ mol}^{-1}} = 44.87 \times 10^{-30} \text{ m}^3$$

In the unit cell of NaCl, there are eight Na$^+$ ions at the corners of the cube, each shared between eight cubes, and six Na$^+$ ions at the face centers, each shared between two cubes. Thus, per unit cell, there are $\frac{8}{8} + \frac{6}{2} = 4$ Na$^+$ ions. There is an equal number of Cl^- ions, and therefore four NaCl units per unit cell ($Z = 4$). The volume of the unit cell is, therefore, $4 \times 44.87 \times 10^{-30} \text{ m}^3 = 179.48 \times 10^{-30} \text{ m}^3$. The interplanar spacing for the 200 planes is $d_{200} = \frac{1}{2}a = \frac{1}{2}(179.48 \times 10^{-30} \text{ m}^3)^{1/3} = 2.82 \times 10^{-10} \text{ m} = 282 \text{ pm}$. When this value and the observed diffraction angle are put into the Bragg equation, the wavelength of the Pd–$K_{\alpha 1}$ X ray is $\lambda = 2(282) \sin 5°58' = 58.5 \text{ pm}$.

Example 28.6 Fluorite has a face-centered cubic structure with 4 CaF_2 groups in the unit cell. At 25°C the (111) reflection with X rays of $\lambda = 154.2$ pm occurs at $\theta = 14.18°$. Calculate the length of the unit cell and the density of fluorite at 25°C.

From $\lambda = 2d \sin \theta$ and $d_{111} = a_0/\sqrt{3}$,
$a_0 = \sqrt{3} \, \lambda/2 \sin \theta = \sqrt{3} \, (0.1542 \times 10^{-9} \text{ m})/2(0.2450) = 0.5451 \times 10^{-9} \text{ m}$
$\rho = 4M/La_0^3 = 4(78.08)(10^{-3} \text{ kg mol}^{-1})/(6.02 \times 10^{23} \text{ mol}^{-1})(0.5451 \times 10^{-9} \text{ m})^3$
$= 3202 \text{ kg/m}^3 \ [3.202 \text{ g/cm}^3]$

28.10   The Powder Method

The simplest technique for obtaining X-ray diffraction data is the powder method, first used by Debye and Scherrer. Instead of a single crystal with a definite orientation to the X-ray beam, the sample is a mass of finely divided crystals with random orientations. The experimental arrangement is illustrated in Fig. 28.14. The powder is contained in a thin-walled glass capillary, or deposited on a glass fiber; polycrystalline metals are studied in the form of fine wires. The sample is rotated in the beam to average over the orientations of the crystallites.

Out of many random orientations of the little crystals, there will be some at the Bragg angle for X-ray "reflection" from each set of planes. The direction of the reflected beam is limited only by the requirement that angle of reflection equal angle of incidence. Thus, if the incident angle is θ, the reflected beam makes an angle 2θ with the direction of the incident beam, as shown in Fig. 28.14(a). This angle 2θ may itself be oriented in various directions around the central beam direction, corresponding to various orientations of the individual crystallites. For each set of planes, therefore, the reflected beams outline a cone of scattered radiation. This cone, intersecting a cylindrical film surrounding the specimen, gives rise to the observed lines. On a flat plate film, the recorded pattern consists of a series of concentric circles. Figure 28.14(c) shows the Debye–Scherrer (powder) diagrams obtained from several types of cubic crystal structures.

After one obtains a powder diagram, the next step is to index the lines, assigning each to the responsible set of planes. The distance x of each line from the central spot is measured, usually by halving the distance between the two reflections on either side of the center. If the film radius is r, the circumference $2\pi r$ corresponds to a scattering angle of 360°. Then, $2\theta/360° = x/2\pi r$. The interplanar spacing is calculated from θ and Eq. (28.4).

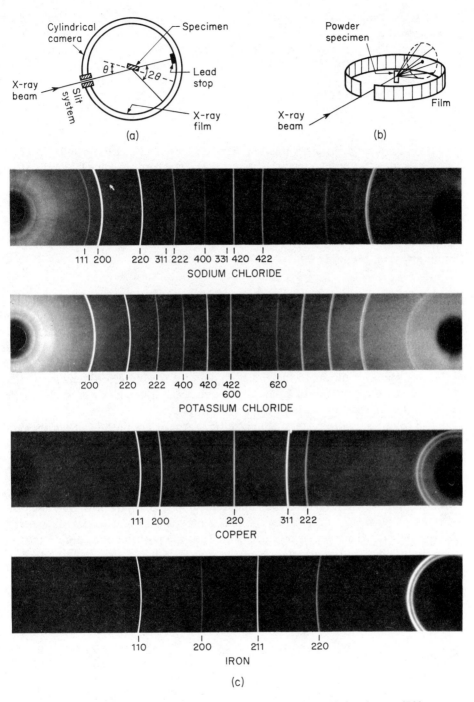

FIGURE 28.14 X-ray diffraction by the powder method. [Arthur Lessor, IBM Research Laboratory, Poughkeepsie, N.Y.]

Example 28.7 Use the 420 line in Debye–Scherrer picture of NaCl in Fig. 28.14 to calculate the wavelength of X rays used to obtain the picture. The length of unit cell of NaCl structure is 564 pm.

The spacing of 420 planes is

$$d_{420} = \frac{a_0}{\sqrt{h^2 + k^2 + \ell^2}} = \frac{564}{\sqrt{20}} \text{ pm} = 126 \text{ pm}$$

By measuring the picture with a centimeter scale we find that line 420 is 6.10 cm from center, the distance between centers is 11.7 cm, corresponding to $2\theta = 180°$, so that line 420 is at $2\theta = (6.10/11.7)180° = 93.8°$, or $\theta = 46.9°$, $\sin\theta = 0.730$. Then $\lambda = 2d \sin\theta = 2(126)(0.730) = 184$ pm.

28.11   Rotating Crystal Methods

From the early work on simple inorganic structures, X-ray crystallography developed rapidly in its applications to complex mineral structures and to both small and large organic molecules, including proteins and nucleic acids. Today X-ray crystallography is unique in its ability to give the detailed three-dimensional structure of very large organic molecules. The fact that crystals are used in the structure determination is merely a means to the end of finding the structure of the molecule. If a chemist can produce crystals of a pure organic compound, the X-ray crystallographer can determine its structure, often within a few days.

The rotating-single-crystal method, in one form or another, has been the most widely used technique for precise structure investigations. The crystal, perhaps a needle 1 mm long and 0.5 mm wide, is mounted with a definite axis perpendicular to the beam, which bathes it in X-radiation. The film may be held in a cylindrical camera. As the crystal is rotated slowly during the course of the exposure, successive planes pass through the orientation corresponding to the Bragg condition. In the Weissenberg and other moving-camera methods, the film is moved back and forth with a period synchronized with the rotation of the crystal. Thus the position of a spot on the film immediately indicates the orientation of the crystal at which the spot was formed. An example is shown in Fig. 28.15, taken from a crystal of an enzyme, lysozyme. The spots are indexed and their intensities are measured. The data thus obtained are the raw materials for the determination of the crystal structure.

28.12   The Structure Factor

The reconstruction of a crystal structure from the intensities of the various X-ray diffraction maxima is analogous to the formation of an image by a microscope. According to Abbe's theory of the microscope, the objective gathers various orders

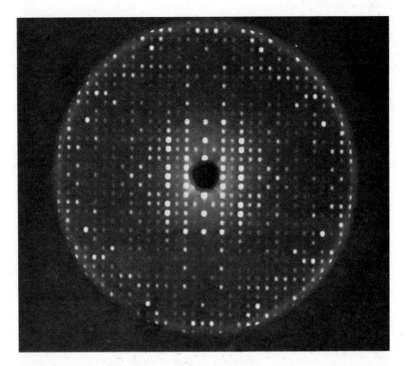

FIGURE 28.15 X-ray diffraction picture of a monoclinic crystal of lysozyme iodide [L. K. Steinrauf, Indiana University]. From the positions and intensities of the diffraction maxima in such pictures, the crystal structure, and hence the molecular structure of lysozyme was determined by D. C. Philips and coworkers in London. Lysozyme was the second protein and first enzyme to have its structure determined by X-ray crystallography. The molecule contains 1950 atoms. [See *Sci. Am.,* Nov. 1966]

of light rays diffracted by the specimen and resynthesizes them into an image. This synthesis is possible because two conditions are fulfilled in the optical case: The phase relationships between the various orders of diffracted light waves are preserved at all times, and optical glass is available to focus and form an image with radiation having the wavelengths of visible light. Electron beams can be focused with electrostatic and magnetic lenses, but we have no lenses for forming X-ray images. Also, the way in which the diffraction data are obtained (one by one) means that all phase information is lost. The essential problem in determining a crystal structure is to regain this lost information and to resynthesize the structure from the amplitudes and phases of the diffracted waves.

We shall return to this problem, but first let us see how the intensities of the various spots in an X-ray picture are governed by the crystal structure. The Bragg relation fixes the angle of scattering in terms of the interplanar spacings, which are determined by the arrangement of points in the crystal lattice. In an actual structure, each lattice point is replaced by a group of atoms. The arrangement and composition of this group are the primary factors that control the intensity of the scattered X rays, once the Bragg condition has been satisfied.

As an example, consider in Fig. 28.16(a) a structure formed by replacing each point in a body-centered orthorhombic lattice by two atoms (e.g., a diatomic molecule). If a set of planes is drawn through the black atoms, another parallel but slightly displaced set can be drawn through the white atoms. When the Bragg condition is met, as in Fig. 28.16(b), reflections from all the black atoms are in phase, and reflections from all the white atoms are in phase. The radiation scattered from the blacks is slightly out of phase with that from the whites, so that the resultant amplitude, and therefore intensity, is diminished by interference.

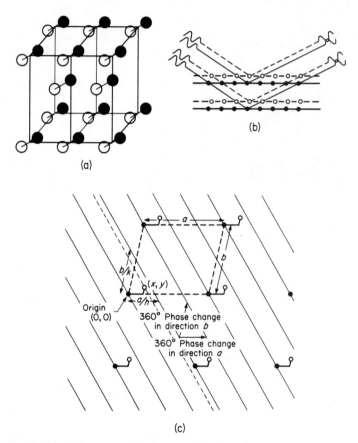

FIGURE 28.16 X-ray scattering from a set of planes in a structure consisting of molecules AB. Derivation of the structure factor.

The problem is to obtain a general expression for the phase difference. An enlarged view of a section through the structure is shown in Fig. 28.16(c), with the black atoms at the corners of a unit cell with sides a and b, and the whites at displaced positions. The coordinates of a black atom may be taken as $(0, 0)$ and those of a white as (x, y). A set of planes hk is shown, for which the Bragg condition is fulfilled; these are actually the 32 planes in the figure. Now the spacings a/h along a and b/k along b correspond to positions from which scattering differs in phase by exactly 360° or 2π radians, i.e., scattering from these positions is exactly in phase. The difference in phase between these planes and those going through the white atoms is proportional to the displacement of the white atoms. The phase difference ϕ_x for displacement x in the a direction is $x/(a/h) = \phi_x/2\pi$, or $\phi_x = 2\pi h(x/a)$. The total phase difference for displacement in both a and b directions becomes

$$\phi_x + \phi_y = 2\pi\left(h\frac{x}{a} + k\frac{y}{b}\right)$$

By extension to three dimensions, the total phase change that an atom at (xyz) in the unit cell contributes to the plane $(hk\ell)$ is

$$\phi = 2\pi\left(\frac{hx}{a} + \frac{ky}{b} + \frac{\ell z}{c}\right) \tag{28.7}$$

Superposition of waves of different amplitude and phase can be accomplished by vectorial addition. If f_1 and f_2 are the amplitudes of the waves scattered by atoms (1) and (2), and ϕ_1 and ϕ_2 are the phases, the resultant amplitude is $F = f_1 e^{i\phi_1} + f_2 e^{i\phi_2}$. For all the atoms in a unit cell,

$$F = \sum_j f_j e^{i\phi_j} \tag{28.8}$$

When the phase ϕ_j from Eq. (28.7) is introduced, an expression is obtained for the resultant amplitude of the waves scattered from the $hk\ell$ planes by all the atoms in a unit cell:

$$F(hk\ell) = \sum_j f_j e^{2\pi i(hx_j/a + ky_j/b + \ell z_j/c)} \tag{28.9}$$

The expression $F(hk\ell)$ is called the *structure factor* of the crystal. Its value is determined by the exponential terms, which depend on the positions of the atoms, and by the *atomic scattering factors* f_j, which depend on the number and distribution of electrons in the atoms and on the scattering angle θ. Structure-factor expressions have been tabulated for all space groups.

The intensity of scattered radiation is proportional to the absolute value of the amplitude squared, $|F(hk\ell)|^2$. Thus, once the Bragg condition is satisfied for a set of planes $hk\ell$, the structure factor allows us to calculate the intensity of X-ray scattering from these planes. The relation between intensity and structure factor includes a number of physical terms for which explicit formulas are available.

Example 28.8 Calculate $F(hk\ell)$ for the 100 planes in a face-centered cubic structure, e.g., that of gold. In this structure, there are four atoms in the unit cell ($Z = 4$), which may be assigned coordinates $(x/a, y/b, z/c)$ as follows: $(0\,0\,0)$, $(\frac{1}{2}\,\frac{1}{2}\,0)$, $(\frac{1}{2}\,0\,\frac{1}{2})$ and $(0\,\frac{1}{2}\,\frac{1}{2})$.

From Eq. (28.9), $F(100) = f_{\mathrm{Au}}(e^{2\pi i.0} + e^{2\pi i.1/2} + e^{2\pi i.1/2} + e^{2\pi i.0}) = f_{\mathrm{Au}}(2 + 2e^{\pi i})$ $= 0$, since $e^{\pi i} = \cos \pi + i \sin \pi = -1$. Thus the structure factor vanishes and the intensity of scattering from the 100 planes is zero. This is almost a trivial case, since inspection of the face-centered cubic structure reveals that there is an equivalent set of planes interleaved midway between the 100 planes, so that the resultant amplitude of the scattered X rays must be reduced to zero by interference. In all cases, however, the structure factor determines the scattering intensity from any set of planes $hk\ell$ in the crystal structure.

X rays are scattered by the electrons of the atoms in crystals. Thus it is rather artificial to represent a crystal structure as an array of atoms located at points (x, y, z). A continuous distribution of electron density $\rho(x, y, z)$ would be a more realistic model. The expression for the structure factor in Eq. (28.9), given as a sum over discrete atoms, then becomes an integral over the continuous distribution of electron density.

$$F(hk\ell) = \int_0^a \int_0^b \int_0^c \rho(x, y, z)e^{2\pi i(hx/a+ky/b+\ell z/c)} \, dx \, dy \, dz \qquad (28.10)$$

Equation (28.10) holds for unit cells with orthogonal axes. A slight modification would be necessary for others, involving a change in the variables x, y, z to a new set of axes.

A crystal structure can be generated by taking a unit cell and repeating it an infinite number of times in three directions of space. The crystal structure is thus a function that is periodic in three dimensions. Therefore, the electron density $\rho(x, y, z)$ is a function with the periodicity of the lattice, and it can be written as a Fourier series in three dimensions (recall that $e^{i\theta} = \cos\theta + i\sin\theta$):

$$\rho(x, y, z) = \sum\sum\sum A(pqr)e^{+2\pi i(px/a+qy/b+rz/c)} \qquad (28.11)$$

To evaluate the Fourier coefficient $A(pqr)$, we substitute Eq. (28.11) into Eq. (28.10), obtaining

$$F(hk\ell) = \int_0^a \int_0^b \int_0^c \sum\sum\sum A(pqr)\exp\left[2\pi i\left(\frac{hx}{a} + \frac{ky}{b} + \frac{\ell z}{c}\right)\right]$$
$$\times \exp\left[2\pi i\left(\frac{px}{a} + \frac{qy}{b} + \frac{rz}{c}\right)\right] dx \, dy \, dz \qquad (28.12)$$

The integrals of the complex exponential functions over a complete period always vanish, so that the only term that remains in Eq. (28.12) is that for which $p = -h$, $q = -k$, and $r = -\ell$, which leads to

$$F(hk\ell) = \int_0^a \int_0^b \int_0^c A(\bar{h}\bar{k}\bar{\ell}) \, dx \, dy \, dz = VA(\bar{h}\bar{k}\bar{\ell})$$

where V is the volume of the unit cell. When this value of the Fourier coefficient is put into Eq. (28.11), we obtain

$$\rho(x, y, z) = \frac{1}{V}\sum\sum\sum F(hk\ell)\exp\left[-2\pi i\left(\frac{hx}{a} + \frac{ky}{b} + \frac{\ell z}{c}\right)\right] \qquad (28.13)$$

The summation is carried out over all values of h, k, ℓ, so that there is one term for each set of planes $hk\ell$, and hence for each spot on the X-ray diffraction diagram.

This remarkable equation summarizes the entire problem of a structure determination, since the crystal structure is represented by the function $\rho(xyz)$ in Eq. (28. 13). Positions of individual atoms are peaks in the electron density function ρ, with heights proportional to the atomic number (number of electrons). If we knew the $F(hk\ell)$'s we could immediately draw the crystal structure. All we know, however, are the intensities of the diffraction maxima, and these are proportional to $|F(hk\ell)|^2$, the square of the absolute value of F. As stated earlier, we know the amplitudes but we have lost the phase information in taking the X-ray pattern.

In one method of solution, we assume a trial structure and calculate the intensities. If the assumed structure is even approximately correct, the most intense observed reflections should have large calculated intensities. We then compute the Fourier series by taking the *observed F*'s for these reflections with the *calculated phases*. If we are on the right track, the graph of the Fourier summation should give new positions of the atoms, from which more of the phases can be correctly determined. As more terms are included in the Fourier synthesis, the resolution of the structure improves, just as the resolution of a microscope improves with objectives that catch more orders of diffracted light.

Sometimes the phases can be readily determined if one atom that is much heavier than any of the others (and so has many more electrons) occupies a known position in the structure. The heavy atom often makes a large contribution to a given structure factor, allowing the phase of that $F(hk\ell)$ to be assigned. An example of a structure determined in this way is the cobalt compound vitamin B_{12}. Part of the Fourier synthesis of the electron density of this molecule is shown in Fig. 28.17.

FIGURE 28.17 X-ray analysis gives this electron-density map of the central portion of the molecule of vitamin B_{12} (cobalamin). Two of the four pyrrole residues are linked directly to each other, and the other links are through —CH_2— bridges. The central peak of high density is due to Co(II), which is coordinated to the four nitrogens. [W. L. Bragg, *The Development of X-ray Analysis* (London: G. Bell & Sons Ltd., and Bell & Hyman, 1974)]

The number of terms that must be included in the Fourier summation to achieve a given resolution increases rapidly with the number of atomic positions that need to be determined in the unit cell. For a protein having a molecular weight of 2×10^4, about 10^4 Fourier terms would be required to achieve a resolution of 200 pm.

28.14 Neutron Diffraction

As well as X rays, beams of electrons and neutrons give diffraction patterns when scattered from a regular array of atoms. Because of their negative charge and consequently low penetrating power, electron beams are mostly useful for surfaces and thin films. Neutrons, however, have a high penetrating power and a number of special advantages for crystallography.

Wavelength is related to mass and speed by the Broglie equation $\lambda = h/mv$.

Thus a neutron with a speed of 3.9×10^3 m s^{-1} (kinetic energy 0.08 eV compared to $kT = 0.026$ eV at 300 K) has a wavelength of 100 pm.

The diffraction of X rays is caused by the orbital electrons of the atoms in the material through which they pass; the atomic nuclei contribute practically nothing to the X-ray scattering. The diffraction of neutrons, on the other hand, is primarily caused by two other effects: (1) nuclear scattering due to interaction of the neutrons with the atomic nuclei, and (2) magnetic scattering due to interaction of the magnetic moments of the neutrons with permanent magnetic moments of atoms or ions.

X-ray diffraction by heavy atoms greatly outweighs that by light atoms. This effect has been valuable in determining the phases of experimental structure factors $F(hk\ell)$. On the other hand, the positions of hydrogen atoms are difficult or impossible to pinpoint from X-ray data. Neutron scattering by H and D nuclei is especially intense, however, and thus neutron diffraction is ideally suited for locating these nuclei in a structure.

An example of the powers of the neutron-diffraction method can be seen in the structure of sucrose in Fig. 28.18. All the hydrogens have been precisely located. The atoms are depicted as ellipsoids whose axial lengths and orientations represent the mean thermal vibrations of the respective nuclei. The vibrations of the nuclei affect the intensity of scattering and hence the shapes of the diffraction maxima, and these data can be used to determine the amplitudes of the thermal vibrations. When the H atoms are engaged in hydrogen bonding their motions are restricted. The exceptionally large vibration of the H atom at O-4 is due to the fact that it is not hydrogen bonded.

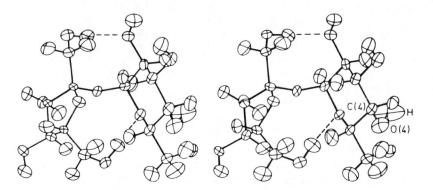

FIGURE 28.18 Stereoscopic view of the structure of sucrose as determined by neutron diffraction. [G. M. Brown and H. A. Levy, *Acta Cryst. B29*, 793 (1973)]

Problems

1. Draw a primitive two-dimensional square lattice and indicate the sets of lines with Miller indices (10), (20), (11), and (31).

2. Draw an alternative primitive lattice through the same points as in (1) and indicate the lines (10), (20), (11) and (31), in the new lattice.

3. The two possible tetragonal Bravais lattices are primitive P and body-centered I. Show that the end-centered lattice C is not a distinct type.

4. At 25°C orthorhombic sulfur has unit cell dimensions $a_0 = 1.0465$, $b_0 = 1.2866$, $c_0 = 2.4486$ nm. and density $\rho = 2067$ kg m^{-3}. Calculate the number of S atoms in the unit cell.

5. Ni crystallizes in the fcc structure with $a_0 = 352$ pm and NiO in the NaCl structure with $a_0 = 418$ pm. Calculate the closest distance between Ni nuclei in the two structures. How would you explain the result?

6. For the unit cell of orthorhombic sulfur at 25°C (Problem 4), calculate the spacing of the planes (111), (101), (011), and (110).

7. Glycylglycine forms monoclinic crystals. At 25°C, $a = 1793$ pm, $b = 462$ pm, $c = 1709$ pm, $\beta = 125°10'$. There are eight molecules in each unit cell. What is the density of the crystal?

8. The only metal known to crystallize in a simple cubic structure is polonium. Sketch the X-ray diffraction powder pattern to be expected from Po. Compare it with that from a fcc structure.

9. The density of Po is 9.15 g cm^{-3} at 25°C. Calculate the Bragg angle for diffraction from the planes (234) with X rays of $\lambda = 153.9$ pm (Cu–K$_\alpha$).

10. What is the plane of closest packing in the bcc structure? At 25°C Fe has a structure based on a bcc lattice with a basis 1Fe at (000), $a_0 = 286.4$ pm. What is the closest distance between Fe atoms?

11. Argon crystallizes in a fcc structure. At 20 K, $a_0 = 543$ pm. What is the effective diameter of an argon atom?

12. MgO has the NaCl structure and density $\rho = 3.650$ g cm^{-3} at 25°C. Calculate the values of $\sin \theta/\lambda$ for which X-ray scattering occurs from the planes 100, 110, 111, 200, 300 and 221, if it does occur.

13. The diamond structure is based on a fcc lattice with a basis (000, $\frac{1}{4}\frac{1}{4}\frac{1}{4}$). Derive an expression for the structure factor $F(hk\ell)$ for diamond structure. What are the first six maxima observed in a Debye–Sherrer pattern from diamond, and what are their approximate relative intensities?

14. A simple cubic crystal with $a_0 = 360.7$ pm is studied with X rays of $\lambda = 143.6$ pm. What are the smallest and largest possible Bragg angles for X-ray "reflection"?

15. At 1183 K iron undergoes a transition from a bcc to a fcc structure. The radius of the atom is the same in the two structures. Calculate the ratio of the densities of iron in the two structures.

16. With X rays of $\lambda = 153.7$ pm, the diffraction maximum from the 111 planes in Al is at a Bragg angle of 19.2°. Al is fcc with $\rho = 2.699$ g cm^{-3}, $M = 26.98$ g mol^{-1}. Calculate from these data a value for Avogadro's number L.

17. The structure of fluorite (CaF$_2$) can be based on a fcc lattice with a basis of (Ca^{2+} at 000 and 2F$^-$ at $\frac{1}{4}\frac{1}{4}\frac{1}{4}$ and $\frac{1}{4}\frac{1}{4}\frac{3}{4}$). Sketch the structure. Write the structure factor for the planes 100, 110, 111, 200, 220, and 222, in terms of atomic scattering factors f_{Ca} and f_F.

18. The following lines are observed in the Debye–Scherrer pattern of a cubic crystal of Ba(NO$_3$)$_2$ with X rays at $\lambda = 154$ pm. $\theta° = 9.48, 10.95, 12.25, 13.48, 15.65, 18.38, 19.25, 22.45, 24.50, 25.18, 27.88,$ and 29.65. Index the lines and deduce the lattice type of the crystal. Intensities of lines are $vs, m, w, w, m, m, s, m, m, w, w, m$, respectively.

19. Look up the crystal structure of Ba(NO$_3$)$_2$ in *Structure Reports* and calculate the structure factors for the "reflections" indexed in Problem 18. How well do they explain the observed intensities?

29

The Solid State

We live in a world of solid-state devices: lasers, phosphors, photographs, ferrites, and transistors. The strength of engineering materials, such as alloys and ceramics, depends on deviations from purity and regularity in the solid state. The design of new catalysts for chemical processes and better electrodes for conversion of energy requires an understanding of solid-state theory and an ability to apply it to complex industrial problems. This chapter introduces some basic concepts of the solid state, but a chemist who plans to work in this field should study in depth both solid-state chemistry and solid-state physics.

Two theoretical approaches to the nature of the chemical bond in molecules were described in Chapter 22, the valence-bond method and the molecular-orbital method. For studying the nature of bonding in solids, methods closely related to each of the two basic models for molecules are available. In one case, a crystal structure is pictured as an array of regularly spaced atoms, each possessing electrons used to form bonds with neighboring atoms. These bonds may be ionic, covalent, or intermediate in type. Extending throughout three dimensions, they hold the crystal together. The alternative approach is to place the nuclei at fixed positions and then to pour the electron cement into the periodic array of nuclear bricks. Both these methods yield useful and distinctive results, displaying complementary aspects of the nature of the solid state. The first method, growing out of the valence-bond theory, is the *bond model* of the solid state. The second method, an extension of the method of molecular orbitals, is called, for reasons to appear later, the *band model* of the solid state. The bond model is sometimes called the *tight binding approximation*, and the band model, the *collective electron approximation*.

A rough classification of the bonds that occur in solids is as follows.

1. *Van der Waals bonds.* These result from the forces between atoms that are responsible for the *a* term in the van der Waals equation. Crystals held together in this way are sometimes called *molecular crystals.* Examples are nitrogen, carbon tetrachloride, and benzene. The molecules pack together as closely as their sizes and shapes allow. The binding between molecules in van der Waals structures is due to a combination of factors, such as dipole–dipole and dipole–polarization interactions, and the quantum mechanical dispersion forces first elucidated by F. London, which are often the principal component. The theory of these forces is discussed in Chapter 30.

2. *Ionic bonds.* In the NaCl crystal, the Coulombic interaction between oppositely charged ions leads to a regular three-dimensional structure; each positively charged Na^+ ion is surrounded by six negatively charged Cl^- ions, and each Cl^- is surrounded by six Na^+. There are no distinct NaCl molecules in a crystal.

X-ray diffraction data can be used to calculate the electron density in crystals. Results for NaCl and LiF are shown as contour maps in Fig. 29.1. In the case of NaCl, the electron-density contours for each ion are almost spherically symmetric. In the case of LiF, however, the smaller Li^+ ion has polarized the F^- ion and drawn some electron density into the region between the two ions.

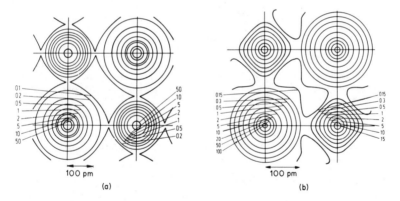

FIGURE 29.1   Map of electron density in (a) LiF; (b) NaCl. [H. Witte and E. Wölfel, *Z. Phys. Chem. N.F. 3*, 296 (1955)]

3. *Covalent bonds.* These bonds are the result of the sharing of electrons by atoms. When extended through three dimensions, they lead to a variety of crystal structures, depending on the number of electrons available for bond formation.

An example is the diamond structure in Fig. 29.2. This structure is based on a face-centered cubic lattice. The basis consists of two atoms, one at (0 0 0), the other at $(\frac{1}{4}\frac{1}{4}\frac{1}{4})$. Each atom is surrounded tetrahedrally by four equidistant neighbors. This

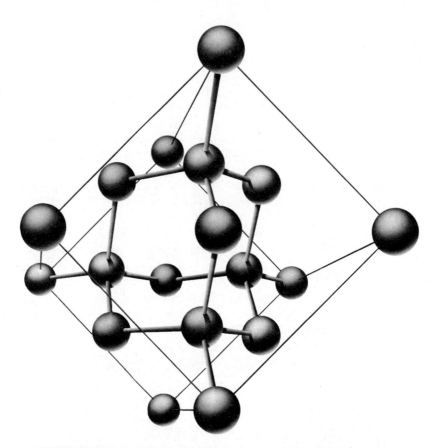

FIGURE 29.2 Cubic unit cell of the diamond structure. If the four atoms arranged tetrahedrally about a central atom are of a different species, the structure would be that of zinc blende (ZnS). [G. H. Wannier, *Solid State Theory*, (Cambridge, Cambridge University Press, 1959); drawn by F. M. Thayer]

arrangement constitutes a three-dimensional polymer of carbon atoms connected by tetrahedrally oriented bonds. The configuration of the carbon bonds in diamond is similar to that in aliphatic compounds such as ethane. Germanium, silicon, and gray tin also crystallize in the diamond structure. A similar structure is assumed by compounds such as ZnS (in the zincblende structure), AgI, AlP, and SiC. In these structures, each atom is surrounded by four unlike atoms situated at the corners of a regular tetrahedron. In every case, the binding is primarily covalent. The structure can occur whenever the number of valence-shell electrons is four times the number of atoms; it is not necessary that each atom provide the same number of valence electrons.

Atoms with a valence of only 2 cannot form isotropic three-dimensional structures. Consequently, we find interesting structures such as that of selenium and tellurium, which consist of endless chains of atoms extending through the crystal, the individual chains being held together by much weaker forces. Elements like arsenic and antimony, which in their compounds display a covalence of 3, tend to crystallize in structures containing layers or sheets of atoms.

Example 29.1 X-ray data show that the length of the unit cell in diamond is $a_0 = 356.7$ pm. Calculate the length of the C—C bond and the C—C—C angle θ in diamond.

The length of the bond is calculated from the general formula for the distance between two points in cartesian coordinates: $d = \sqrt{(x_1 - x_2)^2 + (y_1 - y_2)^2 + (z_1 - z_2)^2}$. A pair of nearest-neighbor atoms are those at (000) and $(\frac{1}{4}\frac{1}{4}\frac{1}{4})$. Thus $d = \sqrt{(\frac{1}{4})^2 + (\frac{1}{4})^2 + (\frac{1}{4})^2}\,a_0 = (\sqrt{3}/4)a_0 = 154.5$ pm. From Fig. (29.1), $\sin \theta/2 = (\frac{1}{4})\sqrt{2}\,a_0/(\frac{1}{4})\sqrt{3}\,a_0 = \sqrt{2}/\sqrt{3} = 0.8165$. $\theta/2 = 54.74°$, $\theta = 109.47°$.

---

4. *Bonds of intermediate type.* An ion is polarized when its electron distribution is distorted by the presence of an oppositely charged ion. The larger an ion, the more readily it is polarized, and the smaller an ion, the more intense its electric field and the greater its polarizing power. Usually, therefore, the smaller cations polarize the larger anions. Even apart from the effect of size, cations are less polarizable than anions because their net positive charge tends to hold their electrons in place. The structure of an ion is also important: Rare-gas cations, such as K^+, have less polarizing power than transition cations, such as Ag^+, because their positive nuclei are more effectively shielded.

The effect of polarization can be seen in the structures of the silver halides. AgF, AgCl, and AgBr have the ionic rock-salt structure, but as the anion becomes larger, it becomes more strongly polarized by the small Ag^+ ion. Finally, in AgI the binding has little ionic character and the crystal has the zincblende structure.

5. *Hydrogen bonds.* The hydrogen bond is important in many crystal structures, e.g., inorganic and organic acids, salt hydrates, ice. The structure of ordinary ice I is shown in Fig. 29.3. Each oxygen is surrounded tetrahedrally by four nearest neighbors at a distance of 276 pm. The hydrogen bonds between the oxygens result in a very open structure.

6. *Metallic bonds.* The metallic bond is related to the ordinary covalant electron-pair bond. Each atom in a metal forms covalent bonds by sharing electrons with its nearest neighbors, but the number of orbitals available for bond formation exceeds the number of electron pairs available to fill them. The empty orbitals permit a ready flow of electrons under the influence of an applied field, leading to metallic conductivity.

29.2 Closest Packing of Spheres

Spheres of uniform size can be packed in different ways to yield the minimum fraction of empty space, 0.259. Two simple closest packings are the basis of many crystal structures.

There is only one way to make a closest packed layer of uniform spheres. It is shown in Fig. 29.4(a). Note the hexagonal symmetry of the arrangement. Each

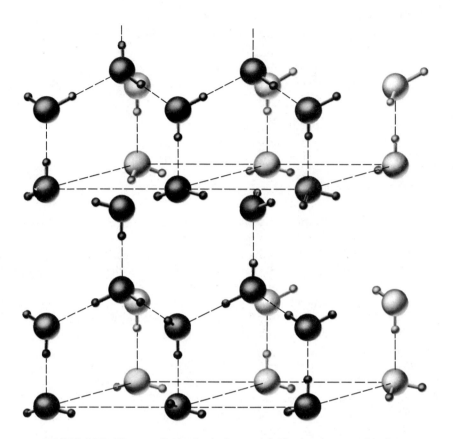

FIGURE 29.3 Water molecules in the ice crystal. The arrangement of hydrogen bonds is arbitrary; there is one proton along each O–O axis, closer to one or the other of the two O atoms. [Linus Pauling, *The Nature of the Chemical Bond*, 3rd ed. Copyright © 1960 by Cornell University. Used by permission of the publisher, Cornell University Press]

sphere has six nearest neighbors with centers arrayed at the vertices of a regular hexagon.

Now notice the triangular void spaces in the closest packed layer. There are two sets of these triangular spaces, one with the vertices of the triangles pointing up and one with the vertices pointing down. We can call these sites *B* and *C*, respectively, and the positions of the centers of the spheres in the first layer, we call sites *A*. Now suppose that we wish to place a second closest packed layer on top of the first. If we place some spheres on the *B* sites, we cannot also place some on the *C* sites if we wish to maintain closest packing. Therefore, we can pack the second layer by nesting spheres either all in *B* sites or all in *C* sites. Suppose that we choose, therefore, *AB* as sites for the first two layers [Fig. 29.4(b).] This is not really a

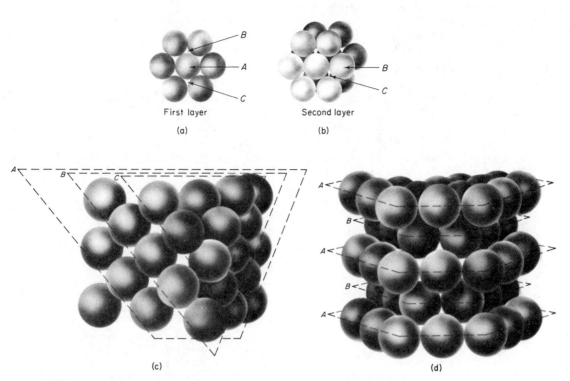

First layer
(a)

Second layer
(b)

(c)

(d)

FIGURE 29.4 Structures that arise from closest packing of uniform spheres: (a) a close-packed layer of spheres; (b) a second layer resting on the positions B of the first layer; (c) cubic closest packing, the closest packed layers are the 111 planes in a fcc structure; (d) hexagonal closest packing, the closest packed layers are the 001 planes.

different structure from *AC*, since *AB* can be converted to *AC* by 180° rotation of the double layers.

When we come to a third layer, however, a real choice is available. We can either use *C* sites for a closest packed third layer, or go back and place the third layer exactly above the first layer, that is, use *A* sites. The two possible structures that result are designated as *ABC* and *ABA* [Fig. 29.4(c) and (d).] Both these structures have ideal closest packing. Each sphere touches 12 neighbors, six in the same plane, three above and three below.

These two structures were worked out about 1883 by Barlow. The structure *ABAB* . . . can be based on a hexagonal unit cell containing two atoms (basis: 000, $\frac{1}{3}\frac{2}{3}\frac{1}{2}$). It is called the *hexagonal closest packed* (hcp) *structure*. The closest packed layers are normal to the hexagonal *c* axis. The structure *ABCABC* . . . can be based on a face-centered cubic lattice and is called the *cubic closest packed* (ccp) *structure*. The closest packed layers are the 111 planes in the structure, normal to the [111] direction which is the threefold symmetry axis of the cube.

Example 29.2 Calculate the fractional void volume in the ccp and hcp structures of hard spheres.

For spheres of radius r, the face diagonal of the cubic unit cell is $4r$ (Fig. 29.4), so that the side of unit cell is $2\sqrt{2}\,r$ and its volume $16\sqrt{2}\,r^3$. The unit cell contains four spheres each of volume $\frac{4}{3}\pi r^3$, so that the fractional void volume for ccp is $(16\sqrt{2}\,r^3 - \frac{16}{3}\pi r^3)/16\sqrt{2}\,r^3 = 1 - (\pi/3\sqrt{2}) = 0.2594$.

For hcp the volume V of the unit cell is the area A of the base $\times$ the height c. $A = (2r)(\sqrt{3}\,r) = 2\sqrt{3}\,r^2$. The height c is $2(4 - \frac{4}{3})^{1/2}r = (4\sqrt{2}/\sqrt{3})r$. Hence $V = 8\sqrt{2}\,r^3$. The cell contains two atoms, with hard-sphere volume $2(\frac{4}{3}\pi r^3)$. Hence the fractional void volume is $(8\sqrt{2}\,r^3 - \frac{8}{3}\pi r^3)/8\sqrt{2}\,r^3 = 1 - \pi/3\sqrt{2} = 0.2594$ as for ccp.

Most metals crystallize in either the ccp or hcp structure. Some metal structures are summarized in Table 29.1. Many metals are polymorphic; i.e., they have different structures depending on temperature and pressure.

TABLE 29.1
CRYSTAL STRUCTURES OF METALS

Cubic closest packed	Ag, Al, Au, αCa, βCo, Cu, γFe, Sr, Th, Ni, Pb, Pt
Hexagonal closest packed	αBe, γCa, Cd, αCo, βCr, Mg, αZr, Os, αTi, Zn
Body-centered cubic	Ba, αCr, Cs, αFe, K, Li, Mo, Na, βZr, Ta, βTi, V
Simple-cubic	Po
Rhombohedral, layers	Bi, As, Sb
Body-centered tetragonal	βSn, βGe
Face-centered tetragonal	γMn, In

29.3 Electron-Gas Theory of Metals

The cohesive energy of metals, and their high electric and thermal conductivities, remained a mystery until after the discovery of the electron in 1895. Drude in 1905 suggested that a piece of metal is like a box containing a gas of mobile electrons. If an electric field is applied, the negative electrons flow up the gradient of potential; i.e., an electric current occurs. As in the case of ionic conductivity, the electronic conductivity $\kappa = C|ze|u$, where C is the number concentration of charge carriers, $|ze|$ the absolute value of their charge, and u their mobility. The conductivity of metals can be explained if it is assumed that all the valence electrons are included in C, and, furthermore, that the mobility u is so high that the electrons move freely over hundreds of atomic distances without being appreciably deflected by collisions with nuclei or other electrons. In other words, the term *electron gas* is no mere figure of speech; electrons in metals really appear to have kinetic properties similar to those of gas molecules.

There was, however, a serious objection to the Drude theory. If electrons really behave like gas molecules, they should take up kinetic energy when a metal is heated. In accord with the principle of equipartition of energy (Chapter 3) the translational energy of a mole of electrons should be $\frac{3}{2}LkT$, where L and k are the Avogadro and Boltzmann constants, giving an electronic contribution to the heat capacity of $C_V = (\partial U/\partial T)_V = \frac{3}{2}Lk = \frac{3}{2}R$ per mole. Experimentally, however, there is no electronic heat capacity of any such magnitude. In fact, the heat capacities of many metals at ordinary temperatures are close to the value given by the rule of Dulong and Petit, $C_V = 3Lk$ per mole, which is accounted for by the $3L$ degrees of vibrational freedom alone. At temperatures near absolute zero a small electronic heat capacity can be detected but the large value predicted by the Drude theory is nowhere to be found.

29.4 Quantum Statistics

The solution to the problem of the missing electronic heat capacity was found by Sommerfeld in 1928. In Chapter 20 we outlined the quantum mechanical theory for the motion of a particle in a three-dimensional box. We were then considering particles without spin and the Pauli Exclusion Principle was not applied. Electrons are elementary particles of spin $s = \frac{1}{2}$, and the Pauli Principle allows only two electrons to be placed in any energy level ϵ_i. Therefore, even in the limit of absolute zero, there must be a wide spread of occupied energy levels. All the lowest states are filled with pairs of electrons of opposite spin until some maximum energy level ϵ_F is reached. If we draw the distribution function, the probability $p(\epsilon)$ of filling a level as a function of energy ϵ of the level, we find the result shown in the dashed curve in Fig. 29.5: $p(\epsilon) = 1$ until we reach ϵ_F, after which $p(\epsilon) = 0$.

This $p(\epsilon)$ is an example of a *Fermi–Dirac distribution function*. It is the function to be expected when elementary particles are distributed in translational energy levels with the requirement that they obey the Pauli Exclusion Principle. At any temperature above 0 K, some electrons move into higher energy levels, and at temperatures still small compared to ϵ_F/k, the distribution function has the appearance of the solid

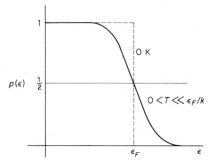

FIGURE 29.5 The Fermi–Dirac function for probability that an electron occupies a given energy level in a metal.

curve in Fig. 29.5. The Fermi–Dirac function is thus completely different from the Maxwell–Boltzmann distribution function which is the basis of the classical equipartition of energy.

The mathematical expression for the Fermi–Dirac distribution function is

$$p(\epsilon) = \frac{1}{1 + e^{(\epsilon - \epsilon_F)/kT}} \tag{29.1}$$

Note that when $p(\epsilon) = \frac{1}{2}$, $\epsilon = \epsilon_F$, where ϵ_F is called the *Fermi energy*. In metals, the Fermi energy acts as an effective cutoff level for allowed energies of electrons.

As long as $\epsilon_F \gg kT$, the $p(\epsilon)$ function has the general form shown in Fig. 29.5. If $\epsilon_F \sim kT$, the distribution would become similar to the Boltzmann one. But, at 1000 K, $kT = 0.086$ eV, and for sodium, $\epsilon_F = 3.12$ eV, a typical value for a metal. Thus the electron gas in metals follows the quantum mechanical (Fermi–Dirac) distribution law at all accessible temperatures.

The Fermi–Dirac function of Eq. (29.1) gives the probability that a state of energy ϵ is occupied by an electron. The number of energy states that are available for occupancy can be obtained from the expression for the number of translational energy states with an energy not greater than some value ϵ,

$$N(\epsilon) = 2 \left(\frac{1}{8}\right) \frac{4\pi}{3} \left(\frac{8m_e\epsilon}{h^2}\right)^{3/2} V \tag{29.2}$$

The factor of 2 compared to Eq. (12.11) comes from the fact that each state can hold two electrons of opposite spin.

Example 29.3 The density of Na at 0 K is about 10^3 kg m^{-3}. Calculate the Fermi level ϵ_F for Na at 0 K from the free-electron-theory model.

From Eq. (29.2), $\epsilon_F = (h^2/8m_e)(3N/\pi V)^{2/3}$. The molar volume of Na is

$$M/\rho = 23.0 \times 10^{-3} \text{ kg mol}^{-1}/10^3 \text{ kg m}^{-3} = 23.0 \times 10^{-6} \text{ m}^3 \text{ mol}^{-1}$$

Hence

$$N/V = 6.02 \times 10^{23} \text{ mol}^{-1}/23.0 \times 10^{-6} \text{ m}^3 \text{ mol}^{-1} = 2.62 \times 10^{28} \text{ m}^{-3}$$

and

$$\epsilon = \frac{(6.63 \times 10^{-34} \text{ J s})^2}{8 \times 9.11 \times 10^{-31} \text{ kg}} \left(\frac{3 \times 2.62 \times 10^{28} \text{ m}^{-3}}{3.14}\right)^{2/3}$$

$$= 5.16 \times 10^{-19} \text{ J} = 3.22 \text{ eV}$$

We can now see why the electron gas does not contribute to the heat capacity of metals. As a metal is heated, its electrons cannot take up energy unless they can move into somewhat higher energy levels, but most electrons are buried deep in the Fermi sea, and there are no empty levels nearby into which they can move. Only the relatively few electrons at the top of the distribution can find empty levels above them. These electrons, in the *Maxwellian tail* of the Fermi–Dirac distribution, are the only ones that can contribute to the heat capacity.

How, then, can all the free electrons contribute to the conductivity, since to do so they must take energy from the electric field and move to higher levels? In this case the electric field **E** exerts a force $-e\mathbf{E}$ on all the electrons simultaneously. As a result of the applied field, all energy levels are shifted to higher values.

29.5 Cohesive Energy of Metals

The cohesion of metals is caused by electrostatic attraction between the positive cores of metal atoms and the negative fluid of mobile electrons. A quantitative treatment of the problem involves solution of the Schrödinger equation for the energy of many electrons in a periodic electric field specified by the crystal structure of the metal.

Consider a simplified model of a one-dimensional structure. For concreteness, the nuclei are taken to be those of sodium, with a charge of $+11$. The position of each nucleus represents a deep potential-energy well for the electrons. If these wells are far apart, the electrons all fall into fixed positions on the sodium nuclei, giving rise to the $1s^2\,2s^2\,2p^6\,3s^1$ configurations of isolated sodium atoms. In the metal, however, the potential wells are neither far apart nor infinitely deep. The electrons can tunnel through the barriers, and we are thus no longer concerned with energy levels of individual sodium atoms but with levels of the crystal as a whole. The Pauli Principle tells us that no more than two electrons can occupy exactly the same energy level. Instead of the sharp $3s$ energy levels in isolated sodium atoms, there is a $3s$ *band* of closely packed energy levels in sodium metal.

The topmost band actually broadens sufficiently to overlap the peaks of the potential-energy barriers, so that electrons in the top energy levels can move quite freely throughout the crystal structure. According to this idealized model, in which the nuclei are always situated at the points of a perfectly periodic lattice, there would be no resistance to the flow of an electric current. The actual resistance is caused by departures from perfect periodicity. One important loss of periodicity results from thermal vibrations of the atomic nuclei. As would be expected from this effect, the resistance of metals increases with temperature.

What of magnesium with its two $3s$ electrons and therefore completely filled $3s$ band? Why is it not an insulator instead of a metal? More detailed calculations show that in such cases, the $3p$ band is low enough to overlap the top of the $3s$ band, providing a large number of available empty levels. Actually, this overlap occurs for the alkali metals also. The way in which the $3s$ and $3p$ bands in sodium broaden as the atoms are brought together is shown in Fig. 29.6. The interatomic distance in sodium at 100 kPa and 298 K is $R_e = 380$ pm. At this distance, there is no longer any gap between the $3s$ and $3p$ bands. In the case of diamond, on the other hand, the quantum mechanical calculations indicate a large energy gap between the filled *valence band* and the empty *conduction band* at the experimental R_e of 154 pm. To excite an electron from the strong covalent C—C bond in diamond requires a lot of energy.

Thus conductors are characterized either by partial filling of bands or by overlapping of the topmost bands. Insulators have completely filled lower bands with a wide energy gap between the topmost filled band and the lowest empty band. These models are represented in Fig. 29.7.

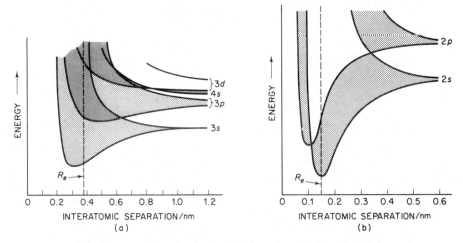

FIGURE 29.6 Quantum-mechanical calculations show the formation of electronic energy bands as atoms are brought together to form a crystal: plots of energy vs. internuclear distance for (a) sodium, (b) diamond.

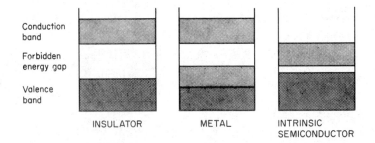

FIGURE 29.7 Schematic band models of solids classified according to their electronic properties.

29.6 Intrinsic Semiconductors

Semiconductors have resistivities which are intermediate between those of typical metals and typical insulators, and which decrease with rising temperature. The electrical properties of insulators and semiconductors depend on the magnitude of the energy spacing (*band gap*) between a filled valence band and a higher empty conduction band. One way to determine the gap is to measure the wavelength at which optical absorption begins in the crystal, the so-called *absorption edge*. The energy at the onset of absorption corresponds to the transfer of an electron from the top of the filled valence band to the bottom of the conduction band. Values of the band gap ϵ for crystals with the diamond structure are C, 5.33 eV; Si, 1.14 eV; Ge, 0.67 eV; Sn (gray), 0.08 eV.

The ratio of the number of electrons thermally excited to the conduction band to the number in the valence band is given by a Boltzmann factor $e^{-\epsilon/2kT}$. [Show how the factor 2 occurs because the electron leaves a *hole* in the valence band when it jumps into the conduction band.] For diamond, ϵ is so high that electrons rarely reach the conduction band by thermal excitation; thus the crystal is a good insulator. In the cases of Si and Ge, however, there will be an appreciable number of conduction electrons produced by thermal excitation from the valence band. These crystals are typical *intrinsic semiconductors*.

Electrons in a completely filled valence band make no contribution to the conductivity. As soon as holes are formed in the valence band, however, the remaining electrons find empty states available and can then contribute to the conductivity. A hole in a band of negative electrons is effectively a point of positive charge. The jump of an electron into a hole is equivalent to the jump of a positive charge into the position vacated by the electron. We can therefore treat the motion of electrons in an almost filled band as if the holes were positive charges moving in an almost empty hole band.

Electrons e^- and holes h^+, besides having opposite charges, have different mobilities, u_e and u_h. If C_e and C_h are the concentrations of electrons and holes, the electrical conductivity becomes

$$\kappa = e(C_e u_e + C_h u_h) \tag{29.3}$$

where e is the protonic charge. In an intrinsic semiconductor, $C_e = C_h$.

We can draw an analogy between an intrinsic semiconductor like Si or Ge and a weakly ionizing solvent like water.

$$H_2O \rightleftharpoons H^+ + OH^-, \qquad K_w = [H^+][OH^-]$$
$$Si \rightleftharpoons h^+ + e^-, \qquad K_i = [h^+][e^-]$$

In pure intrinsic silicon,

$$[h^+] = [e^-] = K_i^{1/2} = A(T)e^{-\epsilon/2kT} \tag{29.4}$$

From statistical thermodynamics, $A(T) = 2(2\pi kT/h^2)^{3/2}(m_e m_h)^{3/4}$, where m_e and m_h are the effective masses of electrons and holes, respectively.*

Example 29.4 The band gap in pure germanium is 0.67 eV. The electron and hole mobilities are $u_e = 0.38$ m^2 V^{-1} s^{-1} and $u_h = 0.18$ m^2 V^{-1} s^{-1}. Calculate the conductivity of Ge at 400 K.

From Eq. (29.4) with $m_e = m_h = 9.11 \times 10^{-31}$ kg.

$$A(T) = 2\left[\frac{(2\pi)(1.38 \times 10^{-23} \text{ J K}^{-1})(400 \text{ K})(9.11 \times 10^{-31} \text{ kg})}{(6.63 \times 10^{-34} \text{ J s})^2}\right]^{3/2}$$

$$= 3.85 \times 10^{25} \text{ m}^{-3}$$

$$C_e = C_h = 3.85 \times 10^{25} \text{ m}^{-3} \exp\left(\frac{-0.67 \times 1.60 \times 10^{-19} \text{ J}}{2 \times 1.38 \times 10^{-23} \times 400 \text{ J}}\right)$$

$$= 3.85 \times 10^{25} \exp(-9.71) = 2.34 \times 10^{21} \text{ m}^{-3}.$$

From Eq. (29.3),

$$\kappa = (1.60 \times 10^{-19} \text{ C})(2.34 \times 10^{21} \text{ m}^{-3})(0.38 + 0.18 \text{ m}^2/\text{V s})$$

$$= 2.1 \times 10^2 \ \Omega^{-1} \text{ m}^{-1}$$

(In comparison, for copper, $\kappa \sim 10^7 \ \Omega^{-1}$ m^{-1}.)

---

*Derivation: C. Kittel, *Introduction to Solid State Physics*, 5th ed. (New York: John Wiley & Sons, Inc., 1976), p. 313.

Figure 29.8 depicts the doping of silicon. When [as in Fig. 29.8(a)] a P atom is substituted for a Si atom, four of the valence electrons of P can enter the valence band, but the fifth electron must enter some level of higher energy. In fact, this energy state is only 0.012 eV below the conduction band. Therefore, the bound electrons in the *impurity levels* can readily be thermally excited into the conduction band. The doped semiconductor will have a greatly enhanced conductivity compared to pure intrinsic silicon. A semiconductor such as Si doped with P is called *n-type*, because the majority current carriers are negatively charged (electrons). An atom such as P or As, which can give electrons to the conduction band, is called a *donor*, and the extra energy levels just below the conduction band are *donor levels*.

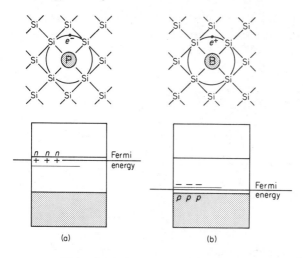

FIGURE 29.8 (a) Doping of silicon with electron donors (for example, P) leads to *n*-type silicon with electrons in the conduction band. (b) Doping of silicon with electron receptors (for example, B) leads to *p*-type silicon with positive holes in the valence band.

If the doping atom, boron for example, has fewer valence electrons than silicon, the schematic structure looks like Fig. 29.8(b). There is a hole or missing electron in the tetrahedral bonds about the B atom, creating a new level within the band gap. In the case of B in Si, these impurity levels are only 0.01 eV above the top of the valence band. Electrons can readily jump from the top of the valence band to fill such *acceptor levels*. Silicon doped with boron thus becomes a *p-type* semiconductor, because the majority current carriers are positively charged holes.

29.8 Ionic Crystals

The binding in many inorganic crystals is predominantly ionic. Since Coulombic forces between ions are undirected, the sizes of the ions play an important role in determining the final structure. Several attempts have been made to calculate from X-ray crystallographic data a consistent set of ionic radii, from which the internuclear distances in ionic crystals can be estimated. The first table, given by V. M. Goldschmidt in 1926, was modified by Pauling. These radii are listed in Table 29.2.

TABLE 29.2

IONIC CRYSTAL RADII (pm)

Li+	60	Na+	95	K+	133	Rb+	148	Cs+	169
Be²+	31	Mg²+	65	Ca²+	99	Sr²+	113	Ba²+	135
B³+	20	Al³+	50	Sc³+	81	Y³+	93	La³+	115
C⁴+	15	Si⁴+	41	Ti⁴+	68	Zr⁴+	80	Ce⁴+	101
O²⁻	140	S²⁻	184	Cr⁶+	52	Mo⁶+	62		
F⁻	136	Cl⁻	181	Cu+	96	Ag+	126	Au+	137
				Zn²+	74	Cd²+	97	Hg²+	110
				Se²⁻	198	Te²⁻	221	Tl³+	95
				Br⁻	195	I⁻	216		

Consider ionic crystals having the general formula CA. They may be classified according to the *coordination number* of the ions, the number of ions of opposite charge surrounding a given ion. The CsCl structure [Fig. 29.9(a)] has eightfold coordination. The NaCl structure [Fig 29. 9.(b)] has sixfold coordination. Although zincblende is itself covalent, there are a few ionic crystals, e.g., BeO [Fig. 29.9(c)] with this structure, which has fourfold coordination. The coordination number of a structure is determined primarily by the number of larger ions, usually anions, that can be packed around a smaller ion, usually the cation. It should therefore depend upon the radius ratio, of cation to anion, r_C/r_A. A critical radius ratio is obtained when the anions packed around a cation are in contact with both the cation and with one another.

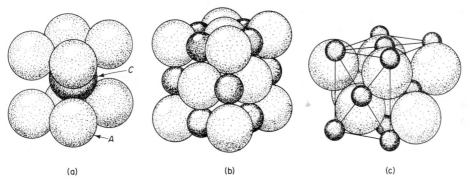

(a)   (b)   (c)

FIGURE 29.9 Three important crystal structures of type CA: (a) the unit cube of the CsCl structure; (b) the unit cube of the NaCl structure (the larger spheres represent Cl⁻ ions and the smaller spheres, Na+ ions); (c) ZnS (zincblende) structure. [R. W. G. Wyckoff, *Crystal Structures* (New York: John Wiley & Sons, Inc., 1963)]

Example 29.5 What is the critical radius ratio for the CsCl structure?

From Fig. 29.9(b) for anions of radius r_A, the side of the cubic unit cell $= 2r_A$. The diameter of the cation $2r_C$ then equals the length of the cube diagonal minus $2r_A$, or $r_C = (\sqrt{3} - 1)r_A$. The critical radius ratio is $(\sqrt{3} - 1) = 0.732$.

In Eq. (22.21) the potential energy between a pair of singly charged ions is given as the sum of an attractive term and a repulsive term,

$$U = \frac{-e^2}{4\pi\epsilon_0 r} + Be^{-r/c}$$

For NaCl, $B = 3.0 \times 10^{-16}$ J and $c = 3.3 \times 10^{-10}$ m. To calculate the electrostatic energy of a crystal, we take one ion as a center and calculate its electrostatic interaction with all the other ions in the crystal. The electrostatic interaction has a long range, so that this calculation requires a summation over many ions out to quite a large distance from the center. The repulsive interaction, however, falls off rapidly with distance, and its calculation for nearest neighbors of the central ion usually suffices.

Let us consider the computation of the total electrostatic attractive energy. We take the attractive term due to nearest neighbors, the repulsive term due to next nearest neighbors, the attractive term due to third nearest neighbors, and so on. The distance r_j from the central ion to any other ion j in a surrounding shell can be expressed in terms of the nearest neighbor distance r_0 in the crystal structure as $r_j = p_j r_0$. The total electrostatic energy is

$$\epsilon_M = \sum_j \frac{\pm e^2}{4\pi\epsilon_0 r_j} = \frac{-e^2}{4\pi\epsilon_0 r_0} \sum (\pm) p_j^{-1} = \frac{-\alpha e^2}{4\pi\epsilon_0 r_0} \tag{29.5}$$

The energy ϵ_M is called the *Madelung energy* and the constant α is called the *Madelung constant*, after the inventor of this theory. In the sum in Eq. (29.5), if the central ion is $+$, the terms are $+$ for negative ions j and $-$ for positive ions j.

Consider the crystal structure of NaCl as an example, starting with a central Na$^+$ ion. The 6 nearest Cl$^-$ ions are at r_0 or $p_1 = 1$. Then there are 12 Na$^+$ ions at $p_2 = \sqrt{2}$, giving a term $-12/\sqrt{2}$ in the sum; next there are 8 Cl$^-$ ions at $p_3 = \sqrt{3}$, and so on. The first few terms in the sum give

$$\alpha = \frac{6}{1} - \frac{12}{\sqrt{2}} + \frac{8}{\sqrt{3}} - \frac{6}{2} + \cdots = 6.000 - 8.484 + 4.620 - 3.000 + \cdots$$

The series is converging slowly. Madelung made the first summation by a clever grouping of $+$ and $-$ terms to obtain $\alpha = 1.7476$. The constants for several crystal structures are given in Table 29.3.

TABLE 29.3
MADELUNG CONSTANTS BASED ON
THE NEAREST CATION–ANION DISTANCE

Crystal structure	α
NaCl	1.7476
CsCl	1.7627
Zincblende (ZnS)	1.6381
Fluorite (CaF$_2$)	3.0388
Rutile (TiO$_2$)	4.816
Corundum (Al$_2$O$_3$)	25.0312

When the repulsive interaction with the six nearest neighbors is added to the electrostatic interaction in Eq. (29.5), an expression for the crystal energy per NaCl unit is obtained.

$$\epsilon_c = \frac{-\alpha e^2}{4\pi\epsilon_0 r_0} + 6Be^{-r_0/c} \qquad (29.6)$$

The repulsive energy usually amounts to only about 10% of the Madelung energy.

All crystals of a given structure have the same Madelung constant. To calculate $-\epsilon_M$, we need only to know r_0 the distance of closest ionic approach for the particular substance.

Example 29.6 KCl, NaCl, and MgO all have the NaCl structure, with $r_0 = 314, 282,$ and 210 pm, respectively. Calculate their Madelung energies per mole.

$$-E_M = -L\epsilon_M = \frac{L\alpha e^2}{4\pi\epsilon_0 r_0}$$

$$= \frac{(6.022 \times 10^{23}\ \text{mol}^{-1})(1.7476)(1.602 \times 10^{-19}\ \text{C})^2}{4\pi(8.854 \times 10^{-12}\ \text{J}^{-1}\ \text{C}^2\ \text{m}^{-1})}\frac{1}{r_0}$$

$$= 2.43 \times 10^{-4}\ (1/r_0)\ \text{J mol}^{-1}$$

$$\text{KCl} -E_M = (2.43 \times 10^{-4})(314 \times 10^{-12})^{-1} = 774\,000\ \text{J mol}^{-1}$$

$$= 774\ \text{kJ mol}^{-1}$$

$$\text{NaCl} -E_M = (2.43 \times 10^{-4})(282 \times 10^{-12})^{-1} = 862\,000\ \text{J mol}^{-1}$$

$$= 862\ \text{kJ mol}^{-1}$$

$$\text{MgO} -E_M = 4(2.43 \times 10^{-4})(210 \times 10^{-12})^{-1} = 4\,630\,000\ \text{J mol}^{-1}$$

$$= 4630\ \text{kJ mol}^{-1}$$

(Note that in case of MgO α for NaCl must be multiplied by the product of the charge numbers $z_1 z_2$.)

Multiple ionic charges greatly increase the Madelung energies of crystals—we have here an explanation of the great stability of many oxide crystals. For example, crystalline thorium oxide, ThO_2, with Th^{4+} and O^{2-} ions, has the fluorite structure. Its Madelung energy has the enormous value of $-23\,100\ \text{kJ mol}^{-1}$.

29.10 Crystal Energies

We have so far neglected a fairly important contribution to the total cohesive energy of the crystal. The ion Na^+ is like a neon atom, and Cl^- is like an argon atom, except for their net charges. We know that at low enough temperatures even these inert gases condense to liquids and freeze to solids. The intermolecular attractions between such neutral atoms and molecules are called van der Waals or London forces. Many organic crystals are held together entirely by such forces. In the case of interactions of ions like Na^+ and Cl^- the London forces are still important even if overshadowed by the much larger electrostatic interaction. Table 29.4 lists a few of the

TABLE 29.4

CONTRIBUTIONS TO CRYSTAL ENERGIES $-\Delta U_c$
OF ALKALI HALIDES (KJ MOL^{-1})

Crystal	Madelung	Repulsive	London	Total	Experimental (Born–Haber)[a]
LiF	1195	−185	16	1026	1021
NaF	1038	−148	19	909	912
NaCl	855	−98	22	779	762
NaBr	807	−86	23	744	724
NaI	745	−72	26	699	678
KCl	766	−90	30	706	694
RbCl	736	−83	33	686	670
CsCl	680	−74	49	655	627

[a]The zero-point vibrational energies are included in these values.

alkali halides with the calculated contributions from each of the three sources of conesive energy mentioned so far: Madelung energy, repulsive energy, and London energy. The London energies in ionic crystals are by no means negligible and, in fact, may approach the magnitude of the repulsive energies.

Experimental values for crystal energies are also included in Table 29.4. These can be obtained by a well-known thermochemical method called the Born–Haber cycle, provided that we know the standard enthalpy of formation ΔH_f° of the crystal-line compound and other thermochemical quantities. The cycle goes as follows:

$$Na(c) + \tfrac{1}{2}Cl_2(g) = NaCl(c) \qquad \Delta U_f^\circ \text{ (standard energy of formation)}$$
$$Na(g) = Na(c) \qquad -\Delta U_s \text{ (sublimation)}$$
$$Cl(g) = \tfrac{1}{2}Cl_2(g) \qquad -\tfrac{1}{2}D_0 \text{ (energy of dissociation per atom)}$$
$$Na^+(g) + e^- = Na(g) \qquad -I \text{ (ionization energy)}$$
$$\underline{Cl^-(g) = Cl(g) + e^- \qquad A \text{ (electron affinity)}}$$
$$Na^+(g) + Cl^-(g) = NaCl(c) \qquad \Delta U_c$$

Therefore,

$$-\Delta U_c = \Delta U_f^\circ - \Delta U_s - \tfrac{1}{2}D_0 - I + A \qquad (29.7)$$

It would be nice to be able to say that the theory of crystal energy as outlined now allows us to predict the crystal structure of a substance in its most stable state. We then could explain, for example, why cesium chloride has a different structure from sodium chloride. However, the net energy differences between possible alternative crystal structures are usually quite small (40 kJ or so) and the theoretical calculations of the energy terms are often not sufficiently precise to pick out with certainty the most stable structure. Thus the Madelung energy of sodium chloride in a cesium chloride–type structure would be only 5 kJ higher than in the actual structure, supposing that the same interionic distance was maintained.

In 1896, Roberts-Austen, an English metallurgist, showed that gold diffuses faster in lead at 300°C than sodium chloride diffuses in water at 15°C. This is one example of the surprising ease with which atoms can sometimes move about in the solid state. It was difficult to believe that atoms or ions could move easily in solids by changing places with one another: The activation energy for such a process would be too high. More reasonable mechanisms were suggested by I. Frenkel in 1926 and by W. Schottky in 1930. They proposed models for what are now called *point defects* in crystals. Various point defects are illustrated in Fig. 29.10.

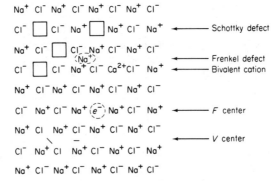

FIGURE 29.10 Various kinds of intrinsic point defects in alkali–halide crystals.

The Schottky defect consists of a pair of vacancies of opposite signs. The Frenkel defect consists of an ion that has moved to an interstitial site in the crystal structure plus the vacancy left behind. Frenkel and Schottky defects are called *intrinsic defects*. They do not alter the exact stoichiometry of a crystal. They provide mechanisms through which atoms and ions can move within a crystal, either by jumping from an occupied site into a vacancy or by jumping from one interstitial site to the next.

We can calculate the concentrations of point defects from simple statistical considerations. It costs an energy ϵ to make a defect, but entropy S is gained owing to the disorder associated with the entropy of mixing of the defects with the occupied lattice sites. If N defects are distributed among the total of N_0 crystal sites, the entropy of mixing is $S = k \ln W = k \ln N_0!/(N_0 - N)!\,N!$. If ϵ is the increase in energy per defect, the change in the Helmholtz free energy is $\Delta A = \Delta U - T\Delta S = N\epsilon - kT \ln N_0!/(N_0 - N)!\,N!$. At equilibrium, $(\partial\,\Delta A/\partial N)_T = 0$. Applying the Stirling formula ($\ln X! = X \ln X - X$), we find that

$$\ln \frac{N}{N_0 - N} = \frac{\epsilon}{kT}$$

and thus if $N \ll N_0$,

$$N = N_0\,e^{-\epsilon/kT} \tag{29.8}$$

As an example, if ϵ is about 1 eV and T is 1000 K, $N/N_0 \approx 10^{-5}$. For a pair of vacancies, the expression for the number of ways of forming the defect is squared and for

Schottky defects,

$$N = N_0 e^{-\epsilon/2kT} \qquad (29.9)$$

For Frenkel defects, if N_0' is the number of interstitial sites,

$$N = (N_0 N_0')^{1/2} e^{-\epsilon/2kT} \qquad (29.10)$$

Example 29.7 The energy of formation of a Schottky defect in NaCl is about 2.0 eV and that for a Frenkel defect about 3.0 eV. Estimate the mole fraction of these defects in a crystal of NaCl at 1300 K. 1 eV $= 1.602 \times 10^{-19}$ J.

Schottky defects:

$$N/N_0 = e^{-\epsilon/2kT} = \exp\left[-(2.0)(1.60 \times 10^{-19})/(2)(1.38 \times 10^{-23})(1300)\right]$$
$$= 1.3 \times 10^{-4}$$

Frenkel defects: There are eight interstitial sites for Na$^+$ ions in the NaCl structure (Fig. 29.9(b)). (Only the smaller Na$^+$ ions form Frenkel defects.) Thus $N_0' = 2N_0$ in Eq. (29.10).

$$N/N_0 = \sqrt{2}\, e^{-\epsilon/2kT} = \sqrt{2} \exp\left[-(3.0)(1.60 \times 10^{-19})/(2)(1.38 \times 10^{-23})(1300)\right]$$
$$= 2.2 \times 10^{-6}$$

If a NaCl crystal is heated in Na vapor it acquires a deep yellow color; KCl crystals heated in K vapor acquire a magenta color. The color centers are called F-centers. When NaCl takes up extra Na, there is an excess of occupied sodium sites in the crystal so that some Cl$^-$ sites are vacant. There are six Na$^+$ sites adjacent to a vacant Cl$^-$ site. The extra electron on the excess Na atom can be shared between all six Na$^+$ ions to give a delocalized electron at the vacant Cl$^-$ site. This electron is similar to the delocalized electrons in a box described in Section 20.16. Light is absorbed as the electron makes a transition from its ground state to an excited state.

29.12 Linear Defects—Dislocations

The problem for metals under stress is not why they are so strong, but why they are so weak. The calculated elastic limit of a perfect crystal is 10^2 to 10^4 times that actually observed. There must be some imperfections or defects in actual metal crystals that cause them to deform plastically under quite small loads.

A solution to this problem was worked out independently in 1934 by Taylor, Orowan, and Polanyi. Crystals contain linear defects called *dislocations*. These defects have been compared to rucks in carpets. We all have pulled carpets over floors and know that there are two ways of doing it. One can take hold of one end and tug, or one can make a ruck in one end of the carpet and gently edge it to the other end. For a big heavy carpet, the second way involves less effort. The dislocation most like a ruck is the *edge dislocation* shown in Fig. 29.11, which represents a model of crystal structure viewed along a dislocation line. The dislocation line is perpendicular to the

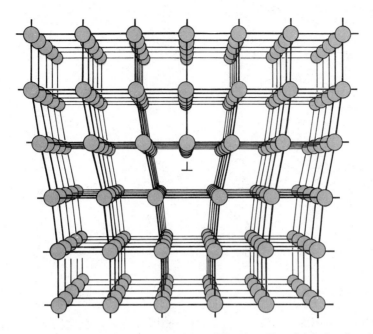

FIGURE 29.11 The structure of an edge dislocation. The dislocation line is normal to the plane of the paper at point marked ⊥. The dislocation is the result of inserting an extra half-plane of atoms in the upper half of this crystal structure.

plane represented by the plane of the paper. The presence of the dislocation allows the crystal to deform readily under the influence of a shear stress. The atoms are displaced in the *slip plane* that includes the dislocation line. Thus the dislocation can move across the crystal from one side to the other, the result being a displacement of the top half of the crystal relative to the bottom half.

The other basic kind of dislocation is the *screw dislocation*. You can visualize this defect by cutting a rubber stopper parallel to its axis, and then pushing on one end so as to create jog at the other end. If you suppose that initially the stopper contained atoms at regular lattice points, the results of the deformation would be to convert the parallel planes of atoms normal to the axis into a kind of spiral ramp. Such a displacement of the atoms constitutes a screw dislocation; the dislocation line is along the axis of the stopper. A model of the emergence of a screw dislocation at the surface of a crystal is shown in Fig. 29.12.

29.13 Effects Due to Dislocations

Cohesive forces between atoms in a crystal offer little resistance to the gliding motion of dislocations, but a crystal without dislocations would approach the theoretical strength of a perfect crystal. Thin metallic whiskers can be grown virtually free of dislocations. Pure iron whiskers have been obtained with tensile strengths up to 1.4×10^{12} N m^{-2}, compared to a maximum of 3×10^{11} for the strongest steel wire.

FIGURE 29.12 An emergent screw dislocation at the surface of a crystal. Atoms or molecules are denoted by small cubes.

How can the addition of a small percentage of an alloying element (carbon in iron, copper in aluminum, as examples) so greatly increase the mechanical strength of a metal? Since the metal deforms through movements of dislocations, anything that hinders such movements increases the resistance of the metal to deformation. A foreign atom introduced into the structure of a metal tends to reside at a position of minimum free energy. Since positions adjacent to dislocations are sites of lower free energy, foreign atoms tend to segregate at dislocations. In this way, they tend to stabilize dislocations by locking them into place. Thus alloying elements can increase the strength of metals by rendering their dislocations less mobile.

Dislocations also provide preferred sites within crystals for chemical reactions and physical changes (such as phase transformation, precipitation, or etching). The point of emergence of a dislocation at the surface of a crystal is a site of enhanced chemical reactivity, a fact often revealed by the pattern of etch pits formed at the surface. We can measure the number of dislocations per unit area by counting such etch pits. Values range from 10^5 m^{-2} in the best silver and germanium crystals (small crystals can be grown practically free of dislocations) to 10^{16} m^{-2} in a severely deformed metal.

When a crystal is etched, the easiest place to remove an atom from the surface is at a dislocation, since here the crystal structure is not perfect. But, when a crystal grows from the vapor or melt, the easiest place to deposit each new atom or molecule is also at the site of a dislocation. In Section 18.4, we saw that spontaneous nucleation of a new phase in a homogeneous system is extremely improbable; and this improbability extends also to the deposition of an atom onto a perfect crystal face. In brief, such a process is improbable because there is a loss of entropy in the condensation process, but there is no compensating decrease in energy for the first few atoms in a new layer, since they have no neighbors to bind them. If there is an emergent screw dislocation at the surface, however, a new atom can readily deposit at the edge of the developing screw. This mechanism of crystal growth was discovered in 1949 by F. C. Frank. An experimental example is shown in Fig. 29.13.

FIGURE 29.13 A double spiral at the surface of a crystal of silicon carbide, centered on the site of emergence of a pair of screw dislocations. Photomicrograph 300 ×. [W. F. Knippenberg, Philips Research Laboratory]

Problems

1. What is the plane of closest packing in the bcc structure? What is the number of iron atoms per unit area in this plane at 25°C where $a_0 = 286.1$ pm?

2. What is the void volume in a bcc structure in which the corner atoms are in contact with the central atom?

3. Consider a fcc structure based on cubic closest packing of spheres of unit radius. What is the largest radius of a sphere that could fit into the interstices of the ccp structure?

4. Tridymite, the high-temperature form of SiO_2, is hexagonal with $a_0 = 503$, $c_0 = 822$ pm. There are four SiO_2 units per unit cell.

   Si: $(\frac{1}{3} \frac{2}{3} \pm z)(\frac{2}{3} \frac{1}{3} \frac{1}{2} \pm z)$ $z = \frac{3}{16}$

   O: two at $(\frac{1}{3} \frac{2}{3} 0)(\frac{2}{3} \frac{1}{3} \frac{1}{2})$

   six at $(\frac{1}{2} 0 \frac{1}{4})(0 \frac{1}{2} \frac{1}{4})(\frac{1}{2} \frac{1}{2} \frac{1}{4})(\frac{1}{2} 0 \frac{3}{4})(0 \frac{1}{2} \frac{3}{4})(\frac{1}{2} \frac{1}{2} \frac{3}{4})$

   Draw the projection of this structure on the *ab* plane, showing all the atoms in one unit cell. Calculate the distance from a Si atom to each of the two types of O bonded to it.

5. Solid Cl_2 at 113 K has a structure based on an orthorhombic unit cell with $a_0 = 629$, $b_0 = 450$, $c_0 = 821$ pm. There are 8 Cl atoms in the unit cell, located at $\pm[0, y, z; \frac{1}{2}, y, \frac{1}{2} - z; \frac{1}{2}, \frac{1}{2} + y, z; 0, \frac{1}{2} + y, \frac{1}{2} - z]$ with $y = 0.1173$, $z = 0.1016$. Draw the unit cell and calculate the Cl—Cl bond length and the closest distance of approach between Cl_2 molecules. [Note: For example, $-(\frac{1}{2}, \frac{1}{2} + y, z) \equiv (-\frac{1}{2}, -\frac{1}{2} - y, -z) = (\frac{1}{2}, \frac{1}{2} - y, 1 - z)$].

6. From Eq. (29.2) show that the average energy of an electron in the valence band (conduction band) is $\bar{\epsilon} = \frac{3}{5}\epsilon_F$. Hence calculate the total translational kinetic energy of the conduction electrons in Na at 0 K. (Data in Example 29.3.)

7. Calculate the Fermi energy ϵ_F of the conduction electrons in a gold crystal, assuming one conduction electron per atom. Take $\rho(Au) = 19.32$ g cm^{-3}.

8. The electronic heat capacity of graphite is $C_P = 3.3 \times 10^{-6}T$. Calculate the ΔS of graphite between 10^{-6} and 10^{-3} K.

9. From Eq. (29.6) show that for a NaCl crystal structure with equilibrium interionic distance r_o,

$$\epsilon_c = \frac{-\alpha e^2}{4\pi\epsilon_o r_o}\left(1 - \frac{c}{r_o}\right)$$

From data in Table 29.4, calculate the constant c for KCl and NaCl, with $r_o = 314$ and 282 pm, respectively.

10. If the change in ϵ_F of Na with T is due entirely to the expansion of the metal with T, calculate ϵ_F at 300 K and compare with value at 0 K in Example 29.3. $\alpha = V^{-1}(\partial V/\partial T)_P = 7.1 \times 10^{-5}$ K^{-1}.

11. The Fermi energy of gold is 5.51 eV. Plot the Fermi–Dirac distribution function p vs. ϵ for gold at $T = 100$ K and $T = 1000$ K.

12. A crystal of pure Ge is doped with Sb to the extent of 1.65×10^{22} m^{-3} Sb atoms. If all the Sb is ionized, calculate C_e and C_h and the conductivity of the doped crystal at 400 K. (Use data for pure Ge in Example 29.4.)

13. What are the critical radius ratios r_C/r_A for the NaCl and cubic ZnS structures? What structures would you predict for AgCl and AgI? What are the actual structures? Any comment?

14. Calculate the Madelung energy of CsCl for which the unit-cell dimension $a_0 = 362$ pm. Suppose that CsCl crystallized in the NaCl structure with the same value of a_0. What would its Madelung energy be?

15. Calculate the Madelung constant for a line of alternating $+$ and $-$ charges with a separation a.

16. NH$_4$F crystallizes in the zinc-blende structure with $r_o = 263$ pm. The crystal energy is 757 kJ mol^{-1}, the electron affinity of F is 3.45 eV, and the ionization potential of H is 13.6 eV. For NH$_4$F(c), $\Delta H_f(298$ K$) = 469$ kJ mol^{-1}. Calculate the proton affinity of NH$_3$ (i.e., ΔU for NH$_3 + H^+ = NH_4^+$).

17. MgO has the NaCl structure with $a_0 = 421.3$ pm at 25°C. The ΔH(sublimation) of Mg is 149.0 kJ mol^{-1}, and Mg $\longrightarrow$ Mg$^{2+} + 2e$ has $\Delta H = 2179$ kJ mol^{-1}. The $\Delta H_f^\circ(MgO) = -601.5$ kJ mol^{-1}. Calculate the ΔH° of O $+ 2e \longrightarrow$ O^{2-}.

18. The unit cell of α Fe has $a_0 = 286.1$ pm at 25°C. The coefficient of linear expansion of Fe from 25 to 906°C is 1.20×10^{-5} K^{-1}. At 906°C (1 atm) αFe $\longrightarrow \gamma$Fe (fcc) with $\Delta v = 0.001$ cm^3 g^{-1}. Calculate the nearest distance apart of Fe atoms in αFe and in γFe at the transition point.

19. Suppose that two Ni^{2+} ions in NiO are replaced by a Li$^+$ ion and a Ni^{3+} ion. Calculate the change in Madelung energy caused by this substitution (when the new ions are well separated).

30

The Liquid State and Intermolecular Forces

Because the extremes of total chaos and perfect order are both relatively simple to treat mathematically, the theories of gases and of crystals are well advanced. The liquid state, a compromise between order and disorder, has so far defied a comprehensive theoretical treatment. In an ideal gas, molecules move independently and interactions between them are neglected. The energy of the perfect gas is simply the sum of the energies of individual molecules, their internal energies plus their translational kinetic energies; there is no intermolecular potential energy. It is therefore possible to write down a partition function such as that in Eq. (12.10) from which all equilibrium properties of the gas are readily derived. In a crystalline solid, the molecules, atoms, or ions vibrate about equilibrium positions to which they are held by strong forces. In this case, too, an adequate partition function, such as that in Eq. (12.35), can be obtained. In a liquid, on the other hand, the situation is much harder to define. The cohesive forces are sufficiently strong to lead to a condensed state, but not strong enough to exclude a considerable translational energy for individual molecules. The thermal motions introduce some disorder into the liquid without completely destroying the regularity of its structure. It is therefore difficult to evaluate the partition function for a liquid.

Liquids, like crystals, can be classified according to the kinds of cohesive forces that hold them together. There are ionic liquids such as molten salts, metallic liquids consisting of ions and mobile electrons, liquids such as water held together mainly by hydrogen bonds, and molecular liquids in which cohesion is due to van der Waals forces between molecules. Many liquids fall into this last group, and even when other forces are present, the van der Waals contribution may be large. The origin of these forces will be considered in this chapter.

671

Consider a liquid in equilibrium with a solid at some temperature and pressure. The molar enthalpy and molar entropy of the liquid are both higher than those of the solid. The stable phase is determined by the difference in molar Gibbs free energy, $G^l - G^s = \Delta G_f = \Delta H_f - T\,\Delta S_f$. With increasing temperature, the greater randomness of the liquid, and hence its greater entropy, finally make the $T\,\Delta S_f$ term large enough to overcome the ΔH_f term, so that the crystal melts when the condition $\Delta G_f = 0$ is reached: $T(S^l - S^s) = H^l - H^s$.

The sharpness of the melting point is noteworthy: There is no continuous gradation of properties between liquid and crystal. The sharp transition is due to the rigorous geometrical requirements that must be fulfilled by a crystal structure. It is not possible to introduce small regions of disorder into a crystal without seriously disturbing the structure over a long range.

J. D. Bernal made several experiments in which molecules were modeled by ball bearings, and the packing patterns studied after vigorous shaking. Walton and Woodruff coated the ball bearings with oil to simulate the attractive forces between molecules, and provided random thermal motions by shaking the bearings in a tray. A snapshot of a typical pattern is shown in Fig. 30.1. This pattern is like a large-scale version of a two-dimensional liquid of spherical molecules, e.g., argon. Note the failure to achieve closest packing and the occurrence of holes in the structure of nearly "molecular" size. In three dimensions, the number of contacts between ball bearings is about 9 compared to 12 in a close-packed solid structure.

Oily ball bearings in a vibrating tray may be considered as a sort of analog

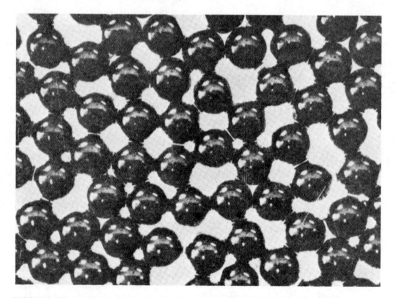

FIGURE 30.1 Molecules in the liquid state simulated by oily ball bearings shaken in a tray. [A. J. Walton and A. G. Woodruff, *Contemp. Phys. 10*, 59 (1969)]

computer for the study of two-dimensional liquids. Calculations with powerful digital computers have also been applied to these problems by the methods of *molecular dynamics*. A collection of molecules is taken with certain initial positions and velocities. A law of intermolecular force is chosen and the forces on each molecule due to all the other molecules are calculated. Then Newton's equation of motion, $\mathbf{F} = m\mathbf{a} = md^2\mathbf{r}/dt^2$, is integrated numerically for each molecule. The trajectories of the molecules are plotted to give a dynamic picture of how their positions change with time.

Figure 30.2 shows the results of such calculations by Wainwright and Alder. The interactions between 32 hard-sphere molecules were calculated. The pictures show the projection onto one face of the container of the motions of the centers of some of the molecules over a short interval of time. The initial condition was chosen by assigning to all the molecules a certain constant speed, which would determine an initial "temperature." The left set of trajectories corresponds to a "temperature" just below a "melting point" and the right set to one just above a "melting point." Actually,

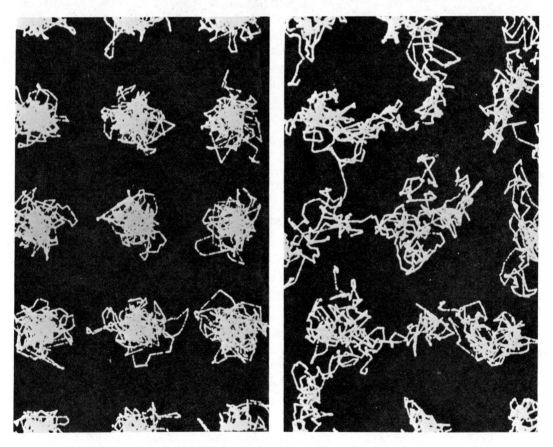

FIGURE 30.2 Calculated trajectories of a set of hard-sphere molecules: (a) below a "melting point"; (b) above a "melting point."

in this calculation the "melting point" was not sharp but it was confined to a narrow range of initial "temperatures." The significance of this work is that even a simple hard-sphere model of 32 atoms can yield a molecular-dynamic picture that reproduces quite well the phenomenon of melting. Much of the detailed theoretical work on liquid structure has dealt with the behavior of hard-sphere models: They are believed to be capable of reproducing most of the interesting properties of liquids.

30.2 X-Ray Diffraction of Liquid Structures

If a liquid were completely amorphous, i.e., without any regularity of structure, it would give a continuous scattering of X rays without maxima or minima. This is not the case. A typical pattern, obtained from liquid mercury, is shown in Fig. 30.3(a), as a microphotometer tracing of a photographic recording of the diffraction pattern. Several intensity maxima appear, having positions that correspond to some of the larger interplanar spacings in the crystal structure. The fact that only a few maxima are observed in diffraction patterns from liquids is in accord with a structure that has short-range order and increasing disorder at longer range. The Bragg condition, $n\lambda = 2d \sin \theta$, shows that as interplanar spacing d goes down, scattering angle θ goes up. To obtain maxima corresponding to smaller interplanar spacings or higher orders of diffraction, the long-range order of the crystal must be present.

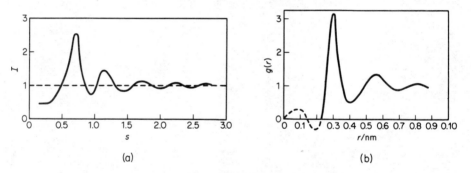

(a)   (b)

FIGURE 30.3 X-ray diffraction of liquid mercury at 20°C: (a) photometric trace of diffraction pattern; (b) radial distribution function calculated from (a).

The arrangement of atoms in a monatomic liquid like mercury is described by introducing a number density function $\rho(r)$. Taking the center of one atom as origin, this $\rho(r)$ gives the probability of finding the center of another atom at the end of a vector of length r drawn from the origin. The chance of finding another atom between a distance r and $r + dr$, irrespective of angular orientation, is the radial distribution function $4\pi r^2 \rho(r) \, dr$ (cf. page 469). The intensity of scattered X-radiation is given by the expression

$$I(\theta) \propto \int_0^\infty 4\pi r^2 \rho(r) \frac{\sin sr}{sr} \, dr \qquad \text{with} \qquad s = \frac{4\pi}{\lambda} \sin \frac{\theta}{2} \qquad (30.1)$$

By an application of Fourier's integral theorem, the integral in Eq. (30.1) can be inverted, yielding

$$g(r) = 4\pi r^2 \rho(r) \propto \frac{2}{\lambda} \int_0^\infty I(\theta) \frac{\sin sr}{sr} d\theta \qquad (30.2)$$

With this relationship, we can calculate a radial distribution function, such as that plotted in Fig. 30.3(b), from an experimental scattering curve, such as that in Fig. 30.3(a).

In the plot of $4\pi r^2 \rho(r)$ vs. r in Fig. 30.3(b) the regular coordination in the liquid-mercury structure is clearly evident, but the fact that maxima in the curve are rapidly damped out at larger interatomic distances indicates that deviations from an ordered arrangement become greater as one travels outward from any centrally chosen atom. In general, liquid metals have approximately close-packed structures quite similar to those of solid metals, but with interatomic spacings expanded by about 5%. The number of nearest neighbors in a close-packed structure is 12. In liquid sodium, each atom is found to have on average 10 nearest neighbors.

30.3 Liquid Water

One of the most difficult structures to decipher is that of water, but its great importance has led to many experimental and theoretical investigations. The first extensive X-ray diffraction study of water was by Morgan and Warren in 1938. Narten and Levy in 1971 were able to apply improved techniques. The scattering by X rays is due to the electrons in atoms, and therefore X rays primarily "see" the oxygen atoms in water. Consequently, the molecule H_2O appears to X rays as virtually spherically symmetrical. Figure 30.4 shows results from the X-ray diffraction of water over a range of temperatures. At higher temperatures, the long-range order becomes less evident, and at 200°C it disappears beyond 600 pm. There is a sharp peak in $g(r)$ at 290 pm due primarily to nearest neighbors of the central molecule, and when $g(r)$ is integrated over the volume element $4\pi r^2\, dr$ in this shell, the number of nearest neighbors is calculated to be 4.4 at all temperatures from 4 to 200°C. These results indicate that the coordination in liquid water is approximately tetrahedral, as we know to be the case in ice I (Fig. 29.3). The peaks in the $g(r)$ function for water at 450 to 530 pm and at 640 to 780 pm are also in good accord with the tetrahedral arrangement.

The small but distinct peak at 350 pm cannot be explained by the tetrahedral structure. The structure of ice I has six interstitial sites at a distance of 348 pm from the central molecule. It was therefore suggested that when ice melts, some of the water molecules move from their tetrahedral sites into these interstitial sites, thus accounting for the peak in $g(r)$ at 350 pm. There is a contraction in volume of about 9% when ice melts. The X-ray data indicate that occupancy of the interstitial sites increases from 45% at 4°C to 57% at 200°C.

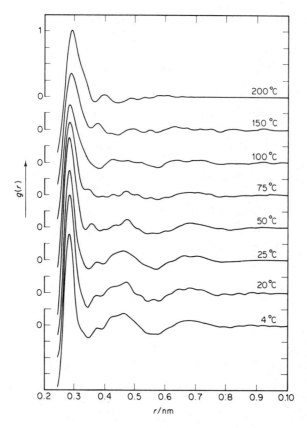

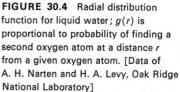

FIGURE 30.4 Radial distribution function for liquid water; $g(r)$ is proportional to probability of finding a second oxygen atom at a distance r from a given oxygen atom. [Data of A. H. Narten and H. A. Levy, Oak Ridge National Laboratory]

One of the most interesting features of the structure of water, however, is the angular correlation between molecules, as determined by the pattern of hydrogen bonds. X-ray diffraction cannot provide this information. Neutron diffraction and particularly electron diffraction studies can, however, yield pertinent data. It is necessary to use heavy water, D_2O, because neutron scattering from protons is mainly incoherent owing to nuclear-spin effects. The neutron-diffraction results indicate that water has a less ordered structure than previously believed.

Although assemblies of hard spheres may be able to reproduce many general features of the behavior of liquids, they cannot explain all the structure of liquids. Thus, in the case of water, it is necessary to include interactions of the many-body type; i.e., one cannot simply add pairwise interactions. A computed structure from this kind of model is shown in Fig. 30.5. All hydrogen bonds are shown that have energies greater than 8.4 kJ mol^{-1}. One must not get the impression that water is a static structure—Fig. 30.5 is simply an instantaneous picture of a rapidly changing dynamic pattern, in which the molecules are continually agitated by thermal motion and the network of hydrogen bonds flickers and varies with great rapidity. Experimental work on the structure of water lends no support at all to a model that contains regions of icelike structure interspersed in a more random arrangement.

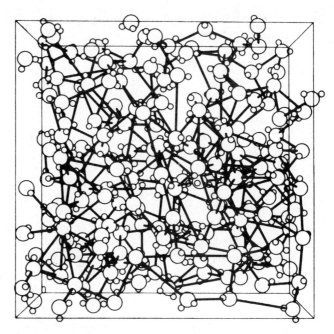

FIGURE 30.5 Pattern of hydrogen bonds in liquid water as indicated by energy computations. [Reprinted by permission from P. Barnes, J. L. Finney, J. D. Nicholas, and J. E. Quinn, *Nature* 282, 459 (1979). Copyright © 1979 Macmillan Journals Limited.]

30.4 Cohesion of Liquids—Internal Pressure

Whatever the model chosen for the liquid state, the cohesive forces are of primary importance. Ignoring, for the time being, the origin of these forces, we can estimate their magnitudes from thermodynamic consideration of the *internal pressure, P_i*.

We recall from Eq. (8.14) that $P_i = (\partial U/\partial V)_T = T(\partial P/\partial T)_V - P$. In the case of an ideal gas, the internal pressure is zero because intermolecular forces are absent. In the case of an imperfect gas, $(\partial U/\partial V)_T$ becomes appreciable, and in the case of a liquid it may become much greater than the external pressure.

Example 30.1   At 273 K, the expansivity of mercury is $\alpha = 17.8 \times 10^{-5}$ K^{-1} at 100 kPa and 15.3×10^{-5} at 7×10^5 kPa, and the compressibility is $\beta = 38.8 \times 10^{-12}$ Pa^{-1} at 100 kPa and 31.8×10^{-12} Pa^{-1} at 7×10^5 kPa. Calculate the internal pressure P_i of mercury at 273 K and the two pressures cited.

From Eq. (2.25), $(\partial P/\partial T)_V = \alpha/\beta$ so that $P_i = T\alpha/\beta - P$.
   At 100 kPa:
$$P_i = (273 \text{ K})\left(\frac{17.8 \times 10^{-5} \text{ K}^{-1}}{38.8 \times 10^{-12} \text{ Pa}^{-1}}\right) - 10^5 \text{ Pa} = 1250 \text{ MPa}$$

At 7×10^5 kPa:
$$P_i = (273 \text{ K})\left(\frac{15.3 \times 10^{-5} \text{ K}^{-1}}{31.8 \times 10^{-12} \text{ Pa}^{-1}}\right) - 7 \times 10^8 \text{ Pa} = 610 \text{ MPa}$$

The internal pressure is the resultant of the forces of attraction and the forces of repulsion between the molecules in a liquid. It therefore depends markedly on the volume V, and thus on the external pressure P. This effect is shown in the following data for diethyl ether at 298 K:

P (MPa)	20	80	200	500	700	900	1100
P_i (MPa)	275	280	250	200	4	-150	-420

For moderate increases in P, P_i decreases only slightly, but as P exceeds 500 MPa, P_i begins to decrease rapidly and goes to large negative values as the liquid is further compressed. This behavior reflects on a larger scale the law of force between individual molecules; at high compressions, the repulsive forces become predominant.

Joel Hildebrand was the first to point out the significance of the internal pressures of liquids in determining solubility relationships. A solution of two liquids that differ considerably in P_i usually exhibits considerable positive deviation from ideality, that is, a tendency toward lowered mutual solubility.

30.5 Intermolecular Forces

In the van der Waals model, molecules behave as hard spheres with forces of attraction between them. A "hard sphere" means that at some distance between molecular centers, an infinite repulsive force suddenly occurs between a pair of molecules. Since a molecule has a relatively open structure of positive nuclei and electrons, the hard-sphere model is not realistic. In reality, both attractive and repulsive forces between molecules vary with distance. The repulsive forces have a much *shorter range* than the attractive forces; i.e., they decrease more rapidly with separation r, but they do not occur suddenly as in the hard-sphere model. At long distances the intermolecular force is a net attraction, but at short distances it is a net repulsion.

Experimental and theoretical studies both indicate that the attractive forces fall off approximately as the seventh power of the separation between molecular centers:

$$F_a = -k_a r^{-7} \tag{30.3}$$

Note that, although force is a vector quantity, we are considering simply the force along the direction between two molecules. An attractive force has a negative sign, since if we take the center of one molecule as the origin $r = 0$, the force is opposite in direction to the vector from $r = 0$ to the second molecule.

The repulsive force has a shorter range and can be represented approximately as

$$F_r = k_r r^{-13} \tag{30.4}$$

The total net intermolecular force is the sum of the attractive and the repulsive terms,

$$F = F_a + F_r = -k_a r^{-7} + k_r r^{-13} \tag{30.5}$$

Instead of the forces themselves, it is often more convenient to consider the corresponding intermolecular potential energies. The potential energy U is defined by

$$dU = -F\,dr \tag{30.6}$$

We suppose one molecule to be at an infinite distance from the other one and bring it up to a separation r. This process is represented mathematically by the integration of Eq. (30.6),

$$\int_{\infty}^{r} dU = -\int_{\infty}^{r} F\,dr' \tag{30.7}$$

At infinite separation between molecules, the intermolecular potential energy is zero. Thus Eq. (30.7) gives

$$U(r) - U(\infty) = U(r) = -\int_{\infty}^{r} F\,dr'$$

We introduce the intermolecular force from Eq. (30.5), and carry out the integration,

$$U(r) = \frac{k_r r^{-12}}{12} - \frac{k_a r^{-6}}{6} \tag{30.8}$$

Figure 30.6 is a plot of the repulsive and attractive parts of the potential energy from Eq. (30.8) and of the sum of the two terms, $U(r)$.

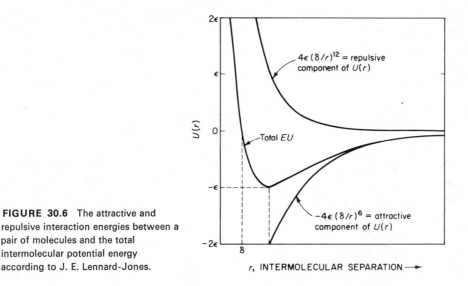

FIGURE 30.6 The attractive and repulsive interaction energies between a pair of molecules and the total intermolecular potential energy according to J. E. Lennard-Jones.

The intermolecular potential energy in Eq. (30.8) is usually written in a somewhat different form,

$$U(r) = 4\epsilon\left[\left(\frac{\delta}{r}\right)^{12} - \left(\frac{\delta}{r}\right)^{6}\right] \tag{30.9}$$

As shown in Fig. 30.6, δ is one of the values of r for which $U(r) = 0$ (the other being $r = \infty$), and ϵ is the energy at the minimum in the potential-energy curve. [Please prove these statements.] The value of ϵ is often called the "depth of the potential well." Values of δ and ϵ for various gases are summarized in Table 30.1.

CONSTANTS FOR LENNARD-JONES 6-12
INTERMOLECULAR POTENTIAL OF EQ. (30.9)
OBTAINED FROM SECOND VIRIAL COEFFICIENTS

Gas	ϵ $(10^{-22}$ J)	δ (pm)
He	1.41	256
Ne	4.92	275
Ar	16.5	341
Kr	23.6	360
Xe	30.5	410
H_2	5.11	293
N_2	13.1	370
O_2	16.2	358
CO_2	26.1	449
CH_4	20.5	382
C_2H_4	26.8	522

The potential in Eq. (30.9) is called the *Lennard-Jones 6-12 potential* after the English physicist who first made extensive use of it in the theory of imperfect gases. Values for δ and ϵ can be calculated from experimentally determined values of the virial coefficients. Figure 30.7 shows the Lennard-Jones potential energies for several gases.

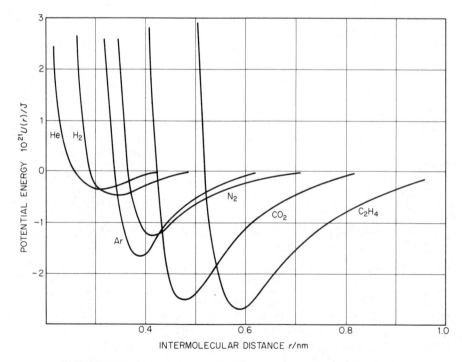

FIGURE 30.7 Lennard-Jones potential-energy curves for several molecules.

All forces between atoms and molecules are electric in origin. They are ultimately based on Coulomb's Law of attraction between unlike and repulsion between like charges. We usually describe the potential energies rather than the forces themselves. We can then distinguish the following classes of intermolecular and interionic potential energy:

1. The Coulombic energy of interaction between ions with net charges, leading to a long-range attraction or repulsion, with $U \propto r^{-1}$.
2. The energy of interaction between permanent dipoles, with $U \propto r^{-6}$.
3. The energy of interaction between an ion and a dipole induced by it in another molecule, with $U \propto r^{-4}$.
4. The energy of interaction between a permanent dipole and a dipole induced by it in another molecule, with $U \propto r^{-6}$.
5. The potential energy between nonpolar atoms or molecules, such as the inert gases, with $U \propto r^{-6}$.
6. The overlap energy arising from interaction of the nuclei and electrons of one molecule with those of another. The overlap leads to repulsion at very close intermolecular separations, with $U \propto r^{-9}$ to r^{-12}.

The van der Waals attractions between molecules without net charges must arise from interactions belonging to classes 2, 4, and 5.

The first attempt to explain intermolecular forces was made by W. H. Keesom in 1912, on the basis of an interaction between permanent dipoles. Two dipoles in rapid thermal motion may sometimes be oriented so as to attract each other, sometimes so as to repel each other. On average, they are somewhat closer together in attractive configurations, and consequently there is a net attractive energy. This energy is calculated to be

$$U_d = -\frac{2\mu_1^2\mu_2^2}{3(4\pi\epsilon_0)^2 kTr^6} \tag{30.10}$$

where μ is the dipole moment and r is the distance between the centers of the dipoles. The calculated dependence of the interaction energy on r^{-6} is confirmed by experiments with molecular beams. The Keesom theory, of course, is not an adequate general explanation of van der Waals forces, since attractive forces exist between molecules with no permanent dipole moment.

Debye, in 1920, extended the dipole theory to take into account the *induction effect*. A permanent dipole induces a dipole in another molecule and a mutual attraction results. This interaction depends on the polarizabilities α of the molecules, and leads to a formula,

$$U_i = -\frac{\alpha_2\mu_1^2 + \alpha_1\mu_2^2}{(4\pi\epsilon_0)^2 r^6} \tag{30.11}$$

The induction effect is quite small and does not help to explain the forces between nonpolar molecules.

The primary cause of the attractive forces between nonpolar molecules is the *dispersion interaction*. An interpretation can be given as follows. The positive nucleus in an atom is surrounded by a "cloud" of negative charge. Although the time average of this charge distribution is spherically symmetrical, at any instant it is somewhat distorted. Consider a hydrogen atom, for example, in which the electron is sometimes on one side of the proton, sometimes on the other. The instantaneous charge distribution of any atom would reveal a little dipole with a certain orientation. An instant later, the orientation would be different, and so on, so that over any macroscopic period of time the instantaneous dipole moments average to zero. We should not think that these "snapshot dipoles" interact with those of other molecules to produce an attractive potential. This cannot happen, because repulsion occurs just as often as attraction; there is not enough time for the instantaneous dipoles to line up with one another. There is, however, an interaction between snapshot dipoles and the polarization they produce. Each instantaneous dipole induces an appropriately oriented dipole moment in neighboring atoms or molecules, and these moments interact with the original to produce an instantaneous attraction.

The quantum mechanical treatment of this dispersion interaction was given by Fritz London in 1930, and the resultant forces are now often called *London forces*, whereas the term *van der Waals forces* is kept for the sum of all nonionic forces between molecules. If we suppose that the instantaneous dipoles are oscillating with a frequency ν_0, the potential of the London interaction is calculated* to be

$$U_L = -\frac{3h\nu_0\alpha^2}{4(4\pi\epsilon_0)^2r^6} \tag{30.12}$$

A formula for the interaction of two different molecules A and B is

$$U_L = -\frac{3I_AI_B}{2(I_A+I_B)}\frac{\alpha_A\alpha_B}{(4\pi\epsilon_0)^2r^6} \tag{30.13}$$

Here I_A and I_B are the first ionization energies of the molecules, α_A and α_B are their average polarizabilities. Equation (30.13) is related to (30.12) by the substitution of I/h for the characteristic frequency ν_0.

Example 30.2 Xenon has an ionization energy $I_1 = 12.13$ eV and a polarizability given by $\alpha/4\pi\epsilon_0 = 4.00 \times 10^{-30}$ m^3. Calculate the London energy of a pair of xenon atoms at a distance of 350 pm (approximately the distance apart of the atoms in xenon crystals at 100 K).

From Eq. (30.13),
$$-U_L = \frac{3(12.3 \text{ eV})(1.60 \times 10^{-19} \text{ J/eV})(4.00 \times 10^{-30} \text{ m}^3)^2}{4(350 \times 10^{-12} \text{ m})^6}$$
$$= 1.28 \times 10^{-20} \text{ J} \quad \text{(per mole: 7700 J)}$$

The important contributions to the potential energy of intermolecular attraction that we have listed all display an r^{-6} dependence. Numerical values are given in Table 30.2 for a number of gases.

*For a simple quantum mechanical derivation of Eq. (30.12), see W. Kauzmann, *Quantum Chemistry* (New York: Academic Press, Inc., 1957), p. 507.

TABLE 30.2

TABLE 30.2
CALCULATED CONTRIBUTIONS TO INTERMOLECULAR POTENTIAL ENERGIES

Molecule	Dipole moment, $10^{30}\mu$ (C m)	Polarizability, $10^{30}\alpha/4\pi\epsilon_0$, (m^3)	Ionization Energy, $h\nu_0$ (eV)	Coefficients of r^{-6}		
				Orientation[a] $\dfrac{\frac{2}{3}\mu^4/kT}{(4\pi\epsilon_0)^2}$	Induction[a] $\dfrac{2\mu^2\alpha}{(4\pi\epsilon_0)^2}$	Dispersion[a] $\dfrac{\frac{3}{4}\alpha^2 h\nu_0}{(4\pi\epsilon_0)^2}$
CO	0.40	1.99	14.3	0.0034	0.057	67.5
HI	1.27	5.4	12	0.35	1.68	382
HBr	2.60	3.58	13.3	6.2	4.05	176
HCl	3.44	2.63	13.7	18.6	5.4	105
HN$_3$	5.00	2.21	16	84	10	93
H$_2$O	6.14	1.43	18	190	10	47
He	0	0.20	24.5	0	0	1.2
Ar	0	1.63	15.4	0	0	52
Xe	0	4.00	11.5	0	0	217

[a]Units of J m$^6 \times 10^{-19}$.

The expression for intermolecular potential energy must include also a repulsive term, the *overlap energy*, which becomes appreciable at very close distances. Thus the net interaction can be written in the form of the Lennard-Jones equation (30.9).

30.7 Equation of State and Intermolecular Forces

The calculation of an equation of state for a substance reduces to calculating its partition function Z [Eq. (12.2)]. From Z, the Helmholtz free energy A is immediately derivable, and hence the pressure, $P = -(\partial A/\partial V)_T$. To determine the partition function, $Z = \sum e^{-E_t/kT}$, the energy levels of the system must be known. In the cases of ideal gases and crystals, we can use energy levels for individual constituents of the system, such as molecules or oscillators, ignoring interactions between them. In the cases of nonideal gases and liquids, such a simplification is not possible.

Let us consider the problem of the equation of state of a nonideal monatomic gas. We shall follow a delightful derivation given by Landau and Lifshitz.*

The energy is written as the sum of kinetic and potential energies:

$$\epsilon(p, q) = \sum_{j=1}^{3N} \frac{p_j^2}{2m} + U(q_1 q_2 \cdots q_{3N})$$

Here p and q stand for generalized momentum and position coordinates, respectively. For N atoms there would be $3N$ each of p's and q's. The kinetic energy is obtained simply by summing over the independent kinetic energies of all the atoms, but the potential energy is, in general, a function of all the coordinates.

The partition function is taken as an integral over all the p's and q's instead of

*L. S. Landau and E. M. Lifshitz, *Statistical Physics* (Oxford: Pergamon Press Ltd., 1959), p. 219.

a summation over discrete states,

$$Z = \int e^{-E(p,q)} \, d\tau \tag{30.14}$$

where $d\tau = dp_1 \cdots dp_{3N} \, dq_1 \cdots dq_{3N}$ includes all the differentials of momentum and position. The integral in Eq. (30.14) is the product of a kinetic-energy term and a potential-energy term. The latter can be written as

$$Q = \int \cdots \int e^{-U/kT} \, dV_1 \, dV_2 \cdots dV_N \tag{30.15}$$

where the *configuration integral* Q is now taken over the volume elements of each atom ($dq_1 \, dq_2 \, dq_3 = dV_1$, etc.)

For a perfect gas, the intermolecular potential energy $U = 0$, so that Q becomes simply V^N. The kinetic-energy term is the same for perfect and imperfect gases. Thus we can write

$$A = -kT \ln Z = A^P - kT \ln \frac{1}{V^N} \int \cdots \int e^{-U/kT} \, dV_1 \cdots dV_N$$

where A^P is the Helmholtz free energy for a perfect gas. We add and subtract 1 from the integrand,

$$A = A^P - kT \ln \left[1 + \frac{1}{V^N} \int \cdots \int (e^{-U/kT} - 1) \, dV_1 \cdots dV_N \right] \tag{30.16}$$

We now assume that only interactions between pairs of molecules need be considered. An interacting pair can be chosen from among N atoms in $\frac{1}{2}N(N-1)$ different ways. We assume that the density of the gas is low enough so that the interaction integral in Eq. (30.16) is that for a single pair times the number of ways of making a pair. If the energy of interaction of the pair is U_{12}, the integral in Eq. (30.16) becomes

$$\frac{N(N-1)}{2} \int \cdots \int (e^{-U_{12}/kT} - 1) \, dV_1 \cdots dV_N$$

Since U_{12} depends only on the coordinates of two atoms, we can integrate over all the others, obtaining V^{N-2}. Since N is large, we can replace $N(N-1)$ by N^2, to obtain

$$\frac{N^2 V^{N-2}}{2} \iint (e^{-U_{12}/kT} - 1) \, dV_1 \, dV_2$$

We introduce this expression for the integral into Eq. (30.16), to find

$$A = A^P - kT \ln \left[1 + \frac{N^2}{2V^2} \iint (e^{-U_{12}/kT} - 1) \, dV_1 \, dV_2 \right] \tag{30.17}$$

We now use the fact that for $x \ll 1$, $\ln(1 + x) \approx x$, so that Eq. (30.17) becomes

$$A = A^P - \frac{kTN^2}{2V^2} \iint (e^{-U_{12}/kT} - 1) \, dV_1 \, dV_2 \tag{30.18}$$

Instead of the coordinates of the two atoms, we can introduce the coordinates of their center of mass and their relative coordinates. Integration over the relative coordinates gives V, so that Eq. (30.18) simplifies to

$$A = A^P - \frac{kTN^2}{2V} \int (e^{-U_{12}/kT} - 1)\, dV \tag{30.19}$$

This expression is usually written as

$$A = A^P + \frac{N^2 kT}{V} B(T) \tag{30.20}$$

where

$$B(T) = \frac{1}{2} \int (1 - e^{-U_{12}/kT})\, dV \tag{30.21}$$

The pressure is $P = -(\partial A/\partial V)_T$, so that

$$P = \frac{NkT}{V}\left[1 + \frac{NB(T)}{V}\right] \tag{30.22}$$

Since the perfect gas pressure is $P^P = NkT/V$, this derivation has given us an expression for the second virial coefficient $B(T)$ in terms of the potential energy of interaction U_{12} between a pair of molecules. This is a lovely result, but it applies only to gases with small deviations from ideality. For dense gases, and especially for liquids, further development of the theory depends on evaluation of the configuration integral Q for more general interactions. For dense gases, a series development of Q is possible, paralleling the empirical expressions for the third, fourth, and higher virial coefficients. For liquids, however, the series does not converge, and this avenue of theoretical progress is blocked.

Example 30.3 Calculate the second virial coefficient for a gas composed of hard-sphere molecules.

The hard-sphere potential is $U(r) = \infty,\ r < \sigma;\ U(r) = 0,\ r > \sigma$. From Eq. (30.21), with $dV = 4\pi r^2\, dr$,

$$B(T) = \frac{1}{2} \int_0^\sigma (1 - e^{-\infty/kT}) 4\pi r^2\, dr + \frac{1}{2} \int_\sigma^\infty (1 - e^{-0/kT}) 4\pi r^2\, dr$$

$$= \frac{1}{2} \int_0^\sigma 4\pi r^2\, dr = 2\pi\sigma^3/3$$

Note that $B(T)$ is four times the volume of the hard-sphere molecules and is independent of temperature. [Equation (30.21) cannot be integrated analytically for realistic potential functions such as that of Lennard-Jones, but the calculation is easy to do with a pocket calculator.]

30.8 Theory of Liquids

Faced with the unsolved mathematical problems of the general configuration integral, theoreticians have been left with three other approaches to the theory of the liquid state:

1. The construction of simplified models from which the configuration integral can be evaluated, and comparsion of results from these models with experiment.

2. Attempts to calculate the radial distribution function, from which, for certain simple liquids (argon, for example) thermodynamic properties can be evaluated.

3. Numerical calculations with large digital computers.

Since 1933, Eyring and his coworkers have sought a simple model for the structure of liquids which would avoid the intricacies of the general statistical theory. They began by considering the *free volume* of a liquid, the void space that is not occupied by rigid molecular volumes. In a liquid at ordinary temperature and pressure, about 3% of the volume is free volume. We can deduce this figure, for example, from the studies of Bridgman on the compressibility β. As long as compression consists essentially in squeezing free volume out of the liquid, β remains relatively high. When the free volume is exhausted, β drops precipitously. The Eyring model for a liquid is shown in Fig. 30.8. The vapor is mostly void space with a few molecules moving at random. The liquid is mostly filled space with a few vacancies moving at random.

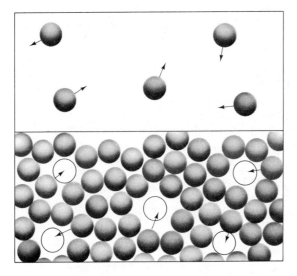

FIGURE 30.8 According to the Eyring model, vacancies in a liquid behave like molecules in a gas.

As the temperature is raised, the concentration of molecules in the vapor increases and the concentration of vacancies in the liquid also increases. Thus, as the vapor density increases, the liquid density decreases, until they become equal at the critical point. The average density ρ_{av} of liquid and vapor in equilibrium should remain approximately constant. Actually, there is a slight linear decrease with temperature, so that

$$\rho_{av} = \rho_0 - aT \tag{30.23}$$

where ρ_0 and a are constants characteristic of each substance. This relation was discovered experimentally by L. Cailletet and E. Mathias in 1886 and is called the

Law of Rectilinear Diameters. Examples are shown in Fig. 30.9, where the data for helium, argon, and ether are plotted in terms of reduced variables to bring them onto the same scale.

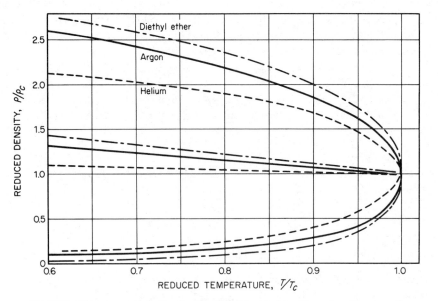

FIGURE 30.9 Law of rectilinear diameters. The straight lines give the mean of the reduced densities (ρ/ρ_c) of liquid and vapor phases in equilibrium as functions of temperature.

30.9   Viscosity of Liquids

Consider in Fig. 30.10 a stationary plate I and a moving plate II with a layer of liquid of thickness d between them. The upper plate II is displaced relative to the lower plate I at a constant velocity v_x as the result of an applied force F_x. The necessary force is proportional to the area α between the plates and the gradient of velocity v_x/d between them. Thus we can write

$$F_x = \eta(\alpha v_x/d) \tag{30.24}$$

The proportionality factor η is called the *viscosity* of the liquid. This is Newton's Law of Viscous Flow, and η is the Newtonian or dynamic viscosity. The law is often followed but there are many exceptions called *non-Newtonian* flow processes. The dimensions of η are those of (force)(time)/(area) $[m\, t^{-1}\, \ell^{-1}]$. In SI, therefore, the unit of viscosity is Pa s (kg s^{-1} m^{-1}). An older unit, the *poise*, equals 10^{-1} Pa s.

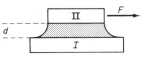

FIGURE 30.10   System for definition of Newtonian viscosity.

In Newtonian flow, one layer of liquid moves smoothly over another. Such flow is called *laminar* or *streamline*. An important case of streamline flow is the flow through pipes or tubes when the diameter of the tube is large compared to the mean free path in the fluid. The measurement of flow through tubes has been the basis for many experimental determinations of viscosity coefficients.

The theory of the process was worked out by J. L. Poiseuille in 1844. Consider an incompressible fluid flowing through a tube with length ℓ and a circular cross section of radius R. The fluid at the walls of the tube is assumed to be stagnant, and the rate of flow increases to a maximum at the center of the tube (see Fig. 30.11).

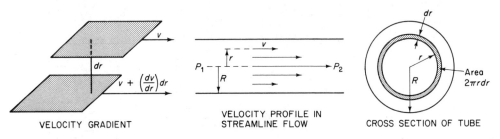

VELOCITY GRADIENT   VELOCITY PROFILE IN   CROSS SECTION OF TUBE
                           STREAMLINE FLOW

FIGURE 30.11 Aspects of streamline fluid flow considered in derivation of the Poiseuille equation.

Let v be the linear velocity at any distance r from the axis of the tube. A cylinder of fluid of radius r experiences a viscous drag given by Eq. (30.24) as $F_r = -\eta 2\pi r \ell \, dv/dr$. For steady flow, this force must be exactly balanced by the force driving the fluid in this cylinder through the tube. Since pressure is force per unit area, the driving force is $F_r = \pi r^2 (P_1 - P_2)$, where P_1 is the fore pressure and P_2 the back pressure. Thus, for steady flow,

$$-\eta 2\pi r \ell \frac{dv}{dr} = \pi r^2 (P_1 - P_2) \quad \text{and} \quad dv = -\frac{r}{2\eta \ell}(P_1 - P_2)\, dr$$

On integration, we obtain $v = -(P_1 - P_2)r^2/4\eta\ell + \text{const.}$ According to our hypothesis, $v = 0$ when $r = R$; this boundary condition determines the integration constant, so that we obtain finally

$$v = \frac{(P_1 - P_2)}{4\eta\ell}(R^2 - r^2) \tag{30.25}$$

The total volume of fluid flowing through the tube in unit time is calculated by integrating over each element of cross-sectional area, given by $2\pi r \, dr$. Thus

$$\frac{dV}{dt} = \int_0^R 2\pi r v \, dr = \frac{\pi (P_1 - P_2) R^4}{8\ell\eta} \tag{30.26}$$

This is the *Poiseuille equation*. It was derived for an incompressible fluid and therefore may be satisfactorily applied to liquids but not to gases. For a gas, the volume depends on the pressure. The average pressure along the tube is $(P_1 + P_2)/2$.

If P_0 is the pressure at which the volume is measured, the equation becomes

$$\frac{dV}{dt} = \frac{\pi(P_1 - P_2)R^4}{8\ell\eta}\frac{P_1 + P_2}{2P_0} = \frac{\pi(P_1^2 - P_2^2)R^4}{16\ell\eta P_0} \qquad (30.27)$$

By measuring the volume rate of flow through a tube of known dimensions, we can determine the viscosity η of a gas or liquid. A simple device for measuring the viscosity of a liquid is the Ostwald viscometer, shown in Fig. 30.12. The liquid is drawn into the volume A from the reservoir B, and the time required for it to flow out under the force of gravity from between the calibration marks (a) and (b) is measured. Usually the viscosity of the liquid is determined by comparison with a standard such as water: $\eta_1/\eta_2 = \rho_1 t_1/\rho_2 t_2$.

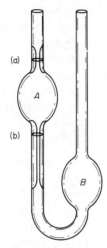

FIGURE 30.12 An Ostwald viscometer.

Another useful viscometer is the Höppler type, based on the Stokes formula for the frictional resistance of a spherical body falling through a fluid, $F = 6\pi\eta r v$. By measuring the rate of fall in the liquid (terminal velocity v) of metal spheres of known radius r and mass m, the viscosity can be calculated, since the magnitude of the force F is equal to $(m - m_0)g$, where m_0 is the mass of liquid displaced by the ball. Thus

$$\eta = \frac{F}{6\pi r v} = \frac{(m - m_0)g}{6\pi r v} \qquad (30.28)$$

Example 30.4 The viscosity of a molten glass at 1050 K was determined by timing the fall of a platinum sphere 1.00 cm in diameter through the melt. The density ρ_0 of the glass was 3.54×10^3 kg m^{-3} and that of platinum, $\rho = 21.4 \times 10^3$ kg m^{-3}. In 25.0 min the sphere fell 1.00 cm. What was the viscosity of the glass?

The Stokes formula [Eq. (30.28)] is applied.

$$\eta = \frac{(m - m_0)g}{6\pi r v} = \frac{\frac{4}{3}\pi r^3(\rho - \rho_0)g}{6\pi r v} = \frac{2}{9}r^2\frac{(\rho - \rho_0)g}{v}$$

$$= \frac{2}{9}\frac{(0.500 \times 10^{-2} \text{ m})^2[(21.4 - 3.54) 10^3 \text{ kg m}^{-3}](9.807 \text{ m s}^{-2})}{6.667 \times 10^{-6} \text{ m s}^{-1}}$$

$$= 1.46 \times 10^5 \text{ kg s}^{-1} \text{ m}^{-1} \text{ [Pa s]}$$

The hydrodynamic theories for liquid flow and gas flow are very similar. The kinetic-molecular mechanisms differ, however, as we might suspect from the different ways in which gas and liquid viscosities depend on temperature and pressure. For a gas, viscosity increases with temperature and is practically independent of pressure. For a liquid, viscosity increases with pressure and decreases with increasing temperature.

The viscosity of a liquid is strongly influenced by its molecular structure. For example, polar liquids have much higher viscosities than nonpolar ones. At 300 K, aniline has six times the viscosity of benzene. Glycerol with three hydroxyl groups has over 300 times the viscosity of n-propanol at 300 K.

In both gases and liquids each layer of flowing fluid exerts a viscous drag on adjacent layers. This viscous force is associated with a net transfer of momentum from the more rapidly moving layer to the more slowly moving one. In a gas, the momentum is transferred by the actual flights of molecules between the layers and the intermolecular collisions at the ends of the free paths of these flights. In a liquid, by contrast, the momentum transfer is due to intermolecular attractive forces between the molecules, which cause a frictional drag between the moving layers.

The viscosity coefficient for a liquid may be written as

$$\eta = A \exp\left(\frac{\Delta E_{vis}^{\ddagger}}{RT}\right) \qquad (30.29)$$

The quantity $\Delta E_{vis}^{\ddagger}$ is the energy barrier that must be overcome before the elementary flow process can occur. It is expressed per mole of liquid. The term $\exp\left(-\Delta E_{vis}^{\ddagger}/RT\right)$ can then be explained as a Boltzmann factor giving the fraction of molecules having enough energy to surmount the barrier. Thus $\Delta E_{vis}^{\ddagger}$ is an activation energy for the rate process of viscous flow.

When the viscosity of a liquid over a range of temperatures is plotted as log η vs. T^{-1}, a linear graph is usually obtained. Examples are shown in Fig. 30.13. The activation energies $\Delta E_{vis}^{\ddagger}$ are usually about 1/3 to 1/4 the enthalpies of vaporization. Liquids with hydrogen-bonded structures often exhibit a deviation from the linear log η vs. T^{-1} plot at higher T, in the direction of lowered η. [How would you explain this effect?]

Example 30.5 From the data in Fig. 30.13, calculate $\Delta E_{vis}^{\ddagger}$ for benzene and compare it with ΔH_v of benzene at its boiling point.

Slope is $0.490/0.0800 \times 10^{-3}$ K^{-1} = 613 K. $\Delta E_{vis}^{\ddagger}/R$ = (613 K)(2.303) (conversion of slope from log to ln). $\Delta E_{vis}^{\ddagger}$ = 11 700 J mol^{-1}; ΔH_v = 34 700 J mol^{-1}.

30.11 Viscosity of Polymer Solutions

The viscosities of solutions of polymers depend on the sizes and shapes of the molecules in solution. In 1906, Einstein derived an equation for rigid spherical particles in dilute solutions,

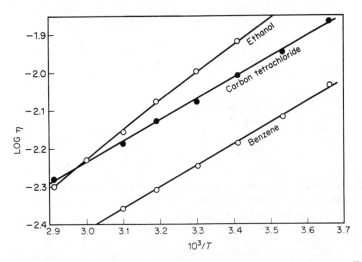

FIGURE 30.13 Temperature dependence of viscosity η plotted as log η vs. T^{-1}. The data for carbon tetrachloride and benzene fall on straight lines, but data for ethanol at higher temperatures deviate from a straight line.

$$\lim_{\phi \to 0} \left[\frac{(\eta/\eta_0) - 1}{\phi} \right] = 2.5 \tag{30.30}$$

Here η is the viscosity of the solution and η_0 that of pure solvent; ϕ is the volume fraction of solution that is occupied by solute molecules. The fraction η/η_0 is called the *viscosity ratio*. The limit as $\phi \to 0$ is obtained by plotting $(\eta/\eta_0 - 1)/\phi$ vs. ϕ and extrapolating to $\phi = 0$.

Since ϕ is difficult to measure, we often use the expression

$$[\eta] = \lim_{C' \to 0} \left[\frac{(\eta/\eta_0) - 1}{C'} \right] \tag{30.31}$$

where C' is the mass of polymer per unit volume of solution. $[\eta]$ is called the *limiting viscosity* [formerly, *intrinsic viscosity*].

In 1932, Staudinger proposed the semiempirical relation $[\eta] = KM$, where M is the molar mass and K is a constant. Mark and Houwink gave a more general relation

$$[\eta] = KM^\alpha \tag{30.32}$$

This relation predicts that log $[\eta]$ is a linear function of log $[M]$. Some data for solutions of proteins are plotted in Fig. 30.14.

The value of α depends on the shape of the polymer, but for polymers of the same shape, for example random coils, the value of $[\eta]$ provides a way to estimate the molar mass. For solid spheres $[\eta]$ is independent of M, i.e., $\alpha = 0$ in Eq. (30.32). It has been found experimentally that many native proteins, regardless of molar mass, have the same $[\eta]$. This result is one of the best pieces of evidence that globular proteins in solution are compact, almost spherical molecules.

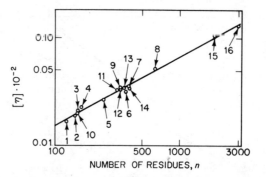

FIGURE 30.14 Limiting viscosities $[\eta]$ plotted vs. number of amino-acid residues in the protein for 16 proteins in the form of random coils. Scales are logarithmic. (1) ribonuclease; (2) hemoglobin; (3) myoglobin; (4) lactoglobin; (5) chymotrypsinogen; (6) pepsinogen; (7) aldolase; (8) bovine serum albumin; (9) glyceraldehyde-3-phosphate dehydrogenase; (10) methemoglobin; (11) lactate dehydrogenase; (12) enolase; (13) alcohol dehydrogenase; (14) ovalbumin; (15) heavy subunit of myosin; (16) "51A" protein. [Reprinted by permission from A. H. Reisner and J. Rowe, *Nature* **222**, 558 (1969). Copyright © 1969 Macmillan Journals Limited.]

Problems

1. For CS_2 at $P = P°$:

T (K)	193	233	273	313
$10^4\alpha$ (K^{-1})	10.57	11.04	11.55	12.11
$10^9\beta$ (Pa^{-1})	0.467	0.612	0.820	1.051

Calculate the internal pressures P_i at the temperatures cited.

2. H_2O has a dipole moment of magnitude $\mu = 6.18 \times 10^{-30}$ C m. At what distance of separation would the average dipole–dipole interaction between two H_2O molecules equal kT at 273 K?

3. For H_2O the polarizability is given by $\alpha/4\pi\epsilon_0 = 1.48 \times 10^{-30}$ m^3. At the separation found in Problem 2, what would be the average dipole–dipole and dipole–induced dipole intermolecular potential energies between two H_2O molecules?

4. Values of $\alpha/4\pi\epsilon_0$ and of I are:

Molecule	He	Ne	Ar	Kr	Xe
10^{30} $(\alpha/4\pi\epsilon_0)$ (m^3)	0.204	0.392	1.630	2.465	4.008
I (eV)	24.45	21.47	15.96	13.94	12.13

Calculate the London interaction energies between atoms of He, Ne, Ar, Kr, Xe and also those of Ar with the others. Plot vs. R.

5. The sublimation enthalpies ΔH_s of Ar and Xe are 7.74 and 16.02 kJ mol^{-1}, respectively. From calculation in Problem 4, at what interatomic distances would the London energies equal the ΔH_s?

6. For interaction of an ion of charge Q and a neutral molecule of polarizability α, $U_i = -\alpha Q^2/2(4\pi\epsilon_0)^2 r^4$. An Ar$^+$ ion and an Ar atom associate to form an Ar$_2^+$ molecule ion. If Ar$^+$ and Ar have a repulsive interaction, $U = Br^{-9}$ with $B = 6.18 \times 10^{-106}$ J m^9, calculate the internuclear distance R_e and the dissociation energy D_e of Ar$_2^+$.

7. From the statistical formulas of Chapter 12, calculate the extent of dissociation of Ar$_2^+$ as a function of T from 10 to 100 K.

8. It is proposed to evacuate continuously a large vessel to a pressure of 0.10 Pa or lower with a mechanical vacuum pump. The vessel has a continuous inward leak of air from the atmosphere equivalent to a capillary 1.0-cm long and 10^{-1}-mm diameter. The viscosity of air at 20°C is 1.82×10^{-5} kg m^{-1} s^{-1}. The pump has a rated capacity of 6.0 m^3/min at 0.10 Pa. Is it suitable for the proposed use?

9. For N$_2$ the intermolecular energy can be represented by $U = Ar^{-6} + Br^{-9}$ with $A = 1.99 \times 10^{-77}$ J m^6 and $B = 9.90 \times 10^{-106}$ J m^9. Estimate the nearest approach of centers of two N$_2$ molecules in head-on collisions with relative energies equal to $10^{-1}kT$, kT, and $10kT$.

10. Evaluate the second virial coefficient for N$_2$ at 300 K from the data in Problem 9. It will be necessary to do a numerical integration of Eq. (30.21). See Texas Instruments SR-52 Math Programs or similar source.

11. Consider a mole of N$_2$ at 200 K and 10 MPa. Calculate the average distance apart of the N$_2$ molecules. At this distance what would be the Lennard-Jones intermolecular energy per mole of N$_2$?

12. For several gases listed in Tables 2.3 and 30.1, compare the Lennard-Jones parameter δ with the hard-sphere diameters calculated from van der Waals b.

13. The viscosity of mercury:

T (K)	273	393	513	633
$10^4\eta$ (Pa s)	16.9	11.9	9.85	8.85

Calculate ΔE_{vis}^{+}, and compare it to ΔH_v of Hg at its boiling point (60.0 kJ mol^{-1}).

14. A cylindrical tank of benzene 2.0 m high and 1.0 m diameter is connected via a 2.0-m vertical length of 5-mm inside diameter tubing to a reaction vessel at a lower level. How long will it take to empty the tank into the vessel at 300 K? The viscosity of benzene is 5.64×10^{-4} kg m^{-1} s^{-1} and its density is 875 kg m^{-3}. (Neglect the time to empty the tubing.)

15. Petroleum with $\eta = 8.0 \times 10^{-2}$ Pa s (kg m^{-1} s^{-1}) at 350 K, is pumped through a pipeline 1000 km long at a rate of 6.0 km^3 h^{-1}. The pipe has a diameter of 1.0 m. What is the minimum power necessary to drive the petroleum through the pipeline?

16. The viscosities η relative to pure solvent (%) of a fraction of polystyrene $\bar{M}_N = 280\,000$ g mol^{-1} dissolved in tetralin at 20°C were:

g polystyrene/100 cm^3 soln	0.01	0.025	0.05	0.10	0.25
η/η_0	1.05	1.12	1.25	1.59	2.70

Calculate the limiting viscosity $[\eta]$ of the polystyrene. From the Staudinger equation, estimate the relative viscosity of a solution of 0.10 g of polystyrene of $\bar{M}_N = 500\,000$ in the same solvent at 20°C.

17. For β-amylose in dimethylsulfoxide at 25°C, the constants of Eq. (30.32) are $K = 3.6 \times 10^{-4}$ and $\alpha = 0.58$. Calculate the molar mass $\bar{M}$ of a sample of β-amylose from the data:

$g/100$ cm^3	0.15	0.20	0.30	0.40
η/η_0	1.11	1.14	1.23	1.30

18. The viscosity of water from 0° to 100°C is represented with better than 1% accuracy by

$$\log \frac{\eta_{20}}{\eta_t} = \frac{1.37023[t - 20] + 8.36 \times 10^{-4}[t - 20]^2}{109 + t}$$

where t is Celsius temperature and η is in kg m^{-1} s^{-1}, and $\eta_{20} = 0.001002$. Use this equation to make a plot of $\ln \eta$ vs. T^{-1} and comment on the result in terms of $\Delta E_{vis}^{\ddagger}$.

Tables

TABLE A.1
THERMODYNAMIC DATA FOR SUBSTANCES IN THEIR
STANDARD STATES AT 298.15 K[a]

Substance	ΔH_f° (kJ mol^{-1})	S° (J K^{-1} mol^{-1})	ΔG_f° (kJ mol^{-1})	C_P° (J K^{-1} mol^{-1})
Ag(c)	0.0	42.7	0.0	25.5
AgBr(c)	−100.4	107.1	−96.9	52.4
AgCl(c)	−127.1	96.2	−109.8	50.8
AgNO$_3$(c)	−124.5	141	−33.6	93.1
Al(c)	0:0	28.3	0.0	24.3
Al$_2$O$_3$(c)	−1676	50.9	−1583	79.0
B(c)	0.0	5.86	0.0	12.0
B(g)	570	143	530	20.8
B$_2$H$_6$(g)	31.4	233	82.3	56.4
B$_2$O$_3$(c)	−1273	54.0	−1194	62.3
BN(c)	−251	14.8	−225	12.4
Ba(c)	0.0	62.8	0.0	—
BaCl$_2$(c)	−859	124	−811	83.5
BaO	−554	70.4	−526	45.3
Be(c)	0.0	9.54	0.0	17.8
Be(g)	320	136	290	20.8

[a]Data are mainly from G. N. Lewis and M. Randall, *Thermodynamics*, 2nd ed. revised by K. S. Pitzer and L. Brewer (New York: McGraw Hill Book Company, 1961), with amended tables through courtesy of Professors Pitzer and Brewer; and O. Kubaschewski and C. B. Alcock, *Metallurgical Thermochemistry*, 5th ed. (Oxford: Pergamon Press, 1979).

TABLE A.1 (cont.)

Substance	ΔH_f° (kJ mol^{-1})	S° (J K^{-1} mol^{-1})	ΔG_f° (kJ mol^{-1})	C_P° (J K^{-1} mol^{-1})
BeO(c)	-610	14.1	-580	25.4
Br$_2$(l)	0.0	152	0.0	—
Br$_2$(g)	30.90	245.3	3.08	36.0
C(graphite)	0.0	5.73	0.0	8.64
C(g)	716.7	157.99	671.3	20.8
CO(g)	-110.41	197.91	-137.15	29.1
CO$_2$(g)	-393.51	213.64	-394.38	37.1
CS$_2$(ℓ)	89.58	152	65.0	77.0
Ca(c)	0.0	41.4	0.0	26.3
CaF$_2$(c)	-1220	68.9	-1168	67.0
CaCl$_2$(c)	-796	104.6	-748	72.8
CaO(c)	-635	39.7	-604	43.1
Ca(OH)$_2$(c)	-986	83.4	-898	87.5
CaC$_2$(c)	-59.0	70.7	-64.3	62.4
CaCO$_3$(c)	-1205	91.7	-1127	81.9
Cl$_2$(g)	0.0	223	0.0	33.9
Co(c)	0.0	30.0	0.0	24.6
CoO(c)	-239	53.0	-215	52.7
Cr(c)	0.0	23.6	0.0	23.2
Cr$_2$O$_3$(c)	-1140	81.2	-1060	104.5
Cu(c)	0.0	33.1	0.0	24.5
Cu$_2$O(c)	-171	92.4	-141	69.9
CuO(c)	-156	42.6	-128	44.4
CuS(c)	-52.3	66.5	-52.8	47.7
CuSO$_4$(c)	-770	113	-672	101
CuSO$_4 \cdot$5H$_2$O(c)	-2280	305	-1880	281
F$_2$(g)	0.0	203	0.0	31.5
Fe(c)	0.0	27.3	0.0	25.2
Fe$_{0.947}$O(c)	-266	57.5	-244	35.9
Fe$_2$O$_3$(c)	-824	87.4	-742	105
Fe$_3$O$_4$(c)	-1119	146	-1016	152
FeS(c)	-100	60.3	-105	54.6
Ge(c)	0.0	31.1	0.0	23.3
GeH$_4$(g)	90.8	217	113	—
H$_2$(g)	0.0	130.6	0.0	28.8
H(g)	217.97	114.6	203.27	20.8
HBr(g)	-36.23	198.6	-53.3	29.1
HCl(g)	-92.31	186.8	-95.29	29.1
HF(g)	-271.1	173.7	-273.2	28.5
HI(g)	26.5	206.4	1.87	29.2
HCN(g)	134.7	202	124.2	35.9
HNO$_3$(ℓ)	-173.2	156	-80.0	110
H$_2$O(ℓ)	-285.84	69.44	-237.04	75.3
H$_2$O(g)	-241.83	188.72	-228.59	33.6
H$_2$O$_2$(ℓ)	-187.6	109.5	-120.2	—
H$_2$S(g)	-20.6	206	-33.6	34.0
Hg(ℓ)	0.0	75.9	0.0	27.8

Substance	ΔH_f° (kJ mol^{-1})	S° (J K^{-1} mol^{-1})	ΔG_f° (kJ mol^{-1})	C_P° (J K^{-1} mol^{-1})
Hg(g)	60.8	174.8	31.9	20.8
HgCl(c)	−132.6	96.2	−105.4	50.9
I$_2$(c)	0.0	117	0.0	55.0
I$_2$(g)	62.42	261	19.5	36.9
K(c)	0.0	64.6	0.0	29.2
K(g)	89.2	160	60.8	20.8
KCl(c)	−436.7	82.5	−409	51.5
KI(c)	−328	106.4	−323	52.0
Li(c)	0.0	29.3	0.0	23.6
Li(g)	159	139	126	20.8
LiF(c)	−615	35.6	−586	44.7
LiCl(c)	−408	59.3	−384	50.2
LiH(c)	−90.0	20.1	−67.8	—
Mg(c)	0.0	32.7	0.0	23.9
MgCl$_2$(c)	−641.4	89.5	−592	71.3
MgSO$_4$(c)	−1285	91.6	−1171	98
Mn(c)	0.0	32.0	0.0	26.3
MnO$_2$(c)	−520	53.1	−465	54.0
N$_2$(g)	0.0	191.5	0.0	29.1
N(g)	472.7	153.2	455.6	20.8
NH$_3$(g)	−45.9	192.5	−16.3	35.7
NH$_4$Cl(c)	−315	94.6	−203	84.1
N$_2$O(g)	82.05	220	104	38.6
NO(g)	90.8	211	87.0	29.9
NO$_2$(g)	33	240.5	51.1	37.9
N$_2$O$_5$(g)	11.3	346	118.0	—
Na(c)	0.0	51.2	0.0	28.4
Na(g)	106.7	153.6	76.2	20.8
Na$_2$(g)	142	230	104	—
NaCl(c)	−411	72.1	−384	49.7
NaBr(c)	−361	86.8	−349	52.3
NaOH(c)	−427	64.4	−381	80.3
NaNO$_3$(c)	−467	116	−366	93.1
Na$_2$CO$_3$(c)	−1130	136	−1050	110.5
Na$_2$SO$_4$	−1380	149.5	−1262	128
Na$_2$SO$_4$·10H$_2$O(c)	−4320	593	−3640	587
Ni(c)	0.0	29.9	0.0	25.8
NiO(c)	−241	38.1	−213	44.1
NiS(c)	−94.1	52.9	−91.4	46.9
O$_2$(g)	0.0	205	0.0	29.4
O(g)	249.2	161.0	231.8	20.8
P(c, white)	17.4	41.1	11.9	23.2
P(c, red)	0.0	22.8	0.0	—
P(g)	334	163	292	20.8
PCl$_3$(g)	−271	312	−257	88.2
PCl$_5$(g)	−350	364	−285	148

TABLE A.1 *(cont.)*

Substance	ΔH_f° (kJ mol^{-1})	S° (J K^{-1} mol^{-1})	ΔG_f° (kJ mol^{-1})	C_P° (J K^{-1} mol^{-1})
PH$_3$(g)	+22.8	210	+25.4	46.8
P$_2$O$_5$(c)	-1475	114	-1343	123.4
Pb(c)	0.0	65.1	0.0	26.8
PbO(red)	-219.0	66.5	-189.0	54.2
PbO(yellow)	-217.3	68.7	-187.8	46.2
PbS(c)	-100.4	91.2	-98.7	49.5
Pu(c)	0.0	55.2	0.0	—
PuO$_2$(c)	-1060	68.4	-1003	69.8
S(orthorh.)	0.0	31.9	0.0	22.6
S(g)	279	167.7	237	20.8
SF$_6$(g)	-1220	293	-1116	—
SO$_2$(g)	-296.8	249	-300	31.8
SO$_3$(g)	-395.7	256	-370	50.6
Si(c)	0.0	18.8	0.0	19.9
Si(g)	456	168	412	20.8
SiC(c)	-73.2	16.6	70.8	26.3
SiCl$_4$(g)	-663	331	-623	112
SiO$_2$(α quartz)	-911	41.5	-857	44.4
Sn(white)	0.0	51.4	0.0	26.3
Sn(gray)	1.97	44.1	4.15	—
SnCl$_4$(g)	-529	259	-458	98.4
SnO(c)	-286	56.5	-257	44.4
Ti(c)	0.0	30.6	0.0	25.0
TiO$_2$(rutile)	-940	50.3	-890	55.0
U(c)	0.0	50.3	0.0	27.8
UF$_4$(c)	-1900	151.7	-1810	116
UF$_6$(g)	-2140	378	-2060	129
UO$_2$(c)	-1080	77.8	-1040	63.7
U$_3$O$_8$(c)	-3570	282	-3265	237
Zn(c)	0.0	41.6	0.0	25.1
ZnO(c)	-348	43.6	-318	40.3
ZnS(hex)	-206	57.7	-201	55.7

TABLE A.2

GIBBS FREE ENERGIES BASED ON H_0°

	$-(G^\circ - H_0^\circ)/T$ (J K^{-1} mol^{-1})					$H_{298}^\circ - H_0^\circ$ (kJ mol^{-1})	ΔH_0° (kJ mol^{-1})
	298 K	500 K	1000 K	1500 K	2000 K		
Elements							
Br$_2$ (g)	212.8	230.1	254.4	269.1	279.6	9.73	45.7
C (g)	136.1	147.3	162.0	170.6	176.6	6.53	711.3
C (graphite)	2.23	4.85	11.6	17.5	22.5	1.05	0
Cl$_2$	192.2	208.6	231.9	246.2	256.6	9.18	0
F$_2$	173.1	188.7	214.0	224.8	235.0	8.83	0
H$_2$	102.2	116.9	137.0	148.9	157.6	8.47	0
I$_2$ (g)	226.7	244.6	269.4	284.3	295.1	10.12	65.5
N$_2$	162.4	177.5	197.9	210.4	219.6	8.67	0
O$_2$	176.0	191.1	212.1	225.1	234.7	8.66	0
S (g)	145.4	157.1	172.7	181.7	188.0	6.66	274.8
Various compounds							
Cl$_2$O	228.7	250.1	282.8	303.5	318.8	11.7	89.5
HF	144.9	159.8	179.9	191.9	200.6	8.60	−271.1
HCl	157.8	172.8	193.1	205.4	214.3	8.64	−92.1
HBr	169.6	184.6	205.0	217.4	226.5	8.65	−28.6
HI	177.4	192.4	213.0	225.5	234.8	8.66	28.7
H$_2$O (g)	155.5	172.8	196.7	211.7	223.1	9.91	−238.9
H$_2$S	172.3	189.7	214.6	230.8	243.1	9.98	−17.7
NH$_3$	159.0	176.9	203.5	222.1	236.9	10.06	−38.9
NO	179.8	195.6	217.0	230.0	239.5	9.18	90.4
N$_2$O	187.7	205.4	233.0	251.6	265.9	9.59	85.5
NO$_2$	205.7	224.0	251.4	269.5	283.0	10.31	35.9
SO$_2$	212.6	231.7	260.6	279.6	294.0	10.54	−294.3
SO$_3$	217.4	239.5	276.0	301.1	320.5	11.70	−390.0
Gaseous carbon compounds[a]							
CO	168.4	183.5	204.1	216.6	225.9	8.673	−113.80
CO$_2$	182.3	199.5	226.4	244.7	258.8	9.364	−393.17
COS	198.3	216.7	245.8	265.3	279.9	9.92	−137
CS$_2$	202.0	221.9	253.2	273.8	289.1	10.07	116.6
CH$_4$	152.5	170.5	199.4	221.1	238.9	10.03	−66.82
CH$_3$Cl	199.5	218.8	251.1	275.2	294.9	10.41	−78.03
CH$_2$Cl$_2$	230.5	252.8	291.0	314.5	340.3	11.86	−89.0
CHCl$_3$	248.1	275.3	321.1	352.7	377.1	14.18	−98.3
CCl$_4$	251.9	285.3	340.2	376.7	404.0	17.24	−93.8
COCl$_2$	240.6	265.0	304.6	331.1	351.1	12.87	−217.1
CF$_4$	218.7	243.7	288.0	319.7	344.3	12.73	−927.9
CH$_3$OH	201.4	222.3	257.7	—	—	11.43	−190.2
CH$_2$O	185.1	203.1	230.6	250.2	266.0	10.01	−104.7
HCOOH	212.2	232.6	267.7	293.6	314.4	10.88	−371.4

[a]From G. N. Lewis and M. Randall, *Thermodynamics*, 2nd ed. revised by K. S. Pitzer and L. Brewer (New York: McGraw Hill Book Co., 1961) through courtesy of Professors Pitzer and Brewer.

TABLE A.2 *(cont.)*

	$-(G° - H_0°)/T$ (J K^{-1} mol^{-1})					$H_{298}° - H_0°$ (kJ mol^{-1})	$\Delta H_0°$ (kJ mol^{-1})
	298 K	500 K	1000 K	1500 K	2000 K		
HCN	170.8	187.7	213.4	230.7	244.0	9.234	136 ± 8
CN	173.5	188.5	209.2	221.8	231.0	8.673	456 ± 12
C$_2$H$_2$	167.3	186.2	217.6	239.5	256.6	10.01	227.3
C$_2$H$_4$	184.0	203.7	239.1	266.7	289.7	10.52	60.73
C$_2$H$_6$	189.4	212.4	255.7	290.6	—	11.95	−69.12
C$_2$H$_5$OH	235.0	262.6	314.8	356.1	—	14.2	−217.4
CH$_3$CHO	221.1	245.5	288.8	—	—	12.84	−155.4
CH$_3$COOH	236.4	264.6	317.6	357.1	—	13.75	−418.3
CH$_3$CN	202.9	225.8	265.5	—	—	12.10	94.47
CH$_3$NC	204.3	228 1	268.7	—	—	12.66	1.56
C$_3$H$_6$	221.5	248.2	299.4	340.7	—	13.54	35.4
C$_3$H$_8$	220.6	250.2	310.0	359.2	—	14.69	−81.50
(CH$_3$)$_2$CO	240.2	272.1	331.5	378.8	—	16.27	−201
n-C$_4$H$_{10}$	244.9	284.1	362.3	426.6	—	19.44	−99.04
iso-C$_4$H$_{10}$	234.6	271.8	348.9	412.7	—	17.89	−106.5
n-C$_5$H$_{12}$	270.0	317.7	413.7	492.5	—	23.55	−114.3
iso-C$_5$H$_{12}$	269.3	315.0	409.9	488.6	—	22.15	−120.2
neo-C$_5$H$_{12}$	235.8	280.5	376.1	455.7	—	21.04	−133.4
C$_6$H$_6$	221.5	252.0	320.4	378.4	—	14.23	100.4
cyc-C$_6$H$_{12}$	238.8	277.8	371.3	455.2	—	17.73	−83.72

TABLE A.3
THERMODYNAMIC PROPERTIES OF SUBSTANCES IN
STANDARD STATE ($a = 1$) IN AQUEOUS SOLUTION

Species in solution	$\Delta H_f°$ (kJ mol^{-1})	$S°$ (J K^{-1} mol^{-1})	$\Delta G_f°$ (kJ mol^{-1})
Ag$^+$	105.9	73.93	77.11
Ag(NH$_3$)$_2^{2+}$	−111.8	242	−17.4
Al^{3+}	−525	−313	−481
Be^{2+}	−389	−230	−356.5
Br$^-$	−120.9	80.71	−102.8
Ca^{2+}	−543	−55.2	−553
CO$_2$	−412.9	121.3	−386.2
CO$_3^{2-}$	−676	−53.1	−528
Cl$^-$	−167.4	55.2	−131.2
ClO$_4^-$	−131.4	182.0	−8.0
Cu^{2+}	64.4	−98.7	65.0
Cu(NH$_3$)$_4^{2+}$	−334	807	−256
Cr^{3+}	−256	−307.5	−215.5
Cr$_2$O$_7^{2-}$	−1461	214	−1257
CrO$_4^{2-}$	−894	38.5	−737
F$^-$	−329.1	−9.6	−276.5

TABLE A.3 (cont.)

Species in solution	ΔH_f° (kJ mol^{-1})	S° (J K^{-1} mol^{-1})	ΔG_f° (kJ mol^{-1})
Fe^{2+}	-87.9	-113.4	-84.9
Fe^{3+}	-47.7	-293	-10.6
H^+	0.0	0.0	0.0
H_3O^+	-285.9	70.0	-237.2
H_3BO_3	-1068	160	-963
$H_2BO_3^-$	-1054	30.5	-910
HCl	-167.4	55.2	-131.2
H_2CO_3	-699	191	-623
HCO_3^-	-691	95.0	-587
HNO_3	-206.6	146.4	-110.6
H_3PO_4	-1290	176.1	-1147
$H_2PO_4^-$	-1303	89.1	-1135
HPO_4^{2-}	-1299	-36.0	-1094
H_2S	-39.3	122	-27.4
HS^-	-17.7	61.1	12.6
H_2SO_4	-907.5	17.1	-742
HSO_4^-	-885.8	126.9	-752.9
I^-	-55.94	109.4	-51.67
I_2	20.9	—	16.44
I_3^-	-51.9	174	-51.5
K^+	-251.2	102.5	-282.3
Li^+	-278.4	14.2	-293.8
Mg^{2+}	-462.0	-118	-456.0
Mn^{2+}	-218.8	-84	-223.4
MnO_4^-	-518	190	-425
NH_3	-80.8	110	-26.6
NH_4^+	-132.8	112.8	-79.5
NO_3^-	-206.6	146	-110.6
Na^+	-239.7	60.2	-261.9
OH^-	-230.0	-10.54	-157.3
PO_4^{3-}	-1284	-218	-1026
Pb^{2+}	1.63	21.3	-24.3
S^{2-}	35.8	-26.8	26.88
SO_3^{2-}	-635	-29	-486
SO_4^{2-}	-907.5	17.1	-742
Zn^{2+}	-152.4	-106.5	-147.2

TABLE A.4

DEBYE HEAT CAPACITY FUNCTION, $C_V/3R$
AS A FUNCTION OF Θ_D/T[a]

Θ_D/T	0.0	0.1	0.2	0.3	0.4
0.0	1.0000	0.9995	0.9980	0.9955	0.9920
1.0	0.9517	0.9420	0.9315	0.9203	0.9085
2.0	0.8254	0.8100	0.7943	0.7784	0.7622
3.0	0.6628	0.6461	0.6296	0.6132	0.5968
4.0	0.5031	0.4883	0.4738	0.4595	0.4456
5.0	0.3686	0.3569	0.3455	0.3345	0.3237
6.0	0.2656	0.2569	0.2486	0.2405	0.2326
7.0	0.1909	0.1847	0.1788	0.1730	0.1675
8.0	0.1382	0.1339	0.1297	0.1257	0.1219
9.0	0.1015	0.09847	0.09558	0.09280	0.09011
10.0	0.07582	0.07372	0.07169	0.06973	0.06783
11.0	0.05773	0.05624	0.05479	0.05339	0.05204
12.0	0.04478	0.04370	0.04265	0.04164	0.04066
13.0	0.03535	0.03455	0.03378	0.03303	0.03230
14.0	0.02835	0.02776	0.02718	0.02661	0.02607
15.0	0.02307	0.02262	0.02218	0.02174	0.02132

Θ_D/T	0.5	0.6	0.7	0.8	0.9	1.0
0.0	0.9876	0.9822	0.9759	0.9687	0.9606	0.9517
1.0	0.8960	0.8828	0.8692	0.8550	0.8404	0.8254
2.0	0.7459	0.7294	0.7128	0.6961	0.6794	0.6628
3.0	0.5807	0.5647	0.5490	0.5334	0.5181	0.5031
4.0	0.4320	0.4187	0.4057	0.3930	0.3807	0.3686
5.0	0.3133	0.3031	0.2933	0.2838	0.2745	0.2656
6.0	0.2251	0.2177	0.2107	0.2038	0.1972	0.1909
7.0	0.1622	0.1570	0.1521	0.1473	0.1426	0.1382
8.0	0.1182	0.1146	0.1111	0.1078	0.1046	0.1015
9.0	0.08751	0.08500	0.08259	0.08025	0.07800	0.07582
10.0	0.06600	0.06424	0.06253	0.06087	0.05928	0.05773
11.0	0.05073	0.04946	0.04823	0.04705	0.04590	0.04478
12.0	0.03970	0.03878	0.03788	0.03701	0.03617	0.03535
13.0	0.03160	0.03091	0.03024	0.02959	0.02896	0.02835
14.0	0.02553	0.02501	0.02451	0.02402	0.02354	0.02307
15.0	0.02092	0.02052	0.02013	0.01975	0.01938	0.01902

[a]From K. S. Pitzer, *Quantum Chemistry* (Englewood Cliffs, N.J.: Prentice-Hall, Inc., 1963).

Index

Absolute zero, 137, 232
  entropy at, 138, 269
Absorbance, 604
Absorption edge, 658
Absorption spectra, 580, 603
Absorptivity, 604
Acetylene vibrations, 601
Acid catalysis, 311–13
Actinometry, 613
Action potential, 399
Activated complex, 282, 335
  theory, 338
Activation energy, 301, 331
  in catalysis, 310
Activation entropy, 427
Activation free energy, 340, 427
Activity, 198, 204, 366
  ion, 367
  mean ionic, 367
  standard state for, 366
  vapor pressure, 206
  water, 207
Activity coefficient, 198, 368
  Debye-Hückel, 374
Adiabatic demagnetization, 233
Adiabatic expansion, 128
Adsorption, 319–22, 412
  coefficient, 320
  isotherm, 319, 412

Alder, B. J., 673
Alkali-halide:
  crystals, 664
  molecules, 510
Allotropy, 225
Amino acids, 211
Ammonia:
  inversion, 584
  synthesis, 203
Amount of substance, 3
Ampere, 3
Amplitude, wave, 45, 437
Andrews, T., 19
Anesthesia, 184
Angular momentum, 460, 465
  molecular, 491
  orbital, 465, 476
  photon, 555
  quantization, 468
  spin, 473
Anharmonicity, 588, 606
Antisymmetry, 474
Arrhenius, S., 357
  ionization theory, 356
  rate equation, 302
Atomic orbitals, 467, 480, 482
Atomic spectra, 459
Atomization enthalpy, 112

Average:
  energy, 48
  ensemble, 249
  speed, 18
Average-value theorem, 88
Avogadro constant, 4, 74
Austenite, 242
Axis, symmetry,
Azeotrope, 215

Bacon, F., 98
Bacteriophage, 622
Balmer, J. J., 459
Balmer series, 461
Band gap, 658
Band model, 648, 657
Band spectra, 606
Barlow, W., 636
Barometric formula, 73
Base catalysis, 311
Basis, 629
Beer-Lambert law, 604
Benzene m.o.'s, 512
Bernal, J. D., 672
Berthelot equation, 32
Berthelot, M., 114
Berzelius, J. J., 309
B.E.T. Theory, 322

Binary solution, 176
  distillation, 187
Bjerrum, N., 459
Black, J., 102
Bloch, F., 537
Bockris, J. O'M., 425
Bodenstein, M., 297
Body-centered:
  lattice, 627
  structure, 629
Bohr:
  atomic theory, 460
  magneton, 535, 552
  radius, 461
Bohr, N., 460, 535
Boiling point:
  diagram, 214
  pressure dependence, 154
Boltzmann:
  constant, 40, 135
  distribution, 71–76
  equation, 526
  factor, 79, 542, 554
Boltzmann, L., 135
Bond:
  angles, 513
  covalent, 487–512
  ionic, 508
  enthalpy, 111–14
  length, 513
  order, 487
Bonding orbital, 501
Born-Haber cycle, 664
Born, M., 443
Born-Oppenheimer approximation,
    488, 578, 607
Bouguer, P., 604
Bovey, F., 540
Boyle, R., 13
  law, 13
Bragg condition, 674
Bragg equation, 633
Bragg, W. H., 632
Bragg, W. L., 632, 645
Bravais lattice, 626
Bridgman, P. W., 229, 686
Broglie equation, 57, 441
Broglie, L. de, 57, 440
Brønsted acid, 324
Brønsted, J. N., 314
Brunauer-Emmett-Teller isotherm,
    322, 326
Bubble-cap column, 187
Butler-Volmer equation, 428

Cailletet, L., 18, 686
Calorie, 98
Calorimeter, 104
  heat capacity, 48
Candela, 3
Canonical ensemble, 249
Capacitance, 521
Capillarity, 407
Carbon spectra, 606
Carbonyl compounds, 611
Carbon dioxide:
  laser, 593
  normal vibrations, 46, 591
  phase diagram, 224
  spectrum, I R, 591
Carbon monoxide:
  moment of inertia, 42
  vibrations, 44
  energy levels, 579
  spectrum, 56
Carboxypeptidase, 317
Carnot, S., 127
  cycle, 128
  theorem, 129
Cast iron, 244
Catalysis, 309
  acid-base, 311
  enzyme, 314
  homogeneous, 310
  surface, 314
Catalytic constant, 312
Cell, electrochemical, 378
  classification, 382
  concentration, 395
  reversible, 386
Cell membrane, 417
Center of mass, 38
Center of symmetry, 562
Central-field approximation, 474
Chain reaction, 297–301, 307,
    618
Character, group, 571, 574
Chemical bond, 486–517
Chemical dynamics, 341
Chemical potential, 166, 199
  diffusion and, 365
  equilibrium and, 168, 223
  solutions and, 205
Chemical shift, 544–49
Chemiluminescence, 609
Chemisorption, 319
Chromophore, 611
Clapeyron-Clausius equation,
    151

Class:
  crystal, 628
  group theory, 561, 571
Clausius-Mossotti equation, 526
Clausius, R., 118, 151, 129, 354
Closest packing, 651
Cobalamin, 645
Colligative properties, 190
Collision, resonant, 593
Collision theory, 328
Colloids, 404, 418
Color centers, 666
Component, 28, 205
Composition measures, 173
Composition-pressure diagram, 185
Composition-temperature diagram,
    188
Compound formation, 239
Compressibility, 25
  factor, 23
Concentration, 15, 174
  cell, 382
  polarization, 430
Conductance, 354
Conduction band, 657
Conductivity, 354
  ionic, 362
  metals, 654
  molar, 355
  water, 355
Configuration integral, 684
Configuration interaction, 496
Consecutive reactions, 289
Conservation:
  angular momentum, 465, 581
  energy, 34
  laws, 37
Consolute temperature, 216
Contact:
  angle, 407
  metal, 381
  potential, 381
  time, 294
Continuity of states, 19
Coordination number, 661
Correlation diagram, 502
Correlation energy, 482
Correspondence principle, 61
Corresponding states, 22, 201
Cosmology, 131
Cotton, F. A., 572
Coulombic integral, 515
Coulomb law, 357
Coulometer, 353

Coupling of angular momenta, 479
Cousteau, J., 183
Covalent bond:
  H_2^+, 492
  H_2, 493
Cratic, 219
Critical:
  composition, 216
  micelle concentration, 415
  point data, 18
  solution temperature, 216
Crystal:
  classes, 628
  energy, 663
  faces, 624
  growth, 668
  planes, 624
  radii, ionic, 661
  statistical thermodynamics, 266
  structure, Fourier synthesis, 644
  systems, 625
Crystallography, first law, 623
Cubic closest packing, 653
Curie, P., 535
Cyclopropane, isomerization, 345

d'Alembert, J., 34
Dalton law, partial pressure, 16,
  186
Davydov mechanism, 620
Davy, H., 352
Debye equation, heat capacity, 267
Debye-Hückel theory, 370–73
Debye length, 373, 420
Debye, P., 348, 370, 527, 638, 681
Debye-Scherrer method, 638
Defects, point, 665
Degeneracy, 60, 72, 455
Degree of freedom, 29, 38, 224
Delocalization, 512
Demagnetization, adiabatic, 233
Density:
  function, 86
  gas, 13
Detergent, 416
Dewar, G., 232
Diamagnetism, 533
Diamond structure, 650
Diathermic wall, 249
Diatomic molecule, 38, 596
Dielectric constant, 357
Dieterici equation, 32
Differential, exact, 99, 119

Diffraction:
  electron, 57
  neutron, 645, 676
  x-ray, 631
Diffusion, 132, 364
Diffusion coefficient, 333, 365
Diffusion overpotential, 432
Dinitrogen pentoxide
  decomposition, 284
Dipole:
  -dipole interaction, 681
  electric, 522
  moment, 521, 528, 564
  nuclear magnetic, 538
Dirac, P.A.M., 473
Dislocation, 666
Disorder, entropy and, 135
Dispersion interaction, 682
Dissipative process, 296
Dissociation energy:
  ionic molecule, 509
  molecular, 265
  spectroscopic, 589
Distillation, 187, 217
Distortion polarization, 526
Distribution function, 86
D.L.V.O. Theory, 420
Donnan equilibrium, 399
Donnan, F. G., 399
Donor, electron, 660
Double layer, 420
Drude theory, 654
Dulong and Petit rule, 655
Dunn, T., 608
du Noüy, L., 423

Eddington, A. S., 131
Eigenfunction, 452
Eigen, M., 291
Eigenvalue, 452
Einstein, A., 580, 612, 690
Einstein unit, 612
Electric double layer, 420, 426, 431
Electric field:
  effective, 524
Electric potentials, 392
Electrochemical cell, 378
  classification, 383
  equilibrium condition, 384
  sign convention, 383
  standard, 386
Electrochemical equivalence, 352
Electrochemical potential, 380
Electrochemical rate processes, 424

Electrode potential, 390–92
  standard, 390
Electrodes, 378, 382
  diffusion at, 432
  kinetics at, 424
Electromotive force (emf):
  calculation, 393
  definition, 385
  equilibrium constant, 394
  Nernst equation, 389
  solubility product, 394
  standard, 389
  temperature coefficient, 388
  thermodynamics, 387
Electrolytes, 352
Electrolytic cell, 378
Electromagnetic spectrum, 578
Electron affinity, 487, 664
Electron density:
  H_2O, 512
  NaCl, 643
Electron diffraction, 58
Electronegativity, 506
Electron gas, 654
Electronic partition function,
  267
Electronic spectra, 603
  transitions, 608
Electron microscopy, 58
Electron paramagnetic resonance,
  552
Electron spin, 473
Electron volt, 9
Emission:
  spontaneous, 580
  stimulated, 580
Enantiotropic change, 225
Energy, 92
  activation, 301, 337
  conservation, 35
  crystal, 663
  electronic, 268, 603–12
  internal, 36, 100
  level diagram, 579, 593
  levels, atomic, 483
  mechanical, 35
  molecular, 34, 40, 48
  quantization, 448
  reaction, 103
  rotational, 40, 257
  statistical calculation, 250
  transfer, 620
  translational, 441
  vibrational, 43, 262

Ensemble, 248
  average, 249
Enthalpy, 100
  activation, 340
  atomization, 112
  bond, 111–13
  formation, 107–9
  fusion, 227
  mixing, 219
  partial molar, 178
  phase changes, 102
  reaction, 105–9
  statistical formula, 252
  temperature variation, 109
  vaporization, 227
  water, 152
Entropy, 118, 135
  activation, 340
  arrow of time, 131
  Boltzmann formula, 136
  equilibrium, 144
  fusion, 227
  mixing, 131, 181
  probability, 133–36
  reaction, 140
  residual, 270
  rotational, 257
  statistical formula, 251
  third law, 139
  universe, 127
  vaporization, 227
  vibrational, 262
Enzyme:
  catalysis, 314
  structure, 317
Equation of state, 12, 148, 251,
    683
  thermodynamic, 149
  van der Waals, 23
  virial, 20
Equilibrium, 144–68
  dynamic, 145
  electrochemical, 384
  entropy and, 144
  phase, 223
  state, 8
Equilibrium constant:
  activities, 210
  ammonia synthesis, 203
  cell emf, 394
  concentrations, 210
  Gibbs free energy, 159
  mole fractions, 210
  partial pressures, 5
  pressure dependence, 165

statistical theory, 270
  temperature dependence, 163
Equipartition principle, 47
Ernst, R. R., 551
Error function, 86
Ethane, internal rotations, 586
Ethanol, nmr spectrum, 545
Euler rule, 149
Eutectic, 188, 237
  diagram, 236
Eutectoid, 242
Ewald, P. P., 631
Exact differential, 99
Exchange:
  current, 380, 425
  interaction, 477
  rate, 560
Excited states, 611, 615
Exclusion principle, 474, 483
Expansivity, 25
Expectation value, 450
Explosions, 300
Extensive property, 9
Extent of reaction, 7, 278
Extinction coefficient, 604
Eyring, H., 335, 686

Face-centered cell, 626
Faraday, 352
Faraday, M., 352
Faujastite, 325
Fermi-Dirac statistics, 655
Fermi, E.:
  contact interaction, 549
  level, 381, 656
Fick, R., 365
  diffusion laws, 365
Film balance, 412
First law of thermodynamics, 92
  catalysis and, 309
  mathematical statements, 98–99
  second law combined, 120
First-order reaction, 280
  consecutive, 289
  flow system, 294
  half life, 284
  rate equation, 283
Flash photolysis, 619
Flow systems, 293
Fluid-mosaic membrane, 417
Fluid state, 19
Fluorescence, 615
  quenching, 616
  sensitized, 620
Fluorite, 638

Flux, 365
  density, 533
Flygare, W. M., 515
Force, 5, 92–97
Force constant, 43, 65, 592, 606
Förster, T., 620
Fourier, J. B.:
  series, 644
    in crystallography, 644
  theorem, 675
  transform, 551
Fractional distillation, 187
Franck-Condon principle, 607, 609,
    617
Franck, J., 607
Free-electron model, 453
Free-energy functions, 146 (See
    Gibbs-Helmholtz)
Free induction decay, 552
Free radicals:
  chain reactions, 299
  epr spectra, 556
Free volume, 686
Freezing-point depression, 188
Frenkel defects, 665
Freundlich isotherm, 321
Frequency factor, 302
Friedrich, W., 631
Fuel cell, 433
Fugacity, 198
  chemical potential, 199
  coefficient, 198, 201
  corresponding states, 201
  equilibrium constant, 202
Fusion:
  entropy, 123
  thermodynamic data, 227

Galvanic cell, 378 (See
    Electrochemical cell)
Gas, 11–25
  constant, 12
  ideal, 12, 96, 252
  real, 198
  solid reactions, 165
Gaussian density function, 86
Gay-Lussac law, 13
Geophysical Laboratory, 230
G factor:
  electronic, 552
  nuclear, 538
Giauque, W., 232
Gibbs adsorption isotherm, 412
Gibbs-Duhem equation, 177, 208

Gibbs free energy, 146, 150, 167
    activation, 340
    electrostatic, 370
    equilibrium, 149–51, 159
    extent of reaction, 157
    ideal gas, 157
    partial molar, 166, 181, 205
    pressure dependence, 157, 199
    reaction, 159
    standard state, 155
    temperature dependence, 161
Gibbs-Helmholtz equation, 162,
    388
Gibbs, J. W., 27, 146, 166, 248
Glide plane, 631
Goniometer, 623
Goldman, D., 398
Gouy-Chapman double layer, 421
Graham T., 404
Grotthuss, C. J., 354, 612
Group:
    definition, 561
    representation, 567–71
    G, u classification, 491
Guldberg-Waage principle, 145,
    288

Habit, crystal, 623
Half cell, 378
Half life, 284, 286
Hall process, 388
Hamiltonian operator, 441, 452
Hard-sphere potential, 685
Harmonic oscillator, 44, 64, 263
    Schrödinger equation, 586
Harris, R. K., 560
Hartree, D., 482
Hartree-Fock method, 482
Haüy, R. J., 624
Heat, 98
    molecular picture, 121
Heat capacity, 84, 264
    constant pressure, 101
    constant volume, 48
    Debye theory, 267
    Einstein theory, 267
    gases, 50
    entropy from, 136
    solids, 266
    temperature dependence, 264
Heat conduction, 124
Heat engine, 127
Heisenberg, W., 436
    uncertainty principle, 443

Helium:
    atom, 481
    liquid, 234
    spectrum, 475
Helmholtz, H. V., 146
Helmholtz free energy, 146, 167,
    251, 683
    equation of state, 148, 251
    statistical, 251
Henderson, L. J., 127
Henry, W., 183
Henry's law, 182, 213
Hermite polynomial, 587
Herzberg, G., 617
Herzfeld, K., 299
Hess, G. H., 106
Heteronuclear diatomics, 505
Hexagonal closest packing, 653
High-pressure systems, 230
Hildebrand, J., 678
Hole conduction, 658
Homonuclear diatomics, 505
Hooke, R., 43, 623
Hooke's law, 43
Hund, F., 479, 612
Hückel, E., 370, 515
Hydrocarbons in water, 219
Hydrogen atom:
    energy levels, 461
    orbitals, 471
    Schrodinger equation, 463
Hydrogen bond, 651
Hydrogen bromide, 297
Hydrogen chloride, 582
Hydrogen iodide, 289, 590
Hydrogen ions, 364, 431
Hydrogen molecule:
    electron density, 497
    energy, 496
    potential-energy curve, 493
Hydrogen molecule ion, 488
Hydrophobic interaction, 651
Hyperfine interaction, 558

Ice, crystal structure, 652
Ice skating, 153
Icosahedron, 565
Ideal gas (See Gas, ideal)
Ideal solution, 179–81
Indicator diagram, 94
Integrating factor, 119
Intensive property, 9
Intermolecular energy, 679, 683
Intermolecular forces, 23, 678–85
Internal pressure, 149, 677

Internal rotation, 586
Interplanar spacing, 628
Ionic:
    atmosphere, 371
    bond, 508
    character, 507
    conductivity, 362
    crystals, 660–62
    enthalpy, 109
    strength, 370
Ionization potential, 460, 463
Ionization theory, 356
Iron-carbon phase diagram, 244
Iron, crystal structure, 628
Irreducible representation, 570
Isometric, 14
Isopiestic method, 208
Isopleth, 237
Isotherm, 14
Isotopic exchange, 274

Jablonski diagram, 615
Joule-Thomson coefficient, 169

Kammerlingh-Onnes, K., 232
Kasha, M., 483, 609
Kauzmann, W. J., 682
Keesom, W. H., 681
Kelvin, 3
Kelvin (See Thomson, W.)
Kelvin equation, 409
Kinetic energy:
    gas, 17
    hydrogen molecule, 493
Kirkwood, J. G., 527
Kittel, C., 659
Knipping, P., 631
Kohlrausch, F., 355

Lambda transition, 235
Lambert, J. H., 604
Laminar flow, 688
Langevin, P., 527, 535
Landau, L. S., 420, 683
Langmuir adsorption isotherm, 319
Langmuir, I., 319, 412
Laplacian, 464
Larmor frequency, 541
Laser, 592, 612
Latent heat, 102
Latter, R., 483
Lattice, 626, 630
Laue method, 631
Laue, M. V., 623, 631
LeChatelier principle, 163

Lennard-Jones, J. E., 679
Lessor, A., 639
Levy, H. A., 675
Lewis, G. N., 138, 487
Light:
　absorption, 580
　scattering, 594
Lindemann, F. A., 343
Lineweaver-Burk equation, 316
Lipscomb, W. N., 317
Liquid:
　cohesion, 673, 677
　disorder, 672
　hard sphere model, 673
　theory, 671–87
　x-ray diffraction, 674
Liquid crystals, 227
Liquid junction, 386
　potential, 395
Lithium hydride, electron density, 506
London, F., 682
London forces, 663, 682
Lorentz-Lorenz equation, 530
Luminescence, 615
Lyman series, 461
Lynden-Bell, R. M., 560

Madelung constant, 662
Madelung energy, 662–64
Magic — T, 554
Magnetic:
　field, 533
　induction, 533
　resonance, 537
　susceptibility, 533
Magnetogyric ratio, 538
Magneton:
　Bohr, 535, 552
　nuclear, 538
Manganese, crystal structure, 630
Mao, H. K., 230
Marcelin, J., 335
Mark, H., 691
Maser, 584
Mass average, 419
Maxwell, J. C., 89
　distribution function, 349
　equation, 86
　relations, 147, 530
McMillan-Mayer equation, 193
McQuarrie, D. A., 525
Mean free path, 330
Mean-value theorem, 88
Melting, 227

Melting point, 239
Membrane potential, 397
Mercury, x-ray diffraction, 674
Metals:
　cohesion, 657
　crystal structure, 654
　electronic structure, 654
Metastable state, 151
Methane symmetry, 562
Micelle, 415
Michaelis, L.:
　constant, 315
Michaelis-Menten equation, 315
Microcanonical ensemble, 248
Microstate, 135
Microwave spectra, 583
Miller index, 624
Mirror plane, 560
Mobility, ionic, 358, 362
Mohorovičić discontinuity, 231
Molality, 175
Molarity, 174
Mole, 4
Molecular beams, 342
Molecular dynamics, 673
Molecularity, 281
Molecular orbitals, 491, 494, 611
　benzene, 514
　H_2, 494
　H_2^+, 491
　homonuclear diatomics, 499
　LCAO approximation, 491
　N_2, 517
　O_2, 504
Molecular speed, 18, 85, 90
Molecular structures, 513
Mole fraction, 15
Moment of inertia, 40, 258
Monatomic gas, 254
Moore, C. A., 611
Morse, P. M., 589
Morse function, 589, 607
Mulliken, R., 507
Mullite, 239
Multiplication table, group, 566
Multiplicity, 476

Nematic liquid crystals, 228
Nernst, W., 138, 618
　diffusion layer, 432
　equation, 389
　potential, 398
Nerve conduction, 399
Neutron diffraction, 645
Newton, 5

Newton, I., 53, 687
Newtonian viscosity, 687
Nitrogen, molecular orbitals, 503
Nitrous oxide entropy, 139
Noncrossing rule, 502
Normal coordinates, 597
Normal density function, 86
Normal distribution, 87
Normalization, 444, 588
Normal mode, vibration, 46, 591, 598
Norrish, R., 619
Nuclear g factor, 538
Nuclear magnetic resonance, 539
　dynamic, 549
　Fourier transform, 551
Nuclear magneton, 538
Nucleation, 411, 668
Number average, 419

Observable, 450
Ocean, energy, 130
Octahedron, 573
Ohm's law, 353
Oil and water, 219
Onsager, L., 527
Operator, 451
Optical activity, 565
Optical density, 604
Orbital, 467
　atomic, 480
　benzene, 514
　bonding, 501
　molecular, 494
　pi, 501
　sigma, 501
Order of reaction, 280
Orthogonality, 445
Oscillator strength, 605
Osmotic pressure, 190–93
　polymer solutions, 193
　sucrose solutions, 191
Ostwald, W., 309, 312
　viscometer
Overbeek, T., 420
Overpotential, 426, 431
Overtone, 588
Oxygen:
　dissociation, 611
　excited states, 609
　molecular orbitals, 504
Ozone decomposition, 311

Paracrystalline state, 227
Paramagnetism, 533

Partial molar quantity, 175–78
Partial molar volume, 175
Partial pressure, 15
Particle in box, 59, 442
Partition function, 251
  electronic, 267
  ensemble, 250
  molecular, 72, 256
  rotation, 256, 259
  translation, 254
  vibration, 261
Pascal, 5
Pauli exclusion principle, 473, 655
Pauling, L., 113, 506, 652
Pauli, W., 474, 483
P branch, 66, 81, 590
Pearlite, 252
Peritectic, 240
Permittivity, relative, 520
  frequency dependence, 529
Perrin, J., 74, 620
pH, 396
Phase, 27
Phase change, 102
Phase equilibrium, 151
Phase rule, 29, 223, 225
Phosphorescence, 615
Photochemistry, 612–21
  secondary processes, 618
Photoelectron spectroscopy, 516
Photoionization, 516
Photon, 54, 555
Photosensitization, 614
Photosynthesis, 620
Planck constant, 54
Planck, M., 54
Plane wave, 54
Plate, theoretical, 187
Poggendorf method, 385
Point defect, 665
Point group, 561–66, 631
Poise, 687
Poiseuille equation, 688
Polanyi, M., 335, 666
Polar coordinates, 464
Polarizability, 523, 682
  tensor, 524
Polarization, 426, 521, 651
  molar, 526
Polymerization kinetics, 307, 323
Polymer solutions, 193, 418
Polymer structures, 323
Pople, J. A., 513
Porter, G., 619
Potential energy, 36

Potential-energy curve, 489
  harmonic oscillator, 46
  H_2, 493
  H_2^+, 489
  NaC1, 508
  O_2, 610
Potential-energy surface, 335
Potential, intermolecular, 679
Potentiometer, 385
Powder method, 638, 640
Predissociation, 617
Pre-exponential factor, 302
Pressure, 5, 16
Probability, 135, 250
  entropy and, 135, 251
  mixtures, 133
Pseudo-first-order reaction, 287
Purcell, E. M., 537

Q branch, 590
Quadrupole, 537
Quantum mechanics, 53, 61,
  436–56
Quantum number, 59, 447, 467
  angular momentum, 465
  azimuthal, 465
  inner, 477
  magnetic, 468
  principal, 460, 465, 468
  rotational, 62, 81
  vibrational, 64
Quantum statistics, 450
Quantum yield, 613
Quenching cross-section, 616

R-branch, 66, 590
Radial distribution function, 469,
  675
Radial wave function,
Radius ratio, 661
Raman, C. V., 595
Raman spectra, 595
Raoult, F. M., 180
Raoult's law, 179, 213
  deviations, 214
Rate constant, 278
  reduced, 281
  temperature dependence, 301
Rate law, 280
Rate of reaction, 277
Rational intercepts law, 624
Reaction:
  consecutive, 289
  coordinate, 336
  diffusion controlled, 348

elementary, 281
flow system, 293
mechanism, 282
opposing, 288
parallel, 290
solution, 346
Reactor-stirred flow, 295
Real gas, 199
Rectolinear diameters, law, 687
Reduced mass, 41, 329
Reduced rate constants, 281
Reduced state variables, 22, 201
Refraction, molar, 530
Refractive index, 529
Relaxation:
  chemical, 291
  spin-lattice, 542
  spin-spin, 543
Relaxation time, nmr:
  longitudinal, 543
  transverse, 544
Representation, group, 570
Resistivity, 354
Resonance energy, 516
Resonance integral, 515
Resonance, magnetic, 537, 552
  (See also Nuclear magnetic
  resonance)
Resonant collision, 593
Reversible heat, 119
Reversible process, 95
Rice, F. O., 299
Rigid rotor, 41, 581
Ring current, 546
Ritchie, G. L., 533
Ritz, W., 460
Rocksalt structure, 637
Rotary-reflection axis, 562
Rotational constant, 67, 591
Rotational quantum number, 62
Rotation, molecular, 39
  axis, 560
  energy, 62
  internal, 586
Russell-Saunders coupling, 478
Rutherford, E., 460
Rydberg constant, 459, 462

Sackur-Tetrode equation, 255
Salt bridge, 387
Scale factor, 481
Schoenflies notation, 563
Schottky, W., 665
Schrödinger, E., 436
  wave equation, 436, 440, 587

Schumann-Runge bands, 610
Screw axis, 631
Second law of thermodynamics, 120, 130, 140
Second-order rate equation, 285
Secular equation, 492, 515
Sedimentation equilibrium, 74
Selection rule, 66
Self-consistent field, 482
Semiconductor, 658–60
Shielding constant, 544
SI units, 3
Silicon, doping, 660
Silicon nitride, spectrum,
Similarity transformation, 561
Slip plane, 667
Smectic state, 228
Sodium chloride:
    crystal structure, 633–37
    molecule, 508
Sodium dodecyl sulfate, 404
Sodium pump, 400
Solid solution, 240
Solid state, 11, 26
Sol, monodisperse, 419
Solubility:
    curve, 188
    gases in liquids, 182
    gases in water, 183
    liquids in liquids, 215
    product constant, 394
    temperature dependence, 189
Solutions:
    dilute, 188
    ideal, 173
    nonideal, 204
    solid, 240
    solids in liquids, 188
Sommerfeld, A., 631
Sorensen, 396
Space group, 631
Spectra:
    absorption, 61
    atomic, 459
    carbon monoxide, 55
    electronic, 579, 603
    infrared, 577
    isotope effects, 584
    microwave, 583
    optical, 577
    photoelectron, 516
    Raman, 594
    rotational, 581
    vibrational, 589

Spectral terms, 460
Spectrofluorimeter, 615
Spectrometer:
    epr, 554
    microwave, 583
    nmr, 540
    x-ray, 632
Spectroscopic data, molecules, 596
Spectroscopy, 54
    photoelectron, 516
Speed of light, 54
Spherical top, 260
Spin:
    electron, 535
    nuclear, 537
Spin density, 556
Spin-orbit interaction, 477
Spin-spin coupling, 547
Standard cell, 386
Standard Gibbs free energy, 156
Standard hydrogen electrode, 391
Standard state, 107, 205, 367
    gas, 199
    Gibbs free energy, 155
    solution components, 205
Standing wave, 59
Stark effect, 584
Stark-Einstein principle, 612
State function, 9, 99
Stationary state, 8
Statistical thermodynamics, 247–76
    formulas, 252
Statistical weight, 60, 72, 256, 268
Statistics, quantum, 655
Staudinger equation, 691
Steady state, 294, 298, 345
Steel, 242
Steinfeld, J. I., 609
Stensen, N., 623
Stereoregular polymers, 323
Steric factor, 334
Stern double layer, 421
Stern-Gerlach experiment, 473
Stern-Volmer equation, 616
Stoichiometric number, 6
Stokes equation, 689
Stopped-flow method, 279
Streamline flow, 688
Streitweiser, A., 506
Structure factor, 640–43
Subgroup, 551
Sucrose structure, 646
Sullivan, J. H., 295
Superconductivity, 234

Supercooled liquid, 126
Superfluidity, 233
Surface catalysis, 318
Surface film, 412–14
Surface pressure, 413
Surface tension, 97, 405, 434
    solutions, 411
Surfactant, 415
Susceptibility:
    electric, 522
    magnetic, 533
Swan bands, 606
Symmetry, 37
    element, 563
    operation, 559, 562
Symmetric top, 260
Systems, 7
    one component, 224
    two component, 236
Syzszkowski equation, 422

Tafel equation, 429
Tammann, G., 229
Temperature, 3, 71, 82
Tensile strength, 667
Tension, 97
Tensor, polarizability, 524
Tesla, 538
Tetrahedral anvil, 229
Tetrahedral group, 562, 575
Thermochemistry, 114
Thermodynamics, 35, 118
    cell reactions, 387
    equation of state, 149
    first law, 92
    functions, 147
    phase equilibrium, 151
    second law, 118
    statistical, 247
    temperature, 120
    third law, 138, 269
Thermopile, 613
Third law entropies, 139
Third law of thermodynamics, 138, 269
Thomsen, J., 114, 356
Thomson, G. P., 58
Thomson, W., 128, 130, 409
Tie-line, 185
Tin, allotropy, 225
T-jump, 291
Townes, C., 584
Transfer coefficient, 428

Transference number (*See* Transport number)
Transition moment, 581, 595, 605
Translation, 40
Translational:
  energy, 40, 58
  partition function, 254
  wavefunction, 446
Transmittance, 604
Transport number, 358
  Hittorf method, 359
  moving-boundary method, 360
Triplet state, 614
Tunnel effect, 455, 585
Tyrosine, UV absorption, 605

Uncertainty principle, 449
Unimolecular reactions, 343–46
Unit cell, 626
United atom, 502
Updike, J., 321
Uyeda, N., 58

Vacancies:
  crystal, 665
  liquid state, 686
Valence band, 657
Valence-bond method, 499
Valence theory, 487
van der Waals:
  bonds, 649
  constants, 19
  equation, 23
  forces, 681
van der Waals, J. H., 23
van't Hoff equation, 163, 193
van't Hoff i factor, 357

van't Hoff, J. H., 357
van Vleck paramagnetism, 535
Vaporization, thermodynamic data, 227
Vapor pressure, 25
  droplets, 409
  liquids, 155
  osmotic pressure and, 191
  temperature dependence, 153
Variance, 29
Variation method, 480
  H_2^+, 491
Vector model of atom, 477
Vector, symmetry transformation, 565
Vibrational energy, 43, 64
Vibrational partition function, 261
Vibration, normal modes, 596
Virial coefficient, 21, 193, 685
Virial equation, 20
Virial theorem, 497
Viscometer, 689
Viscosity, 333
  activation energy, 690
  intrinsic, 691
  limiting, 691
  liquid, 687
  Newtonian, 687
  polymer solutions, 690
  temperature dependence, 690
Void volume, 654

Wahl, A. C., 502
Walton, A. J., 672
Wannier, G. H., 650
Warren, B. E., 675

Water:
  dipole moment, 532
  dissociation, 363
  hydrogen bonds, 677
  liquid structure, 675
  molecular structure, 509
  phase diagram, 230
  x-ray diffraction, 675
Watt, J., 127
Wave equation, 438
Wavefunction, 436, 438, 447, 467, 469, 474
Wave guide, 554
Wave mechanics, 57, 436
Wave number, 54, 61
Weston standard cell, 386
Wheatstone bridge, 354
Whewell, W., 352
Whisker, metal, 667
Wilhelmy, L., 278
Witte, H., 649
Work, 14, 92, 97
  maximum, 148
  reversible, 95, 119
Wyckoff, R. W., 661

X-ray, 623
  crystallography, 631
  intensity, 641
  wavelength, 632

Young, T., 405
Young-Laplace equation, 406

Zawidski, J. V., 213
Zeolite, 325
Zero-point energy, 64, 449
Ziegler, K., 323

Electronic Configurations of the Elements in Their Ground States

Z	Element	Configuration	State	Z	Element	Configuration	State
1	H	$1s$	$^2S_{1/2}$	54	Xe	$[\text{Kr}]4d^{10}5s^25p^6$	$^1S_0$
2	He	$1s^2$	$^1S_0$	55	Cs	$[\text{Xe}]6s$	$^2S_{1/2}$
3	Li	$1s^22s$	$^2S_{1/2}$	56	Ba	$[\text{Xe}]6s^2$	$^1S_0$
4	Be	$1s^22s^2$	$^1S_0$	57	La	$[\text{Xe}]5d6s^2$	$^2D_{3/2}$
5	B	$1s^22s^22p$	$^2P_{1/2}$	58	Ce	$[\text{Xe}]4f5d6s^2$	$^1G_4$
6	C	$1s^22s^22p^2$	$^3P_0$	59	Pr	$[\text{Xe}]4f^36s^2$	$^4I_{9/2}$
7	N	$1s^22s^22p^3$	$^4S_{3/2}$	60	Nd	$[\text{Xe}]4f^46s^2$	$^5I_4$
8	O	$1s^22s^22p^4$	$^3P_2$	61	Pm	$[\text{Xe}]4f^56s^2$	$^6H_{5/2}$
9	F	$1s^22s^22p^5$	$^2P_{3/2}$	62	Sm	$[\text{Xe}]4f^66s^2$	$^7F_0$
10	Ne	$1s^22s^22p^6$	$^1S_0$	63	Eu	$[\text{Xe}]4f^76s^2$	$^8S_{7/2}$
11	Na	$[\text{Ne}]3s$	$^2S_{1/2}$	64	Gd	$[\text{Xe}]4f^75d6s^2$	$^9D_2$
12	Mg	$[\text{Ne}]3s^2$	$^1S_0$	65	Tb	$[\text{Xe}]4f^96s^2$	$^6H_{15/2}$
13	Al	$[\text{Ne}]3s^23p$	$^2P_{1/2}$	66	Dy	$[\text{Xe}]4f^{10}6s^2$	$^5I_8$
14	Si	$[\text{Ne}]3s^23p^2$	$^3P_0$	67	Ho	$[\text{Xe}]4f^{11}6s^2$	$^4I_{15/2}$
15	P	$[\text{Ne}]3s^23p^3$	$^4S_{3/2}$	68	Er	$[\text{Xe}]4f^{12}6s^2$	$^3H_6$
16	S	$[\text{Ne}]3s^23p^4$	$^3P_2$	69	Tm	$[\text{Xe}]4f^{13}6s^2$	$^2F_{7/2}$
17	Cl	$[\text{Ne}]3s^23p^5$	$^2P_{3/2}$	70	Yb	$[\text{Xe}]4f^{14}6s^2$	$^1S_0$
18	Ar	$[\text{Ne}]3s^23p^6$	$^1S_0$	71	Lu	$[\text{Xe}]4f^{14}5d6s^2$	$^2D_{3/2}$
19	K	$[\text{Ar}]4s$	$^2S_{1/2}$	72	Hf	$[\text{Xe}]4f^{14}5d^26s^2$	$^3F_2$
20	Ca	$[\text{Ar}]4s^2$	$^1S_0$	73	Ta	$[\text{Xe}]4f^{14}5d^36s^2$	$^4F_{3/2}$
21	Sc	$[\text{Ar}]3d4s^2$	$^2D_{3/2}$	74	W	$[\text{Xe}]4f^{14}5d^46s^2$	$^5D_0$
22	Ti	$[\text{Ar}]3d^24s^2$	$^3F_2$	75	Re	$[\text{Xe}]4f^{14}5d^56s^2$	$^6S_{5/2}$
23	V	$[\text{Ar}]3d^34s^2$	$^4F_{3/2}$	76	Os	$[\text{Xe}]4f^{14}5d^66s^2$	$^5D_4$
24	Cr	$[\text{Ar}]3d^54s$	$^7S_3$	77	Ir	$[\text{Xe}]4f^{14}5d^76s^2$	$^4F_{9/2}$
25	Mn	$[\text{Ar}]3d^54s^2$	$^6S_{5/2}$	78	Pt	$[\text{Xe}]4f^{14}5d^96s$	$^3D_3$
26	Fe	$[\text{Ar}]3d^64s^2$	$^5D_4$	79	Au	$[\text{Xe}]4f^{14}5d^{10}6s$	$^2S_{1/2}$
27	Co	$[\text{Ar}]3d^74s^2$	$^4F_{9/2}$	80	Hg	$[\text{Xe}]4f^{14}5d^{10}6s^2$	$^1S_0$
28	Ni	$[\text{Ar}]3d^84s^2$	$^3F_4$	81	Tl	$[\text{Xe}]4f^{14}5d^{10}6s^26p$	$^2P_{1/2}$
29	Cu	$[\text{Ar}]3d^{10}4s$	$^2S_{1/2}$	82	Pb	$[\text{Xe}]4f^{14}5d^{10}6s^26p^2$	$^3P_0$
30	Zn	$[\text{Ar}]3d^{10}4s^2$	$^1S_0$	83	Bi	$[\text{Xe}]4f^{14}5d^{10}6s^26p^3$	$^4F_{3/2}$
31	Ga	$[\text{Ar}]3d^{10}4s^24p$	$^2P_{1/2}$	84	Po	$[\text{Xe}]4f^{14}5d^{10}6s^26p^4$	$^3P_2$
32	Ge	$[\text{Ar}]3d^{10}4s^24p^2$	$^3P_0$	85	At	$[\text{Xe}]4f^{14}5d^{10}6s^26p^5$	$^2P_{3/2}$
33	As	$[\text{Ar}]3d^{10}4s^24p^3$	$^4S_{3/2}$	86	Rn	$[\text{Xe}]4f^{14}5d^{10}6s^26p^6$	$^1S_0$
34	Se	$[\text{Ar}]3d^{10}4s^24p^4$	$^3P_2$	87	Fr	$[\text{Rn}]7s$	$^2S_{1/2}$
35	Br	$[\text{Ar}]3d^{10}4s^24p^5$	$^2P_{3/2}$	88	Ra	$[\text{Rn}]7s^2$	$^1S_0$
36	Kr	$[\text{Ar}]3d^{10}4s^24p^6$	$^1S_0$	89	Ac	$[\text{Rn}]6d7s^2$	$^2D_{3/2}$
37	Rb	$[\text{Kr}]5s$	$^2S_{1/2}$	90	Th	$[\text{Rn}]6d^27s^2$	$^3F_2$
38	Sr	$[\text{Kr}]5s^2$	$^1S_0$	91	Pa	$[\text{Rn}]5f^26d7s^2$	$^4K_{11/2}$
39	Y	$[\text{Kr}]4d5s^2$	$^2D_{3/2}$	92	U	$[\text{Rn}]5f^36d7s^2$	$^5L_6$
40	Zr	$[\text{Kr}]4d^25s^2$	$^3F_2$	93	Np	$[\text{Rn}]5f^46d7s^2$	$^6L_{11/2}$
41	Nb	$[\text{Kr}]4d^45s$	$^6D_{1/2}$	94	Pu	$[\text{Rn}]5f^67s^2$	$^7F_0$
42	Mo	$[\text{Kr}]4d^55s$	$^7S_3$	95	Am	$[\text{Rn}]5f^77s^2$	$^8S_{7/2}$
43	Tc	$[\text{Kr}]4d^55s^2$	$^6S_{5/2}$	96	Cm	$[\text{Rn}]5f^76d7s^2$	$^9D_2$
44	Ru	$[\text{Kr}]4d^75s$	$^5F_5$	97	Bk	$[\text{Rn}]5f^86d7s^2$ (or $5f^97s^2$)	
45	Rh	$[\text{Kr}]4d^85s$	$^4F_{9/2}$	98	Cf	$[\text{Rn}]5f^96d7s^2$ (or $5f^{10}7s^2$)	
46	Pd	$[\text{Kr}]4d^{10}$	$^1S_0$	99	Es	$[\text{Rn}]5f^{10}6d7s^2$ (or $5f^{11}7s^2$)	
47	Ag	$[\text{Kr}]4d^{10}5s$	$^2S_{1/2}$	100	Fm	$[\text{Rn}]5f^{11}6d7s^2$ (or $5f^{12}7s^2$)	
48	Cd	$[\text{Kr}]4d^{10}5s^2$	$^1S_{1/2}$	101	Md	$[\text{Rn}]5f^{12}6d7s^2$ (or $5f^{13}7s^2$)	
49	In	$[\text{Kr}]4d^{10}5s^25p$	$^2P_{1/2}$	102	No	$[\text{Rn}]5f^{13}6d7s^2$ (or $5f^{14}7s^2$)	
50	Sn	$[\text{Kr}]4d^{10}5s^25p^2$	$^3P_0$	103	Lw	$[\text{Rn}]5f^{14}6d7s^2$	
51	Sb	$[\text{Kr}]4d^{10}5s^25p^3$	$^6D_{1/2}$	104	Ku	$[\text{Rn}]5f^{14}6d^27s^2$	
52	Te	$[\text{Kr}]4d^{10}5s^25p^4$	$^3P_2$	105	Ha	$[\text{Rn}]5f^{14}6d^37s^2$	
53	I	$[\text{Kr}]4d^{10}5s^25p^5$	$^2P_{3/2}$	106		$[\text{Rn}]5f^{14}6d^47s^2$	